WORLD RAILWAY ACCIDENTS 1961-1975

Francis Voisey

ISBN: 9781916201118

This book is dedicated to signalman Brian Stone, who first awakened my lifelong interest in railways way back in 1966.

Thanks to the staff of the Exeter Reference Library and the University of Exeter Library for their assistance while I was researching this book, to my friend June Nightingale for encouraging me to go ahead and get it finished when my enthusiasm was waning a little, and to my publishing team, without whose guidance and assistance it is doubtful if I would have been able to navigate the mysteries of getting my first book into print.

CONTENTS

INTRODUCTION 1
CHAPTER 1 DERAILMENTS: TRACK DEFECTS 3
CHAPTER 2 DERAILMENTS - BROKEN TYRES, BROKEN AXLES AND OTHER VEHICLE DEFECTS 22
CHAPTER 3 OTHER DERAILMENTS 40
CHAPTER 4 REAR COLLISIONS 91
CHAPTER 5 SIDELONG, HEAD-ON, DOUBLE AND FOULING COLLISIONS ... 167
CHAPTER 6 SINGLE LINE COLLISIONS 241
CHAPTER 7 RUNAWAY TRAINS AND VEHICLES 299
CHAPTER 8 BRIDGE FAILURES 327
CHAPTER 9 OPENING BRIDGE ACCIDENTS 334
CHAPTER 10 TRAIN ACCIDENTS INVOLVING BRIDGES 336
CHAPTER 11 TUNNEL ACCIDENTS 346
CHAPTER 12 LEVEL CROSSING COLLISIONS 376
CHAPTER 13 RAIL/RAIL CROSSING COLLISIONS 542
CHAPTER 14 FIRES 548
CHAPTER 15 CHEMICAL EXPLOSIONS 577
CHAPTER 16 BOILER EXPLOSIONS 590
CHAPTER 17 VANDALISM 592
CHAPTER 18 SABOTAGE 605
CHAPTER 19 MISCELLANEOUS ACCIDENTS 635
INDEX 716
TABLE OF MAJOR RAILWAY ACCIDENTS 1961-1975 739
ABOUT THE AUTHOR 754

INTRODUCTION

This book is the seventh of what is hoped will be a series detailing railway accidents worldwide. A good many books have been written over the years about railway accidents, some concentrating on single disasters, others covering accidents in a given country or geographical area. Accidents in countries such as Great Britain, Germany, the Netherlands, the United States, Australia, New Zealand and so on have been documented comprehensively in one or more books but no attempt has been made previously to detail every significant accident worldwide. This is clearly an ambitious project, but six volumes dealing with the earlier years having been completed, this book will attempt to detail the accidents that occurred between 1961 and 1975. Previous volumes have covered the periods 1811-80, 1881-1900, 1901-15,1916-30,1931-45 and 1946-60. Further volumes are planned to cover the remainder of the 20th century and the opening years of the 21st. As before, an attempt will be made to include all accidents involving five or more fatalities to train passengers. A good number of less serious accidents considered of interest have also been included. The accidents have been grouped into chapters according to their primary cause and are dealt with, in general, chronologically within each chapter.

The gathering of information about accidents in distant parts of the world has not been easy, though it is much easier nowadays than when this project began in the days before the internet. Outside of Great Britain and Ireland few countries seem to have had a formal body charged with investigating all serious railway accidents until recent years, though one or two major disasters such as the Bussey Bridge failure in the United States and the Baptiste Creek collision in Canada were the subject of full formal investigations even in the early days. Otherwise reliance has largely had to be placed on contemporary newspaper reports and previously published works on the subject, though the author has tried as far as possible to keep to primary sources. If the book appears to be biased towards accidents in Great Britain and the United States this is probably partly because these are the countries about which information on accidents has been easiest to obtain, though these were also in any case the countries where many of the most serious accidents did occur.

Contemporary spellings of place names have been used where appropriate and for the many American accidents, terms such as railroad, engineer, caboose and so on have been used instead of railway, driver, brake van etc. By this date the 24-hour clock was in widespread use, except in the United States. The 12- and 24-hour clocks will both be used depending upon the custom in the country concerned. Distances likewise will be given in miles and yards/feet or kilometres and metres depending on the practice in the country concerned.

By 1961 railways had spread to almost all parts of the world, though some countries only had very small systems or perhaps just one line.

The names and the boundaries of countries have undergone many changes over the years. The names of a number of countries and of various places have also changed. The names of locations in use at the time an accident occurred have been given, together with present names where appropriate, as for example Ürsküb (now Skopje).

CHAPTER 1

DERAILMENTS: TRACK DEFECTS

The tracks over which trains run are obviously a vital part of the railway and any problems with the track are likely to lead to the train being derailed. By 1961 the technology involved in laying railway track had much advanced since the early days, but train speeds and weights had also much increased. Accidents occurred from time to time due to failure to keep the track in a good state of maintenance, insufficient allowance being made for hot weather expanding the rails, damage caused to the track from floodwaters and so on. Also included in this chapter, though not strictly due to defects in the track, are accidents due to trains having become derailed on open points. Broken rails were another hazard that caused mishaps and worse from time to time, as exemplified by the Hither Green disaster in England towards the end of 1967.

An express train consisting of an engine, a luggage van and 14 carriages was passing through Castelfranco Veneto, Italy, at 100 kilometres an hour, when its sixth vehicle became derailed on a set of points, followed by the remaining nine carriages. Twenty-nine passengers lost their lives and 76 were injured.

A Ministerial Inquiry found that the accident had been caused by a faulty set of points; one of the first sets of a new design constructed by the Ferrovie dello Stato in its own workshops. The accident reportedly occurred at an unstated date in March 1961, though there seems to be no reference to it in newspapers between late February and early April 1961, and it has only proved possible to find one reference to it on the internet. [1]

A Pennsylvania Railroad Baseball Special, carrying fans from the Harrisburg area to a Phillies game in Philadelphia, met with disaster at Steelton, Pennsylvania, United States, on Saturday 28th July 1962. The nine-coach train had left Harrisburg at 5.02 p.m., two minutes late, and had only travelled about four miles when it left the track next to the Bethlehem Steel Co plant, at a point where the line ran alongside the Susquehanna River. The last three carriages of the train fell down a 25-foot embankment to the river, two ending up on their sides and the third coming to rest upright, all three of them ending up in the river, which was, however, only a couple of feet deep at this point; the third was

---

[1] http://www.macchinistisicuri.info/ms/inciferr/index.php#1961

minus its roof. Nineteen passengers lost their lives in the disaster and about 100 were injured. Two empty coaches, reserved for passengers boarding later in the journey at Lancaster and Coatesville, were also derailed but remained on the track.

A newspaper wrote, 'Aboard {the train} were baseball fans – parents and children, including some Little Leaguers. The talk was of batting averages and pitchers, of Clemente and Skinner, Gonzalez, and Groat. Youngsters fidgeted with the excitement of seeing the Pittsburgh Pirates and the Philadelphia Phillies under the lights. The ballpark was less than 100 miles and two hours away. They didn't know they would never hear the call "play ball."

'Several miles away, David Wilbert stepped from the open hearth furnace building of the Bethlehem Steel Co. The bright sunlight forced him to squint at the few pleasure boats on the Susquehanna River. It was 5.8 p.m. and Wilbert idly watched a crew working on the railroad tracks about 200 feet from the building. Then he noticed the electric powered train and several cars passing into view from behind a long "crane run" building.

'Horror rooted him to the spot. One of the cars suddenly tore loose from the car before it and bowled over a high tension pole. The car, dragging two others, bulldozed two more poles. Sounds of steel on steel and splintering wood pierced the summer day and Wilbert saw the cars somersault down a 25 foot embankment into the river. They stood in about two feet of water. Shocked from his momentary stupor the smelter foreman ran to the office and notified the telephone operator, setting a massive rescue operation in motion.

'W.K. Besteder, of New Cumberland, Pa., a veteran conductor on the Pennsy, was riding in the next to last car. "It was a real happy car-ful. The kids were laughing and joking with each other and everybody was feeling fine. All of a sudden there was a bump and I was thrown from where I was standing in the aisle over to the right side."

'John V. Miller, 22, of Newark, Del., a recent bridegroom who is a graduate student at the University of Delaware, is a Los Angeles Dodgers fan. But he was headed for Philadelphia for the game with his brother Harry, 17, their cousin Robert Walter, 15, of Johnstown, and a friend Donald S. Harlacher, 16, of Dillsburg. Miller said they had boarded the train about 4.30 p.m. and were talking baseball. "We were wondering where we were when somebody noticed the steel mill and we knew we were in Steelton. It felt like we were switching tracks. It (the train) jumped and seemed to get more violent. The car rolled over and never seemed to stop."

'Within minutes after Wilbert raised the alarm, the air was split by sirens as police cars, ambulances and rescue units raced for the river. Pleasure boats on the river sped toward the wreck. On a nearby highway, 27 members of a U.S. Navy drill team, bound to their base at Bainbridge, Md., from a parade stopped and aided the rescue. One of them was Michael Pierce, of Buffalo, New York. Pierce and his shipmates found an eight-year-old boy pinned in the wreckage by one leg. The youngster was Robert Kennedy, a third grader en route to the ball game with his father, Robert, 30, from their home in Harrisburg. When the sailors reached Bobby he was crying and lying atop his father. The elder Kennedy was dead. Pierce called Bobby "the bravest little kid I ever saw." To keep the boy's mind off what was happening, Pierce said he talked about Mickey Mantle. Bobby asked him baseball questions until they managed to free him. Working in teams, Pierce, and the other sailors, mostly in their teens, placed the dead and injured on litters.

Holding the stretchers with one hand and guide ropes with the other, they struggled up the steep, sloping embankment. Then returned for others.

'On the river, state police worked from boats. Other rescuers waded in the river pulling victims through open windows and doors. On the embankment above the river, Roman Catholic priests moved solemnly imparting a final absolution over the bodies and trying to comfort the injured. Above the hundreds who gathered to give aid, a helicopter from nearby Olmsted Air Force base hovered. Skindivers with acetylene torches cut through the twisted steel to free those pinned inside the cars.

'One of the first to reach the scene was Jack Murray, security officer at Harrisburg Hospital. "It was like a bad dream. There were bodies in the river and parts of bodies, some decapitated," he said. "Most of the people who may have had an arm outside the window or were falling through the window when the cars hit – they are all dead."

'The injured were rushed to Harrisburg Polyclinic hospital in nearby Harrisburg. An emergency plan was put in effect and Harrisburg Hospital quickly added 35 cots to the 24 beds it had available. Both hospitals sent out urgent calls for help and it came fast. Some doctors rushed to the hospital in sports clothes, one wore Bermuda shorts. A woman administrative employee showed up in a bathing suit. Harrisburg Hospital estimated it had 60 doctors on hand in a short time. Among them was C.P. Fowler, a 71-year-old retired physician who volunteered. Hospital officials said they had called for 15 blood donors and within 20 minutes had offers from 22 volunteers. All means of transportation were used to rush the injured to the hospitals. One victim was brought by an Air Force helicopter pilot who set his "chopper" down on a main thoroughfare in front of the hospital while police stopped traffic.'

Engineer J. F. Shue stated afterwards that he had been travelling at between 65 and 70 miles an hour at the time of the derailment, below the maximum permitted speed of 75 miles an hour at that point.

Investigations into the cause of the derailment were opened by the railroad company, the Pennsylvania Public Utilities Commission (PUC) and the Interstate Commerce Commission (ICC), while soon after the accident railroad officials stated that it had apparently resulted from the track being out of alignment. [2]

Railroad officials, testifying before the Public Utilities Commission, blamed the derailment on poor judgement on the part of Crew Foreman R.D. Fritz. Pennsylvania Railroad Assistant Chief Engineer of Maintenance of Way Ernest Schultz blamed Schultz for not ensuring that track anchors had been replaced at the end of the day's work on Saturday. Schultz told the ICC that he had left the anchors because he had not wished to authorise overtime, but he believed that the track was safe for the running of trains over the weekend.

James Newell, the vice president of the railroad company, blamed the derailment on "lateral (sidewise) deflection of the track due to the fact that there were not enough rail anchors, and abrupt changes in the temperature," adding that, "We hold the supervision responsible for this and the lack of good judgement" (on the part of Schultz).

ICC investigator Earl Delaney found various other problems with the track such as insufficient ballast, improperly spaced rails, and loose tie plates, in addition to the

[2] *Lebanon Daily News*, Lebanon, Pennsylvania, 30th July 1962

missing rail anchors, and in conflict with the PUC the ICC blamed a general failure of maintenance for the derailment. [3] [4]

Poland suffered a major disaster on Tuesday 9th October 1962. Two international expresses were involved in the accident. One was an express from Sofia, Bulgaria, via Belgrade, Yugoslavia, Budapest, Romania, and Zebrzydowice, Poland, to Warsaw and the other was the *Chopin* express from Warsaw to Wien, Austria, with through coaches between Moscow, USSR, and Roma, Italy. At about 21:00, and only a few minutes after leaving the station at Piotrków Trybunalski, Poland, the Sofia express became derailed. Its last three carriages detached from the remainder of the train and ended up foul of the opposite running line, while several other coaches also left the rails. The *Chopin* express was close at hand, running in the opposite direction. Railwaymen attempted to warn its driver, Wieslaw Kolarczyk, but it was foggy and he only saw their hand signals in time to reduce speed from 105 to 90 kilometres an hour.

Sixteen coaches were derailed in all and the casualties in the derailment and collision were heavy. It was officially stated that 34 passengers had died and 67 were injured, though the true figures are believed to have been higher. Some, possibly the majority, of the fatalities were in a Hungarian sleeping car at the front of the *Chopin* express. The rescue operation was not well coordinated, for example it was an hour before ambulances from Piotrków, only 7 kilometres away, reached the scene. The injured were taken to hospitals in Piotrków, Tomaszów Mazowiecki, Rawa Mazowiecka, Radomsko and Lódz. According to contemporary local newspapers 60 injured were admitted to Piotrków hospital alone. A London newspaper added that three English passengers were on the Sofia train. They were Mrs Edith Ann Crowhurst, a London businesswoman dealing in scientific instruments, Mr Charles Wragg, of Crawley, Sussex; and Mr Mike Twomey, of West Drayton, Middlesex. Mrs Crowhurst was slightly hurt, the two men were unhurt. Mr Wragg later told a reporter, "We had just ordered some food and had returned to our coach when we seemed to become airborne. We grabbed anything there was to hang on to." [5]

After a three-day investigation the initial derailment was attributed to a broken rail. Made, it was stated, of unsuitable steel, the rail had broken in several places over a length of five metres.[6] The collision that followed was, of course, unavoidable.

Half of a 14-coach passenger train became derailed on the night of Tuesday 26th February 1963 in what was described as 'rugged bush country' in north-western Argentina, probably between Apóstoles and Pindapoy stations on the Buenos Aires – Posadas main line of the Ferrocarril General Urquiza division of the Ferrocarriles Argentinos. Seven people died and 20 were injured. The derailment was ascribed to subsidence of the rain-softened track bed. [7]

---

[3] Ibid, 12th September 1962

[4] *The Era*, Bradford, Pennsylvania, United States, 12th September 1962

[5] *The Times*, London, England, 11th October 1962

[6] Wikipedia – Railway disaster near Piotrków Trybunalski

[7] *Corpus Christi Times*, Corpus Christi, Texas, United States, 27th February 1963

A major catastrophe occurred in Japan on Saturday 9th November 1963. It began with the derailment of the 43rd waggon of a long freight train travelling on the Down freight line between Tsurumi and Shin-Koyashu stations on the Tōkaidō main line of Japanese National Railways. The derailed waggon and two following ones fouled the Up main line in front of a 12-coach electric multiple unit (EMU) passenger train bound for Tokyo, while another 12-coach EMU passenger train from Tokyo to Kurihama was passing on the Down passenger line. The three front cars of the Tokyo-bound train were derailed in turn and struck the fifth and sixth cars of the Kurihama-bound train. No fewer than 162 people died in the derailment and double collision and another 120 were injured. [8]

A reporter for an American newspaper wrote, 'Railway authorities said that 142 people were killed – one of them American – and that at least 173 persons were injured, many seriously. Scores of other victims were feared pinned in the mass of twisted steel scattered along the Tokyo – Yokohoma main line. The dead American was identified as William Scott, 29, a student at Tokyo's International Christian University. A friend of Scott's in Tokyo said the dead student was from Colorado Springs, Colo.

'Rescue workers operating by powerful floodlights were dismantling the two shattered trains with welding torches, spurred on by screams of "Taskette, Taskette" (Help, Help) and "Itai," the Japanese word for pain, coming from the wreckage. A sickening parade of bodies borne on stretchers still was emerging from the nightmare of the wreckage hours after the triple crash on a six line speedway where trains zip through at 70 to 100 miles an hour, sometimes on a one-train-a-minute schedule.

'It was the second major disaster to hit Japan in one day. An explosion deep in a coal mine in Omuta, on the southernmost island of Kyushu, killed at least 171 miners and trapped an estimated 200 others who were feared dead in clouds of poisonous gas.

'It was Japan's second major rail disaster in 19 months. In May 1962, a train crash at Mikawashima station, near Tokyo, snuffed out 160 lives.

'Railway authorities were unable to explain the Saturday night triple crash. They said the freight train was derailed along the six-mile speedway from unknown causes and denied previous reports that it hit a dump truck. The light freight cars scattered across the roadbed and a Tokyo-bound train smashed into the wreckage. It caromed off the rails and across the other tracks and plunged into the fourth and fifth cars of a Yokosuka-bound passenger train. The accident took place near the railway station at Tsurumi City, about ten miles from Yokohoma. Two of the six parallel sets of tracks are reserved for freight and two for Japanese national railway trains plying between Tokyo and Yokohoma. The other two are used by a private suburban line.' [9]

Another newspaper report, clearly written slightly earlier, added, 'The line is heavily travelled by U.S. naval men in Japan because it connects Tokyo with the big navy base in Yokohoma, but rescue workers operating in drenching rain punctuated by flashes of lightning said no Americans were found in the wreckage.' [10]

With parliamentary elections due in less than a fortnight Socialist and Democratic Socialist politicians used the railway and colliery disasters to attack the Conservative

---

[8] Webpage – Japanese Railway Safety and the Technology of the Day by Masao Saito

[9] *The Daily Plainsman*, Huron, South Dakota, United States, 10th November 1963

[10] *Grand Prairie News Texan*, Grande Prairie, Texas, United States, 10th November 1963

government of Premier Hayato Ikeda, alleging that the disasters had occurred due to the government's one-sided policy of protecting the monopolistic capitalist classes at the expense of workers' lives. Meanwhile the president of Japanese National Railways, Reisuke Ishida, stated that he would accept full responsibility for the Tsurumi disaster and tendered his resignation. [11]

An investigation was immediately launched into the cause of the initial derailment but no answers were quickly forthcoming. The railway company stated that the speed of the freight train (60 kilometres an hour) had not been excessive, while no problems had been detected by investigators with the flanges, axles, springs, couplings, or air brakes of the waggon, or with the permanent way. It took five years of research by the Railway Technology Research Institute, running similar rolling stock on a test track, before it was finally able to state that the derailment had resulted from a combination of rail cross-section wear and track geometry, with waggon wheelset design and tyre profile. It was urged that traditional methods of track inspection were inadequate to cope with the increased train speeds and frequencies that had followed electrification of the line. Following the accident new track inspection methods using vehicles employing dynamic inspection methods and data collection were introduced and the regulations relating to rolling stock, track, signalling, and operation were updated. [12]

Passenger train P 1221 from Potsdam to Berlin Schönefeld was travelling at 90 kilometres an hour at Waβmannsdorf, East Germany, at 13:00 on Thursday 24th June 1965, when its two drivers felt it sway in an abnormal manner and applied the brakes in emergency. The rear six coaches of the train were derailed and the last four became detached from the rest of the train. Nine passengers lost their lives, 10 were seriously hurt and a number were slightly hurt. The derailment was a result of track defects. [13]

The Franco-Ethiopian Railway Company announced that a passenger train bound for Djibouti in French Somaliland had been derailed at a river crossing 40 kilometres east of Diredawa in eastern Ethiopia on Friday 1st April 1966. Heavy rains had washed away the track. Nine passengers died and 30 were injured. The accident was said to have been the worst in the 40-year history of the company. [14]

Five carriages of a passenger train were derailed near Dzialdowo, Poland, about 160 kilometres north of Warsaw, on Monday 3rd July 1967. Seven people died and 22 more were seriously injured. The accident was variously blamed on a broken rail and distortion of the track due to the heat. [15] [16]

Ten years after the disaster at St John's, Lewisham, and only about a mile away, the

[11] *The Hillsdale Daily News*, Hillsdale, Michigan, United States, 11th November 1963
[12] Webpage – Japanese Railway Safety and the Technology of the Day by Masao Saito
[13] *Katastrophen der deutschen Bahnen Teil II* by H.J. Ritzau, pg. 296
[14] *Independent*, Long Beach, California, United States, 2nd April 1966.
[15] *Lubbock Avalanche-Journal*, Lubbock, Texas, United States, 4th July 1967
[16] *The Des Moines Register*, Des Moines, Iowa, United States, 4th July 1967

most serious accident since the 1957 collision occurred on British Railways, Southern Region, on the evening of Guy Fawkes Day, Sunday 5th November 1967, at Hither Green, England. The 19:43 Hastings to London Charing Cross Up express consisted of two six-coach diesel electric sets, in charge of Goods Guard J. Gray with Motorman D.S. Purves. It was travelling under clear signals at about 70 miles an hour when it was derailed by a broken rail while running on the Up Fast line at 21:16. The rail broke beneath the leading wheels of the third coach of the train, which continued in a derailed state for a quarter mile until a diamond crossing was reached, where that coach, the second coach and the fourth to twelfth coaches became completely derailed. The second to fifth coaches overturned and were very badly damaged. The train had been well-filled with passengers standing in the corridors and 49 of them lost their lives while 78 were injured, 27 seriously.

The emergency services were immediately advised and police were on the scene in five minutes the first ambulance arrived a minute later and the fire brigade two minutes after that. In all 33 ambulances and 28 fire engines attended the scene. The first casualty reached hospital at 21:34 and the last at 01:00 on Monday. The Women's Voluntary Service, The Salvation Army, other organisations, and local residents all provided assistance.

All lines were blocked by the accident, the Fast Lines by the derailed train and the Slow Lines for the clearance work. Much disruption was thereby caused, with trains having to be cancelled, terminated short of their normal destinations, or diverted. The Slow Lines were reopened for the Tuesday morning rush hour, then closed again, but reopened on the Tuesday afternoon; the Fast Lines were reopened on the Wednesday morning.

'"My God, we're crashing," a woman passenger shouted as the railway carriage began jolting and swaying. In the same compartment was Miss Shirley Ward, aged 21, a secretary, of North Foreland, Broadstairs, Kent. "The lights went out before we turned over," she said. "We felt ourselves travelling along on the carriage's side. Everyone was clutching on to one another. There were some terrible screams." Miss Ward was uninjured.

'Another passenger, Mr David Prince, aged 22, who was with his fiancée in the third from last coach said: "It ran off the rails and bumped along the sleepers. There were flames and sparks outside the window. Then we ground to a halt. The last carriages were still upright but the third was tilted over at an angle of 35 degrees. The rest was shambles. There were at least six coaches on their side."

'In one of the worst-hit coaches was Mr Ray Moore, aged 19, a medical student from Hastings. He said, "Stones came chipping up from the track as we went along this stretch which is usually a pretty fast one. Then it rocked and swayed and went over on its side. At least three people in my compartment were killed."

'Mr Phillip Gorin, a Lewisham councillor who works at the Hither Green Continental Freight Depot, said: "I have never seen anything like it. It is horrible."

'Mrs Joan Mowbray, of Sandhurst Road, Catford, was walking under the railway bridge as the train approached: "What I saw was a carriage that had been ripped open and people were lying on the grass," she said. "They had been thrown out of the train. Then I heard people crying and screaming." Mrs Mowbray was taken to hospital suffering from shock.

'Mr Ronald Williams, of Hither Green Lane, about 30 yards from where the crash happened, said he and his wife put the noise down to fireworks. "We went outside to have a look and then we saw the train all lying on its side. We could hear people shouting and I did what I could to help. I suppose I must have helped carry at least a dozen bodies from the coaches."

'Mr Ronald Humphrey, aged 42, a G.P.O. worker, also heard a bang and tearing noises. Outside, he joined in the rescue: "We started giving first aid to people who had crawled from the wreckage and gave them cups of tea. We then organized the neighbours into providing help and called all the major emergency services."

'A young doctor who volunteered his services said: "I climbed down into a shattered compartment. It was chaotic, carnage. I felt completely helpless. The rescuers, the firemen and ambulance men were doing the really valuable work. There was one woman who was trapped. She had been standing in the corridor when the coach turned over and the compartment was on top of her. They got her free and she seemed to be completely unhurt. It was quite miraculous."'

Motorman Purves told the Ministry of Transport inquiry into the disaster that driving trains on the Up Fast line it was the normal practice to apply the Westinghouse brakes between Grove Park and Hither Green, "near the beginning of the concrete sleepers." As he had been about to do so he felt a slight drag on the train. Soon afterwards the drag had become severe, there had been much snatching, then there was a severe snatch followed by a "terrific bang." The brakes had come on and his train had stopped. He realised that his train had become divided but not that it had been derailed, until he looked back. He had then protected the Down lines and contacted the Hither Green signalman.

Guard Gray stated that the run had been normal until after Grove Park he had seen a lot of debris flying past the windows of the sixth coach. Realising that something serious was happening he had lain down on the floor and advised the passengers in the coach to do likewise. There had been no time to apply the brakes. After the accident he had walked back along the line accompanied by two police officers and they found the broken rail.

Several passengers also gave evidence at the inquiry. Mr G. Deans, a company director, had ridden the line for ten years. He said that he had joined the train at Battle. The run to Hither Green had been normal but "on that line coaches always swayed a lot and I have always had an uneasy feeling that they were going too fast." Near Hither Green the coach had begun to sway more than usual and there had been a sound as if the train was running over splintered glass. The rocking then became even more severe, then there was a crash, the lights had gone out and the coach overturned on its right-hand side. He escaped unhurt. He repeated that it was usual for the trains to "sway and rock all over the place," and that he had considered that the train was going too fast, though no faster than usual.

Mr J. Danks, an advertising executive and a regular weekend traveller, had noticed nothing unusual before the derailment, though he agreed that the train had been rocking.

Mr J.J. Brennan, a clerk, was another regular passenger who had joined the train at Tunbridge Wells. He related that the coaches of the train always swayed considerably but about two minutes and two miles before the derailment the swaying had appeared to get worse and the train seemed to be going faster than usual. He had been in a toilet just before the accident and had difficulty opening its sliding door to return to his seat.

Mr J.C. Reitchel, the Vice-Chairman of Uckfield Rural District Council, was not a passenger on the train but said that he had travelled several times a week over the line between Wadhurst and London for 30 years. He and other passengers he had spoken to all thought that the trains went too fast. He had noted especially bad points for swaying between Grove Park and Hither Green on the Up Fast line and approaching Sevenoaks on the Down line.

District Engineer D.L. Carden told the inquiry that mechanised track maintenance had been introduced over the section involved in the derailment in June 1966 using two Plassermatic tamping machines and one lining machine. An ultrasonic ganger examined the rail ends of all rails on running lines once a year, with more frequent inspections of areas causing concern. A patrolman inspected the track every Monday, Wednesday and Friday and a sub-inspector examined the track once a fortnight. He (Mr Carden) walked most of the track once a year, as did the chief permanent way inspector at a different time.

Responding to the discovery by the inspecting officer after the disaster of numerous track defects in the vicinity of the derailment Mr Carden replied, "Well, sir, I have also inspected the tracks since the derailment, from the point of derailment towards Grove Park, and the tracks have been examined by my Chief Inspector from Orpington onwards since the derailment. From what I personally saw at the site and in the stretch between the point of derailment and Grove Park, I must agree that in some places the maintenance was poor. This was particularly applicable to the points and crossings as far as fastenings were concerned. The top in so far as the rails themselves was reasonably good but I must admit the fastenings and bolts, 'T' bolts etc, quite a number of which were loose, were even missing or pulled out. I wouldn't regard the general standard as being very good at all; from what I saw it should have been higher." He added that the permanent way staff should have taken "some further action from that they did." The maximum speed at the site of the derailment was reduced from 90 to 60 miles an hour immediately after the accident. It appeared that the Matisa track-recording car had never been run over the Up Fast line between Grove Park and Hither Green, having twice been diverted onto the Up Slow line at Grove Park.

Asked if he considered the increased train speeds, which had come with the new timetable in July, had had a more punishing effect on the track Mr Carden replied that he did not think they had. Having initially said that he thought the number of permanent way staff was sufficient, Mr Carden later stated that he thought the quantity of staff was slightly deficient.

Mr Carden went on to give a very detailed account of his examination of the broken rail. "I first examined as closely as I could the fractured surface of the parent rail. This was quite evident to me to be a very fresh break, almost to the bolt holes, with no signs of metal flaws or fatigue cracking. It was also evident from the shiny batter of the running face of the fracture that already a number of wheels had pounded over the gap in the rail surface. I observed the complete fracture in the fishplate on the four foot side but since this appeared only as a crack, I could not tell if this was a fresh break at the time. The concrete sleeper at the running on end under the 6 ft rail seating was completely shattered with the rubber pad, punctured in the centre, still in its relative position. The sleeper beds were full of ballast and there were slight signs of wetness. What I mean by this, bearing in mind that it had been raining for some time, there were signs of wetness on the vertical faces of the sleepers, but not on the top faces. There were no signs of slurry on the top but the conductor rail in the vicinity certainly showed some signs of splashing mud. I thought, although I could not be sure, that this mud had spurted from the sleepers in the vicinity of the joint. I then checked if the sleepers were in any way 'hanging' by sounding the beds. Of course I could not do this on the concrete sleeper adjacent to the 6 ft rail because this was shattered, but I sounded those joint sleepers by hammering them with my foot, my heel, which I often do, and this I did also to two or three sleepers on either side of those joint sleepers (including the timber sleeper), and every one of those seemed to be to sound and solid. Then with the help of my own lamp I eyed the line and level of both rails from the approach direction, from a point beyond 140 points, and I was quite satisfied with the line and top. I also checked the fish bolts at that time, too, by kicking each nut with my toe, which had a steel cap on it and they were all solid, on both joints, very tight. Subsequently I personally arranged for that joint to be removed by cutting, complete, without touching the fish bolts, and I personally undid these bolts myself in the dry. Two officers from the Research Department, Derby, were present at the time and they both can confirm that each of those bolts was tight."

Chief Permanent Way Inspector T. Davenport stated that he was responsible for 700 miles of track and so could not possibly inspect all of it carefully, thus he had concentrated on what he regarded as the worst areas, which did not include the vicinity of Hither Green. He had not noticed that the fracture had occurred at a point where the rail was supported by one wooden and one concrete sleeper. He was aware that a broken rail and five defective rails had been found in the Up Fast line between Elmstead Woods and Grove Park on 23rd October by the ultrasonic ganger, which he attributed to the track being old; it was due for renewal in 1968. He had not thought that the Up Fast line between Grove Park and Hither Green needed a special inspection after the October discovery of problems a little further south. The top of the Up Fast line had appeared satisfactory when viewed during a cab ride along the Down Fast line a month earlier, while he had been busy dealing with track problems elsewhere. He did not think that the line through Hither Green was suitable for a speed of 90 miles an hour.

Permanent Way Inspector A.L.J. Wright testified that continuous welded rail (c.w.r.) had been laid in the Hither Green area in February 1967 with two short closure rails in the vicinity of points No. 140. The end concrete sleeper at the joint between the 30 ft and 34 ft closure rails (the location of the broken rail) had later been found to be cracked and was replaced by a wooden sleeper in at the end of June 1967. After the derailment he had found "pumping" at the site, indicating a low joint. He stated that he had written to the District Engineer in mid-July complaining of difficulties in keeping up with the track maintenance due to lack of staff. He said that he had realised there were more trains with the new July timetable but had not immediately been aware that they were running at 90 miles an hour.

Permanent Way Sub-Inspector P.G. Dawes stated that the track in the area of the broken rail had needed more than the usual amount of maintenance. He attributed this to the age of the track but pressed by the inspecting officer, he agreed that he had not been satisfied with the formation of the track at any point between Grove Park and Hither Green. Further questioned as to his supervision of Patrolman H.I. Suttle, Dawes said that the patrolman was conscientious and took an interest in his job and he was surprised that the patrolman had failed to spot the problems at the point where the rail broke. He suggested that the fittings might have deteriorated between the patrolman's inspection on Friday and the derailment on Sunday.

Patrolman Suttle testified that he had examined the track at the point of the derailment between 09:30 and 10:00 on Friday 3rd. He had not noticed a fatigue crack in the inner fishplate where the rail had later broken. He said that he had noticed some joints on the countryside (Grove Park) side of points No. 140 were pumping and would have verbally informed Sub-Inspector Dawes had he met him on the track. He added that there had been so many defects due to lack of maintenance and shortage of materials that it would have needed a foolscap sheet every day to report them. After the accident, though, he had reported all defects.

Mobile Ganger P.J. Cartland spoke of changing the cracked concrete sleeper at the point of derailment with a wooden one on 30th June. The wooden replacement sleeper was shallower than the concrete one, but there had been ample clean and dry ballast to pack beneath it. On 1st November he had also replaced a nearby concrete sleeper with a wooden one. He had not noticed any "pumping" at the broken rail joint.

Ultrasound Ganger D.V. Gateland spoke of finding two defective rails in the Up Fast between Grove Park and Hither Green on 20th June, neither near the derailment site. He was also the person who had found the five defective rails between Elmstead Woods and Grove Park in late October. All had been at the running off ends of joints, as, he added, such defects almost invariably were.

Chief Civil Engineer A.H. Cantrell stated that he had been unaware of any location where there was one concrete and one wooden sleeper at a joint. He did not like this arrangement and had issued instructions that the concrete sleepers at such locations were to be at once replaced with wooden ones. He said that the number of rail fractures on the Southern Region was on the increase and was the subject of special investigation.

Asked about the effect of higher train speeds on the track Engineer Cantrell replied that he had considered that the increased speeds would have had some effect on the track, mostly on points and crossings, but it would have taken some time to judge the extent of the damage caused.

Mr Carden had represented to him that they needed more track staff and he was working to increase the number of staff and also the amount of ultrasonic track testing. He was under pressure, he said, from higher up, to save money, but not at the expense of safety.

Chief Civil Engineer A. Paterson of the British Railways Board explained how the number of permanent way staff needed was calculated, taking into account various factors. He then outlined progress made in the manufacture of rails to prevent fracturing and in the development of ultrasonic testing. He pointed out that the incidence of broken rails on British Railways was lower than that of France, Belgium, or Italy, to name three European countries, presumably chosen more or less at random.

In view of passengers' statements of trains running too fast Divisional Movements Manager J.F. Rogers was asked what steps had been taken to prevent trains running at excessive speed. He stated that various speed tests were carried out, and in 1967 of 2,259 trains checked only 121 had been found to have been travelling too fast, the greater number of which had only been going slightly too fast. None of 19 trains tested between Chislehurst and Hither Green on 16th August had exceeded the speed limit.

Comparative tests of train riding were carried out on the Eastern and Southern Regions, which showed that the riding of trains on the Eastern Region lines was uniformly superior to that on the Southern Region, though the riding of trains on the latter region was acceptable from a safety point of view.

The findings of the Railways Research Department at Derby, who examined the broken rail, were detailed. The department found that:

The chemical composition and mechanical properties of the steel in the rail conformed to B.S. 11 of 1959, to which the rail was ordered, and it was adequately free of harmful trace elements.

The failure was initiated by a fatigue crack, which had started at the upper surface of the first bolt hole and extended upwards at an angle of about 45 degrees for ⅛ in. The crack had formed under high stress, multiple initiation fatigue conditions. It had multiple initiation points, denoting that initiation had occurred under relatively large cyclic stresses compared to those expected from a well-supported joint. It was unlikely that this crack would have been detected by the type of ultrasonic testing equipment in use.

The fatigue crack had initiated a brittle fracture on approximately a 45-degree plane, which extended upwards to and through the rail head and downwards to the rail end and had caused the detachment of a portion of the rail including 5½ ins. of the rail head. The brittle fracture had occurred, probably, in a single loading event.

There was a fatigue crack extending from the top edge of the inside fishplate opposite the end of the running-on rail. The remainder of the fracture had propagated from that crack downwards throughout the section of the fishplate by brittle fracture.

Examination of the fishing surfaces of the rails and the fishplates indicated that the joint had been "working" immediately prior to failure.

There was adequate evidence from examination of the ballast that pumping conditions had existed at the joint, especially under the rail that had failed, but these were not considered as severe.

It was deduced that differential deformation of the ballast had occurred as a result of the replacement of the running-off concrete sleeper by a shallower timber sleeper, and this was considered to be the principal defect in the joint support conditions.

Concluding his report the inspecting officer, Colonel D. McMullen, stated that there could be no doubt that the initial derailment had resulted from the fracture of a rail at a joint as a train had passed over it. The rail that had broken was one of two closure rails between a length of c.w.r. and a set of trailing points, the fracture having occurred at the "running-on" end of the right-hand rail. The concrete "running-off" joint sleeper had been replaced by a wooden sleeper. The unsatisfactory support condition of the joint had led to its excessive "working."

Explaining the cause of the broken rail in detail Colonel McMullen wrote, 'The evidence that a joint had been "working" excessively is incontrovertible... and the abrasion of the original "running-off" concrete sleeper rail seat where the pad was missing is evidence that it must have been "working" for some considerable time before the sleeper had cracked... There had been a depth of only 2-4 in. of clean ballast under both the concrete sleepers and both the ballast beds would have been well compacted. When the "running-off" concrete sleeper was replaced by a wooden one, ballast was added to fill the gap left by the deeper concrete sleeper. This newly formed bed was 4-6 in. or more deep, and it became compacted under load. When not under load the stiffness of the rail and the joint would have caused the wood sleeper to "hang" slightly above the ballast bed and give the appearance of a good "top" to the track. On account of the absence of the pad, the concrete sleeper had not given its full support to the rail, which had also not been properly supported by the wood sleeper which replaced the concrete one. The rail had therefore been acting as a continuous girder spanning the gap between the second sleeper from the joint at the "running-off" end of the rail and the well consolidated concrete joint sleeper at the "running-on" end, and the joint was well towards the latter end; it would thus have "worked" even though the fish bolts were tight, and I accept the evidence of Mr. Carden, the District Engineer, that they were so. The "running-on" rail end would have been subjected to heavy impact which would have caused stresses much higher than would have obtained in an evenly packed joint. The high stresses were applied very frequently under the wheels of the large number of trains that run on that line and initiated the fatigue crack in the rail end. The "working" of the joint also induced excessive stresses in the upper surface of the fishplate and either initiated the fatigue crack or, if it already existed, accelerated its propagation. These fatigue cracks led to the brittle fractures in the rail and fishplate.'

It was not clear if the rubber pad beneath the concrete sleeper had been missing when the track was laid in February 1967 or had worked out later. If it had never been there Patrolman Suttle would probably not have noticed its absence. Had it worked out later he should have noticed it but he could be given the benefit of the doubt on this. Nor did the inspecting officer blame P.W. Inspector Wright for substituting a wooden for a concrete sleeper, but the cracking of the sleeper should have alerted the inspector to the fact that the pad was missing. Nor would the "working" of the joint under load, even if noticed by Sub-Inspector Dawes or Patrolman Suttle, have especially alarmed them. Colonel McMullen did not therefore consider either Wright, Dawes, or Suttle responsible for the unsatisfactory condition of the joint, nor did he blame Mobile Ganger Gartland, who had probably packed the new wooden sleeper well when it was installed, or Ultrasonic Ganger Gateland, whose equipment would not have detected the fatigue crack in June.

While not blaming Patrolman Suttle for failing to notice the missing pad or the cracked fishplate, the inspecting officer noted that Suttle had missed a large number of other track defects in the weeks leading up to the accident. He had clearly not been carrying out his duties properly, had not been fully alive to his responsibilities and had not been adequately supervised by Sub-Inspector Dawes. The inspecting officer was not impressed by the attitude of these two men that although the track was not good, its standard was adequate and its maintenance acceptable. Nor was Colonel McMullen happy with the comment made by Chief P.W. Inspector Davenport that the track "could have been better, there was room for improvement" or with the view of District Engineer Carden that the plain track was not too bad, though some of the points and crossings needed attention. The inspecting officer remarked, 'In my view. the attitude of the permanent way staff on the ground towards the standard of track maintenance generally stemmed from that of Mr. Davenport and Mr. Carden, and indeed from Mr. Cantrell, all of whom were, I think, prepared to accept too low a standard of maintenance.' Given that Chief P.W. Inspector Davenport did not consider that the track in the Grove Park – Hither Green area was fit for train speeds of 90 it was, wrote the inspecting officer, highly irresponsible of him not to have stated his views to the District Engineer, while he and Mr Carden should have kept a closer eye on the condition of the track. Chief Civil Engineer Cantrell also came into criticism in this respect.

It was surprising that District Engineer Carden had not become aware of the rail fracture discovered on 23rd October until after the accident and that no further action had been taken regarding the several problems revealed on that day.

The inspecting officer stated that he did not consider Messrs Davenport, Carden and Cantrell responsible for the unsatisfactory condition of the rail joint though he did hold them responsible for permitting the increased speed of 90 miles an hour through the area given the unsatisfactory state of track maintenance.

Turning to remarks and recommendations the inspecting officer noted that most rail fractures occurred at joints. It was unfortunate that the Southern Region had continued to use closure rails on c.w.r. sections of track, a practice largely abandoned by the other regions. It was doubtful if the rail fracture could have been foreseen but the general standard of track maintenance at Hither Green had been poor. Immediate action had been taken after the accident by the Chief Civil Engineer to deal with the three factors directly leading to the broken rail – the inadequacy of ballast under the joint, the lack of a rubber pad under a concrete sleeper and the replacement of a concrete sleeper with a wooden one.

Further measures taken included an increase in track patrolling, an increase in the number of permanent way staff and the introduction of improved pulse-echo rail-testing equipment. Improved types of rails, more resistant to fracture, were also being introduced.

Research was in progress regarding the heavy battering rail ends received from electric multiple-unit trains with axle-hung traction motors and smaller wheels than steam locomotives, especially if the condition of the track was not good and with increased train speeds. The subject was a complex one. The short-term solution must be the replacement of jointed track with c.w.r., while a longer-term solution would be the redesign of new rolling stock with traction motors suspended from the sprung portion of bogies, as was done almost universally on the Continent.

Express train D136 from Leipzig to Düsseldorf was derailed between Leiferde and Meinersen, West Germany, on Monday 28th July 1969. Just after the train passed over a bridge its driver felt a jerk, then the air brake pressure began to drop and he realised that something must be wrong. Looking back he then saw a cloud of dust towards the rear of his train, applied his brake fully and the train came to a stand near the Meinersen Einfahrhauptsignal (Home signal).

The rear coach of the train collided with a bridge abutment, had its bogies torn off and then overturned onto its left side. The two coaches next in front were also derailed. Five passengers died and 23 were injured.

A train conductor related, "When I was in the rear portion of the penultimate coach, I suddenly noticed repeated powerful jerks. The coach lurched and leaned to the right and then to the left five or six times." A second conductor further up the train noticed nothing wrong before the train stopped but several passengers reported that they had.

Because the cause of the derailment was not immediately obvious a careful and thorough examination of the wreckage and the derailment site was made by an expert. His conclusion was that insufficient expansion joints, exacerbated by the disturbance of the track due to tamping operations, had caused the track to warp on a hot day. The summary of his report read:

'As it must be assumed that the expansion gaps had not been corrected prior to the tamping operation, it cannot be proved that the tamping itself caused the fault. If the fault had arisen in a track which had not been freshly tamped, then insufficient expansion joints would have been put down as the cause of the derailment.

'The air temperature and hence the rail temperature on the day of the accident had hardly been any higher than those of the previous week – a summary of the temperatures from an official agency was not available to me, but it appears that it will be presented by the Federal Railway Directorate with the results of their investigation – and if there was no distortion of the track which had not been tamped during the previous week, it must be objectively deduced on the basis of the on the spot examination and an overview of the "Joint Gap Measurements," that insufficient expansion gaps at least promoted the track warping.

'The foreman of the Meinersen maintenance department is responsible for measuring the expansion gaps and checking the results, which should also have been presented to the directors of the Hannover 2 works office, but this has apparently not been done since 1966.' [17]

Express train D 47, which ran non-stop between Hannover and Hamburg, was derailed at No. 3 points at Celle station, West Germany, at 12:02 on Friday 5th June 1970, the rear seven carriages of the train leaving the track. Five passengers died and another 41 were injured, six seriously. Damage was put at 900,000 Deutsche Mark.

During the investigation of the derailment the driver of train E 1573, which had passed the derailment site at 10:02 travelling at 135 kilometres an hour, stated he had noticed some swaying at that point, but nothing he considered unusual or dangerous.

There had been two drivers and a senior inspector on the engine of train D 47. The first driver stated that at the points in question he had seen the track ahead was distorted in a serpentine manner and exclaimed, "That looks bad!" Neither the second driver nor the inspector would admit to hearing this comment or having noticed any problem with the track, but all three agreed that at this point the engine lurched violently, so that some timetables had fallen out of their holders and other loose objects had been thrown about. The engine had then steadied itself again. As to exactly where the lurch had occurred the first driver thought it was at No. 3 points, the second driver was unsure exactly where and the inspector named a different point just before a bridge.

The sequence of the accident was reconstructed by the examining officer. The engine and leading five carriages of the train had managed to negotiate the defective spot in the track and the trailing bogie of the sixth carriage had been the first to leave the rails. Marks on the track showed that as the train continued its journey, further bogies had been forced off the line, whereby the track had been so damaged that the derailment of the remainder of the rear carriages had followed. The train had then become divided behind the eighth coach and the last three coaches had come to a stand, after the brakes had come on automatically by the separation of the brake pipes. The sixth, seventh and eighth coaches had continued further along the line and the trailing bogie of the sixth coach had been further derailed at No. 34 points. The seventh and eighth coaches had then been thrown towards the opposite track, No. 35 points had pushed the bogies of the eighth

---

[17] *Katastrophen der deutschen Bahnen Teil I* by H.J. Ritzau, pgs. 86-101

coach to the right and its body, pulled by the seventh coach, had collided with a post supporting the overhead line equipment. The leading bogie of the sixth coach and both bogies of the fifth had been derailed on No. 37 points. The dead and injured passengers had been riding in the eighth coach.

On 6th July the central office of the railway in Minden reported:

'The Celle accident clearly resulted from an hitherto unknown type of track deformation…due to the conditions prevailing at the point of the derailment with a set of points on a curve the vehicles of the train were obviously subject to strong lateral jolts here, especially at high speed. The cause of the track having given away laterally can only be that the vehicles of the train involved had exerted greater lateral forces on the track than it was able to withstand and had moved sideways beneath the axles of the train. Existing small track defects had apparently increased under the movement of the fast running train which had begun to hunt (oscillate). This is confirmed by the fact that the oscillation of the train vehicles at a speed of 155 kilometres an hour, corresponded to the variations in the track levels. It is not known why this condition occurred and further detailed research appears necessary.'

The official in charge of the accident investigation was not convinced that on well-ballasted and freshly worked-on track the derailment could be due simply to the presence of two sets of points on a curve, uneven track surfaces and poor work execution and considered that the vehicles of the train must have been somehow also to blame, so he requested the assistance of a second expert, the two producing a detailed report in April 1971.

In part their conclusions read, 'The lack of clear signs of wheels climbing the rails proves that the cause of the derailment was not excessive horizontal forces, Y, but rather insufficient wheel loads Q. In addition to the above described variation in track levels, there was also spatial torsion, which further increased the load on the wheels' and, 'the form of the track shifting differs from the cause and form of track displacement and track warping. The track shifting occurred at the point where the vertical and horizontal forces were repeatedly applied, beginning at a weak point (an initial directional error, plus a weakening of the resistance to transverse and longitudinal displacement after overhauling work). The initial amplitude of the vibration, the wavelength of which was determined by the dynamic forces acting on the track, became progressively greater and a resonant-like build-up of vibrations occurred.

'It was no longer possible to determine whether the track had been entirely stable on the day of the accident…there had been a great variation in rail temperatures during the days preceding the derailment, sometimes as much as 28 degrees C. Investigations should be carried out to determine whether the stress in tracks and points is changed during maintenance work.'

A separate report by the railway added that the Kershaw type of track machine used in overhauling the line should not have been employed, because it had a motorised flywheel and no edge compactors, that improvements were being made to the Class 112 locomotives used on the train involved in the accident, and that the ultimate aim was to have a completely new design of locomotives and carriages which would impose less stress on the track. [18]

(It is hoped that the above account makes some sense – the author has only a limited understanding of both German and of the technicalities of track maintenance.)

Another serious derailment occurred in West Germany on Tuesday 18th May 1971 at Kellmünz. Prior to the day of the accident slight problems had been found with the track west of the station and the decision had been taken to deal with them. Thus on the morning of the accident a permanent way gang were working on the track, doing preparatory work prior to the use of a tamping machine. At 11:50 the gang stopped work for their dinner break. Eleven trains had passed the site done during the morning without any problem, as did train E 2156, travelling at 112 kilometres an hour, at 12:25. However, when express train D 1013 from Obertsdorf to Dortmund arrived at 12:53, disaster struck.

Train D 1013 consisted of a V class locomotive weighing 78 tonnes, four passenger coaches, a dining car, a passenger coach, a combined luggage van and passenger coach and three passenger coaches, ten vehicles in all. It was travelling at 130 kilometres an hour at the time of the accident and carrying 160 passengers. Just its rear two carriages left the rails, going first to the left and then to the right. The rear coach had its side torn away when it collided with a signal post and it was in this vehicle that the fatalities and most of the injuries occurred. In all, five passengers lost their lives and 46 were hurt.

The first warning the driver of the train had of the accident was a loss of speed. Assuming there was an engine problem he had shut it down, then the speed of the train suddenly dropped rapidly and he made an emergency brake application.

The conductor had been in the dining car as the train passed through Kellmünz station and had felt a severe shock. Looking out of the window and seeing a cloud of dust, he pulled the communication cord.

Another (assistant?) conductor in the combined luggage and passenger coach felt a slight and then a severe jolt, the second of which threw him to the side of the carriage, and he also applied the brakes, while a retired railwayman in the rear coach likewise attempted to do so but found that the brakes were already applied.

The conductor informed the fahrdienstleiter (signalman) at Kellmünz station of the accident by telephone at 12:56, advising him that ambulances were required and that all lines were blocked. At 14:30 the front part of the train continued the few kilometres to the next station at Altenstadt, where its driver, who was suffering from shock, was relieved. At 16:53 it continued at reduced speed to Ulm, as the conductor had noticed that the dining car was swaying badly, even at normal speed. The dining car was there examined and found to have suffered damage in the accident.

The accident was the subject of very detailed investigation of a highly technical nature. The final conclusion was that on a hot day the track had been distorted by the heat, this being the result of it having been disturbed to too great a degree by the track work, which should either have been postponed until a cooler day or alternatively a temporary speed restriction should have been imposed. In addition the foreman of the

[18] Ibid, pgs. 119-130

permanent way gang should have checked the rail temperature with a thermometer. Instead he had merely judged the temperature by touching the rails with his hand. He duly appeared in court charged with causing death and injury through negligence and endangering passenger safety, was found guilty and given a year's suspended prison sentence, plus ordered to pay court costs. It seems that, to some degree at any rate, he was made a scapegoat for the accident. [19]

A Kolobrzeg – Warszawa Glówna passenger train was derailed near Bydgoszcz, Poland, on the Ślesin – Zielonczyn line, at about 12:20 on Saturday 3rd June 1973, again apparently due to buckling of the rails during hot weather. The rear coach of the train overturned and was dragged a short way along the track by the Ty 51 Class engine hauling the train before it could be stopped. Twelve passengers died and another 26 were injured. [20]

Five unauthorised passengers reportedly died and three people were hurt in the derailment of a freight train near Encarnación de Diaz, Mexico, in Jalisco state, on Monday 8th January 1973. The accident reportedly resulted from inadequate protection of work on the track. On the same day a passenger train and a freight train collided head-on near La Piedad, killing three people, including both drivers, and injuring 20. In this accident the freight train driver had apparently left Paty station without authority. [21]

First reports stated that 22 passengers had died and about 100 had been injured as a result of the derailment of Rede Ferroviária Federal *Expresso Campista* passenger train (No. 2-415) at kilometre 121, on the Caixa curve, near Cesário Alvim, Brazil, just after 09:30 on Friday 2nd February 1973. By Sunday the number of known dead had been corrected to eight, two of whom remained unidentified because they carried no documents, being only poor rural people who did not carry identity cards. Also by Sunday only nine passengers were badly enough hurt to be still detained in hospital.

Train No. 2-415 had left Barão de Mauá at 05:25 for Cachoeiro de Itapemirim and thence Campos and was carrying 207 passengers at the time of the accident. It was travelling at over 100 kilometres an hour at the time of the derailment, which began with its rear coach leaving the rails, supposedly because it was the lightest vehicle of the train, dragging the six coaches in front of it off the track in turn, the train covering another 200 metres before it stopped. The derailed carriages reportedly fell into a lineside ravine. The locomotive and three leading carriages were not derailed, so Driver Sebastião Nogueira da Silva was subsequently able to uncouple them from the derailed vehicles and use them to evacuate the injured passengers. Even then, because of the isolated location of the accident, it was three hours before assistance reached the scene, access to which was also difficult. When firefighters from Niteróand and São Gonçalo did arrive they found that they were unable to properly search the wreckage pending the arrival of a breakdown train to lift the overturned carriages, and the crane only reached the scene at nightfall.

---

[19] Ibid, pgs. 139-156

[20] Wikipedia – Railway accidents in Poland

[21] *Wichita Daily Times*, Wichita, Texas, United States, 9th January 1973

The injured were taken to Darcy Vargas hospital in Rio Bonito. The Rio state Health Secretary arranged for an additional 30 doctors and 40 nurses to be sent to the hospital to cope with the influx of casualties, to which were added relatives who had gone there in search of their loved ones who had been on the train. The railway company meanwhile sent 200 staff to the site of the derailment to assist in the rescue work, remove the wreckage and clear the line. There were reports of fairly widespread looting, including, for example, one person who had taken soft drinks from the dining car and was selling them at the scene, until police appeared and arrested him.

There was some doubt as to the cause of the accident, about which Driver da Silva and Conductor José Vicente Sousa were shortly to be interviewed when the newspaper report of it was published, but it seems likely that it resulted from heat distortion of the track (Brazil being in the southern hemisphere, February was of course mid-summer). The railway company blamed the derailment on the failure of the rails to expand due to 'rusty screws,' presumably meaning those securing the fishplates. On the other hand, according to a newspaper report, the police raised the possibility of an act of sabotage, based on the discovery of a ticket on the body of passenger Jorcellino Armando Conveição. Quite why the finding of a ticket should lead to a suspicion of sabotage the newspaper did not explain.

A railway official told a newspaper that the number of deaths in the accident would have been halved had all the carriages of the train been of steel construction instead of some of them still being old wooden ones, adding that only the *Nocturno Campista* overnight train between Rio de Janeiro and Cachoeiro de Itapemirim had all-steel coaches. The same newspaper stated that the blockage of the line would cause little inconvenience as most people preferred to take the bus anyway, while most of the goods traffic was carried on lorries. [22] [23] [24]

[22] *Jornal do Brasil*, Rio de Janeiro, Brazil, 4th February 1973

[23] *O Estado de S.P.*, São Paulo, Brazil, 3rd February 1973

[24] https://www-estacoesferroviarias-com-br.translate.goog/efl_rj_litoral/cesario.htm?_x_tr_sch=http&_x_tr_sl=pt&_x_tr_tl=en&_x_tr_hl=en&_x_tr_pto=sc

CHAPTER 2

DERAILMENTS - BROKEN TYRES, BROKEN AXLES AND OTHER VEHICLE DEFECTS

A frequent cause of derailments of trains other than problems with the track was defects in the trains themselves, such as broken tyres, wheels, and axles. Such accidents were not as frequent by the late 20th century as in the early days of railways, but they still happened quite often and were to a fair degree unavoidable.

On Wednesday 1st February 1961 the Paris – Trieste direttissimo which was due to leave Milano Centrale at 13:18 had only covered about 20 kilometres when, at 13:48, it was derailed at a set of facing points less than two kilometres from Cassano d'Adda, Italy. The express consisted of a locomotive, a luggage van, four second-class carriages, two first-class carriages, a first- and second-class composite, a buffet car, and a mail van. Only the rear three coaches of the train left the rails. The composite coach remained attached to the main portion of the train and was dragged along for about two kilometres before the train came to a stand, despite a passenger having pulled the communication cord. The two rear vehicles, however, became detached and after running about 100 metres were thrown against the side of the cutting through which the double-track line ran at this point. The buffet car overturned on its left-hand side in the derailment and was then crushed by the following mail van. The train was not booked to stop at Cassano d'Adda and was running at 110 kilometres an hour at the time of the accident. Fourteen passengers lost their lives in the accident and another 25 were more or less seriously injured and taken to Cassano hospital.

The work of rescuing the passengers began shortly after the derailment and involved local railwaymen, the Cassano and Milano fire brigades, ambulances from Milano, the railway police and carabinieri, the Red Cross and the Green Cross. A number of victims were trapped in the wreckage and oxy-acetylene equipment had to be used to release them. Darkness fell before the rescue operation could be completed and floodlights powered by generators had to be set up. Eleven corpses were eventually discovered and three of the injured died after admission to hospital. Some of the bodies were badly mangled.

A passenger who was injured and did not wish to give his name for fear of alarming his relatives, told a newspaper reporter that he had been in the penultimate vehicle of the train. He had suffered injuries to his head and left arm. He said that he had been standing in the corridor at the time of the crash, having given up his seat to an elderly woman. The whole thing had only lasted a few seconds. There had been a severe jolt which threw him to the floor. When he had picked himself up he realised that the coach had become detached from the rest of the train and was heading towards side of the cutting. He had closed his eyes and tried to cover his face with his hands. The next thing that he knew two strong arms were pulling him out of the wreckage through a window. His arm had been crushed by falling luggage; he was not sure how.

Another passenger, Giuseppe Nazzucchelli, from Monte Isola, said that he had been sat near the door of his compartment, when the coach had tilted and overturned. Fortunately, he had been on the corridor side, which was not crushed against the wall of the cutting. Only he and two others, of the eight who had been in his compartment had, he said, escaped. The other five occupants had disappeared in the wreckage.

Signora Dolores Picco, 31, from Milano, told the reporter she had likewise been sat next to the door. "For the passengers who were sat next to the window it was horrible. I saw them all lying motionless, covered in blood, after the accident, and I fear that none of them survived." [25]

Signora Bersano, the mother of Antonio Bersano, one of those who had died, kept repeating to relatives that it was her fault her son had lost his life. They had both been passengers on the train. A few minutes before the train left the rails her son had suggested to her that they swop places in the coach, as she was sat in a draught. They had done so; she had survived and he had died.

An engaged Milano couple, Mariuccia Morandini, 23, and Orvedo Dotti, 27, also lost their lives. The morning of the accident Mariuccia, a nurse, had learnt that her mother was dying in her home town of Carpenedolo. She had telephoned her fiancé at the shop where he worked but he had been out. When he returned to the shop, he was given the message and the two of them just had time to catch the fateful train.

Two days after the disaster the newspaper was able to include a statement from Ferrovie dello Stato Deputy General Manager Mario Cirillo, speaking on behalf of General Manager Rissone, who was away. He stated that the speed at which the train had been travelling at the time had been well within the permitted limit on that stretch of the line. Similarly, the points at which the derailment had occurred were of a modern type, having not long been installed, and were controlled from Cassano d'Adda. The points and signals were in perfect condition. The coach third from the end of the train, first to become derailed, was fairly new, dating from 1950. It had been properly maintained and was due for its next service on 29th February. The penultimate coach, that which had overturned and been wrecked, dated from 1948, and it was due for its next service on 13th June. Both coaches, like those in the remainder of the train, were of the Stella type, authorised to run at speeds up to 120 kilometres an hour.

Signor Cirillo went on to deny that the passenger trains were running too fast for the rolling stock of which they were composed, pointing out that train speeds in general were no higher than they had been 25 years earlier.

Engineer Forte stated that the points had probably opened beneath the train because they struck a foreign body. It appeared that a piece of metal had become detached from

[25] *La Stampa*, Torino, Italy, 2nd February 1961

the undercarriage of the third from last vehicle of the train and caught the stretcher bar of the facing points, forcing them open. The coach that had suffered the most damage, although 13 years old, had been overhauled in the railway workshops at Napoli only the previous year. It would be the subject of a detailed examination. It should normally be impossible for the points to be opened beneath a train.

The line past the scene of the accident was reopened at 17:40 the following day, with a severe speed restriction for 400 metres where the work of restoration was still in progress. [26]

The *Cisalpin* international express of the Société nationale des chemins de fer français, inaugurated the year before and running between Paris Lyon and Milano Centrale, met with disaster on the evening of Friday 5th October 1962. It was the Milano – Paris service, which had left Lausanne, Switzerland, at 18:12, and was due in Paris at 22:50, which was involved in the accident. The events leading up to disaster reportedly began with the derailment of the tank waggon of a freight train travelling in the opposite direction, which had suffered a broken axle. The derailed waggon ended up leaning over at an angle of 45 degrees and foul of the opposite running line at Combe aux Loup, France, about 4 kilometres on the Paris side of Montbard, and rather nearer to Saint-Rémy.

The front of the *Cisalpin* collided with the derailed waggon, was thrown across the line and demolished a former crossing keeper's house in its path. The next two cars of the express were derailed in turn, continuing their journey for several hundred metres along the ballast and then overturning at the side of the line. The overhead line equipment was also torn down. Nine people died and 12 were injured, three seriously. Fortunately, many of the occupants of the train were still in the restaurant car, marshalled third in the train, and escaped with slight bruises.[27]

A second report on the accident stated that the crew of the freight train had time to stop the *Aquilon* passenger train from Paris to Marseille, which had been following them, but not the *Cisalpin.* What exactly had happened in the cab of the latter train in the last moments before the accident it was impossible to say, for its French driver and the Swiss technician of the Chemins de fer de fédéraux suisses, who had been accompanying the driver, were both killed, but a violent brake application had apparently been made immediately before the crash.

The leading vehicle of the train had consisted of the driving cab, a luggage compartment, and a passenger compartment. Its second vehicle was the power car, the first of its type, capable of using six different European electrification systems. Then came the restaurant car and two further passenger cars. The fact that the power car was the second in the train, doubtless reduced the number of passenger fatalities and injuries, as did the fact that none of the windows were broken and no one was ejected from the train.

Following the accident help soon came from inhabitants of Saint-Rémy-les-Montbard, firemen, police, and ambulances, together with the implementation of the ORSEC emergency plan. [28]

---

[26] Ibid, 3rd February 1961

[27] *Gazette de Lausanne*, Lausanne, Switzerland, 6th October 1962

[28] Ibid, 8th October 1962

The next day it was reported that Driver Caron and Assistant Driver Vignier of the freight train had been charged with negligent homicide in connection with the accident. [29]

Five days after the disaster the facts of what had happened became clearer. The derailed waggon, which had led to the collision, was one of seven to become detached from the train and left behind. The *Aquilon* passenger train had been stopped, according to this later report, not by the crew of the freight train but because its driver had been warned by seeing the red tail lamp at the rear of the last of the detached waggons. The front part of the freight train, meanwhile, had continued past Montbard to the next station at Nogent-les-Montbard, where it had been shunted into a siding. By then the *Cisalpin* had passed on the opposite running line, travelling at 130 kilometres an hour. The driver of the freight train stated that the brake pressure dropped when his train had become divided and it had stopped. He had then sent his assistant back along the train to investigate. Reaching the end of the train the assistant had been surprised to find that the brake pipe of what had then become the rear waggon was hanging down, and there was no tail lamp. He had not known how many waggons there should be in the train and hence had not realised that there were now less, nor had he apparently attached any significance to the missing tail lamp. He had simply resealed the brake pipe and returned to the engine, the vacuum had been recreated and the train had restarted.[30] (An accident had happened in very similar circumstances at Lunan Bay in Scotland in 1958, though with less serious consequences.) It was not mentioned in the newspaper reports but presumably there must also have been a signalling error, in reporting the line clear for the *Aquilon* although the freight train had arrived at Nogent-les-Montbard incomplete. Possibly by this date the line was equipped with automatic signalling and the *Aquilon* had been stopped not because its driver had seen the tail lamp of the rear waggon, as the newspaper alleged, but by the protecting signal being held at danger by the detached waggons still occupying the track circuits ahead of it.

The international express from La Paz, Bolivia, to Buenos Aires, Argentina, was derailed about a kilometre from Tahona, Argentina, on the General Belgrano division of the Ferrocarriles Argentinos in Tucumán Province, on Wednesday 21st November 1962. Nine people died and 20 were injured. [31] The derailment occurred while the train was rounding a curve at high speed and was blamed on a broken axle on the locomotive. [32]

At about 7:45 a.m. on Saturday 17th October 1964 the *Talaimannar Express* was derailed at Wilwatte, Ceylon, between Ambepussa and Mirigama and about 30 miles from Colombo. This was a night mail train between Colombo and Talaimannar Pier in northern Ceylon, from where there was a ferry service to Dhanuskodi in India. The train consisted of a General Motors M2 Class diesel locomotive, eight passenger bogies (carriages), of which two were brake composites, 14 four-wheeled waggons, and one bogie low-sided waggon. In the derailment the engine and nine of the waggons broke away from the remainder of the train and stopped some distance down the line, while one

---

29 Ibid, 9th October 1962

30 Ibid, 10th October 1962

31 *Clarin*, Buenos Aires, Argentina, 22nd November 1962

32 *The Lovington Daily Leader*, Lovington, New Mexico, United States, 22nd November 1962

of the brake composites and three passenger bogies were reportedly completely destroyed. An early report in an Indian newspaper spoke of 21 passengers being killed and 180 injured, 56 of them seriously.

Immediately upon news of the accident reaching Colombo four ambulances, blood transfusions and a team of doctors were despatched to the scene. Over 300 Rover Scout Leaders, who were attending a mini jamboree about 400 yards away, also went to the scene with first-aid kits and water bottles. The seriously injured passengers were taken to Colombo General and others were treated at other nearby hospitals. The limbs of some of the victims had to be amputated to free them from the wreckage. Mr Anil Moonesinghe, the Minister of Communications, and other ministry officials made their way to the scene as did the General Manager of the Ceylon Government Railway. [33]

On Monday 19th October it was stated that the number of dead had risen to 25 and that the government had announced a public inquiry into the cause of the disaster and how to prevent a recurrence. [34]

The inquiry determined that the derailment had resulted from a mechanical defect in the couplings connecting the bogies. [35]

Seven passengers died and 45 were injured, 15 seriously, in an unlucky accident near Thirsk, England, on British Railways, Eastern Region, on the afternoon of Monday 31st July 1967. Two trains were involved in the accident, the 12:00 Down express passenger from London King's Cross to Edinburgh Waverly and the 02:40 Down freight train from Cliffe to Uddingston.

The passenger train consisted of a Class DP 2 prototype diesel electric locomotive and 13 coaches, in charge of Passenger Guard William Anderson, with Driver John Evans and Secondman Dennis Smith. The freight train consisted of a type 4 diesel electric locomotive, 26 loaded Cemflo bulk cement waggons and two brake vans, in charge of Goods Guard H.P. Wake and Driver J. Walker.

The main line through Thirsk was quadruple tracked. The freight train was travelling on the Down Slow line at its maximum authorised speed of 45 miles an hour, when the rear axle of its 12th waggon became derailed towards the cess on plain track. Shortly afterwards the vacuum pipe between the 11th and 12th waggons fractured and the brakes were applied, stopping the front of the train 470 yards down the line. Meanwhile the 13th to 20th waggons in the train had become derailed in turn and fallen down the embankment, while the 23rd waggon had slewed around foul of the Down Fast line. Approaching on that line, travelling under clear signals at 80 miles an hour, was the express. Driver Evans saw a cloud of dust 600 yards in front of him and then the derailed waggon. [36] He told a newspaper reporter, "I just shut my eyes at the last moment and hoped for the best. It would probably have been worse if it had been head-on." [37] Although he had immediately made a full brake application when he had seen the obstruction the express was still travelling at 50 miles an hour when it struck the cement waggon.

---

[33] *The Sunday Standard*, Madras, India, 18th October 1964

[34] *The Indian Express*, Madras, India, 20th October 1964

[35] http://archives.dailynews.lk/2001/09/11/fea04.html

[36] Ministry of Transport report

[37] *The Times*, London, England, 1st August 1967

The engine and seven leading coaches of the express were all derailed and badly damaged and Driver Evans and Fireman Smith had lucky escapes. Worst damaged were the leading coach, which had its left side torn away, and the next two coaches.

A farm worker who witnessed the accident telephoned the emergency services. The pilot of a Jet Provost aircraft on a training flight also witnessed the derailment and informed his control room at R.A.F. Topcliffe by radio, they in turn making a 999 call to the police. Finally Driver Walker of the freight train had informed the Thirsk signalman of the accident from a signal post telephone and the guard of the express had telephoned Pilmoor signal box from another signal post telephone. The derailment occurred at about 15:17. The first ambulance reached the rather inaccessible site at 15:40 and the last of the injured had been removed by 16:30. The Fast lines and Down Slow line were blocked but single-line working was introduced over the Up Slow line. Later Down trains were diverted over the recently closed Harrogate – Northallerton line and sleeper trains via Carlisle.

Driver Walker told the Ministry of Transport inquiry into the accident that he was well aware of the 45 mile an hour speed restriction on the Cemflo waggons and had been careful not to exceed that. His journey had been uneventful with clear signals until near Thirsk he had noticed the brakes rubbing and the vacuum pressure falling and quickly realised that something was wrong. Looking back, he saw that his train had become derailed and when his engine had stopped he had gone to the telephone at signal 19 BS and reported what had happened to the Thirsk signalman.

Goods Guard Wake, riding in the rear brake van, testified that he always looked out continuously when he was in charge of Cemflo waggons because they often oscillated badly. The last time he had been able to see along the train had been the left-hand curve at Sessay Wood, about 3½ miles short of the derailment point. Between signals D 19 S and D 19 BS he had seen a bulge in the middle of the train and almost at once 9 or 10 waggons had rolled down the embankment. Seeing that the Down Fast line was also obstructed he had seized detonators and a flag and started to run back along the line, waving his flag. When he had covered about 100 yards the express passed him, with its brakes hard on.

Driver Evans of the express testified that he had had a normal run from York until, approaching signal D 19, he found that his view seemed to be somewhat obscured by dust or haze and had closed his controller. Passing that signal he had realised that there was a cloud of dust, then had seen the derailed waggons and applied his brakes in emergency. Coming closer he had seen Guard Wake waving his arms but there was nothing more that he could do.

Secondman Smith confirmed his driver's evidence.

Passenger Guard Anderson said that he had felt his train's brakes coming on hard and the collision had thrown him across his van. After the accident he reported it to the Pilmoor signalman from a signal post telephone.

Signalman M. Fraiser at Pilmoor stated that he had noted the characteristic lateral oscillation of the waggons of the cement train but that was nothing out of the ordinary. He had received the *Obstruction Danger* signal from Thirsk at 15:18 and Guard Anderson had telephoned him about a minute later, when he contacted Control, but they already knew of the accident from the Thirsk signalman.

Signalman J. Russell at Thirsk said that Driver Walker had telephoned him at 15:18 from signal D19 BS. He had replaced his Up signals, stopping an Up express, sent the *Obstruction Danger* signal to Pilmoor and Northallerton boxes, and informed Control.

Stationmaster J. Reed had also been in Thirsk box. He said he had notified the station staff to call out ambulances, doctors, the fire service, and the police, then proceeded to the accident scene.

Mr R. Burns, the farm worker who had been operating a potato harvester, spoke of witnessing the accident. He put the interval of time between the initial derailment and the subsequent collision at no more than a minute. He had then run to his home a quarter mile away to ring the police.

Two Electricity Board workers, Messrs H.I. Jackson, and J.W. Young, who were interested in railways, timed the trains as passing Pilmoor at 15:12 and 15:15. Mr Young did not consider that the cement trains ran through Pilmoor at excessive speed, rarely more than 40 miles an hour. He had noticed an unusual movement in the leading axle of a waggon about midway along the train. Mr Jackson did not notice this.

Mr A.P. Mann, sat in his car near Pilmoor, said he had seen the cement train pass at about 40 miles an hour and a waggon near its rear oscillating severely from side to side.

Area Assistant District Engineer J.R. Stockdale stated that the line was inspected three times a week by a patrolman and he himself had walked it ten days before the derailment. He considered that the track was in a generally satisfactory condition.

Ganger (Acting Sub-Inspector) R. Watson testified that he had last examined the line on 21st July and the last work done on the Down Slow had been at the end of April when some low joints had been attended to.

Permanent Way Inspector T.E. Watson said he had walked the Down Slow on 4th July when he had found a few low joints but nothing needing the attention of the ganger being drawn to it. The line, he thought, was safe for a maximum speed of 60 miles an hour.

Sub-Inspector R. Coldwell described running a track-recording trolley over the Down Slow line on 13th July. Nothing of significance had been detected in the Thirsk area.

Carriage and Waggon Examiner H. Portus at Peterborough said that his examination of the cement train had taken 15 to 20 minutes. Two waggons with broken springs he had "red carded" but he had found no other problems.

Carriage and Waggon Examiner J.W. Hathaway at Doncaster said that he had "red carded" one waggon with a broken suspension spring.

Carriage and Waggon Examiner T.H. Walker at Skelton Sidings, York, stated that he had found no problems with the cement train.

Technical Assistant (Waggons) G.A. Appleby stated that he had carried out a very full examination of the waggons after the derailment. He explained that the waggon first to leave the track, Cemflo No. LA 23, the 12th vehicle of the train, he had found to have signs of heavy wear inside the axleguards, indicating that it had been running for some time with worn and extended suspension links, allowing excessive lateral oscillation. The links, however, were not excessively worn and appeared to have been recently changed.

Scientific Officer J.B.C. McCann agreed that the 12th waggon in the train had been the first to derail. He found the flange of its left trailing wheel to be severely worn and notably more worn than the right. This would have caused the waggon to oscillate laterally ('hunt') and could have been a major factor in the derailment. New suspension links would reduce the hunting but worn links or the oiling or greasing of the links would increase the tendency to hunt.

The Down Slow line was also subject to a detailed examination after the accident, which disclosed slight defects such as minor kinks, gauge variations and a few loose or missing keys, but nothing of any great significance.

Tests as to how far a similar train would run after its vacuum pipe had been severed confirmed that the accident train had not been travelling at more than 45 miles an hour.

The cause(s) of the derailment not having been firmly established, further tests were carried out by the Research Department at Derby.

Concluding his report, the inspecting officer, Colonel D. McMullen, wrote that the derailment had been the direct result of the bad riding characteristics of Cemflo bulk cement waggon No. LA 23 as a result of excessive lateral oscillation, the waggon having become derailed at a point where there were minor track irregularities, which were nevertheless within the specified tolerances of the line concerned. The type of suspension of the waggon, though used extensively on other types of waggons and on the Continent, was not suitable for the cement waggons.

The cement train was being driven within the speed limit and Driver Walker was in no way responsible for the derailment. After the accident he and Guard Wake had taken immediate and correct action but had no time to warn the express. Driver Evans and his secondman had also done what they could to minimise the effects of the collision.

Since the accident the maximum speed of Cemflo waggons had been reduced to 35 miles an hour.

Turning to remarks and recommendations the inspecting officer noted that the reduction of the maximum authorised speed at which Cemflo waggons were allowed to run would be highly restrictive from a traffic point of view and could not be regarded as a satisfactory long-term solution. Modifications to the suspension and buffers of the waggons were being tried in an attempt to alleviate the problem. In addition there needed to be a greater awareness in permanent way staff of the danger of minor track defects and the need to recognise and rectify them as soon as possible. [38]

The *Treno del Sol* overnight direttissimo from Palermo to Torino, with 200 passengers, was derailed at Gioia Taura, Italy, at about 17:08 on Wednesday 22nd July 1970. The train was running about 40 minutes late due to difficulties with the ferry from Messina in Sicily to Villa San Giovanni on the mainland. It was travelling at about 80 to 90 kilometres an hour when the rear 12 carriages of the 17 in the train left the rails. Two sleeping cars, one carrying 50 Sicilian pilgrims on their way to the shrine at Lourdes, in France, overturned. Six people lost their lives and about 70 were injured, 32 of whom were taken to hospital in Palmi and 14 to Taurianova hospital.

A newspaper reporter managed to interview a number of the survivors of the accident when the engine and front coaches of the train eventually arrived at Torino. Calogera

[38] Ministry of Transport report

Schifano, 18, said, "We were sitting in a carriage in the middle of the train. We were passing through Gioia Taura station at full speed. Suddenly we heard the coaches crashing against one another. I grabbed my mother and pulled her out through a door. There was a lot of smoke, like a fog, and injured people were screaming. There were some missing a leg or an arm. After about five minutes the first fire engine appeared and they worked to get the victims out of the rear carriages, which had overturned. As each body was removed it was covered with a blanket. It seemed to me that all the dead were children, but I cannot remember much, only ambulances coming and going and a helicopter flying very low. I do not know if we were there for an hour or half a day. Eventually we were told to board another train and we set off again."

Calogera's father, Francesco, 62, suffered a blow to his head and was wearing bandages. All he remembered was hearing what sounded like a bomb going off. He began crying, "Thank you God, for saving me," and was still giving thanks to the Almighty when interviewed by the reporter.

Travelling in the same coach as the Scifanos was Calogera Pesce, 65, coming from Catania to visit his only son, Rosario, a Fiat employee, whose fifth daughter, Maria Grazia, had just been born. His son had waited all morning for him at the station (the train had finally arrived at Porta Nuova station at 15:30, 6 hours and 42 minutes late). Sig Pesce had jumped from a coach window after the derailment, breaking his ankle and hitting his head. When he got off the train he was unable to walk and was uttering incoherent sentences. Taken in a wheelchair to a room at the station he was examined by Dr Zagaria, who ordered his removal to hospital.

Loredano and Piero Panzanaro had been visiting an aunt in Palermo. Making their way along the line after the accident they had stumbled upon the mangled body of a young man. Sig Panzanaro was worried about his 19-year-old wife, who was expecting their first child.

Widow Maria Barraca, 55, had come from Marsala with her son Carmelo, to see her sister Rosa, who was also a widow, for the first time in seven years. The nightmare journey, as the reporter described it, was the first time that her son had been on a train.

Angelo Tufiano had come from Corleone to attend the wedding of his son Stefano, who was getting married on the following Sunday in Venaria (part of Torino).

Nunzio Giungrande, 32, had come from Monreale in Sicily to visit his youngest son Antonio who was a patient at the Molinette hospital in Torino. "Seeing the suffering," he said, "I almost forgot that I was coming to Torino to visit the hospital."

The derailment coincided with the beginning of the Reggio revolt, violent disturbances that had begun in July after the town of Catanzaro was chosen as the regional centre of Calabria region, in place of the considerably larger town of Reggio Calabria. There was a strong suspicion at first that the derailment had resulted from an act of sabotage, especially as there had been other attacks on the railway by the protesters, for example a railwayman walking the line between Villa San Giovanni and Cannitello had discovered 15 sticks of gelignite attached to a fuse, placed near a pylon supporting the overhead line equipment. The fuse had been lit to set off the explosives but had gone out after burning for 15 centimetres. Police officers searching the tracks had found another ten sticks of gelignite at a different point.

Two investigations were begun into the derailment, one under the direction of Dr

Paulo Scopelliti, the Palmi public prosecutor, the other under the direction of the Ministry of Transport. Driver Giovanni Bilardi, 41, and Assistant Driver Antonio Romeo, 31, were interrogated at length, but the newspaper noted it was not considered that they had been responsible for the accident; the questioning was merely to establish all the facts relative to the derailment. The train had been running under clear signals and at less than the maximum permitted speed.

In view of the suspicion of sabotage Deputy Police Chief Catenacci and the Police Commissioner visited the scene and an examination was carried out by a forensic team. However, no trace of any explosive device was found, while the loosening of some of the bolts securing the rails to the sleepers was attributed to the force exerted on the track by the derailed vehicles. Also another train had passed over the same spot just a few minutes earlier without any problem. Following the examination it was theorised that the actual cause of the derailment had been an undetected problem with the undercarriage of the fifth coach of the train, which had dropped, fouled, and spread the track, resulting the derailment of all following vehicles. [39]

According to one newspaper report there were seven fatalities in the accident, possibly one of the seriously injured passengers later died. [40]

Amtrack train No. 1, otherwise the southbound *City of New Orleans*, was derailed on Saturday 10th June 1971 on Illinois Central Railroad tracks at Tonti, near Salem, Illinois, United States. The train consisted of diesel units Nos. 4031, 4109, 4106 and 2024, a baggage car, two coaches, a diner counter car, a lounge car, six coaches, a coach food car and three coaches, in charge of Conductor Carroll Fletcher with Engineer Lacy Haney and an unnamed fireman. It had started from Chicago Central station at 8 a.m., and shortly afterwards it was found that the locomotives were lacking power. The fireman found that this was because the two engines in the centre units and the No. 2 engine of unit No. 4071 were not loading. He made adjustments and was then able to get the two engines of the centre units to work but not that of unit No. 4071. The train reached Champaign at 10.44 a.m., 34 minutes late. Here the crews were changed and the new crew was informed of the problem, but leaving there the speed of the train apparently increased normally upon the engineer opening the throttle and the new fireman concluded that the problem had disappeared. The train left Effingham, Illinois at 11.53 a.m. with 211 passengers, passed Edgewood, the last open office, about 24½ miles north of the point of the accident, at 12.05 p.m., and was travelling at 90 miles an hour as it approached Tonti. The crew of a northbound freight train that passed the passenger train near Edgewood had noticed nothing wrong with it. As the train was reducing speed at Tonti for a signal ahead and was passing a crossover there, Engineer Haney noticed a bump and then realised that the engine was derailed. As he went to apply the brakes in emergency the locomotive turned over on its right side and slid along the ground for 350 feet. The engineer and his fireman were then able to evacuate the locomotive, seeing as they did so that the second engine had caught fire. The fire later involved the leading engine as well.

Behind the engines all the coaches of the train were derailed. The first six cars were

---

[39] *La Stampa*, Torino, Italy, 23rd and 24th July 1970

[40] *L'Unita*, Roma, Italy, 24th July 1970

badly damaged, the next four were extensively damaged and the last four less so. Six people were killed by being thrown from the train through coach windows, three were crushed or thrown from the ends of cars, one was killed by a crosstie (sleeper), which had penetrated a window, and another died from being thrown about inside the car. There were thus 11 fatalities in all, while in addition 163 passengers and railwaymen were injured. [41] Conductor Fletcher was among those who lost their lives.

Various passengers and others gave their accounts of the accident to newspaper reporters. '"Everything was spinning, turning 180 degrees, and I had no idea what was happening," said Kwoc Koe, of Cairo {Illinois}, who was not injured. "I have no idea how many people were killed in my car."

'"I was sitting, talking to a conductor," said Mrs Gertrude Harris, 28, of 6805, S. Normal Blvd., Chicago. And then, I think, the train put on its brakes, and the conductor screamed to hold on, and then I saw a lot of dust. That's all I can remember." Mrs Harris is among the 52 survivors hospitalized in this town's {Salem's} 40 bed Memorial Hospital. At least 44 more are in hospitals in towns near here or in St Louis.

'"I was in the back of the train," said Jeffrey Couchman, 17, of 195, Schiller St., Elmhurst. "My car didn't tip over, so I didn't think it was too bad until I started seeing people coming from other cars with blood all over them."

'However not everyone was unaware of what was happening. "I knew the car was tipping over and we were crashing," said William Black, 75, of 1110, 53d St., Chicago, a retired Illinois Central Railroad janitor who previously was aboard a train involved in a minor wreck. "I've got high blood pressure so I just held on and didn't worry about it," said Black, who was treated for a minor head injury and released.

'Melvin Maxey, 63, a farmer, had just finished lunch and walked outside his house when he heard the crash of the train. "My boy and I jumped in the old truck as soon as we called an ambulance," Maxey said. "People were climbing out of the cars and lying mangled on the rails and under the cars. Everybody was screaming. Everybody was hysterical." Maxey and his son were the first to arrive at the accident scene. Maxey's wife Pansy, 60, arrived at the crash scene a few minutes later. "I saw two little kids – about 4 and 6 – who came running down the road crying their eyes out," she said. "They wanted their mommy and daddy. The last I saw of them was when the sheriff took them in a car and said he'd try and find their mommy and daddy." "There were other kids too," Maxey said. I ran across a field to direct the ambulances and fire trucks in. The engine was on fire and the flames were jumping 15 feet in the air."

'After the rescue work was under way, Maxey and his wife went home. Later, they returned to stand at the edge of a cornfield and watched workers beginning to clear the wreckage. "It's a horrible thing," he said.

'Nearby a woman's black floppy brim hat rested atop one of the steps leading into one of the twisted railroad cars. The neat lettering on the steps warned, "Watch Your Step."' [42]

Information on the accident was received in the town of Salem at 12.27 p.m. and broadcast over the town's communication system. Emergency vehicles were at once despatched to the scene and nearby communities were alerted to provide additional

---

[41] National Transportation Safety Board report

[42] *Chicago Tribune*, Chicago, Illinois, United States, 11th June 1971

assistance. The hospital was also alerted to prepare to receive the injured. A portable field hospital unit was made available and set up in the high school as an emergency station where the less badly hurt were treated. All injured were removed from the scene within two hours.

The accident was the subject of an inquiry by the National Transportation Safety Board (NTSB), which by this date had taken over the investigation of American railway accidents from the Interstate Commerce Commission. An examination of locomotive No. 4071 showed that the leading wheels of its rear truck (bogie) had flat spots about 10¾ inches long, which had caused false flanges to develop. The tracks north of the site of the derailment were then examined and various marks were found as far back as Edgewood.

It emerged that locomotive No. 4071 had formed part of train No. 1 on Tuesday 6th June and shortly after the train had left Chicago its No. 2 engine failed to supply electrical power and had been shut down. At the train's destination, New Orleans, it was determined that the auxiliary generator was defective and the engine was returned to Chicago with the No. 2 engine still shut down, for it to be serviced there. The power reverser for the No. 2 power circuit was also locked in the neutral position at New Orleans.

The conclusions of the NTSB report had 24 parts:

Locomotive No. 4071 had operated from New Orleans to Chicago on 7th October with the No. 2 motor shut down and with the power reverser locked in the neutral position. The motor was tagged.

The daily locomotive inspection report indicated that the power reverser was blocked.

The motor was repaired on 10th June but the reverser was left blocked.

Tests on the repaired motor did not detect the blocked reverser.

Tests on the locomotive prior to its leaving Chicago likewise did not detect the blocked reverser.

Because the locomotive left Chicago with the power reverser still blocked in the neutral position, the wheel slip detection device was inoperative and unable to indicate sliding wheels on its rear truck.

Operation of the locomotive without a wheel slide protection device violated Code of Federal Regulations (CFR) 230.201 (d).

The engineers were unable to detect if the warning devices were operative.

The engine crew of the accident train were changed at Champaign and the new crew were informed of the condition of No. 2 motor circuit.

The load meter in the locomotive cab only indicated the amount of current being supplied to one motor circuit.

The engineer and fireman with their knowledge and appliances could not be expected to determine whether power was being supplied to all traction motors.

The train was operated according to the rules between Champaign and Tonti.

The leading wheels of the rear truck of locomotive No. 4071 probably locked during the stop at Effingham and slid continuously over the 38¾ miles onwards to Tonti.

Loosening of a locking plate in a traction motor armature bearing permitted excessive lateral movement of the armature.

The excessive lateral movement caused overheating of the bearings, deterioration of the lubrication and eventual failure.

The reconditioning of the armature did not contribute to the failure of the bearings in the accident.

Regulation CFR 230.262 permitted a unit to remain in service with a defective motor cut out, nullifying the wheel slip indicator, and was in conflict with CFR 230.201 (d) requiring an indication of slipping or sliding wheels.

Neither the Federal Regulations nor the Recommended Standards of The Association of American Railroads (AAR) provided for the maintenance and operation of traction motors and other main locomotive component parts.

The high compressive forces generated between the locomotive and the cars during the derailment caused the cars to jackknife.

Almost all of the 43 large observation-style windows on the lower sides of the five overturned coaches had been broken either due to impact with the ground or due to the weight of passengers or luggage.

Six people who had fallen through the broken windows were fatally hurt.

The basic structures of the cars had withstood the stresses of the accident well but they had 'inadequate design crashworthiness' with sharp edges, hard surfaces, and poorly secured seats and furniture, while the large observation windows were a critical weakness destroying the crash integrity of the structure.

The use of propane as a fuel in the cars created a serious fire hazard and could have led to a catastrophic accident.

'The people of Salem provided an outstanding example of what can be accomplished by preplanning and practising for emergencies.'

The NTSB found that the probable cause of the accident was 'the displacement of the east stock rail of the southbound main track by the false flange on the left hand wheel on the leading axle of the rear truck of the locomotive unit 4031. The wheel slid flat when the traction-motor armature bearings failed and locked the driving wheels. Failure to detect the sliding wheels was caused by the dispatch of the locomotive with an inoperative wheel slip indicator.'

The NTSB recommended that:

The Federal Railroad Administration (FRA) review CFR 230 to reconcile the conflict between sections 201 (d) and 262; that it require a practical terminal test to indicate whether a locomotive complied with the Federal regulations before being brought into service and that it require the use of warning systems other than wheel slip indicators to protect the operation of locomotives and a means for testing the warning systems.

The FRA establish near future safety standards to deal with the serious problem of passengers being ejected through large car windows.

The FRA promulgate regulations for passenger cars to minimize sources of direct impact injury.

Amtrak meanwhile correct impact injury causing features as cars were renovated or rebuilt. Purchase specifications for new cars should include designs to reduce impact injuries, prevent passenger ejection through windows and make provision for the escape of passengers when the end doors of overturned cars were blocked.

The AAR evaluate the requirements for the maintenance and overhaul of locomotive unit traction motors and other major components and establish standards and recommended practices for such maintenance.

The Office of Civil Defence of the United States Army review the plans made by the town of Salem, Illinois, for emergency situations and consider advising associated State and county Civil Defence Agencies of their contents and results in this instance. [43]

A Burlington Northern Railroad freight train from Minot, North Dakota, to Duluth, Minnesota, and consisting of a locomotive, 72 loaded cars and 12 empty cars, was derailed at about 5 p.m. a mile and a half north of Larimore, North Dakota, United States, on Monday 28th August 1972. Sixty-two of the cars left the track, scattering grain, coal, lumber, and farm implements along the wayside. Riding in an empty boxcar of the train were six transients, five of whom were killed and the sixth who was critically injured was taken to Grand Forks Hospital. Eyewitnesses said that the brakes had been slammed on a moment before the middle part of the train was derailed. None of the five train crew was hurt and they later returned to Grand Forks. The engineer of the train was Ed Thygeson.

Four of those who died were quickly identified as Dominick Cavanagh and Roy Hasset of Devil's Lake, George Fasset of Fort Totten, and Raymond Richott, 27, of Walhalla; all being in North Dakota state.[44] The following day the fifth fatality was identified as Edward Delonais, of Belcourt, also of North Dakota. The injured man was Roman Cavanagh, of Devil's Lake, brother of Dominick.

Divisional Superintendent P.B. Rasmusson stated that the wheel of one of the cars had broken, causing the derailment. [45]

The 17:18 British Railways, Western Region, Down express passenger train from London Paddington to Oxford, 1A82, was only eight minutes into its journey, having started 11 minutes late, when it was derailed at 17:37 at West Ealing, England, on the evening of Wednesday 19th December 1973. The whole train left the track and the fourth to sixth coaches were badly damaged. Ten of the 650 passengers on the train were killed and 94 were injured, 53 of whom were taken to hospital.

The emergency services were called two minutes after the accident by the signalman in Old Oak Common panel box when he saw the track circuit indications for all the lines east of West Ealing station light up and the points lights flashing, from which his supervisor had deduced that there must have been a derailment. The police arrived at 17:44, the first ambulance at 17:46 and the fire brigade at 17:48. Fourteen ambulances took the injured to the King Edward Memorial Hospital in Ealing and the Hillingdon Hospital. All four lines were blocked, causing extensive disruption to traffic. Most main line trains were terminated at Reading. The Up and Down Relief lines were reopened at

[43] National Transportation Safety Board report

[44] *Winona Daily News*, Winona, Minnesota, United States, 29th August 1972

[45] Ibid, 30th August 1972

18:45 the next day, the main lines not until the evening of 28th December, with Christmas intervening. [46]

'Three hours after the crash firemen were cutting with hydraulic and compressed air equipment to reach the bodies in the fourth and fifth carriages. Working by arc light in a light drizzle they cut into the two overturned carriages, which completely straddled the four lines. Firemen who arrived just after the crash found hundreds of passengers standing on the embankments on each side of the track. People living in houses nearby said they first knew of the accident when passengers knocked at their back windows to ask for help.

'Mrs T. Banks, who lives in Craven Avenue, said, "We were watching the news when we heard a screech and a bump. The first thing we knew was people knocking at the windows to use the telephone. When we went outside we saw people clambering up the grass embankment."

'Mr Ray Hennessy, from Woodley, near Reading, said: "The train started to rumble and the track disappeared. I realised what was happening. It was very, very, lumpy. There was no panic in the coach I was in."

'Another passenger, Mr John Carroll, of Wokingham, said: "There was a sudden braking and then a jumping all over the place. It was all a matter of seconds and the next thing the coach was on its side in the rubble." Two of the front three carriages had formed themselves into a V shape with a third squeezed in the middle. "I was trying to help people out of the carriage in front and most got out. But some wanted a ladder to get down as it was a 17ft drop. We got a ladder out of the guard's van."

'Mrs Pat Bowden, of The Meadway, Reading, who escaped with slight injuries, estimated that there were 70 or 80 people in her saloon carriage. "I was just standing there when suddenly the train seemed to start to brake. Then it started to judder and all of a sudden I could see the rails flying past the windows where I was standing. I realized that we were off the rails. Then the carriage began to rock from side to side. The lights went off and everybody was being thrown all over the place."' [47]

The train consisted of Western class diesel hydraulic locomotive No. 1007 *Western Talisman*, two second-class coaches, a third-class coach, two first- and second-class composite coaches, a first- and second-class brake composite coach and five second-class coaches, in charge of Conductor Guard J. Wells with Driver T.D. Owen and Secondman R.P. Woolnough.

Driver Owen told the Department of the Environment inquiry that he had taken over the locomotive at Old Oak Common on the day of the accident and had prepared it for service, carrying out the series of checks set down in the Driver's Manual, only three of which related to checking items outside of the locomotive – the air, vacuum, and heater hoses; that no cables were attached and the main lighting switch. There was no reference to checking the battery box doors. He had then worked the empty coaching stock from Old Oak Common to Paddington to form the Oxford train. After leaving Paddington he had received green signals all the way. His first warning of a problem had been feeling a lurch as his train passed under a bridge at the west end of Ealing Broadway station.

[46] Department of the Environment report
[47] *The Times*, London, England, 20th December 1973

Attributing it to a bad rail joint he had resolved to report it when he reached Reading. The next signal had also been green but suddenly he felt a thumping from the rear of the engine, it had slowly tipped up in the air, fallen over on its right-hand side and slid along the track till it came to rest. Recovering himself after the accident he had made his way to a lineside telephone and contacted the signalman, who had told him that all the lines were protected.

Secondman Woodnough confirmed his driver's evidence. In retrospect he thought that the bump at Ealing had been the battery box door of the locomotive striking the platform ramp. He had telephoned the signalman from signal OO 27.

Guard Wells stated that he had been making his way through the sixth coach towards the front of his train when the derailment occurred. He had been able to return to the brake compartment for his handlamp and some track circuit operating clips. He had then attempted to place a clip on the Up Main line but the rails had been splayed apart by the derailment. Instead he had telephoned Old Oak Common panel box from signal OO 134.

Mr R.V.C. Pearson, an assistant in the Passenger Manager's Office at Paddington, had been a passenger in the third coach. He said there had been a very loud metallic noise at Ealing Broadway, then 12 or 15 seconds later the coach had begun to sway and had overturned. With difficulty he had got out of the coach and realising the seriousness of the accident made his way via a private garden to an office on Uxbridge Road, from where he had been to report the accident by telephone to Mr. C.R. Postle, the Divisional Movements Manager, and he had remained there as a link between the railway control and the accident site.

Messrs R. Pope and W. Cattermole, British Railways Board employees who were passengers on the train, also made their way to the signal post telephone where the secondman was speaking to the signalman, and spoke to the signalman.

Signalman C.F. Duffin at Old Oak Common panel testified that he had had no occasion to use the Down Main to Down Relief crossover at which the derailment occurred since he had come on duty. When numerous track circuits had suddenly shown occupied and points lights started flashing at about 17:35, although only train 1A82 was in the area, he had at first suspected a power failure, but then the secondman had spoken to him from the signal post telephone, followed by Mr Pope, whom he knew, and he realised that there had been a derailment.

Panel Supervisor O.G. Hawkins said that when all the track circuits had become illuminated he suspected a derailment and instructed Signalman Duffin to place the protecting signals at danger, while he called Ealing Broadway station for more information.

Chief Signal and Telecommunications Technician D.L. Turner chanced to be on a following train stopped at Ealing Broadway. He related that he had made his way to the accident location on foot. The points machine at No. 807 points, the site of the derailment, he had found to be extensively damaged but the points machine itself was in the fully normal position and bolted. The light weather shield over the hole where the points rodding emerged from the machine had apparently been torn off by a violent blow and carried 20 yards down the track.

Traction Inspector A.E. Holder reached the scene at about 18:35. He stated that he had found the switch blades of the points almost in the reverse position and the points machine damaged. A little further on he had found a battered blue steel object which appeared to have come from a locomotive. Examining the overturned locomotive he had found that one of its battery box doors was missing and realised that was what the battered object must be.

Chief Inspector J.L. James of the Civil Engineer's Department said he had heard of the derailment on the BBC television news and had gone to the scene. He found that the fourth battery box door from the leading end of the locomotive had been torn from its hinges and had met Inspector Holder, who showed him the damaged object.

Chief Mechanical and Electrical Engineer S.R.D. Power stated that although from the circumstantial evidence it was virtually certain that the derailment had been caused by the detached battery box door striking the point rods, he had had this confirmed by an examination of paint particles at the British Railways London Area Laboratory.

Old Oak Common Area Maintenance Engineer B.M. Jordan explained that when a locomotive was undergoing a casual repair that would cause some drain on its battery, it was usual to put the battery on charge so the engine would go back into service with a fully charged battery. A battery door box would need to be opened to check the specific gravity of one or two cells to monitor the level of charge. It was usual to leave the doors open as long as the battery was charging.

Shopman C.O. Pitter stated that he had opened the battery door box at the A end of engine No. 1007 on the night of 18th December to check the specific gravity of the electrolyte. The battery had still been charging and the battery box door open when he left duty in the early morning. Various other staff at Old Oak Common depot gave evidence. Only one of them, Shopman F.A. Barret, remembered seeing the battery door box open on the morning shift on 19th October. He said that in order for him to move a scaffold to clean the locomotive someone had closed the battery box door and removed the charging leads.

Beginning the conclusions of his report Lieut. Col. I.K.A. McNaughton, the inspecting officer, stated that the immediate cause of the derailment was not in doubt. The rearmost battery box door on the nearside of the locomotive had been unlocked with its pear-shaped safety catch secured in the raised position and so was ineffective. Shortly after leaving Paddington, as a result of vibration, the battery box door had fallen open. It protruded about 12 inches outside the loading gauge and had struck a number of lineside objects, including the platform ramp at Ealing Broadway, which had damaged it and forced it into a position in which it was about two feet outside of the running rail and six inches below rail level. In this state it had then struck the rods of No. 807 points, forcing the switch blades open and the train, travelling at about 70 miles an hour, had been derailed.

A properly locked battery box door could not have worked open by itself and there was no evidence that anyone had interfered with it after the engine had left Old Oak Common. It could not, therefore, have been properly closed when the locomotive left the depot. The evidence as to who had closed the battery box door in the depot, and when, was contradictory but was in any case unimportant. What was important was that there was no effective system at the depot to check that all roof hatches, body side doors, battery box doors and so on were closed and secured befoe the engine was put into service. Assistant Shift Supervisor G. Abbas at the depot must share some responsibility for this. It was not part of Driver Owens' duties to check the battery box doors and he could not be held responsible for failing to notice that it was not properly secured.

Turning to remarks and regulations the inspecting officer noted that the most fundamental way to prevent a similar accident in the future was to have a specific person designated to carry out a series of safety checks after a casual repair and this had now been done.

The pear drop safety catch was also being redesigned.

Protection of the point machine, situated over two feet outside of the running rail, could, the inspecting officer added, not be justified by the fortuitous circumstances of the accident. [48]

---

[48] Department of the Environment report

CHAPTER 3

OTHER DERAILMENTS

Other chapters in this book cover derailments caused by track defects, defects with the trains themselves and those caused deliberately by acts of vandalism or sabotage. This chapter will deal with derailments due to other causes such as excessive speed, collisions with animals and other accidental obstructions, traction shocks and so on, together with those for which no cause was given.

A brief report in a single American newspaper mentioned the derailment of a train at Valdivia, Chile, on the night of Wednesday 11th January 1961. Thirty soldiers reportedly died in the accident, including 11 officers. [49]

A special Pennsylvania Railroad train, carrying racegoers to a meeting at Bowie, Maryland, United States, met with an accident close to its destination at about 12.58 p.m. on Tuesday 2nd February 1961. The accident occurred about 2½ miles north of Bowie station at the point where a branch just over 2 miles long left the Baltimore – Washington D.C. main line at Del block station, the line curving around sharply to the left as it entered the branch. The maximum speed of passenger trains on the main line was 50 miles an hour and on the branch 30 miles an hour, but over the turnout onto the branch itself trains were limited to 15 miles an hour.

Passenger train Extra No. 5893 South, the race special, consisted of diesel electric units Nos. 5893 and 5890 coupled in multiple control, four coaches, a dining car, a café car and five coaches, in charge of an unnamed conductor with Engineer Howard Horner and Fireman William Wyatt. It had left Philadelphia 30th Street station at 10.28 a.m. after a terminal brake test had been made, called at Chester, Pennsylvania; Wilmington, Delaware; and Elkton, Aberdeen, and Baltimore, Maryland; and departed from Baltimore, 22½ miles from Del, at 12.29 p.m. Seven miles north of Del it was switched from track No. 3 to track No. 2 and at Odenton, 4½ miles from Del, it was switched from track No. 2 to track No. 1 (track No. 1 was the only one from which the branch could be reached).

Engineer Horner told the Interstate Commerce Commission (ICC) inquiry into the accident that he had sounded the horn of his engine for a level crossing with Anderson Road about 3,300 feet north of the junction points at Del. Immediately afterwards he had

---

[49] *The Gettysburg Times*, Gettysburg, Pennsylvania, United States, 16th January 1961

made a service brake application of about 8 pounds, with no apparent effect, followed a few seconds later by a further brake pipe reduction of 12 pounds, equivalent to a full service application. There had still been no apparent effect and he had shouted a warning to his fireman, acknowledged with his horn a yellow flag signal from the Del operator indicating that the points were set for the branch, then made an emergency brake application, at a point he said was about 1,500 feet from the junction. Other evidence, the ICC noted, indicated that the emergency application had only been made at the Patuxent river bridge, 600 feet from the junction points.

Fireman Wyatt testified that shortly before the river bridge he had called to his engineer that they were close to Del and were going too fast and he (the fireman) had reached for the emergency brake valve but the engineer had acknowledged his warning and applied the brake himself. The fireman put the speed at 45 to 60 miles an hour at this point. Both he and the engineer thought that the brakes had failed because the brake pipe had been frozen (the temperature was 16 degrees Fahrenheit {-9 Centigrade} at Odenton at 12 noon).

The conductor and flagman were too badly hurt to give evidence at the inquiry. The front brakeman had been speaking to a passenger and could give no useful evidence. The swing brakeman had only noticed the emergency brake application when the dining car was close to the bridge (750 to 1,000 feet from the junction).

The Del operator said he had been advised by Odenton that the train had passed there at 12.52 p.m. When he had heard the warning whistle for the level crossing he left his box with his signal flags and proceeded to the junction points, 30 feet away. He had shown his yellow flag as soon as the engine of the train had come into view (2,000 to 2,250 feet away) and it had been acknowledged. He had then exhibited a green flag to indicate that the line was clear to Track block station on the branch, but this signal was not acknowledged. Returning to his post he had become concerned that the train was not slowing up. As he had seen the engine begin to overturn when it entered the curve he had run away eastward from the track to save himself from being struck. He had seen no indication of braking.

Both engines were derailed, overturned, and were considerably damaged. Behind them the leading seven coaches of the train were derailed and the first six were badly damaged. The remaining four were slightly damaged. Five passengers and a dining car employee were killed and 232 passengers, 6 railwaymen, 4 dining car employees and a news vendor were injured. [50]

'"All we felt was two sharp jerks," said dining car steward Eugene B Webster. "One big sudden jerk, then a small one."

'One of the passengers who escaped injury, Edward F. Janus, 48, of Philadelphia, said he thought at first the train was just making a sudden stop. "There was one jolt after another and the car was rocking like a ship at sea," he said, after climbing from the last derailed car.' [51]

The ICC noted that an examination of the wreckage indicated the train had 'entered the turnout materially in excess of the safe speed and at or near the theoretical

---

[50] Interstate Commerce Commission report

[51] *The Daily Mail*, Hagerstown, Maryland, United States, 3rd February 1961

overturning speed of the diesel-electric units.'

No problem was found with the tracks at Del. Examination of the leading engine showed the throttle closed and the emergency brake applied but the sanding controls were "off." The angle cocks visible were all open and there were indications of hard braking on the brake shoes of all the coaches. No defect in the brakes could be found.

The tachograph from the trailing diesel locomotive was examined and showed normal braking between Philadelphia and Baltimore. South of Baltimore speed had increased to 70 miles an hour at first, had fallen to 20 miles an hour for Vern interlocking and to 15 miles an hour at Odenton. Increasing beyond there to 63, it had fallen to 57 miles an hour on the rising gradient to Anderson Crossing, then on the falling gradient beyond the crossing it had risen again to about 62 miles an hour, a speed which had been maintained to the Patuxent River bridge, less than 1,500 feet from the junction points. Finally the tape showed a marked deceleration, evidently resulting from the emergency brake application, which had slightly reduced the speed of the train to about 55 miles an hour at the point of derailment.

Engineer Horner was qualified to drive over the line between Philadelphia and Washington and had done so in both directions on numerous occasions since 1st March 1960.

According to the regulations and timetable, the ICC report continued, the speed of the train should not have exceeded 50 miles an hour and should have been reduced to 15 miles an hour at Del block post. The train brakes had functioned properly at Odenton and on being applied in emergency. The fact that the manual sand release controls on the locomotive had been found in the "off" position after the accident showed that the sand found on the rails close to the point of derailment must have been due to the automatic application of the brakes. These findings indicated that the train would have been under proper control had it approached the junction at a lower speed and had the brakes been applied at a reasonable distance from the junction. The ICC concluded that:

(1) The locomotive and train-brake apparatus were functioning properly at the time the accident occurred.

(2) The train was being operated materially in excess of the maximum authorised speed on the track over which it was moving and braking action was taken too late to make a safe approach to the branch turnout.

(3) The accident was caused by the train's entering the turnout at an excessive rate of speed, coupled with tardy brake application. [52]

The engine, tender and four leading bogies of passenger train No. 2 TF from Tundla Junction to Farrukhabad was derailed between Mainpuri and Bhongaon stations, Uttar Pradesh, India, on the Northern Railway, at 10:35 on Sunday 29th October 1961. Sixteen passengers, the driver and second fireman of the train were killed outright and a pointsman, another railwayman and two passengers died in Mainpuri Hospital. Fourteen passengers and three railwaymen were seriously injured, and 43 passengers and two railwaymen were slightly hurt. All but one of the passengers killed had been Ceylonese

[52] Interstate Commerce Commission report

Buddhist pilgrims, who had been on their way to a pilgrim centre in the Farrukhabad area riding in the second-class bogie next to the engine. Thirty other monks, the remainder of the party of 45, were among those injured.

Civil and police officials rushed to the scene of the accident, about six miles from Mainpuri, upon receiving news of it and the injured were taken to Mainpuri hospital. It was hard to obtain the names of the dead and injured because of language difficulties.

Mr Shah Nawaz Khan, Union Deputy for Railways, proceeded to the accident scene from Delhi. Mr C.B. Gupta, Chief Minister of Uttar Pradesh, announced that he had instructed the authorities of the Agra Medical College to provide medical relief to the sufferers and Mr Aluwihare, the Ceylonese High Commissioner in India, said that he had arranged for two members of the commission to proceed to the location of the derailment by car.

Guard B.K. Asthvana told the inquiry into the accident that his train had left Mainpuri Kacheri at 10:28. Immediately before the accident he had felt a slight jerk, had gone to the window of his van and then been thrown down by a more violent jerk. Seeing the extent of the accident he had gone to Crossing No. 6 nearby to get assistance from the crossing keeper in protecting the train.

Gateman Kanauji at Crossing No. 6 said that seeing the train approaching, he had closed and padlocked his gates. As the train had come closer he had seen that it was running much faster than usual. When the train was two and a half telegraph poles away he had seen the engine lurching violently, there had been some loud noises, and the engine and some coaches capsized just short of the crossing. He had seen nothing wrong with the track before the train appeared and had given the train no flag signals.

Gangman Kamaruddin stated that the train passed him and his gang going faster than usual. Shortly afterwards he had heard a loud noise. The next thing was that a villager sent by the gateman had appeared and told him of the derailment, whereupon he and his gang had gone to the scene. When he had passed the scene earlier there had been nothing wrong with the track.

Stationmaster R.P. Gautam at Mainpuri mentioned reports of rough running at different points.

Cowherd Basta Singh, who had been grazing his cattle near the line, said that the train had been going faster than any he had previously seen and he had witnessed the accident. No gangmen had been working at the site of the derailment; they had been 4 or 5 furlongs (800 or 1,000 metres) nearer Mainpuri. He had seen the gateman closing his gates; he had not seen him showing any flags.

Pedestrian Bankey Lal said he had been near the crossing. He agreed that the train had been going much faster than usual. The crossing gates were closed and the gateman had not exhibited any flag signals.

Various passengers on the train gave evidence and spoke of jerks preceding the derailment. Most seemed to think that the train had been travelling at its usual speed.

Driver E.S. Chatterjee of train No. 4ATF, which had left Mainpuri Kacheri at 06:07, and Driver Iqbal Hussain of a goods train that passed the derailment site at 06:55 stated that they had noticed no problems with the track near the crossing.

Traffic Inspector S.L. Kackar testified that he had issued some chits for rough running during the previous year but none in the current year.

Gateman Moharman, who had been on duty at the crossing until 06:00, stated that there had been nothing unusual in the running of the four trains that passed during the night and there had been nothing wrong with the track.

Booking Clerk-in-Charge Jagannath Prasad at Mainpuri Kacheri stated that the train had left there at 10:32 by the station clock.

Assistant Engineer Thakur Singh and Permanent Way Inspector K.N. Verna spoke of the drivers of Class AWD and CWD engines exceeding the speed limit through the section.

First Fireman Gurbachan Singh of train 2TF stated that the train had proceeded at the normal 25 to 30 miles an hour after leaving Mainpuri and he had not noticed any jerks or lurching of the engine. Approaching Crossing No. 6 he had just finished firing and was observing through the look-out window when a jerk from the front of the engine had thrown him against the window. He received burns and other injuries and could not say what had happened in the cab. As they neared the crossing his driver had his hand on the brake handle but he did not know if the brakes had actually been applied.

Divisional Engineer Madan Gopal stated that some drivers had exceeded the speed limit and he had warned them or reported them to the divisional mechanical engineer.

Divisional Mechanical Engineer J.L. Soni stated that he had warned or censured drivers reported to him as speeding by the divisional engineer.

Discussing the causes of the accident Additional Commissioner A.C. Khastgir noted that there were four possible causes for the derailment:

(a) Tampering with the track

(b) Track defects or obstruction

(c) Defect in the train equipment

(d) Excessive speed

(a) The track had been noted to be in good condition by an assistant commercial superintendent and the gateman shortly before the derailment. No unauthorised person or permanent way staff had been seen in the area after that and it was hardly conceivable that persons could have interfered with the track in broad daylight in full view of the gateman. Tampering with the track could be ruled out.

(b) Slight defects found along the whole stretch of the line by a motor trolley inspection would not have been sufficient to cause a derailment. If the gross damage to the track found after the derailment had existed before it had occurred it would have been obvious to the gangman, a keyman and the gateman. There was no evidence of the train driver having braked suddenly because the crossing gates had been opened in front of him or of any obstruction on the line.

(c) There was no evidence of any defect in the engine or rolling stock which could have caused a derailment.

(d) The guard unfortunately did not note the time of the derailment but a passenger who appeared to be a reliable witness had timed it at 10:35, which could be accepted as correct. The guard stated that his train had left Mainpuri Kacheri at 10:28, the booking clerk stated 10:32. The latter figure must be rejected as it would imply that the train had taken only 4 minutes, start to stop, to cover the 3.9 miles to the derailment site; the 10:28 figure could be accepted. Calculations as to the train speed based on such short distances and time intervals were not reliable but there was other evidence that the train had been going too fast.

Various witnesses who had been on the train or observed it from the lineside stated that it had been running at 40 miles an hour or more. In addition drivers had been warned on several occasions about exceeding speed limits and track recorders also showed trains running at up to 50% in excess of the maximum permissible speed. The statements of some of the train passengers that the speed had been normal might be because the train was regularly driven too fast.

The manner in which the engine had embedded itself in the ballast and the damage caused to the track were also indicative of high speed.

The additional commissioner decided that the train must have been travelling at about 45 miles an hour when it left the track.

Reconstructing the likely sequence of events leading up to the derailment the additional commissioner deduced that after passing Isan Bridge, just outside of Mainpuri, the driver must have increased speed. Gathering speed the train had started lurching, slightly distorting the track at Km 1267.28 and Km 1267.93. Further on, just beyond the culvert at Km 1268.63, further increased speed had caused violent lurching and a severe track distortion. At this point the driver, awakening to the danger, had closed the regulator and applied the brakes but the left Bissel wheel had mounted the rail, causing the jerk described by First Fireman Singh, and total derailment had followed.

The additional commissioner concluded that the high speed of the train had resulted in excessive flange forces and violent oscillation of the train, distorting the track, and causing the derailment. The sudden brake application had then caused the engine to capsize and the following four coaches to telescope. The deceased Driver Bhagita must be held responsible for the accident. [53]

Diretto train No. 152 from Lecce and Bari to Milano was crowded with possibly over 1,000 passengers when it arrived at the station at Castel Bolognese, Italy, at 01:56 on the morning of Thursday 8th March 1962. It was in charge of Conductor Gino Bizzarri from Ancona, with Driver Ennio Covacci of Ancona and Assistant Driver Otello Manzi from Civitanova Marche, and consisted of an electric locomotive (No. E428.217), two luggage vans and 12 carriages. Its passengers were largely soldiers returning from leave and men from the poorer areas of Southern Italy, who were going to look for better employment either in Lombardy or abroad in France, Germany, or Switzerland; the daily service was nicknamed the Train of Hope. As it was the early hours of the morning most of the passengers would have been asleep or trying to sleep. The train was due in Milano at 06:27 and was running about five minutes late when it reached Castel Bolognese.

---

[53] Government of India Ministry of Tourism and Civil Aviation (Commission of Railway Safety): Report On the Derailment of 2TF Down passenger train between Mainpuri and Bhongaon Stations (Northern Railway) on 29th October 1961

Work had been in progress on renewal of the northbound track at Castel Bolognese since 27th February, as a result of which trains were being diverted via a facing crossover 350 metres from the station platforms onto the southbound line through the station. The speed of trains through the crossover was limited to 30 kilometres an hour but the diretto arrived travelling at 102 kilometres an hour. As a result its engine and six leading vehicles were derailed and overturned. Damage to the overhead line equipment from the derailment caused the circuit breakers to operate and cut the current so the added danger of electrocution was fortunately absent but fourteen passengers died [54] and at least 86 were more or less seriously injured, with another 40 or 50 slightly hurt. The scene after the accident was likened to a war zone, with ambulances and other vehicles going to and fro with injured people and army, carabinieri and police vehicles. The station access roads were closed off and only cars carrying those involved in the rescue work were admitted. Pedestrians were also excluded unless they had a specific reason to be there. Even so a large proportion of the population went to the scene. The gutted carriages were said to have reminded an observer of the wartime bombing raids and there was said to be more devastation than there had been in the war.

Living alongside the railway in Via Parini No. 2 were Maria Dal Prato Sangiorgi and her husband Lino, both aged 58. The noise of the derailment, which was estimated to have gone on in total for only seven or eight seconds, woke the woman up and she shouted to her husband, "Lino, it's the end of the world." A derailed luggage van smashed up against their house, cracking its wall. Alarmed for the safety of her two young children Mrs Sangiorgi had run to their bedroom, to find them, very frightened, looking out of the window towards the railway.

As the cries of men, women and children emanated from the wrecked coaches Signor Sangiorgi, having realised what had happened, quickly dressed and made his way to the scene, where he was able to assist Stationmaster Giovanni Tarlazzi, Head Shunter Romano Corelli and other railwaymen. The rescue work began in almost total darkness, with the only light being provided by matches, torches, and from the headlight of the derailed engine, which remained on. Signor Sangiori, who was a lorry driver, loaded some of the injured into his vehicle to take them to hospital, while the stationmaster alerted Dr Bagnaresi, who lived nearby, and the latter in turn telephoned the Red Cross offices in Castel Bolognese, Feanza, Imola and Lugo, for ambulances. Within half an hour many rescuers had reached the scene but the work was poorly co-ordinated and there were complaints about the lack of lighting, stretchers, and rescue equipment, while some of the surviving passengers threatened to attack the railwaymen whom they believed were responsible for the disaster.

One of the first to be taken to Faenza hospital was 24-year-old Carmel Pavan from Venezia, who was returning from her honeymoon and was in despair because she could not find her husband. He survived, and they were reunited. Others were less fortunate. Fioravente Madonna, 36, from Chieti, told a reporter, "I had been sitting alongside my brother-in-law, Giuseppe Pasquale. We were going to look for work in France. Suddenly I felt a terrific blow and the next thing that I knew I was lying with my head down on a sleeper, my coach having overturned and been wrecked. I looked for Giuseppe and found

[54] https://www-bardellarotaia-it.translate.goog/viewtopic.php?f=28&t=774&start=105&_x_tr_sch=http&_x_tr_sl=it&_x_tr_tl=en&_x_tr_hl=en&_x_tr_pto=sc

him dead by my side."

A rail torn from the track had penetrated a carriage, the fourth vehicle in the train, passing through a corridor and several compartments (it was thus what on early United States railroads was termed a "snakehead"). It was in this coach that the fatalities in the accident occurred. The rail struck a woman passenger in the neck and grazed the nine-month-old baby she was holding in her arms. The woman, Annamaria Ventura, 22, from Rocco Morice in Pescara Province, was killed, but her baby son, a 4-year-old child, and her husband all survived. The husband was trapped in the wreckage and called for rescuers to save his children. Romano Scardovi, a Rimini footballer who lived in Castel Bolognese, rescued the children, and took them to the local hospital. When the husband was later released he taken to Imola hospital, where his children were later transferred to join him, thanks to the efforts of the Ravenna prefect.

The morning after the accident several senior officials went to the scene of the disaster including General Camilleri, commander of the Emilia police; General Simonetti of the Bologna brigade of the carabinieri; Dr Petta, the Ravenna prosecutor of the Republic; Dr Albanese, the Attorney General of the Bolgna Court of Appeal and Engineer Robertthen, the departmental director of the Bologna railways. By the afternoon the bodies of all the dead had been recovered, together with the passengers' personal property and the contents of the luggage vans, which included thousands of pairs of women's shoes destined for customers in Northern Italy from a factory at Porto S. Elpidio in Ascoli Piceno, large numbers of poultry in cages, fish, rabbits, and truffles. Then a powerful crane and 200 workers had begun to remove the wreckage and repair the damage to the track.

News of the accident was announced in the Camera dei Deputani (the Chamber of Deputies – the lower house of the Italian parliament) and the members expressed their condolences with the victims and their families. It was stated that Undersecretary Cappugi and Engineer Rissone, the general director of the Ferrovie dello Stato, were on their way to Castel Bolognese to open an investigation, while Prime Minister Fanfani stated that the government had a plan for the reorganisation and upgrading of the railway system over the following five years, which would involve an expenditure of 800 billion lira.

It emerged that the procedure for warning train drivers of a temporary speed restriction such as that involved in the disaster had been altered a year previously. Prior to that date the signals in the vicinity of the restriction were maintained at Danger until the train had been brought to a stand and the train had then been authorised to proceed through the restricted area by the stationmaster. In the revised procedure the signals were no longer maintained at Danger but instead the train drivers were issued with a special form advising them of the restriction. Driver Covacci had duly been issued with the relevant form at Rimini station less than a quarter of an hour earlier but apparently had then forgotten about it until his engine jolted violently as it reached the crossover. He had then applied the brakes in emergency but by then, of course, it was too late. Coming from the Faenza (Ravenna) direction there were three signals approaching Castel Bolognese, situated 2 kilometres, 1 kilometre 200 metres, and 1,130 metres in the rear of the crossover. All had been clear for the train. In addition there was a large lineside warning sign with white (?) numerals on a black background, displaying '30/30', advising drivers of the restriction and situated 1 kilometre 500 metres from the crossover.

The driver, assistant driver and conductor were all placed under arrest on multiple charges of manslaughter and causing injury. They were not taken to prison as they were all patients in Castel Bolognese hospital, where they were guarded by carabinieri.

A possible factor Driver Covacci's error was the presence of a third person in the engine cab. This was Electrical Technician Lodovico Fiumana from Forli, who worked for the Implanti elettrici segnalamentali (Signalling Electrical Systems – a branch of the Ferrovie dello Stato). He stated, "I had come from Forli upon the orders of my superiors to check the voltage values and current intensity of the locomotive over the section between Castel Bolognese and Imola, where work had recently been undertaken on the electrical system. I exchanged a few words with the two drivers, whom I did not know, when I joined the engine, and then began to work with my instruments. I was sat behind Covacci and Manzi in the cab and did not speak to them as we reached Castel Bolognese." The two drivers likewise denied that they had been distracted by the technician. They admitted that they had received an M40 form at Rimini station containing the information regarding the speed restriction. It appears that they might have been misled by the clear signals and forgotten that the system had changed, while the warning sign was apparently unlit except by the locomotive headlight. [55]

Driver Covacci was later convicted of manslaughter and sentenced to five years and eight months' imprisonment but did not lose his job. Assistant Driver Manzi was not charged with any offence. The method of warning drivers of temporary speed restrictions was changed after the disaster so that the signals no longer showed clear at a location where a restriction was in force. [56]

A crowded passenger train jumped the rails at Iswepe, South Africa, in the eastern Transvaal, on Wednesday 21st March 1962 and five coaches were telescoped. Ten 'non-whites,' as they were referred to in the language of the time, lost their lives and 52 'non-whites' were injured. The fireman of the train was reportedly critically ill with burns. [57]

Eight or nine passengers died and 35 were injured in the derailment of a train on Sunday 1st April 1962 between Pas Viejo and Pas Centro in the western part of Nicaragua. [58] All those who lost their lives had been standing on the rear platform of one of the coaches. [59]

Thirty-two people died and 54 were injured on Sunday 22nd July 1962 when a passenger train was derailed between Bucharest and Mogosoia, Romania. The Ministry of Transport announced that the accident had apparently resulted from excessive speed on a curve, that it was being investigated, and that all legal steps would be taken to punish those responsible for it. [60]

---

[55] *La Stampa*, Torino, Italy, 9th and 10th March 1962
[56] Wikipedia – Castel Bolognese train disaster
[57] *The Times*, London, England, 22nd March 1962
[58] *The Idaho Free Press*, Nampa, Idaho, United States, 4th April 1962
[59] *The Republican Courier*, Findlay, Ohio, United States, 3rd April 1962
[60] *The Times*, London, England, 26th July 1962

Nine people died and 80 were injured in the derailment of suburban train PM-4 from Rio Acima to Belo Horizante at the Curva do Boiadeiro, near the station at Horto Florestal, Brazil. Alfredo Luis de Freitas, the driver of the train, was allegedly drunk, having consumed a quantity of alcohol at Cacamo Furquim station. In his inebriated state, it was said, he had threatened to kill any passenger who remonstrated with him, then proceeded to drive the train 30 kilometres an hour faster than he should have done, so that it left the rails on the curve.

After the accident passenger Valdomiro Patrocimo carried out a citizen's arrest on the driver and handed him over to Stationmaster Artur Braga at Horto Florestal, but in the face of threats from some of the other passengers to lynch the culprit, the stationmaster allowed the driver to flee.

The injured passengers were taken to Belo Horizonte Pronto Socorro Hospital. [61]

A passenger train was derailed at Holfontein, South Africa, not far from Kroonstad in the Orange Free State, on Tuesday 2nd October 1962. An initial report spoke of 12 fatalities and 30 injured, while according to a second report the number of dead had risen to 17 and injured to 40. The dead were all Africans, the injured both Africans and whites. About 200 children, who had presumably been in a different part of the train, escaped injury. [62] [63]

Twenty-three or twenty-four people died when a Belgrade – Skopje passenger train was derailed as it was entering the station at Kabanovce, Yugoslavia, early on Monday 5th November 1962, The engine and a guards van left the rails and the following passenger coach was telescoped into the guards van. At the time of the first reports on the accident rescuers were still struggling to free victims trapped underneath two overturned coaches. About 17 other passengers were injured. The cause of the derailment was not immediately clear. [64] [65]

Twenty-one people died and 40 were injured in the derailment of a train near Surabaya, East Java, Indonesia, on Saturday 13th July 1963. [66]

Twenty-one people died and 108 were injured when a night train jumped the track and reportedly plunged into a ravine at an unstated location in West Java, Indonesia, on Thursday 30th April 1964. [67]

On Sunday 26th July 1964, Portugal was to suffer the worst railway accident in its

[61] *Jornal do Brasil*, Rio de Janeiro, Brazil, 19th September 1962
[62] *The Times*, London, England, 3rd and 4th October 1962
[63] *Joplin News Herald*, Joplin, Missouri, United States, 3rd October 1962
[64] *The Times*, London, England, 6th November 1962
[65] *Wichita Falls Times*, Wichita Falls, Texas, United States, 5th November 1962
[66] *The Gettysburg Times*, Gettysburg, Pennsylvania, United States, 15th July 1963
[67] *Reno Evening Gazette*, Reno, Nevada, United States, 2nd May 1964

history. On summer weekends many people escaped the city of Porto to spend time on one of the beaches north of the city at resorts such as Espinho, Vila do Conde and Póvoa de Varzim. On Sunday evenings they took the train back into Porto. Usually a train left Póvoa de Varzim at 21:10, followed by a railcar and trailer at 21:30, but on the evening in question the trains were running out of course, and diesel railcar No. 9309 with its trailer was first. By now it was dark and the townspeople, having enjoyed their day at the beach, were keen to get home. Few of them wished to wait for the second train, which was only 12 minutes behind, thus they all piled into the railcar and its trailer, so that the train was already overcrowded when it left Póvoa de Varzim, while more passengers got on at Pedras Rubras and Crestins. Ticket Inspector Orlando Pereira Mendonça, who suffered a broken spine in the accident, estimated that there had been 200 passengers packed into the trailer alone, though it only had a capacity of 112 people (68 sitting and 44 standing).

Not long after leaving the halt at Crestins, as the train was descending a steep gradient and negotiating a sharp curve, Driver Manuel Ferreira suddenly felt the load of his train lighten dramatically, the front of his train shot forward and he heard a crash behind him. The brakes were applied and the train stopped a short distance down the line. The trailer had left the rails on a curve, smashed into one of the granite piers of Carvalhas Bridge, and had been demolished. The accident happened near Custóias, nine kilometres short of Porto, at about 22:20. [68] The train had been due at Porto Trinidade at 22:21.

A cyclist chanced to witness the accident and the sound of it was also heard by nearby residents, so the emergency services were quickly alerted. The rescue operation was difficult though, not only because of the darkness but because at the point of the derailment the railway ran at a lower level than an adjoining road.

A reporter wrote: 'The first rescuers were local residents and the occupants of cars on the road. They had to grope about in the darkness among the twisted metal of the wrecked coach to try to drag out the victims. Some of the injured were carried on the back of their rescuers up the steep slope, or on improvised stretchers. The bodies of the dead were hauled up on ropes. Many people were killed outright, and others were suffocated by those who fell on them. Many had lost limbs, and some were unrecognizable. Children's beach toys and picnic baskets were collected in a pile at the side of the wreckage. Firemen worked through the night sawing through metal with acetylene lamps to find victims. Scores of volunteers have besieged Oporto hospitals offering blood for the injured, while every available doctor and nurse was mobilized.' [69]

The injured were taken to the Santa António, São João, Militar and the Casa de Saúde da Boavista hospitals in Porto; the bodies of the dead were taken to mortuary at Santa António, which was overwhelmed by the large number of them. The last body was not recovered until 10:30 the next morning. Initially it was stated that there were over 100 dead and 170 injured.

As might be expected the disaster caused a sensation in the country, especially in and around Porto. It received extensive coverage in the national newspaper and reporters managed to conduct brief interviews with several of the surviving passengers. An unnamed man spoke of lighting a match to see and walking over the bodies of the dead.

---

[68] Wikipedia – Desastre ferroviário de Custóias

[69] *The Times*, London, England, 28th July 1964

He found his wife under a bench and took her in his arms (the report does not make it clear if she was dead or merely injured). Later at the hospital he was reunited with his little girl, whom someone else had brought there. He was clearly also worried about how the accident would affect him afterwards, telling the reporter, "I'm a metallurgist. How long am I going to be in hospital? Who is going to look after my house? I shall be blamed for getting into an overcrowded carriage but no one forbade me from doing so."

Sr. Jorge Victor Guimãraes was rather luckier as he had been riding in the front car of the train. He said that suddenly all the passengers had been thrown over to the right. There had been a loud noise and something had shattered. The train had screeched to a halt and there had been shouts and screams in the darkness behind him. "The second car has gone!" [70]

Not everyone on the train had been returning from the beach. Sr. João Martinho, a fisherman, and his wife, Sra. Maria de Luz Rocha, a fruit seller, had been on their way to Porto to visit a son who was in military service. They were somewhat hurt in the accident, abandoned their visit to their son and instead walked back along the main road through the night the 25 kilometres to their home near Vila do Conde. Many other uninjured and slightly injured passengers did likewise. [71] Meanwhile in the aftermath of the disaster the local population offered meals, money and other assistance to strangers who had been in the accident, as did various organisations. On the morning after the accident a mother who had emerged from the mortuary carrying the white coffin of her little daughter was surrounded by people offering her help.

A Minister of Communications stated that a commision of inquiry had been set up to investigate the disaster, composed of senior officials from the railway company, the Caminhos de Ferro Portugueses, together with officials from the Direcção-Geral dos Transportes Terrestres (the body set up in 1951 to regulate road and rail transport). The minister stressed that the inquiry would have nothing to do with the determination of civil or criminal responsibility, which would be determined by the Public Prosecutor's Office and the judicial authorities.

The train involved in the accident had been built in the Netherlands and was designed to run on metre-gauge lines, such as that between Póvoa and Porto. It was of steel construction and its motor unit was 19.5 metres long, weighed 37 tonnes tare and 43 tonne gross, with 12 seats and 10 standing places in first class and 32 seats and 16 standing places in second class, according to a newspaper report (quite why anyone would want to pay extra to travel first class, just to stand anyway, is not clear). The trailer was the same length, had a tare weight of 22 tonnes and a gross weight when full of 31 tonnes, with 86 seats and 44 standing places. At the time of the accident, however, the calculated laden weight of the trailer, given the number of excess passengers, was over 20 tonnes, giving a gross weight of over 42 tonnes.

There had been no apparent defect in the train itself before the accident and afterwards it could be towed to Porto Boavista, though not without becoming derailed again en route between Senhora da Hora and Boavista. It was believed that the excess weight of the trailer had prevented its bogies rotating as they should have for the trailer to negotiate the curve, hence it had become derailed, separating at the same time from the leading car.

[70] *Diário de Lisboa*, Lisboa, Portugal, 27th July 1974

[71] Ibid, 28th July 1964

What had made the accident so serious was the chance proximity of the bridge pier (as in the March 1951 derailment at Doncaster, England, described in the previous volume of this series). There were also questions as to the speed of the train, the maximum speed around the curve being limited to 55 kilometres an hour. Some of the surviving passengers spoke of high speeds during the journey but Driver Ferreira insisted that he had not exceeded the speed limit. [72]

Various figures have been quoted for the total number of fatalities in the disaster but it would appear from contemporary newspaper reports that the correct figure is probably 88, the final two deaths being those of passengers who succumbed to their injuries on about Thursday 30th. No further deaths had been reported by Wednesday 5th August, by which date only a handful of patients remained in hospital. [73]

It is hard to see how anyone could have been held responsible for the disaster. There seems to have been no proof that the train was travelling too fast. Perhaps the station staff at Póvoa de Varzim and elsewhere should have made more effort to prevent too many people boarding the train, though that would probably have been difficult.

A passenger train was derailed about five kilometres from the mining town of Randfontein, South Africa, at about 20:00 on the evening of Wednesday 29th July 1964. The two electric locomotives and six wooden coaches of the train left the rails, three telescoping one another. Firemen had to use cutting equipment to free passengers from the wreckage. Some reports stated that 21 passengers were killed and 36 were injured, and that 16 of the passengers, including two babies, were white and five were black. According to another report a railway spokesman stated that only ten white passengers had lost their lives [74] and a third that 18 bodies had been recovered. [75] The train had left Randfontein station a few minutes earlier.

The police and Mr J.P. Hugo, general manager of the railway, stated that there was no immediate evidence of sabotage as a cause of the derailment, though Nationalist groups had carried out bombings in the area. [76] [77]

In March 1965 Driver Nicholas Erasmus was sentenced to 12 months' imprisonment at Johannesburg after pleading guilty to a charge of culpable homicide, probably in connection with this accident. [78]

Express train No. 94, otherwise the *Nordpilen*, met with an accident at Alby, Sweden, on the evening of Saturday 5th September 1965. The train consisted of a Statens Järnvägar Ma class electric locomotive, a mail van, a baggage van, two sleeping cars, two sitting cars, a restaurant car, two sleeping cars and seven sleeping cars. The line from Ramsjö via Alby to Ånge was double track with a maximum authorised speed of 100 kilometres

[72] Ibid, 29th and 30th July 1964
[73] Ibid, 31st July 1964
[74] *The Kingston Daily Freeman*, Kingston, New York, United States, 30th July 1964
[75] *The Republican-Courier*, Findlay, Ohio, United States, 30th July 1964
[76] *The Times*, London, England, 30th July 1964
[77] *Nieuwsblad van het Noorden*, Groningen, The Netherlands, 30th July 1964
[78] *The Times*, London, England, 30th March 1965

an hour but on the evening of the accident relaying operations were in progress on the Up line between Alby and Ånge and there was single-line working over the Down line between those stations. Speed over the crossover from the Up to the Down line at the north end of Alby station was limited to 40 kilometres an hour. The line between Ljusdal, south of Ramsjō, and Ånge was remotely controlled from the latter station and Alby station was unstaffed at night.

The engine crew of train No. 94 from Stockholm had been changed at Bollnäs. The train approached Alby at the regulation 100 kilometres an hour with the driver and his assistant repeating the signal indications to one another, as per the regulations. They observed that the *utfartsförsignalen* (distant signal for the exit or starting signal) was displaying a flashing green and a steady yellow (indicating caution) and the *utfartssignalen* (exit or starting signal) was displaying two steady greens (proceed at not more than 40 kilometres an hour). The line fell sharply approaching Alby, so that an early and fairly forceful brake application was necessary to bring the speed of a Down train down to 40 kilometres an hour at the exit from the station. The crew of the *Nordpilen* either failed to apply the brakes at all or at any rate did not do so early enough, with the result that the train was still travelling at not less than 80 or 90 kilometres an hour when it reached the crossover. The rear bogie of the fourth vehicle of the train was probably the first to leave the rails and pulled the coaches in front and behind with it. In all, nine of the coaches of the train left the rails. The coupling between the second and third vehicles of the train broke during the derailment and the engine and leading two vehicles came to a stand, on the line, 750 metres further on. Five vehicles at the rear of the train also escaped derailment but the middle coaches ended up spreadeagled across the track. Eight passengers lost their lives in the accident, all but one of whom had been in the sleeping cars, which were the third and fourth vehicles on their sides, and which overturned with their side corridors uppermost and their sleeping berths against the ground. The eighth fatality was a passenger in the sitting car marshalled fifth in the train. Another 31 passengers were injured, three seriously. Those who were injured included two occupants of a lineside house, which was struck by one of the derailed coaches.

The derailment occurred at 22:48. Five minutes after the accident, the SOS centre in Sundsvall was advised by a passenger on the train by telephone from a house nearby and at about the same time the train crew advised the control at Ånge of the derailment and they in turn also contacted Sundsvall. Ambulances, firemen and police were sent to the scene from Ånge and elsewhere and local railwaymen and other inhabitants were also quickly on the scene. A breakdown train was also readied at Ånge but could not reach the accident site until a steam locomotive brought the electric locomotive and leading two coaches of the accident train there. The rescue operation went on in darkness and increasingly thick fog. A woman procured all the torches that could be found in a nearby store and fortunately the battery-operated lighting of the train coaches mostly remained on.

There had been 400 passengers on the train, most of whom were asleep at the time of the accident and they were helped off the train dressed in their nightclothes. Locals provided them with clothes while they were awaiting evacuation by ambulances, busses, and taxis. A rescuer related later, “When I was able to climb up onto one of the overturned coaches I could hear people desperately pounding on the coach walls below. I saw passengers lying about who had been thrown on top of one another, suitcases, bedding, and broken glass lying in the mud and injured with luggage piled up on them.”

The last of the injured and trapped passengers was released from the wreckage at 02:30 and the work of clearing the line could then begin.

The line was reopened for traffic at 13:57 on Sunday with a 10 kilometres an hour speed restriction past the accident site.

It was fortunate that the accident had occurred in a populated area so that help was quickly on hand from people nearby, and also fortunate that there chanced to have been both a doctor and a nurse amongst the train passengers. [79]

How the engine crew had accounted for their failure to reduce speed for the crossover was not stated.

Twelve people lost their lives and another 39 were injured on Monday 1st February 1965 in the derailment of a freight train carrying timber near Sukabumi, West Java, Indonesia. The victims were presumably unauthorised passengers who were riding atop the timber. The lumber slipped off the waggons and in doing so derailed the train. [80]

A major disaster on the Rede Ferroviária Federal, on a line of the former Estrada de Central do Brasil, at Comendador Soares, Brazil, on the evening of Sunday 28th March 1965, involved crowded passenger train M-279 from Don Pedro II station in Rio de Janeiro, bound for Taireta, and freight train KP-2 from São Paulo. The trains were passing one another when the freight train became derailed, following which the passenger train was either struck by or ran into the derailed waggons. A newspaper reported that there had been a deafening roar followed by dramatic scenes with men, women, and children running in all directions, screaming for help, while among the twisted ironwork and mutilated corpses dozens of people were groaning.

Ambulances from Carlos Chagas and Getúilo Vargas hospitals took the injured to nearby Nova Iguaçu. They were assisted by residents of Comendador Soares, police and soldiers from Paracambi, while firemen from the Camphino barracks removed the bodies and the seriously injured from the wreckage. Several of the badly injured died on their way to hospital. The drivers of the two trains, whom the newspaper seemed to consider somehow responsible, which seems doubtful at any rate, both disappeared after the accident, doubtless because they feared indefinite detention if they were arrested within 48 hours under an obscure Brazilian law. An initial report spoke of 19 dead and about 33 injured. [81]

Three coaches of a crowded passenger train carrying 1,500 black workers home to Kwa Mashu Township were derailed at Effingham Junction, Natal, South Africa, 10 miles north of Durban and about two miles from its destination, at 6 p.m. on Monday 4th October 1965. Initial reports spoke of 100 passengers killed and several hundred injured.

'A white railroad employee told from his hospital bed Tuesday how he watched frenzied Negroes kill his friend minutes after the worst train crash in South African

[79] Statens Järnvägar: Redogōrelse fōr tägmissōdet I Alby den 5.9.1964
[80] *The Independent*, Pasadena, California, 2nd February 1965
[81] *Folha de S. Paulo*, São Paulo, Brazil, 29th March 1965

history killed 81 Negroes and injured 130 Monday night.

'Franz Jagles, 34-year-old station foreman, was the only white passenger aboard the packed suburban train, part of which was derailed as it approached Effingham Road. Jagles told reporters he was sitting in the next-to-last car next to some Negroes. "I looked around. All around me were bodies. A chap who had his head cut off was lying next to me, and not far from me a woman sat on a rail holding half the body of her child."

'Then, Jagles said, he saw his white friend, signalman Wally Hartslief, running from his signal box to help; a group of Negroes started screaming and rushed at Hartslief. "They were kicking and hitting. I saw Wally throw his hands up and I knew it was the end."

'Jagles said he managed to get clear of the wreckage and set out on a nightmare crawl through undergrowth to escape the crowd. A Negro night watchman at the quarry telephoned for police and ambulances.' [82]

A broadcast for doctors, nurses and medical equipment was sent over the national radio network and over 40 ambulances went to the scene. [83] [84]

The day after the accident the number of fatalities was stated to be 81, though it was believed that the final figure would be higher. An official inquiry into the accident had been opened. [85]

It has not been possible to uncover the results of the inquiry into the derailment, which presumably was able to determine its cause. The final number of dead has been quoted as 91 with 56 injured. [86]

At about noon on Monday 14th February 1966 a passenger train descending what was described as a 'zig-zag hill section of track' and travelling at speed was derailed shortly after leaving Shwenyaung station in central Burma. Seven coaches crashed into a deep valley. Helicopters took the injured to hospitals in Kalaw, the nearest town, and Rangoon. Twenty-three passengers died and 30 were critically injured according to reports reaching Rangoon. Six women and a child were said to be among the dead. The train was crowded with delegates who had been returning from celebrations marking the union of Burma's states. [87] [88] [89]

At least nine people died and 14 were injured when a Philippine National Railways passenger train was derailed about 160 miles south of Manila on Monday 4th April 1966. Retired General Alfredo Santos, the manager of the railway, believed that the final number of victims would be considerably higher. He said, "We have received only initial

[82] *The Citizen*, Prince George, British Columbia, Canada, 6th October 1965
[83] *The Daily Colonist*, Victoria, British Columbia, Canada, 5th October 1965
[84] *The Daily Gleaner*, Kingston, Jamaica, 5th October 1965
[85] *The Times*, London, England, 6th October 1965
[86] https://www.heraldscotland.com/news/12681976.63-killed-as-south-african-commuter-train-jumps-rails/
[87] *The Lethbridge Herald*, Lethbridge, Alberta, Canada, 14th February 1966
[88] *The Salt Lake Tribune*, Salt Lake City, Utah, United States, 15th February 1966
[89] *Press-Telegram*, Long Beach, California, United States, 15th February 1966

reports so far from the scene. I presume that the casualty will increase as soon as we get reports." He added that the train had 10 or 12 coaches, each with a seating capacity of about 76, and there were several hundred passengers on it at the time of the accident. It had departed from Manila at 5.50 a.m. and been on its way to Legaspi in southern Luzon Province, when it left the rails at 1.20 p.m., at Sumulong in Quezon Province. A baggage car, a kitchen car, two first-class and two third-class carriages had become derailed. Railway officials had sent doctors, nurses, and an investigating team to the scene and the governor of the province was sending a helicopter loaded with medical supplies. The cause of the derailment was not immediately clear. [90]

The rear coach of a Montargis – Nevers train was derailed about 100 metres from the station at Cosney-sur-Loire, France, on the evening of Friday 21st October 1966. The train had just left the station. The derailed coach was dragged some distance before the train could be brought to a stand, when cries for assistance were heard from it. A rescue operation was begun at once, the rescuers releasing the passengers under the glare of electric lights powered by generators and waiting ambulances took them away. Ten passengers died and 14 were injured. An investigation was opened into the accident. [91]

At least 13 people were killed and 20 injured on Saturday 17th December 1966 when a passenger train was derailed near Khandvara station, India. [92] Another report put the location of the accident as near Surat, stated that there were 24 injured, three coaches were overturned and three more were derailed. [93]

Six people died and a number were presumably injured in the derailment of a passenger train near Frankfurt Oder, East Germany, on Monday 19th December 1966. [94]

Five passengers and a driver's assistant lost their lives and others were injured in the derailment of an 'accelerated' train hauled by steam locomotive No. Pt 47-73 near Rzepin, Poland, on the line between there and Drzeńsko on Thursday 29th December 1966. The train driver, unfamiliar with the line, had attempted to negotiate a crossover at too high a speed. [95]

A serious derailment occurred on the Căile Ferate Române (Romanian State Railway) at Frunzeasca, Romania, probably on Sunday 12th March 1967, though another source states that the accident occurred two years earlier, in March 1965. Frunzeasca was a crossing place on the single line equipped with semaphore signalling, which appears to have followed German practice. Two trains were indirectly involved in the accident, No. 612, an

---

[90] *Pacific Stars and Stripes*, Tokyo, Japan, 6th April 1966
[91] *Gazette de Lausanne*, Lausanne, Switzerland, 22nd-23rd October 1966
[92] *The Brownsville Herald*, Brownsville, Texas, United States, 18th December 1966
[93] *The Lowell Sun*, Lowell, Massachusetts, United States, 17th December 1966
[94] *Katastrophen der deutschen Bahnen Teil II* by H.J. Ritzau, pg. 296
[95] Wikipedia – Railway Accidents in Poland

Iaşi – Bucharest accelerated passenger train, drawn by Pacific class steam locomotive No. 231.067, and a freight train. The passenger train normally passed through the station on the main line, travelling at full speed, the freight train being routed onto the loop line, but on the day of the accident the freight train was too long to fit in the loop and so Stationmaster Stroe Tiberiu decided to keep the freight train on the main line and instead route the passenger train onto the loop line, an operation which required it to pass through the connection from the main line to the loop at considerably reduced speed. Normally the driver of the passenger train would have been advised of such a diversion in advance, via a traffic order, but the decision was presumably taken at the last moment, too late for this to be done, probably when the freight train arrived and it was seen how long it was.

Approaching Frunzeasca station the driver of the passenger train received a clear signal at the distant signal, despite the route being set for the loop line, and so continued at full speed. At the equivalent of the home signal both its arms were raised, indicating that he needed to reduce speed for the diverging route, but he failed to see the signal quickly enough to reduce speed sufficiently to negotiate the facing connection to the loop line. The engine and leading four coaches left the rails and were overturned and the following two coaches were derailed. Ten people died and 24 were seriously hurt (another source speaks of 50 dead and 80 injured).

Following the accident the signalling was modified with the provision of an additional arm at the distant signal to indicate to the driver if the line was set for a diverging route, while from 1973 radio communication began to be provided between stations and locomotives to enable drivers to be informed of matters such as late diversions. [96] [97]

At least six people, most of them children, died and 131 were injured in the derailment of a Philippine National Railways passenger train about 290 miles south of Manila on Tuesday 30th May 1967. There had been about 800 passengers on the train when it left Manila on Monday night but fortunately it had been near the end of its journey, with only two stops to go, when the accident occurred at Locsin, the Philippines, in Albay Province. Three of the ten coaches overturned and three others left the rails. Railway officials sent doctors and nurses to the scene and the injured were taken to clinics and hospitals. [98]

On the afternoon of Saturday 30th December 1967 passenger train No. 62 was derailed at Kilometre 16 of the Cali – Buenaventura line of the Ferrocarriles Nacionales de Colombia near the town of Cordoba, Colombia. Seven people lost their lives and 37 were injured, five of whom were seriously enough hurt to be admitted to the Santa Helena Hospital in Cali. There were some fears a couple of days later that other victims might still be trapped in the wreckage, as a few people thought to have been on the train had not returned home. The derailment had occurred on a curve, and was believed to have resulted from excessive speed, though a definite cause awaited the results of an official investigation. [99]

---

[96] https://www.railnet.ro/viewtopic.php?f=119&t=4119
[97] https://forum.lokomotiv.ro/threads/noi-carti-feroviare.6786/
[98] *Pacific Stars and Stripes*, Tokyo, Japan, 1st June 1967
[99] *El Tiempe*, Bogotá, Colombia, 2nd January 1968

Six coaches of a passenger train were derailed between Quincieux and Saint-Germains-au-Mont d'Or, France, on the afternoon of Monday 1st July 1968. It was the rear part of the train that left the tracks on a curve and the first three of the derailed coaches, including the dining car, piled up on top of one another, while one coach ended up on an embankment. Five people were initially reported to have lost their lives and 30 to have been injured. One of the injured probably died later. Many of those on the train were children. The plight of the injured was made worse by the great heat reigning in the Rhône Valley. One theory as to the cause of the derailment was that it had been caused by a broken coupling. [100]

A horrific accident occurred at Langlaagte, South Africa, in the suburbs of Johannesburg, on Monday 17th February 1969. An engine was shunting seven tanks waggons loaded with petrol on a siding when two of them derailed, fouling a running line in the path of a passenger train carrying African workers from the township of Orlando into Johannesburg. Sixteen thousand gallons of petrol were spilt onto six passenger coaches, the petrol caught fire and enveloped the coaches while the remaining tank waggons were probably also involved in the fire. Flames reportedly shot 100 feet into the air and there was an explosion that could be heard a mile away. Witnesses reported seeing "human torches" fleeing from the train, while burning petrol spread over an area of half a square mile. The coaches and the track were buckled by the heat and smoke from the fire was visible throughout the western suburbs. The heat also damaged the overhead line equipment, stranding thousands of commuters to the West Rand.

The only figure published as to the number of dead was 13, but it seems likely that many more of the 300 workers on the train must have perished in the following days, as some of the injured reportedly suffered 95% burns. White ticket examiners S.H. Roets (46) and Johan Viljoen (23) were found afterwards wandering along the track, 300 yards from the scene. Both were badly burnt. Mr Roets' wife collapsed from shock upon seeing him. The couple had three daughters of school age. Miss Martie Cronje, fiancée of Mr Viljoen, was not permitted to see him in his severe condition.

Fire engines and ambulances rushed to the accident scene.

Mr B.J. Schoeman, the Minister of Transport, promised to issue a statement regarding the accident the following day. [101] [102]

A passenger train packed with May Day trippers was derailed near Sofia, Bulgaria, on Thursday 1st May 1969. Twenty-six people lost their lives and 40 were injured, many seriously. [103]

Late editions of a London newspaper on Wednesday 7th May 1969 reported the

---

[100] *Journal de Genève*, Geneva, Switzerland, 2nd July 1968

[101] *The Rhodesia Herald*, Salisbury, Rhodesia, 18th February 1969

[102] *Austin Daily Herald*, Austin, Minnesota, United States, 18th February 1969

[103] *The Holland Evening Sentinel*, Holland, Michigan, United States, 2nd May 1969

derailment of a London – Aberdeen a quarter of a mile from the station at Morpeth, England, on British Railways, Eastern Region, in which one person had died. The following day it was stated that the accident had been considerably more serious than first reported and in fact six people had died: five passengers and a travelling ticket inspector.[104]

Describing the accident a newspaper reporter wrote, 'A doctor, aged 27, who was injured when the overnight Aberdonian Express train crashed at Morpeth, Northumberland, worked until he was exhausted trying to help other injured passengers.

'Dr. John A. Habeshaw, a lecturer in pathology at Edinburgh University, was returning home from a business trip to London. He gave pain-killing injections to the more seriously injured passengers. Eventually police persuaded him to go to hospital to receive treatment for his own bruises and shock.

'The train from King's Cross to Aberdeen was derailed about a quarter of a mile from Morpeth station, as reported in later editions of *The Times* this morning. Six people were killed and 103 injured.

'The train's 11 coaches were scattered in many directions. The diesel electric locomotive continued along the track for several hundred yards before it came to rest in Morpeth station. Two coaches shot up the embankment and stopped with their wings resting on the wrecked carriages below. A van coupled to the engine was smashed.

'Heavy lifting gear was soon in use, and immediately after the alarm an emergency arrangement for the area was operated; crews of policemen, firemen, doctors, ambulance workers, and civil defence and W.R.V.S. workers went to the scene. More than 100 firemen worked in the dark and later with the aid of flares to free trapped people. It was a wet, moonless night. About 84 passengers were able to walk away from the wreckage and later boarded another train. The injured were taken by convoys of ambulances to hospitals at Ashington, Morpeth and Newcastle.

'Mr John Rinfret, aged 38, a company director, of West Ashby, Horncastle, from his bedside in Newcastle Infirmary, said he was in his pyjamas fast asleep in a bottom bunk when he was awakened by the sways and jolts of the carriage, which eventually came to a stop. All was in darkness, he said, but there was no panic.

'He managed to find an overcoat, a shoe, and a slipper. "We just waited silently until someone came. I felt a trickle on my forehead but I wasn't much hurt. Eventually we all got out very quickly when there reports that propane gas was leaking from a cylinder and we all hoped that no one would strike a match." He was taken to a house nearby.

'Twenty-seven injured people were taken to Ashington Hospital but only six were detained; 23 were admitted to Newcastle Infirmary, but none was seriously hurt. Nineteen went to Newcastle General Hospital, but only 13 were detained, six with severe injuries. Two were critically ill today. They were Mr Brian Coppard, aged 19, of Redbourn, near St Albans, and Mr James Mein, aged 40, of Bartongate Avenue, Edinburgh.

'Northumberland police said tonight they had identified three of the dead, They were Mr Samuel Elliot, aged 38, of Brighton Road, Edinburgh, a travelling ticket collector; Mr Malcolm Henderson, of 229 Signal Squadron, Berwick, and Mr P.S. Harrison, of

[104] *The Times*, London, England, 7th May 1969

Eastfield Avenue, Melton Mowbray. The other dead were two elderly women and a middle-aged man.

'David Wilsworth writes from Morpeth: The British Railways inquiry will have to look at three possible factors contributing to the disaster; the speed of the train, the condition of the track and the possibility of an obstruction on the line.

'There is a 40 m.p.h. limit on the bend which curves to the right approaching Morpeth station from the south station from the south. There is little evidence that the bend has been particularly dangerous in the past. The last accident there was at least 60 years ago.

'A railway gang was working at full speed today to lay a new section of track over several hundred yards.

'It was regarded as significant that the locomotive, the Northumberland Fusilier, was not derailed, as it probably would have been had it hit an obstruction.

'Mr Joseph Douds and his wife, Ida, both aged 78, live in a small cottage about 10 yards down a bank from the railway line. One of the carriages mounted the bank and was saved from plunging on to their cottage by a derelict signal box. Mrs Douds said: "I have dreaded that a train would crash here since we moved in three years ago. I have said many times that the old signal box should be pulled down. It was a real eyesore but last night it saved us."

'The main up and down lines at Morpeth were opened at 10.30 p.m. and trains were running normally, British Railways said late last night.' [105]

Involved in the derailment was *The Aberdonian*, the 19:40 Down passenger train from London Kings Cross to Aberdeen, 1S60, which consisted of Class 55 Deltic diesel electric locomotive No. 9011, a brake van, a first-class sleeping car, a convertible (first- or second-class) sleeping car, a first- and second-class composite coach, three second-class coaches, a brake van, a first-class sleeping car, a convertible sleeping car and a brake composite coach, in charge of Guard R. Hewitt with Driver Leslie Byers and Secondman C. Graham.

The Department of the Environment report into the accident noted that the line between Newcastle and Morpeth was equipped with continuous track circuiting and four-aspect colourlight signals, with Track Circuit Block working using single-stroke block bells between the signal boxes. The last signal box before Morpeth coming from Newcastle was at Stannington, 2½ miles away.

From Dudley, 8½ miles from Morpeth, the line was undulating; it fell at 1 in 350, 1 in 286 and 1 in 170 for 1¾ miles approaching Morpeth. For 12 miles to Stannington there were a series of gentle curves, followed beyond Stannington by a stretch of almost straight line leading to the right-hand Morpeth Curve.

Between Newcastle and Morpeth there were permanent speed restrictions of 15 and 25 miles an hour at Newcastle itself, 45 miles an hour at Heaton South Junction (Heaton Curve), 2 miles from Newcastle; 75 miles an hour over Plessey Viaduct, 12 miles from Newcastle; and 40 miles an hour at Morpeth Curve. In addition there was a 20 miles an hour temporary speed restriction due to a bridge replacement at Cramlington, 10½ miles from Newcastle. Each of the permanent speed restrictions was indicated by a 6 ft high

[105] Ibid, 8th March 1969

cutout sign, painted yellow and showing the permissible speed. The signs were not illuminated. Fishtail warning boards lit by two oil lamps marked the temporary speed restriction.

Plessey Viaduct, over the River Blyth and 4½ miles short of Morpeth, was an easily recognisable landmark, immediately beyond signal D12. Next came signal D12B and, on the right-hand side of the line, Stannington signal box, which should also have been easily located. Also approaching the box on the left-hand side were the large and conspicuous buildings of Netherton Training School. Further on, the gate-box at Clifton Crossing was closed and unlit at night and so not a good landmark but signal D15 came into the driver's view at this point, as well as a row of sodium street lights along the Morpeth to Tynemouth road. It was at this point that the first brake application for the Morpeth Curve should have been made, to reduce the speed of a Down express to about 60 miles an hour. Two hundred yards beyond signal D15 was an old water tank and the street lights along the A1 road on the southern outskirts of Morpeth then came into view. Passing the water tank signal M80 came into view and the speed of the train should here have been reduced to 45 miles an hour as the train neared Morpeth Curve.

Permanent Way Supervisor D. Mellen was not available for the Department of the Environment inquiry but told the railway inquiry that the track at Morpeth had been inspected by a patrolman the day before the accident and he was confident that it had been in good condition. He had not observed any cases of speeding and none were reported to him.

Area Assistant to the Civil Engineer P.G. Rawling described his examination of the track at 04:15 on the day of the accident and his noting the marks of the derailment. The track had been otherwise in good condition and there were no signs of any obstruction.

Assistant Divisional Civil Engineer T.F. Grisdale stated that the track had been examined using a Neptune track-recording trolley on 24th March and had been found satisfactory, with only minor defects. He agreed that there was no evidence of any obstruction.

Carriage and Waggon Examiner S.F. Bunn stated that the train had been in good order when it left Hornsey Carriage Sidings to form the train after minor defects had been corrected.

Signalman T. Davidson at Stannington said that *The Aberdonian* had been described to him by Dudley box at 01:17, he had passed its description on to Morpeth at 01:22 and it passed him at 01:25, running at its usual speed. At about 01:29 one of his telephones had rung but he could get no reply from any of them, then he found that none of his track circuits showed *Occupied,* although a freightliner had been described from Dudley, and all his signal lights were out. Assuming that there must have been a total power failure, when the secondman of the freightliner had come to his box at 01:37 he instructed him to proceed to Morpeth at extreme caution under Block Regulation 25 (relating to the signalling of trains during the failure of block instruments or bells). The first that he had known of the derailment was a telephone message from an Up line signal at 01:48, made by a railwayman who had been a passenger on the derailed train. He had at once advised Control and the Dudley signalman. Later the guard of the freightliner had also telephoned him.

Signalman T.M. Halbert at Morpeth stated that *The Aberdonian* had been described to him at 01:23 and he had cleared his signals for it and an Up express. As it approached he had heard a "terrific bang," all his track circuits had shown *Occupied* and the point and signal indications towards Stannington had gone out. He had gone to replace his Up signals to Danger but found they were already red (because the derailed train had cut the signal cables). He could see the engine of *The Aberdonian* at the station platform and its secondman had then come to his box and advised him of the derailment. He found that he could contact neither Stannington box nor Control and his box lad had been on the way to alert the assistant station manager, but was able to stop a police panda car, whose occupants contacted the emergency services by wireless.

Guard Hewitt said that he had examined the train at King's Cross before departure. About 17 minutes had been lost at Grantham and the train had left Newcastle 7 minutes late. As far as he knew the speed restrictions between Newcastle and Morpeth had been obeyed. Morpeth had been approached at the usual speed and he had been standing up, checking the registered mail loaded at Newcastle, when suddenly, without any brake application, the coach he had been riding in (the leading brake van) had swung to the left and he found himself on the floor, with parcels on top of him. Getting up and unable to find his detonators he had given someone a track circuit operating clip to put down on the Up line, then gone to a nearby house to ask its occupants to alert the station, the police, and ambulances. He had then gone to assist the passengers, assuming (incorrectly, the accident report noted) that the Down line was fully protected.

Secondman Graham testified that he and Driver Byers had taken over the engine of *The Aberdonian* at Newcastle, where his driver had been handed a *Notice* (commonly referred to as a *Form 1*, though that term is not used in the accident report) when he signed on, requiring an explanation as to why he had lost four minutes when working a train between Edinburgh and Grantshouse a fortnight earlier. They had discussed the notice and Byers was puzzled as to how he lost the time but had not appeared worried. The speed restrictions as far as Morpeth had been obeyed but he did not remember a preliminary brake application for the Morpeth Curve. It was only when close to the start of the curve and the brakes had not been applied that he (the secondman) realised that they were going too fast. Although he had been reluctant to interfere with the way his driver was handling the locomotive he had then begun to move across the cab to do so, when he was flung to the floor as, he thought, they had entered the curve. He did not remember any brake application. As he had struggled to his feet his driver said something like, "I lost my bearings… I shouldn't have been thinking about that report."

Driver Byers stated that he had a normal, restful day before the accident. He had nothing to drink, was familiar with Deltic locomotives and knew the very line well. He had taken over the engine at 00:52 and left Newcastle at 01:06, a few minutes later. The signals had all been green, the Automatic Warning System (AWS) had been working correctly and he had adhered to all the speed restrictions on the way to Morpeth. He thought that he had closed the controller nearing Morpeth Curve and his mind must have turned to the notice. He then recalled adverse signals at Cockburnspath (as the reason for his having lost time on the previous trip). Suddenly, he had been "brought around and made the full brake application." The speedometer had been, "just on the line of 80, possibly 82 m.p.h." His engine had "rode the curve very well" and he thought the remainder of the train had also done so until, having stopped, he looked back and saw the wreckage behind him.

Sending his secondman to the signal box, he had gone back to protect the rear of the train, when the freightliner slowly approached and its driver told him that he had had no warning of the obstruction. Driver Byers had, remarked the inspecting officer, Colonel J.R.H. Robertson, frankly and freely admitted that he entered the curve at over 80 miles an hour. The driver had received a similar notice about a month previously regarding the loss of two minutes. The inspecting officer considered that such notices tended to unsettle Driver Byers more than they should have. Driver Byers repeated that after passing Stannington box he had been puzzling over how he had lost the time until his secondman, moving across the cab of the engine, belatedly reminded him that he was at Morpeth Curve, and he had applied the brakes fully. He said that he had nothing else on his mind, had been quite fit and had a happy and contented home life.

Colonel Robertson concluded that there had been nothing wrong with the track or train and that the sole cause of the derailment was the grossly excessive speed at which the train had entered the curve. Driver Byers had become distracted and had relaxed his customary concentration somewhere between Stannington signal box and Clifton Crossing, not being brought to his senses until too late. Secondman Graham had trusted his driver and been naturally reluctant to interfere; when he had decided to act there was very little time left. Driver Byers alone was responsible for the derailment. He was very seriously at fault for taking the notice too much to heart and dwelling on it at the expense of concentrating on his driving.

Turning to recommendations the inspecting officer stated that in his view the drivers of trains on high-speed lines – which he defined as those with speeds of 75 miles an hour and higher – required a special and advance warning of severe speed restrictions on open line. He regarded severe restrictions as ones requiring a reduction to not more than two-thirds of the line speed, for example on a 90 m.p.h. line to 60 m.p.h. or less. The problem, he stressed, would only get worse with higher train speeds. He suggested the use in such cases of an illuminated warning board together with a permanent magnet similar to an AWS inductor, which would give a driver an audible warning of the restriction, such as was received at a distant signal at caution.

Considerable discussion had followed, the inspecting officer continued, regarding problems such as whether the warning would be needed at facing junctions or at facing loops on high-speed single lines. It was estimated that about 550 locations would be involved, entailing considerable expense. The proposal was eventually accepted by the British Railways Board, the illuminated warning boards to be provided at locations with line speeds of 75 m.p.h. or more where there was a permanent restriction of speed to 33⅓% or less of the approach speed, the warning board being preceded by an AWS permanent inductor 200 yards in the rear of it. The warning board was to consist of a circular disc 750mm in diameter painted black with a 40mm wide border, showing the restriction in 320mm numerals. By 1st November 1972, 47 of some 500 restrictions had been provided. [106]

Down Mixed passenger train No. 156 from Kaduna Junction to Igunmale was derailed at about 17:15 at mile 379½ of the eastern line near Langalanga, Nigeria, on Tuesday 17th February 1970. The train had been carrying about 1,000 Muslim pilgrims on their way home at the end of the first day of the Eid-al-Kabir festival when it became derailed

[106] Department of the Environment report

approaching a curve, when the diesel engine and four coaches fell down an embankment. The results were disastrous; some accounts spoke of 150 or even 400 deaths. Another report, which stated that there had been 105 deaths, was perhaps closer to the truth. [107] Eighty-one bodies were soon recovered from the wreckage. [108]

After the accident a passenger ran three miles to Mada village to raise the alarm. The injured were taken by rail and road to hospitals at Keffi, Alushi, and Lafia, with rescuers working through the night to free victims from the wreckage, working by the light of the headlights of cars and lorries. [109]

Many years later, victims of the disaster told their stories. Kachalla Waya said, "I just heard a loud sound and found myself outside the train with people screaming and others shouting the name of God. I had sharp cuts on my forehead. I had a cut on my tongue and lost a tooth. I also sustained injuries in some other parts of my body but did not lose consciousness. I was taken to hospital in Alushi, near Akwanga. Unfortunately the train fireman, who was taken to hospital with me, died the next day."

He was in hospital three months and nearly 50 years on was still suffering from the effects of the accident. "Anytime I want to cut my hair I feel severe pains in my forehead, especially where I was injured. I also experience joint pains," he said.

Monday Kpaji was a witness to the disaster. He related, "I was close by the sharp bend where the train derailed. The engine of the train and the front wagons first derailed. I took part in the rescue operations before government officials arrived. I still remember how we evacuated many victims. I remember how many people were screaming for help with blood rushing from their heads and other parts of their body. I can still remember how I and my late friend dragged out one victim with head injuries through the window. We were able to rescue many before fire broke out and we also witnessed how some victims died slowly because goods and wagons trapped them. In such cases we were helpless as there was nothing we could do.

"Doctors from Alushi came and provided medical services to the victims at the spot. I witnessed how doctors amputated victims to rescue them from the debris because debris or goods buried most of them. It was the doctors that cut either the legs or hands trapped by iron to save victims. The doctors really worked hard to save many lives."

According to Kpaij, government officials were overwhelmed with the carnage but evacuated as many victims as they could. However, he explained that they dug a deep trench near the accident scene and buried the dead and pieces of flesh that littered the site. [110]

An especially sad case among those who survived the disaster was that of a man named Malam Adamu. He was 29 in 1970 and preparing for his wedding. He had intended to take a taxi to Port Harcourt but none had been available. Instead a friend who worked on the railway had given him a free ticket for the train. As a result of the accident he suffered a catastrophic spinal injury, which left him unable to walk. For the next 35 years he was a bedridden patient in the Ahmadu Bello Teaching Hospital in Kaduna, then

---

[107] *The Canberra Times*, Canberra, Australian Capital Territory, 19th February 1970
[108] *Las Vegas Sun*, Las Vegas, Nevada, United States, 18th February 1970
[109] *The Times*, London, England, 18th February 1970
[110] https://dailytrust.com/langa-langa-nigerias-unforgettable-train-disaster-48-years-after/

when that was moved he was transferred to the Yusuf Dantsoho Memorial Hospital in Tudun Wada, Kaduna, where he remained a patient for a further eight years at the time of the report. His fiancée had married someone else after his accident, some of his family had abandoned him, and he was now blind, but he still thanked Allah for saving his life in the accident.

The accident was the subject of a full inquiry, [111] which was unable to determine with certainty the cause of the derailment. Nigerian Railway Corporation District Superintendent Alhaji Ja'afaru Makarfi, speaking in 2013, blamed it on the excessive speed of the train, stating that its driver had been drunk. [112]

Heavy rainfall for 24 hours in West Germany led to two separate train accidents on Monday 11th May 1970, both of which involved fatalities. In the more serious of the two, at Beckingen on the Saarbrücken – Trier line, railbus E 1855 ran into the landslide, was derailed fouling the opposite running line, and its rear coach was struck by passenger train P 4864 and thrown down an embankment. Five people died and three were injured.

In the other accident, at Odernheim, on the line between Homburg and Bad Kreuznach, train Pto 2784, a three-car diesel multiple unit, was derailed at a point where the foundations of the track had been washed away. Two passengers, brothers, died and nine passengers were injured, three seriously. [113] [114]

A passenger train from Oran, Algeria, to Oujda, Morocco, was derailed on the afternoon of Sunday 16th August 1970 at Zahana, Algeria, 30 kilometres from Oran. A mail and luggage van and a passenger coach became detached from the train and overturned. Ten people lost their lives. [115] [116]

According to what was probably the most accurate account 13 people, two of them children, died when a coach of a Cairo – Aswan express passenger train was derailed and overturned early on Saturday 22nd August 1970 at Tamouth, Egypt (officially at this date the United Arab Republic). [117] Earlier unconfirmed reports had spoken of 24 deaths in the accident 25 miles south of Cairo. [118] Another 53 people were injured.

West Germany suffered a major disaster on Tuesday 9th February 1971 when the München – Zurich Trans Europe Express (TEE) Bavaria was derailed near Aitrang with the loss of 28 lives.

---

[111] https://gazettes.africa/archive/ng/1970/ng-government-gazette-supplement-dated-1970-04-02-no-18-part-b.pdf
[112] https://dailytrust.com/meet-kaduna-man-who-s-been-on-hospital-beds-for-43-years/
[113] *Katastrophen der deutschen Bahnen Teil II* by H.J. Ritzau, pg. 144
[114] *European Stars and Stripes*, Kaiserstlautern, Germany, 22nd September 1971
[115] *Journal de Genève*, Genève, Switzerland, 17th August 1970
[116] *Lubbock Avalanche Journal*, Lubbock, Texas, United States, 17th August 1970
[117] *The Daily Gleaner*, Kingston, Jamaica, 24th August 1970
[118] *Wisconsin State Journal*, Madison, Wisconsin, United States, 23rd August 1970

On the evening of the day of the accident the *fahrdienstleiter* (signalman) at Aitrang watched the TEE pass at about 18:44 from his ground-floor office and realised that it was travelling faster than usual, though it appeared its brakes had been applied, if somewhat belatedly. The next train he was expecting from the opposite direction, at 18:49, was diesel multiple unit No. 2513 from Kempten, due to terminate at Aitrang at 18:49. At 18:46 he saw from the red lights displayed on his illuminated track diagram that the line from Kempten was showing occupied, although the Kempten train was not yet due, and realising that something was wrong he immediately placed all his signals at Danger. Unfortunately train No. 2513 appears to have been too close at hand for its motorman to be warned by the signals and it collided with the derailed coaches of the TEE. Due to the peculiar local acoustic conditions the *fahrdienstleiter* heard nothing of the derailment and collision, which had happened only about 100 metres from him. There was thus a delay of about 10 minutes before the emergency services were alerted to the accident.

There had only been a total of 53 passengers on the train and almost all of its occupants who were not killed were injured to a greater or lesser degree. The high proportion of casualties in the accident was partly explained by the fact that the windows of the SBB (Schweizerische Bundesbahnen – the Swiss Federal Railways) rolling stock were not constructed from laminated glass but only from toughened glass, which crumbled when subjected to a hard impact, meaning that a number of passengers were thrown through the windows when the coaches overturned and were then crushed beneath them. In addition the fittings in the dining car were not properly secured and the mirrors were made of ordinary glass. The Swiss rolling stock was said not to have met the safety standards of equivalent German rolling stock of the 1930s.

The disaster had clearly resulted from the failure of Driver Rahn to reduce speed for the curve beyond Aitrang station for reasons that could not be determined, as he lost his life in the accident, as did the Swiss technician who was accompanying him. The train conductor survived but could throw no light on the accident. A post mortem was conducted on the body of the driver but found nothing, such as an indication of a heart attack, that could have explained his mistake, while the approach to the curve was sufficiently distinctive to mean that there was no possibility of his having mistaken his location. He had, probably, made a brake application at the last moment, but by then it was far too late. A suggestion by a Professor Hans Thoma, published in a technical magazine, that ice had formed in the brake pipe lines, causing the brakes to fail at the crucial moment, was dismissed by the Deutsche Bundesbahn, who stated that there had been no problem with the train's brakes. Like other excessive speed derailments before and since, the accident remains unexplained. [119]

Just over six months after the Aitrang disaster another accident of similar magnitude due to excessive speed on a curve occurred at Rheinweiler, when the *Schweiz Express,* train D 370, which had left Basel, Switzerland, at 13:08 for København, Denmark, was derailed on a sharp curve south of Rheinweiler station at about 13:22 on Wednesday 21st July 1971. As a result of the derailment the driver and 20 passengers on the train lost their lives and 124 people were injured. In addition the derailed vehicles collided with four lineside dwellings, killing two people in them, and injuring another eight. Property

[119] *Katastrophen der deutschen Bahnen Teil I* by H.J. Ritzau, pgs. 157-177

damage was estimated at DM 5,000,000.

The train consisted of a Class 103 electric locomotive, a through Danske Statsbaner coach and seven Deutsche Bundesbahn coaches. The engine turned through 180 degrees and fell into a pit, the leading coach probably had its roof torn off by a mast supporting the overhead line equipment (OLE), and the second coach collided with the concrete base of the OLE mast and then demolished a house. The train then separated behind the second coach and the next four went down an embankment. The seventh coach was derailed but did not go down the embankment and the eighth coach was only partly derailed.

The line from Basel northwards through the Rhine Valley had considerable curvature and in the 20 kilometres between Basel Bad and Rheinweiler ten changes of speed were required. Starting from Basel the train was permitted to run at 100 kilometres an hour (km/h), thence at 120 km/h, dropping to 105 km/h after only 4.3 kilometres for 800 metres. Speed could then be increased to 120 km/h for 8.2 kilometres. At Istein speed was limited to 80/85 km/h for 3.5 kilometres, then to 105 km/h past Kleinkims, then 120 km/h for 2 kilometres. At kilometre 250.4 speed had to be reduced to 75 km/h for the curves, which began with the one south of Rheinweiler station. As far as Kleinkims (km 252.4) the restrictions had been adhered to but beyond there the train's speed had increased to 140 km/h instead of only 120, which had been maintained, with no apparent brake application, prior to the derailment between km 250.3 and km 250.1.

The engine was equipped with *Automatische Fahr und Bremssteuerung* (AFB) (Automatic Driver and Brake Control), *Induktive Zugsicherung* (Indusi) intermittent train signalling to ensure compliance with lineside signals and *Sicherheitsfahrschaltung* (Sifa, equivalent to a dead man's handle), none of which had prevented the accident.

Three experts examined the cause of the disaster. There was disagreement as to whether the driver had the AFB switched on at the time of the accident but it was thought that it probably had been for several reasons:

(a) The driver had been instructed to use it during his training,

(b) The driver had previously stated that he valued the AFB,

(c) The position of the engine controls after the derailment, and

(d) The regularity of the train running when AFB was in use.

The principal puzzle in the accident was why the train had accelerated to 140 km/h at km 252.4 and why, when the AFB indicated that it had done so, the driver had not intervened and applied the brakes. (The train speed as recorded on the Indusi tape and the condition of the tyres of the train wheels indicated that the brakes had not been applied at all, nor had power been shut off.)

Three theories were advanced to explain the accident:

(a) The driver had mistaken his whereabouts and panicked when he realised that he was approaching a sharp curve.

(b) The driver had suffered a momentary loss of consciousness due to some medical condition.

(c) There had been a malfunction of the AFB, which had taken the driver by surprise and he had not reacted in time.

Against theory (a) was the fact that the driver had travelled over the route for years and was very familiar with it.

Against theory (b) was the fact that the driver had apparently been in good health, was well-balanced and had apparently been behaving perfectly normally before the accident. Also no sign of disease was detected in a post mortem examination, though this presumably would not have been able to detect a condition such as transient global amnesia or a transient ischaemic attack (mini stroke). Again, it was presumably possible that the driver had just let his attention wander, thinking about his wife's birthday, his forthcoming holiday, or whatever.

Theory (c) was a possibility as other drivers had reported problems with the AFB with deviations in the indicated speeds up to +45 and -50 km/h reported. If the driver had noticed that the train was overspeeding due to a fault with the AFB he should, of course, have had ample time to apply the brakes, but perhaps at this point, faced with an unexpected malfunction, he had panicked. It might have been significant that on the day of the disaster he had only been driving a Class 103 locomotive equipped with AFB for the third time unaccompanied by an instructor. If the driver had set the AFB speed to 120 km/h after Kleinkims, but the AFB had wrongly set it higher, it might have taken him some time to grasp the situation due to the very quiet running of the electric locomotive, and he might have tried to reset the AFB before applying the brakes.

Following the disaster the Deutsche Bundesbahn introduced Punktförmige Zugbeeinflussung (PZB, punctual train control) which used track magnets to apply the brakes if the driver failed to reduce speed for a curve. The interval at which the Sifa came into action to apply the brakes if the driver failed to operate it by pressing a button was also reduced from 55 to 30 seconds. [120] [121]

At least 22 passengers died and another 76 were injured when a mixed train bound for Touggourt was derailed 12 kilometres south of Biskra, Algeria, on Monday 20th September 1971. Twenty-two of the 40 vehicles of the train left the rails. The victims were mostly Bedouin nomads from the Tell region returning to their winter pastures with their herds of sheep and goats. The injured were taken to local hospitals. [122] [123]

Twenty-one people died and 37 were injured, some seriously, when the engine and four coaches of a passenger train were derailed in Hindupur station, north of Bangalore, India, in Mysore state (since 1st November 1973 renamed Karnataka state), India, on Wednesday 26th April 1972. [124] [125]

A return excursion train from Margate to Kentish Town, carrying mainly railwaymen and their families, became derailed on a curve at Eltham (Well Hall), England, on British

[120] Ibid, 178-194

[121] Wikipedia – Eisenbahnunfall von Rheinwheiler

[122] *European Stars and Stripes*, Kaisertlautern, Germany, 13th May 1970

[123] *Gazette de Lausanne*, Lausanne, Switzerland, 22nd September 1971

[124] *The Times*, London, England, 27th April 1972

[125] *The Lincoln Star*, Lincoln, Nebraska, United States, 27th April 1972

Railways, Southern Region, on the evening of Sunday 11th June 1972. Three passengers and the driver were killed more or less outright and 125 passengers and the secondman were injured, of whom 40 were detained in hospital. A fourth passenger died of her injuries in mid-August and a fifth of his injuries in November. [126]

A newspaper reported the next day, 'Four people, including the driver, died and 115 were injured when a railmen's excursion train crashed off the rails near Eltham Well Hall station in south-east London last night. The train was bringing several hundred passengers, many of them children, back from a day trip to Margate when it left the track on one of the tightest bends on the Southern Region, about 400 yards on the Kidbrooke side of the station. No other train was involved. The bend has a 20 m.p.h. speed limit.

'The passengers were members of the Kentish Town Railwaymen's Club and their families and friends.

'The driver's mate was pulled out of the tangled cab alive, but critically ill, two-and-a-half hours after the crash. He was carried by stretcher with a drip-feed to an ambulance. As firemen and police cut through the wreckage with oxy-acetylene burners, doctors giving him pain-killing injections. Other firemen held up the shattered side of the cab with a ladder as the man was dragged clear and ambulance men used their ties to make tourniquets.

'A team of Army surgeons and nurses from the Royal Herbert Military Hospital, Shooters Hill, were called to the accident. As teams of doctors and nurses stood by to hèlp the badly injured, residents brewed tea for the shocked passengers and brought out hot water to help bathe their wounds.

'On the track beside the train belongings of the passengers, handbags, clothing, and fairground straw hats. Among them was a small child in a pushchair who was hauled from the wreckage. Another, unhurt, still clutched a piece of rock.

'Police detained three youths at the scene of the crash suspected of looting scattered belongings of the returning day trippers.

'After the crash only one coach was left standing properly on the line. The engine was derailed, the first coach after the engine was on its side, then two jack-knifed, three coaches derailed and the last upright.

'The smashed coaches were scattered for 100 yards on either side of the track. Heavy sleepers were splintered to the size of firewood.

'Dazed and injured passengers were wandering around the area stopping drivers and asking to be taken to hospital. Police called by loud-hailer to local residents asking if they could drive the not-too-badly injured to hospital.

'Of the 38 injured taken to the Brook General Hospital at Shooters Hill, two were dead on arrival. One was a man and the other a woman. A hospital spokesman said, "I believe 10 will be detained."

'A railway breakdown train was brought to the scene and workers began the process of removing the wreckage. Twisted fragments of steel, carriage seats and splintered sleepers covered the ground by the wreck. There were several sets of upturned bogies which had sheared completely off the carriages. A large part of the roof of one upturned

[126] Department of the Environment report

carriage was broken open by the end of another carriage. The cab of the engine was smashed in and partly buried in the earth.

'Gregory Turner, aged 15, of Froissart Road, Eltham, was one of the first at the scene. He was walking along the path next to the railway track when the crash happened. He said: "I heard the crash and looked over and saw a lot of smoke and heard people screaming. A man ran down the bank and said there had been a crash and could we go and get help. My mate tried to get the driver out. We were pulling his body but his legs were just staying there. We saw a woman lying on the ground with some debris on her back. I think she must have been thrown from the train. She asked us if we could move the stuff from her back and help her to stand up. We moved the stuff but when we tried to stand her we saw her leg was badly injured. We called the ambulance men and they took over. Then some men were passing young children down to us from the carriages and we were passing them to the firemen who took them to the ambulancemen."

'The down line from London will be open this evening between 4 p.m. and 7 p.m. only to cope with peak traffic.'

In a separate piece the newspaper carried interviews with passengers and rescuers. 'Mr John Costen said: "There was an impact then we careered on and started to keel over. I was thrown over and then tumbled out of the coach. There were people trapped. I helped to get a baby out but I couldn't get the mother out."

'Mrs Emily Jones, of Kentish Town, comforted a tearful girl of 13 who had lost her father in the confusion. The girl repeated: "He was singing, he was singing." Mrs Jones explained: "Everyone in the compartment was singing songs when the crash happened."

'Mrs Violet North, who lives opposite the crash scene, said a shocked girl aged nine came to her home looking for her family who were on the train. Mrs North said: "A couple brought the little girl to the door. They said they had found her walking around in the street looking for a phone box to try and find out what had happened to her mother, father, and little brother. She was with a little boy of three who was in the same carriage."

'Mr Peter Solari, a British Rail salvage engineer, said, "It is a miracle that most of the children survived."

'Mr John Bengs, aged 44, a slater, of Churchill Road, Kentish Town, said: "There was a bump and then another bump and we stopped with a terrific jolt. Our coach tilted over. I got out with my brother Edward and a friend and helped pull six people out. One of them was my eight-year-old nephew Jason Bengs. He had lost a leg and was covered in blood."

'One of the first rescuers on the scene, Mr Paul Levie, said, "I heard a woman crying out for her baby and saw a man wandering around with his jacket sleeve ripped to shreds and his arm missing. There were bodies lying everywhere."

'Mr John Duffy, aged 39, an insurance agent who was a passenger, said: "Suddenly there was a loud bang and the next thing I knew the coach we were in was overtaking the engine. The engine finished up on its side. I don't know how many other coaches toppled over. It happened so fast. Suddenly there was chaos. People were being flung all over the place. It was a miracle that more people were not killed."

'Mr William Hill, a St John Ambulance Brigade area staff officer, was bitter about the number of people who swarmed to the scene.' [127]

The 20:05 from Margate to Kentish Town train consisted of Brush Type 4 diesel electric locomotive No. 1630, a second-class brake composite coach, a second-class coach, a first- and second-class composite coach, three second-class coaches, a first- and second-class composite coach, two second-class coaches and a second-class brake composite coach, in charge of Guard Henry Atterbury with Driver Robert Wilsdon and Secondman Paul Stokes.

Eltham (Well Hall) station lay between Eltham Park and Kidbrooke stations on the line between Dartford and Blackheath via Bexleyheath (the Bexleyheath line), exactly 9 miles from London Charing Cross. Beyond it on the Dartford side Eltham Park station was 9 miles 41 chains from Charing Cross and Falconwood station was 10 miles 28 chains from Charing Cross. Between Margate and Dartford there were permanent speed restrictions at Faversham, Rochester Bridge Junction, Strood, Gravesend and Dartford itself. Beyond Dartford station there were speed restrictions at Dartford Junction, Crayford Creek Junction and Perry Street Junction. From Perry Street Junction the line was more or less straight as far as Eltham (Well Hall). It climbed steeply at first, was level for a mile past Bexleyheath station then climbed again to a summit beyond Falconwood station, whence it fell steeply at 1 in 65 to Eltham (Well Hall). On the Charing Cross side of Eltham (Well Hall) station was the 12 chain Well Hall right-hand curve, over which the speed of Up trains was restricted to 20 miles an hour. The restriction was marked by the standard cut-out yellow sign.

The Department of the Environment inspecting officer, Colonel J.R.H. Robertson, rode over the line 12 days after the accident and commented that it had a number of easily recognisable landmarks, particularly the station and footbridge at Falconwood.

The line was fully track circuited with 4-aspect colourlight signals and worked by Track Circuit Block from Dartford and St Johns (Lewisham) boxes.

Permanent Way Supervisor P.G. Dawes told the railway inquiry that the track at the point of derailment had been in a generally good state of repair. It was not considered necessary to interview him again for the Department of the Environment inquiry.

Leading Trackman A. Sandy told the inquiry that he had last examined the track on Friday, when it had been in good condition with a good ballast shoulder.

Leading Trackman A. Corri stated that he had patrolled the track on the day of the accident, Trackman Sandy having been on leave that day. He had noticed five or six rail keys out which he had driven home but nothing else.

Divisional Civil Engineer J.L. Newing stated that a Matisa track-recording trolley had been run over the line on 22nd April 1972. He was of the opinion that the track had not been in a very good fettle and in some need of attention but that it was safe for the permitted line speed of 60 miles an hour and certainly for the 20 miles an hour around the Well Hall curve.

Assistant Civil Engineer K.C. Haysom spoke of examining the site after the accident. From what he had found he was of the opinion that the engine had rolled off the rails rather than climbing the high rail into derailment.

---

[127] *The Times*, London, England, 12th June 1972

Driver J.E. Wilkinson testified that he had taken over the engine from its London Midland Region driver at Cricklewood for the outward journey of the excursion. He had make a brake test on taking over and had experienced a number of signal checks between Longhedge Junction and Hither Green, through following a goods train, and the engine brakes had worked satisfactorily.

The inspecting officer in his report on the accident then detailed the evidence at the opening of the Coroner's inquest on 14th June, where it had emerged that Driver Wilsdon must have consumed a considerable quantity of alcohol in the hours before the accident.

Dr J.M. Cameron, a distinguished pathologist, told the inquiry that he had taken blood and urine samples from the body of the driver at about 14:20 on the day after the accident. These were analysed at the London Hospital Medical College and indicated a blood alcohol level of 278 milligrams/100 millilitres and a urine alcohol level of 326 milligrams/millilitre. There was no evidence of bacterial contamination of the samples and these levels of alcohol would have been the levels present in the driver's body at the time of death, equivalent to about 8½ pints of beer or 8½ tots of 70% proof spirits. The difference in levels between the blood and urine samples indicated that the driver's body had still been absorbing alcohol at the time of his death and that he might have consumed it within the previous hour. No other drugs were detected and there was no indication of a heart attack.

Mr R. Farmer, the driver's father-in-law, said that his son-in-law drank a fair amount when off duty, but never to his knowledge whisky. His daughter said that had her husband been obviously drunk when he was setting off for work she would not have let him go. Mr Farmer agreed that his son-in-law sometimes drank to excess when off duty.

Driver Wilsdon's brother, Mr W. Wilsdon, said that he, the driver, and John Wilsdon, another brother, had gone by car to the Cherry Orchard public house in Rainham at lunchtime on the day of the accident. He could recall the driver drinking two pints of light and bitter and a half pint of light ale. He denied that the driver had afterwards drunk any sherry and said that his brother could "carry" a lot of beer so that eight pints would have been "well within his capacity." John Wilsdon agreed.

Driver P.S. Stevens, acting as motive power supervisor at Hither Green, said that Driver Wilsdon had telephoned the depot to book on duty to work the excursion train. His voice and manner had seemed perfectly normal. He said that it was the practice for drivers to book on by telephone when they lived some way from the depot and agreed that the only check on the driver's condition in such circumstances was his manner of speech, though he pointed out that the driver would likely meet some supervisor during his course of duty. Secondman Stokes had booked on in person at the depot.

Leading Railman F.T. Fleming at Rainham said that he had spoken to Driver Wilsdon at 17:30 on the evening of the accident; he assumed that the driver was catching the 17:45 train to Ramsgate (where the stock of the excursion was stabled during the day). The driver had been wearing his ordinary clothes rather than his uniform; his voice had appeared normal.

Secondman Stokes, aged 18, stated that he had known Driver Wilsdon for about 18 months and seen him occasionally in public houses. He thought that the driver "could hold a fair bit" of alcohol. He had signed on at Hither Green and taken a train to Ramsgate, where he met Driver Wilsdon, who "was still dressed in his holiday togs."

They had talked and the driver's speech appeared normal, but "he smelt something pretty strong," and he had asked Wilsdon, "Christ, what have you been doing?" The driver replied that he had "had some beer at dinner time" and that later he had, "ended up going somewhere and drinking some sherry." In another statement the secondman said that Wilsdon had told him, "that during the afternoon he had drunk a bottle of sherry with his brothers after he had been out drinking with them."

Continuing, Secondman Stokes said that after a further chat with Driver Wilsdon at Ramsgate the latter had suggested that they had time to go to the Railway Staff Association Club there before taking out the train. They had duly done so and had each consumed three pints of light ale and bitter. He remembered little of what had happened after leaving Ramsgate.

Motive Power Supervisor W.C.T. Reeves at Ramsgate spoke of seeing a railwayman whom he had assumed was Driver Wilsdon but was apparently Secondman Stokes.

Steward J. Buckett remembered serving the driver and secondman at the Staff Association Club but could not recall them having three pints each, nor had he served any miniature bottles of spirits that evening.

Station Supervisor K.R. Henderson at Ramsgate said that he had seen Driver Wilsdon and Secondman Stokes between about 19:15 and 19:30 on the evening of the accident. He had noticed nothing out of the ordinary about them.

Guard Attlebury stated that he had signed on at London Bridge at 15:15 and travelled as a passenger to Ramsgate, where he met his driver, who had seemed quite normal. Told of the driver's alcohol level he replied, "It puzzles me. I just can't make it out, sir, just can't make it out at all...I just can't believe it." He said that his train had stopped normally at Margate where the excursionists, mostly railwaymen, had boarded it; none of them had offered him a drink. The booked departure time was 20:05 but the train had not started until 20:13. The station staff had told him they were waiting for the driver.

Station Manager B.L.H. Arundell, normally at Herne Bay but on duty at Margate on the evening in question, said that after blowing his whistle for the train to start at 20:05 it had not done so and going to the engine he had found that there was no one there. Going back along the train he had seen the driver and secondman, neither of whom were in uniform, and had told them to hurry up as the train was already late. Driver Wilsdon had replied, "All right, we'll make it up before we get there." Returning to the engine the station manager said that he had seen two small beer bottles there, which he "assumed the organisers had given to the driver, one each for the driver and secondman, which is practically normal on these occasions." He did not think that excursion passengers usually gave engine crews any more than that.

Guard Atterbury said that the excursion train had been stopped by signals at Sittingbourne and Rainham on the return journey. The speed of the train around the curves between Gillingham and Chatham had "seemed to be a little bit excessive." His driver had braked, "it was a heavy one," between those two places and "seemed to get the train under control for going round to Rochester and Strood." The speed restrictions around Dartford had been reasonably observed but approaching Eltham (Well Hall) he had thought that his driver "was going a bit excessive, over and what he should have done" so he made two brief brake applications, "splashes is what we call it," to draw his driver's attention to the high speed at that point. He did not hear or feel any brake application made by his driver and the next thing he knew was the "terrific banging and that was when the accident happened." He could not tell if his driver had shut off power but was positive that there had been no brake application. Asked by the inspecting officer if he had considered stopping the train himself he replied that he had an experienced driver who had braked well previously on the journey, and by the time he had made the two "splashes" and waited for a response, it had been too late.

Signalman F.D. Obee at Rainham stated that as the excursion train had been four minutes late leaving Sittingbourne he had opened the gates at his level crossing after a previous train passed to let two buses across, stopping the excursion train momentarily at his Up home signal. Although he had then cleared his Up signals the train stopped again in the station and its driver had got on the telephone and asked him if the previous train was to precede him all the way. No, he (Obee) had replied, it was to be looped at Gillingham. "I'm booked to pass it at Newington," the driver had told him, and that he (the signalman) should read his weekly notices. Signalman Obee said that he had been surprised by the query and did not agree but had not argued. The driver had not been aggressive and his speech had been clear.

Leading Railman Fleming at Rainham said he had heard Driver Wilsdon speaking to Signalman Obee using the telephone in the ticket collector's box. The driver had chosen the correct one of the two telephones there and his speech sounded clear and not slurred.

Secondman Stokes stated that he could now remember very hazily what had happened at Rainham and "just the last few seconds" before the derailment. At Rainham his driver had remarked regarding the train ahead, which he thought was why they had been checked by signals, there. "He's a slow bastard." His driver had appeared annoyed but the remark had not been made aggressively. At Eltham (Well Hall) he "saw the bend coming" and his driver had cried out, in a frightened manner, "f.....g hell." There had just been time to brace himself before the engine "tipped over." He did not recall any brake application made there by the driver. His statement to reporters the day after the accident that the brakes of the engine had failed had, he explained, only been made to protect his driver, whom he had not at that time known was dead, and he had had no intention of alleging a brake failure at the official inquiry. He put the speed just before the derailment at 60 m.p.h.

Traffic Regulator H.L. Hickmott in Dartford box said that when he had taken over the excursion train had been at Dartford Junction and was running under clear signals. The excursion train in front had already reached Falconwood. The first he had known of the derailment was at 21:35 when all the track circuit lights on the panel had momentarily shown *Occupied.* Shortly afterwards the leading railman at Eltham Well Hall advised him of the derailment. He had told the signalman to replace a controlled Down signal at Falconwood to Danger, arranged with the Lewisham control room for the current to be kept off, dialled 999 for the emergency services and the signalman sent the *Obstruction Danger* signal to St Johns box.

Leading Railman Ralph Akehurst at Eltham Well Hall testified that he had been tidying the Up platform when something attracted his attention, and he had seen the excursion train approaching. Its engine had been "wobbling about" and thinking that it was moving much faster than usual, and that its driver might have forgotten the speed restriction, he dropped his broom and waved his arms to the driver, but there had been no response and no signs of braking. He could not estimate its speed but it had been far faster than the previous excursion train 20 minutes earlier. He had been able to make out details of the people on the earlier train but not on this one. He could see the driver, whom he thought was a big man slouched over, but told that Wilsdon was a small man he said that he must have been "standing up over his controls…leaning forward." Due to the debris and dust he had seen little of the derailment itself but he had immediately realised what had happened, telephoned Dartford panel, and had gone back and protected the Up line with detonators.

Booking Clerk W.R.A. Thorley at Eltham Well Hall spoke of hearing the excursion train coming and seeing it "flash past" from the ticket office at a speed he estimated as 65 m.p.h.

Mr W.R. Lasseter, a retired station manager who lived in the station house at Eltham Well Hall, spoke of seeing the excursion train passing Hall from his sitting room and saying to himself, "Blimey, you're belting." He put the speed of the train at 50 m.p.h. He could not hear if the brakes were applied above the sound of his television.

Concluding his report Colonel Robertson stated that the evidence in general was that the speed of the train at the moment of derailment had been about 65 miles an hour, the overturning speed at the start of the curve. There was no evidence that the brakes of the train had been applied at all by the driver for the curve and the guard's two very brief "splashes" would have been too short to have provided any braking effect. The brakes were in good order and there was no evidence of any brake failure. He accepted that Guard Atterbury had attempted to alert his driver; the decision of a guard to take over the control of a train was a difficult one to make and the guard was in no way responsible for the accident.

There could be no doubt as to the high levels of alcohol in the driver's blood and urine. What was in doubt was when Driver Wilsdon had consumed the alcohol, a point of importance as the earlier he had drunk it the more likely it was that his condition would have been spotted by an alert supervisor. The evidence was to some extent conflicting but Colonel Robertson finally concluded that it was fairly certain that Driver Wilsdon had consumed 2½ pints of beer between 12:55 and 14:05 and three pints of bitter between 19:05 and 19:55; plus, possibly, a bottle of sherry between 14:30 and 16:30 and up to a quarter bottle of spirits between 20:15 and 21:30 (while driving the train). It seemed odd that none of the supervisors had noticed any change in the driver's manner but a questioning by Mr Boardman, General Secretary of the National Federation of Licensed Victuallers, of a cross-section of licensees found that they all stated categorically that customers who were regular heavy drinkers (as Driver Wilsdon apparently also was) could after 10 pints continue to converse intelligibly, play darts with their customary efficiency and in general appear in full possession of their normal faculties. The experience of the licensees did not, however, extend to how capable such customers would be of driving a vehicle. It seemed therefore that the supervisors could not fairly be criticised for having failed to detect that Driver Walsdon had been drinking.

Professor R.D. Teare of the Department of Forensic Medicine at St George's Hospital Medical School was of the opinion that the beer, bitter and sherry consumed by Driver Wilsden, plus a quarter of a bottle of spirits in the last 75 minutes of his life, would just have accounted for the levels of alcohol found in his remains. Reference books, the professor said, all agreed that the three stages of alcohol intoxication were excitement, confusion and stupor and that the driver's behaviour at Rainham suggested that he had reached the second stage "which could be identified as irascibility due to alcohol intoxication, though his speech and gait still appeared normal."

It seemed likely, said the inspecting officer, that by the time his train had reached Well Hall curve, Driver Wilsdon had progressed to confusion so that he had forgotten all about it.

Secondman Stokes had on his own admission, drunk three pints of beer while on duty at Ramsgate, in direct disobedience of the then Rule 3(v) of the 1950 British Railways rule book, which stated 'employees MUST NOT consume intoxicating liquor while on duty.' This was disgraceful, but Wilsden had clearly been the ring leader and Stokes would have needed a stronger character than he had, not to have followed his driver's lead.

Driver Wilsdon alone was to blame for the accident.

Going on to remarks and recommendations, Colonel Robertson noted that Driver Wilsdon had joined the railway at the end of 1953. In December 1961 he had been dismissed after a conviction for causing malicious damage and assaulting a police officer while drunk. Following a successful appeal he had been reinstated as a fireman and only seven days after his reinstatement had been automatically promoted to driver. On 29th March 1969 he had again been convicted of a drinking offence but due to a combination of circumstances this second offence had not been reviewed until 19th June, when he was given a warning.

The inspecting officer recommended that the British Railways Board should re-examine their disciplinary procedures 'with a view to a formal withholding of promotion, for a term appropriate to the original offence, as the reduction in punishment when a passed secondman, or some other railwayman in a similar position as regards promotion, is reinstated on appeal but still cannot be absolved from serious blame.'

Rule A1.2.2 of the October 1972 rule book stated that employees "must not report for duty under the influence of intoxicating liquor or dangerous drugs or consume such whilst on duty." These rules, the inspecting officer considered, were sensible and sufficient.

A suggestion had been made by a Member of Parliament that every train driver should be tested for alcohol before going on duty. An equivalent suggestion regarding the random testing of motorists had caused an outcry from the public, motoring organisations and the press and could only be justified if there was a serious problem of drinking amongst train drivers, which there was not. The last serious accident in Great Britain involving a suggestion that the driver had been drinking had been at Wormit in Scotland in 1955, previous to which the last case had been at Donemana in Ireland in 1913. In addition there had been a collision near Todmorden, England, in August 1967, not the subject of a full inquiry, in which the driver had undoubtedly been drinking.

A check made by the British Railways Board regarding cases of drivers being under the influence of alcohol during the previous three years showed that:

There had been no train accidents prior to that Eltham (Well Hall) caused by drivers being under the influence of alcohol.

a. Seven drivers had been stopped from taking up duty.
b. Fourteen drivers on duty but not actually driving had been charged with being under the influence of alcohol.

c. Twenty-one drivers had been removed from the driving cab while on duty.

d. Booking on by telephone, unless part of an official arrangement, should, Colonel Robertson recommended, be forbidden.

The Bexleyheath line was not a high-speed line and therefore did not qualify for the sort of advanced warning notices that had been suggested after the Morpeth derailment (see previous). All but one of the regular Up trains stopped at Eltham (Well Hall) station anyway and would thus approach the curve at low speed. It was thus only to this one regular train, plus excursion trains and suchlike, for which there was any need for a special warning. The inspecting officer suggested that, as a special case, an approach control be installed on signal A479 at the beginning of the curve, such that it was held at red, and the preceding signals at yellow and double yellow, until its track circuit had been occupied for such a time that the driver had reduced the speed of his train to 20 miles an hour in response to it. Such a control was subsequently installed, with a corresponding control on the appropriate Down line signal approaching the curve. [128]

Eleven people died and 27 were injured when a Port Elizabeth – Johannesburg passenger train was derailed near Norvalspont, South Africa, on Monday 3rd July 1972. Two engines, a steam waggon (?) and eight coaches left the track, five coaches went down a 22 ft embankment and two coaches reserved for African passengers were reportedly ripped apart. 'A spokesman for the railways ruled out sabotage.' [129] [130]

Most initial reports of a derailment of a nine-coach passenger train on the evening of Friday 29th September 1972 at Rust Stasie, near Malmesbury, South Africa, put the number of deaths at 36 with 140 injured, [131] though a report in a London, England newspaper spoke of 100 possible fatalities. Major J.P. Villiers, Regional Officer at Cape Town station, who was directing the rescue work, told a reporter that it was impossible to make an accurate guess as to how many had been killed. He added, "There are many dead under the heap of rubble. We'll be busy through much of the day digging out the occupants of the first few coaches. It could be anything up to 100 or more." [132]

A further report put the number of dead at 48 and injured at 144. Most of the passengers on the train had been African labourers, returning from Cape Town to their homes at Bitterfontein for the weekend. They had been in the three front coaches that had plunged down a gorge after the derailment and all the dead were Africans. Two coaches at the rear of the train reserved for white passengers were largely undamaged. [133]

On Saturday night the Transport Ministry announced that 61 people were still in hospital, including one white passenger and three white crewmen. About 30 of the injured were children. A team of 40 African labourers were busy clearing the wreckage.

The cause of the derailment was not known, though it was noted that three derailments

---

[128] Department of the Environment report

[129] *The Times*, London, England, 4th July 1972

[130] *The Modesto Bee*, Modesto, California, United States, 7th July 1972

[131] *Iowa City Press Citizen*, Iowa City, Iowa, United States, 30th September 1972

[132] *The Times*, London, England, 2nd October 1972

[133] *Sunday Journal and Star*, Lincoln, Nebraska, United States, 1st October 1972

earlier in the year had all resulted from sabotage. Driver Fanie van Niekerk, clearly suffering from post-traumatic amnesia, said that he could remember nothing of the accident except being thrown about in his cab. [134]

A 15-coach passenger train was derailed on Boxing Day, Tuesday 26th December 1972 at a point 482 miles from Lagos, between Maiduguri and Jos in northern Nigeria. Six people lost their lives and four others were seriously hurt. A special train carrying relief material and medical supplies was sent to the scene. [135]

A five-coach passenger train No. 252 from Annaba to Tebessa, near the border with Tunisia, was derailed at 21:15 on Thursday 1st February 1973 at Medjez-Sfa station, Algeria. At least 35 people died and 51 were injured. Many casualties were trapped in the wreckage and 15 hours after the accident rescuers were still striving to release them. [136] [137] The cause of the derailment was not immediately clear, though a defect in the diesel locomotive pulling the train was one theory. [138]

First reports stated that ten people had died and 17 were hurt in the derailment of a passenger train from Patras to Athens near Megara, Greece, on the night of Friday 16th February 1973. [139] Other reports gave 12 dead and 25 injured, [140] though a police report the following day mentioned only four dead and 36 injured. [141] The cause of the derailment was not stated. Army units worked under floodlights to free passengers trapped in the wreckage.

The *Anatolian Express*, bound for Istanbul, ran into a landslide and was derailed shortly after leaving the station at Elazig, Turkey, at 06:05 on Wednesday 27th February 1974. Rescuers recovered ten bodies from the wreckage and the same number of people were reportedly hurt. [142]

Up mixed narrow-gauge train No. 662 was derailed between Sillipur and Birpur stations, India, on the Central Railway, on Saturday 9th March 1974. The accident was not a major disaster but five passengers died, three were seriously injured and 41 were slightly hurt.

Train No. 662 consisted of Class NH/5 2-8-2 steam locomotive No. 813, seven waggons and, behind them, seven coaches. It was not fitted with automatic brakes, either

[134] *Star News*, Pasadena, California, United States, 1st October 1972
[135] *The Charleston Gazette*, Charleston, West Virginia, United States, 28th December 1972
[136] *The Herald*, Provo, Utah, United States, 2nd February 1973
[137] *Gazette de Lausanne*, Lausanne, Switzerland, 3rd February 1973
[138] *The Lethbridge Herald*, Lethbridge, Alberta, Canada, 3rd February 1973
[139] *The Times*, London, England, 17th February 1973
[140] *Sunday News-Globe*, Amarillo, Texas, United States, 18th February 1973
[141] *The New York Times*, New York, New York, United States, 18th February 1973
[142] *The Tribune*, Terre Haute, Indiana, United States, 27th February 1974

air or vacuum, there were just steam and vacuum brakes on the engine and a hand brake in the guard's van. Just the thirteenth and fourteenth coaches left the rails in the accident and went off to the left, the thirteenth coach was damaged and the fourteenth was 'broken into shambles.'

Driver Ahmed Ali told the inquiry into the accident that the run to Sillipur had been uneventful. At Sillipur some passengers had climbed up onto the roofs of the coaches. The guard had ordered them down but on rounding curves beyond there he had seen that there were still passengers on the roofs, so stopped his train while they were brought down. Restarting, he had reached a falling gradient (1 in 80 for 720 metres, followed by 1 in 100 for 600 metres and 1 in 80 for another 1,680 metres), where he had shut off steam and applied the brakes. Midway down the gradient he had found that the brakes were not working and reversed his engine. The train had gradually slowed down when the fireman, under his instructions, had got off the engine and applied the hand brakes on the leading two waggons. After the train had stopped some passengers told him that the rear two vehicles had become detached, derailed, and capsized, and that people were hurt. He stated that the brakes had been working properly and he had not requested assistance from the guard in braking the train, nor had he applied the hand brake on the engine as it was not in working order. The whistle code to the guard for brakes was either three short or three long whistles, he was not sure at first which. (It was three short, repeated several times.) He had not been trying to make up time.

Leading Fireman Nur Mohomad stated that the train had got out of control in a cutting about midway down the falling gradient. Reversing his engine the driver had gradually brought the train under control; he assisting by applying the hand brakes on the first two waggons. He thought his driver had whistled to the guard for brakes but was not sure.

Second Fireman Abdul Salam said that he had been busy breaking coal when the train had gathered speed. He had seen the first fireman alight and apply brakes and saw that two coaches had capsized.

Guard L.D. Mishra spoke of the train stopping at Km 1361/1, quickly restarting, then gathering speed. He had applied the hand brake intermittently, for fear that it would jam, but the speed had been too high for it to be effective. His van had swayed and derailed, he had been thrown to the floor and had lost consciousness. He put the speed of his train at much more than 30 kilometres an hour. He had not heard the driver whistling for brakes.

Fitter Jagannath at Gwalior stated that he had examined the engine brakes earlier on the day of the accident and found them to be very poor. He had adjusted them.

Permanent Way Inspector S. C. Maheshwari at Sheopur Kalan attributed defects in the track present after the accident to damage caused by the derailment.

Additional Commissioner D.V. Divgi estimated that the derailment had occurred at 10:20.

The driver mentioned that the speed of his train had been excessive, then incongruously had put it at 25 to 30 kilometres an hour when the maximum permissive speed was 30 km/h. The guard and passengers also thought that the speed had been excessive. It had probably been over 40 km/h.

There had been no serious track or vehicle defects but changes of versine and superelevation of the track could have been a factor in the derailment. The commissioner wrote: 'There is no evidence to indicate any particular dynamic characteristics of the moving vehicles. However, it can be expected that with the train rolling down at a high speed and with the engine brakes trying to retard the motion there would be severe wagging of the tail tending to cause heavy lurches at the trailing portion of the train. Added to this the intermittent application of the brakes by the guard would cause further jerks. In these circumstances a coincidence of a jerk towards the left would easily worsen the condition of low stability inherent in the narrow rail gauge {only 2 foot} and the high centre of gravity of the rolling stock further jeopardised by the static parameters discussed earlier {the track versine and superelevation changes, depressions, and cant deficiencies}, each of which was conducive to the mounting of the left (inner) wheel, at speeds in excess of 30 Km/h. The perhaps accounts for the very peculiar nature of the derailment in which the inner wheel mounted the rail.'

The brake gear in the rear portion of the engine tender had not been fitted due to want of spares, while the front brake assembly of the tender, though fitted, had not functioned because it was too slack. Hence only the brakes of the engine had been available. The district engineer did not consider this to be a serious matter as other engines did not have tender brakes anyway but the deficiency of brake power had handicapped the driver's ability to recover from an unsafe high-speed situation.

The driver had not whistled for the guard's brakes, being under the wrong impression that the whistle was only to be used if the train had parted.

The driver had not used the hand brake on the engine, which in any case was usually jammed and too stiff to operate.

It was unfortunate that there had been no speedometer on the train.

The overall maximum permitted line speed was 30 km/h but speed over the Kunnoo bridge, just beyond the point of derailment, was limited to 5 km/h, requiring trains to approach it at no more than 12 to 15 Km/h. A Caution board would be useful there to assist drivers in controlling their speed.

The additional commissioner concluded that the derailment was the responsibility of Driver Ali, though the absence of tender brakes and of a speedometer were extenuating circumstances. The driver had subsequently stood trial at the Indore magistrates court and was acquitted.

The additional commissioner made seven remarks and recommendations.

(1) The matters of the engine being allowed to leave the shed without the slackness of the front tender brakes being attended to on the day of the accident and of the rear tender brakes not having being fitted in the previous month should be taken up with the staff involved.

(2) Engines lacking tender brakes should have them fitted.

(3) The speed or loads of trains worked by engines not fitted with tender brakes should be restricted so as to enable them to stop within about 450 metres on a 1 in 80 falling gradient.

(4) The fitting of engine speedometers should be expedited.

(5) The knowledge of drivers and guards regarding when guards should assist with train braking should be kept up-to-date.

(6) Norms for variations in versines and superelevation should be laid down.

(7) The permanent way maintenance should be tightened up. [143]

At 21:00 on Sunday 4th August 1974 a Caen – Rennes passenger train was derailed on the approach to the station at Dol-de-Bretagne, France. The train consisted of a diesel locomotive, a van and five carriages and was carrying about 350 passengers, many of them soldiers returning from weekend leave. The accident occurred on a curve over which speed was restricted to 40 kilometres an hour, despite which the train entered the curve at a speed of 120 kilometres an hour, with the result that the locomotive and all but perhaps the rear vehicle of the train left the rails and overturned. Ten people died, including the driver and conductor, and another 30 were injured and taken to hospitals in Dol-de-Bretagne, Combourg and Saint-Malo, though only 14 remained in hospital by the following day.

The statements of surviving passengers indicated they had more or less anticipated an accident. One passenger said, "The train was going very fast. There was a burning smell {due to a last-minute brake application?} and, all of a sudden, the conductor had appeared, shouting, 'Stop, stop! We must stop.'" Another passenger related, "The express was going so fast that all the metalwork was vibrating and the conductor lost his balance as he was checking the tickets." A third passenger commented that the train had overshot the previous station at Pontorson and had to set back.

The accident unfortunately proved to be a repetition of that at Eltham (Well Hall), England, just over two years earlier (see above). A post mortem on the driver showed a blood alcohol concentration of 3.55 g/l (355 mg/dl). The legal limit for driving a motor vehicle was 50 mg/dl. [144] [145]

Forty years after the disaster a French couple from Bayeux spoke of the personal tragedy they had suffered as a result of the accident. Their son Patrick had celebrated his 20th birthday with them that Sunday, leaving them at 19:05 to catch the train back to his barracks. At the completion of his military service he had hoped to begin work as an accountant. The next morning an aunt had informed them of the accident. Unable to obtain any news of their son they had driven to Dol-de-Bretagne, where a police officer informed them that their son's name was on a list of the dead. A gendarme had accompanied them to the chapel ardente at the town hall where they found their son's body, and had been given his wallet and identity card, though his money and watch were missing. Ten days after the disaster they had received a cheque from the Société nationale des chemins de fer français for moral damages, which was ripped up and thrown away. What they wanted was not the money but to know who was responsible for the disaster. After a long legal fight and three investigations they had finally had their answer in 1978, when the medical officer who had been in charge of the dead driver was held responsible for the accident. It appeared that there had been a previous incident on the Paris – Caen

---

[143] Government of India Ministry of Tourism and Civil Aviation (Commission of Railway Safety): Report On the Derailment of No. 662 Up Mixed Train between Sillipur and Birpur Stations on Sheopur Kalan – Gwalior Narrow Gauge Section of Jhansi Division on Central Railway on 9th March 1974

[144] https://fr.wikipedia.org/wiki/Liste_des_accidents_ferroviaires_en_France_au_XXe_si%C3%A8cle#cite_note-648

[145] *Le Monde*, Paris, France, 4th August 1974

main line in 1968 when the driver had overrun signals, presumably when intoxicated, but he had been permitted to continue driving, though relegated to operating trains on secondary lines such as that between Caen and Rennes. [146]

A major catastrophe occurred in Yugoslavia on the night of Friday 30th August 1974. Train No. 10410, a Belgrade – Dortmund express, was due in Zagreb from Vincovci at 19:45. It was carrying about 400 passengers, mostly *gastarbeiters* (guest workers) going for better paid work in West Germany and accompanied by their families, including many children, though there were also Greek, Turkish and West German passengers on the train. It was almost three hours late when it reached Zagreb at 22:33. As it entered the station on track IIA, overrunning a red signal, all nine of its carriages became derailed on a curve and were spreadeagled across the track. Some of the coaches rammed the end of the platform and others demolished some of the posts supporting the overhead line equipment. Some passengers were crushed between the coaches; others were electrocuted.

Passenger Divna Mijalovic told a reporter from her hospital bed, "At least two hours before Zagreb I told my husband, 'We are going to jump the tracks,' but he said that maybe trains go very fast these days. Say what you like but the driver must have been either drunk or crazy." [147] Other passengers noted that an hour before the accident the train had not slowed down to pass through Ludina and Noveselec stations and that it had been leaning dangerously.

'Nearly 24 hours after the train crashed Friday night, rescue workers were still digging through the tangle of overturned rail cars, scattered luggage, and dismembered bodies. The acrid smell of smoke, burnt rubber, and death hung heavy in Zagreb station.

'The morgue at Zagreb, Yugoslavia's second largest city, with a metropolitan area population of over half a million, was unable to cope with all the dead. Ambulances carried scores of bodies to morgues in other cities.'

'"The crash felt like an earthquake," one of the survivors recalled. "There was a child in my compartment crying. I grabbed his hand and pulled him out with me."' [148]

Initially it was reported that over 170 people had died in the accident. Later the figure was reduced to 121, while the final figure was apparently 153, with about the same number of injured. It soon became clear that the express had entered the station at a grossly excessive speed and the police quickly arrested Driver Nikola Knežević, Assistant Driver Stjepan Varga, a signalman and a station official. Those arrested had to submit blood samples for alcohol testing at a hospital, all of which were negative. [149] The drivers alleged that the brakes of the train had failed but these were found to be in order after the accident.

President Tito ordered a day of national mourning and personally intervened to see that it was observed. Becoming aware that a concert was being broadcast on Zagreb

---

[146] https://www.ouest-france.fr/normandie/le-drame-de-bretigny-les-renvoie-au-4-aout-1974-1431287

[147] *Galveston Daily News*, Galveston, Texas, United States, 1st September 1974

[148] *Telegraph Herald*, Dubuque, Iowa, United States, 1st September 1974

[149] *South Mississippi Sun*, Biloxi, Mississippi, United States, 2nd September 1974

television he ordered that the broadcast be stopped and it was. [150]

A subsequent investigation found that the train had been travelling at 104 kilometres an hour as it had entered the station at a point where speed was restricted to 40 kilometres an hour.

The two railwaymen at the station who had been taken into custody were subsequently exonerated but Driver Kneževič and Assistant Driver Varga were convicted of having caused the disaster, though with mitigating circumstances, and they were sentenced to 15 and 8 years' imprisonment respectively. It appears that extreme tiredness, rather than drunkenness, was the cause of their mistake, they having been reportedly on duty for 52 hours at the time of the derailment. [151] Why they had been working such long hours was not stated.

A serious derailment occurred in Portugal on the afternoon of Friday 27th December 1974. Passenger train No. 15 was due to leave the station at Santa Apolónia in Lisboa at 15:17 for Campanhza (Porto). At 15:22, just as it was setting off its rear coach, a second-class 'double carriage' carrying about 150 passengers, left the track on a set of points as passenger train No. 130 from Entrocamento was entering the station on the opposite running line and the engine of that train struck the derailed vehicle. Nine passengers died and 66 were injured, some seriously, as a result of the derailment and collision. [152]

Passengers and the driver of train No. 130 told their stories to newspaper reporters. Many of those on train No. 15 were troops going to spend the New Year with their families. Soldier José Joaquim Lopez declared that it was a miracle he had not been among the dead. He related, "I suddenly found myself with my left leg stuck underneath a seat. There were many people on top of me. I started screaming for people to let me out, but some could not move because they were dead." [153]

Soldier José Ferreira de Araújo said, "When I realised what had happened I jumped out of the carriage in which I had been riding and rushed to the one which had been wrecked. When I heard the screams and moans coming from it I did not hesitate. I broke a coach window with a piece of iron and jumped into it. It was a horror which faced me. There were destroyed and crumpled bodies and others who were still alive who were screaming to be released. I began to pull out the injured and dead. As I was removing one of the casualties from a tangle of twisted metal I was hurt myself." He showed the reporter his bandaged arm.

Another passenger, Vitaliano Araújo de Castro, had been on train No. 15, bound for Coimbra. He related, "I had been going to get into the rear coach but I found that it was very crowded, so I walked to the front of the train instead, which was obviously very fortunate, as it turned out. When I became aware of the magnitude of the accident I ran to the adjoining street and stopped cars so as to get the injured to hospital as quickly as possible. Strangely the police, or at least some of them, did not want private cars entering the station area and transporting the injured, saying that that was the job of the ambulances. Surely, in an emergency such as this, they should make use of every means available?"

---

[150] *The Times*, London, England, 2nd September 1974
[151] Wikipedia – Zagreb train disaster
[152] *Diário de Lisboa*, Lisboa, Portugal, 28th December 1974
[153] *The Times*, London, England, 28th December 1974

The unnamed driver of train No. 130 said, "I was already passing train No. 15 when it occurred to me that there was something wrong with it. Its rear carriage was rocking a lot, then it left the rails altogether, hit the catenary and toppled over right in front of my engine. I jammed on the brakes but there was nothing that I could do to avoid hitting it."

With the accident happening in a metropolitan area help was soon at hand. Local radio stations broadcast an appeal for doctors and ambulances and police sealed off roads near the station to allow access for emergency vehicles through the evening rush-hour traffic already slowed by a heavy fog. Firemen, police, the military, and railwaymen worked to free the injured. The police and military also had to deal with a large crowd of spectators who soon gathered at the scene.

Work to clear the line began at 17:00 and it could be reopened the next morning. The Santa Apolónia stationmaster, Goncalor Correia, told reporters that 14 trains had had to be cancelled because of the accident.

Two more minor accidents added to the woes of the Caminhos de Ferro Portugueses in the following days. On 28th December the engine and four waggons of a freight train were derailed near Lavre station, on the Setil – Vendas Novas line, blocking it until the evening. Then on the night of 29th December an express on the Norte line was stopped near Póvoa de Santa Iria after a tree had fallen onto the catenary. The current had to be cut off, single-line working introduced and the train arrived at Lisboa two hours late.

The cause of the derailment was not immediately clear. Presumably either the coach involved was defective or there was a problem with the points, though a newspaper speculated whether it had really been an accident at all. [154]

A train from the Mediterranean port of Damietta to Cairo was derailed at Qalyūb, Egypt, 10 miles north of Cairo, on Saturday 18th January 1975. It was very overcrowded, with some passengers riding on the locomotive because there was no room elsewhere. As a result of the derailment the locomotive overturned twice and the six carriages that followed it into derailment were partially telescoped, though none overturned. According to the police 27 people died in the accident and another 52 were injured.

The train was being diverted onto a secondary line at Qalyūb to give priority to a following Alexandria – Cairo express. According to the Mustafa el Sheikh, the Assistant Interior Minister, the accident happened because the driver of the Damietta train failed to reduce speed for diversion. [155]

At least 47 people died in a derailment at Sidi Yahia du Rharb, Morocco, 23 kilometres north of Kenitra, on Thursday 22nd May 1975. About 150 other passengers on the Rabat – Tangier passenger train were injured, while about 300 survived unhurt and were able to return home. [156] The engine and three of the seven carriages were reportedly destroyed when the train left the track 200 metres from the station while travelling at 100 kilometres an hour. [157] The investigation into the cause of the accident was hampered by

[154] *Diário de Lisboa*, Lisboa, Portugal, 28th and 30th December 1974

[155] *Sunday News-Globe*, Amarillo, Texas, United States, 19th January 1975

[156] *Bucks County Courier Times*, Levittown, Pennsylvania, United States, 23rd May 1975

[157] *Daily Review*, Hayward, California, United States, 23rd May 1975

the fact that the train driver was among the fatalities, while the conductor, although he survived, was seriously injured and in a coma. [158]

News of an early-morning train crash in the Midlands at Nuneaton, England, just reached the London national newspapers in time to be included in them the same day, Friday 6th June 1975, though only brief details were available. A paper was able to state that least 50 people were injured and an unknown number were trapped when a London to Glasgow sleeper train had become derailed. British Railways, London Midland Region, stated that all four lines through the station were understood to be blocked. [159]

The train involved in the accident was the 23:30 Down night sleeper train from London Euston to Glasgow Central, consisting of two Class 86 electric locomotives (one not under power), a passenger brake van, five sleeping cars, a buffet car, seven sleeping cars and a passenger brake van, in the order stated. It was in charge of Guard E.A. Shore with Driver John McKay and Senior Secondman Leslie Norman. The start from London Euston had been on time and all went well until the train reached Kings Langley, 28 minutes later, where the engine became a total failure. Another engine was eventually provided to take the train, including its failed engine, forward, and the train eventually restarted at 00:58, now 75 minutes late. The run between Kings Langley and Nuneaton was made at the maximum permissible speed but the train was still 66 minutes late as it approached there on the Down Fast line, travelling at 80 miles an hour. A mile short of the station a gas-lit warning board, the lights of which had gone out, was provided to remind the driver of a temporary speed restriction of 20 miles an hour through the station. Despite this, the train failed to reduce speed at all and was totally derailed in the station, apart from its rear van. The front coaches were badly damaged and two passengers and two sleeping car attendants were killed outright, while two more passengers died later from their injuries. Another 36 people were more or less injured. All the lines through the station were blocked and there was extensive damage to both the signalling and overhead line equipment. The Slow lines were not reopened until late on Tuesday 10th June and the Fast lines not until the afternoon of Thursday 12th June. [160]

A fuller report on the accident appeared in the newspapers on Saturday. 'The Euston-Glasgow sleeper express crash early yesterday, which killed five people and injured 40, happened on a stretch of track where a temporary 20 mph speed limit was in force. But there were indications that the train may have been travelling much faster than that when it left the rails at Trent Valley station, Nuneaton, Warwickshire…The train was carrying 101 passengers, including Mr Peart, Minister of Agriculture, Fisheries and Food and seven British Rail staff…Mr Peart was travelling to Workington for a referendum count. After his rescue, the police searched his compartment for government papers. One of his first visitors, Mr Leslie Huckfield, Labour MP for Nuneaton, said Mr Peart woke up to find the compartment very compressed. Mr Peart later described how he was thrown out of his bunk and trapped. He said: "One of the first things I knew was the attendant, Mr George Norman, who knows we well, coming up and saying: 'hang on, Mr Peart, and we will get you out.' Shortly afterwards two firemen arrived and they brought me out

---

[158] *The Daily Report*, Oxnard-Upland, California, United States, 23rd May 1975
[159] *The Times*, London, England, 6th June 1975
[160] Department of Transport report

through the shattered window and took me to hospital. Although I felt shaken up, I was all right and wanted to leave. But the hospital suggested that I should stay on for a time. Mr Peart, whose condition was comfortable in Manor Hospital, Nuneaton, added: I am going back to London; but I do not want to go back by train.'" [161]

The normal maximum permissible speed through Nuneaton station was 100 miles an hour but since midnight on Saturday/Sunday 24th/25th May extensive remodelling and rationalisation of the track at the south end of Nuneaton station had been in progress, hence the need for the temporary speed restriction. The restriction was detailed in a weekly engineering notice issued to all train crews, the notice specifying the location of the restriction, the lines affected, the mileage, the restriction, the reason for it, the date and time that it commenced and, where known, the time and date that it would be lifted. The notice was headed:

TEMPORARY SPEED RESTRICTIONS

(Until further notice unless otherwise specified)

In addition to the engineering notices the commencement of the speed restrictions was indicated by a lineside square or circular speed indicator, showing the maximum permissible speed in black numerals on a white background. Preceding the speed indicator at a distance specified in the Rule Book should have been a warning board consisting of a horizontal arrow-shaped yellow board surmounted by a panel showing the maximum permitted speed in white numerals on a black background. There were also termination indicators, showing a white T on a black background, at the end of the restrictions. The speed and termination indicators were constantly lit and the warning board indicators and two lamps in their arrow-shaped boards must be lit after dark or during fog or falling snow.

The Department of Transport report on the accident noted that with the disappearance of many signal boxes long stretches of the railway had become relatively featureless but this was not the case on the approach to Nuneaton with the number of tracks changing from four to three and back to four, the adjacent Oxford Canal, various overbridges, and a large lineside works with several tall chimneys.

Assistant Permanent Way Supervisor P.R. Batkin told the inquiry into the accident that the work at the south end of Nuneaton station involved the removal of points and crossings, their temporary replacement by plain track and the relaying of the track on a new alignment. In connection with this work relatively sharp curves were temporarily required between the old and new track alignments, hence the need for the speed restriction, which he expected to be in place for about four months.

Assistant Permanent Way Inspector W.S. Evans stated that the warning boards had been placed at 95 miles 36 chains rather than at the 95½ milepost, thus four chains (88 yards) too close, because of the likelihood of vandalism at the 95½ milepost. The warning board for the Down Fast line was lit by gas supplied by two large (42lb) cylinders. The correct procedure was to have both cylinders turned on, an automatic valve would then change the supply to the second cylinder when the pressure dropped; one cylinder, burning day and night, should last between a week and ten days. In the Nuneaton area, however, since the gas-lit boards had been introduced eight or nine years previously, the procedure had been to

---

[161] *The Times*, London, England, 7th June 1975

have just one cylinder in use at a time, the patrolman carrying out a regular inspection and changing the cylinders when necessary. Inspector Evans could not recall a previous case of the lamps having been extinguished because the gas had run out.

The patrolman responsible for inspecting the warning board was Leading Trackman D. Costello. He stated that previous to the accident he had been unaware that the correct procedure was to have both gas cylinders on all the time, rather than manually switching from one to the other. He had checked the warning board at 09:30 and again on 13:30 on Wednesday 4th June, when both lamps had been burning well and his next patrol would have been on the morning of Friday 6th (thus after the accident). In his nine months as Nuneaton Patrolman he could not remember a light of a warning board going out, though sometimes deteriorated mantles had resulted in poor lights.

Track Chargeman Harold Merrick, Costello's immediate superior, said that the lights of the warning board had been fine at 11:00 on Thursday 5th June. He, too, thought that the 'one cylinder' method was standard practice and could not recall a previous case of the warning board lights going out. After the accident he had examined the warning board at about 05:00, accompanied by Assistant P.W. Section Supervisor Simpson, and found that all three lights were extinguished, the mantle of the light behind the 20 mph speed indicator being broken.

The drivers of 15 trains that had passed on the Down Fast line between 22:53 on 5th June and 00:58 were interviewed. Six said that the speed indication light at the warning board had been out, four that it had been dim, four that it had been well lit and one had not noticed. All agreed that the two 'bull's eye' lights in the arrow-shaped board had been lit, some saying that they had been dim. None of those who reported the speed indication light as out had reported it in accordance with Rule T 25.5.

The evidence of the crews of three further trains that had passed before the accident was detailed in depth. Train 1P54, the 23:45 London Euston to Barrow sleeper, had passed at 01:12. Driver C.W. Ireson testified that he had approached Nuneaton travelling at 90 miles an hour. He had known exactly where the warning board was. Its '20' indication had been unlit and the two 'bull's eye' lights had been dim. He was aware of Rule T 25.5 but had not reported the matter.

Driver William Sharpes of train 1S15, the 23:30 Kensington (Olympia) to Stirling Motorail, said that when he had approached the warning board at 01:40 all of its lamps were out, and he had remarked to his secondman, "It's out." He had caught a glimpse of the board itself. He knew that the track work was going to be a long job and had at no time considered that the restriction had been lifted. He knew that he should have reported the lights out and did not know why he had not done so; perhaps it was because he had been running late. He had learnt of the accident on arrival at Crewe.

Senior Secondman F.S. Roddy of train 1S15 agreed that the warning board lights had been out and that his driver had remarked upon the fact. He likewise excused their not reporting the problem on their being late, having had a diesel locomotive pulling the train which could not equal the performance of the usual electric one.

Driver Robert Turner, in charge of the engine of train 6F72, the 22:20 Willesden to Runcorn freight train, alleged at the public inquiry that when his train had reached Nuneaton at 01:45 all three lights of the warning board had been brightly lit. Later he withdrew his statement, saying that he had been bending down to look in his bag and missed seeing the warning board entirely.

Guard D.L. Simcoe, accompanying Driver Turner, pleaded that he had been making tea as the train neared Nuneaton and had not seen anything.

Driver McKay testified that he had received the weekly engineering notice, which he had studied and which he kept in front of him in the cab of his engine. He had driven the same train the previous night and had obeyed the restriction. On the night of the accident his secondman, who was a qualified driver, had driven the train until the failure at Kings Langley. After that, he (McKay) had taken over. He had then driven at the maximum speed of 90 miles an hour in order to try and regain some time. Approaching Nuneaton he had shut off power as usual at Bulkington, reducing speed to 80 miles an hour on the approach to where the warning board had been. Seeing the board for the Down Slow line but nothing of that for the Down Fast line on which he was running, he had been convinced that the restriction must have been lifted and released his brakes. It was only when he had seen the '20' board at the start of the restriction itself that he realised the restriction must be still in force. He had then applied the brakes in emergency and the derailment occurred moments later. He added that he had not at that time received the weekly engineering notices for the period from the 7th to the 13th (which confirmed that the restriction was still in force) and had thought that the work might have been finished early. Unlike Driver Sharpes, he had been unaware of how big a job it was. He had been well, properly rested and had nothing to drink on the day before the accident. He had been keen to regain as much time as he could.

Secondman Norman agreed with his driver's evidence. He said that he, too, had failed to see the Down Fast line warning board and had assumed that the restriction was lifted. He agreed that in accordance with Rule T 21.1 should the restriction have been lifted early the warning board should have been left in place with the higher speed displayed.

Guard Shore stated that his train had been driven "pretty fast" after Kings Langley. He had been aware of the speed restriction and could have applied the brakes had he realised that they were running into danger, but had not realised how close they were to Nuneaton in time to do so.

Regulator J. Marlow and Signalman G.E. Young spoke of witnessing the derailment from Nuneaton panel box, calling the emergency services, and arranging for the power to be switched off. They could remember numerous reports over the years of the lights of warning boards being extinguished, though never all three lights at once.

Concluding his report into the accident the inspecting officer, Major C.F. Rose, noted that the derailment had occurred because the train attempted to negotiate a speed restriction of 20 miles an hour at 80 miles an hour, well in excess of the maximum possible safe speed of 45 miles an hour. Information regarding the restriction had been correctly published in the engineering notices. The lineside signs had been correctly positioned but the propane gas supply to the warning board had not been set up correctly. The light in the '20' sign of the warning board had gone out at about 22:30 on 5th June, doubtless due to failure of its mantle. The other two lights of the warning board had dimmed at about 01:10 on 6th June and had gone out altogether at some time between 01:10 and 01:40 because the gas had run out. A number of drivers had seen that the lights were out but none had reported it.

Major Rose did not believe that Driver McKay, a very experienced driver, could have made a conscious decision that the speed restriction had been lifted merely because he had failed to see the warning board, when it had been in force the day before, the engineering notice had stated that it would remain in force 'until further notice' and he was aware that if the restriction had been lifted early the warning board should have been left in place, showing the higher permitted speed. It seemed more likely that deprived of the reminder of the lights of the warning board he had briefly forgotten all about the restriction, as had his secondman, or else the thoughts of the latter had been elsewhere. Guard Shore had not been in a position to know exactly where his train was.

Driver McKay was therefore primarily responsible for the accident but others were also to blame, namely Secondman Norman, who had not assisted his driver, the permanent way staff who had misused or condoned the misuse of the warning board gas equipment, and the crews of the previous trains who had failed to report that the lights of the warning board were out.

A survey of the failures of lights of warning and speed indicator boards in the Stoke division of the London Midland Region during the previous year showed 72 failures from various causes such as high winds, blocked gas pipes and vandalism. The last previous accident due to failure to observe a temporary speed restriction had occurred at Cheadle Hulme in 1964.

An inquiry into the technical departments of French, German, Swiss, Belgian, Dutch, Italian, Swedish, and Japanese railways showed that all used some type of printed notice to inform drivers of restrictions, plus lineside warning boards. In addition, the French, German, Swiss and Belgian railways used the equivalent of the British automatic warning system (AWS) at the warning boards. The Italian and Japanese railways warned their drivers by advanced cab signalling systems and the Dutch and Swedish railways were working towards similar measures.

Following the accident improved gas mantles were being introduced along with improved designs for warning boards, possibly using reflective material. The basic risk remained though and as an additional measure the British Railways Board had decided to install AWS magnets at warning boards. Problems arose where the warning board was near a signal or at a junction where the restriction did not apply to all routes but the matter was being worked on, with plans to install such magnets first on high-speed Inter City lines. [162]

In June 1976 Driver McKay appeared at Birmingham Crown Court charged with the manslaughter of the six people who lost their lives. He pleaded *Not Guilty* while admitting that he had made "a stupid decision" in assuming that the speed restriction had been lifted (he did not say, as the inspecting officer believed, that he had forgotten about it). [163]

He was acquitted. [164]

An accident occurred in the suburbs of Rio de Janeiro, Brazil, in the early evening of Thursday 17th July 1975. Commuter train UM-59, of the Rede Ferroviária Federal (RFF), which had left Don Pedro II station at 18:27 for Japeri, was derailed on the Linha Auxiliar (Auxillary Line) close to Magno station at 18:40. One of the coaches of the

[162] Department of Transport report
[163] *The Times*, London, England, 11th June 1976
[164] Wikipedia – Nuneaton Rail Crash

crowded train struck and damaged the Império Serrano sambo school (a dancing, marching, and drumming club). Another two coaches collided with the supports of the overhead line equipment (44,000 volts) and also the signalling cables (4,000 volts). Initial reports estimated 20 dead and about 220 injured. The casualties were taken to the Carlos Chagas Hospital and the Souza Aguiar Hospital; some later being transferred from the first-named hospital to the Clinico Nossa Senhora de Lourdes, which had an agreement with the RFF. The line was completely blocked by the accident and trains were diverted and a replacement bus service was set up. Later traffic was restored past the accident site using diesel traction pending the restoration of the overhead line equipment. Quite apart from the deaths and injuries the accident was an expensive one for the railway in financial terms, for example Señor Irani dos Santos Ferreira, the president of the sambo school, demanded compensation from the railway amounting to 1,500,000 cruzeiros novo, saying that rehearsals at his school had had to be postponed until further notice due to the damage caused by the derailed train.

The derailment occurred on a curve and first suggestions were that it had been due to the excessive speed of the train, though there was also the possibility of a brake failure. Several surviving passengers said the train had been travelling too fast earlier in the journey, so that while passing through Tomás Coelho station a number of them had begun to panic. The motorman of the electric train, Vilmar Valentin Limasse, suffered serious chest injuries in the accident and was confined to hospital, where he was due to be questioned when he had sufficiently recovered from his injuries.

According to one newspaper 15 passengers died in the derailment but a possibly more reliable source states that there were only 13 fatalities, including two passengers who had died from their injuries later. [165]

The day after the accident the Brazilian President, Ernesto Geisel, called a meeting with the Ministers of Justice and Transport; Colonel Milton Gonçalves, President of the RFF; and others to discuss the disaster and the general question of railway safety in the Rio area. Colonel Gonçalves told the president that the railway needed more staff and resources. At the end of the meeting the president suggested the establishment of an emergency programme to improve the safety of the railway and told Colonel Gonçalves that he hoped everything would be done to avoid further such accidents. The colonel replied that everything possible was being done, but sometimes accidents were unavoidable. [166]

On Tuesday 12th August 1975 omnibus train No. 2745 from Santiago de Compostela to Vigo consisted of a pair of two-car diesel multiple-unit sets. It was apparently running late when it was derailed on a curve at Bamio, Spain, at 17:04 in the evening. The three leading vehicles of the train left the track. The motorman and five passengers (all from the same Madrid family) lost their lives and about 40 passengers were hurt. There had been about 200 people on the train. [167] [168] [169]

---

[165] *Jornal do Brasil*, Rio de Janeiro, Brazil, 19th and 21st July 1975

[166] *Folha de S. Paulo*, 18th and 19th July 1975

[167] Facebook page – Palco de Estrelas – Daniel Seijas. Orquestras, historias e incendios

[168] Federación Castellano Manchega de Amigos del Ferrocarril

[169] *Morning Journal*, Daytona Beach, Florida, United States, 13th August 1975

CHAPTER 4

REAR COLLISIONS

The risks associated with one train running into the back of another had been low in the formative years of the railways when trains were few and speeds not very high, but as the years progressed train speeds and traffic intensity often grew faster than advances in brake power, rolling stock construction and signalling and as a result by the early twentieth century serious rear collisions had become common. This chapter is therefore a long one, even though a great number of less serious accidents have not been included.

On the evening of Sunday 30th April 1961 a local train bringing families home after a day at the seaside collided with the rear of a stationary express from Adana at Kartal, Turkey, not far from Istanbul. Three coaches were reportedly 'smashed to matchwood.' Eleven passengers were killed outright, two or three died later of their injuries and about 100 were injured, half of them seriously. 'Ambulances filled the ferry-boats crossing the Bosphorus as they carried victims of the crash from the Asian side to hospitals on the European side.' [170] A brake failure rather than a driver or signalling error was apparently the cause of the accident. [171]

Holidaymakers bound for the Isle of Man, via the port of Fleetwood, were the victims of an accident at Singleton Bank, England, on British Railways, London Midland Region, on the morning of Sunday 16th July 1961. Six passengers and the driver of the passenger train lost their lives in the accident and 116 passengers were injured. Engineering work had been in progress in the vicinity of the accident from midnight on Saturday until just before the accident and the passenger train ran into the rear of a ballast train due to a signalman's error.

The passenger train was the 8.50 a.m. express from Colne to Fleetwood, a six-car diesel multiple unit, in charge of Passenger Guard G.R. Bradshaw with Motorman Thomas Shaw and carrying about 350 people. The ballast train consisted of a Class 8 2-8-0 steam locomotive, a brake van, 39 open high-sided waggons and a second brake van, in charge of Goods Guard E. Fenton with Passed Fireman J. Dean and an unnamed fireman.

A man out for a walk with his son witnessed the accident, as a newspaper reported.

[170] *The Times*, London, England, 1st May 1961
[171] *Wisconsin State Journal*, Madison, Wisconsin, United States, 2nd May 1961

'Mr Charles Basey, of Knowlsley Crescent, Weeton, an instructional officer with the R.A.F., was out for a Sunday morning walk with his son, Eric, aged 8, in the field at the foot of the embankment where the wrecked train came to rest. "I saw the train coming and I could see there was going to be a crash," he said. "The driver braked for 800 yards but he couldn't stop, and he ploughed into the waggons and leapfrogged into the air. I rushed home with my son and then called the men from the village of Weeton to come and help. We went back to the scene in cars and started to pull the injured out. I managed to get two women out before the police and fire brigades arrived. Passengers were laid all over the field."

'Police, fire engines and convoys of ambulances from Blackpool, the Lancashire county area and 100 men from the R.A.F. station at Weeton went to the scene.

'The crash brought down all the telephone wires over the railway tracks and communications were operated from police radio cars.

'An official casualty list issued by the police tonight showed that 18 of the injured had been detained at Blackpool Victoria Hospital and one at Preston Royal Infirmary. Ninety-three had been treated and discharged at Blackpool and four at Preston.

'A bus driver, Mr Allan Hollings, of Southfleet Avenue, Fleetwood, was on his way to pick up a gang of railwaymen when he arrived at the scene. He turned his bus round and took about 50 sitting cases to hospital.

'Two Lancashire County ambulance men, Mr Malcolm Wright, and Mr Bernard Tudgey, were in their homes nearby when they heard the crash and ran to help drag out the dead and injured and give first-aid. Diesel oil fumes threatened to burst into flames and Mr Wright was overcome by fumes.

'Among the injured were children. In some cases several casualties were from the same family.

'A railway spokesman said that of the passengers fit enough to continue their journey, 59 who were travelling to the Isle of Man were taken by bus to Fleetwood. More than 20 were returned home by ambulance after receiving treatment at Blackpool.' [172]

As a result of the accident the leading car of the passenger train demolished the brake van at the rear of the ballast train, mounted the rear six waggons, then veered to the left and plunged down a 15 ft embankment. The front of the second car was also pulled down the embankment and it came to rest at right angles across the track. The couplings between the coaches held and the third coach was held high in the air by the rear of the second, while the fourth coach was also raised up at its leading end. The rear two coaches remained on the rails. The motorman's cab and first-class compartment at the front of the leading coach were destroyed, the damage to the remainder of the train was progressively less from front to rear. The brake van and three waggons of the ballast train were wrecked and three other waggons were damaged.

The events leading up to the accident might be said to have begun at midnight. Both of the lines between Weeton and Singleton signal boxes, on the Preston and Fleetwood sides respectively of Singleton Bank signal box, the exact site of the collision, were scheduled to be closed at 12.01 a.m. that Sunday for re-ballasting and re-sleepering.

[172] *The Times*, London, England, 17th July 1961

Singleton Bank signal box, a simple block post, was switched out of use, though its telephone could be used. Possession of the Down line was to be given up at 4.15 a.m., with single-line working over it from that time until possession of the Up line was also to be given up and normal working resumed at 11 a.m. The engineering operations were in charge of Permanent Way Inspector J.N. Wilkinson. The operating arrangements were in charge of District Signalman and Traffic Inspector J. Redmond and Assistant District Signalman and Traffic Inspector M.A. Bush.

In connection with the engineering work a ballast cleaner was used on the Up line at Singleton Bank and a Matisa tamper worked from Weeton towards Singleton Bank, also on the Up line. Two ballast trains were also in use. One, with sleepers, was working near Weeton, the other, for the spoil, stood on the Down line opposite the ballast cleaner, from which a conveyor belt fed the spoil into its waggons. Inspector Wilkinson mostly remained at Weeton, while Sub-Inspector T. Dagger was with the ballast cleaner. District Inspector Redmond made the arrangements at Weeton and Assistant District Inspector Bush did likewise at Singleton and later also acted as pilotman over the single line.

Of the many irregularities that led to the accident the first two occurred right at the beginning of the possession. Inspector Redmond arrived at Weeton at 11.30 p.m. on Saturday and after an Up freight train had passed arranged with Assistant Inspector Bush for the tamping machine to be sent from Singleton on the Up line, followed by the ballast cleaner. He arranged for two track circuits at Weeton to be disconnected and took possession of the Up and Down lines at midnight, recording that he had done so in the train register book, though not using the correct wording as designated in the General Appendix. This was, perhaps, the first minor transgression. Four minutes before the possession had been taken a Down freight train was accepted by the Weeton signalman from Kirkham North Junction box and duly signalled to Singleton box, although the line was supposedly in the engineer's possession, and passed through the section, clearing the latter box at 12.09 a.m. At the Ministry of Transport inquiry Inspector Redmond pleaded that possession of the Down line had not in fact been taken until 12.10 a.m., not at 12 midnight as recorded in the Weeton train register book. Perhaps, but more irregular working was to follow.

Following the freight train the two ballast trains arrived from sidings at Kirkham. The first had two engines, the front one of which was bound for Fleetwood. This train entered the Weeton – Singleton section on the Down line at 1.02 a.m., having been offered and accepted on the block instrument (again despite the line supposedly being in the engineer's possession) and was signalled as a light engine (which it was not). The front engine was detached at Singleton Bank and reached Singleton at 1.15 a.m., whereupon the Singleton signalman (H. Monks) gave the *Train out of Section* signal to Weeton, despite the ballast train remaining in the section.

The ballast train returned to Weeton at 4.15 a.m. without a Wrong Line Order form being issued as per the regulations. Single-line working was then opened up at 4.34 a.m. over the Down line. At the Weeton end, according to Rule 200(a), a handsignalman should have been stationed at the Weeton Down distant signal to protect the beginning and end of the single line. There was a shortage of competent handsignalmen, so none was appointed. Inspector Redmond considered that this was safe, having cautioned the Weeton signalman not to accept a Down train from Kirkham North unless the pilotman was present at Weeton.

Meanwhile at Singleton Assistant Inspector Bush failed arrange for a lineman to disconnect track circuit No. 5654 on the Down Line in accordance with Rule 198 (c). The effect of this was that whenever an Up train used the Down line during the single-line working occupation of the track circuit put the needle of the Down line Weeton – Singleton block instrument to *Train on Line*. This being the case the block instruments should have been treated as having failed and the section operated under Block Regulation 25 (working during failure of block instruments or bells). Instead the assistant inspector condoned the signalman irregularly replacing the indication of the block instrument to *Normal* each time an Up train passed through the section over the Down line.

A number of trains then passed over the temporary single line in each direction. Between 7.30 and 8.15 a.m. Inspector Redmond began to arrange for the disposal of the different engineering trains. The ballast and sleeper train left Weeton for Singleton on the Down line at 8.30 a.m., followed by the ballast spoil train at 8.40. a.m., the arrangement being that Assistant Inspector Bush would use the latter train to withdraw the single-line working. Signalman F.M. Whiteside at Weeton recorded in his train register 'Single Line Working withdrawn at 7.35 a.m.' Inspector Redmond, however, had arranged for the spoil train to stand on the Down line opposite the ballast cleaner until the work was complete. Signalman Whiteside instructed Sub-Ganger D. Semple to place three detonators on the line at Weeton to protect the ballast train, though Semple should have been working under the instructions of Goods Guard Fenton of the ballast train rather than those of the signalman. The ballast train thus stood in the section without being properly protected for an hour and a half (until it was struck by the passenger train) in contravention of Rule No. 216. Inspectors Wilkinson and Redmond considered that this rule did not apply because single-line working was still in force until pilotman Bush had withdrawn it at the Singleton end as well.

Just before 10 a.m. Inspector Redmond at Weeton gathered from Assistant Inspector Bush at Singleton Bank that the work of ballast cleaning was almost complete and he instructed the latter to withdraw the single-line working at the Singleton end when he reached there as pilotman with the second ballast (spoil) train on the Down line. The ballast cleaner and Matisa tamper would return to Singleton on the Up line, all three engineering trains would then continue on the Down line to Poulton (on the Down or Fleetwood side of Singleton) and normal working would be resumed. These arrangements, like the earlier ones, were made using the omnibus telephone (linked to all the local boxes) and both Inspector Redmond and Assistant Inspector Bush assumed that Signalman Whiteside at Weeton and Signalman Derek Harrison at Singleton would be listening in to the conversations, as was normal practice. (Signalman Harrison had relieved Signalman Monks at Singleton at 6 a.m., Signalman Whiteside, on the other hand, was on duty from 11.30 p.m. on Saturday until after the time of the accident.)

At about 10.15 a.m. Assistant Inspector Bush telephoned Signalman Harrison from Singleton Bank box and in the inspector's words told him, "We are ready to leave on the Down line in two or three minutes."

Signalman Harrison had not repeated the message but had merely replied, "Right, Mr Bush."

Earlier, at 9.38 a.m. Signalman Harrison had spoken to Assistant Inspector Bush about two light engines that had arrived at Singleton from Poulton on the Up line. The first engine had to be crossed over to the Down line at Singleton to allow the second to enter the "dead" Up line, which it did at 9.55 a.m. Signalman Harrison told the inquiry into the accident that he did not always listen to all conversations on the omnibus telephone, had not done so on that morning, and had therefore been unaware of the arrangements for giving up the single-line working. Between 10.10 a.m. and 10.25 a.m., Signalman Harrison continued, Assistant Inspector Bush had telephoned him from Singleton Bank and, as he (Harrison) understood it, had told him that the spoil train was coming to Singleton on the Up line. He had then assumed that there had been a change of plan and only later realised that in that case, Signalman Whiteside at Weeton should have cancelled the spoil (ballast) train which he (Harrison) had accepted an hour and a half earlier on the Down line.

According to Signalman Harrison at 10.21 a.m., while his Down line block instrument from Weeton was still at *Train on Line* for the spoil train Signalman Whiteside had offered him the Colne express. He had not accepted it but had got on the telephone to Weeton asking why the express had been offered with his block at *Train on Line*. Signalman Whiteside had replied, "Why is your indicator at *Train on Line*? Come on, Derek, straighten up your needle and give me a *Line Clear*." The Weeton signalman had appeared to be in a hurry and anxious not to delay the express and had, Harrison thought, also said that the spoil train was coming back on the Up line. With this, Harrison said, he had unpegged his block instrument and accepted the express from Weeton at 10.22 a.m. He thought that the spoil train must have been crossed over to the Up line at Weeton out of the way of the express. He admitted to the inspecting officer, Brigadier C.A. Langley, that he should not have replaced his block instrument to *Normal* merely on the basis of a telephone conversation and without having received a cancelling signal, but he was only a young and inexperienced signalman, whereas Whiteside was a much more senior man. The inspecting officer concluded that Signalman Harrison had become thoroughly confused with the different telephone messages passing to and fro. That Harrison had been sure the ballast train was coming back on the Up line was shown by the fact that he had instructed his handsignalman to remove the detonators from the Up line, to allow the ballast train to come down to his box as soon as the express had passed.

Signalman Whiteside's version of the events leading to the acceptance of the express was somewhat different. He maintained that at 10.17 a.m. he had heard via the omnibus telephone Assistant Inspector Bush at Singleton Bank telling Signalman Harrison that the ballast cleaner would be coming back to Singleton on the Up line and that he (Bush) would be coming back to there with the spoil train on the Down line. He (Whiteside) had then spoken to Harrison for about five minutes about a light engine at Singleton.

At the end of this conversation, thinking that the spoil train must by then have covered the mile and a half from Singleton Bank to Singleton, he had asked Harrison, "How are you doing on the Down?"

Harrison had replied, "I am clear on the Down."

He had repeated the question, received the same reply. Harrison had given the *Train out of Section* signal for the spoil train, and accepted the express. Whiteside contended that he had not been pressing Harrison and could not explain why he had thought it necessary to twice ask Harrison over the telephone if the line was clear instead of relying on the indication of his block instrument.

The express having been accepted, Whiteside instructed Sub-Ganger Semple to pick up the detonators he had laid on the Down line, showing why the rules required that Semple should have been working under the direction of Guard Fenton, who knew that the spoil train was still in the section, rather than under Signalman Whiteside, who did not. The Colne express had passed Weeton at 10.26 a.m. travelling at about 50 miles an hour. Shortly afterwards Whiteside was advised of the collision by a ganger from a signal post telephone, called the ambulances and fire brigade and sent the *Obstruction Danger* signal to Bradkirk box, which had by then been switched in between Weeton and Kirkham North Junction. It could not be sent to Singleton because the accident had brought the telegraph wires down.

The reason the spoil train had taken longer than anticipated to proceed to Singleton was that although it was started promptly it had no sooner done so when it had been stopped again by a hand signal from Inspector Dagger, who had seen that one of the side doors of a waggon had come undone. It took a couple of minutes to rectify the problem and the train was just about to restart from Singleton Bank when it was run into by the express.

Brigadier Langley concluded that 'slack working, by senior as well as junior members of the railway staff, the contravention of the Block Regulations on several occasions, the misinterpretation of Railway Rules and the misunderstanding of telephone conversations were major factors affecting the issue.'

The fundamental cause was aptly summarised by the jury in their verdict given at the inquest on the victims, which was concluded at Preston on 23rd August. The jury found that the seven persons died from injuries caused when the passenger train in which they were travelling came into collision with a stationary ballast train at Weeton, due to the misunderstanding of and the noncompliance with the Signalling Regulations and Rules laid down by British Railways. The jury returned a verdict of misadventure in each case.

The inspecting officer then proceeded to outline the various misinterpretations, contraventions of the rules and misunderstanding which had led to the disaster.

Inspector Redmond had, the inspecting officer noted, carried out the spirit but not the letter of his instructions in taking the engineering possession. Having done so he had then allowed a Down freight train to pass through the section. This was a comparatively minor mistake of making a premature entry in the train register.

An hour later a light engine forming part of a ballast train had been signalled through the section on the block instruments despite the line being in the engineer's possession. This was highly irregular. The ballast train had been wrongly described as a light engine and *Train out of Section* had been given back by Singleton when the ballast train was still in the section. The inspecting officer commented, 'It is not surprising that other block irregularities occurred when such misuse of the block instruments was authorised by a senior inspector.'

Wrong Line Order forms had not been issued by any of the Weeton or Singleton signalmen and neither Inspector Redmond nor Assistant Inspector Bush had done anything about this.

When single-line working was begun no handsignalman had been appointed at the Weeton Down distant signal. Inspector Redmond's instructions that Down trains were not to be accepted by Weeton, unless the pilotman was there, was a safe one but was not in accordance with Rule No. 200, which directed that a handsignalman should always be appointed.

There was no objection to the telephone conversations made during the possession and these were in fact necessary but the assumption by the inspectors that all the signalmen would invariably be listening in to the conversations on the omnibus circuit was sure to lead to misunderstandings.

The use by Inspector Redmond of the spoil train to withdraw single-line working after it had been stopped in the section for a prolonged period (8.40 a.m. to 10.30 a.m.) was not against the rules but was unwise.

Inspector Redmond had not arranged for the ballast train to be protected in accordance with Rule No. 216 because he considered that single-line working was still in force, even though it was noted as having been withdrawn at the Weeton end at 7.35 a.m. Had the ballast train been protected by handsignalmen, working to the instructions of the guard, at a ¼ mile and ¾ mile distances, as per the rule, the collision would have been prevented. The inspector's view was, however, held by other officials, so he could not fairly be criticised for it.

Inspector Redmond had stayed at Weeton the whole night rather than going to Singleton to see how the young and inexperienced signalman there was getting on. In general, 'He must take the responsibility for allowing slack work to go unchecked, block working to be carried out irregularly, and important telephone messages to be passed with little attempt to check that the right recipients heard, let alone understood them.'

Assistant Inspector Bush was seriously at fault for having failed to have the track circuit at Singleton disconnected and for condoning the signalman replacing the Down block instrument to *Normal* every time an Up train went over the Down line. The same thing had apparently been done the previous Sunday. The bad habit Signalman Harrison had thereby acquired had doubtless influenced his actions later. 'Misuse of block instruments is a dangerous practice which must not be allowed,' the inspecting officer added.

There was a conflict of evidence regarding the acceptance of the Colne express. Signalman Whiteside's version was partly confirmed by the evidence of Sub-Ganger Semple, who was, however, an unsatisfactory witness, and had been coached by the signalman as to what he was to say. Signalman Harrison's version was partly confirmed by the evidence of Lengthman A.J.P. Terras and Passed Fireman K.S. Craven of the light engine, both of whom had been in Singleton box at the time and both of whom were good witnesses. Brigadier Langley said that he accepted neither signalman's story. He was satisfied that Whiteside had not told Harrison that the spoil train was coming back on the Up line, but that he thought that the spoil train must have passed Singleton and Harrison had forgotten it, and he had then pressed Harrison to clear the block and accept the express. Harrison, on the other hand, had been convinced there had been a change of plan, misunderstood the telephone message from Assistant Inspector Bush and confused the ballast (spoil) train and the ballast cleaner. The pressing by a more senior signalman to clear the line he had taken as a confirmation of the change of plan.

Signalman Harrison frankly admitted a grave breach of the block regulations in clearing out the ballast train without having seen it pass his box or alternatively having received the cancelling signal. He must therefore bear the main responsibility for the collision.

Signalman Whiteside's actions also called for criticism. Having wrongly assumed that the ballast train must have covered the short distance to Singleton he had, no doubt, in spite of his denials, and in his anxiety to give the express a clear run, unduly pressed his colleague to clear his block. Whether he had actually asked for the express with the block still at *Train on Line* was open to doubt, possibly repeated impatient rings of the block bell by him had been taken as a request for *Line Clear*.

It was clear that Motorman Shaw had had his train under control, had been driving within the speed limit and had been properly alert. His prompt braking upon seeing the obstruction had lessened the severity of the collision, for which he had been in no way responsible.

Moving on to remarks and recommendations the inspecting officer noted the integral all-welded steel construction of the multiple-unit express had stood up well to the violence of the collision.

The misunderstanding of telephone messages had led to accidents before. It was of the utmost importance that such messages relating to signalling were properly understood and repeated. The assumption by two inspectors that everyone concerned was listening to their telephone messages on the omnibus circuit was astonishing. The system of giving messages and recommendations by telephone should be reviewed and more comprehensive instructions issued.

The Rules and Regulations had grown in length and complexity throughout the years; many of them stemming from lessons learnt from past accidents and some from recommendations made by the inspecting officers. The problem had been discussed with the Officers of the British Transport Commission and they had agreed to review these Rules and Regulations to see whether they could be simplified. This would by no means be an easy task. In doing so the aim should be to set out in clear and simple language those rules which must be obeyed implicitly and to provide a comprehensive series of instructions and codes of practice to cover the various abnormal and unusual contingencies that could arise.

Finally, Brigadier Langley wrote, 'I would draw attention to the lack of leadership displayed by a senior supervisor on this occasion. Supervisors of all ranks and grades should set a high standard of conduct and discipline and take a personal interest in their staff. By so doing, they will do much to increase safety and improve efficiency on which the future of the railways so largely depends.' [173]

A simple act of forgetfulness on the part of a signalman led to a horrific accident on the night of Thursday 5th October 1961. At 22:34 starting signal F was cleared at Hamburg Berliner Tor station, West Germany, for S-Bahn train No. PS 3819 to depart for Bergedorf, by Fahrdienstleiter Alfred Messer. Shortly afterwards, too late to do anything to avert the accident, the fahrdienstleiter remembered that an engineering train, loaded with massive I girders for the construction of a new bridge, was still occupying the section. Filled with passengers returning home from the cinema or theatre the electric train quickly gathered speed, and it was moving at an estimated 70 kilometres an hour when it collided with the engineering train at 22:38 between Berliner Tor and Rothenburgsort stations. Its motorman, possibly its conductor, and many of the passengers in the leading coach of the train were killed instantly when two of the massive girders penetrated some 13 metres into the train, while other passengers were trapped in

[173] Ministry of Transport report

the wreckage. Rescuers were soon on the scene but the collision had occurred on a high embankment, making access difficult. Firemen and police worked in the darkness to free the injured, assisted by uninjured passengers, including women and girls, while doctors performed several amputations on trapped victims. No less than 150 firemen attended the scene. The rescue work continued until 05:00.

Passenger Heike Rackza was one of the lucky ones. She had intended to board the leading coach, then when she had seen how full it was had instead got into the middle part of the train. She lost her hat and a shoe in the accident, but that was all. In the dark she did not see the dead and injured and did not realise the seriousness of what had happened. "They told us that there had been an accident and we should walk back along the line to Berliner Tor," she related later. It was only when she read the newspaper the following morning that she became aware of the extent of the disaster.

In all, 28 people died in the disaster and about 100 were injured.

Fahrdienstleiter Messer and his subordinate, Assistentenwärter Harald Kruse, 28, were subsequently charged with manslaughter through negligence, causing injury through negligence and endangering the safety of people travelling on the railway. Messer was found guilty and awarded a suspended sentence of a year's imprisonment; Kruse was acquitted. It was thought that a distraction caused by having to complete paperwork might have led the fahrdienstleiter to make his mistake, while the Deutsche Bundesbahn was criticised for not having installed automatic block working on the line for reasons of economy.

Messer suffered from extreme guilt for having caused the disaster for the remainder of his life. [174] [175]

Thursday 31st May 1962 was Ascension Day, and in Italy for many people it was also the start of a four-day holiday, thus *accelerato* train No. 1391, from Milano to Genova and Ventimiglia, was very crowded with families going on holiday to the Ligurian Riviera, students returning home, Sardinians en route to Genova to embark for the island, soldiers, workers returning home, and foreign tourists. The train consisted of an electric locomotive and 12 coaches. The rear coach consisted of two long compartments with a central corridor and had a seating capacity of 90 passengers; on the day of the disaster every seat was occupied and another 40 passengers were standing in the corridor. The train started from Milano Centrale at 00:45, a minute late, and after stopping at numerous stations it arrived at Voghera, Italy, at 02:22, now running 15 minutes late. It was due in Genova Brignole at 05:22. At this date the Milano-Voghera line was electrified with a DC system, while the line on to Genova was electrified with a three-phase AC system, with the result that all trains had to change engines at Voghera. This necessarily took a few minutes. At 02:30 the signalman in A cabin cleared his signals for the train to depart and at 02:34 the official in charge of the station, seeing that the station work was complete, signalled the train away. A minute later, at 02:35, as the *accelerato* was about

---

174 https://www.ndr.de/geschichte/schauplaetze/5-Oktober-1961-28-Menschen-sterben-in-der-Hamburger-S-Bahn,sbahnunglueck100.html

175 https://www.abendblatt.de/hamburg/article233470539/28-menschen-sterben-bei-hamburgs-wohl-schwersten-zugunglueck.html

to restart, it was run into from behind. [176] [177]

Freight train No. 8151 had left Lecco at 00:02 for Milano Rogoredo and Arquata Scrivia, in charge of Conductor Mario Antonangeli with Driver Lanfranco Pigiani, 28, and Assistant Driver Soriano Fabbri, 22. The driver and his assistant were well familiar with the line. The train consisted of a Class 626 electric locomotive and 39 empty cement waggons. There was no brake van; the conductor rode in the rear cab of the engine. The approach to Voghera was fully protected by colourlight signals, the first of which, 2,400 metres from the station, was displaying a yellow caution indication when the freight train reached it. The next signal, displaying red, was 1,200 metres from the station. Six hundred metres closer to the station again was cabin C, which on the morning in question was manned by Signalman Franco Sarchi, while shunters Lislo Barbieri and Guglielmo Balbi were also in the cabin. The three men watched the freight train coming and realised that it was still moving at full speed despite the adverse signals. They ran from the cabin, shouting at the top of their voices and waving red lamps at the advancing train, but to no effect. According to a newspaper account Shunter Barbieri had a narrow escape from being run down in his attempts to stop the train. The brakes of the freight train were apparently applied at the last moment; examination of the tachograph of the engine showed that its speed had been slightly reduced, from 72 to 60 kilometres an hour, in the last few hundred metres before it struck the rear of the passenger train. Prior to that the train had travelled at between 70 and 75 kilometres an hour all the way from Milano, apart from a brief reduction of speed at Pavia.

The results of the collision were disastrous, especially on the rear coach of the passenger train, into which the engine of the freight train embedded itself. The engine of the colliding train was little damaged, though ten freight waggons were more or less wrecked. Driver Pigiani and Assistant Driver Fabbri had fled into the interior of the locomotive at the last moment and escaped almost unhurt, as did their conductor. The first two alighted from their engine, saw the scene of devastation and fled. The conductor made his way to the station.

"We were chatting in Via Cavour," a Voghera witness said. "It is 300 or 400 metres to the station as the crow flies but the roar that we heard shook the windows of all the buildings in the street."

A night watchman came running from the station and stopped the first person he came across and shouted, "Disaster, disaster, quick, call help."

Assistance was quickly at hand and soon there were 15 doctors on the scene from the local hospital, including Professor Bergonzoli, in charge of the hospital, nurses, firemen from Voghera and later also Pavia and Alessandria, railway and town police, carabinieri, priests and nuns. The firemen used oxy-acetylene torches to cut through the steel wreckage as doctors administered transfusions of blood and plasma, gave morphine injections, and performed amputations to free trapped victims. Professor Castelani performed an amputation and then, doubtless overcome by the horror of it all, he fainted. A toddler was rescued from the roof of the rear carriage, crying but apparently little hurt. It was not clear how she had escaped. Perhaps her mother and father had protected her.

[176] Wikipedia – Voghera train crash

[177] https://www.ferrovie.info/index.php/it/34-correva-l-anno/14231-ferrovie-31-maggio-1962-lo-schianto-di-voghera

But where were they? She was too young to explain and was taken to the hospital where she cried for a long time, then fell asleep, clutching a celluloid doll that the nuns had given her. In the evening she would be identified as Daniela Gritti (see below).

The spectacle, a newspaper reporter wrote, became even more chilling when dawn broke. All around were scattered pieces of metal, shards of glass, gutted seats, and bloodstains. The passengers could be seen to have been people of modest means, wearing working-class clothes, the clothes of people on their way to the seaside for a few days or returning to their families after a week working away from home. Scattered around were fibre bags, a bottle of wine, a sports magazine, and a picture album to keep the children quiet.

By 03:30 there were known to be 15 dead, by 07:00 the number had risen to 21, by 09:30 to 32, by 10:30 to 46, and a little later to 50. By the end of the rescue operation 62 dead would be counted, and two more victims would succumb later. The bodies of many of the victims were unrecognisable, and by noon only ten had been identified. Men and women arrived at the station from Milano and other places in Lombardy, searching for missing friends and relatives. About another 60 passengers were more or less hurt. "We have not had such a terrible day since the bombings of August 1944," locals said. A number of the passengers on the *accelerato* who were not hurt were too upset to continue their journeys on another train.

Among the passengers in the rear coach had been ten Sardinian miners, returning from work in Germany. One of them, named Savarore Lupino, survived the disaster. Later he would be asked to identify the bodies of his brother Giuseppe, aged 56, and his nephew Angelo, aged 18.

Another passenger was 29-year-old Aldo Masolo, a plumber from Dezio, who had saved up for a month for his holiday. Unable to get a seat he had been one of those standing in the corridor of the rear coach. When the train had arrived at Voghera he had made his way to the toilet to freshen up, to wash his hands and comb his hair, and he was there at the moment of the collision. The impact trapped him between the washbasin and the toilet door, fracturing his legs and pelvis. He was trapped in the wreckage for a long time, in agony at first, though later doctors were able to administer morphine, while firemen fought to release him. Later a friar, Father Barison, arrived, and stayed with the victim for over two hours, comforting him as he grew weaker.

At 10:30 someone exclaimed, "He's dead, there's nothing more we can do."

"No, no, he's alive," the friar protested, feeling an extremely weak pulse. At 10:45 he could finally be released and Father Barison blessed him. Five minutes later Aldo Masolo died in an ambulance.

The coffins of the victims in the mortuary were all numbered. Coffin No. 33 was marked Atenzi Giuseppe. It contained the body of a 24-year-old, a Sardinian who had been working in Milano for a short time. He had been on the way to Genova to meet his brother who had come over from the island. His fiancée, described as a small and frightened Sardinian, read of the disaster in a Milano newspaper. His name was not on the first list of dead but she had a premonition and travelled to Voghera. At the first-class waiting room in Voghera station there was a notice: 'Information for victim's relatives.' She went in, looked lost, and could not at first summon the courage to ask. Then in a quiet voice she told a Red Cross nurse, "I'm looking for my fiancée."

"What's his name?"

"Atenzi Pino." They scrolled down a list of names and then said to the girl, "We'll have to accompany you to the mortuary." Someone had to hold her up as she collapsed in a near faint. The girl's name as Catterina Pizadri, and she and her fiancée were shortly to be married.

Perhaps especially sad was the story of little Daniela Gritti, the two-year-old girl mentioned earlier. She was indeed only slightly hurt in the accident, but in the hospital she repeatedly asked for her mother. She did not understand that her mother, Annunziata, had died in the accident, as had her father, Enrico; her five-year-old sister, Agostina; an aunt, an uncle, and a cousin. Another uncle, Luigi Crispiatico, her mother's brother, came to the hospital at 19:00 and was able to identify the little girl in cot No. 47 as Daniela and say that her father was Enrico Gritti. She recognised him and smiled. He explained that he had taken his seven relatives in his car to the station to catch the train but had himself stayed behind in Milano, though they had teased him for not coming with them. He recalled that all his relatives on the train had been in the same compartment of the last coach, that the train had been absolutely full, spoke of the frantic rush of people to get seats, the happiness of those who had succeeded, and the disappointment of others. What became of little Daniela is not known, but there were many people who were anxious to adopt her.

The Treasury Minister, Roberto Tremelloni, arrived at the station at 14:35, looking pale and deeply disturbed. He examined the wreckage, asked how it had happened, enquired about the victims and the drivers. Asked to make a statement he said, "Tragedies like this happen in every country, but it's very sad that people have died. Now it's just a question of finding who is responsible." He then went to the hospital to visit the injured, and upon leaving declared, "All over Italy people will be crying over this terrible disaster."

Another official who visited the scene of the accident was the public prosecutor, Dr Manfra, who upon learning what had happened issued a warrant for the arrest of the driver and his assistant. About an hour after the accident the two railwaymen had gone to the house of Dr Brigliadori, in Via Piana, where the doctor had advised them to consult a lawyer and then surrender themselves to the police. This they had done, and they had appeared at the police station, accompanied by a lawyer, just after 10:00. From there they had been taken to prison, where they were questioned in the late afternoon by the public prosecutor, being subject to a further interrogation early the following morning. The men were confined separately. They were said to be in despair, spending their first night walking up and down their cells. Initially, Driver Pigiani apparently admitted having fallen asleep in the moments before the collision; later he alleged that he had only received a caution signal, and his statement was supported by Assistant Driver Fabbri. But his train had in any case to stop in the station to change engines, and it had been clearly travelling much too fast to do that. Driver Pigiani had a good record and was also qualified to drive steam locomotives. He was a married man and the father of a little girl who was a few months old. Assistant Driver Fabbri was single. The charges that the two men faced were those of causing a railway accident (penalty 2 to 10 years) and manslaughter (maximum penalty 12 years). The investigation by the judiciary, which was of course separate from the internal railway investigation, was expected to take a long time, perhaps several months, as all the railwaymen involved would have to be examined, the technical aspects would have to be gone into by independent experts, the financial situations of the families of the victims would be assessed for the purposes of determining compensation, and so on.

Conductor Arvangeli testified that when the collision had occurred he believed that they were still approaching the station, and he stated, "When I heard the tremendous crash I was thrown against the side of my cab. I thought that we had collided with a lorry on a level crossing, I never realised that we had actually entered Voghera station."

Engineer Severo Rissone, the general director of the Ferrovie dello Stato (FS), ordered an investigation of the disaster on behalf of the railway, entrusting it to Engineers Fasoli, Rossetti, and De Nicola, while Engineer Fienca and Doctor Branca, deputy general directors, and Undersecretary for Transport Angrisani went to Voghera. The undersecretary arranged for an immediate payment of 100,000 lira to the relatives of the dead and to those of the 23 injured admitted to hospital.

The freight engine was removed to the goods yard at 17:00, still partially covered by the remains of the rear carriage. During the evening the damaged overhead line equipment was repaired and at midnight, 21½ hours after the disaster, normal working was resumed. [178]

Driver Pigiani and Assistant Driver Fabbri duly stood trial for causing the disaster. This finally took place in early December 1963. Their defence counsel, lawyers Guizzi, Rovatti and Venturi, argued that irregularities on the part Conductor Antonangeli and Stationmaster Francesco Sinopoli had contributed to the disaster. The stationmaster was presumably not actually on duty in the early hours of the morning but was expected to 'carry the can' for his subordinate who had been in charge of the station at the time of the accident. The defence argued that the freight train should have been routed onto another track rather than track 3, that occupied by the passenger train, or at least the driver of the freight should have been warned at the previous station that he was to be routed onto an occupied track. To the author this is a fatuous argument; the basic duty of a train driver is to obey signals and it is these that tell him that the line is clear or occupied. Conductor Arvangeli was blamed, perhaps with more justification, for not having taken any action to apply the brakes when he found that the train was approaching Voghera station at too high a speed, though guards in general were usually very reluctant to intervene in such situations. At any rate the court appears to have decided that there were insufficient grounds to put either the conductor or the stationmaster on trial and only the case against the driver and assistant driver was proceeded with.

The public prosecutor argued that Driver Pigiani and Assistant Driver Fabbri were equally responsible for overrunning signals and for entering the station at a grossly excessive speed of 67 kilometres an hour, when the maximum permissive speed was 30 kilometres an hour. The attempt by the defence to implicate the stationmaster and conductor in causing the accident was, Prosecutor Oronzo declared, an admission of guilt on the part of the defendants.

For the defence lawyer Rovatti argued that the numerous railway accidents that had occurred in recent years demonstrated a serious problem in the entire organisation of the state railway, hence it was impossible to attribute the responsibility for accidents to individual railwaymen. He asked the court to acquit Driver Pigiani, due to if nothing else, insufficient evidence.

---

[178] *La Stampa*, Torino, Italy, 1st, 2nd and 3rd June 1962

Lawyer Venturi argued that the main proof of the railwaymen's innocence was that the results of the internal investigation by the FS, although not made public, had apparently been in their favour.

Lawyer Guizzi claimed that it was completely unacceptable that the experts appointed to investigate the disaster had not even bothered to examine the signals involved (it hardly seems that this can be correct; the working of the signals would surely have been thoroughly checked after the accident, but this is what the lawyer claimed, according to the newspaper report). He also argued that in any case Assistant Driver Fabbri should be acquitted, as he was only a subordinate to the driver.

After an adjournment to consider the case the President of the Court (there was no jury) announced the verdict. Driver Pigiani and Assistant Driver Fabbri were both found guilty of multiple manslaughter and of causing a railway disaster. As there were extenuating circumstances they were both sentenced to 14 years' imprisonment, with one year condoned. The defence counsel immediately announced their intention to launch an appeal.

As lawyer Guizzi had argued, it does seem strange that Assistant Driver Fabbri was punished just as severely as his driver. [179]

Buena Vista station, Guatemala, was the scene of a major disaster in the early hours of Thursday 20th December 1962. Overnight passenger train No. 16 from Tecún Umán to Guatemala City was actually a mixed train. It consisted of locomotive No. 180, first- and second-class passenger carriages and some freight waggons, in charge of Conductor Gabriel Godínez Campos, with Driver Silvino Orellana and Fireman Ramón Casasola. It had left Tecún Umán at 18:00 the previous evening and arrived late in Buena Vista, where it had to attach a cattle waggon.

Following the mixed train was a freight train consisting of locomotive No. 604 and 35 vans loaded with bananas, in charge of Conductor César Hugo Barrios y Barrios, with Driver Diego Cabrera and Fireman Octavio Cárdenas.

At 02:20, while train No. 16 was still stopped in the station, it was run into from behind by the freight train, leaving 22 passengers dead and another 45 injured.

A newspaper report on the accident included the stories of several passengers. Gonzalo de Jesús Aragón Rosales, who had been waiting for the passenger train at the station, told of seeing the freight train suddenly arrive, of witnessing the collision, of seeing the bloodied faces of the injured passengers and hearing their cries of pain.

Passenger Adela Solares de Iglesias, a Mexican national, was interviewed at the Hospital Nacional de Escuintla, where the injured and the bodies of the dead had been taken. She was sat in a wheelchair, presumably as a result of injuries sustained in the accident, and apparently suffering some sort of nervous breakdown, perhaps due to severe shock. She said that she had come to visit her nephews who lived in Guatemala over Christmas.

Dolores Morales complained of not being able to move.

Brothers Marco Tulio Fernando and German Fong, who were injured, said that they had been returning to Guatemala City from Tapachula. Their cousin Miriam Spigaris had

[179] Ibid, 8th and 10th December 1963

been accompanying them. He had been reading some jokes when they had seen a bright light rapidly approaching them (the headlight of the freight engine) and then there had been a violent crash. Many of their fellow passengers had been trapped in the wreckage but they were able to get out.

There was presumably no signalling at the station, for Driver Cabrera of the freight train blamed the accident on the conductor of the mixed train not having gone back far enough to protect his train by exhibiting a red light, so that he was unable to pull up in time. [180]

Late December 1962 saw the beginning of what in England became known as the Big Freeze, with abnormally low temperatures lasting for three months. At Crewe, on Boxing Day, Wednesday 26th December 1962, frozen points in the station area led to a backup of trains arriving from the north. The 4.45 p.m. British Railways, London Midland Region, Up express train from Liverpool Lime Street to Birmingham New Street was thus stopped by signals between Winsford and Coppenhall Junction, and it was just restarting upon a signal being cleared for it when it was run into from behind by the 1.30 Up express from Glasgow Central to London Euston, the *Mid-day Scot*. Two coaches of the Birmingham train were telescoped and 18 passengers lost their lives, whilst 33 passengers and the guard of the Birmingham train were badly enough hurt to be taken to hospital and many other passengers were slightly hurt or suffered from shock. [181]

A newspaper described the aftermath of the accident. 'Fifteen fire engines went to the scene with 60 firemen with full rescue equipment. With the help of spotlights they worked within inches of 25,000-volt overhead electric wires. Oxy-acetylene equipment was used to free the dead and injured.

'First aid was given in the bare, frozen fields under the headlights of cars.

'A dozen doctors and 20 ambulances were at the scene. Others were held up by ice on the roads. A shuttle service of ambulances took the injured to Crewe Memorial Hospital and Northwich Infirmary. Crewe Memorial Hospital said that they had 25 casualties, three of them dead and four seriously injured. The dead included two children.

'A loose box at Lea Hall Farm, near the crash, was used as a mortuary. The bodies there included those of four children. Mrs F. Wilson was playing with her son Brian at the farm when a ticket collector from the Scottish train came in and asked her to telephone for help. Because her telephone was out of order her husband Frank ran a mile across fields to Dairy House Farm, from where the alarm was given out.

'Injured were laid in fields by the line and at 9 p.m. a second mortuary was opened in the farm buildings.

'Two casualties taken to Northwich Infirmary were Mr Timothy Brivett, aged 24, labourer, of Archibald Street, Gloucester, and Corporal Dennis Norman Osborn, aged 30, of Innesworth R.A.F. Station, Gloucester. Both were suffering from minor injuries, bruising and shock and were detained for the night. Corporal Osborn and Mr Brivett were both in the Pullman coach which apparently suffered the most damage. Corporal Osborn,

---

[180] https://www.prensalibre.com/hemeroteca/1962-accidente-de-tren-deja-22-muertos-en-escuintla/

[181] Ministry of Transport report

who was accompanied by his wife and two children, a girl aged four and a boy of three, said in hospital that he had been spending Christmas with relatives in Liverpool. "The train had been stopped for, I think, about 40 minutes when the crash happened," he said. "The coach bounced, the lights went out and all the luggage shot on top of us." He understood that his wife and children had been sent to another hospital but were apparently all right.

'Five hours after the crash, floodlights were brought up to light the track as engineers swarmed on an overhead electric gantry to hang slings down to the last carriage of the Liverpool – Birmingham train, which was telescoped on to the top of the second carriage from the end. The carriage was stripped to its chassis for half its length by the impact of the rear coach, and the verge of the track was covered with smashed seats and wrecked upholstery. The Scottish train, which was a diesel, was backed a short distance from the wreckage. Trains which had been waiting began to filter through at 11 p.m.

'Mr Arthur John Ross, of Edinburgh Drive, Short Heath, Willenhall, Staffordshire, who had been spending Christmas with relatives in Waterloo, Liverpool, and was a passenger in the third coach of the Liverpool train, said: "I heard a terrible smash, windows were shattered, there was glass everywhere and all the passengers in my compartment were hurled off their seats. Although I was dazed I could hear children screaming and cries for help."

'Another passenger in the coach, Mr Percival Johnson, aged 48, of Richmond Road, Sedgley, near Wolverhampton, said: "I have been spending Christmas in a guest house at Southport. I was chatting to a woman in the compartment when suddenly there was a loud crash. I was hurled from my seat. The screams and cries of children were dreadful. Christmas toys and clothes were strewn all about."

'Mr George Stanley Tulloch, of Chace Guild House, Coventry, who had been visiting his sister in West Kirby, said: "I was in the first compartment, and we had been waiting there for half an hour. I was having a cup of tea when suddenly there was a terrific crash and everything shattered before me. I am trained in first aid and went down to see what I could do. I was dazed by the jolt but not knocked out. What I want to know is why we have been waiting five and a half hours. It was one and a half hours before any assistance came."

'Mr Harold Shiff, aged 26, of Russell Road, Liverpool, was in the coach that was hit. "We just sat there like rocks after we felt the crash," he said. "Then I threw a table through a window. There was very little panic apart from a few women screaming."

'Mr Daniel Russell, of West Street, Crewe, was searching for his daughter Jane, aged two, when the crash happened. His daughter was in the corridor, and his wife was away with their other two children at the toilets. "All the lights went out," he said, "and all I could find was people lying on the floor. "I threw Jane out of the train. Help was a long time coming. I counted the dead and dying before the first police car arrived. It was over three hours before I saw my wife and children, but luckily they were safe."' [182]

The Birmingham train consisted of an electric locomotive and eight coaches, in charge of Guard R. Wilkinson with Driver George Francis William Hedgcock and Passed Fireman F.W. Kelly.

---

[182] *The Times*, London, England, 27th December 1962

The *Mid-day Scot* consisted of a Class 4 diesel locomotive and 13 coaches, in charge of Guard K. McRae with Driver John Russell and Fireman Victor McCallum.

The accident occurred only about two miles off the 1948 Winsford disaster described in the previous volume of this series. Since that time the line had been resignalled and Minshull Vernon box had been abolished, though the boxes at Winsford Station and Coppenhall Junction remained. North of Winsford Station box there were only two tracks but from there southwards there were four tracks. The entry to the Up Fast and Up Slow lines at Winsford Station box was controlled by signals WS 28 and WS 29. The next four subsequent pairs of signals were Nos. 114 and 116, 110 and 112, 106 and 108, and 102 and 104, the lower numbered signals in each case applying to the Up Fast line and the higher to the Up Slow line. The line was fully track circuited, these four sets of signals were all automatic four-aspect colourlights, and each pair of signals was carried on a cantilever structure over the tracks. There was a telephone provided at each cantilever post. The accident occurred almost directly opposite the cantilever carrying automatic signals Nos. 101 and 103 for the Down Fast and Down Slow lines. Each automatic signal could display green, single yellow, double yellow or red, depending on the occupancy of the line ahead. The signal post telephones had a selective ringing code and to call the signalman a switch had to be turned against a spring, then released. If the switch did not properly return to its starting position no code was sent, nor could a call be made on any other telephone on the same circuit. The Up line telephones were connected to Coppenhall Junction box and the Down line telephones to Winsford Station box.

The Birmingham train passed Winsford Station box at 5.26 p.m. Signalman Herbert Sutton there said that its tail lamp had been lit and showing an average light. The train stopped at signal No. 114, which was red, and Driver Hedgcock rang Relief Signalman Cecil Jackson at Coppenhall Junction. He was told to wait until the signal cleared, which he did. He then proceeded, after some delay, to signal No. 110, which was also red, and again tried to ring the signalman. Unfortunately he had failed to properly restore the telephone to its receiver at signal No. 114 and so all the Up line signal telephones were put out of order. Passed Fireman Kelly then managed to ring Winsford Station from Down Slow line signal No. 103 opposite. He was told that the Up line signal telephones were out of order and that his train must wait until signal No. 110 cleared in turn. Again the train did so, the signal eventually cleared to yellow, the brakes of the train were released and it had just restarted when it was run into by the *Mid-day Scot*.

The Winsford Junction and Winsford Station signalmen recorded the *Mid-day Scot* as having passed them at 5.50 p.m. and 5.51 p.m. respectively. Signal No. 114 was displaying a red aspect because the Birmingham train was stopped ahead at signal No. 110. First Fireman McCallum and then Driver Russell attempted to use the signal post telephone but, of course, they could not get it to work. Driver Russell told the Ministry of Transport inquiry that he was aware of a supplementary local instruction that in such circumstances the train crew should attempt to use another signal post telephone but he explained, "The telephone on the Down line seemed pretty far away. I said (to Fireman McCallum) we have wasted enough time in the section, the road looks clear ahead to signal No. 110, and we will make for there." He said that he could see signal No. 110 at red and signal No. 112 for the Up Slow line at yellow. In the darkness he could not see the Birmingham train stopped in front of him. After having stopped six or seven minutes he had restarted the train and proceeded, he said, at a speed he put at five or six miles an hour, opening, and closing the controller to maintain a steady speed. Halfway between the two signals he had seen signal No. 110 change to yellow but had not increased speed as, he explained, he had been going to try and contact the signalman again from that signal post telephone. Almost at once his fireman had shouted, "Stop, John!" when he applied the brakes and released the dead man's handle. At about a coach length he had seen a "black object" in front of him. He estimated the speed of his train on impact at only two or three miles an hour. He added that the cab of his engine had been unlit and he had been alternately looking ahead and glancing at his speedometer, the needle of which had just been moving slightly and not registering any speed. The inspecting officer, Colonel D. McMullen, pointed out that if Driver Russell's train had been travelling at five miles an hour it would have taken nine minutes from signal No. 114 to the point of collision. The collision had occurred at 6.01 p.m., only ten minutes after passing Winsford Station box. At least two minutes must have been occupied in travelling from Winsford Station box to signal No. 114 and five or six minutes had been spent attempting to telephone Coppenhall box from signal No. 114. Only two or three minutes, therefore, could have been occupied in travelling from signal No. 114 to the point of collision, implying an average speed of from 15 to 22 miles an hour. Driver Russell was adamant, however, that he had not been going that fast, despite the inspecting officer explaining his calculations, nor would he admit to having opened his controller when he had seen signal No. 110 go to yellow. Russell was 63 years old, stated that he was in good health, had good eyesight, had been a driver for 26 years and did not drink, nor had he been talking to his fireman at the time.

Fireman McCallum confirmed his driver's account of standing at signal No. 114 for five or six minutes, of having unsuccessfully attempted to use the signal post telephone there, and of their having then gone forward at five or six miles an hour. He thought that they had been more than halfway to signal No. 110 when it changed to yellow, then he had seen the outline of the standing train and warned his driver; he had not seen the tail lamp of the Birmingham train until the last moment. When he had seen that there was a train in front of them he thought they would not be able to pull up short of it and sounded the horn. He confirmed that there had been no conversation in the cab. The heater had been on and the cab windows had been closed. He explained his not having seen the tail lamp or the train ahead by saying that he had wondered if there had been a signal failure due to a broken rail or a buffer which had fallen from a train and thus had been closely studying the track. It was pointed out to him that the stationary train would have obscured the view of signal No. 110 at a distance of 100 yards but he insisted that he had only seen the train ahead when it had been a coach length away.

Guard McRae booked the stop of the *Mid-day Scot* at signal No. 114 as having lasted five minutes and said that the train had then restarted in "a normal manner." He timed the train as restarting at 6 p.m. and the time of the collision at 6.03 p.m. He thought the train had been travelling at 10 miles after restarting but agreed that it could have been more.

Guard Wilkinson of the Birmingham train was described as having had a remarkable escape. He recalled his train being stopped at signal No. 110 for ten minutes, the signal clearing to yellow, the brakes being released and his train restarting. At that moment he had heard the whistle of the *Mid-day Scot* and the collision had occurred. He remembered nothing more.

Driver Hedgcock agreed that he had just been on the point of restarting his train when the collision had taken place. Realising that his train had been run into from behind he had instructed his fireman to go forward and protect the Down lines and telephoned the Winsford Station signalman from Down Fast line signal No. 101 within three minutes of the collision. Knowing that his train was well filled and having seen arcing and telescoping he had realised that the situation was serious and believed that he had conveyed that fact to Signalman Sutton, having told him, "Get ambulances and assistance at once."

Passed Fireman Kelly testified that having protected the Down lines by putting down detonators as instructed by his driver he had telephoned Winsford Station box from Down Slow line signal No. 107, 800 yards ahead of his train. At this point he had only known that there had been a collision, and not how serious it was. He thought, though he was not sure, that this was the first notice that the signalman had had of the accident. Returning to his train and after speaking to his driver he had phoned the signalman again from Down Slow line signal No. 103. He had informed the signalman that the Down lines were clear but that people were trapped. The signalman had replied that help was on the way. He put the time of this second call between 6.20 p.m. and 6.25 p.m.

Guard McRae stated that he had gone back to protect the Up lines and telephoned the Winsford Station signalman from signal WS 46 at 6.20 p.m. He, too, thought that this was the first news that the signalman had had of the accident.

Travelling Ticket Collector D. Mulhearn said that seeing the extent of the damage and realising that help was required without delay he had made his way across some fields to a farm where he thought there would be a telephone, not realising that there was a lineside telephone nearby. The telephone at the farm had, by chance, been out of order but someone had gone from the farm on a motorcycle (not run, as per the newspaper report) to another one to raise the alarm.

Relief Signalman Jackson at Coppenhall stated that he had received the *Obstruction Danger* from Winsford Station box at 6.05 p.m. Signalman Sutton had then spoken to him on the telephone and said that the *Mid-day Scot* had run into the back of the Birmingham train and that two or three people were injured – he was sure that this was the message. He had protected the Down lines with detonators, which had taken seven or eight minutes, then telephoned Control, taking two or three minutes more to get through. He had repeated Signalman Sutton's message to Control but, assuming that the Winsford Station signalman was taking care of the relief arrangements, he had not asked Control to take any action.

Signalman Sutton at Winsford Station stated that someone from the Birmingham train, he thought the fireman, had telephoned him from signal No. 103 and said that the train had been run into, people were trapped and assistance was required. Upon this he had replaced his Down signals, sent the *Obstruction Danger* signal to Coppenhall at 6.05 p.m., then told the signalman there over the telephone that two or three people were hurt. He then had to deal with a telephone call from signal No. 27 and at about 6.10 p.m. he telephoned Porter/Booking Clerk D.A. Leigh at his station to ask him to call the ambulance and the police (there was no public telephone in his box). He did not know exactly where the accident was but had assumed that it was in the vicinity of the former Minshull Vernon station. He had not thought that the accident was serious. He had then had many telephone calls to deal with and sent a sub-ganger to notify the stationmaster, who reached his box at 6.30 p.m. At that time there had been a further telephone call from signal No. 103 with a message that "bodies and people are all over the track," but even then he had not realised the accident was a serious one, assuming that this merely meant that passengers were alighting from the trains. He admitted that it might have been later than he stated that he had spoken to the porter.

Porter/Booking Clerk Leigh said that he had had some passengers in the waiting room for the 6.19 p.m. Down local train. When it had not appeared they asked where it was and he, not knowing of the accident, telephoned Coppenhall Junction. It was only sometime later that Signalman Sutton advised him of the accident, when he had dialled 999. He had not gathered from the signalman that the accident was serious but that an ambulance was required somewhere about the old Minshull Vernon station. According to the Northwich postmaster the 999 call was made at 6.34 p.m.

Stationmaster T.O. Staley said that the message from the sub-ganger relayed that there had been an accident involving one passenger train running into another, that no one had been killed but one or two were hurt. He had quickly put on his overcoat and boots and made his way on foot to the station, about half a mile away. When he had reached the station the porter had been calling an ambulance but was told that more information about the location of the accident was required. He had gone to the signal box after attempting to contact control, tried to find the exact location of the collision, and called more emergency services. Signalman Sutton had told him the gist of the second telephone call from signal No. 103 but he, too, failed to appreciate the seriousness of the accident. It was only when he had reached the scene just after 7 p.m. that he found how bad it was. By then two ambulances and Winsford Fire Brigade were there, having been alerted by the 999 call from Dairy House Farm.

Tests showed that the tail lamp of the Birmingham train was first visible from a distance of 400 yards, though dim and it was not nearly as conspicuous as the bright light of signal No. 114.

The results of the accident had been made much more severe by the fracture of the Buckeye couplings between the seventh and eighth coaches of the Birmingham train, which had allowed the eighth coach to mount and telescope the seventh. Subsequently the couplings were metallurgically examined and the steel of the pivot pin of the coupling at the front end of the eighth coach was found not to be up to specification.

The speedometers of the engine of the *Mid-day Scot* were found to read correctly within one or two miles an hour.

Concluding his report, Colonel McMullen stated that he had no doubt that the collision would not have occurred had the Up line telephones not been out of order, but it would not have happened anyway had Driver Russell obeyed the Rules. There was no doubt that the driver had assumed, without justification, that the line was clear to the next signal and that, to save time, he had disregarded the modification to Rule 55 (h) (i) requiring him in such circumstances to attempt to use a signal post telephone on the Down side.

From the time calculations (see above) it was clear that the *average* speed of the *Mid-day Scot* after passing signal No. 114 must have been at least 15 miles an hour. The extent of the impact damage indicated a closing speed of at least 20 miles an hour.

It was possible that Driver Russell had looked at the speedometer and misread it but more likely that he had not looked at it at all, but had estimated his speed, as he had been accustomed to do on steam locomotives. The enclosed cab of a diesel locomotive, the lack of a relationship between the speed of the engine and the speed of the power unit and the smoothness of the ride on continuously welded track were doubtless factors and it was clear that the driver had seriously misjudged the speed of his train.

Despite Driver Russell's statement it was likely that he had accelerated when he saw signal No. 110 clear to yellow, as he thought for his train, and unlikely that he would have stopped again to attempt to telephone the signalman when he was so close to Crewe station.

It was understandable that neither Driver Russell nor Fireman McCallum had seen the tail lamp of the Birmingham train, overpowered as it had been by the light of the signal, but hard to see why they had not noticed the silhouette of the train against the signal until they were a coach length (22 yards) from it. An emergency brake application when the silhouette had first become visible might have prevented the collision. It could only be supposed that neither of them had been sufficiently alert. Driver Russell must therefore bear the full responsibility for the collision, though he had been without the assistance he could have expected from his fireman. (The contrast of the inspecting officer's conclusion, that Driver Russell alone was wholly responsible for the collision, with the ruling after the Voghera accident in Italy, described above, where Driver Pigiani and Assistant Driver Fabbri were held to be equally responsible, will be noted.)

The first message received by Signalman Sutton at about 6.03 p.m. must have been from Driver Hedgcock. The latter was an excellent witness and there was no doubt that he had realised the seriousness of the accident and conveyed his impression to the signalman. It was difficult to understand how Signalman Sutton had failed to appreciate the urgency of the situation, even when a further telephone call had mentioned "bodies and people all over the line." He no doubt had not rung the porter until 6.30 p.m., when the emergency services had been promptly called but there was uncertainty as to the location of the accident. The emergency services had all responded promptly when they were called.

Turning to remarks and recommendations the inspecting officer noted that the accident had happened on a line with the most modern signalling and the Automatic Warning System of train control. It had occurred in the exact circumstances the rules had been framed to prevent. The 'stop and proceed' rule required the driver to proceed with extreme caution. The judgement of speed, especially in difficult conditions, such as at night, was not easy in a diesel or electric locomotive, and reference to the speedometer was essential. The training of drivers in slow driving, by day or by night, was recommended, perhaps using simulators.

That part of the rule dealing with drivers passing signals at danger, worded "can see or ascertain that the line is clear to the next stop signal," was open to misinterpretation and the rule should be expanded.

The last line of defence against a rear collision, an oil-lit tail lamp, had changed little for a century. It could be easily overpowered by a modern colourlight signal. Experiments with flashing electric tail lamps should be pursued, or an alternative solution, such as the red electrically lit screens used on some Southern Region electric multiple units, should be considered.

On multiple tracks the provision of separate telephones for each line on different circuits was recommended.

A "10" figure should be provided on speedometers.

The use of powerful headlights on locomotives had been suggested but they would be no use on curved track and they might affect the sighting of signals by the drivers of trains on adjoining lines.

The use of wireless telephones on trains would be extremely expensive, would involve technical difficulties and 'was not, in practically any country, considered a suitable means of regulating the movement on trains on running lines, on account of the large number of trains and signal boxes involved and of the consequent difficulty of ensuring that a message is received and understood by the right person.'

The suggestion that stationmasters should have telephones in their residences so that they could be called in an emergency was a good one and should be adopted. Clearly even in 1962 many stationmasters did not have telephones in their homes.

Although in the present case had the driver of the *Mid-day Scot* used a Down line telephone the accident would not have happened, the use of such telephones attached to signals applying to running lines going in the opposite direction and linked to different signal boxes was liable to cause confusion. The local instruction to do so had now been withdrawn.

Staff should be reminded of the existence of separate electrification telephones, one of which could have been used.

The failure of the Buckeye couplings had been entirely due to the severity of the blow but steps were being taken to remove pivot pins made from an incorrect grade of steel.

With modern signalling, signal boxes were often 20 or more miles apart. Although the signalman had known the number of the signal from which a telephone call was made he had not known its location. The locations of signals should be marked on the signal box diagram or on large-scale maps in the control rooms of the emergency services. [183]

At the inquest on the victims of the accident at the end of January 1963 the jury returned a verdict of accidental death. They added a rider that the British Transport Commission should redraft Rule 55 permitting drivers to pass red signals in certain circumstances, that all red signals should be stop signals (!), that all trains should carry two tail lamps and that radio links should be considered between engines and key signal boxes. [184]

A careless mistake by a railwayman led to a rear collision in Spain at 05:57 on Sunday 24th February 1963. The *El Shangai* express from Madrid to Barcelona was preceded that morning by a freight train. When the latter arrived at Almudévar station, Stationmaster José Manuel Vallejo Gárate promptly reported it back over the telephone to Tardienta station and accepted the express. It was only belatedly that he then discovered the freight train had arrived incomplete; its 11 rearmost waggons had become detached and been left in the section. He immediately notified Tardienta but was told that the express was on its way. The accident was somewhat less serious than it might have been thanks to Porter Bruno Martínez, who, hurrying along the snow-covered track on his bicycle, was in time to give some warning to Driver José Grané of the express, who was able to reduce his

[183] Ministry of Transport report

[184] *The Times*, London, England, 1st February 1963

speed from 90 to 30 kilometres an hour. Even then the collision was a very violent one; five passengers lost their lives and 43 were injured, some seriously. The injured were taken in ambulances to Huesca, while the uninjured passengers continued their journeys in the same train, which was taken back to Tardienta and resumed its journey via an alternative route.

The passengers who lost their lives were Manuel Díaz Díaz, of Benamocarra; Marcelino Penas Mato, from Arzúa; Julia Trebull Díaz and her two-year-old son; and a young woman who was later identified as 20-year-old Gloria Váquez Castro, from La Puebla de Brollen. Gloria was working as a servant in Barcelona and had been returning from attending the baptism of a nephew. Driver Grané and Fireman Felipe Ulibarre of the express escaped injury, as did Conductor Andrés Fernández Trapote of the freight train.

Stationmaster Gárate was subsequently arrested. [185] [186]

Traffic on the railways at times such as Christmas is always heavier than at other times of the year and in Hungary in 1964 it was especially heavy as bus services had been disrupted due to heavy snow, forcing passengers to switch to trains instead. Thus it was that a passenger train involved in an accident on Christmas Eve, Tuesday 24th December 1963, was more than usually crowded. The driver of a passenger train overran at least one stop signal in thick fog and collided with the rear of a freight train near Szolnok. An initial report stated that 32 people had died and 36 were injured. By the following day the number of fatalities had risen to 45.

Following the accident the engine crew of the passenger train were arrested and the driver was subsequently sentenced to 11 years' imprisonment for negligence, his fireman receiving a lighter sentence of three years' imprisonment. [187]

First reports stated that at least 40 people had died in a rear collision at Jajinci, Yugoslavia, 15 kilometres south of Belgrade, on Saturday 4th January 1964. By noon it was known that there were about 130 injured, a dozen of whom had died in hospitals, while rescuers were still recovering bodies and many of the injured were in a critical condition. A passenger train had been standing in the station there when it was run into by a commuter train from Požarevac, which was more crowded than usual because shops and factories had been closed for three years for the New Year holidays. Eight coaches were crushed. Most of the casualties were in the commuter train, which had been travelling at about 65 kilometres an hour. Its driver said that he had been running under clear signals and, "I could not see the rear coach of the other train because the line is on a slight curve." [188]

The following day the number of dead had risen to 61 and the stationmaster and a signalman at the station, plus the driver and guard of the forward train, had been taken into custody, charged with causing the disaster through negligence. Newspaper reports assumed that the train had stopped because someone had pulled the communication cord

[185] *ABC*, Madrid, Spain, 26th February 1963
[186] Federación Castellano Manchega de Amigos del Ferrocarril
[187] *Railroad Wrecks* by Edgar A. Haine, pg. 162
[188] *The Lethbridge Herald*, Lethbridge,. Alberta, Canada, 4th January 1964

but it seemed that the real reason was that a brake pipe had separated, causing the brakes to apply automatically. Whatever the cause of the stoppage, the train had been stationary for 20 minutes while its crew attempted to rectify the problem, without it having been protected in the rear, hence the implied negligence. [189]

A serious accident occurred on the main line between Kraków and Oświęcem, Poland, late on Sunday 9th May 1964. The circumstances of the collision are a little unclear from the brief newspaper reports in English and a somewhat longer report in a Polish newspaper but it would appear that a freight train approaching the station at Chełmek became divided and its rear portion was left behind in the section. [190] This, it seems, was not noticed, and a following passenger train, which was mostly carrying factory workers home at the end of their late shift, was permitted to leave the previous station at Libiąż. This passenger train collided with the waggons at about 23:45. Ten passengers and Conductor Władysław Czapla of the passenger train were killed outright and 16 passengers were injured, two of whom later succumbed. The injured were taken to hospitals in Chrzanów and Oświęcem.

The rescue operation was coordinated by the director of the DOKP (Dyrekcja Okręgowych Kolei Państwowych – Directorate of Regional State Railways) T. Mieszkowski. It was not immediately possible to identify all of the victims, some of whom did not have any documents on them. Financial assistance was promised to the victims of the accident and to their dependents. [191]

Initial reports stated that as many as 70 people had lost their lives and over 100 had been injured in a collision Tacotalpa, Mexico, about 80 kilometres from Villahermosa in Tabasco state, on Sunday 20th December 1964. The next day, however, state authorities announced that only 31 people had died and 26 had been seriously injured, while five were reportedly missing. State secretary-general Dr Julian Manzur Ocana added that many of the other injured had not required admission to hospital.

The stationary mixed train from Tenosique to Coatzacoalcos [192] involved in the accident had consisted of an engine, a first-class coach, three second-class coaches and 15 freight waggons loaded with cattle. It had been run into by a freight train consisting of an engine and 20 waggons. The colliding train had reportedly still been travelling at 80 kilometres an hour when it struck the mixed one. One of the second-class coaches of the latter had then caught fire and some of the passengers were said to have burnt to death.

Those injured in the accident included four American tourists. Nils Werner, 75, of Alhambra, California, had to be cut free from the wreckage using oxy-acetylene torches and was taken to hospital in a critical condition. His right arm had to be amputated and he was also in danger of losing a leg. Mrs Werner was also hurt. Colonel Carl and Mrs Margaret Benson, of Fresno, California, suffered leg fractures and other injuries, though they were said to be in good condition the next day. [193] The tourists had been returning to

---

[189] *The Daily Colonist*, Victoria, British Columbia, Canada, 7th January 1964
[190] *The Arizona Republic*, Phoenix, Arizona, United States, 11th May 1964
[191] *Gazeta Krakowska*, Kraków, Poland, 10th May 1964
[192] *The Gettysburg Times*, Gettysburg, Pennsylvania, United States, 22nd December 1964
[193] *Pacific Stars and Stripes*, Tokyo, Japan, 23rd December 1964

Mexico City after visiting the ancient Maya city at Palenque.

A survivor, José Sasso Moravy, said, "I was saved by a miracle. It was horrible. It seemed as if we had stopped at Tacotalpa only about 10 minutes. A tremendous bump threw me against a wall. I fell under part of a broken seat. I saw pieces of bodies flying around me." [194]

Driver Miguel Sanchez Cruz had apparently fallen asleep just before the collision and only awoke (from a microsleep?) too late to brake effectively. It was also foggy at the time of the accident. He fled after the collision and was still being sought several days later. Other members of the train crew were arrested.

The main line of the Ferrocarriles Nacionales de México was reopened on Tuesday. By then eight of the injured had died, increasing the number of dead to 39. [195]

On Saturday 20th February 1965 extra express train No. 3876 from Göteborg to Stockholm ran into the rear of passenger train No. 124, which had stopped at the halt at Skultorp, Sweden. Ten passengers died and 37 were injured, all of whom had been in the front train. Most of the casualties were in the penultimate coach of the train, an old wooden one, brought into use because of the heavy traffic at that time of the year, and crushed by the rear steel coach. There were allegations that the wood of the penultimate coach was rotten and had been painted over to hide the cracks. Whether or not this was true is not known. The driver of the colliding train had overrun a stop signal for reasons that could not be determined. [196]

A work train carrying railway employees stopped for unexplained reasons under the Baba Aleoua Bridge on the outskirts of Tunis, Tunisia, on the evening of Friday 23rd April 1965. It was still stationary when it was run into from behind by passenger train from the nearby coastal resort of Hamman Lif, which was travelling at 50 kilometres an hour. Most of the casualties were in the work train, the rear coaches of which were badly damaged. Initially it was stated that there had been seven fatalities and the toll later rose to nine, with 101 injured, 41 of whom were admitted to hospital. [197] [198] [199] [200]

A rear collision early on the morning of Saturday 28th August 1965 at Pont D'Héry, France, involved two international expresses from Italy. The first train was No. 562, the *Lombardie Express,* from Milano to Paris Lyon, where it was due at 08:33. The second train was No. 536, also from Milano to Paris Lyon, where it was due at 09:25. Five kilometres short of Pont D'Héry the driver of the *Lombardie Express* had noticed an unspecified mechanical problem, which had caused his train to lose time until eventually it stopped out of course at Pont D'Héry for an examination to be made. The second,

---

[194] *The Brownsville Herald*, Brownsville, Texas, United States, 22nd December 1964
[195] Ibid, 23rd December 1964
[196] https://www.ekeving.se/hi/SJdok/SakTj/1969.pdf
[197] *Kokomo Morning Times*, Kokomo, Indiana, United States, 24th April 1965
[198] *Panama City News*, Panama City, Florida, United States, 24th April 1965
[199] *The Troy Record*, Troy, New York, United States, 24th April 1965
[200] *Albuquerque Journal*, Albuquerque, New Mexico, United States, 25th April 1965

sleeper train was, by the time the stop was made, running 20 minutes behind the first. The first train was just on the point of restarting at about 04:40 when the second appeared around the curve on the approach to the station and a violent collision ensued. Despite the low speed of the second train its great weight led to serious results. In the collision the rear coach of the *Lombardie Express* turned completely around and three metres of it was destroyed, the front of the leading engine of the two pulling train No. 536 rose into the air and then fell to the left side of the track and the front coaches of the second train were derailed. Most of the victims of the accident were in the rear coach of the front train, a second-class sitting coach added at the last moment to cope with the large number of passengers. Passengers sleeping in the couchette coach further along the train thought that they were having nightmares until splinters of glass rained onto them from broken windows. Some were said to have been knocked out by falling pieces of luggage.

A rescue operation was immediately begun. A crossing keeper who lived close by and two *Wagons-Lits* employees were the first on scene. Two doctors who chanced to be passengers on the *Lombardie Express* were also quickly in attendance. Eight dead were soon recovered. [201] The implementation of the ORSEC emergency plan was considered but the prefect of the Jura department decided that this was not necessary. The injured were taken to hospitals in Mouchard, Salins-les-Bains and Champagnole, while a *chapelle ardente* (chapel of rest) was set up in the Pont D'Héry communal school. Twelve passengers were eventually found to have died in the accident and about 30 were hurt. Most of the victims were French though some were Italian workers going to jobs in France. Eight of the dead were children. [202] [203]

Driver Charles Neyrat of the colliding train and Assistant Driver Faivre escaped with slight injuries. They were arrested and after an interrogation by the *juge d'instruction* (examining magistrate) the driver was charged with involuntary homicide. Said to be distraught and overwhelmed by what had happened, he was placed under medical supervision at Besançon prison, where he had been taken.

From the initial investigation it appeared that Driver Neyrat should first have received a yellow *signal de ralentissement* followed after 1,500 by a red *carré d'arrét* semaphore stop signal. It appeared that at night, with visibility further obscured by rain and fog, he had probably missed the caution signal altogether. According to the tachograph recovered from the leading engine of train No. 536 the brakes had been belatedly applied 800 metres from the point of collision, and this agreed with the statements of surviving passengers. The speed of the train had thus been reduced to 10 kilometres an hour at the moment of collision. The driver claimed that his brakes had not responded quickly or fully when he applied them. The newspaper report noted that brakes were to be expertly examined to determine if they had been working properly or not. The composition of the *Lombardie Express* also had to be considered; it was normal practice to place a van at the rear of the train rather than the passenger coach, which had been attached in that position at the last moment in Milano. Had a van been at the rear of the train the consequences of the accident would probably have been much less serious. [204]

---

[201] *Gazette de Lausanne*, Lausanne, Switzerland, 30th August 1965
[202] *The Times*, London, England, 30th August 1965
[203] *Le Monde*, Paris, France, 30th August 1965
[204] *Gazette de Lausanne*, Lausanne, Switzerland, 31st August 1965

A freight train overran a stop signal at Praha-Hloubětin station, Czechoslovakia, on Tuesday 21st September 1965 and collided with the rear of a passenger train. Fourteen people were killed or fatally injured and about 90 were hurt, of whom 24 were taken to hospital and the others were treated on the spot. [205] [206]

Ten, or according to some reports 12, passengers lost their lives during the evening rush hour on Monday 30th May 1966 in a collision between two passenger trains carrying large numbers of 'non-whites' home to their segregated townships on the outskirts of Johannesburg, South Africa. Unofficial estimates were that about 100 others were injured. [207] [208] The first train was stationary in Dube station when the second ran into its rear. Most of the dead and injured were in the rear two coaches of the front train. A crowd of about 2,000 Africans quickly assembled and helped remove some of the victims from the trains, while nine ambulances and five fire engines were soon on the scene from the suburb of Brixton. [209]

In the latest of a series of railway disasters in India and less than a week after a major catastrophe near Matunga, Bombay, train No. 4, the Down *Almedabad Express* bound for Delhi collided with the rear of stationary goods train No. 872 Down in the station at Ladpura, near Ajmer in Rajasthan state, at about 07:48 on Sunday 19th June 1966. [210] Eleven people were killed outright, two died en route to hospital and another two later. The heaviest casualties were in a sleeping car containing military personnel, women, and children. [211]

Minister of Railways S.K. Patil ordered an inquiry into the accident and demanded a meeting with the chief operating superintendents, mechanical engineers and signalling and telecommunications engineers from all eight railway zones.[212]

The express consisted of a Class Y.P. locomotive and 14 bogie coaches in charge of Guard Suraj Narain with Driver S.S. Gupta, First Fireman Ganga Ram, and Second Fireman R.S. Ramdan. The goods train was likewise hauled by a Class Y.P. locomotive and had 57 waggons and a guards van, with a W.D. locomotive at its rear, and was in charge of Guard S.L. Sharma with Driver Ram Lall and an unnamed fireman/firemen.

Ladpura station was situated on the double-track line between Madar and Kishangarh and there were loop limes through the station. A signal cabin worked the points and signals but the Siemens General Electric double-line block instruments were placed in the stationmaster's office, with the stationmaster controlling slots on the Home and Starting signals in each direction, which could not therefore be cleared until he had released them.

The first person whose evidence appeared in the report on the accident was Cabinman

205 *The Daily Gleaner*, Kingston, Jamaica, 22nd September 1965
206 Wikipedia – Liste von Eisenbahnunfällen in Tschechien
207 *Independent*, Long Beach, California, United States, 31st May 1966
208 *Nevada State Journal*, Reno, Nevada, United States, 31st May 1966
209 *The Daily Colonist*, Victoria, British Columbia, Canada, 31st May 1966
210 *Nevada State Journal*, Reno, Nevada, United States, 20th June 1966
211 *Wisconsin State Journal*, Madison, Wisconsin, United States, 20th June 1966
212 *The Times*, London, England, 20th June 1966

Amar Chand, who stated that when he had come on duty at 07:00, No. 872 goods train was already stopped on the Down Main line. No. 20 Down passenger train had been received on the Down Loop line, after which he reset his points for the Main line. Some time later his Down Advanced Starter had been released by the stationmaster, who told him that the goods train was departing. About 10 or 12 minutes later the stationmaster had advised him that the express would be received on the Down Main line and had released his Down Main Home signal. He said that he had placed a collar on the lever of the Down Main Home signal but removed it after clearing the Advanced Starter. It had not occurred to him to remind the stationmaster that the goods train was still on the main line. He said, "I confess my mistake. I am repentant for what I did. I have nothing to say." After the collision he had fainted, only being revived half an hour later.

Stationmaster Chhuttan Lal Jain stated that after the express left Madar he had telephoned the gateman at No. 39 Crossing to close the gates, had given the cabinman the slot on the Down Loop line, then had been busy on the telephone, when the driver of the goods train had rushed up to him asking what he was doing, and the collision had occurred. In response to questions from Additional Commissioner P.B. Aibara he admitted having spent the previous day and night at Ajmer without permission. He said that he had given the lock-up key of the control frame (which worked the slides slotting the signals) to Assistant Stationmaster B. Stephen 'about three hours after the accident'. The assistant stationmaster, on the other hand, stated that he had been given the lock-up key at about 08:15. The stationmaster admitted that he had not verified the route of the express with the cabinman, had not observed the Starting signal before giving the release and had not checked that the correct Home signal had been cleared for the express. He denied a suggestion by the additional commissioner that after the accident he replaced No. 1 slot for the Down Main Home and pulled No. 2 slot for the Down Loop Home instead, to conceal his mistake.

Assistant Stationmaster B. Stephen testified that he had been off duty at the time of the accident. Rushing from his quarters upon hearing the sound of the collision he had found the stationmaster "much perplexed and nervous." On the orders of two officers he had taken over control of the stationmaster's office and the instruments had been sealed in the presence of said officers.

Assistant Stationmaster M.L. Yadav stated that he had informed the stationmaster when the latter had taken over at 07:15 that the goods train was standing on the Down Main line and the stop collar was on the slide control instrument for the Down Main line. Upon hearing the noise of the collision he had gone to help the sufferers and then to the signal cabin. There he found Cabinman Chand in tears and the latter had said that he did not understand how the express had been accepted onto an occupied line.

Pointsman Devi Singh stated that on the instruction of the stationmaster he had told the driver of the goods train that the express was to have priority. He had then seen that the signals for the main line had been lowered for the express and rushed to the station office to tell the stationmaster that the wrong signals had been cleared. It had appeared to him that the stationmaster lost his nerves after the accident and was in a dazed state.

Assistant Stationmaster Ganpat Singh, at Gegal Akhri, the next station down the line, testified that the goods train had never been offered to him by Ladpura.

Gateman Ram Singh at level crossing No. 39, near the entrance to the station, stated,

"I casually glanced at the signals and saw that the Down Main line Home signal was lowered and also the Down Outer and Warner signals were in the lowered position. I ran inside my gate hut to inform the S.M. on telephone, but nobody attended the telephone at the station. I left the telephone and taking my fog signals, I ran towards the train with a red flag in my hand. As the train had come very close, I could not place the detonators on the line but continued showing the red flag but the driver had passed me and then I saw that it had dashed against the goods train."

Gang Mate Bhairon Ghasi spoke of having seen the signal lowered for the goods train to leave, then of seeing that the Down Home, Outer Home, and Warner signals were also lowered and that the express was approaching. He said that he had warned the men of his gang, the crew of the W.D. engine, and the guard of the goods train, then had run towards the express waving a flag.

Driver M.L. Sharma of the W.D. engine spoke of seeing that the signals were lowered for the goods train to leave but the lead engine had not moved, then the gangman pointed out that the Down Main line Home signal was lowered and he too (the driver) had run back showing a red flag.

Guard Sharma said that he had been returning to his van after checking his train when he had seen that the signals were lowered for the express, then picking up detonators and a red flag from his van he had followed the W.D. engine driver.

Driver Lall of the goods train testified that having seen the Starting and Advanced Starting signals lowered he had been ready to start, when the pointsman had told him that priority was to be given to the express. Enquiring why, in that case, the signals had been lowered for him he had been told they would be raised again, then had felt the bump of the collision.

Guard Narain of the express said that the signals had been clear for his train and he had known nothing of anything being wrong until he felt the strong jerk of the collision.

Driver Gupta of the express stated that the signals had been lowered for him to run through on the main line. Passing the Home signal he had seen several men waving red flags, braked, and reduced speed to 10 or 15 miles an hour before the collision. His two firemen concurred with his statement.

Additional Commissioner Aibara concluded that the express had not been travelling at excessive speed and there had been no problem with the condition of the express locomotive or rolling stock. Nor had tests shown anything wrong with the signalling installation.

The weight of the evidence was against the assertion of the stationmaster that he had given the slot for the Down Loop line signal. He had released the signal without ensuring that the points were set for the loop line and had subsequently failed to check that the correct Home and Starting signals had been cleared.

As for the cabinman, he had been well aware that the goods train was standing on the main line but had failed to use a lever collar, had failed to inform the stationmaster that a release had been given for the wrong signal and had cleared that signal for the express.

The driver of the express had done his utmost to avoid the collision and the gangmate, driver of the W.D. engine and guard of the goods train had done their best to warn him. The guard of the goods train and the gateman were, however, criticised by the additional commissioner for not having been sufficiently vigilant, while the marshalling of the express by Deputy Yard Master R.M. Gandhi, with a wooden-bodied coach placed second in the train, was also criticised for having caused a greater number of casualties.

The stationmaster had previously been disciplined for "wrong exchange of tokens" and "irregular manipulation of block instruments." He was subsequently taken to court, presumably charged with manslaughter, but was acquitted.

With regards to recommendations, one made earlier in the year that track circuiting be installed in station areas was already being implemented by the Western Railway. The only other recommendation was a modification of the duties of the guards of trains stopped at stations. [213]

An electric train carrying African workmen from Soweto township into Johannesburg ran into the back of a similar train at Croesus station, South Africa, early on Monday 1st August 1966. [214] A statement from the railway headquarters announced that 'two non-white passenger trains' had collided, and 'one coach was partially derailed and another came to rest on top of the coach immediately in front of it. Immediately after the collision five coaches caught fire.' Six passengers died in the collision and no less than 297 were injured, most of them probably only slightly, as only 15 were detained in Baragwanath Hospital.

Violence erupted after the accident. The Johannesburg correspondent for an English newspaper wrote, 'An angry mob immediately started stoning one of the white train drivers, Mr B.C. van Tonder. He hid in one of the coaches, which Africans set alight. He escaped back to the cab and the stoning continued. Sergeant A.J. Jordaan, of the Railway Police, got into the cab and shielded Mr van Tonder with his body. He then kept the mob at bay by firing shots. Sergeant Jordaan is in hospital with head wounds.

'Five fire engines and 20 ambulances were at the scene of the crash within five minutes. The injured were taken to Baragwanath Hospital, which was later cordoned off to prevent relatives of the injured from entering.

'After a railway accident at Effingham, near Durban, on October 4 last, in which 89 Africans were killed, a white signalman was stoned to death.

'Dozens of police were sent to the scene, and some shots were fired at the mob. Two African women were reported to have been injured.

'Another train standing empty at a station nearby was set on fire by angry Africans.' [215]

Some reports said that one of the rioters had been shot dead by the police. At least one other white railwayman was hurt, he was named as G.D.E. Theron, possibly the guard of the train. [216]

---

[213] Government of India Ministry of Tourism and Civil Aviation (Commission of Railway Safety): Report on the Collision between No. 4 Down Express Train and No. 872 Down Goods Train at Laipur Station, Jaipur Division of Western Railway at about 07.48 hours on 19-6-66

[214] *Sandusky Register*, Sandusky, Ohio, United States, 1st August 1966

[215] *The Times*, London, England, 2nd August 1966

[216] *The Lethbridge Herald*, Lethbridge, Alberta, Canada, 1st August 1966

A serious rear collision occurred in Denmark on Thursday 10th August 1967 at Marslev, rather less than 5 kilometres east of Odense. Two express trains were involved, both bound for København. The first train, No. 38, was the *Sydwestjyden*, formed of trains from Struer and Sønderborg, which had been combined into one train at Fredericia. This train had a booked stop at Odense and had restarted from there and proceeded eastwards again when it had come to a stand due to an engine problem. Behind it was train No. 40, the *Nordjyden* express, which was not booked to stop at Odense. Both trains were timetabled to connect with the ferry across the Storebælt to København, which left Nyborg at 10:22.

By chance the *Sydwestjyden* came to a stand with its front 35 metres in advance of *mellembloksignal* (intermediate block signal) AM 1244 and its rear 114 metres behind the signal. In this position it was 20 metres short of the treadle that would have restored the signal to Danger. The previous signal, AM 1264, two kilometres to the rear of signal AM 1244, was held at Danger. The *Nordjyden* duly came to a stand at this signal and its driver contacted the signalling centre at Odense from the signal post telephone. He should, of course, have been advised to wait at that signal, because the previous train was still in the section ahead, but in some manner the signalling staff at Odense misread or failed to read the *togtidsskriver* (train time recorder), which would have indicated the reason that the signal was being held at Danger. Instead they advised the driver to proceed cautiously to the next signal, at a speed, according to the regulations, not exceeding 30 kilometres an hour.

The situation was then analogous to that in the 1962 accident between Winsford and Coppenhall Junction (see above). The driver of the *Nordjyden* started off cautiously but when his train was still some distance from signal AM 1244 he could see it showing a double green *kør igennem* (literally 'drive through' – that is Clear) indication and increased his speed. Trees, a curve in the line, and heavy rain, meant that he failed to notice the train ahead until he was close to it. His train was still travelling at 70 to 80 kilometres an hour at the moment of the collision and several coaches of the forward train were badly damaged. Eleven passengers were killed, 17 were seriously injured and 19 less badly hurt.

Large numbers of people rushed to see the accident from Odense and neighbouring towns and traffic on Hovedvej 1 to Odense, swollen by spectators, reportedly caused problems for the police for several days after the accident. [217]

Almost two years after the accident, in May 1969, a commission was set up to investigate it, headed by Supreme Court Judge Henrik Tamm. [218] It found that a number of chance factors had played a part, such as the fact that the *Sydwestjyden* had come to a stand just in the rear of signal AM 1244, the fact that the driver of the latter train had failed to report his position to the Odense signalling centre quite soon enough, the torrential rain that had obscured the view of the *Nordjyden* and so on. It held no one responsible. [219]

---

217 https://www.toglyde.dk/ulykke.htm

218 https://da.wikipedia.org/wiki/Lyntogsulykken

219 https://odenseleksikon.wordpress.com/lyntogsulykken-1967/

Two trains collided in the Kruger National Park, South Africa, on New Year's Day, Monday 1st January 1968. Mixed train No. 1565 consisted of diesel locomotive No. 32-108, five goods waggons, four passenger coaches and a guard's van. It left Randspruit station at 12:49, had descended a stretch of track with a 1 in 67 gradient and a series of curves and reached a level stretch when its brakes came on automatically and it came to a stand at 13:01. The secondman went back to investigate the cause and met the tranship porter, who said he had been informed by some passengers that the vacuum pipes had come apart between the first and second passenger carriages. The two men were on the point of recoupling the pipes when they became aware of a goods train bearing down upon then and had to jump clear.

Following the mixed train was goods train No. 1509, consisting of diesel engines Nos. 32-088 and 32-097 and a load of 1,424 tonnes for 92 axles (thus 46 two-axled waggons). This train had left Randspruit at 13:05 and collided with the mixed train eight kilometres from there at 13:12. As a result of the collision the guard's van and four wooden coaches of the mixed train were reportedly completely wrecked. Twelve passengers and the driver of the goods train were killed and 49 passengers and two railwaymen were injured.

After the accident the driver of the mixed train took his engine forward to the next interloop (crossing place) at Bume, from where he reported the accident to the station foreman at Komatipoort at 13:50. He refused to take his engine on to Komatipoort (about 30 kilometres further on) until it had been examined by a fitter, while as he was in possession of the tablet for the single-line section pilot working could not be opened up until the tablet had been brought to Komatipoort. It was arranged to send a fitter to Bume as quickly as possible.

Meanwhile one of the two guards of the goods train had made his way back to Randspruit, getting a lift part of the way, and reported the accident there at 13:40, but could not state its extent.

The emergency services and breakdown cranes were called out at 14:40. A doctor arrived at the site at 14:45, after losing his way, and the Nelspruit breakdown crew reached there at 18:00. The Komatipoort stationmaster reached Bume by road about the same time as a diesel fitter. He decided to couple the engine of the mixed train to the underailed coaches of its train and take them to Komatipoort, from where the uninjured passengers were forwarded by bus. Breakdown trains from Komatipoort and Waterval Boven eventually arrived at the scene at 23:42.

The line was worked by the Van Schoor train tablet system, which allowed for permissive working, with trains being allowed to follow one another through a section. Because there was also a danger from wild animals in the national park, special operating instructions were in force. A minimum interval of 30 minutes by day and 60 minutes by night was required between trains when all interloops were open, unless absolute block working was in use, increased to 120 minutes if an interloop was closed. Guards were not required to protect their trains at night due to possible wild animal attacks. (A game ranger who came across the accident soon after it had happened arranged for park staff to protect the site from animals during the rescue work and clearance operation.)

It emerged at the inquiry into the accident that the guard of the goods train, No. 1509, had no watch (it had been handed in for repair). He alleged that the station foreman had given him a hand signal to depart and had entered the departure time in his journal as

13:19. The train had left with a permissive train tablet and a T 389 written warning that train No. 1565 was in front. The foreman said that he had been busy setting the points for train No. 1566 at the other end of the station and had given no signal to the guard. After lunch he had entered 13:19 as the departure time for train No. 1509, though he did not know what time it actually left. The guard and a ticket collector on train No. 1566 confirmed that train No. 1509 had been pulling out at the same as their train had been at the other end of the station and the board of inquiry concluded that train No. 1509 actually left at 13:05, thus only 16 minutes after train No. 1565, instead of the required minimum of 30 minutes.

Given that the collision occurred at 13:12, train No. 1509 had covered the eight kilometres from Randspruit in seven minutes, at an average speed of over 68 kilometres an hour. The maximum permissible speed was 48 kilometres an hour.

The board of inquiry determined that there had been multiple irregularities on the part of various employees:

(a) The driver of train No. 1509 had left Randspruit without obtaining the "train may depart" signal from the station foreman.

(b) The driver had exceeded the maximum permissible speed of 48 kilometres an hour (reduced from the previous 56 kilometres an hour due to the condition of the track).

(c) The guard of train No. 1509 had given his driver the departure signal without obtaining the "train may depart" signal from the Randspruit station foreman.

(d) The guard had not checked for a danger signal before giving the second right-away signal.

(e) The guard had taken no action when it was apparent that his train was exceeding its maximum authorised speed.

(f) The guard had entered an incorrect departure time in his journal.

(g) The Randspruit station foreman had not attempted to stop train No. 1509 leaving before time.

(h) The station foreman had knowingly entered an incorrect departure time for the train, aware that it left earlier.

(i) The station foreman had gone to lunch when a goods train was ready to depart.

(j) The driver of train No. 1565 had not checked that his secondman had replaced the vacuum pipes, which had fallen off.

(k) The driver had allowed a spare driver to ride in his cab.

(l) The spare driver had ridden in the cab of the engine of train No. 1565, knowing that it was against the regulations.

(m) The guard of train No. 1565 had not recorded the time that his train had stopped in the section.

(n) The guard had not protected his train in accordance with the regulations (he did go back some distance, putting down detonators at 430 and 520 metres to its rear, which was not far enough, though given the excessive speed of the goods train, its weight, and the falling gradient, it was concluded that even if protection had been out the regulation distance the collision would still have occurred, but the speed at impact would, of course, have been less).

(o) The guard had not remained at the point of protection but had started back towards his train.

Disciplinary action was taken against all railwaymen found to be at fault.

The writer of an article on the accident had added that the driver of train No. 1509 was known as a "cowboy," with reputation of driving too fast. It was also possible that the driver and guard had been in a hurry to reach Komatipoort in time to get ready to go to an evening dance. They would not, of course, have been expecting to find the mixed train stopped in the section. [220]

Eighteen people were killed more or less outright when an electric train ran into the rear of a passenger train near Allahabad, Uttar Pradesh, India, on the evening of Saturday 30th March 1968. (Allahabad was renamed Prayagraj in October 2018.) About 24 others were injured, two of whom died in hospital on Monday. [221] [222] [223]

The Howrah-bound Janata Express from Madras ran into the rear of a stationary goods train in Anaparthy station, India, at 01:48 on Wednesday 8th May 1968. Seven people reportedly died in the accident on the South Central Railway and 30 were slightly hurt. The engine and leading parcels van of the express were derailed and overturned, as did the guard's van and rear four waggons of the goods train. Three of those who died were quickly identified as Driver Williams, Fireman Bhimayya and Fireman Johnson of the express. The first two were killed outright and Fireman Johnson died after admission to Rajamundry civil hospital from severe burns. All seven of those who lost their lives were apparently railwaymen, one of the others being the guard of the goods train.

The line was not blocked, traffic being maintained with single-line working. The express resumed its journey at 06:12 with all the passengers, including those slightly injured. The railway company stated that "ex-gratia payments will be made as found necessary."

The accident was the subject of a heated discussion in parliament, with calls for the resignation of the ministers concerned. [224]

A rear collision occurred between Lilulah and Howrah stations, India, on the Eastern Railway, on the afternoon of Thursday 20th June 1968. The multiple-track line concerned was worked on the Absolute Block System with four-aspect automatic and semi-automatic colourlight signals. Train No. 336, a Down Rampur Hat passenger train, consisted of a WG class steam locomotive and nine vehicles. It left Lilulah under clear signals and stopped at semi-automatic signal No. 72/72A/100 because of a drop in steam

[220] http://africawild-forum.com/viewtopic.php?t=10980
[221] *Lubbock Avalanche-Journal*, Lubbock, Texas, United States, 1st April 1968
[222] *The Timers-Reporter*, Dover, Ohio, United States, 1st April 1968
[223] *Laurel Leader-Call*, Laurel, Mississippi, United States, 3rd April 1968
[224] *The Indian Express*, Madras, India, 9th May 1968

pressure. Adequate pressure then having been obtained it restarted, stopped for about two minutes at signal No. 1A-2, which was red, restarted, and ran into the rear train No. M 140, the Burdwan local, an eight-car electric multiple unit (EMU), 1,581 feet further on. Five people died and 41 were injured, five seriously.

Driver Deonarayan of train No. 336 told the inquiry into the accident that his train had left Burdwan six minutes late at 09:21. There had been no problem with its brakes. After stopping at automatic signal No. 1A-2 on Down Main line No. 1 he had been looking out for any obstruction when the left steam gauge glass burst. At the same moment his first fireman had exclaimed "Obstruction ahead," whereupon he had applied the brakes, but the collision had been unavoidable. He insisted that despite the damage caused and the fact that the EMU had been pushed forward 110 feet, he had only been moving at 5 miles an hour at the moment of impact. Although the EMU should have been visible for 850 feet and the gauge glass had burst when he had been only a few feet from the forward train he still blamed the broken glass for restricting visibility.

First Fireman Satya Narayan testified that he had seen the rear of the EMU from about 25 feet away from the left side of the engine. The gauge glass burst just before, when the engine had been about 40 feet away, and his driver had closed the steam cock before applying the brakes. He agreed that the speed at impact had been 5 or 6 miles an hour. He could not explain why he had failed to see the EMU earlier.

Second Fireman P.C. Nandy said that he had been busy breaking coal on the tender and had been taken by surprise by the collision.

Guard G.C. Day of train No. 336 put the time of its leaving Burdwan at 09:26. According to him his train had only stopped a minute at signal No. 1A-2 and the accident had occurred at 14:45 when his train had been moving at 10 to 15 kilometres an hour (6 to 9 miles an hour – the accident reports seem to use both miles and kilometres at different points in the same accounts). There had been about 300 passengers in his train.

Motorman S.P. Banerjee of train No. M 140 said that he had been detained 10 or 12 minutes at signal No. 2 on Down Main line No. 1 controlled by Howrah Route Relay cabin and had just stated moving again when there had been a bump in the rear of his train.

Motorman A.S. Sarkar of train No. M 142 stated that he had stopped at automatic signal No. 2A-2 on Down Mail line No. 2 as train No. 336 was stopped at signal No. 1A-2. The other train had restarted just before he had. His train had been moving at 7 to 8 miles an hour and the other train slightly faster at the time of the collision, when the front of his train had been alongside the second coach from the rear of the other.

Guard S.R. Chowdhury of train No. M 140 said that he jumped off when he had seen train No. 336 approaching at speed and had then lost consciousness. He had not considered it necessary to protect his train in the rear as it had not met with an accident or obstruction, or failed, and its detention of over 5 minutes had not come under exceptional circumstances as per General Rule No. 280. There had been about 500 passengers on his train.

Cabin Assistant Stationmaster A.N. Roy in Sorting Yard cabin stated that train No. 336 had cleared semi-automatic signal No. 72/72A/100 at about 14:40 according to the indications of the panel in his cabin and had stopped at signal No. 1A-2 for 1½ or 2 minutes.

Senior Cabin Assistant Stationmaster S.P. Chowdhury in Howrah Route Relay cabin testified that train No. M 140 had reached signal No. 2 at 14:32. The signal had been taken off at 14:45, it had not cleared the signal, and the same minute he had been advised of the collision.

Additional Commissioner G.S. Pandor noted that the collision had occurred at 14:45 on a bright sunny day with clear visibility and as train No. M 140 had just started to move. The lines were fully track circuited with modern colourlight signalling but this could not prevent accidents resulting from failure to obey the Stop and Proceed with Caution rules.

The additional commissioner decided that full responsibility for the collision rested with Driver Deonarayan, who had failed to exercise the necessary caution after passing signal No. 1A-2 at danger, while First Fireman Narayan had failed to keep a good lookout.

The additional commissioner recommended improvements in the training of motormen and checks to ensure that drivers did not take less than the minimum time between signals after passing an automatic signal at danger. [225]

A rear collision at Gangpur station, India, on the Eastern Railway, on Sunday 18th August 1968 resulted in the death of seven people, one of them the motorman of the colliding train, and the injury of seven more, one seriously.

The events that led up to the accident began with the theft soon after midnight of underground signalling cables between Burdwan and Gangpur, which resulted in a total failure of communications over the almost 7-kilometre section between the two stations. The Burdwan stationmaster then decided to work traffic in accordance with the rules regarding the operation of trains over a double line during a total block failure, which involved all trains being issued with caution orders, and not exceeding 15 kilometres an hour on the straight and 5 kilometres an hour on curves or when visibility was restricted. Trains were also required to stop at the first stop signal at the end of the affected section and sound four long whistles.

Down electric goods train No. C 281 was the 12th train to pass through the quadruple-track section under the block failure regulations. It had been stopped at the Gangpur Down Home signal for the Howrah – Bardamman Chord (HBC) line for five minutes when it was run into from behind by train No. C 166, a six-coach electric multiple unit (EMU). The motorman's cab and leading third-class compartment of the EMU were wrecked and the brake van at the rear of the goods train was overridden by the waggon loaded with buffaloes in front of it.

Driver Lalta Prasad of the goods train told the inquiry into the accident that he had been properly cautioned before leaving Burdwan. Approaching Gangpur he found that the signals had already been cleared for his train, regardless of which he had stopped at the first Home signal and whistled four times, as per the rules. After three minutes he had sent the assistant driver to the station to clarify if the signals had been cleared for his

[225] Government of India Ministry of Tourism and Civil Aviation (Commission of Railway Safety): Report on the Collision between No. 336 Down Rampur Hat Passenger Train and M 140 Down Burdwan Local Train between Liluah and Howrah Stations of the Eastern Railway on 20th June 1968

train. Shortly afterwards the traction current had been cut off, he felt a bump and his guard had come forward and told him that there had been a collision. According to him the Gangpur signals had been clearly visible.

Guard Biswanath of the goods train confirmed that the Gangpur signals had already been off when they reached them and said that he had wondered why his driver had stopped even so. He had heard another train coming which he assumed must be on the Down Main line but soon realised that it was on the HBC line, whereupon he had immediately shown a red hand signal to it. There had been no response and he had jumped out to the left and shouted. The EMU had still been travelling at about 40 miles an hour when it struck his train at 07:05.

Guard S.C. Moulik of train No. C 166 testified that his train had left Burdwan at 06:55 after being cautioned. Speed through the section had been 15 miles an hour to the Gangpur Down Distant, after which his train had accelerated to 20 or 25 miles an hour. When this had happened he sounded the bell code for the motorman to slow down, nothing had happened, and he had been about to apply the emergency brake when the collision occurred. Although he had been aware that his motorman had driven through the section at a higher speed than that set out in the caution order (15 km/h or 10 m/h) he pleaded that he had been confused with the time laid down in the block working manual (25 km/h or 15 m/h), also that the speedometer had been out of order. (Another motorman stated that there had been mothing wrong with the speedometer and it was found to be working after the accident.)

Note: As noted in the report on the Lilulah accident above, this one switches between miles and kilometres at different points.

Platform Assistant Yard Manager S.R. Bhownick at Burdwan stated that after being informed by the East cabin of the total communications failure the regulations for the working of trains during such a failure had been brought into use. He personally had ensured that all trains, including No. C 166, were properly cautioned.

Assistant Stationmaster L.M. Das at Gangpur stated that he had begun working under the regulations for operating traffic during a total communications failure at 00:15. He maintained that he had not cleared the Down Home signal for any train after this time until such train had stopped at the signal and whistled, despite being confronted with the guard of train No. C 166 and Motorman Pritam Singh of train No. M 132 Down, who both stated that he had cleared his signals prematurely.

Traffic Inspector P.B. Sarkar did not think the signal had been cleared before the stoppage of the goods train but could not say that the signals had not been cleared too early for some trains.

Driver J. Mahato, off duty, had been riding in train No. C 166. He said that he had met the motorman before leaving Burdwan and the latter seemed cheerful. Shortly after leaving Burdwan the speed of the train had increased to 30 or 35 miles an hour and been maintained. Visibility had been fair and he considered that the goods train could have been at a distance of 1,200 feet. Just after the collision the Gangpur HBC Home signal had been yellow.

Additional Commissioner G.S. Pandor put the time of the collision at 07:05. He considered that train No. C 166 had been travelling at about 35 miles an hour when it hit the goods train.

It was clear that train No. C 166 had been driven well in excess of the speed laid down in the caution order, though even at 35 miles an hour the motorman should have been able to stop short of the goods train had he been vigilant. The guard had been aware that his train had been travelling too fast since it left Burdwan but had taken no action until it was too late, so he was partly responsible for the collision. By clearing the Home signal too hastily the Gangpur assistant stationmaster would have caused the associated Distant signal to clear to green, instead of remaining at yellow, which might have caused the motorman of train No. C 166 to reduce speed.

The additional commissioner found that deceased Motorman J.P. Singh had driven at excessive speed and without great caution in contravention of General Rules Nos. 89(b), 122, and 163 (a) (i) and the caution order and he was thus fully responsible for the accident.

The guard, although fully aware that his train was going too fast, had failed to intervene in time to prevent the collision as demanded by General Rules Nos. 121 and 126.

The only recommendation made by the additional commissioner concerned the gradual provision of roller bearings for medical vans, one of which had been delayed getting to the accident scene because it had developed an overheated axle box ('hot-box'). [226]

The pulling of the communication cord by a woman passenger because another woman had fainted was the prelude to a rear collision in Greece on the night of Monday 30th September 1968. Both trains were bound for Athens from Kyparissia and they were stated to have been running ten minutes apart. They were jammed with passengers returning to Athens after having voted in the national constitutional referendum in towns and villages of the Peloponnesus.[227] The first train stopped because of the pulling of the emergency alarm near Derveni, on the Patras – Athens line, in a gorge 50 kilometres north-west of the Corinth Canal and about 160 kilometres from the Greek capital. When the train stopped its conductor went back with a red lantern to protect it in the rear but he was passed by the second train on a curve and the driver of the engine of the latter failed to see the warning. Whether the line was worked on some sort of block system, as would seem likely, was not stated. If so, there must clearly also have been a failure in the signalling for the second train to have been where it was. The coaches of the forward train were said to have been piled up two and three deep in the gorge as a result of the collision, at a location inaccessible to ambulances. Thirty-four people were killed and 90 were injured according to some reports; others spoke of only 10 or 12 fatalities. [228] Troops assisted with the rescue work.

Mr Stylianos Pattakos, Deputy Prime Minister and Minister of the Interior, flew to the scene by helicopter to supervise the rescue operation. [229]

Hotel worker Ioannis Triantafyllou, who had been returning to Athens on the second

---

[226] Government of India Ministry of Tourism and Civil Aviation (Commission of Railway Safety): Report on the Rear-End Collision Between EC 281 Down Goods train and C 166 Down Burdwan Local (EMU) train at Gangpur Station Eastern Railway on 18th August 1968

[227] *The Bridgeport Telegram*, Bridgeport, Connecticut, United States, 1st October 1968

[228] *The Brazil Daily Times*, Brazil, Indiana, United States, 1st October 1968

[229] *The Canberra Times*, Canberra, Australian Capital Territory, Australia, 2nd October 1968

train after voting at Messalonghi, told a reporter, "It was like a giant hand shaking our train, like an earthquake shaking the land." He added, "We were crowded like sardines in a can. In the first train things were not any better. It was a gory scene, with metal torn and human bodies cut in pieces and blood all over." [230]

Just before noon on Friday 31st January 1969 the *Blue Dragon* express of Korean National Railways (KNR) was nearing Chonan station, South Korea, about 95 kilometres south of Seoul, at "normal" speed, when it ran into the rear of local train No. 102. The rear coach of the local train was all but destroyed and pushed up into the air by the colliding train.

South Korea was suffering its most severe snowfall for 47 years at the time of the accident and 18 inches, which choked all surrounding roads, hampered rescue operations. At first all that was evident was the damage to the rolling stock, until rescuers found bloodstains mixed with the wreckage. It then transpired that the train had been packed with students heading for Seoul to take university entrance examinations and by Sunday it was known that there were 41 dead and 108 injured, 58 of them seriously hurt.

An hour after the accident news of it reached United States Army Colonel Eric Kobbe, commander of the 23rd Direct Support Group at Camp Humphreys, 40 kilometres away, and he immediately left for the scene by helicopter, accompanied by three doctors, an engineer, and First Lieutenant John Forker, who related, "One passenger car was locked into the other. It had just fallen through the wooden roof of the (local) passenger train. There were people still in the train – they were dead. One engine was derailed. There was one Korean military-type ambulance hauling the injured away. Col Kobbe took charge of the rescue work and stayed on the scene all night."

Six of the most seriously injured were sent back to Camp Humphreys in the helicopter where they were treated at the 545th Medical Dispensary. By Saturday they had been transferred to a civilian hospital. Ft. Lt. Forker accompanied the helicopter on its return flight to organise a larger rescue party, and a convoy of ambulances, wreckers, generators with lights and lorries filled with blankets left by road for the scene with a 30-man volunteer party of corpsmen and engineers.

Col. Kobbe told a reporter, "When we arrived it was mass confusion. But someone had already gotten the injured out of the train and into the town's seven local hospitals. We walked to the hospitals, through knee-deep snow, carrying the doctors' bags. At the first hospital there were 15 patients, all lying on the floor. There was only one doctor to work on them.

"The (army) doctors started their sorting work, and we moved out six of the most seriously hurt right away, on the helicopter. The weather closed in even more, and there weren't any more rescue flights. The doctors did a great job. They saved more than one life by going around to the hospitals, sorting the wounded, stopping the major bleeding, directing emergency treatment for those that needed it. The local hospitals were doing a damn fine job, but there just weren't enough doctors in them to take care of everyone."

Kobbe was on the scene for 11 hours. "After about six hours, the doctors had finished the sorting, and my ambulances and a hospital train had both arrived, so we could get the

[230] *The Daily Report*, Ontario-Upland, California, United States, 1st October 1968

men out. It was pretty cold out there. It was snowing off and on, and there wasn't any hot coffee, or anything fancy. We did manage to get blankets out there before it got too cold. The local people were helping out as much as they could, and we got all of the bodies out of the train to Seoul by 10.45 p.m."

KNR director Yi Hun Sop reached the scene in the evening and announced that bereavement money would be paid to the families of those who had died as soon as possible, while the injured would all receive medical care until they were completely recovered.

The driver of the express, who survived, admitted that he had failed to see a red signal shortly before the collision. It was speculated that the falling snow might have obscured the signal, though, of course, the normal ruling on any railway was that a signal not visible where one should be displayed must be treated as a danger signal. Possibly in the blizzard the driver had also failed to realise where he was and had still been looking out for the signal after he had passed it. The local train was running an hour late and had stopped to give precedence to a southbound one.

The 65 medical group, based at Hongsan, sent 35 staff and six ambulances to Yongsan station in Seoul to help unload the dead and injured when the train reached there late in the evening, a job which took them five hours. [231]

Like the Greek accident near Derveni, above, a major catastrophe on the South Eastern Railway in India on the night of Monday 14th July 1969 had its origin in a passenger pulling the communication cord. The passenger train, which was run into by a following freight train while stationary, was crammed with Hindu pilgrims who were on their way to a festival in the town of Puri, in Orissa state. Eighty-six passengers were killed and about 200 were injured, six seriously. [232]

Train No. 398, the Up Asansol-Puri passenger, had left Jaipur-Keonjhar station at 21:22 with token No. 16 for Jakhapura. It had only got as far as a point between the Starting and Advance Starting signals of the South cabin when it was stopped by the pulling of the alarm chain. It was stationary for 10 or 12 minutes while the brakes were released, before restarting. Meanwhile, the South cabin leverman, not having realised that the passenger train had stopped between his signals, had released the slot for the Up Main Home signal. The driver of No. D 513 Up Diesel Goods train thus entered the station under clear signals and seeing the Starting and Advanced Starting signals off, not having been replaced behind the passenger train, assumed that he had a clear run through the station. The rearmost coach of the passenger train was completely smashed in the collision.

Assistant Stationmaster B.B. Das at Jaipur alleged that when the passenger train left the station he had seen the Starting signal return to danger and had given the *Train Entering Section* signal to Jakhapura. About 21:28 he had been advised by the South cabin leverman that the passenger train had passed and had exchanged the private number for the departure of No. 513 Up Goods, replaced the slide for the Up Advanced Starting signal to normal and obtained the slot from the South cabin for the Up Home signal. He had then pulled his slide and the North cabin leverman had cleared the Home signal. The

[231] *Pacific Stars and Stripes*, Tokyo, Japan, 2nd February 1969
[232] *News Journal*, Mansfield, Ohio, United States, 15th July 1969

Up Warner had not been cleared but the goods train nevertheless passed through the station at high speed and the South cabin leverman had shortly informed him of the collision.

Leverman Subal Naik stated that after the passenger train had started the assistant stationmaster advised him that No. D 513 Up Goods was approaching from Baitarani Road and would be received on Line No. 2. The passenger train had then stopped. The assistant stationmaster had asked him for the slot for the goods train and after the passenger train had restarted and passed the Starting signal, he had replaced that signal to danger and given the assistant stationmaster the slot on the Home signal. Before the passenger train had passed the Advanced Starter he had seen the goods train approaching at high speed and tried to stop it by replacing the slot on the Home signal but the goods train had passed it. He had then seen that the Advanced Starter had returned to danger, doubtless by the assistant stationmaster having operated his control slide, and he had frantically tried to get the assistant stationmaster to clear the Advanced Starter again, had shown a red light to the goods train and had tried to divert the goods train onto a siding but the brake van of the passenger train had been on the points. He accepted that he had given the slot for the goods train prematurely and pleaded that he had done so under instruction from the assistant stationmaster.

Leverman Sheik Nabi Bux in the North cabin said that the tail lamp of the passenger train had been burning brightly. After exchanging private numbers and the slot having been released he cleared his Up Home and Outer signals for the goods train but not the Warner, despite which the goods train had passed him at high speed.

Token Porter Padmanah Parida confirmed that the tail lamp of the passenger train had been showing a good light and described showing a danger signal to the goods train.

Assistant Stationmaster B. Adhikari, acting as stationmaster, said that he had been in his office when he heard the goods train passing at high speed and the station staff shouting at it.

Driver G.C. Das of the passenger train testified that after starting from the station his train had been stopped near the South cabin by the pulling of the alarm cord. He had sent his first fireman back to check and rectify the problem. Vacuum had been restored and he had restarted after about 5 or 6 minutes when his train had been run into.

First Fireman S.S. Panda stated that on going back after the alarm chain had been pulled he had seen that the Starting signal was still off with 2 or 3 coaches behind it. The brakeman had told him that the vacuum disc had been reset and returning to the engine he had seen that the Advanced Starter was still green. The train had restarted after 9 or 10 minutes, then it had been run into. The alarm cord had only been pulled once.

Second Fireman N.K. Dhal thought that the train had stopped about 8 minutes.

Traffic Brakesman A.K. Raa put the stop from the pulling of the alarm cord at 5 or 6 minutes. He saw the Advanced Starter at danger after the collision.

Traffic Brakesman B. Latchayya agreed that the train had only stopped once after leaving the station. It had been moving at about 5 miles per hour when it was run into.

Driver Chitta Bhanumurty of the goods train agreed that the Jaipur Warner had been ‘on’ but said that the Outer and Home signals had been clear. He had brought the speed down to about 23 miles an hour, then had seen that the Starter and Advanced Starter were green and opened the throttle. He had expected to get the token either in front of the station building or at the end of the platform. He had not received it and at the same time he had seen what looked like a bogie ahead, when he had applied the emergency and independent brakes. After the collision he had gone straight to the South cabin, whereupon the cabinman had replaced the Starting signal and remarked that it was the assistant stationmaster's fault. He (the driver) had not seen the tail lamp of the passenger train, only the outline of the bogie.

Diesel Assistant B. Peter confirmed that all the Jaipur signals other than the Warner had been 'off.' He had thrown out the *Line clear* for the previous section and had been looking for the token porter but could not find him, whereupon he had shouted to his driver that the *Line clear* was not there. The driver had applied the brake and told him to look out for the *Line clear,* which was sometimes handed over at the end of the platform. The train had been moving at about 15 miles an hour or a little more at the moment of impact.

Guard Abdul Satar of the goods train spoke of having seen the Starter and Advanced Starter lowered and exchanging signals with the cabinman, though there had been no one to do so with at the station. Then the brakes had come on and there had been the sudden jerk of the collision.

The inquiry into the disaster was conducted by Additional Commissioner Arya Bhushan. He fixed the time of the accident at 21:36. The speed of the passenger train he put at 5 or 6 kilometres an hour and of the goods train at about 30 kilometres an hour. The accounts of the assistant stationmaster and the South cabin leverman were, he noted, contradictory, but from other evidence from disinterested parties it was clear that the passenger train had only stopped once and that the Starter and Advanced Starter had not been replaced to danger before the Home signal had been cleared for the goods train.

Having been informed by the South cabin leverman that the passenger train had gone, the assistant stationmaster had been within his rights to accept the goods train. It was unlikely that he could have seen the tail lamp of the passenger train but he should have observed that the Starter and Advanced Starter were still 'Off' and should have put back the slide for the Advanced Starter before clearing the Home signal.

The driver of the goods train had been misled by seeing the Starter and Advanced Starter clear for him but should have braked earlier when he had not been given the token for the forward section.

The ideal positioning of the token pickup apparatus was discussed and it was noted that the rules regarding the use of a torch at night to illuminate the token delivery area were not being strictly observed.

The presence of a wooden-bodied TLR coach at the rear of the passenger train had led to more casualties.

Track circuits on the reception lines would have prevented the accident.

There had been no interlocking between the main line slot lever and the Starting signal.

Additional Commissioner Bhushan concluded that Leverman Naik in the South cabin was principally responsible for the accident, having advised the assistant stationmaster that the passenger train had left when it was still between Starting and Advanced Starting signals, in contravention of the station working rules.

Assistant Stationmaster Das was at fault for failing to put back the control slide for the Advanced Starting signal before releasing the control on the Home signal and Driver Bhanumurty was not entirely free from blame either for not having been more alert when he had not seen the token porter in the usual place, and given that the Warner signal had been against him.

The additional commissioner recommended that General Rule 10 (s) be amended to warn drivers to proceed cautiously if the Warner signal was 'on' even if the following signals at the station were clear.

His other recommendation was for the introduction of track circuiting in all station yards on the main line. [233]

The Tanjung News Agency reported that an accident had occurred on the Belgrade – Zagreb main line on Saturday 25th October 1969. On a stretch of line equipped with automatic signalling, the *Yugoslavia Express*, bound for Zagreb, had run into the rear of a slower moving freight train at Okucani station, Yugoslavia. It stated that at least 15 people had been killed and about 50 injured, although reliable information on the number of dead was still unavailable. The engine and leading three coaches of the express were reportedly destroyed in the collision. The injured were rushed to hospitals in Gradiška, Požega, Dubica and Zagreb. As the line was completely blocked trains were diverted via Sunja, Doboj or Banja Luka. [234] [235]

Twelve people died and 15 were injured when a Sofia – Viding express ran into the rear of a freight train in the station at Oreshetz in north-western Bulgaria, probably late on Friday 2nd January 1970. Visibility was stated to have been poor at the time. [236] A few days later it was stated that the driver of the express and a signalman would shortly be put on trial for causing the accident through negligence. [237]

By far the most serious accident in Argentinian railway history occurred on the former Ferrocarril General Bartholmé Mitre Buenos Aires Retiro Mitre – Rosario Norte line of the Ferrocarriles Argentinos on the late evening of Sunday 1st February 1970. In a familiar scenario the first train stopped out of course in mid-section due to a problem with its locomotive and a following train was admitted into the occupied section due to a signalling error.

Train No. 3832, *El Zarateño*, had left the port and resort of Zárate at 18:48 for the Argentinian capital. It consisted of an engine and five única (second-class?) coaches in charge of Conductor Pedro Pérez with Driver Victor Ceroli and Fireman Molina, and was

---

[233] Government of India Ministry of Tourism and Civil Aviation (Commission of Railway Safety): Report on the Collision of No. D 513 Up Diesel Goods Train with No. 398 Up Asansol – Puri Passenger Train at Jaipur-Keonjhar Road Station on the East Coast Main Line of the South Eastern Railway at 21:36 hours on 14th July 1969

[234] *Pacific Stars and Stripes*, Tokyo, Japan, 27th October 1969

[235] *Bridgeport Sunday Post*, Bridgeport, Connecticut, United States, 26th October 1969

[236] *The Argus*, Fremont-Newark, California, United States, 5th January 1970

[237] *European Stars and Stripes*, Darmstadt, Germany, 5th and 9th January 1970

carrying 1,090 passengers, returning home at the end of a restful Sunday. This train had duly stopped at Campana, Otamendi, Rio Luján, Escobar, Maschwitz and Benavidez stations, but about a kilometre beyond the latter, at kilometre 36, in an uninhabited area of marshland, it came to a stand due to a problem with the fuel injector of the diesel locomotive. [238] When it did so its passengers naturally looked out of the windows to see what was wrong, while Conductor Pérez got down and made his way forward to the engine. "Only a small problem," Driver Ceroli had assured him. "I will get it fixed in a couple of minutes and we can be off again." The conductor realised later that his action in going forward to speak to his driver had saved his life.

Mixed train No. 1016, *El Mixto*, consisted of engines Nos. 6369 and 5232, eight passenger coaches and two car-carrying waggons, with Conductor Fernando Faccio and Drivers Alfredo Amoroso and Juan Diozkez. It was carrying 70 first-class passengers, 179 única-class passengers and 11 who had a cabin, 260 in all. It had started from Tucumán on Saturday and had left Rosario, where there had been a crew change, at 15:30 on Sunday afternoon. According to the working timetable it should have passed through Zárate at 18:28, thus 20 minutes in front of *El Zarateño*, but this day it was running 48 minutes late at that point. When it reached Benavidez the signals were clear and it continued its journey, travelling at somewhat over 100 kilometres an hour. Some of the passengers of *El Zarateño* saw the headlights of it coming in time to jump off but most did not. Driver Amoroso, on the leading engine of *El Mixto* saw the train stopped ahead only as he exited the curve. He immediately applied the brakes in emergency but the distance was far too short to be able to do more than slightly minimise the effects of the collision. The accident probably happened at 20:05 (some accounts state 20:15 but this appears to be incorrect). The train had been due to pass through Benavidez at 19:11.

The two locomotives of *El Mixto* embedded themselves in the two rear coaches of *El Zarateño,* which were destroyed, while the remainder of the train was pushed forward some distance by the force of the impact. Four coaches of the colliding train were derailed.

Driver Amoroso told a reporter, "Barely two minutes must have elapsed from my passing Benavidez station to reaching the scene of the accident, since it was traveling at a speed of nearly 105 kilometres an hour. Leaving the curve beyond the station, I was able to see a train that seemed, I thought at once, to be on my own track. I applied the brakes about 350 metres before the crash, thus reducing the speed to almost half of what it had been. Despite this," he added, "the impact was terrible."

Conductor Faccio of *El Mixto* had been completely unaware of what was about to happen. He said, "Suddenly we heard a terrible explosion, and we were thrown to the ground. When our train stopped, after pushing the local train about 300 metres, the spectacle was terrifying. On the sides of the tracks were the remains of the last coach and dozens of mutilated corpses. The confusion lasted a long time, until first aid arrived and something effective could be done for the injured."

Driver Ceroli of *El Zarateño* said, "With my fireman, Molina, I left Benavidez station at 19:48, that is, at the timetabled hour. It took us eight and a half minutes to get to the place where the horrendous tragedy would later occur, because the locomotive was malfunctioning. Faced with this difficulty, I brought the train to a smooth stop, and it took us three and a half minutes to fix the problem there. As we were about to set off again I felt

[238] *The Times*, London, England, 3rd February 1970

an impact. At first I thought that it was just the repaired engine starting with a jerk. I never imagined that we had been hit by another train. The fact that I was squatting down caused the impact to throw me forward and that was when I sustained the injury to my forehead, when I collided heavily with the control panel of the locomotive. The fact that I did not lose consciousness allowed me to realise what happened," he explained, "and I collaborated with the first aid work, despite having lost a considerable amount of blood. Given the magnitude of the events, I told the fireman to detach the engine from the train, so we could go as quickly as possible to General Pacheco station to request assistance."

Surviving passengers also told their stories. Oscar Amadeo de Pino, 19, suffered a broken right ankle, cuts, and bruises. He related, "...we had boarded the train at Campana station. We managed to get seats in the last carriage. My parents were in the back seats and I was in the front. The train remained stopped for quite a long time. Suddenly the people who had been standing began to run forward. They were the ones who had noticed the proximity of the other train. I did not notice it. Afterwards I tried to work out what was happening, I heard an explosion and thought that everything was coming down on me. Those who had run forward were flying through the air or being pinned between the last two carriages. At no time did I lose consciousness, but I was stunned, having no idea of the catastrophe that had actually happened. I only remember the heart-breaking and plain screams of children."

Jaime Gómez, 16, said, "When I left the sugarcane field to search for my friends, I saw several mutilated bodies near the tracks and a girl running from one place to another asking for help to find her father. I found my friends a few hundred metres from the accident site and we embraced without saying a word; two of them had in fact momentarily lost their voices. We left the place in the direction of the capital on a lorry which had stopped on the road."

Beatrice Saravia declared, "I was riding in the first carriage and to pass the time I had been reading a Saturday newspaper, so I did not pay much attention to the fact that the train had stopped. I just heard a man say, 'It seems like it's the engine.' What happened next is very confusing, since when I regained consciousness I was lying on a field where there were many people in the same position, who seemed to be dead. Despite the fact that my legs were shaking and that I felt very sick, I got up and began to follow people who were moving away from the place along the road. I saw what was left of the train, as well as many scenes of pain, something that I will never forget all my life."

Despite the relatively remote location of the accident the sound of it was heard by a group of young men playing football in a nearby field and by some workers in a beef factory. Meanwhile some of the passengers had begun to make their way along the line towards General Pacheco station. After a while they came to a dwelling, that of the Rodriguez family, where they were able to raise the alarm, and someone set off from the house on horseback to alert the authorities.

The rescue operation was a long and complex one. Doctors and nurses, firemen, police from Tigre and San Martin, the Argentine Naval Prefecture, the Army, and local inhabitants all took part. There were few telephones in the area and radio amateurs played a prominent part in spreading news of the catastrophe and in co-ordinating the rescue work. Several journalists also went to the scene, including one from *ASI Magazine*, a journal with a reputation for sensationalism. Sure enough the magazine published a number of sensational pictures of the accident until, on 5th February, its printing, publication, and distribution was banned by the authorities and copies of it were seized.

Soon after the accident the sun had set and the onset of darkness increased the difficulty of the rescue workers, though electric generators were set up to provide lighting. The initial hours of the rescue were said to have been desperate ones and an influx of relatives and friends, desperate for information, further hindered matters. A local train bound for Zárate that had been stopped at General Pacheco station was used as a rescue train, circulating between there and the accident site with rescuers and injured. The first injured were taken to the Hospital Vecinal in Pacheco but this was soon overwhelmed with the number of casualties (600 were said to have been more or less hurt). Then the injured were taken to local hospitals and some of the more badly hurt to Buenos Aires. A day of mourning was declared throughout the province.

Whilst many people committed acts of heroism others attempted to take advantage of the disaster and there were incidents of looting. One individual was apprehended who had procured a fireman's helmet and jacket in order to merge in with the rescuers and steal jewellery and personal effects. In all, ten people were arrested for theft.

At about 04:00 it began to rain hard, adding to the difficulties. Fire engines, ambulances and private cars had accessed the scene across country from Route 9. With the rain a number of them became bogged down and tractors had to be used to drag them clear, while the Army took control of the area and the police enforced a cordon to keep out looters, sightseers, and relatives of the victims alike.

Provincial governor Saturnino Llorente arrived at the scene at 12:40 by helicopter, followed ten minutes later by the presidential press secretary on another helicopter. President Juan Carlos Organia later arrived by train but reportedly did not come very close, preferring to observe the site through binoculars. Then, it was said, he went to visit some of the injured in a field hospital set up nearby but shocked by their appearance he had quickly left again.

By the following day a large number of bodies had been recovered but an unknown number of dead remained in the two fused coaches. The bodies of some victims were never claimed because the whole family had perished in the collision. The two most badly damaged coaches were later thrown into a lineside ditch, presumably using a crane, to be cut up and removed later. According to one report the wreckage was then burnt, with bodies still inside, in order, it was said, to prevent the outbreak of an epidemic. It was alleged, with how much truth it is impossible to say, that the real motive was to reduce the amount of insurance money the government would have to pay out, that the official figure of 236 dead was far too low and that the real number of dead was over 500. [239]

Sabotage was quickly ruled out as a cause of the catastrophe, which was soon attributed to a mistake made by the staff at Benavidez station, who were arrested. It appeared that the signals there, having been cleared for *El Zarateño* had not been replaced to Danger behind it. The crew of *El Mixto* had assumed that *El Zarateño,* which they knew must be in front of them, had been diverted onto another line at General Pacheco station.

[239] https://noticiasconenfoque.com.ar/nota/9428/1970-el-accidente-ferroviario-de-benavidez-la-mayor-catastrofe-ferroviaria-ocurrida-en-nuestro-pais

According to passengers Juan Sosa and Alfredo Gonzalez, who had arrived at Benavidez station that Sunday evening to take a train to Escobar, Porter Signalman Máximo Blanco had appeared to be very nervous. Later they discovered that a few hours earlier Blanco had been the subject of an assault, which explained his nervousness, especially as he had previously been the victim of a robbery in which his life had been threatened. At 20:04, the witnesses said, Blanco had been in the toilet when the Tucumán express had gone by. Hearing this train go by, he had rushed out, gesticulating and holding his head, and had shouted to a labourer, "Has General Pacheco reported the previous train back?"

"No," was the reply.

At that moment, Sosa and Gonzalez added, they had anticipated that something serious was about to happen. [240]

Ten people lost their lives on Thursday 12th February 1970 in a collision involving two passenger trains about three kilometres north of Hsinchu, Taiwan. An express train was stopped at a signal waiting to enter a station when another train ran into its rear, derailing and overturning its rear two coaches. Six people were killed outright, including Driver Peng Ching-chu of the colliding train, and four others died after admission to hospital. Another 47 were injured. The railway promised to pay the medical expenses of the injured and compensation to the families of the victims. [241] One source stated that a faulty signal was to blame for the collision. [242]

According to first reports at least 13 people reportedly died and more than 85 were injured when train No. 1, the Down Madras – Mangalore Mail, ran into the rear of train No. 19, the Down Madras – Cochin Mail, at 20:10 on the night of Saturday 31st October 1970, in the station at Perambur, India. All the dead and most of the injured were in the Cochin train, the rear two bogies of which were badly damaged. The rear coach, a first- and third-class coach, was twisted so badly in the impact that rescue squads with oxy-acetylene flames were still trying to cut it open at midnight, nearly four hours after the collision. It was feared that the death toll would rise when all the passengers trapped in the rear coach were released. The engine and three coaches of the Mangalore train were derailed.

Neither train was booked to stop at Perambur. The Cochin Mail, which had left Madras at 19:30, had, as in previous rear collisions, had to stop due to engine trouble. The Mangalore train had started half an hour behind it.

The Southern Railway Public Relations Officer stated that 63 injured had been admitted to Perambur Railway Hospital and 20 to the General Hospital.

The Senior Deputy General Manager, Chief Commercial Superintendent, Chief Medical Officer, and Divisional Superintendent of the railway went by road to the scene to organise the rescue work, while a relief medical van was sent from Madras. The General Manager was flying back to Madras from Trivandrum, while Chief Minister Karunanidhi, Industries Minister Madhavan, Food Minister Shanmugham and Inspector

---

[240] http://www.generalpachecoweb.com.ar/elbaul_notas.php?nota=accidenteferroviario
[241] *Pacific Stars and Stripes*, Tokyo, Japan, 14th February 1970
[242] *The Press-Courier*, Oxnard, California, United States, 12th February 1970

General of Police Mahadevan also hurried to Perambur. Doctors and nurses from the general hospital worked at the scene and the hospital blood bank remained open all night. As is often the case a large crowd of onlookers hampered rescue operations.

Chief Minister Karunanidhi offered his condolences to the friends and relatives of the dead and injured. The cause of the collision was not immediately clear. [243]

By the following day the number of dead was stated to have risen to 16 with the death of three of the injured, while the number of those more or less hurt had increased to 108. Those slightly hurt included film star Ravichandram, who had been en route to Salem for a shooting engagement. Those injured who were well enough to return home were given free passes and reservations. The undamaged coaches of the Cochin Mail and those of the Mangalore Mail went forward at 02:05 on Sunday. Other stranded passengers were catered for by attaching extra coaches to Sunday's West Coast Express and the Kerala Express.

Driver S. Chinnappan of the Mangalore Mail, who with his two firemen had jumped from his engine immediately before the collision, was arrested at the Perambur General Hospital, though it was not at that point clear if he had been overrunning signals or if the signals had been wrongly cleared for him.

Ex-gratia payments were arranged by the Southern Railway for the injured and the families of the dead. Meanwhile the inquiry into the cause of the collision was begun and Chief Guard Abdul Hamoed of the Cochin Mail stated that he had been waiting for the green signal for his train to start when he heard the sound of another train approaching. At first he had thought that it must be on another line, then realised to his horror that it was on the same line as his train. He had then rushed towards it but had neither a lamp nor a flag with him.

A passenger who had been riding in a first-class compartment, and whose face was bandaged, said that he had been sat reading a book when he heard a loud noise and almost at once something had hit him in the face. He had been told that he was taken from the compartment unconscious after the window bars had been cut. [244] (Indian trains have barred windows, mainly to prevent thefts from the coaches while the trains are stopped at stations.)

Several days after the disaster the number of fatalities was stated to have only been 14 and not 16 as earlier reported. [245] The final death toll was 15, with 80 injured, 22 seriously.

Driver N.C. Lakshmanan of No. 19 Mail told the inquiry into the accident that he had had problems with the diesel electric engine of his train en route to Perambur, so had to stop at Basin Bridge and again before Veysarpadi. At Perambur there had been trouble once more and he had stopped short of the Down Fast Starting signal and walked back to the station to report the problem. While he had been there his train was run into.

Diesel Assistant R. Raju of train No. 19 confirmed that the signals at Veysarpadi had been clear for their train, they had stopped at the Starting signal at 20:05, and the collision had occurred at 20:09.

Guard A.S. Abdul Hameed stated that train No. 19 had stopped for seven minutes at Basin Bridge, again near Veysarpadi and then run very slowly onwards to Perambur.

[243] *The Sunday Standard*, Madras, India, 1st November 1970

[244] Ibid, 2nd November 1970

[245] Ibid, 4th November 1970

Signal No. F 56 approaching Perambur had been green, going to red after the engine had passed it. When his train had stopped at Perambur he had been aware that train No. 1, due to leave Madras Central at 20:00, must be catching them up, and had kept looking behind and displayed a red hand lamp to the rear. He had then seen train No. 1 approaching at speed and alighting from his brake van had fallen, extinguishing his hand lamp. Undeterred, he had run back with his arms raised, shouting. Although according to the rules he was meant to protect his train in the rear after an unscheduled stop of five minutes or more, he had not done so at Basin Bridge or at Perambur.

Assistant Stationmaster T.G. Balaraman at Perambur testified that train No. 19 had stopped on the Down Fast line at 20:05, although the Down Starting signal No. F 60 had been green. Its driver had immediately come to his office to report the failure of his engine and had been speaking on the telephone to the power controller when they heard the sound of the collision at 20:09. Half an hour after the collision he had seen that Signal No. F 56 was displaying a red aspect; signal No. F 60 was still green. The night had been clear and dark.

Driver Chinnappan of train No. 1 stated that the brakes, headlight, and speedometer of his engine had all been on order. He had left Madras Central at 20:00 and been routed onto the Down Fast line at Veysarpadi. The next three signals on the Down Fast line had all been green but the second signal had changed from green to yellow to red, then back to yellow and then to green several times and suspecting that something was wrong he had reduced speed. Seeing an obstruction ahead he had then applied the vacuum brake and his first fireman the steam brake. In reply to questioning by Additional Commissioner H.S. Hart he said that he had not mentioned the flickering signals immediately after the collision because he had been suffering from shock; he had not encountered such a thing before and he had not seen the tail lamp of the train in front.

First Fireman P. Velappan of train No. 1 confirmed that they had had green signals after Veysarpadi. He had seen a dim red light at the rear of train No. 19; no side lamps had been burning and he had not seen any danger signal. He had never seen the colourlights flickering. His driver had appeared normal and had not been intoxicated.

Additional First Fireman E.V. Selvaraj of train No. 1 agreed that the engine brakes had been in order and his driver had not been intoxicated. He said that the first and third automatic signals after Veysarpadi had been green but he had not seen the second as he had been assisting with the firing of the WP class steam locomotive. His driver had suddenly shouted that there was danger, when he had seen the dim red tail lamp of train No. 19 about 2½ bogies ahead. It was he who had applied the steam brakes.

Second Fireman P. Nachimuthu said he had seen two green lights beyond Veysarpadi but he was not sure if they had been on the Fast or Slow lines.

Guard V.R. Sankaran said that he had not looked out for any signals after Veysarpadi.

As might be expected extensive tests of the signalling were carried out after the accident.

Additional Commissioner Hart accepted that the collision had occurred at 20:09. Train No. 19 had been stationary and the speed of train No. 1 had been 50 kilometres an hour at the moment of impact, according to the speed recorder. The evidence of all the witnesses other than the driver of train No. 1 was that the electric tail lamp of train No. 19 had been lit. It was clear that the signalling had been working correctly prior to the arrival of train

No. 19 and there was no evidence of any tampering with it. The additional commissioner concluded that all the evidence pointed to train No. 1 having passed automatic signals Nos. 159 and F 56 at danger without stopping and that there had been nothing wrong with the brakes of the train.

Driver Chinnappan was, the additional commissioner concluded, directly responsible for the collision because he had failed to stop at signals 159 and F 56. The failure of Firemen Velappan and Selvaraj to correctly identify the aspects of the signals, which must have been red, indicated that they had not been keeping a good lookout as required by the rules.

As train No. 19 had only been stopped for four minutes at Perambur Guard Hameed could not be held responsible for having failed to protect it in the rear.

The eyesight of Driver Chinnappan had deteriorated such that it was poor even using strong glasses, though his colour vision was normal. This might have been a factor in the accident.

The additional commissioner made two main recommendations and no less than 14 additional observations and recommendations. His main recommendations were:

(1) The interval for the retesting of the vision of drivers over 45 should be reduced from a year to 6 months.

(2) Drivers should be tested on their knowledge of a road they had not been over for a month and required to relearn a road they had not been over for 6 months.

The additional observations and recommendations were:

(1) Since all Down trains negotiated a turn-out at Veysarpadi automatic signals Nos. 62 and 107 should only display double or single yellow and not green.

(2) The layout at Veysarpadi should be re-aligned to eliminate the turn-outs.

(3) The emergency crossover at Veysarpadi was in poor condition and should either be put into a better state or removed.

(4) The trough flooring of through girder bridge No. 27 between Veysarpadi and Perambur should be scraped and painted every three years and the ballast needed screening.

(5) Several drivers had refused to work with First Fireman E.V. Selvaraj, whom they considered unreliable. This should be looked into.

(6) The accident had obstructed the Up Fast line but as the wreckage had not contacted the rails, Up Fast line signal No 191 had continued to show a green aspect, despite which the line had not been protected.

(7) Track circuit test cards were not being maintained.

(8) Adequate insulation joints should be supplied.

(9) Signalling equipment should be tested and maintained fortnightly as per the manual.

(10) It was desirable that track-feed resistance adjustments should not be carried out by staff below the rank of assistant signal inspector and they should be recorded in a special register.

(11) Track relays should work to the specified rating, requiring a higher value of track feed resistance.

(12) A higher grade of signal maintainers should be provided.

(13) The design of Integral Coach Factory carriages, one of which had buckled in the accident, should be modified and the ends of carriages strengthened.

(14) Train No. 1 had taken the turn-out at Veysarpadi at 32 kilometres an hour, thus exceeding the permanent speed restriction there of 15 kilometres an hour. [246]

The same day as the disaster in India, Saturday 31st October 1970, a rear collision occurred in South Africa, at Dube station, near Johannesburg, the location of a previous bad accident at the end of May 1966 (see above). Two passenger trains carrying black workers from Soweto township into Johannesburg were involved in the crash, which happened just before 7 a.m. The first train was stationary in the station when the second ran into it. Fourteen passengers died in the accident and about 230 were injured, mostly slightly.

After the accident about 3,000 Africans gathered at the scene to protest about the high number of accidents on the line, in which 65 people had died and nearly 900 had been injured in less than two years. This was, of course, also during the apartheid era, and the protests were doubtless not just about the poor safety record of the railway but about the gross injustices of the apartheid system in general. Ambulance men who were struggling to release the (white) driver of the colliding train, who was trapped in his cab, were stoned by protesters, who cried: "Why always the Bantus? Why not a white train? We are tired of our people getting killed." A police riot squad equipped with machine guns, rifles and dogs was called in to disperse the protesters. Apart from the driver the only whites on the train were the conductor and the ticket collector, both of whom were slightly hurt. The injured had all been taken to hospital by noon.

'Whistles, boos, and shouts broke from hundreds of onlookers when they saw a railway policeman throw a black news photographer to the ground. Tempers flared again when a police dog attacked a black, tearing his clothes and grazing his leg. [247]

'A white official in the standing train said he heard a grinding sound just before the impact. "I was thrown to the floor and I heard screaming coming from the passengers," he said.

'Another uninjured passenger, Alfred Mkwedi, said after the collision he jumped from the train he was in and ran forward to where two coaches had been torn apart. "There I found screaming, blood-splattered people," he said.

'A passenger in the wreck, John Ratsoma, who suffered a serious leg injury, said: "I was dreaming happily. Suddenly my world was blown apart. I thought somebody had put a bomb in the first carriage, the bang was so big. People everywhere screamed as the train seemed to jump off the track. All I could think of was to get out. People everywhere were

[246] Government of India Ministry of Tourism and Civil Aviation (Commission of Railway Safety): Report on the Collision between No. 1 Down Madras-Mangalore Mail and No. 19 Down Madras Cochin Mail at Perambur Station of Southern Railway on 31st October 1970

[247] *Gazette Telegraph*, Colorado Springs, Colorado, United States, 2nd November 1970

wriggling from the windows like mince coming out of a sausage machine." [248]

'Johannesburg mayor Sam Moss called for an immediate judicial inquiry into the disaster, saying: "I am deeply concerned with the frequency with which these accidents have been happening in Soweto. The railway system there is totally inadequate to cope with the huge population and something must be done about the situation and soon."'

Meanwhile, 'The opposition United Party's transport spokesman, Marius Steyn, also criticised congestion on the line and added: "We have warned the government in Parliament about this again and again."' [249]

The Cape Town reporter for a British newspaper wrote, 'Stones and bottles were thrown at a large force of railway police who arrived with dogs and firearms. The angry Africans refused to go home for several hours, in spite of appeals from their own leaders, the police, and the Mayor of Johannesburg.

'The Soweto railway, one of the most heavily used stretches in southern Africa, is the main transport link for thousands of Africans who work in Johannesburg and live in Soweto, which has a population of more than 750,000. During the rush hour, trains often travel at extremely short intervals, with twice as many people on board as they were built to carry.

'The Government, which recently announced that a large sum of money would be spent in the near future on improving the line, will open an inquiry into the crash tomorrow.' [250]

A severe thunderstorm with heavy rain was one of the factors in a rear collision in Portugal early on Friday 11th December 1970. A mail train consisting of 16 coaches left Santa Apollonia station in Lisboa just after midnight on the Norte line for Porto. [251] It had only travelled about 12 kilometres when it came to a stand in the section between São João de Talha and Póvoa de Santa Iria. The reason for its having stopped out of course was stated later by its driver to have been the failure of its headlight, in conditions of very poor visibility due to the heavy rain, which, he had considered, would have made it dangerous to continue without being able to see the track ahead.

A few minutes behind the mail train was a 12-coach local train from Lisboa Rossio to Santarém. [252] For unknown reasons this train overran the signals that should have protected the mail train and ran into the rear of it, all but destroying the rear car of the mail train and the motor unit at the head of the colliding train. There were about 400 people on the two trains, of whom 13 were killed and about 35 were injured. The motorman of the local train only apparently realised the danger at the last moment, when he had made an emergency brake application. [253] Motorman António do Espírito Santo Tavares and Conductor José Canoso of the local train were amongst those who lost their lives, so no explanation of his disregard of the signals on a straight stretch of track was

[248] *Winnipeg Free Press*, Winnipeg, Manitoba, Canada, 3rd November 1970
[249] *The Anderson Herald*, Anderson, Indiana, United States, 4th November 1970
[250] *The Times*, London, England, 2nd November 1970
[251] Ibid, 12th December 1970
[252] *ABC*, Madrid, Spain, 12th December 1970
[253] *The Troy Record*, Troy, New York, United States, 12th December 1970

obtainable. A wrong-side failure of the signalling was quickly ruled out.

Two security guards at the Robbialac factory in São João de Talha witnessed the accident and raised the alarm. One of them, Sebastião Pedro Roque, said, "At first we thought that it {the sound of the collision} was thunder. It was a powerful thunderstorm. But then the lights of the second train suddenly went out and almost at once we began to hear screams."

The accident happened at an easily accessible location and close to the Portuguese capital, so the rescue work was quickly in hand, with doctors, nurses, firemen, railwaymen, and police soon at the spot. Soldiers going home on leave who had been passengers on the mail train also helped in the rescue operations, while other assistance was given by 50 men of the Guarda Nacional Republicana from Santa Barbara and shipyard workers from Lisnave. The soldiers who were travelling as passengers on the mail train had been returning home after two years of fighting against an independence movement in Portuguese Guinea (which eventually became independent in 1974 as Guinea-Bisseau). Floodlights were set up to illuminate the area for the rescuers. A relief train was also sent to the scene. Nine of the dead and 28 injured were taken to the Hospital de Santa Maria and four dead and the other injured were taken to the Hospital de São José.

The usual sad stories featured in newspaper accounts of the accident. Some newspaper reporters, having visited the site of the collision, then made their way to Santa Maria hospital. Here they were approached by an older gentleman with grey hair and a moustache, who asked them, "Forgive me, you people should know, where is our daughter?"

His wife, in tears and humble, added, "I have a presentiment, please don't lie. Where is our daughter? If she were alive, she would have already contacted us by now. She had long hair and was wearing black trousers and a short dark coat. She used to travel in the first coach of the suburban train."

"She's 23 years old. Her name is Estrida Pereira Pinto," her husband said.

"If something bad had happened to your daughter you would surely have been told by now," the reporters reassured the worried couple. Sadly Estrida's body had at that point already been in a refrigerated drawer at the mortuary for three hours. She was the only passenger to lose her life in the front coach of the colliding train.

It was hoped to bring one line back into use at 13:00 but clearance of the wrecked trains from the other line was a longer task. It was initially proposed to use a powerful diesel locomotive, sent from Encontramento, to pull the wreckage apart, but efforts to do this were abandoned after half an hour because the operation was causing too much damage to the permanent way. Instead work began to cut up the wreckage into manageable pieces with oxy-acetylene equipment. Before this was done passengers' possessions were recovered; a reporter wrote that these included a notebook belonging to a student at the Instituto Industrial de Lisboa and a bloodstained pack of playing cards – mostly the suit of clubs.

Various officials, including the Secretary of State for Transport, Engineer Oliveira Martins, later visited the site. The minister said that from the condition of the wreckage it appeared that at the moment of impact the local train must still have been travelling at 95 to 100 kilometres an hour. Caminhos de Ferro Portugueses (CP) engineers, on the other hand, pointed out that the stock of the local train had been much lighter than that of the mail train so the collision speed was probably not as high as the minister suggested. A CP inspector suggested to Rui Sanches, the Minister of Communications, who was also present, that the accident had clearly been due to human error, but he was interrupted by Engineer Martins, who pointed out that such a conclusion could only be reached after the completion of the inquiry into the disaster. [254] The officials later went on to visit the injured in the two hospitals.

Both lines were back in operation at 08:40 the following day. About a hundred local and long-distance trains had been affected, or about 25,000 passengers. Sixteen buses were used to transfer passengers between Sacavém and Póvoa de Santa Iria stations, while freight trains were stopped at various points along the line. By the day after the accident only eight injured patients remained in the Hospital de Santa Maria and only one in the Hospital de São José – a 15-year-old boy suffering from severe shock. [255]

On the evening of Eerste Pinksterdag (Whit Sunday), 30th May 1971, there were slight delays to trains outside of Amsterdam Central station. Due to the number of passengers alighting and boarding a local train from Utrecht was stopped longer than usual in Amsterdam Amstel station. As a result train No. 296, an international express from Genova, Italy, was stopped by signals behind it, while behind this train was an intercity working from Maastricht, train No. 862. Protecting the Genova express, signal No. 512 stood at Danger. [256] The driver of the Genova express had just contacted the signalling centre using a signal post telephone to enquire about the delay and was returning to his cab when his train was run into from behind by the intercity working from Maastricht at a point between Duivendrecht and Amsterdam Amstel and close a viaduct over the Sparkerweg. The collision was a violent one and the rearmost of the five coaches of the Genova express was described as having mounted the engine of the Maastricht train 'like a hood,' while the dining car and the coach marshalled third and fourth in the train were also considerable damaged. Damage to the colliding train, on the other hand, was largely limited to the engine and broken coach windows, while its driver escaped with a slight head injury and shock. Seeing the impending collision he had released the 'dead man's handle,' then vacated his cab and gone back into the interior of his engine, where he had crouched down to await the inevitable impact. The collision occurred at about 18:55. The Genova express had been due in Amsterdam Central at 18:52 and the Maastricht train was due there at 19:02.

At the moment the accident happened 24-year-old fireman R.J.G. Neering had been visiting his parents at their home in Duivendrecht. Glancing out of the window of their living room he had seen the stationary Genova express and then became witness to the actual collision, so he was able at once to telephone the local fire brigade and police. The Duivendrecht, Ouder-Amstel and Amsterdam fire brigades, police, and the Amsterdam GGD (Gemeentelijke Gezondgeidsfienst – Municipal Health Service) were thus quickly on the scene. Motorists took some of the injured to hospital and a relief train from Amsterdam Central brought uninjured passengers from both trains there. The Salvation Army and local people provided coffee, soup, and hot drinks for the rescuers. Less useful

---

[254] *Diário de Lisboa*, Lisboa, Portugal, 11th December 1970
[255] Ibid, 12th December 1970
[256] *Spoorwegongevallen in Nederland 1839-1993* by R.T. Jongerius, pg. 177

was the appearance of large numbers of sightseers ('rubberneckers') whose parked cars sometimes hindered the progress of fire engines and ambulances. The release of the dead and injured passengers from the rear coach of the Genova express was no easy task, even with the use of oxy-acetylene torches, and the last two passengers, both deceased, were still trapped at 23:00.

Travelling in the fatal last coach of the Genova train had been a party of American students from the University of Maryland, on a tour of Europe. Two of them, Maureen Priddy, 17, from La Vale, Maryland; and Elvin Ed Thomas, 18, from Texas, were amongst those who lost their lives, while several other students were injured. Of the other three passengers who died, two were soon identified as Tietske Scheffer, 21, a typist from Lelystad, the Netherlands, and her fiancée, P.F. Bissen, 23, from Luxembourg. The body of the fifth victim was badly mutilated and was not immediately identified. Another 33 passengers were more or less hurt.

Fifty men worked throughout the night by the light of emergency generators to clear the line. Meanwhile some trains were diverted via Hilversum and the Nederlandse Spoorwegen (NS) arranged for a shuttle service between Amsterdam and Utrecht. At 05:00 on Monday morning the rear Italian coach, which had suffered most in the accident, was pulled free from the engine that had struck it and deposited on the embankment. The colliding locomotive, although considerably damaged, was, it was thought, repairable. Fifty metres of track was also damaged but the line was partially reopened to traffic at 05:55, with trains running at reduced speed.

The tachograph of the engine of the Maastricht train indicated that it had been travelling at 90 kilometres an hour when it hit the Genova express. Its driver, who was identified only as Willem H., had apparently run past yellow and red signals without reducing speed at all, until he had made an emergency brake application 250 metres from the point of collision, reducing the speed of the train from 113 to 90 kilometres an hour. The signalling was, of course, subject to exhaustive testing after the accident but no fault could be found with it. Driver H. was 49 years old, had been with the NS for over 25 years and a driver for 17 years. On the day of the accident he had begun his shift at 17:30 by working a train from Amsterdam to Utrecht. After half an hour's break he had taken over command of the Maastricht – Utrecht – Amsterdam – Zandvoort intercity train from a colleague at Utrecht just after 18:30. It was theorised early on that he might have been blinded by the low-lying sun and hence failed to see the signals but he insisted that they had all been green for him.[257 258 259]

Driver H. stood trial in Amsterdam for having caused the accident in April 1972. He pleaded that at the Bijimermeer he had passed a signal, which had been green (it should have been yellow), and at 1,500 metres a second signal (No. 512), which had also been green (it should have been red). Immediately afterwards he had seen another train. "I thought it was an oncoming train. Only when I saw that it was on my line did I apply the brakes in emergency," the driver told Mr B van der Waerden, the president of the court.

Driver A. Jansen of the Genova express stated, "If you start worrying about whether you're on the right line and whether or not the train that you see in front of you is an

[257] *Leeuwarder Courant*, Leeuwarden, The Netherlands, 1st June 1971
[258] *Tubantia*, Enschede, The Netherlands, 1st June 1971
[259] *De Volksrant*, Amsterdam, The Netherlands, 1st June 1971

oncoming one, you have no life left."

"Does this mean that you have to rely completely on the signals?" Counsel A.J. Mondria asked for the defence.

"Yes, you have to," the witness replied.

Driver Blijkerk, testifying for the defence, said, "If you see a train in front of you and you cannot see the three white lights at the front of it you do not immediately become alarmed. The signals of oncoming trains are sometimes only faint due to weak batteries. If you are running under green signals you don't expect to find a train in front of you. By relying on the signal you will go on a long way in such a situation. You realise that there should be no train on your line."

Mr F.H. von Meyenfeldt, speaking for the prosecution, criticised the NS for having already disciplined the driver (who had been given a severe reprimand) stating that it would have been better if they had waited for the outcome of the judicial investigation. There could be little doubt, Mr Meyenfeldt stated, that the signals had been against the Maastricht train. But in any case there had been 1,200 metres of straight track behind the stationary train. At full speed the colliding train should have been able to stop in 675 metres and even allowing for it having taken the driver ten seconds to recover from the shock of seeing a train in front of him, there should have been plenty of time for him to have pulled up. It appeared, therefore, that he had not applied the brakes, or at any rate had only done so far too late. In view of Driver H.'s good record, the fact that he had been deeply affected by the disaster and that he would carry a sense of guilt for the rest of his life, the prosecutor announced that he would not demand a custodial sentence for the defendant, but only a fine of 1,500 guilders.

Mr Mondria pointed out that despite its supposed technical perfection problems had occurred even with the Apollo-16 space module (which had suffered technical problems during its trip to the moon a few days before the trial). It was NS dogma that any failure of the signalling system placed all the signals at danger but two defence witnesses had spoken of 'wrong side' signal failures and signal malfunctions occurred every day. There was therefore always the possibility of an unexplained 'wrong side' failure, while train drivers absolutely depended on the signals.

Driver H., described as a married man with three children, had little to say. He had been reinstated as a driver but only given shunting work. "I've already asked a few times if I can drive normally again," he said. [260]

On 9th May 1972 Driver H. was fined 1,000 guilders. He had maintained throughout that he had had two green signals but the court based its judgement that he had overrun signals on research conducted by Professor De Kroes of Delph on behalf of the NS, which had concluded that the signals had worked perfectly. [261]

Early on Saturday 5th June 1971 passenger train No. 184 Down was standing at the Home signal at Gonda Junction, India, on the North Eastern Railway, when it was run into from behind by passenger train No. 32 Down. Five passengers were killed outright

---

[260] *De Telegraaf*, Amsterdam, The Netherlands, 28th April 1972
[261] *Leeuwarder Courant*, Leeuwarden, The Netherlands, 9th May 1972

and 13 were injured, three fatally and six seriously. Train No. 184 Down consisted of a YP Class engine, three waggons, 12 coaches and three inspection carriages, while train No. 32 Down consisted of another YP Class engine and 16 coaches.

Gonda Junction was a Class B junction station. To its west lay the single-track main line to Gonda Kachahri and a branch line to Bangain. To the east the single-track main line ran to Baruachak with a branch to Subhagpur. The station had four platform lines and there were West and Central cabins concerned with the signalling of trains from Gonda Kachahri. The latter was another Class B station with just a Central cabin. The line between Gonda Kachahri and Gonda Junction was worked on the absolute block system with tokenless block, continuous track circuiting and semaphore signalling.

Driver Bindeshwari told the inquiry into the accident that he had been in charge of the engine pulling train No. 32 Down. The Down Distant at Gonda Kachahri had been green, the Main Home had been yellow and he had seen as he reached the station that the Down Main Starter and Advanced Starter were 'Off.' He had stopped at the station for four minutes and left there at 03:42. The Distant signals at Gate No. 262 A and Gonda Junction had both been unlit and their arms at 90 degrees ('Off'). He had been looking out for the Gonda Junction Home signal when he had seen an obstruction ahead at short range. He alleged that the Starting signals at Gonda Kachahri might have been off due to a defect in the signalling and disputed the statement of the first fireman that the Gonda Junction distant had been lit, the speed of his train implied by the bookings of the guard and the statement of the gateman at Gonda Kachahri that he had overrun the Departure signals there.

First Fireman Hamid Hussain of train No. 32 corroborated the statement of his driver except that he said that the two Distant signals near Gonda Junction had been lit.

Second Fireman B.B. Singh of train No. 32 said that he had been busy breaking coal and had seen no signals other than the Departure signal at Gonda Kachahri, which had been 'Off.' He put the speed of the train at the moment of impact at 30 kilometres an hour (his driver had said 15).

Guard H.K. Mathur of train No. 32 stated that he had signalled his train away from Gonda Kachahri on seeing the Starting signal 'Off.' He had not seen the Advanced Starter. He denied having told the Assistant Operating Superintendent that he had not seen the Starting signal before signalling the train away.

Brakesman H. Lal of train No. 32 said that he had not seen the Departure signals and denied having previously stated that he had seen them 'Off.'

Driver Manohar Lal of train No. 184 testified that all the signals at Gonda Kachahri had been clear and the Gonda Junction signals had all been lit. He had stopped at the Gonda Junction Home which had been 'On' and after waiting there for five minutes had sent his second fireman to the West cabin to find out why he was being held. At about 03:51 there had been a severe bump followed by three jerks. On finding the cause he had sent the first fireman to the West cabin to report the collision.

Guard O.P. Varma of train No. 184 said the tail lamp of his train had been burning well. He put the time of the collision at 03:53.

Relieving Assistant Stationmaster Mohammed Hafiz Khan at Gonda Kachahri stated that train No. 184 had passed through at 03:34 and the Home and Distant signals had then returned to the 'On' position. At 03:35 he had given *Line Clear* to Kathola for train No. 32. At 03:46 he had heard train No. 32 leaving the station despite the signals not having been cleared for it and at the same time the cabinman had rung him to ask where the train was going. He had rushed out to stop it but it was too late. He had then informed Gonda West cabin in the hope that they could stop it at Gate No. 262, then informed Control, then sent for the stationmaster.

Acting Stationmaster S. Dubey at Gonda Kachahri said that on being informed of the accident at 04:10 he had proceeded to the station. He had found all the levers in the cabin normal and the indicator for the Down Advanced Starting signal had been red. In the stationmaster's office the indicators on the tokenless panel were:

Down train on line; switch on TGT position; advanced starter light red; cancellation counter number as last recorded in the train signal register; all slides in SM's control frame normal.

{TGT = Train Going to, SM = Stationmaster}

Cabinman Angnoo at Gonda Kachahri testified that he had replaced all of his signals after the departure of train No. 184. Five minutes later the assistant stationmaster had advised him that train No. 32 would be received on line No. 2 and he had cleared his Reception signals for it. This train had arrived and after a couple of minutes had started away again, although the signals had been against it. On his enquiring of the assistant stationmaster what was happening, the latter had rushed out to try and stop the train, while he, the cabinman, had tried to contact the gateman at Crossing No 264 B near the Down Departure (Starting) signal and had shown a red flag to the train. He had not known of any failure of the signals in ten years.

Gateman Ali Raza at Crossing No. 264 B stated that he had seen the Departure signals go to 'On' after the passing of train No. 184, the tail lamp of which had been burning brightly. He had been outside his gate lodge when train No. 32 had left and thus had not heard the telephone call from the cabinman. The Departure signals had been 'On' after train No. 32 had passed.

Leave Reserve Assistant Stationmaster K.D. Pandey in Gonda West cabin explained that he had been unable to reverse crossover No. 30 after the departure of train No. 17 Up for the reception of train No. 184. The signal maintainer had been sent for at 03:25 and meanwhile train No. 184 had been detained outside the station. At about 03:50 he had been advised by Gonda Kachahri that train No. 32 had entered the block section without authority. Within a minute he had contacted the gateman at Crossing No. 262 A but the train had already passed there. Shortly afterwards, the second fireman of train No. 184 had arrived at his cabin to ask why his train was being detained, followed by the first fireman, reporting the collision.

Gateman Mangal at Crossing No. 262 A stated that he had closed his gates and cleared his Down gate signals for train No. 184. That train had passed, he had opened his gates, then when he had seen the headlights of another Down train he had closed his gates again and cleared his crossing signal. Just after the second train had passed, the West cabin had asked him to try and stop it. Due to a curve in the line he had been unable to see train No. 184 stopped at the Down Home signal ahead. The lights of all his signals had been burning.

A thorough investigation of the signalling was carried out.

Additional Commissioner G.S. Pandor put the time of the collision at 03:53. He concluded that the speed of train No. 32 must have been at least 30 to 35 kilometres an hour. There could be no doubt, he stated, that train No. 184 had entered the Gonda Kachahri – Gonda West block section on proper authority and had been stopped at the Gonda Junction Home signal due to a problem in setting the route ahead. Train No. 32 had been following close behind and received at Gonda Kachahri on Line No. 2. According to the crew of train No. 32 the signals had been clear for them to leave, while according to the station staff there the signals had been at danger. Both were interested parties.

There had been no reported wrong-side failures of the signalling at Gonda Kachahri before the accident. The panel there had failed on 29th May due to heavy rain causing track circuit failures. Odd failures of the Advanced Starting signal at Burhwal, causing it to stick 'Off,' had been due to faulty equipment. Two failures of the Down Distant signal at Gonda Kachahri, one due to excessive rusting and the other to a broken contact, had prevented the arm being returned to the 'On' position. In these circumstances the proving of the signal arm would have prevented *Line Clear* being given for another train.

The tokenless block panel at Gonda Kachahri and assorted equipment had been in good order. There was no indication of the Down Advanced Starting signal having stuck 'Off.' Had it done so the cabinman would not have been able to normalise lever No. 13, which worked it. It would not therefore have been possible for the Starting and Advanced Starting signals to have been 'Off' for train No. 32 with the line ahead occupied. At the same time it was difficult to believe that a driver of long experience, on the verge of retirement, would have ignored both signals.

The additional commissioner outlined what he believed had been the course of events leading to the collision: Having instructed the cabinman to obtain *Line Clear* for train No. 32 as soon as train No. 184 had cleared, the assistant stationmaster at Gonda Kachahri had left his office to go to the brake van of train No. 32 (as stated by three railwaymen but denied by the stationmaster). The cabinman had then left his box, omitting to normalise the Departure (Starting) signals behind train No. 184. The driver of train No. 32, seeing the Starting signal 'Off,' had set off and failed to observe the Advanced Starting signal. Full responsibility for the accident therefore rested upon him. Returning to his office the assistant stationmaster, not realising that anything was wrong, had reported the train as 'In' and 'Out' to Control. He had then asked Gonda West cabin for a private number to complete the record in his train register, when Gonda West had queried why he wanted one as he already had private number 12 for train No. 184. It was only at that point that the assistant stationmaster had realised what had happened and tried to get Gonda West to stop train No. 32 at Crossing No. 262 A.

Additional Commissioner Pandor concluded that Driver Bindeshwari had been guilty of gross negligence in passing the Advanced Starting signal at Gonda Kachahri at danger while Cabinman Angmoo had failed to normalise the levers of the Departure signals after train No. 184 had passed in contravention of General Rule 36 (a).

The assistant commissioner made five recommendations and remarks:

(1) The Starting signals on tokenless block sections should be provided with electrical reversers so that when the Advanced Starting signal automatically returned to the 'On' position the Starting signals did too.

(2) Investigations by the signal inspector into the causes of failures were not being carried out.

(3) All failures in tokenless block sections were being attended to and rectified by the maintainers. It was desirable to issue instructions regarding investigations and the rectification of failures at the appropriate supervisory level.

(4) The normal practice to work a line relay at not more than twice its pick up voltage should be followed.

(5) Trials should be carried out to achieve a minimum value of 4 ohms per kilometre of ballast resistance on sections worked with coded track circuits.

The report added that Driver Bindeshwari and Cabinman Angmoo had stood trial in connection with the accident and had both been acquitted. [262]

Talgo passenger train No. 5475, which had left Malaga at 13:15 and was due in Madrid at 20:55, collided with a freight train between Marmolejo and Arjonilla, Spain, and about three kilometres on the Madrid side Marmolejo, at a point in the line where there was a road underpass and a curve. The accident happened at 16:35 on Saturday 11th September 1971. The Talgo consisted of the *Virgen Peregrina* locomotive, a generator car, four first-class coaches, seven second-class coaches and a rear van, and at the time of the accident was carrying 283 passengers and five railwaymen. (The Virgen Peregrina – Pilgrim Virgin – is a patron saint of the Catholic church.) The freight train consisted of an engine and 32 waggons, the last eight of which were empty tank cars.

The Talgo was travelling at between 120 and 130 kilometres an hour as it passed through Marmolejo station, though its passengers and a railwayman at Marmolejo declared that its driver had made three heavy brake applications between there and the point of collision, reducing the speed of the Talgo to 50 kilometres an hour, while the freight train was moving at 20 kilometres an hour. It was later concluded that the driver had also applied sand to the rails to help brake his train more quickly.

As a result of the collision the engine, generator car and leading first-class of the Talgo ended up beneath the rear two tank cars of the freight train. The next four waggons of the freight train were derailed and three of them were overturned, while the generator car and leading first-class coach of the Talgo were crushed. The Talgo driver, the assistant driver and four first-class passengers were killed; two passengers were gravely hurt, 12 were seriously hurt, seven were less seriously hurt and four escaped with slight injuries.

The collision was blamed on the driver of the Talgo having overrun the stop signal at the Madrid end of Marmolejo station because he had mistakenly read the signal at the Cordoba end of the station as displaying green and yellow (Proceed at reduced speed) whereas it had in fact been displaying yellow (Prepare to stop at next signal). Although the line was signalled by Centralised Traffic Control it was not yet equipped with AFSA [263] (Anucio de Señales y Frenado Automático – Announcement of Signals and Automatic

[262] Government of India Ministry of Tourism and Civil Aviation (Commission of Railway Safety): Report on the Rear-End Collision of No. 32 Down Kanpur Passenger Train with No. 184 Down Passenger Train between Gonda Kachahri and Gonda Jn. Stations of North Eastern Railway on 5th June 1971

[263] Federación Castellano Manchega de Amigos del Ferrocarril

Braking – a system of automatic train protection designed to stop a train should a driver overrun signals, which was being introduced on Spanish railways about this time).[264]

Press reports of an accident in Greece on Thursday 23rd September 1971 were more than usually confused, with reports of either a rear collision or a head-on collision, involving a local train and either an express passenger train or a freight train and resulting in three, four, five, six or seven fatalities. The evening collision occurred at Dekéleia, about 30 kilometres north of Athens. It appears most likely that the diesel-hauled local train ran into the rear of a Salonika – Athens freight train, the engine climbing up onto the rear three waggons of the freight, and that at least six people died and about 25 were injured. The injured were taken to hospital in Athens.[265] [266] [267] [268]

Brazil suffered a major disaster at about 07:47 on the morning of Thursday 8th June 1972. The accident occurred on a former line of the Estrada do Central do Brasil, now part of Rede Ferroviária Federal. Train No. UPE-2 was an electrically hauled service popularly known as the students' train because it mostly carried students on their way from São Paulo to attend university at Moji das Cruzes. It stopped at all stations after Itaquera: Parada 15, Ferraz, Poá, Calman Vianna, Suzana, Jundiapeba, Brás Cubas before Moji das Cruzes. On the day of the accident it came to a stand in the section between Suzano and Jundiapeba stations because there had been a failure of the electricity supply. While it was still stationary, waiting for the power to come back, it was run into by train No. SP-2 from São Paulo to Rio de Janeiro, the *Litorina*, also known as Trem de Baianos because it carried many internal immigrants from Minas Gerais state. (Baiano was a derogatory term used in southern Brazil for people from the northern parts.) Train SP-2, being diesel hauled, had not been affected by the power failure. Twenty-four people, comprising Driver Geraldo Couto, Assistant Driver João Modesto Azevedo, Conductor Ulisses da Silva and 21 passengers, lost their lives and over 100 others were injured, many seriously. The leading mail van of the express was derailed and telescoped between the engine and the coach behind it. Some of the passengers in the express were killed or injured but the majority of the deaths and injuries occurred in the rear two coaches of the students' train.

Firemen from Moji das Cruzes were the first to arrive at the scene and were soon joined by others from São Paulo and from Santa André, by military police and by soldiers from the Tiro de Guerra da Moji das Cruzes. Ambulances, lorries, and private cars took the injured to hospitals and later helicopters flew the most serious cases to the Hospital das Clinicas in São Paulo. Firemen worked for some hours to free all the dead and injured and the work to clear the line went on into the night. Shocked by the accident the residents of Moji das Cruzes helped as best they could with medicines, sheets, sandwiches and soft drinks, the stock of several local chemists running out.

The 40 students in the rear coach of train No. UPE-2 were playing cards or sticks or

---

[264] Wikipedia – Anuncio de Señales y Frenado Automático
[265] *European Stars and Stripes*, Darmstadt, Germany, 25th December 1971
[266] *The Daily Gleaner*, Kingston, Jamaica, 24th December 1971
[267] *The Janesville Gazette*, Janesville, Wisconsin, United States, 24th December 1971
[268] *The Times*, London, England, 24th December 1971

catching up on some sleep. The train had been stopped 15 minutes at kilometre 44 but being used to such things happening none of the students were very concerned. Student José Carlos Cadoso dos Santos, a 19-year-old first-year economics student at Brás Cubas, told a reporter afterwards, "We were yelling, 'Agora vai, agora vai!' (Now go, now go!)" He said that none of the students had been concerned about being late for their classes as most of them came on the train, so their classes would not begin until those coming by train had arrived.

How the accident had likely happened quickly became apparent. It appeared that when train SP-2 had arrived at Suzano it had come to a stand at the starting signal, which, according to another railwayman at the station, Esperidião Jonas de Prado, had been red at that time. The train had only been stopped a couple of minutes when Agent José Carlos da Costa had handed Driver Couto a ticket authorising him to proceed, whereupon the train had started away. Whether or not the signal had also been cleared for the express or whether the driver had proceeded past the signal at red on the basis of the ticket that he had been given, was unclear. Conductor Ulisses da Silva, who survived long enough to be taken to hospital, told fifth-year medical student Jaimie Camargo, who was attending him, that the signal had been green. The line was, of course, not clear at all, and Agent da Costa had not received any telephone message from Jundiapeba that the students' train had reached there. The morning was foggy and travelling at 80 kilometres an hour Driver Couto could have had little warning of the train ahead.

The disaster received very extensive coverage in a São Paulo newspaper. It told the story of Lieutenant Expedito Pereira Lima, commander of the Suzano Military Police, who had arrived to teach a fourth-year law class in Moji and had been surprised to find that none of his 120 students had appeared by 08:00. He had been wondering if he had forgotten that it was a holiday when a colleague informed him of the accident, whereupon he had got into his car and driven to the scene.

Student Beethoven Tavares de Lima Junior had not been killed outright in the collision but trapped in the wreckage and badly hurt. A fellow student, Tasso Rosa Campos, hurried to the nearby works of Hoechst do Brasil, and returning with some cutting equipment worked feverishly to free his mate, but in less than five minutes Beethoven had expired.

By 11:00 all of the known injured passengers and railwaymen had been freed and taken away, though doctors and nurses remained at the scene, while buses picked up workers from the Hoechst factory who had offered to give blood. Meanwhile a breakdown train had arrived and was attempting to pull the engine of the express from the rear coach of the students' train.

At 13:00 a request came from the Hospital das Clinicas for firemen to retrieve the amputated left hand of student Fernando Antonio Duarte Leme, in the hope that surgeons might be able to reattach it. Fernando was left-handed and an excellent violin player. The hand was duly found and rushed to the hospital but the surgeons found that it was badly crushed and re-attachment was not possible. About the same time the father of Beethoven Tavares arrived at Santa Casa de Suzano. Informed of his son's death he refused to believe it and protested that there must have been some mistake.

The body of Driver Couto was finally released at 14:30.

For at least one family there was a relatively happy ending to the disaster. Among the students on the train had been 21-year-old Pedro Henrique Lucas Martins, studying engineering, and known among his family for his beautiful teeth. A news announcement broadcast on a local radio station wrongly stated that he had lost his life in the disaster. Hearing this his family rushed to the scene, travelling at 150 kilometres an hour along the Via Dutra (the main road between São Paulo and Rio de Janeiro). Searching the ambulances they found Pedro Henrique in one of them. Seeing his relatives, he exclaimed, “Hi, I’m here!” He had lost his teeth in the accident and had also sustained a broken collarbone but he was still very much alive.

Señor Alipio Monteiro, head of the public relations department of the railway, issued a statement describing how the disaster had occurred. He stated that train UPE-2 had left the Roosevelt station in São Paulo at 06:50 with about 800 passengers, followed by train SP-2. Going on to describe how the accident had occurred, as detailed above, he added that the line would soon be equipped with Centralised Traffic Control which should, at any rate, prevent a recurrence of the disaster.

A commission of inquiry was quickly opened to investigate the accident, chaired by Engineer Oswaldo Tavolaro, assisted by Engineer Carlos Olipio Nogueira and Lawyer Elizeu Soares da Silva.

José da Costa disappeared after the accident. It was believed that he was being held at a secret location so that he could give evidence at the railway’s internal inquiry. The Suzano police wished to interrogate him in connection with their investigation but the Suzano stationmaster, João Biu, refused to divulge da Costa’s whereabouts to them. Colleagues of José da Costa defended him, putting the blame for the disaster on the dead Driver Couto, who, of course, was no longer able to defend himself.

João Flavio da Costa, brother of José, said that the latter was 37 years old and had worked for the railway for 15 years, starting off in the ticket office before becoming an agent. He did not drink but suffered from severe asthma, for which he had had treatment in a Moji hospital a few days earlier, adding that he had taken over as agent from Jair de Lima Ferreira only on the Monday before the accident. José had a salary of 600 cruzeiros from the railway but was never very well off financially as his wife was disabled, suffering from a nervous condition which required a lot of expensive medicine. He had, his brother added, worked hard between his spells in hospital. [269]

At 00:16 on Friday 27th July 1973 Eastern Railway train No. 21, the Up *North Bihar Express* arrived at Madhupur, India, on line No. 1 and had to wait there for goods train No. JC-9 to clear Nawa Patra block hut. Behind train No. 21, train No. 5, the Up Howrah – Amritsar Mail, was stopped in turn at the Madhupur Outer Home signal at about 00:47. Following train No. 5 was No. 87, the Up *South Bihar Express*, which should therefore have been stopped at Joramow block hut. Instead this train entered the Joramow – Madhupur East block section and collided with the rear of train No. 5 at about 01:40. The rear coach of the forward train was smashed and two other coaches were badly damaged. Eighteen people were killed outright, two more were fatally injured, seven were seriously hurt and 48 were slightly hurt.

Train No. 5 consisted of a diesel engine and 16 coaches. Train No. 87 consisted of WP

[269] *Folha de São Paulo*, São Paulo, Brazil, 9th and 10th June 1972

class steam locomotive No. 7266 and 12 bogies (coaches).

Madhupur was a Class B station with East and West cabins and a slide control frame in the stationmaster's office. The signals at Madhupur and Joramow were lower quadrant semaphores. Absolute block working with Siemens General Electric instruments was in operation between Madankata block hut, Joramow block hut, and Madhupur East cabins.

Driver K.L. Purohit train No. 87 Up told the investigation into the accident that he had observed the Warner (Distant) signal at Joramow to be 'On' at close range and reduced speed. The Home signal had been green, he had seen a green light waved from inside the cabin and the Starter and Advanced Starter signals had also been green. He had then increased speed to 60 kilometres an hour. The Madhupur Warner had been red but he only closed the regulator and braked slightly in view of the rising gradient. Exiting a curve his fireman had warned him of a low red light, which he had also seen, and he had braked fully. Immediately there had been a loud report and a sudden jerk.

In answer to various questions Driver Purohit stated:

(a) Due to the heavy rain he had not been able to see the arms of either Starting signal at Joramow.

(b) His engine headlight had been burning brightly throughout.

(c) Two of his firemen had been repeating the signal aspects; the third had been breaking coal.

(d) At various locations the Warner had not been taken off although the stop signals had been clear.

(e) He had had nothing on his mind on the day of the accident.

(f) He had not noticed the explosion of any detonators before the accident.

(g) Neither he nor the firemen had seen anyone near the Joramow Advanced Starter.

(h) He had not seen the electric tail lamp or side lamps of train No. 5, only a dim tail lamp.

(i) He had been about 200 feet from the tail lamp of train No. 5 when he had seen it.

(j) He had worked train No. 87 since 1968.

(k) He had never seen the Joramow Starter 'Off' and the Advanced Starter red.

(l) His guard said that he had seen a stationary green hand signal from Joramow cabin.

(m) Tests showed that the tail lamp of train No. 5 should have been visible for 752 metres from the fireman's side and for 359 metres from the driver's side but he had not been able to see them from that distance.

(n) The Joramow Starter and Advanced Starter could only be seen after passing the Home.

(o) He did not think that despite the rain he could have mistaken the red light of the Joramow Advanced Starter for that of the Loop Starter.

Leading Fireman Mukdeo Ram of train No. 87 confirmed that he and First Fireman Arjun Prasad had been calling out the signal aspects to their driver. He had seen the green light but not the arm of the Joramow Advanced Starter. Upon his alerting his driver to the tail lamp of train No. 5, the latter had braked at once.

First Fireman Prasad generally corroborated the statement of his driver.

Guard B.N. Bhaduri stated that he had not seen any signals at Joramow, though he had seen a stationary green light from the cabin. He had been completing his records when a severe jerk alerted him to the collision.

Driver Ramanand Kurmi of train No. 5 testified that he had been stopped at Madhupur Outer Home signal at 00:42 and at 01:10 had sent his fireman to the East Cabin to find the cause of the stoppage. The latter had only returned about the time of the collision. The Joramow Warner had been 'On,' showing a red light with a fixed green above it. The Home for the Main Line, the Starter, and the Advanced Starter had been green and the switchman had waved a green light towards him. His vacuum brake had been released while waiting at Madhupur. The side lamp of his train had been burning upon leaving Chittaranjan. The Joramow signals had been clearly visible despite the rain.

Switchman Muhammad Khalil at Joramow stated that his clock had been out of order for the previous 10 days and he had to obtain the times to enter in his train register from either Madhupur East or Madankata cabins, or from Madhupur Control. Madhupur East had given him *Line Clear* for train No. 5 at 00:36 and it had passed him at 00:37. He had then turned all his signals to red. Madankata then asked *Line Clear* for train No. 87 which he had given. He had then lowered his Up Main Home signal and stood at the window. Train No. 87 had run past his Starter and Advanced Starter and kept going, whereupon he had tried to advise Madhupur East cabin on the block phone. This failing, he had told Asansol Control that train No. 87 had run through his signals. The Madhupur East switchman had also been on the phone and exclaimed, "How can the train come? I have not given line clear for the same yet. The train must have stopped ahead, you go out and check-up." Control had said the same thing, so he had gone out and checked as far as the Starter. Returning to tell Control that the train was not there, he had been told of the collision. He denied having shown any hand signal to the driver or guard of train No. 87. He could not give the correct order in which his signals had been replaced after the passage of train No. 5. In addition there was an entry in his train register book in the "Time Train Out of Section" column for train No. 5 from Madhupur East, which he explained by saying that his reliever had told him that he could not go until train No. 5 had cleared the block section, so he entered an estimated time, then realising that the entry would be wrong he had erased it.

Substitute Porter Abdul Latif stated that he had lit the signals at Joramow on the evening of 26th July, then slept in the cabin from 22:00. At 01:30 he had been woken to call up Switchman M.M. Mukherji.

Switchman Mukherji said that he had taken over from Switchman Khalil at 02:00. At the time the signal levers had been in their normal positions and the signal lights had been burning. The Joramow – Madhupur East block instrument had been at *Train on Line*. His colleague had told him that train No. 87 had overrun his signals and collided with train No. 5 outside of Madhupur. It had been drizzling but the signals were easily visible. He denied having asked his colleague to wait until train No. 5 had cleared the section. Instead, immediately after telling him of the accident Switchman Khalil had said that his brain was not working, had given him a sick memo, and had left. Only then had he (Mukherji) noticed the erasure in the train register. The hand lamp in his cabin had been on the window facing inside. There had been no lights near the Advanced Starter which could have confused the driver.

Section Controller A.N. Sen at Asansol explained that goods train JC-9 Up had been stopped at Nawa Patra block hut, train No. 21 in Madhupur station, and train No. 5 at Madhupur Outer Home signal because of a problem with the points at Mathurapur block hut, beyond Nawa Patra. He had warned the switchman at Joramow to stop train No. 87 but at 01:34 the latter had advised him that the train had run past his signals. He had asked the switchman if he had received *Line Clear* from Madhupur East for it and was told that he had not. Learning from the Madhupur East switchman that he had not received the slot for train No. 5, he told the latter that train No. 87 had run past Joramow. At 01:55 he had been advised by Madhupur that the fireman of train No. 87 had reported that his train had run into No. 5. He *had not enquired of the fireman as to the extent of the casualties* (this sentence is in italics in the accident report). Although he had been informed of train No. 87 overrunning signals at 01:34 he had not informed the Deputy Controller or Assistant Operating Superintendent of the imminent collision until he received the report of it at 01:55.

Assistant Signal Inspector R.K. Biswas stated that he had carried out regular inspection at Joramow and investigated seven failures since January; all had been right-side failures. Efforts to deliberately cause a wrong-side failure of the Advanced Starter during a routine inspection, on the day before the accident, had failed, the signal stubbornly returning to Danger.

Divisional Signal and Telecommunications Engineer Harbans Lal at Asansol described tests on the signalling he had performed after the accident.

Switchman B.B. Tanti at Madhupur East stated that train No. 21 had arrived at Madhupur on line No. 1 at 00:11, whereupon he had given *Line Clear* to Joramow for train No. 5, which had left there at 00:36, arriving at his Outer Home signal at 00:47. At about 01:40 he had heard what sounded like the bursting of a cracker but could see nothing. At 01:42 Control had advised him that train No. 87 had passed Joramow, which puzzled him, as he had not given *Line Clear* for it, and he told the Joramow switchman to check that the train was not still at his cabin. At 01:55 he had finally received the slot from the West cabin for train No. 5 and had just been pulling his Home signal when the fireman of train No. 87 had come running up to tell him of the collision. He had then reported the collision to the assistant stationmaster, Asansol Control, and the West cabin. He denied that the Joramow switchman had asked him any times or said anything about his clock not working.

Assistant Engineer A.K. Ganguly spoke of overhearing the driver of train No. 87 asking his guard if the latter had seen a green light from Joramow cabin and the latter answering in the affirmative. The driver had not appeared intoxicated.

Additional Commissioner S.K. Mojumder outlined tests that had been done to check the integrity of the lock and block working. He put the time of the collision at 01:40 and the speed of train No. 87 at about 50 kilometres an hour.

The collision must, the additional commissioner pointed out, have been due either to the Advanced Starter having been wrongly cleared for train No. 87 or to that train having passed said signal at Danger. The passage of train No. 5 would have automatically restored the Advanced Starter to Danger and with the block instrument at Madhupur East at *Train on Line* the signal could not have been lowered again. No signalling staff had been working in the area at the time, no one had been seen in the vicinity of the signal, and overall there was no evidence of an irregular clearing of the signal.

With regard to the Starter signal, on the other hand, there was no electrical control to prevent it from being cleared and the driver and two firemen had all declared that it was 'Off.' Switchman Khalil at Joramow had made fictitious entries in his train register book and displayed ignorance or confusion regarding the manipulation of his levers. The most likely sequence of events, therefore, was that Switchman Khalil had booked the clearance of train No. 5 and the receipt of *Line Clear* for train No. 87 in his register and pulled levers Nos. 9 and 10 (the Starter and Advanced Starter), whereupon the Starter, though not the Advanced Starter had cleared. As train No. 87 had passed his cabin he had given it a green hand signal and reported its passing to Control at 01:32. It was only when Control had queried his receiving *Line Clear* from Madhupur East for train No. 87 that Switchman Khalil had belatedly realised that he had not had it, and that the train had overrun his Advanced Starting signal.

Having seen the Home and Starter signals 'Off' the driver of train No. 87 had apparently taken it for granted that the Advanced Starting signal was also clear. In addition, despite having passed the Madhupur East Warner in the 'On' position the speed of train No. 87 had still been 50 kilometres an hour at the moment of impact, while the driver had failed to observe the tail lamp of train No. 5 as soon as he should have done.

Despite it having been stationary for over 15 minutes Guard Munilal of train No. 5 (who was killed) had failed to protect his train, as per Rules Nos. 125 and 166, with detonators placed 600 and 1,200 metres in its rear, nor had he used fusees; possibly, though, he had not had any fusees, which were in short supply.

The accident could have been averted or its results mitigated if the engine crew of train No. 87 had been more vigilant or if the Joramow switchman had been more vigilant and ensured that his Starter was 'On' before clearing his Home signal for train No. 87 as per Rule 39 (d).

The points failure at Mathurapur block hut had been due to the failure of the substitute porter there to set the points for the Loop line and to transmit the key electronically back to the cabin so that *Line Clear* could be given back to Nawa Patra block hut for train JC-9 Up. These operations required training and a degree of proficiency that could not be expected from an unskilled porter. The qualified pointsman who had had the job until June 1967 had been replaced by a porter (doubtless to save money), which, the commissioner considered, had 'been a injudicious act.' The regular porters had somehow managed but the substitute porter had naturally failed. It was essential that properly trained staff should be employed on jobs involving the safety of train running.

There was a wide scope for improvement in the work of the Madhanpur and Asansol section controllers.

The additional commissioner found that Driver Purohit was primarily responsible for the accident.

Firemen Ram and Prasad were responsible for failing to draw the attention of their driver to the Danger position of the Up Advanced Starting signal at Joramow.

Switchman Khalil at Joramow had contributed to the accident by not making an adequate effort to stop train No. 87 and had made fictitious entries in his train register book.

Guard Munilal had failed to protect his train in the rear and paid for his negligence with his life.

The additional commissioner made five recommendations (numbers 1 to 5 below) and three incidental observations (numbers 6 to 8 below)

(1) The installation of the Automatic Warning System on trunk routes to alert a driver passing a Danger signal and to apply the brakes should he fail to respond should receive the highest possible priority.

(2) Starting signals on double-track lines should be provided with automatic reversers to ensure that they returned to Danger after the passage of a train, like Advanced Starters.

(3) The large-scale manufacture of fusees should be pursued more vigorously.

(4) The side lamps of guards' vans should be properly maintained and guards should ensure that they showed red to the rear before starting.

(5) Suitably trained staff must be employed at stations where the operation of points and locking and the electrical transmission of keys was involved.

(6) Section controllers should be properly trained before being posted as such.

(7) By agreement with his colleague Switchman Khalil at Joramow had been working continuously for over 20 hours. Such long hours of duty must be prevented.

(8) The clock in Joramow cabin had been out of order for 10 days. Action should be taken to prevent such things happening. [270]

The apparent negligence of several railwaymen led to a major disaster on the Kilece – Jędezejów main line in Poland on the night of Monday 27th August 1973. As a freight train was passing Radkowice station the signalman there noted that it lacked a tail lamp. He duly telephoned his colleague at Wolica, the station in the rear, and was assured that the train had been complete when it passed Wolica. This, however, was untrue, the Wolica signalman had in fact dozed off and failed to notice that the train lacked a tail lamp. The automatic application of the brakes upon the division of the train had brought it to a stand in the Wolica – Radkowice block section but the driver and his assistant had just released the brakes and continued, without presumably checking that their train was complete or bothering to inform the Radkowice signalman. In fact, 20 waggons had been left behind and they were shortly run into by a Zakopane – Warszawa Wschodnia passenger train at about 23:53. Sixteen people died and 24 were injured, though a worse disaster was prevented by the crew of the passenger train stopping a train on the opposite line. [271]

On Wednesday 26th September 1973 a goods train was stopped at a signal at Mandini station, South Africa, when it was run into from behind by a following passenger train.

---

[270] Government of India Ministry of Tourism and Civil Aviation (Commission of Railway Safety): Report on the Collision of 87 Up South Bihar Express with the rear of No. 5 Up Howrah – Amritsar Mail at Km 291.767 between Joranmow Block Hut and Madhupur Junction Station of Eastern Railway on 27-7-1973

[271] Wikipedia – Railway disaster near Radkowice

An engine was derailed and another engine and two coaches of the passenger train were badly damaged; one of the coaches of the passenger train was described as having been jack-knifed and bent double. According to a Reuter report there were nine fatalities, all Africans, including three women and a child, eight of whom were killed outright and the ninth having succumbed after being admitted to hospital, with 97 injured. [272] Another source, however, speaks of 17 dead and 40 seriously injured. The driver of the colliding train was thrown through a window and landed on a waggon of wood pulp; he was seriously hurt. Whether the accident had resulted from a signalling error or from the driver of the passenger train overrunning signals was not stated. [273]

In early November 1973 a change in the weather conditions to rain and strong winds resulted in a sudden heavy leaf-fall affecting trees alongside the railway tracks around Guntershausen, West Germany, and the accumulation of wet leaves on the rails made them very slippery. On Monday 5th November 1973 interzonal express train No. D 453 from Mönchengladbach to Leipzig Hbf in East Germany had left Kassel Hbf at 14:18, seven minutes late. At Rengershausen, a couple of kilometres short of Guntershausen, the line fell at 10 ‰ (1 in 100) and then 7 ‰ (about 1 in 143). Leaves on the rails made them especially slippery here and the lack of adhesion caused the anti-slip device on the train to operate repeatedly; in addition the brake pressure fell and could not be recreated, so that the train came to a stand with its last coach 30 metres in the rear of the Guntershausen Einfahrsignal (Home signal). The crew of train No. D 453 then tried to determine why its brakes had come on in emergency. What the engine crew were unaware of, was that because of the steep falling gradients and the short block sections between Rengershausen, an intermediate block post at Buchberg, and Guntershausen, as an additional safety measure an Indusi track magnet had been installed at Buchberg, which applied the brakes of a train in emergency if its speed exceeded 65 kilometres an hour at that point, the Indusi operating independently of the signal aspects. The speed of train No. D 453 had exceeded 65 kilometres an hour and so the Indusi magnet had applied the brakes.

Following train No. D 453 was intercity express No. DC 973 *Ederland* from Kassel to Frankfurt (Main). It was due to start at 14:18 and actually started at 14:22, thus four minutes behind train No. D 453. The train was braked to 25 kilometres an hour at Wilhelmshöhe, because it was overhauling the slower train in front, but its speed had then increased to 119 kilometres an hour at Rengershausen. The Buchberg Vorsignal (distant) was yellow and the brakes were applied there but due to the slippery rails the speed of the train dropped very little. The anti-slip device released the brakes on the coaches at intervals and the train driver did the same thing repeatedly to the engine brakes, in an effort to get better braking. As the train had passed the Buchberg Hauptsignal (stop signal) at Danger, still travelling at 90 kilometres an hour, the Indusi magnet had again initiated an emergency brake application, but sliding on the wet rails meant that the train was still moving at 40 kilometres an hour when it struck the first-class coach at the rear of train No. D 453 at 14:29. Despite the relatively low collision speed the rear coach of the forward train was wrecked and 13 or 14 people died, while 25 others were seriously and 40 slightly injured.

---

[272] *The Times*, London, England, 27th September 1973

[273] https://cdn-s3.sappi.com/s3fs-public/slices/downloads/Paper-Chain-book-The-Story-of-Sappi-p189-220-Chapter-6-1973-1978.pdf

The Rengershausen signalman thought that train No. D 453 had passed him going rather more slowly than usual, while before it had done so he had already been advised of the approach of train No. DC 973, so that he had only been able to clear his signals for the latter train at the last moment. Concerned that this train was catching up the one in front, he had also exhibited a hand signal to its engine crew, without getting any response from them.

Although the driver of train No. DC 973 had not actually broken any rules his driving technique came in for considerable criticism during the investigation into the accident. He knew that train No. D 453 had only left Kassel four minutes in front of him and having been severely checked at Wilhelmshöhe, plus the signals at Rengershausen only having been belatedly cleared for him, he must have realised that he was repeatedly overhauling the other train, despite which he had continued to drive at maximum speed, instead of running a little more slowly as far as Guntershausen, where train No. D 453 would in any case have taken a diverging route. It appeared also, that he had probably been inexperienced regarding the anti-wheel-slip equipment, so that he had not realised how much effect it would have on the train braking, and that, perhaps because he had been distracted by his assistant, he had not applied sand to get better adhesion, other than possibly at the last moment. The investigation concluded that he had failed to use 'distributed attention,' concentrating on one matter to the exclusion of everything else.

Following the disaster Deutsche Bundesbahn imposed a speed limit of 80 kilometres an hour between Rengershausen and Guntershausen, increased the training of drivers in train handling during adverse conditions, and undertook the regular clearing of leaf fall during the autumn months. [274] [275] [276] [277]

What was variously described as an Istanbul – Baghdad express and an Adana – Ankara express came to a stand in mid-section near Kayseri, Turkey, when its engine failed, early on Friday 13th December 1974. While it was stationary it was run into from behind by a freight train. Seven people died and 25 were injured. [278] [279]

India suffered a major disaster at Ultadanga Road station on the Eastern Railway at 19:28 on Wednesday 29th January 1975. Train No. H-185, an Up Habra local, was run into from behind by train No. 43, the Up Darjeeling Mail.

The Habra local train was an eight-coach electric multiple-unit (EMU), in charge of Guard J.K. Paul with Motorman Gopal Sharma. The Darjeeling Mail consisted of WP Class steam locomotive No. 7248 and 13 coaches in charge of Guard T.N. Bhattacharjee, with Driver Mohamad Abbas, Leading Fireman Mohamad Salimuddin and Additional Fireman Raghunath Prasad.

---

[274] *Katastrophen der deutschen Bahnen Teil I* by H.J. Ritzau, pgs. 201-213
[275] https://mx-schroeder.medium.com/fatal-fall-the-1973-guntershausen-germany-train-collision-1eecbb5594ec
[276] https://www.hna.de/lokales/kreis-kassel/zugunglueck-jahren-guntershausen-3201797.html
[277] Wikipedia – Eisenbahnunfall von Guntershausen
[278] *The Times*, London, England, 14th December 1974
[279] *Independent Journal*, San Rafael, California, United States, 13th December 1974

The engine of the mail train penetrated the rear coach of the EMU for almost 8 metres, the roof of the coach caved in and its side panels were sheared off for 13 metres. It was in this coach that almost all the casualties occurred. In all 35 passengers died in the accident, 52 people were seriously hurt and 41 slightly hurt.

Passenger Amal Ghosh was the first witness to give evidence at the subsequent inquiry. He said that he had been riding in the rear coach of the EMU in the compartment next to the guard's lobby, and that the train had started slowly from Sealdah, then gained speed until stopping normally at Ultadanga Road. The train had then slowly moved forward and backwards several times. Suddenly, he had seen a light behind, which appeared to be on the same line, and a person had remarked that it might be the Darjeeling Mail. Someone in the guard's lobby had then switched the rear headlight of the EMU on and off as a warning to the oncoming train. Shouting a warning to other passengers, he had alighted onto the platform. Given his agitated state, he could not estimate the speed of the incoming train but he thought it had been coming quite fast. The tail lamp of the local train had been lit.

Sushanta Adhikari said that he had been standing on the second-floor terrace of his house. He had observed the local train moving forward and backward, had heard a noise, and had seen many passengers alighting from it. Then another train had appeared from behind, whistling and apparently slowing, but it nevertheless struck the local train. Looking at his wristwatch it had been 19:20 (his watch was subsequently found to be three minutes slow). The headlight of the colliding train had been burning brightly; he had not looked for the tail lamp of the local train. He put the speed of the Mail train at 10 to 12 kilometres an hour.

Retired Additional District Judge T.P. Mukherjee stated he had been waiting on the station footbridge when he had seen the local moving to and fro, then a through train coming on the same line. He had heard people shouting, 'Stop! Stop!' and thought that the train would be able to pull up in time, but it had not. He, too, put its speed at 10 to 12 kilometres an hour.

Guard Bhattacharjee of train No. 43 stated that it had stopped at signal No. AD 13 at red from 19:24 to 19:26. The driver had then exchanged signals with him and the train had proceeded at restricted speed, not more than 15 or 20 kilometres an hour, the speed dropping immediately before the collision, which he timed at 19:30. As regards visibility there had been some smoke but signals could be seen clearly.

Guard Paul of the EMU said that his train had left Sealdah Main platform at 18:55, four minutes late. There had been a two-minute power failure at about 19:05. Ultadanga Road station had been reached at 19:14. After a couple of minutes he had given the starting signal and the train had begun to set off, then it had moved forward and backward several times, to the annoyance of the passengers. Checking his tail lamp he had gone forward to meet the motorman, who had told him that he was going to the booking office to tell the traction loco controller that his train was not picking up speed. When the motorman did not quickly reappear he, too, had gone to the booking office. Returning, he had seen the headlight of the Mail train, had shouted for his passengers to jump out and had run towards the oncoming train exhibiting a red light. He put the speed of the Mail at 20 to 25 kilometres an hour.

Motorman Sharma put the time of his train leaving Sealdah Main platform No. 6 at

18:58. He said he had started cautiously under yellow signals and after signal No. K 5 at Kankurgachi speed had been further reduced by a power failure. Though it had soon been restored he could not regain full power, but had somehow reached Ultadanga Road at 19:15. There he had moved the train backwards and forwards to test it but could not locate the fault, so informing the guard, he had gone to the train announcer's room to arrange for some maintenance staff to come and fix the problem. Instead the traction loco controller had told him to proceed to Dum Dum Junction at whatever speed he could manage. He had thus just been setting off when he felt the jerk of the collision.

Driver Abbas of train No. 43 said that he had started from Sealdah at 19:15. Signal No. K 5 at Kankurgachi had been yellow and signal No. AD 13 red. Stopping there two minutes he had then whistled and restarted, the signal remaining red. Approaching the end of a curve his fireman had called out that signal No. AD 15 at Ultadanga Road was yellow, which he had verified. Reaching the station platform there, he had heard people shouting at him to stop, had applied the steam and vacuum brakes, and had almost stopped when the local train backed into his. He had not seen the local train before hearing the shouts. Told that the motorman had said that signal No. AD 15 had been red he insisted that it had been yellow. He alleged that the tail lamp of the local train had been out and added that he had not seen its rear headlight being switched on and off to warn him. He had had no illness or family problems on his mind.

Leading Fireman Salimuddin generally corroborated his driver's statement.

Additional Fireman Prasad stated that his train had started slowly after stopping at signal No. AD 13, with a highest speed of 15 to 20 kilometres an hour. He had seen signal No. AD 15 at yellow.

Section Controller S.R. Roy at Sealdah said that after finding that Train No. H-185 had motor trouble at Ultadanga Road he contacted Sealdah RRI cabin with the thought of diverting train No. 43 to the Up Suburban line but had been informed that it left Sealdah at 19:15 on the Up Main line and had passed Kankurgachi at 19:24. He had been told of the collision at 19:30.

Ticket Collector Tarapada Guha confirmed that the motorman train No. H-185 had used his telephone and stated that he had witnessed train No. 43 approaching and the collision, which he had reported to the section controller.

Additional Commissioner S.K. Mojumder fixed the time of the collision at about 19:28.

The additional commissioner believed that train No. H-185 had been crawling forward at 2 or 3 kilometres an hour and the speed of train No. 43 had not been high, perhaps 12 kilometres an hour.

The signalling had been thoroughly checked after the collision and found to be working perfectly. Signal No. AD 15 could not therefore have been yellow for train No. 43. Either the crew of the engine of the Mail train had mistaken the station lights for the signal, which had been obscured by the stock of the EMU, or they had deliberately misrepresented the facts. Even if the signal had been yellow the driver of train No. 43 should have proceeded with great caution prepared to stop short of any obstruction until he reached it, after having passed signal No. AD 13 at red, as per General Rules Nos. 277 and 279. The statement of Driver Abbas that train No. H-185 had backed into his train was disproved for several reasons. There had been no reason for the front train to do so at that moment after its driver had already been instructed by the controller to continue to

Dum Dum Junction as quickly as he could manage; the heavy damage caused could not have been due to train H-185 setting back at the low speed which it was capable of with its defective motors; the evidence of independent witnesses, passengers and the guard of train No. H-185 was that train No. 43 had run into train No. H-185 and not vice versa; and the damage to the platform shelter caused by the pushing up of the roof of train No. H-185 was clearly due to train No. 43 having moved forward about 35 metres after the collision. It was evident therefore that Driver Abbas had not exercised sufficient caution, moreover he had failed to observe the tail lamp of the EMU and its headlight being switched on and off, as confirmed by independent witnesses.

As train No. H-185 had been detained in Ultadanga Road station for more than five minutes according to General Rule No. 280 Guard Paul should have protected it in the rear with detonators placed at 90 and 180 metres, but he had been expecting the motorman to return at any moment, plus the guards of suburban trains were constantly in fear of being manhandled by inpatient passengers if there were any delays. The guard's failure must be considered in view of these circumstances, while he had not been supplied with any fusees.

Additional Commissioner Mojumder concluded that the accident had resulted from Driver Abbas having failed to proceed with great caution after stopping at signal No. AD 13 at red and that he was therefore mainly responsible for the collision.

Leading Fireman Salimuddin was responsible for failing to keep a good lookout in violation of General Rule No. 122.

Guard Paul was responsible for failing to protect his train in the rear but there had been extenuating circumstances.

The assistant commissioner made seven remarks and recommendations:

(1)(a) Trains of vastly different types such as lightweight EMUs, moderately heavy expresses and heavy goods trains used the same lines where automatic signalling rules permitted trains to pass Danger signals after stopping, prepared to stop short of any obstruction. Protection of trains was to be provided by placing detonators 90 and 180 metres in their rear. Even though the signals were spaced a kilometre apart it was difficult to stop a conventional passenger or goods train short of an obstruction, especially at night or when visibility was restricted, unless it was travelling at very low speed. Drivers mostly tended to speed up and often then came to grief. Furthermore the coaches of EMU trains had very little collision resistance, especially when struck by a heavier train, as in the present accident. Segregation of EMU services in automatic signalling areas was desirable. As such segregation of trains was not practicable in the present circumstances, it was recommended that both ends of EMUs should he provided with reflective material visible to the drivers of following trains at night.

(b) The rules should be amended to restrict the speed of trains passing automatic signals at Danger at night to 8 kilometres an hour.

(2) EMU's should be redesigned to make them more crash resistant.

(3) Signal No. AD 15 should be raised so as to prevent its aspect being obscured by a standing train and the station lights being mistaken for it.

(4) A disturbing feature of the accident was that the engine of train No. 43 had not been fitted with a speed recorder, while the speedometer of train No. H-185 was also out

of order. Special steps should be taken to rectify this situation.

(5) The tail lamp of train No. H-185 had been fitted with glass painted red instead of a proper red lens, and other EMUs were similarly fitted. Such painted glass lamps were considerably less conspicuous and only proper red lenses should be used.

(6) The headlight of train No. 43 did not have the proper fittings and had been out of focus. 'Such a condition in a Mail train engine is to be highly deprecated and immediate action is called for.'

(7) The railway board had indicated that until the design of fusees had been improved they should only be supplied to the guards of ghat and suburban trains. Suburban guards were, however, not all supplied with them. In the present case Guard Paul could have used fusees effectively had he been supplied with any. The subject of supplying guards with fusees had been 'on the anvil' for years and it was high time that the board took positive action in the matter. [280]

A local train from Vinovci in western Yugoslavia to Belgrade was standing in Zemun Polje station, 10 kilometres short of its destination, on Saturday 19th July 1975, when it was run into from behind by a following express. Only one minute was scheduled for the station stop and a guard related afterwards that he had heard the stationmaster order the train away several times but there were over 100 items of luggage to be unloaded which necessarily took time.

One of the passengers on the train, Miladin Trkulja, described how he suddenly heard the stationmaster shout, "Jump out of the train, people. There's another one coming!" Miladin and his girlfriend just had time to climb out through a window and escaped harm. Others were not so lucky; 16 were killed and another 37 were hurt.

A Sarajevo – Belgrade express travelling at 65 kilometres an hour demolished the three rear carriages of the local train, causing the casualties.

Dragan Milovic, heading the team investigating the accident, blamed it on human error, and the stationmaster and the driver of the express were arrested, though it was not immediately clear where the blame for the crash lay. The express was not booked to stop at the station and was authorised to pass through it, with clear signals, at 80 kilometres an hour. [281] [282] Station officials maintained afterwards that the signals had been at Danger to protect the local train, implying that the driver of the express was at fault. He had applied his brakes at the last moment. [283]

In what was presumably a rear collision six people died and 19 were injured when Katowice – Gdynia Glówna Osobowa long-distance passenger train No. 4513 collided with a freight train near Chorzów, Poland, at 02:08 on Monday 1st September 1975.

---

[280] Government of India Ministry of Tourism and Civil Aviation (Commission of Railway Safety): Report on the Collision of No. 43 Up Darjeeling Mail with the rear of No. H 185 Up Habra Local on the Up Main line at Ultadanga Road Station on Eastern Railway on 29th January 1975
[281] *The Joplin Globe*, Joplin, Missouri, United States, 20th July 1975
[282] *Pacific Stars and Stripes*, Tokyo, Japan, 21st July 1975
[283] *Syracuse Herald-American*, Syracuse, New York, United States, 20th July 1975

Electric engine EU-07 194 and several waggons were destroyed. [284]

On the evening of Monday 29th September 1975 the *El Rosarino* passenger train of the Ferrocarriles Argentinos set off from Buenos Aires Retiro station over the Mitre division of the railway en route to Rosario in Santa Fe state. At 20:00 it was stopped by signals 300 metres short of Río Luján station, Argentina, when it was run into from behind by a suburban train from Buenos Aires bound for Escobar and Campana. The luggage van at the rear of the front train was demolished and pushed into the sleeping cars in front of it, three of which were badly damaged. The diesel engine of the local train was wrecked but the remainder of it did not suffer a great deal in the collision. Thirty-two passengers on *El Rosarino* were killed and about 25 people were injured, including the driver, driver's assistant, and conductor of the suburban train; none of the passengers on the latter train were badly hurt.

A witness said, "People were screaming, jumping out of windows and running off into the countryside." [285]

The emergency services were alerted by the Río Luján station staff and assistance was provided by ambulances, police, and firemen from Campana, Escobar and Zarate. At first the rescue work was hampered by the darkness but later the firemen erected portable floodlights. The rescue operation went on through the night, with about a hundred rescuers cutting through the wreckage with oxy-acetylene torches in the usual manner. At about 23:00 the engine of *El Rosarino* proceeded to its destination with the undamaged coaches of the train and the uninjured passengers. The injured passengers were taken to local hospitals and some to Buenos Aires. Doctors in the Argentine capital, who had been in the middle of a 24-hour strike for increased wages, abandoned their industrial action when the injured patients began to arrive.

The accident had apparently resulted from a signalling error. Several hours after the collision three railwaymen were arrested and questioned in an attempt to determine who had been responsible. [286]

Six years after Mexico City had first been provided with an urban rail system a serious disaster occurred on Line 2 of the Sistema de Transporte Collective. It happened just after the end of the morning rush hour, at about 09:40 on Monday 20th October 1975, at Viaducto station, Mexico, on an above-ground part of the line.

Train No. 8 was stopped in Viaducto station, apparently because a passenger had operated the communication cord, as had already happened at Hidalgo, Bellas Artes, Allende, and Pino Suárez stations. Motorman Alfonso Sánchez Martínez secured his train, climbed down from his cab, and prepared to release the brakes of car No. 6, so that his train could proceed. As he did so he overheard the signalman speak to his colleague in charge of train No. 10, behind. "I clearly and perfectly heard the signalman tell the motorman not to move forward but to stop immediately," he said.

Train No. 10 had stopped in Chabacano station, the one before Viaducto, at 09:40. Signalman Rudolfo Luis Flores Guriérrez confirmed afterwards that he had instructed

---

[284] Wikipedia – Railway disaster near Chorzewo

[285] *The Arkansas City Traveler*, Arkansas City, Kansas, United States, 30th September 1972

[286] Wikipedia – Accidente ferroviario de Río Luján de 1975

Motorman Carlos Fernández Sánchez of that train to wait at Chabacano, "but I did not hear a response from the driver, which seemed strange to me because he was an experienced man, with three years' service as a motorman. Since train No. 8 had stopped at Viaducto I allowed train No. 10 to come as far as the Chabacano, rather than holding it at the station behind, because of the number of passengers. Then I had occupied myself with other tasks."

Rather than waiting at Chabacano station, Motorman C.F. Sánchez set off again, gradually increasing speed. The first signals that he came to were clear. As his train climbed the incline to the Miguel Alemán Viaduct the motorman failed to notice the train stopped in Viaducto station. Then as he reached the summit of the climb he did see it. "I saw the train at Viaducto as I was going down the hill and signal No. 1-2 was red. I wanted to stop but I couldn't. The brakes should have come on automatically anyway, but they didn't." The train was now descending the slope and travelling at about 70 kilometres an hour.

At 09:40 train No. 10 ran into the rear of train No. 8. The rear car of the front train and the first car of the colliding train telescoped one another and reared up in the air, striking the roof of the station and remaining in that position. Thirty-one people died, 70 were seriously injured and about 100 were slightly hurt. There had been about 135 passengers on each train.

Motorman C. F. Sánchez was himself seriously injured. He was also blamed for causing the disaster and was taken first to the Lecumberri Penitentiary (presumably to the hospital wing in view of his injuries) and then to the Reclusorio Norte (North Prison). His defence counsel and his union argued that there were inconsistencies and irregularities in the operation of trains on the Metro. President Luis Echeverria Álvarez set up a special commission to investigate the accident, consisting of lawyer Horacio Castellanos Coutiño, Metro director Jorge Espinosa Ulloa and the Secretary General of the Federal District. After a five-day investigation the commission found Motorman C. F. Sánchez wholly to blame for the accident. He declared that the brakes of his train had failed but they were found to be working correctly after the collision. It was stated that he had disabled the automatic train-stop system, had failed to apply the brakes, had not been paying attention to his work and had overrun signals. He was dismissed from his employment and sentenced to 12 years' imprisonment. He was released due to good behaviour in 1986. [287] [288] [289]

Speaking rather more than 30 years after the accident a union official, Héctor Manual Zavala Buclo, alleged that the collision had not been an accident at all but deliberately engineered in an effort to show President Echeverria Álvarez what could happen if some form of automatic train control were not installed, so as to persuade him to agree to its installation and the attendant cost. There had been no intention to sacrifice innocent lives, merely to cause a minor accident to demonstrate the potential dangers, but matters had got out of hand. After the accident, the official continued, there had been a cover-up. The tachograph from train No. 10 had disappeared and people had been bribed and/or threatened to keep quiet about the matter. [290]

---

[287] Wikipedia – Colisión de trenes en el Metro de la Ciudad de México de 1975

[288] *El Universal*, Mexico City, Mexico, 20th October 2008

[289] https://www.excelsior.com.mx/comunidad/asi-fue-el-accidente-del-metro-en-20-de-octubre-de-1975/1562915

[290] *El Universal*, Mexico City, Mexico, 22nd October 2008

CHAPTER 5

SIDELONG, HEAD-ON, DOUBLE AND FOULING COLLISIONS

Trains can come into collision with one another in a variety of ways. Rear collisions, that is ones involving one train running into the back of another, trains meeting one another head-on on a single track and trains colliding on rail/rail level crossings are detailed in other chapters. This chapter will deal with other types of collision such as those involving trains running into one another at junctions, meeting head-on on multiple tracks and colliding with vehicles left in sidings but which end up foul of running lines.

Express train No. 704, which had left Valencia at 23:00 the night before with 360 passengers, and was due in Barcelona at 07:45, was near its destination when it was in collision with freight train No. CB-1, an *especial de circunvalación* train (extra local trip working?) coming from Sants, at L'Hospitalet de Llobregat, Spain, at 07:25 on Monday 9th January 1961. The trains collided violently head-on at a junction, with disastrous results, the wreckage piling up to a height of 10 metres. The collision claimed the lives of the driver of the express and its heating manager, the assistant driver, conductor and a third railwayman on the freight train and 20 passengers on the express.

Express train No. 704 consisted of electric engine No. 7652, a luggage van, a heating car, a first-class coach, a first- and second-class composite coach, a second-class coach, two third-class coaches and a sleeping car, in charge of an unnamed conductor with Driver Fernando Ruiz Noriego and Assistant Driver Nicolás Fernández Fernández. The passenger vehicles were all of metal construction, except for the second-class coach fifth in the train, which was telescoped by the composite coach in front of it and in which all the passenger deaths and most of the injuries occurred. A fire also broke out after the collision. The freight train consisted of an engine and an unstated number of waggons, in charge of Conductor José Maria Pallejá Sangenis with Driver Fernando Benavente Molero and Assistant Driver Antonio Avila Lloret. The other railwaymen who lost their lives were Heating Manager José Gutiérrez Giménez and Brakeman José Luis Maroto Echevarria. Among the passengers who died were nine students and a teacher who were returning to Barcelona at the end of their Christmas holidays. Fortunately wires of the overhead line equipment which were brought down by the collision did not cause any further casualties. Assistant Driver Fernández of the express and Driver Molero of the freight train were injured, as were about 45 passengers.

Travelling on the express in the sleeping car were the players and support staff of RCD Español, a first division Barcelona football team, who were returning from playing a National League Championship game in the Mestalla stadium, Valencia. Ernesto Pons, the coach of the team, described his experiences of the accident. “Suddenly we noticed a very strong jolt, followed by violent braking, and this was repeated two or three times before the train stopped in about seven metres. The sudden stoppage threw us all to the floor, one on top of the other, but we picked ourselves up again without attaching much importance to it, thinking that someone must have pulled the communication cord, stopping the train abruptly. Someone opened a window and looked out, without seeing anything unusual. However, I thought to go and investigate further and got a shock when I saw that we had collided with a freight train and there was an enormous amount of damage.”

The emergency services were quickly alerted and the Barcelona Fire Brigade, the Red Cross, the Civil Guard, and other organisations went to the scene, though the initial help was provided by uninjured passengers, local people, and factory workers. Jamie Fernández, the football team masseur, improvised an emergency first-aid kit and together with a player, Señor Torres, treated many injured. Ernesto Pons added, “We only felt a series of bumps in our part of the train and we had a great shock when we alighted and realised the scale of the accident. All the team, together with our masseur, Jamie Fernández, worked to do what we could to rescue the victims and treat the injured.” The rescue work lasted 10 hours. Some looting apparently occurred, one of the injured losing a wallet containing 50,000 pesetas.

As is often the case, there were stories of lucky survivors. An Alicante couple, Carlos Torra, escaped because at the last moment they decided not to travel and sold their tickets instead. Student Vicente Serradell was thrown out of a window of the second-class coach, while fellow students Juan Palomares, Salvador Vendrell, Vicente Meseguer, Manuel Cortés, Salvador Verdaguer and Vicente Domínguez all should have been riding in the ill-fated coach but there were no second-class tickets left so they had to travel third-class instead. Vicente Ballestero, 28, on his way to England where he hoped to marry his fiancée, Amparo Hernández, was a passenger in the second-class coach. Hearing a noise he quickly fell to the floor and was apparently unhurt, though trapped for a time by an iron bar. [291]

The accident had been caused by Driver Noriego of the express, who had run past two yellow signals without reducing speed and only applied the brakes in emergency in the vicinity of a third, red, signal. At the moment of impact the express was still travelling at an estimated 80 kilometres an hour and the freight train at 30 kilometres an hour. Because he was killed in the disaster it was, of course, impossible to explain his mistake. [292] [293] The express was not booked to stop at L’Hospitalet. [294] [295]

---

[291] http://elpueblodelasfiebres.blogspot.com/2013/02/accidente-ferroviario-en-hospitalet-de.html
[292] https://www.lavanguardia.com/hemeroteca/20101231/54094705238/medio-siglo-del-tragico-accidente-de-tren-a-las-puertas-de-barcelona.html
[293] https://lhospitaletdellobregat.wordpress.com/2011/01/09/accident-de-tren-del-9-de-gener-de-1961/
[294] Federación Castellano Manchega de Amigos del Ferrocarril
[295] *The Times*, London, England, 10th January 1961

A passenger train collided head-on with a goods train at Teltar, India, on the Kishenganj – Katihar section of the North East Frontier Railway, early on Wednesday 8th March 1961. Two coaches of the passenger train were telescoped into one another and five waggons of the goods train were derailed. Eleven passengers lost their lives and 39 were injured, 16 seriously. The latter were taken in a medical van to the Civil Hospital in Katihar.

Mr Jagjivan Ram, the Minister of Railways, and Mr Karnall Singh, Chairman of the Railway Board, flew to the scene of the accident from Delhi. [296]

Fourteen people died and about 50 were said to be seriously hurt in a collision between a passenger train and a freight train in the station at Beska, Yugoslavia, on Wednesday 23rd August 1961. The passenger train was stopped in the station ready to leave for Belgrade when the freight train, which consisted of a locomotive and 49 waggons, collided violently with it in a head-on collision. The freight locomotive penetrated some distance into the passenger train, the leading carriages of which were thrown up into the air and landed on top of one another in a pile of wreckage, joined by the leading waggons of the freight train, all of which were derailed. No trace of the drivers could be found after the accident and it was not known if they were among the dead.

Soldiers stationed nearby assisted in the rescue operation along with firemen, police, and Red Cross workers. The injured were taken to hospitals in Novi Sad and to a military hospital at Petrovaradim and there were radio appeals in Novi Sad and Belgrade for blood donors. There were desperate scenes in Novi Sad amongst the parents of a party of 100 children who had been in the train, leading the authorities to issue a statement to the effect that the children were all safe.

An investigation was launched into the cause of the accident, which was provisionally blamed on the malfunctioning of a set of points. Human error would appear to be another likely possibility. [297]

The Roscoff – Paris express that had left Le Mans at 16:36 collided with an engine at Theil station, France, on the line between Ferté Bernard and Voget-le-Rotrou, at 17:07 on Tuesday 29th August 1961. Ten coaches of the express left the track following the collision, ploughed up the ballast for several hundred metres, and two of them overturned. Initial reports stated that eight passengers had died, 32 were seriously injured and another 30 were less badly hurt. [298] [299]

Six people perished in an accident near Sarajevo, Yugoslavia, on Tuesday 3rd October 1961. A passenger train and a freight train collided. [300]

---

[296] *The Indian Express*, Chittor, India, 9th March 1961

[297] *La Stampa*, Torino, Italy, 24th August 1961

[298] *Gazette de Lausanne*, Geneva, Switzerland, 30th August 1961

[299] *Journal de Genève*, Geneva, Switzerland, 30th August 1961

[300] *The Joplin Globe*, Joplin, Missouri, United States, 4th October 1961

The only information available regarding a railway accident between Ruse and Kaspichen, Bulgaria, on Saturday 6th January 1962, appears to be a report some days later from the Bulgarian news agency, B.T.A., that a commission of inquiry had found it had been due to a gross violation of safety regulations by the staff at two stations, adding that those responsible were to be put on trial. Twelve people were said to have died in the accident and another 26 to have been injured. It is assumed that the accident was a collision. [301]

The worst accident in the history of the Dutch railways occurred on the morning of Monday 8th January 1962 at Harmelen, where the line to Amsterdam via Breukelen left the main line between Rotterdam and Utrecht. By this date a so-called ladder junction was provided here, meaning that instead of passing directly from the main line to Utrecht onto the Amsterdam line via a diamond crossing, trains from Rotterdam bound for Amsterdam had first to cross over to the line from Utrecht and then run along that line in the facing direction for a short distance before diverging onto the Amsterdam line. A common arrangement on modern high-speed railways to avoid a diamond crossing but in the author's opinion a dubious one as it introduces the possibility of a head-on collision where none previously existed. Just such a head-on collision happened at Harmelen on the day in question. A significant feature of the disaster was that there was a thick fog in the area at the time, with visibility restricted to 30 to 150 metres according to different witnesses. [302]

One of the trains involved in this accident was No. 164 from Leeuwarden via Utrecht to Rotterdam. This train arrived at Utrecht C.S. station a few minutes late with engine No. 1104 and 11 carriages. Here it had to reverse direction; engine No. 1104 was detached and engine No. 1131 was attached to what had been the rear carriage for the onward journey. A brake continuity test was carried out and the train left Utrecht C.S. at 09:11, six minutes late, with Driver Van Leer in charge of its engine.

The other train involved in the disaster was train No. 464, a six-car electric multiple-unit from Rotterdam via Gouda to Amsterdam, with Motorman Fictor. It left Woerden station, just short of Harmelen, at 09:15.

The junction at Harmelen was controlled from a small NX panel box (post T) in Woerden. When the trains were running to schedule train No. 164 passed Harmelen from Utrecht at 09:14, train No. 464 went through the junction in the opposite direction at 09:18, and train No. 114 (Grongingen – Utrecht – Den Haag) followed from Utrecht at 09:21, a pattern that repeated hourly. On this morning the late running of train No. 164 gave Signalman Kwakkenbus in Woerden a slight problem: should he let the trains through the junction in their normal sequence, or, given that train No. 164 was slightly late, should he instead let train No. 464 across the junction first? Initially he kept his signals at danger against both trains but a little later he saw from the indicator lights on his panel that train No. 464 was drawing up to signal No. 16, protecting the junction, while train No. 164 was still not indicated as approaching, so he decided to give train No. 464 priority, operated points Nos. 5B, 5A and 3 to let it across the junction and cleared signal No. 16 for it. A moment later a light lit up on the panel and an annunciator buzzer indicated the approach of

[301] *The Times*, London, England, 19th January 1962
[302] http://danger-ahead.railfan.net/accidents/harmelen/home.html

train No. 164, which was still at that moment 2,650 metres in the rear of signal No. 8, protecting the junction on the line from Utrecht. In any case the route for train No. 464 had been set up and could not now be cancelled. Watching the progress of train No. 464 on his panel Kwakkenbus was suddenly surprised to see that all the indicator lights in the vicinity of the junction remained lit. At first he thought that there must be a malfunction but then both he and Technician Clief, who was at the former signal box at Harmelen, heard a terrible crash, though nothing could be seen through the fog. A few moments later Clief informed Kwakkenbus that there had been a collision at the junction.

The trains met one another almost head-on. The engine of train No. 164 was thrown to the right and overturned, while the leading vehicles of each train were compressed into one another, forming a pile of twisted steel. The front part of the second coach of train No. 164 was destroyed and it skidded off to the left, striking and partially demolishing lineside relay room 10. The next three coaches of train No. 164 were derailed, the rear six remained on the line. In train 464 behind the destroyed leading car the side of the second car was completely ripped open along one side and the third car was very badly damaged. Also damaged to a lesser extent were the rear three cars, though they remained on the track. There were an estimated 900 passengers on train No. 164 and 180 on train No. 464. Ninety-one people, including both drivers, were killed more or less outright and 54 were injured, 43 of them seriously. Two of the badly hurt passengers died after reaching hospital.

Upon becoming aware of the collision Signalman Kwakkenbus immediately contacted the central electrical control centre to have the current taken off as well as issuing a warning for people to stay away from the overhead wires. Together with the Stationmaster Van Zevenberg at Woerden he then made his way to the accident scene, Porter Molenaar being left in charge of post T. The police, ambulance and fire services were alerted and the first ambulance arrived at 09:35. The injured were taken to various hospitals and the bodies of the dead were removed to the Central Military Hospital in Utrecht. A chapelle ardente was set up in the Buurkerk church in Utrecht for the bereaved. Radio and television programmes were interrupted to broadcast news of the disaster and Queen Juliana returned prematurely from a winter sports holiday at Lech in Austria. Friday 12th January was declared a national day of mourning.

As soon as all the injured had been removed the job of clearing the line began. The undamaged coaches of the two trains were drawn back to Utrecht and Woerden and the damaged overhead line equipment was removed, to allow the breakdown cranes access. By 02:00 the next morning the track had been cleared of wreckage and work began on repairs to the track. At 08:00 the track was passable at 5 kilometres an hour and at 09:00 work began on the replacement of the overhead line equipment. The lines were re-energized at 16:30 and the first three trains passed the accident site at 30 kilometres an hour. Normal working was resumed at 08:00 the next morning. [303]

A correspondent for an English newspaper interviewed several passengers. 'A 19-year-old airman who was in the express train told reporters: "I was in the corridor between two other passengers. Suddenly there was a tremendous crash. I was thrown forward. I was in one of the coaches which was lying aslant. When I climbed out I could not believe my eyes. The havoc was terrible. There was screaming and moaning which did not seem to be of this world. It was like a battlefield, people running round without

[303] *Spoorwegongevallen in Nederland 1839-1993* by R.T. Jongerius, pgs. 142-9

understanding what they did. Others ran away."

'Mr J.A. de Ries, a 58-year-old union official from Rotterdam, who was a passenger, said, "First I heard an enormous crash and then an ear-splitting rumbling. Carriages telescoped into and over one another. It seemed as if everything was being smashed, including our own carriage. The shock of the crash threw my coat from the hook and it fell over my face, protecting me from glass splinters." He climbed out through a smashed window. The air was ringing with the screams of the dying.

'A 32-year-old Rotterdam office worker, Mr B. Bouman, who was on his way to Utrecht, said: "The train was full, as it always is on Monday morning. I was lucky in this case not to have a seat. I was in the space between the seats and when the collision came I saw others crushed where they sat by twisted steel. They must have been killed instantly. I was in the front carriage and just cannot imagine how I got out alive. I climbed through a smashed window. In some places I saw passengers lying underneath the carriages and wheels. I helped to cut free the injured. Assistance was on the spot very quickly."

'Witnesses described how nurses and doctors moved through the fog tending the injured while soldiers and civil defence workers searched for survivors. An official on the scene said the only clues to the identity of the dead would in many cases be clothing and jewellery.' [304]

The panel at Woerden had only been in use since 28th May 1961, so a catastrophic wrong-side signalling failure as a cause of the disaster was a possibility. Shortly after the collision it was confirmed, though, that signal No. 8 was displaying a red light and signal No. 713, which preceded it, was displaying yellow. The signal lamps were rechecked a few days later to rule out an intermittent fault. Nothing was found. The remainder of the signalling installation at Woerden was also thoroughly checked for faults and all was found to be in order. It was therefore clear that signal No. 713 must have been yellow and signal No. 8 red when train No. 164 had passed them.

Engine No. 1131 was also subject to a comprehensive examination by Nederlandse Spoorwegen (NS) at the request of the public prosecutor and it was determined that its traction motors were switched off and the brakes fully applied. The tachograph tape was removed undamaged from the locomotive and indicated that its speed had been reduced from 125 km/h to 107 km/h immediately before the collision. Tests carried out using a train of the same composition between Utrecht and Arnhem suggested that the engine brakes had been applied 85 to 105 metres before signal No. 8. It appeared therefore that Driver Van Leer had completely missed seeing signal No. 713, whose yellow aspect would have required him to reduce speed to 30 kilometres an hour, but had seen signal No. 8 through the fog at short range and had then applied his brakes in emergency. Driver Van Leer was said to have been in good health and had a reasonable service record.

Various suggestions were made via the newspapers, to the Transport Secretary and elsewhere regarding measures to prevent a recurrence of such an accident, such as increasing the overlap beyond the distant signal, the provision of two yellow signals before the red, or of a yellow and two reds before an obstruction, and so on. A Dutch writer has criticised all these suggestions, pointing out that they would reduce line capacity and that the duty of a train driver is always to obey every signal, adding that a driver who was aware that there was a long overlap after a signal, or that there would be

---

[304] *The Times*, London, England, 9th January 1962

another signal before an obstruction was reached, would be inclined to take risks. The only real solution, the writer suggested, was the provision of either an Automatic Warning System, to alert drivers that they were passing a signal with a restrictive aspect, or better still a system of Automatic Train Control, which would automatically apply the brakes of the train upon it passing a restrictive signal if the driver did not. Various mechanical and later inductive systems had been developed in other European countries such as France, Germany, Great Britain, Italy and so on by this date, which gave intermittent indications at lineside signals, while in the United States on lines with the automatic block system a different system was in use, which displayed signal indications to the driver continuously.

The Spoorwegongevallenraad (Accident Investigation Board), concerned at the number of cases of trains overrunning signals, including of course the Harmelen disaster, agreed that a system of Automatic Train Control (ATC) must be brought into use on the Dutch railways. It left it to the NS to decide whether they wished to opt for an inductive or continuous system, while pointing out that the latter was more expensive and would take longer to introduce than the former. It required that the NS begin the installation of one or the other system not later than 1st January 1964. In the end, despite the reservations of the accident board, the NS chose to go for the continuous system, trials of which had already been in progress in collaboration with an American company.

As a short-term measure, pending the introduction of ATC, the NS took measures to improve the visibility of signals at special danger points, such as opening bridges, junctions and so forth. For some years colourlight signals on plain track had been preceded by reflector boards placed 50, 110 and 170 metres in the rear of the signal (these were described in the 1954 signalling regulations as Signal 251). At special danger points a new set of signals (described as Signal 251A) was now to be installed. At the last signal before the danger point (for example signal No. 8 on the Utrecht line at Harmelen) an extra rectangular white sign with two horizontal black stripes was to be provided, while at the penultimate signal before the danger point (such as No. 713 on the Utrecht line at Harmelen) a similar board with slanting black stripes would be provided. [305]

The misplaced overconfidence of a train driver who was unfamiliar with the stretch of line over which he was taking a train led to a bad accident Tábor, Czechoslovakia, at 00:43 on Monday 29th January 1962. A number of troop trains were being run on this date, including train No. 7726 from Benešov via Tábor to České Budějovice. This train consisted of an engine and 40 vehicles, weighing 763 tonnes, and was in charge of a crew from Praha-Vršovice depot. The crew included the 22-year-old driver, who failed to reveal to his superiors at the depot that he was unfamiliar with the route.

All appears to have gone without any problems until the train reached Tábor, where the driver wished to obtain water for his steam locomotive. Believing that the water crane must be located where he could see the central lighting of the station through the winter fog he proceeded towards that point, running past a signal at danger, and his train was diverted onto track No. 20, which ended in a set of buffer stops. The engine demolished the buffers and the sudden stop caused severe damage to the passenger coaches of the train, which were marshalled directly behind the engine, so they could be heated with

[305] *Spoorwegongevallen in Nederland 1839-1993* by R.T. Jongerius, pgs. 142-9

steam from it. In the collision the coaches were crushed between the engine and the flat waggons at the rear of the train, which were loaded with heavy military equipment, probably tanks. The leading two coaches were telescoped and five were wrecked. The presence of a road overbridge beyond the buffers also likely contributed to the severity of the accident. Eight soldiers and the conductor were killed more or less outright and another 42 soldiers were injured, 15 seriously. Several of the injured subsequently succumbed in Tábor hospital. Damage was estimated at 300,000 Czech koruna. [306]

The driver survived the accident and later found himself in court, the case being tried in a large classroom of the local labour school. He was found guilty of being responsible for the accident and was sentenced to 5½ years' imprisonment for endangering the safety of the passengers on the train.

Seven people reportedly died and another seven were injured in a head-on collision between two passenger trains in the station at Padang Besar, Thailand, on Tuesday 17th April 1962. [307]

Japan suffered a major catastrophe on the evening of Thursday 3rd May 1962. The accident began when Down steam-hauled freight train No. 287 from Tabata to Milo was climbing a 1 in 100 incline at 35 kilometres an hour, approaching the station at Mikawashima, Japan, on a freight line, at 21:36. The freight line ended there with a trailing connection to the Down main line. As the freight train arrived at the junction with the main line the signal protecting the connection was at danger as an electric multiple-unit (EMU) passenger train was approaching on the main line. The driver of the freight train failed to stop at the signal and his train was diverted into a trap siding, at the end of which it collided with the buffer stops. As a result of the collision the locomotive and a tank waggon at the front of the freight train were derailed and fouled the Down main line, where they were almost immediately run into by train No. 2117H, a nine-car EMU, from Ueno to Toride, which was derailed in turn, fouling the Up line.

It was at this point that an omission by the Mikawashima signalmen made what would probably not have been that serious of an accident into one of the worst in Japanese railway history. Their mistake was to fail to immediately alert the next station at Minami Senju that the line was blocked. As a result five or six minutes later train No. 2000H, a six-car Up EMU, arrived at the station, collided with the wreckage fouling the Up line, was derailed in turn and partly fell down an embankment. It also ran down many passengers who had in the meanwhile got down onto the line from the Down EMU. [308]

A newspaper wrote, 'Fifteen years ago 106 Japanese burned to death, trapped in a short-circuited electric train at Sakuragicho station in Yokohoma. Another 92 were seriously injured. The passengers were unable escape because the automatic doors were closed tight. And only the conductor could operate them. For one reason or another he was unable to release the compressed air that holds the doors tightly shut.

'Since then, in every coach, levers have been installed so passengers in an emergency,

[306] https://www.koridory.cz/nehody-v-minulosti-tabor-1962/
[307] *Independent*, Long Beach, California, United States, 18th April 1962
[308] Wikipedia – Mikawashima train crash

open the doors by hand and climb out. Thursday night this device allowing passengers to abandon the coaches added to the tragedy at Mikawashima. As soon as the passenger train stopped after a collision with a freight train causing it to derail, the passengers worked the levers to open the doors and dropped onto the tracks. This was at a place 200 yards out of Mikawashima station and on a 30-ft-high embankment.

'While about 60 passengers stood on the tracks and some tried to climb down the embankment, the incoming train arrived at full speed. It was too late for the engineer to brake and too late for those on the track to escape. "It rained people," a witness related. "It was hell," said another who was watching the scene from the second floor of his house near the rail tracks. "I closed my eyes. It was too terrible." Many were mowed down by the train; others were hurled in the air. Those who got off the rails and tried to jump clear were also slaughtered.

'The incoming train's front coach rammed into the derailed outgoing coaches and plunged over the embankment through the roof of a warehouse, crushing those who had tried to escape with steel or burying them with gravel or earth.

'The careering railroad cars knocked down power lines for the electric commuter trains, setting up a shower of sparks and plunging the accident scene into darkness.

'Mrs Massayo Kamei, 33, a housewife who lives near the station, heard the crash, and rushed to the scene. "Just then," she said, "the second commuter train came rushing in with a screech. There was a blinding flash and a booming crash. I could see bodies flying in all directions."

'Massao Inose, a worker at a trackside factory, and his friend Kenji Kando, were pulling victims from the first passenger train when it was rocked by the second crash. Inose looked on helplessly as his friend was ground under the wreckage of the second commuter train and killed.

'Officials of the Joban rail line said both commuter lines were packed with standing-room passengers taking advantage of Thursday's Constitution Day holiday.

'Several thousand policemen, firemen, and other rescue workers were rushed to the scene along with 86 fire department ambulances. The injured were sped to 20 hospitals in the area. Volunteer doctors and nurses treated many victims on the scene. Firemen hurried to turn on water, shut off because of Tokyo's water shortage, for the hospitals and so the wreckage of the trains could be watered down to prevent a fire.

'Railroad officials tentatively laid the blame for the disaster on the engineer of the freight train. They said they believe his train jumped the tracks after he failed to halt for a stop signal. The engineer was among the injured. When the freight train was hit by the first commuter train the boiler of the steam engine burst, spraying scalding steam over the wreckage.

'In the few moments before the second commuter train struck, many of the passengers managed to crawl from the battered cars. Almost all of the fleeing were killed or injured a few seconds later when the second commuter train piled into the wreckage. Both passenger trains were travelling at about 45 miles an hour when they crashed. The force of the second crash broke and scattered the steel passenger cars as if they were matchsticks, fire officials reported. Many of the victims were trapped in the twisted mass of steel.'

Initial reports put the number of dead at 142, with another 385 injured. [309]

The following day the same newspaper reported, 'Five railroad workers were under arrest and charged with negligence Saturday in connection with the three-train pile up in a north Tokyo suburb Thursday that left at least 155 dead and 385 injured. None of the dead or injured were Westerners, according to Japanese sources.

'Workers toiled in driving rain as they continued clearing away the mass of twisted railroad cars at the wreck scene. Johan rail line officials said they expected the clean-up to be completed and the line restored to normal operation by Saturday afternoon.

'The wreck, the second worst rail disaster in Japanese history, occurred at Mikawashima station about three miles north of downtown Tokyo. It involved a derailed freight train that was crashed into by two commuter trains within about three minutes.

'Officials Saturday said they all the deaths resulted from the crash of the second commuter train. Both commuter trains were crammed to capacity with holiday travellers.

'Under arrest and charged with negligence in the tragedy were the engineer and conductor of the freight, the motorman, and the conductor of the second commuter train, and the signalman at Mikawashima station.

'Police charged trainmen on the freight with failing to stop for a signal, and the motorman and the conductor of the second commuter train with failing to exercise caution to avoid the second crash. The signalman at the station was charged with failing to signal the second commuter train to stop.

'UPI reported that the disaster had touched off a political controversy. Transportation Minister Noborn Saito demanded that the government exercise tighter controls over the Japanese National Railways, a public utility. The opposition Socialist Party was calling for an open debate in the Diet on the operation of the railroads. The railroad workers union charged management with overloading facilities and limiting the number of rail workers despite warnings that safety would be endangered, the UPI said.' [310]

On Tuesday 8th May the newspaper stated, 'The Tokyo District Prosecutor's Office Sunday indicted five railwaymen involved in Thursday's pile-up on the outskirts of Tokyo as the death toll climbed one more to 157. The Japanese National Railways employees are accused of failure to perform their duties at the time of Japan's second worst rail disaster since the end of World War II.

'Some 160 persons are still in serious condition and hospital authorities expect the toll to rise. Thirty-two women, including 25-year-old Aiko Iwasaki, who died Sunday night, and 215 men have died so far from injuries suffered in the crash of two commuter trains and a freight at Mikawashima station on JNR's Joban line.

'Noribumi Minakami, 38, engineer of the freight, is accused of failure to correctly identify signals at the crash site while Makoto Ohashi, 52, signalman at Mikawashima station, is charged for turning signal lights to red near the collision scene. The conductor of the freight, Shoju Kiribara, 33, and the engineer and conductor of one of the two commuter trains, Yukio Haga, 30, and Sadahiro Mitojl, 28, are accused of failure to fire signal rockets to notify approaching trains of the accident.

---

[309] *Pacific Stars and Stripes*, Tokyo, Japan, 4th May 1962
[310] Ibid, 5th May 1962

'In a special session of the Japanese Diet Sunday evening Japanese Prime Minister Hayato Ikada and Transportation Minister Noboro Saito promised to prevent repetition of rail accidents by improving training of public sector employees. The JNR is government operated.' [311]

A 20-year-old college student died of his injuries on Friday 11th and a 35-year-old man died a week later. These were the final deaths in the disaster and brought the number of fatalities up to 160. [312]

At the end of May it was announced that a further two Mikawashima signalmen and the assistant station manager had also been indicted for causing the disaster, bringing the number of accused to nine. A further signalman had been released after investigation. [313]

The inquiry into the accident found that traffic on the line had increased from 80 to 240 trains daily over the previous 30 years. The action of the signalmen in reporting the initial collision to the Tokyo Operation Control Centre was blamed on the rules being out of date. The wording of the contemporary rules that 'trains should not be stopped unless absolutely necessary' was doubtless a factor in the decision not to block the line and probably overruled the common-sense reasoning that other trains must be stopped. Or perhaps, faced with a sudden emergency, the signalmen just lost their heads.

The inquiry made three recommendations:

(a) The rules and regulations should be updated to say that if an accident should occur trains should be stopped at once regardless of the magnitude of the accident.

(b) Automatic train stops and train radios should be installed throughout the JNR system.

(c) Separate passenger and freight lines should be provided between Mikawashima and Minowa signal box, a section of track used by a large number of passenger and freight trains.

The recommendations were all accepted and the improvements were made. Automatic train stops were introduced in April 1966.[314] [315]

Between the mid-1930s and the late 1980s details of railway accidents in the USSR were strictly censored, so that no information at all relating to them appeared. However a collision on Thursday 10th May 1962, between two suburban passenger trains at a junction on the lines to Leningrad and the north on the outskirts of Moscow, occurred close to a block of flats occupied by Western journalists and diplomats. Some details of the accident could not therefore be concealed. According to a policeman several people lost their lives. Western reporters noted that a number of coaches were overturned in what was possibly a sidelong collision, that scores of ambulances had gone to the scene and that thousands of Russians were watching the rescue operation. [316]

---

[311] Ibid, 8th May 1962
[312] Ibid, 12th and 19th May 1962
[313] Ibid, 27th May 1962
[314] Webpage - Japanese Railway Safety and the Technology of the Day by Masao Saito
[315] Wikipedia – Mikawashima train crash
[316] *The Times*, London, England, 11th May 1962

Gross negligence on the part of several railwaymen led to a disaster on Friday 6th July 1962 at Církvice, Czechoslovakia, on the Kolín – Děčín line near Sebuzin in Bohemia. At about 16:00 a steam-hauled ballast train approached the station. It was travelling too fast and its driver overran one or more signals. When he saw that there was a freight train on the line ahead of him he applied the brakes but the brake pipe had not been connected up between the engine and the waggons and no brake continuity test had been performed, so that the brakes were only operative on his locomotive and he was unable to pull up. The waggons derailed in the initial collision fouled the opposite running line and were run into in turn by a passenger train. Fourteen people were killed outright, 22 were seriously injured and 43 were slightly hurt. Two of the injured succumbed in the following days.

The disaster was more or less hushed up by the Communist authorities, receiving only a brief mention in a newspaper report. The railwaymen found responsible for it were given long prison sentences. [317]

At about 21:49 on the night of Saturday 21st July 1962 train No. 6, the Down Amritsar Howrah Mail, collided head-on with train No. 1267 Up Small Quick Transit (SQT) goods train at Dumraon station, India, on the Eastern Railway. As a result 66 people were killed outright, eight were fatally injured, and 74 were hurt.

The mail train consisted of WP Class steam locomotive No. 7027 and 12 coaches, in charge of Guard E.V. Shah with Driver L. Braganza and an unnamed fireman/firemen. The SQT train consisted of XC Class steam locomotive No. 22218, 41 waggons and a guards van in charge of Guard R.P. Ram with Driver Mohammad Siddique and an unnamed fireman/firemen.

Both engines were badly damaged, four coaches of the mail train were interlocked and badly damaged and two others were derailed and damaged. The loco rest van and one waggon of the goods train were damaged.

Dumraon was a Class A station on the double line with Standard III interlocking, lock and block working and East and West cabins. The adjoining cabins were Baruna on the west (Mughalsarai) side and Twining Ganj on the east (Howrah) side.

The SQT train arrived at Dumraon from Twining Ganj on the Up Main line at 20:02. Assistant Stationmaster B.D. Sahai had already made out an order, under the instructions of Section Controller A.P. Gupta, at 19:55, for it to cross to the Down Main line for the loading and unloading of packages. The assistant stationmaster gave the driver of the SQT train a shunting order at 20:12, blocked the Down line and the train was crossed via a crossover at the east cabin. It cleared the Up Main line at 20:35 and was then drawn forward along the Down Main line until it was alongside the station platform, where it stopped at 20:45. There were 105 packages to unload and two to load, an operation expected to take about an hour and unlikely to be completed before the Down Mail was due. All had so far been in accordance with the regulations but neither cabinman had recorded the occupation of the Down Main line in his train register. More importantly Cabinman Ram Autar in the West cabin failed to either set the facing points for the Loop line or to block back to Baruna cabin.

[317] https://ustecky.denik.cz/zpravy_region/ustecanka-vzpomina-na-obeti-vlakoveho-nestesti-v-cirkvicich-20120718.html

At 21:40 Baruna cabin asked the West cabin for *Line Clear* for the mail train and a minute later it was given, despite the fact that the SQT train was still standing at the Down platform. At 21:44 the slot was asked for by Cabinman Autar from Cabinman C.P. Singh in the East cabin and at 21:45 it was given, when Cabinman Autar closed his level crossing gates and cleared his signals for the mail train.

The inquiry into the accident examined the actions of each of the railwaymen involved, beginning with Section Controller Gupta, noting that had the SQT train not been shunted onto the Down line according to his instructions the collision would not have occurred. Subsidiary Rule No. 160 laid down that a train must only be shunted onto the opposite line if absolutely necessary but the section controller explained that he had shunted the SQT train to give preference to trains No. 319 Up passenger and No. 71 Up parcels. His decision was stated to have been fully correct by Divisional Operating Superintendent Shahid Ali Khan. The section controller added that at 19:40 he had instructed the assistant stationmaster and the East cabinman to route Down trains via the Loop line.

Assistant Stationmaster Sahai explained that after the SQT train had been shunted to the Down line he had been occupied attending to the Up passenger and parcels trains. According to the rules he had no responsibility for trains not stopping at his station but he said that about a minute before the collision the guard of the SQT train had pointed out to him that the Down signals had been cleared for a train. He had then rushed to his office and tried to contact the cabinmen but at that moment the collision had occurred.

Driver Siddique of the SQT train stated that the headlight of his engine had been lit but dimmed and the red buffer lights had been burning.

Neither the section controller, the assistant stationmaster nor the driver of the SQT train was found to be responsible for the disaster.

Chairman Bishan Narain, who conducted the inquiry into the disaster, noted that it was unfortunate that criminal cases had been opened against the East and West cabinmen as although they had been assured that any statements made before the inquiry commission would not be made public until after the termination of the court case, it was likely that they might withhold some facts from the inquiry.

Cabinman Autar attempted to explain how he had made his fatal mistake. He gave three explanations (excuses):

(1) When the SQT train had been shunted to the Down line the assistant stationmaster had assured him that it would not be at the platform very long and the line would soon be clear. Thus when the East cabin had given him the slot for the mail train he had assumed that the SQT train had been shunted to the Loop line.

(2) He had been unable to see the headlight of the engine of the SQT train and had not therefore realised that the Down Main line was obstructed.

(3) It was up to the East cabinman to give him the slot for a line which was clear.

The first reason the inquiry commission considered to be 'absurd and frivolous.' The assistant stationmaster had denied giving any such assurance and in any case it would have been up to the cabinman to have ensured that the Down Main line was clear, while had the SQT train been moved to the Loop line via the West cabin it would have been obvious to the cabinman there, or if it had been moved via the East cabin then the assistant stationmaster would have had to exchange private numbers with the West cabinman. The explanation was rejected.

Even if the cabinman could not see the headlight of the SQT train he should have known that it could not leave the Down Main line without his being informed by the assistant stationmaster and his log register showed that the Down Main line was blocked.

There was no specific rule as to whether it was the cabinman offering or giving the slot who specified the line on which the train was to run, though in practice it was the offering cabinman who did so, despite Cabinman Autar stating the opposite.

The commission decided that Cabinman Autar was mainly responsible for the accident by clearing his signals for a line which was occupied.

Cabinman Singh in the East cabin admitted that he had given the slot to the West cabin for the mail train and had cleared the Down Starter for it. He pleaded that at 21:40 he had had repeated requests from the section controller to get the assistant stationmaster, as the departure of train No. 71 was being delayed. That train had then left and he had cleared it back to Twining Ganj cabin, while at the same time the West cabin had asked him for the slot for the mail train. He had protested that the line was blocked but Cabinman Autar had persisted and unable to contact the assistant stationmaster, he had finally given in and given the slot. He admitted that afterwards he had altered entries in his train register and had moved levers to make it appear that he had accepted the mail train on the Down loop line.

The commission of inquiry decided that his efforts to cast all the blame on Cabinman Autar were without validity as according to his log register the Down Main line was occupied by the SQT train, while it was up to the assistant stationmaster to advise him if the SQT train had been moved onto the Down loop line after the exchange of private numbers, not the West cabinman. In addition, he was guilty of attempting to destroy evidence.

The commission of inquiry found that Cabinman Singh was also responsible for the disaster.

Driver Braganza stated that the signals had been clear for him to pass through Dumraon and he had been travelling at 52 miles an hour when he reached the station. He had seen the obstruction from a distance of 150 feet. The commission decided that he could not have avoided the collision even had he seen the obstruction earlier.

Turning to means of preventing future accidents of the same type, the commission noted that the collision should have been prevented if the platform line had been track circuited. Track circuits were, however, unsuitable for lines equipped with steel or cast iron sleepers, as many lines in India were, though 90 large Indian stations had been partly track circuited, including 32 on the Eastern Railway. Wooden or concrete sleepers were being substituted for steel or cast iron ones to enable track circuits to be used.

Various alternatives to track circuits used on foreign railways were considered and discussed in detail, such as electronic treadles, inductive devices at the end of trains, ultrasonic presence detectors and axle counters. It was decided, however, that conventional track circuits with wooden sleepers were still the most effective method of ensuring safety and should be installed, at least on the main lines, at wayside stations such as Dumraon.

With regard to other matters the General and Subsidiary Rules and Station Working Rules should be translated into regional languages for staff who could not read English.

The printing of the Hindi rule book should revert to the better and bolder type used in the 1929 edition, while the 1951 rule book in English also needed reprinting.

The education and training of cabinmen should be improved and they should have refresher courses every three years rather than every five years as at present.

The blocking of lines should be recorded in train registers in capital letters.

The Station Working Rules should specify which of the two cabinmen should designate which line the slot to be requested or given applied to.

The proposal that lever locks should be provided on the Home signals, the key of which remained in the possession of the stationmaster while a line was blocked, was worthy of further consideration. The provision of such locks would cost 12,000 rupees for a station the size of Dumraon.

At stations such as Dumraon where there was a single island platform between the Main lines an extra platform serving the loop line would reduce the possibility of accidents and facilitate traffic.

Special instructions should be issued to staff drawing their attention to the necessity of ensuring that the Main line was not ordinarily blocked.

Specific instructions should be issued to the Control staff to the importance of keeping the Main line clear to the maximum extent possible and that whenever instructions were given by the Control staff for the blocking of the Main line by a train, these should be recorded in writing.

Section 101 of the Indian Railways Act should be amended so as to enhance the existing punishment of two years of rigorous imprisonment to five years. [318]

The second major railway disaster in two days in South Africa occurred on Friday 5th October 1962 at the interloop (crossing place) at Klington, near Potchefstroom. A southbound multiple-unit passenger train from Johannesburg to Klerksdorp had to cross a northbound freight train there. The freight train had already entered the loop at the south end but for unexplained reasons the points at the north end of the crossing place were also set for the loop and the passenger train ran into it from the north end, to collide head-on with the Class 5E locomotive of the stationary freight train. Severe telescoping occurred between the steel leading coach of the passenger train and the wooden second coach, killing 11 passengers and injuring 34.

The following month the motorman of the passenger train appeared in the magistrates' court charged with culpable homicide. He was found to have failed to keep a proper lookout, to have been travelling too fast and to have failed to stop in time. He was fined 300 rand and allowed to pay it off at 25 rand a month as he only earnt 160 rand a month. [319]

Forty-two people reportedly died and over 100 were injured when the Oudh – Tirhut mail train and a passenger train collided head-on at Umeshnagar, India, on the Kastihar – Barauni section of the North Western Railway, at about 04:30 on Friday 4th January 1963.

---

[318] Report of the Commission of Inquiry Dumraon Accident 1962

[319] https://issuu.com/railwaysafrica/docs/ra-feb_2010_lr

The engine of the passenger train had just been getting up steam ready to depart when the mail train collided with it. Five coaches overturned and fire broke out in one of them. The injured were taken to hospital in Khagaria, where residents donated blankets and cots for them. 'Failure of Railway Staff' was blamed for the disaster. [320] [321] [322] [323]

Six men died and eight were injured at 4.46 p.m. on Saturday 21st September 1963 when the track motor car on which they were riding was in a head-on collision with an engine. The accident happened at Havre St Pierre, Quebec, Canada, on the line of the Quebec Iron and Titanium Company. [324]

A collision occurred at Leinhausen, West Germany, just before 07:00 on Monday 25th November 1963, during temporary single-line working. As express train No. E 175 from Bremen to Hannover was passing through the driver of passenger train No. P 1161 from Haste to Hannover, filled with workmen and schoolchildren, having received the 'Right Away' signal, started away from the station, against the Ausfahrsignal (starting signal), and collided with the side of the express. Five or six people died and almost 60 were injured, some seriously. [325] [326]

A disaster occurred at Altamirano station, Argentina, at 03:50 on Wednesday 1st February 1964. Standing in the station on a side track was freight train No. 112, bound for the coastal city and resort of Mar de Plata, consisting of a Class 1500 steam locomotive and a number of waggons. It was awaiting the arrival of train No. 56, the *El Luciérnaga* (The Firefly) express from Mar de Plata, which had left there at 22:45 the previous evening and was bound for Buenos Aires Constitución, with 1,040 passengers, mostly returning holidaymakers. The express was due to arrive in the capital at 05:30. It entered the station under clear signals, travelling at 100 kilometres an hour, but was diverted onto the track occupied by the freight train, with which it collided head-on. As a result of the impact the first three coaches of the express flew into the air and landed 60 metres away, while the diesel engine *Meteor* and the fourth and fifth coaches of the express formed a confused heap of wreckage with the engine of the freight train. One of the coaches of the express caught fire. The engine crew of the freight train were able to jump clear when they saw that the express was heading towards them but Driver Villarreol and the unnamed fireman of the express were among the fatalities. In all there were probably 34 deaths and about 70 injured. [327] [328] (Another source states that the accident occurred at 04:10 and that there were 27 fatalities.) A mother who was able to escape from the burning coach realised her two children had not escaped with her and re-

---

320 *The Times*, London, England, 5th January 1963
321 *The Coshocton Tribune*, Coshocton, Ohio, United States, 4th January 1963
322 *Appeal Democrat*, Marysville-Yuba City, California, United States, 4th January 1963
323 https://eparlib.nic.in/bitstream/123456789/2891/1/lsd_03_04_02-03-1963.pdf
324 *The Daily News*, St John's, Newfoundland, Canada, 23rd September 1963
325 *Katastrophen der deutschen Bahnen Teil II* by H.J. Ritzau, pg. 166
326 https://www.nd-archiv.de/ausgabe/1963-11-26
327 http://museoferroviario.flavam.com/galestaltamirano2.html
328 http://museoferroviario.flavam.com/galestaltamirano2.html

entered the coach to save them. She succeeded in doing so but only at the expense of herself suffering terrible burns, from which she soon succumbed. [329]

The accident was reportedly due to the fact that when Altamirano signalman Alberto Basualdo had earlier reversed the points that had directed the express onto the opposite running line they had failed to respond, although the lever was over in his box, thus freeing the running signals for the main track. Presumably there was no detection on the points to prevent this happening, or else that had failed too.

Only 16 hours before the Altamirano accident 17 people had died in a level crossing accident involving a bus at La Picada, Argentina (see Level Crossing Collisions chapter). [330]

Early on Monday 9th March 1964 (or possibly late on Sunday 8th) the Madras – Howrah express was diverted onto the wrong line and collided head-on with a stationary goods train at Baudpur station, India, on the South Eastern Railway. Twenty people died at the scene, two more succumbed on the way to hospital and 110 were injured. The accident came just over a month after the Kunzru Committee had made a series of recommendations to prevent railway accidents. [331]

It was stated in parliament that the disaster had caused 500,000 rupees' worth of damage to the locomotive and 7,000 rupees' damage to the track. *Ex Gratia* payments of 14,350 rupees had been made to the next of kin of 18 of those killed and 34 of those injured. [332]

At about 09:10 on Monday 31st August 1964 two passenger trains collided at Westervoort, the Netherlands, just outside of Arnhem. One was local Dutch train No. 7763, which had Arnhem at 09:07 for Zevenaar and Winterswijk, and the other was international train No. 767 from Hagen, Germany, to Rotterdam, due in Arnhem at 09:15. At Westervoort the double-track line from Zevenaar became single at the east end of a bridge over the River IJssel. The Dutch train had just crossed the bridge and was coming off the single line when its flank was run into by the German train, which all the evidence suggested had run past stop signal No. 304. [333] The German train consisted of an engine and four coaches; the Dutch train was a five-car multiple unit. As a result of the collision the leading vehicle of the German train partially slid into the locomotive, the second vehicle broke in two under the weight of the leading car of the Dutch train, which climbed over it, while the third and fourth coaches of the German train escaped much damage. Fortunately there had been no one in the leading customs car, the most damaged vehicle of the German train, at the time of the collision, the customs officer having alighted at Zevenaar and two other officials being elsewhere in the train. The first four cars of the Dutch train were derailed and all were damaged more or less. Leaking diesel oil from the Dutch train also started a fire shortly after the collision but the blaze was extinguished by the Arnhem fire brigade before it had caused much damage. Struck by

[329] https://archivoferroviario.com.ar/choque-de-colectivo-a-cm-fiat-en-la-picada/
[330] https://archivoferroviario.com.ar/ferro-biblioteca/accidentes/page/5/
[331] *The Indian Express*, Madras, India, 11th March 1964
[332] https://eparlib.nic.in/bitstream/123456789/55673/1/lsd_03_07_31-03-1964.pdf
[333] *Spoorwegongevallen in Nederland* by R.T. Jongerius, pg. 177

the derailed trains, Westervoort signal box was demolished. Signalman H. Oteman saw what was about to happen and just had time to jump out of a window but he was injured, as well as suffering from shock.

Four passengers were killed more or less outright in the collision. They were Mrs Kulper, aged 50, and her 25-year-old son, both from Zevenaar; Driver H. Bongers from Zwolle, who was travelling as a passenger; and Mr H. Kummelink, aged 50, from Amersfoort. A fifth passenger, Mr M. van Dalen, aged 63, from Barneveld, succumbed on the way to hospital. All those who lost their lives had been passengers on the Dutch train. Ambulances went to the scene from Arnhem, Zevenaar and Doetinchem and 22 injured were taken to Arnhem hospital, including the signalman and both train drivers. Fortunately neither train was very full; according to the German staff there had been 47 passengers on their train and the Dutch train was likewise by no means crowded. When Driver Gerritsen on the Dutch train had seen that the other was coming towards his on a collision course he had released the dead man's handle, causing the brakes to apply, and had run back into the train from his cab, shouting to the passengers to do the same. After the accident he was reportedly seen wandering about the scene, clearly suffering from shock, and asking repeatedly, "How could anyone run past a danger signal?" Signalman Oteman agreed with Driver Gerritsen that the signals had been clear for the latter's train.

Head Conductor Franz Gerling of the German train said later, "Suddenly we received a violent blow, but none of the passengers in my car was hurt. The carriage remained on the rails and there was no panic. I immediately got out and was able to call Arnhem station from a house at the foot of the embankment. Assistance was very quickly on the scene." [334]

One of the first rescuers to arrive was Mr H.J. van Slooten, the manager of the Holland concrete factory, which adjoined the railway. He told a reporter, "I was walking about the factory site when I suddenly heard the Dutch train, which had just come off the bridge, blowing its horn loudly. Then I heard a crash and a cloud of dust enveloped the embankment. I saw a train car slowly rise into the air, overturn, and come down with a crash. I ran to the railway with 20 of my employees. We reached the trains in less than five minutes, to find the red carriage of the Dutch Winterswijk train standing on top of the customs compartment of the German train from Hagen. I found 16-year-old Miss Trang from Zevenaar lying bent in two under a bench. We released her at once and she appeared to be unhurt. Nearby were the bodies of a woman and a young man, both dead. A dazed man, his left hand torn off, sat with the stump of his arm between two seats. My men took the necktie from the dead young man and used it as a tourniquet to stem the bleeding of the victim with the badly injured arm. By this time more rescuers were appearing from all sides."

Another person who was early on the scene was 68-year-old rolling mill driver F. Jansen from Westervoort. He related, "I was just eating a piece of bread. I heard a crash. I thought it was an aircraft. Then my grandson came running, shouting, 'Grandpa, the trains are on top of one another.' I grabbed a ladder and crowbars and ran to the line. Others from Westervoort also rushed to the dike in cars and on bicycles. There were no police there yet, no priests, no one. I rescued a woman who kept shouting, 'My money, my money is still on the train.' Then someone said that others were trapped. I went straight where they were and

---

[334] *Het Parool*, Amsterdam, The Netherlands, 31st August 1964

found two passengers who were imprisoned. One was still alive, the other, I think, was dead. Helped by a blacksmith and others I got the survivor out."

An incident which, despite the circumstances, contained perhaps an element of humour, was the presence near the accident of a butcher, Mr W. Hertterschijt, dressed in his white coat. The driver of a Red Cross car, seeing him, assumed that he must be a doctor, stopped, picked him up and took him to the wreckage, where he helped in the rescue work.

The uninjured passengers had to make their way as best they could down a four-metre-high embankment, at the foot of which waiting buses later took them to Arnhem.

Two passengers who had been riding in one of the rear coaches of the German train were interviewed by a reporter in the lounge of the nearby Snelder Hotel, where they were sat awaiting forward transport. Mrs Ingrid Fuchs from Bochum, Germany, said she had boarded the train at Oberhausen and had been on her way to Rotterdam to meet her husband, whose ship was calling there. "We had already had our passports checked," she said. "We felt the train slowing down and thought that we were coming into Arnhem station, then it stopped with a lot of noise and a series of jerks. When we looked out of the window we saw the terrible devastation." Mrianne Ellert from Rotterdam, who had been visiting relatives in Dinslaken and was on her way home, said she had also looked out of the window. "I saw a badly injured man hanging from the train and a child being carried away. There was no panic in our car, which was not badly damaged. We found that the doors were jammed so we were unable to get out. After about ten minutes we were freed. I preferred to complete my journey by car, so I telephoned my fiancé from the hotel here and he will be coming from Rotterdam to pick me up later. Mrs Fuchs will come with us." [335]

The last of the injured were away from the scene by 10:30. A Dutch breakdown crane was supplemented by a 90-tonne crane from Essen, Germany, while during the afternoon Mr van Aatsen, the Minister of Transport, Public Works and Water Management, went to the site. Operations to separate the two trains were somewhat delayed due to fears that there might still be further victims in the wreckage, but they were finally begun at 19:00 and completed by 20:30. While the line was blocked passengers were transported by bus between Zevenaar and Arnhem. Exactly 24 hours after the accident, at 09:00 on Tuesday, the line reopened, trains being signalled by hand at low speed past the accident site pending the replacement of the signal box.

The internal investigation into the cause of the accident by Nederlandse Spoorwegen (NS) concluded that it was almost certain that the German train had run past a yellow signal 1,500 metres short of the stop signal at Westervoort and had then passed the stop signal itself, still without reducing speed. The Dutch signalling was of course different from that on the German side of the border but Driver Erich Röhr, in charge of the engine of the German train, would have received a thorough training in the foreign signalling from Dutch railwaymen before having been passed to drive trains over the line, so it should have been wholly familiar to him. There had been fears initially that Driver Röhr's injuries might prove fatal. After several days his condition improved but almost a week after the accident he was one of the seven injured who were still in hospital. German officials maintained that international trains always had priority, according to the

[335] *Algemeen Handelsblad*, Amsterdam, The Netherlands, 31st August 1964

regulations, but the NS press office stressed that this only applied to D trains, not local workings such as the Hagen one. In any case, under all circumstances it was the driver's responsibility to obey signals. [336]

An interesting question arose as to who was responsible for the financial burden of the accident. An international agreement stipulated that normally the company upon whose line the accident occurred was responsible, but in this case the fault for the accident clearly lay with the foreign driver. Whether this would mean that the Deutsche Bundesbahn rather than NS was financially responsible was an open question, as such a situation had not arisen before. It is not known how it was resolved.

The accident somewhat marred celebrations of the 125th anniversary of the opening of the first Dutch railway in September 1839. [337]

According to first reports eight people lost their lives in a head-on collision between the Up *Kamrup Express* and the Down *Lucknow Express* on the North East Frontier Railway near Kaithalkurichi, Assam, India, on the evening of Saturday 19th September 1964. Two were railwaymen, and five were said to have been on the first named train and three on the other. The general manager of the railway rushed to the scene from Siliguri together with other railway officials, as did Mr K.K. Phookan, the Deputy Commissioner of the Kamrup District, and Mr P.C. Borbera, the Superintendent of Police. A railway medical van took 18 of the injured to the railway hospital at Maligaon. [338]

A second report indicated that the number of dead had risen to 10 and of injured to 26. [339]

The Antara news agency reported that five people had lost their lives and 11 others were injured in a collision between an express and some goods waggons near Purwokerto, Java, Indonesia, on about Sunday 11th October 1964. [340]

On Saturday 24th October 1964 a sidelong collision occurred at Hastedt, West Germany, on the outskirts of Bremen, West Germany. Deutsche Bundesbahn Class 50.40 steam locomotive No. 50 4027 was running light along a goods line as train No. Lz 18 402. At Hastedt the goods line merged with the Bremen – Osnabrück main line and no flank protection was provided against conflicting movements, other than the signals. The driver of the light engine failed to observe the Forsignal (distant signal) for the goods line and at the junction itself misread the adjoining stop signal, which had been cleared for passenger train No. E 526 on the main line, as applying to his engine, and continued at unabated speed. The passenger train was already passing through the junction when the engine fouled the main line and struck the third coach of train No. E 526. The third and fourth coaches overturned and the third coach struck the central girder of an overhead bridge and was demolished, while the engine was derailed all wheels and overturned

[336] *Limburgsch Dagblad*, Heerlen, The Netherlands, 5th September 1964
[337] *De Volkskrant*, Amsterdam, The Netherlands, 1st and 2nd September 1964
[338] *The Sunday Standard*, Madras, India, 20th September 1964
[339] *The Indian Express*, Madras, India, 21st September 1964
[340] *The Times*, London, England, 13th October 1964

against a bridge pier. Seven people died and 36 were injured. [341]

On Monday 2nd November 1964 an English newspaper carried a brief report of a railway accident at Langhagen in East Germany the previous day, in which it was said that 18 people had died and over 100 had been injured. [342] The accident was in fact considerably more serious than this.

On the evening of Sunday 1st November 1964 through freight train Dg 7913, consisting of engine No. 50 323 and 12 stake waggons loaded with gravel, left the VEB Kies und Betong Werk (Gravel and Concrete Works) at Langhagen for the port of Rostock. The train entered the station slowly at about 20:15, the engine running tender first, and as it passed the fahrdienstleiter's office it reduced speed further, when the shunter, who had been riding on the engine, jumped off, Fireman Georg Wensch had asked the shunter if they could proceed on their way and the shunter had replied that he would have to ask the fahrdienstleiter first. At this point the Driver Kart Krause and his fireman observed that the Ausfahrtsignals (starting signals) at the exit from the station were all at danger. The train then proceeded along track No. 4 and came to a stand near the exit from the station. The shunter returned and told the train crew that their signal was at danger because they must wait for express train D 1193 from Berlin to Rostock Hbf, while Driver Krause also knew that his train was not due to depart until 20:36.

When the freight train was some 200 metres from the Ausfahrtsignal for track No. 4 Fireman Wensch announced that it was clear for their train to come out onto the main line. This signal, on the right hand side of the line, could not be seen from the driver's side. According to his own statement, Driver Krause, not wholly convinced, had then come across the engine cab to check for himself and had confirmed that the signal was off. He had then accelerated to 20 to 25 kilometres an hour until, passing the signal for track No. 4, he had observed that the Ausfahrtsignal for track No. 1 was green. Driver Krause claimed that he had at once applied the brakes in emergency. This was not confirmed by Fireman Wensch and it appeared that only the engine brake had been applied, as it would have been anyway to cross to the main track. As the exit points from track No. 4 were not set for the main line the freight train continued for about 230 metres along the trap siding beyond them and then demolished the steel buffer stops at the end of the siding. The engine and tender ploughed into the earth beyond the buffers, the engine then overturning to the right, and the leading stake waggon reared up into the air and in turn overturned to the left. Express train D 1193, consisting of an engine, a mail van and seven coaches, was passing at that moment, travelling at about 105 kilometres an hour, and it was struck by the derailed stake waggon and its load of gravel. Its third and fourth coaches were slightly damaged, the sides of the fifth, sixth and seventh coaches were ripped out and the eighth coach was demolished. The accident happened at about 20:21.

Thirty-six people were killed outright in the accident, five had died of their injuries in the next two days, and at the time of a report on the collision, made on 3rd November, 74 people were being treated in hospitals at Güstrow, Teterow, Rostock, Neustrelitz and Greifswald, six of whom were in a critical condition. The final number of dead was 44. Another 143 passengers were slightly hurt and went home after treatment. Damage was

[341] *Katastrophen der deutschen Bahnen Teil II* by H.J. Ritzau, pg. 135
[342] *The Times*, London, England, 2nd November 1964

estimated at 1,200,200 Mark der Deutschen Notenbank (East German Marks). The line was blocked until 17:00 on Monday 2nd November.

Driver Krause and Fireman Wensch were found to have been in good health and well rested before they had come on duty on the day of the accident. The driver had previously been found responsible for a head-on collision between two freight trains at Sternberg in 1960 and said that he had also been involved in accidents at Teterow in 1956 and at Adamsdorf in 1962, though neither of these had been his fault.

In March 1965 Driver Krause and Fireman Wensch were found guilty of causing the accident through negligence and were sentenced to 5 and 3½ years' imprisonment respectively. [343]

Following the disaster the East German Council of Ministers announced that the remaining wooden coaches in use on the Deutsche Reichsbahn would be phased out as quickly as possible and that in future only safety glass would be used for coach windows. It took over ten years though for these measures to be fully implemented. Train stops were also installed at stop signals to apply the train brakes if they were passed at danger. [344] [345]

A train and a railcar collided between Moatize and Mutarara, Mozambique, and closer to the former place, on Tuesday 30th March 1965. Nine people reportedly lost their lives in the accident, comprising six Europeans, presumably Portuguese, and three 'indigenous people,' while another seven, three adults and four children, were badly hurt and were taken to Tete hospital. [346]

Fifteen railway workmen died in a head-on collision between their special train and a freight train near Kota, Rajasthan, India, on the night of Saturday 19th June 1965. The accident happened on the main line of the Western Railway between Bombay and New Delhi. [347]

Seven people died in a collision on Tuesday 30th November 1965 between a passenger train and a goods train at Zidani Most, Yugoslavia. [348]

A major catastrophe occurred in Burma at about 10 p.m. on Wednesday 8th December 1965. The location was Kyee Taw, a wayside station five miles north of Toungoo Junction (now Taungoo). The Rangoon (now Yangon) – Mandalay night mail, travelling northwards at speed, collided head-on with the engine of a freight train loaded with sugar cane which had been brought out of a siding onto the main line without the protecting

[343] https://www.nd-archiv.de/ausgabe/1965-03-17
[344] https://www.ddr-im-blick.de/jahrgaenge/jahrgang-1964/report/ursachen-des-eisenbahnunfalles-langhagen-2/
[345] Wikipedia – Eisenbahnunfall von Langhagen
[346] *Diario de Lisbõa*, Lisbõa, Portugal 30th and 31st March 1965
[347] *Winona Daily News*, Winona, Minnesota, United States, 22nd Jun 1965
[348] *The Times*, London, England, 1st December 1965

signal being replaced to Danger. The diesel engine and leading five coaches of the mail train were telescoped and at least 80 people died, many of them women and children, while about 100 were injured. The driver and secondman of the mail train were also among the fatalities, as were five members of the armed military escort accompanying the train as a protection against attacks from Communist guerrillas. There were reports of hysterical passengers fighting to escape from the wreckage in the darkness. Ambulances were rushed to the scene while some of the critically injured were evacuated by helicopter. Five railwaymen were arrested in connection with the disaster. [349] [350] [351] [352] [353]

Only two days after at least 16 Portuguese workers on their way home from abroad for Christmas had lost their lives in a train crash at Villars-de-los-Alamos, Spain (see next chapter), Portugal suffered a railway disaster of its own, between Sintra and Algueirão Mem Martins, on Monday 20th December 1965. The accident apparently resulted from a strange mistake on the part of a train driver.

One of the trains in the accident was passenger working No. 4909, consisting of two three-car electric multiple-unit sets coupled together, which had left Lisboa Rossio station at 17:21. This train had called at Algueirão Mem Martins station and was on its way again when it collided head-on with a freight train before it reached Portela de Sintra station. The freight train had been shunting at Sintra station (the next stop on the country side of Portela de Sintra), but for some unexplained reason it left there and proceeded, on the wrong running line, past Portela de Sintra and on towards Algueirão. The guard of the freight train jumped into his van at its rear and applied his handbrake with little effect. A woman crossing keeper at a point 500 kilometres from Sintra station also vainly tried to stop the train but its driver either did not see her in the darkness or ignored her. The driver of the passenger train, Mário Augusto Esteves, stopped it, sounded his whistle, waved a red light towards the oncoming engine, and then ran back into the passenger compartment to warn the travellers. When the trains met at 17:52 the freight locomotive crushed the leading car of the passenger train and its second car was badly damaged.. Driver Esteves and 17 passengers were killed and about 30 were injured, a number seriously. The casualties would doubtless have been greater had a lot of the those in the front cars of the passenger train not alighted at Algueirão, while passengers for Sintra preferred to ride in the rear cars of the train so that they were closer to the station exit. It was also the first day of the school Christmas holidays, so there were no children travelling.

The work of rescuing the injured, of removing the bodies of the dead and of clearing the wreckage went on through the night with the help of floodlights. Firemen, doctors, nurses, Red Cross volunteers, police, soldiers, railwaymen, and civilians all took part. Various senior officials also went to the scene, including Engineer Espregueira Mendes, the director general of the Caminhos de Ferro Portugueses, who later continued by road to another, less serious accident, which had happened the same day at Praia de Ribatejo.

---

[349] *The Canberra Times*, Canberra, Australian Capital Territory, Australia, 11th December 1965
[350] *The Coshocton Tribune*, Coshocton, Ohio, United States, 9th December 1965
[351] *The Victoria Advocate*, Victoria, Texas, United States, 10th December 1965
[352] *The Register*, Santa Ana, California, United States, 10th December 1965
[353] *The Record Herald*, Waynesboro, Pennsylvania, United States, 9th December 1965

Some of the bodies were severely mutilated and at one point it was believed that there had been 20 fatalities. By 08:45 the next morning the freight engine had been removed and the damaged cars of the passenger train had been thrown to the side of the line, for removal later. Slight damage to the track was then repaired and double-line working was resumed; the first train to use the line after the accident passed at 09:25 en route from Lisboa Rossio to Sintra, followed by a train in the opposite direction at 09:50. Police also had to control a large number of spectators by 09:00, though by then there was little for them to see.

Newspaper reports carried the usual sad stories of relatives searching for their loved ones and of occasional happier outcomes. Alice Nogueira Esteves arrived at Sintra hospital in search of her husband, the driver of the passenger train. Firefighter Manual Jacinta da Silva told the woman that he had taken the driver, still alive, to the Hospital de São José in Lisboa. She set off for the capital. It appears that the fireman was mistaken, for the body of the driver was later identified as one of the unnamed bodies recovered from the wreckage.

Sr. Domingos Gaspar identified the body of his brother, Sr. Francisco Antônio de Jesus Gaspar, a young man of 19 who worked during the day and studied during the evenings. At Domingo's side was his brother-in-law, Sr. Feliciano Quinto, who told reporters that the young man had helped to support a large family. Who would provide for them now?

Sitting on a bench was Maria Teresa, aged 16. She was shown a handkerchief and a comb that she thought belonged to her father, Pedro Francisco de Oliveira Graça. She wanted to go into the mortuary to try and find his body but was prevented. An aunt went in instead and returning declared that Maria's father was not there. The girl, near collapse, was supported by staff. She was shown a yellow jumper, was that his? Yes. Black shoes? Yes. With grooved rubber soles? No. She added that her father had been wearing a brown raincoat. But the bodies of the victims had been stripped of their clothes, which were left in the carriages. "We found no brown trench coats," she was told.

It was now 12 noon and Maria Teresa still did not know if her father was dead or alive. An hour later, her waiting was over. Her father appeared at Sentida, alive. He had been unconscious all night at a first aid station in Benfica. In the morning he had regained consciousness and soon recovered sufficiently to go and look for his family. Words were not enough to describe the explosion of joy when they were reunited, a reporter wrote.

Driver Manuel Moreira of the freight train survived the accident, though he sustained a head injury, a broken collarbone, several broken ribs, and possibly other injuries. His mental state after the accident was said to be such that he could not be interviewed by the investigating authorities. He was, however, questioned in São José hospital by a newspaper reporter, to whom he stated that the Algueirão stationmaster had given him orders to proceed. He had done so and could recall seeing a light in front of him, which became larger and larger, and then there had been a terrific bang and a great confusion of people. He had not realised that he had set off on the wrong running line and denied that he had been tired. At this point he was still unaware of the magnitude of the accident, the facts having been kept from him.

Relief Stationmaster Pires de Almeida, who had been standing in for the regular Sintra stationmaster, was also interviewed by a reporter. He said that everyone knew that Driver

Moreira was at fault. He (the stationmaster) had witnesses: the porters at Algueirão, Signalman José Dias and the inspector who should have accompanied the freight train back to Lisboa but had been left behind when the freight train left unexpectedly. He did not know the inspector other than that his surname was Sousa. The order given to the driver had been handed to him upon his arrival and merely indicated the times his train was due to call at different stations; it did not give the driver authority to leave the station. It was a mystery to him why Driver Moreira had done so. [354] [355]

Unsubstantiated suggestions were made that Driver Moreira had been taken ill in some manner, or that he had had a row with the Algueirão stationmaster (and had set off in spite?)

It is not known whether Driver Moreira recovered sufficiently to be charged with causing death and injury through negligence, as he surely would have been, or, if so, he was convicted.

Five people died and about 20 were injured when an express train from Split collided with a stationary freight train at Novska, Yugoslavia, about 100 kilometres from Zagreb, before dawn on Sunday 9th January 1966. A blizzard was raging at the time. [356] Both engines, a sleeper and couchette car, two coaches and three goods waggons were derailed. [357]Both drivers were among the injured, suggesting a head-on collision. [358]

At 16:00 on Tuesday 31st May 1966 Inspector Alexandru Cainaru at Mogosoia station, Romania, received an order to place an approaching railmotor onto the loop line at his station. At the time he was busy playing draughts with the two signalmen in their cabin and thus distracted he forgot all about the order, so that a head-on collision ensued between the railmotor and accelerated train No. 703, arriving from the opposite direction. Forty-one people lost their lives, 71 were injured and material damage was estimated at 1,114,417 lei (about 190,000 pounds Sterling at 2023 exchange rates). The inspector was sentenced to death for his negligence, subsequently commuted to 25 years' hard labour, and he died in prison. [359]

A major catastrophe occurred near Matunga, India, on the Central Railway, at 06:51 on Monday 13th June 1966. A northbound train to Thane had just left Matunga station and was crossing from track No. 1 to track No. 3 when it collided more or less head-on with a southbound train bound for Bombay Victoria Terminus (renamed Mumbai Chhatrapati Shivaji Terminus 1966 and Mumbai Chhatrapati Shivaji Maharaj Terminus 2017).

The two suburban trains were carrying an estimated 2,500 passengers and it was initially feared that there were 100 dead. The day after the accident it was stated that there were 60 dead and another 135 injured had been detained in hospital. The ladies-

---

[354] https://serradesintra.net/acidente-comboio-em-sintra/
[355] *Diário de Lisboa*, Lisboa, Portugal, 21st, 22nd and 23rd December 1965
[356] *La Stampa*, Torino, Italy, 17th February 1966
[357] *The Daily Gleaner*, Kingston, Jamaica, 10th January 1966
[358] *The Arizona Republic*, Phoenix, Arizona, United States, 10th January 1966
[359] https://www.railnet.ro/viewtopic.php?f=119&t=4119

only compartment at the front of the Bombay train suffered the most and at least 18 women were among the dead. Most of the victims were vegetable vendors and factory workers either returning home at the end of their shifts or otherwise on their way to work. Others were schoolchildren. The two motormen were thrown from their cabs and survived the accident, though injured. They were suspended after the accident pending the findings of an inquiry. Hundreds of police, firemen and doctors were called to the scene of the accident, which happened during monsoon heavy rain. Soon after the accident another suburban train trying to pass the scene on an adjoining line was reportedly stoned by irate locals and its motorman was injured. The Municipal General Hospital in nearby Sion was set aside for casualties and closed to other admissions while 100 doctors were rushed from the Dhore Hospital of the Indian navy. [360]

Property damage was estimated at 200,450 rupees. The railway announced an *ex gratia* payment of 500 rupees to the families of those killed and of 300 rupees to those seriously injured.

There was speculation that the rain might have disrupted the signalling system, though one of the trains overrunning signals seems a more likely explanation. Unfortunately it has not been possible to access the results of the inquiry into the collision. [361] [362] [363] [364] [365]

Modern signalling failed to prevent a serious collision near Birmingham, England, on Tuesday 28th February 1967. A London newspaper gave an account of what happened. 'Six men and three women were killed and 11 others were injured when the Manchester – Coventry passenger express and a diesel shunting engine were in collision near Stechford station, Birmingham, today. The four-coach electric train was thrown onto the adjoining line.

'For two hours police and firemen were removing bodies trapped in the wreckage. The passenger train driver, Mr W.D. Burton, of Handford, Stoke-on-Trent, was killed; six of the dead were from Coventry, one from Nuneaton, and the ninth victim was still unidentified last night.

'The crash was on a stretch of line controlled by a new electronic signal box at New Street Station, said to be the most up to date in Britain. The express left New Street at 3.34 p.m. As it approached Stechford the diesel locomotive was shunting carriages on to a section which runs to the main London line.

'Mr William West, a train driver, was standing by another diesel passenger train in the station. He said: "The diesel locomotive was making a movement on the Aston line. The Birmingham – Coventry electric train was approaching very fast from Birmingham. They were in collision at the junction and the electric train was lifted on to the down line. Everything went quiet after the train had come to a stop. I telephoned for ambulances.

---

[360] *Jacksonville Courier*, Jacksonville, Illinois, United States, 13th June 1966

[361] *The Times*, London, England, 14th and 15th June 1966

[362] *The Daily Independent*, Kannapolis, North Carolina, United States, 13th June 1966

[363] *Brownwood Bulletin*, Brownwood, Texas, United States, 13th June 1966

[364] https://www.freepressjournal.in/cmcm/on-this-day-in-history-june-13-1966-major-rail-accident-occurs-in-mumbai-killing-60-people?fbclid=IwAR1t9_7_dpw61ty_NRf3sH6umR5yfTjk4B3Gm7tIpYVtYwYIptqJrmCLvZ8

[365] https://medium.com/@matildayorke/haunting-memories-352ed6e177ff

For a quarter of an hour we saw no one moving except the people we got out."

The driver of the shunting engine, Mr Harold Garnett, of Church Lane, Nuneaton, said: "It was a matter of seconds. I might have been dead. The electric train just caught my engine a glancing blow and I was thrown across the cab."

'Two of the coaches of the train were thrown on their side. The front of the driver's cab was smashed in, and the roof behind the cab rolled back and telescoped.

'People living in houses backing on to the line helped to free those trapped.

'Mr Garnett was treated at hospital; his fireman was unhurt. Three of the injured people were treated in hospital.

'The fast line is part of the newly electrified stretch of railway between New Street Station and Euston. Hourly services of fast trains are due to start on Monday.

'British Railways said there was no danger from electrification after the crash because the inbuilt safety system would have automatically cut off the current.

'An inquiry into the cause of the accident began tonight. Mr Harold Johnson, chairman and general manager of the London Midland Region of British Railways, travelled to Birmingham last night for an on-the-spot report.' [366]

Three and a half miles from Birmingham New Street, Stechford was the junction of the main line from Birmingham (the Stour Valley line), with the double-track main line from Aston and Bushbury Junction (the Grand Junction line), which was used mainly by freight traffic. Since the simplification of the layout the previous year the Down Grand Junction line diverged from the Down Main/Stour Valley line at the Birmingham end of Stechford station but the Up Grand Junction line ran separately through the station to join the Up Stour Valley/Main line at the Coventry end. There were sidings on both Up and Down sides of the station and mains crossovers on both the Stour Valley (at the Coventry end of the station) and Grand Junction lines. The connections at the Coventry end were worked by Stechford Shunting Frame (formerly Stechford No. 1 signal box), while a crossover between No. 1 siding on the Up side and the Up Grand Junction line and the mains crossover on the Up Grand Junction line were worked respectively by Stechford Sidings Ground Frame and Stechford Crossover Ground Frame. None of these three frames could be operated without the permission of the New Street signalman, who operated release switches for them.

The events that led up to the accident began with the arrival of a ballast train at Stechford from Coventry at about 14:50. This train had to return to Coventry, hence its engine had to run round its waggons. It was initially placed in the Down sidings out of the way of a through train, then crossed over via the Down Stour Valley line and the Down Grand Junction line to the Up Grand Junction line. Running round movements were normally made via No. 1 Up siding but in this case the train was too long to fit between the crossovers at each end of the siding without being divided. After the train had come to a stand on the Up Grand Junction line the head shunter uncoupled its engine, which moved forward again over the mains crossover to the Down Grand Junction line. Having reset the crossover again the shunter then shouted and made a hand signal to the guard of the ballast train, which was misinterpreted by the secondman of the ballast train

[366] *The Times*, London, England, 1st March 1967

locomotive as a signal for the engine to move in the wrong direction, up the Down Grand Junction line. As the engine approached the diamond crossing with the Stour Valley lines the secondman noticed that there was a train standing in the Down platform and realised that something must be wrong. He shouted to his driver, who applied the brakes in emergency, but the engine did not stop before it was slightly foul of the Up Stour Valley line, where it was immediately struck by the Up passenger train, the 13:15 Manchester to Coventry express, a four-car electric multiple unit.

Guard J.W. Smith of the Manchester passenger train told the Ministry of Transport inquiry into the accident that his train had been running on time and travelling at about 65 miles an hour under what he had assumed were clear signals, though he could not see them from his position in the train. He had noticed no brake application.

Driver H.J. Lillystone stated that he had been riding in the cab of the passenger train with Driver Cobb and Driver Burton; the latter man had been driving. He (Lillystone) and Cobb had both been learning the road. The signals had all been green and he put the speed of the train at the moment of the accident at 50 miles an hour. They had seen the light engine moving slowly towards their running line but "we were on top of it before I realised it was foul of the line." Asked if Driver Burton had applied the brakes he replied, "I could not tell you but he would not have stood a chance if he had done so, he was too close to the diesel before we realised it was foul."

Signalman C.H. Curtis was in charge of Stechford Shunt Frame. He said that Head Shunter R.H.J. Tolster had told him, when it had been in the Down sidings, that the ballast train was too long for its engine to run round it in the normal manner. He denied having rebutted a suggestion by the shunter that the train should be divided to allow the run round movement because the train crew would object, because they were in a hurry to get away and such an operation would take longer. He said that the shunter had insisted on the train being taken to the Up Grand Junction line via the Down Stour Valley line rather than via the Coventry end of the station. He maintained that the responsibility for the run round movement rested with the shunter, though he more or less admitted that he knew the planned way of doing things would involve an unauthorised movement up the Down Grand Junction line, as the only other way of running round such a long train would have been to take it along the Grand Junction line to Aston and run round there, something that clearly was not intended. Such a move as was planned he could not remember having been made before. When he had realised that the engine was setting back in the path of the Manchester train, he said, "There was nothing I could do about it." Seeing a large cloud of smoke and dust he had realised that there must have been a collision and had telephoned the booking office and Signalman R.W. Bradbury in New Street panel box.

Head Shunter Tolster testified that on realising that the ballast train was a long one he had suggested to Signalman Curtis that it should be divided, but the signalman had rejected the idea. He described uncoupling the engine of the ballast train after it had been moved to the Up Grand Junction line and the locomotive moving forward onto the Down Grand Junction line. He had then seen the guard coming towards the front (Coventry) end of the train so had shouted and hand signalled for him to return to the rear of his train. Suddenly he had realised that the engine was setting back towards the Stour Valley lines. He had not intended to signal to the engine crew, saying that, without the agreement of the panel signalman, "I would have been mad to have done a thing like that." He knew that he had been planning to make an unauthorised move but said that he would only have done so with the permission of the signalman, as had been done once previously. He admitted that he had not come to a clear understanding with the engine crew over what was to be done.

Booking Clerk H. Hunter said that he had been advised over the telephone by Signalman Curtis of a probable accident and having gone to the platform and seen what had happened he had dialled 999 for the emergency services and confirmed with Control that the overhead line equipment had been isolated.

Driver H.D. Garrett of the ballast train said he had arrived at Stechford at about 15:00 and his train had been put in the Down sidings. After about 25 minutes he had taken his train across to the Up Grand Junction line under the instructions of his guard. The head shunter had then uncoupled his engine and he had moved it forward over the crossover onto the Down Grand Junction line. He had realised that his train was too long to run round as normal via No. 1 siding and that it would have to be an unauthorised move back the Down Grand Junction line, though no one had told him so. He had been relying on Secondman B.K. Sugden to convey any signals from the shunter and his secondman had told him that the crossover had been replaced and to set back, whereupon he had begun to do so. All of a sudden, Sugden had shouted for him to stop and he had applied the brakes. He realised that he was responsible for moving the locomotive but said that he had had complete confidence in his secondman, though he had not worked with him for very long. He had worked at Stechford frequently before the resignalling, when a run round movement could be made without fouling the main lines but had not made such a move since. He realised that it was a wrong direction movement onto the main line without fixed signals but denied being aware that such a movement could only be made after obtaining permission of the signalman, and then only in emergency, adding, “We often go back the way we come in under the new signalling.”

Secondman Sugden spoke of his having seen a hand signal from the shunter, which he said he had been expecting, and which he had interpreted as meaning they were to set back towards the Stour Valley lines. They had been fairly close to the Up Stour Valley line when he had seen the train in the Down platform, the line over which they were to pass, had realised that something must be wrong, and had called out to his driver. He considered that the shunter’s hand signal had been sufficient authority for the move and had not realised that it was an unauthorised one.

Goods Guard C.R. Gooden stated that he had not been aware that his train was too long to run round in the normal manner until it was brought over to the Up side. He had realised then that it would have to do so via the Stour Valley lines and had assumed that this move had been arranged with the New Street panel signalman. He had been on his way to the brake van at the Coventry end of the train to remove the tail lamp when he had seen a hand signal from the shunter, though he had not thought that it was intended for him.

Driver William West of the passenger train at the Down platform said that he had seen the engine of the ballast train moving back along the Down Grand Junction line and had not been able to understand it. The driver of the Manchester train had whistled just before the collision, which he (West) had witnessed.

Signalman Bradbury in New Street panel stated that Signalman Curtis had asked for a release for Stechford Shunt Frame at 15:25, to move the ballast train from the Down sidings via the Down Grand Junction line to the Up Grand Junction line. He had not understood why the move was to be made this way as it was usually made via the Coventry end of the station but had not questioned it. The move having been completed he had cleared his signals for the Up and Down Stour Valley lines. He had not realised that the ballast train had been moved to the Up side and had not been asked to permit a move up the Down Grand Junction line. Had he been, he would not have agreed to it without consulting his regulator, as it was an unauthorised move permitted only in an emergency.

New Street Regulator F.H.A. Perry stated that he would not have permitted such a move, had it been proposed to him.

Colonel D. McMullen, the inspecting officer, concluding his report, stated that the cause of the accident was clear. The engine of the ballast train had made an unauthorised move for which no proper arrangement had been made. There had been no failure of the signalling equipment and Driver Burton, who had lost his life, had in no way been to blame.

The situation had been an unusual one inasmuch as the ballast train had been too long to be run round in the normal manner. Other than dividing the train there had therefore been no way of conducting the run round movement at Stechford other than by a wrong direction movement along the Down Grand Junction and Down Stour Valley lines, a manoeuvre which was specifically prohibited.

Head Shunter Tolster stated that he had suggested to Signalman Curtis that the train be divided. Curtis denied this and at any rate the head shunter had not pressed the matter. Although her had at first denied it Signalman Curtis had clearly been aware that an unauthorised move was to be made. He had not been in a position to order the shunter not to make the move but given his greater age and experience could have attempted to dissuade him. He had not done so; nor had he advised the New Street signalman. Instead he had tacitly agreed to the move; he was therefore partly responsible for the accident.

Head Shunter Tolster was responsible to a considerably greater degree. He knew that he was proposing an unauthorised move and should not even have started it without consulting the New Street signalman. He had no doubt intended to do so and his actions had been influenced by a desire to save time. His misinterpreted hand signal, intended for the guard, and his premature reversal of the Grand Junction lines crossover, which had removed the protection provided to the Stour Valley line, had also doubtless been motivated by a wish to save time. To say the least he had been very imprudent, but more than that he had been negligent in not coming to a clear understanding with the driver as to what movements were to be made. He had, however, given his evidence in a most forthright manner and had not attempted to excuse himself.

Secondman Sugden should not have accepted a hand signal from the shunter, which he agreed had only consisted of two movements of the shunter's arm, without verifying it.

Driver Garratt had not been justified in accepting the word of a secondman with whom he had worked so little and should himself have verified the shunter's signal for such an unusual movement. He should have known that it was an unauthorised movement and he and Sugden were both at fault for not coming to a clear understanding with the traffic staff regarding the unsignalled movement. Both men were also at fault for not complying with the Appendix Instructions which required that the secondman should occupy the leading cab of the locomotive in the direction in which it was moving. Had he been there he would have seen the train in the Down platform a few seconds earlier and the collision might have been averted.

It was clear that Signalman Bradbury knew nothing of the unauthorised move and he was not in any way responsible for the accident.

As regards remarks, the inspecting officer stated that it had been considered that a reverse move over the Down Grand Junction line towards the Stour Valley lines would only be made in emergency and the signalling had not been designed to cover such a move, which was covered instead by detailed instructions contained in the Appendices to the Working Timetables. This, he thought, was completely reasonable. However, as further precautions, the Birmingham end points of the Grand Junction lines crossover had since been disconnected from the ground frame and converted to spring points to prevent unauthorised movements up the Down Grand Junction line, while the track circuit covering the diamond crossover with the Stour Valley lines had been extended along the Down Grand Junction line to the mains crossover.

The question had also been raised, Colonel McMullen continued, as to whether vestibule or compartment stock was safer. The coaches of the Manchester train had been of the vestibule type; they had had steel underframes, steel bodies and Buckeye couplings and had not telescoped but most of the passengers had been riding in the front of the train. He did not consider that they would have been any safer had they been of the compartment type. [367]

Barbara Castle, the then Minister of Transport, was questioned in the House of Commons about the accident. She told MPs that the chief inspecting officer had established that the fact that the line was electrified had nothing to do with the accident, nor had it anything to do with the new signalling. [368]

Damage to equipment of an automatic level crossing by a road vehicle, variously described as a car and a lorry, led in turn to a double collision in the station at Fexhe-Le Haut-Clocher, Belgium, on Wednesday 4th October 1967. The initial level crossing accident at Monmalle, which did not involve a train, damaged both the crossing barriers and the warning lights and the crossing had to be placed under manual control, with trains being cautioned at the preceding stations and given written instructions to approach the crossing at reduced speed. The automatic signalling in the vicinity of the crossing was put out of action, with all signals reverting to red, including those at Fexhe-Le Haut-Clocher station, four kilometres from the crossing.

Three trains were involved in the accident. A Verviers – Oostende and a Leuven – Liège train were in Fexhe-Le Haut-Clocher station at 13:30. The train from Verviers was signalled away but a set of points was in the wrong position and its engine was diverted onto the line occupied by the train from Louven, which was heading in the opposite direction, and struck its rear coach. The colliding engine was diverted to the left as a Liège – Brussels express was passing on the third running line, travelling at 120 kilometres an hour, and struck its fourth coach, which was wrecked. The remainder of the express was derailed.

---

[367] Ministry of Transport report

[368] *The Times*, London, England, 2nd March 1967

The station staff reportedly kept their cool and called the emergency services, including the Red Cross. Ten ambulances ferried the injured to hospital in Liège and surrounding towns. Nine passengers were killed outright, a tenth died in an ambulance and an eleventh after admission to hospital. About 75 others were injured, some very seriously. Those who lost their lives included a five-year-old girl. All those who died were Belgians, most of them from the Liège area.

During the afternoon Prince Albert of Liège and Minister of Transport Bertrand visited the scene. [369] [370]

An unnamed 50-year-old signalman at Fexhe-Le Haut-Clocher was questioned later the same day and then arrested, charged with culpable homicide, having admitted causing the accident. [371]

In the late afternoon of Sunday 28th January 1968 was a major disaster at Fanfa station, Brazil, in the municipality of General Câmara, involving a sidelong collision between what was described as a mixed milk train, which was carrying over 100 passengers, and a freight train consisting of an engine and 23 waggons. The accident occurred on a line of the former Viação Férrea do Rio Grande do Sul, by now part of the Rede Ferroviária Federal. Driver Alres Teixeira of the mixed train alleged that the freight train, coming from Santa Maria, had overrun a stop signal to strike his train, which had been bound for Porto Alegre. Seeing the freight train approaching fast he had increased the speed of his train in an effort to avoid the collision. As a result his engine and the leading luggage van of the mixed train had got clear but the locomotive of the freight train had instead struck and demolished the following passenger coaches. A lorry parked near the station was also wrecked. At least 42 people died and about 50 others were injured, many seriously. All rescue services in the area were mobilized, including the Army, and the injured were taken to hospitals in Montenegro, Novo Hamburgo, São Leopoldo, Triunfo and Porte Alegre. Material damage was estimated at a million cruzeiros.

Driver Emilio Andrade da Salva of the freight train reportedly alleged that his 'accelerator' had jammed so that he was unable to stop. After the accident he apparently committed suicide by throwing himself from the water tower at Fanfa station. Governor Peracchi Barcelos visited the scene and promised measures to prevent a recurrence of the disaster. [372]

On the evening of Tuesday 19th March 1968 a Southern Railway Birar – Hubli passenger train was standing in the station at Yalvigi, India, when the Poona – Bangalore *Deccan Express* collided with it head-on at 10.35 p.m. Five bogies of the trains were reportedly smashed to smithereens and others were telescoped. Early reports spoke of 50 people killed and scores of others injured.

'The toll may go up when the rescue teams reach the bodies trapped inside the tangled mass of the crowded compartments which have been badly smashed up. Reports from the

---

[369] *Leeuwarder Courant*, Leeuwarden, The Netherlands, 5th October 1967
[370] *Niewsblad van het Noorden*, Groningen, The Netherlands, 5th October 1967
[371] *Tubantia*, Enschede, The Netherlands, 5th and 6th October 1967
[372] *Folha de S. Paulo*, São Paulo, Brazil, 29th and 30th January 1968

accident spot at 2 a.m. said the concerned officials were too dazed to report in detail even at that time,' a newspaper wrote.

Special trains left for the scene of the accident from Arsikere, Bangalore, Harihar and Hubli, the Bangalore train leaving there at 1.05 a.m. Another special train from Madras was expected at Bangalore at 6 a.m. with the plan being for the officials to continue to the accident site by air. [373]

When Railway Minister Poonacha announced details of the accident in Parliament the following day there were calls for his resignation from the Opposition. He stated that there would be a judicial review into the accident, conducted by a High Court judge with two assessors. A separate Department Inquiry was to be held by Mr H.S. Hart, Additional Commissioner of Railway Safety. By then the number of dead was officially stated to be 53, with 41 injured, 13 seriously. Unofficial sources put the toll much higher. The confirmed dead included Driver Vishwanath of the express, while Ticket Examiner Nandagopal, Driver H.D. Godayya and Firemen Arumugam (of the express), N. Sannappa and H. Homballayya were among the injured.

The Hubli passenger train was running three and a half hours late. It should have reached Hubli at 10.30 p.m. but its engine failed at Harlhar. Its driver eventually managed to rectify the fault temporarily and it had continued to Yalvigi, where the engine had failed again. It had reached Yalvigi only a few minutes before the express was due there.

'The brunt of the impact of the collision was apparently borne by the two engines, two third-class carriages of the express next to the engine, and two bogies of the passenger train. The first two engines had fallen on their sides, one on the right side of the track and the other on the left. The two sleeper coaches of the express, which were packed to capacity, had been smashed and had capsized. The casualties were principally in these two bogies. The bogies of the passenger had telescoped. Fortunately, however, they were sparsely occupied, and it is believed there were no fatal casualties among their passengers as many of their passengers had come out of their compartments and were moving about the platform when the accident occurred.

'A Hubli businessman, who was travelling on the passenger train, said that the train reached Yalvigi ten minutes before the accident occurred. Immediately the train steamed into the station, he said, some passengers brought out of their compartments an alleged thief whom they marched to the office of the stationmaster. This was followed by a little commotion on the platform. In the meanwhile, the express was steaming towards the station.

'Another passenger who was travelling on the passenger train said the impact was such that he any others were thrown down. There was complete confusion for five minutes, and nobody knew for a while what had happened. He said the wheels of the engine of the express were revolving for some time after it had fallen sideways on the loop line.

'An eye witness mentioned that but for the fact that Vishwanath, the driver of the express engine, had applied the brakes, the impact would have been more serious, resulting in many more passengers being killed. The slowing down of the speed of the train reduced the impact, but still an accident of this kind is one of the worst, involving 100 passengers.

---

[373] *The Indian Express*, Madras, India, 20th March 1968

'Mr Vishwanathan, an advocate of Namakkal in Salem district of Madras state, said: "I felt a sudden jerk. Between my III class three-tier bogie and the engine, there was a two-tier I class bogie. As I managed to come out of my compartment, I found the first-class bogie smashed to smithereens. I heard groans beneath the wreckage."

'Mr Srinath, advocate son of a leading advocate of Bangalore, Mr S.K. Venkatarangaiyenger, who travelled by the express, said the wheels of the turtled engine were still spinning when he came out of his compartment. He paid a handsome tribute to the women passengers of his compartment who maintained perfect calm, to the medical staff who were prompt in their help and two the express driver for his pluck in applying the brakes to prevent an even greater disaster.

'The driver, Vishwanath, it is learnt had planned to join his family in Bangalore for a pilgrimage to Palghat (Kerala) to worship his family deity.

'Twenty-year-old Nalini, whose marriage was to have taken place within the next fortnight, sustained fatal injuries in the accident. She was travelling from Poona to Bangalore with her father, Mr T.R.K. Rao, an engineer in a private concern. The father escaped with a fracture in the left forehead. He was dumbfounded.

'Kashinaih, a constable, who boarded the train at Hubli, and was travelling on duty carrying some murder case exhibits for chemical analysis, could be identified only by his turban. Another constable, who also boarded the train at Hubli and sustained fatal injuries, was however easily identified by his uniform.

'Mr Shahanpurkar, a Circle Inspector of Police, got into a compartment at Hubli with five of his friends. He safely jumped out of the train at the impact but two other Circle Inspectors were injured when they followed suit.

'Mr Nandagopal, a ticket examiner, volunteered to take charge of a two-tier compartment in the night, relieving his colleague, Krishnish, only to die.

'Mr K. Ananthanaryan, a young man returning home on leave from Kuwait, brought great relief to his anxious mother, who was waiting on the platform; tears of joy greeted his safe homecoming but he appeared still bewildered and declined to say anything.

'The smashed bogies had to be opened with crowbars before the injured and the dead could be extricated. The railway authorities received help from the people of the nearby village, who rushed to the site, as well as from passengers of the two trains. In one bogie, a man was noticed with one of his hands caught in a window. The window had to be broken open to release him.

'As the team of railway, police, and other officials started clearing the debris, the fears of heavy casualties came true. One by one the bodies were recovered and the process continued the whole of Wednesday. The injured were rushed to Hubli, Haveri and Harihar hospitals.

'The accident caused disruption of traffic as both the main lines and the through line were blocked. The resumption of traffic is expected only on Wednesday.

'The bodies of 35 persons were taken to the Karnatak Medical Collage Hospital at Hubli and placed in the mortuary for identification. The bodies, in badly mutilated condition, included those of seven children and nine women. None of the bodies had been identified till late Wednesday evening. The hospital authorities said that they had admitted 27 injured persons, three of whom died subsequently.

'The relief train carrying the stranded passengers of the *Deccan Express* reached Bangalore on Wednesday evening.

'Ex-gratia payments have been ordered to be paid through the Divisional Commercial Superintendent Hubli, who is at Yalvigi. On proper identification, the next of kin of the deceased will be paid at the rate of Rs 500 each. As regards passengers who had sustained serious and minor injuries, ex-gratia payment was being arranged through a committee, the rates ranging from Rs 150 to Rs 400.

'All the injured persons have been admitted in the civil hospital at Hubli; the relatives of the injured and also of the deceased are advised that they can travel free to Hubli, according to a spokesman of the PRO office of the Southern Railway at Madras. Enquiry offices have also been set up at the accident spot, Yalvigi; and at Bangalore City and Madras Central.

'The Union Minister of State for Railways, Mr Parimal Ghosh, accompanied by Mr. G.P. Bhalla, Member Mechanical Railway Board, have reached the site of the accident for on-the-spot study of the situation. They left Delhi early this morning by special plane for Belgaum. According to a message from Delhi, five Members of Parliament from Mysore had also left by air for Belgaum along with Mr Parimal Ghosh. They are Mr B. Sankaranand, Mr S.H. Moshin, Mr S.B. Patil, Mr Patil Kotappa and Mr Veerannagowda.'

Assistant Stationmaster M.V. Patil and Pointsman Siddappa were arrested, charged with negligence.[374]

The death toll had increased to 54 by the next day.[375]

Renigunta station, India, on the Southern Railway was the scene of a serious collision on Tuesday 28th May 1968. The stock of train No. 418, the Up Renigunta – Madras passenger train, due to depart at 04:45; and of train No. 93 Down Renigunta – Raipur passenger train, due to depart at 04:50, was stabled on No. 5 road when it was run into by No. DD 27 Down goods train at about 03:40. A number of passengers had already boarded the stabled coaches, 11 of whom were killed outright, two of whom died on the way to hospital and another later. Fifty-one other passengers were injured, 15 seriously.

The goods train consisted of Class WDM-2 diesel locomotive No. 18337, running with its long hood leading, 66 waggons and a van, in charge of Guard T.S. Rangasami Achari with Driver Venkatamuni Naidu.

Renigunta was a Class B junction station with four-metre-gauge lines (roads 1 to 4) and eight broad-gauge lines (roads 5 to 12), of which broad-gauge roads 5 and 7 were track-circuited platform lines. The layout was controlled by A and B cabins. There was a permanent speed restriction from the Broad Gauge Down Outer signal through the station yard of 15 kilometres an hour.

Platform Assistant Stationmaster J.N. Purushottam at Renigunta told the inquiry into the collision that in order to clear Road No. 8 for the reception of the goods train he had placed the rakes of the two passenger trains on Road No. 5, separated by about two bogie lengths. There had been no red lights on the ends of the coaches to protect them. He had

---

374 Ibid, 21st March 1968
375 Ibid, 22nd March 1968

exchanged private numbers with the two cabins informing them that the goods train would go into Road No. 8. Shortly afterwards, at 03:45, the goods train had entered Road No. 5 and collided with the coaches. At about 04:10 the guard of the goods train had told him that his train had stopped at the Outer signal, which had then been cleared. He had never known of a wrong-side failure of the track circuits at Renigunta.

Assistant Stationmaster Bheema Naraslugh in Renigunta A cabin stated that the goods train had entered the section at 03:25 and ten minutes later the platform assistant stationmaster had informed him that the train was to be routed to Road No. 8. Pointsman Thyagarajan had then shouted that the goods train was approaching the Home signal. He (Naraslugh) had displayed a red signal as the train was passing the cabin. No one was to be seen in the driver's cab and the collision had then occurred at 03:40.

Pointsman Thyagarajan testified that he had heard Shunting Jamadar Khadar Sahib and Pointsman M. Ramaraju calling out from the ground nearby and seeing train DD 27 approaching he had shown it a red light, to no avail. As the collision occurred the light on his panel had shown that B cabin had just given the slot for the goods train on Road No. 8. He had not attempted to clear any signals and had never known a signal to clear for an occupied road.

Shunting Jamadar Sahib and Pointsman Ramaraju spoke of seeing the goods train passing the Home signal despite its white backlight being visible (indicating that it was showing red to the driver) and its arm horizontal and shouting.

Signal Khalasi D. Pullah spoke of joining the previous two men in trying to raise the alarm.

Assistant Stationmaster M.B. Ramachandra Rao in Renigunta B cabin stated that after exchanging private numbers with A cabin and as he was giving the slot for the reception of train DD 27 on Road No. 8 he had heard the sound of the collision at 03:40.

Pointsman M. Viswanathan in cabin B confirmed the statement of the assistant stationmaster there.

Driver Naidu said that he had left Arkonam with engine No. 18337 at 23:00 the previous evening and had proceeded to Melpakkam where he had picked up his train. Guard Achari had been with him in the engine cab as far as Melpakkam but there had been no assistant driver. Approaching Renigunta the Warner, Outer and Home signals had all been red. He had stopped at the Outer signal for five or seven minutes, then both it and the Home signal had changed to green, though he could not see the route indication at the Home signal. Near the A cabin two or three men had shouted to him when he had crossed his cab but he could not see anything. Crossing back again he had tripped over his line box and accidentally opened the throttle fully. As he had got up, closing the throttle, and applying the brakes, the collision had occurred, at 03:45. His brakes had been working properly as had the speed indicator; the vigilance device, however, had been inoperative. He had hoped the assistant driver would join him at Melpakkam. He knew that it was against the rules to drive without an assistant driver and had done so to oblige the traffic department, though he had not been specifically asked to do so.

Guard Achari pleaded that he had been unaware that the assistant driver had not joined the train at Melpakkam. He at first alleged that his train had been stopped at the Outer signal at Renigunta for 10 minutes, when he had then seen a green light. It was pointed out to him that the timings in his rough journal book had all been overwritten and under further questioning he admitted that he had, in fact, not seen any of the Renigunta signals, nor had the train stopped outside the station, while it had entered the station at an excessive speed of 30 or 40 kilometres an hour. Also, his driver had smelt of liquor, and after the collision Driver B.V.S. Rao and others had persuaded him to alter the timings in his journal and to say that they had been stopped 10 minutes outside of the station.

Assistant Driver G. Punniah said that the engine had already left Arkonam when he had gone to join it.

Assistant Medical Officer Dr. G. Subramanyam at Renigunta agreed that Driver Naidu had smelt slightly of alcohol at about 04:30, despite which he had not carried out a blood test or tested a stomach wash, nor had he asked any other doctor to do so.

Additional Commissioner H.S. Hart noted that the train had passed Pudi at 03:24 or 03:25 and had probably only taken 15 minutes to cover the 9.56 kilometres to Renigunta, putting the time of the collision at about 03:40.

From the damage caused in the collision the speed of the train on impact must have been 45 or 50 kilometres an hour.

The signalling had been tested after the accident and been found to be working normally. With the points still set for the track circuited Road No. 5, awaiting the slot being given from B cabin for No. 8 road, the Home and Outer signals could not have been cleared. In view of this, the statements of the assistant stationmaster and pointsman in A cabin that they had not cleared any signals, and the statements of the shunting jamadar and pointsman that they had seen the signals at danger, the driver's statement that the signals had been clear must be rejected. The guard's statement that he had seen the signals 'Off' he had later confessed to be untrue.

Driver Naidu had admitted that the brakes of his train had been working, and they were tested and found to be working correctly after the accident. Visibility had been good. The story of the driver having tripped over his line box and accidentally opened his throttle was hardly credible and must be dismissed as an afterthought. The driver was solely and personally responsible for the accident. He was also responsible for operating his locomotive without an assistant in violation of General Rule No. 98 (a) and Subsidiary Rule No. 98 (i) (a). In the absence of any tests to determine the matter it could not be known if he had been intoxicated.

The platform assistant stationmaster stated that it was not usual to protect stabled coaches with red lights when they were on track-circuited platform lines within fixed signals. As the platform lines were on a curve and the train had been approaching at high speed it was unlikely that the red lights would have made any difference, and the assistant stationmaster must therefore be absolved from responsibility for the collision, but the omission of such warning lights should be looked into.

Additional Commissioner Hart made seven recommendations and observations.

(1) The rules regarding alcohol consumption by operating staff should be made more stringent.

(2) Subsidiary Rule 154 (vi) (regarding vehicles containing passengers not being left on running lines for long periods and exhibiting warning lights at night) should be complied with in future.

(3) Vigilance devices should be properly maintained and fitted to all locomotives.

(4) Action should be taken in the matter of some drivers having prevailed upon the guard to make false entries in his journal.

(5) The approaches to platform lines 5 and 7 should at once be thoroughly screened, especially as they were track circuited.

(6) Two rails on Road No. 7 had had holes in their webs. They had been changed.

(7) The train examiner at Jalarpet had only superficially examined the vehicles of the goods train and said that he had been told that incoming trains should be thoroughly examined, with the examination of outgoing trains being expedited. As a result the brake rigging of 26 of the 66 waggons in train DD 27 had been improperly adjusted. Outgoing trains should be thoroughly examined. [376]

At 23:55 on Friday 15th November 1968 No. 2 GD passenger train left Son Nagar station, India, on the Eastern Railway under clear signals and was crossing the double track of the Grand Chord line to reach the single line to the Garwa Road branch, when it was run into in a sidelong collision by No. 27 Up Electric Crack goods train. The fourth to seventh coaches of the passenger train were derailed, the first three capsizing, and the engine and 11 waggons of the goods train were also derailed and capsized. Nine people died and 55 were injured, nine seriously.

Driver Shankar of No. 2 train told the inquiry into the accident that he had picked up No. 17 token for the branch from Son Nagar East cabin and was crossing the Up Main line when the goods train ran into the third bogie (carriage) of his train. He had seen the light of the goods train but had not been able to tell whether it was stationary or moving.

First Fireman Narayan Mandal more or less corroborated his driver's evidence.

Second Fireman Nandji Choudhury said he had been collecting coal from the tender when the first fireman had shouted and the goods train had dashed into them.

Guard J.N. Mishra stated he had whistled for train No. 2 to leave Son Nagar as soon as the starting signal had been cleared and had been stood at the door of his van when a severe jerk had thrown him down and knocked him out.

Assistant Stationmaster D. Kar stated that the power had failed at 23:59 and he had learnt of the collision. Priority was given to passenger trains for the branch over Crack Special goods trains on the Up Grand Chord line, unless directed otherwise by Control and train No. 2 had already been 20 minutes late.

Switchman Ram Chandra Ram in Son Nagar East cabin said that he had given *Line Clear* to Anugraha Narayan Road on the Grand Chord line for train EC/27 and received the *Train Entering Section* signal for it at 23:50. At 23:52 he had given the slot to the West cabin for train No. 2, it had arrived at the station at 23:55 and he had handed the driver the single line token at 23:58. As the passenger train was crossing the main line he saw that the goods train had run past the Up Home signal but could do nothing. He disputed the statements of the driver and assistant driver that he had thrown back the

---

[376] Government of India Ministry of Tourism and Civil Aviation (Commission of Railway Safety): Report on the Collision Between Goods Train No. DD 27 and Passenger Coaches and Waggons pertaining to No. 481 Up Renigunta – Madras Passenger and No. 93 Down Renigunta – Raichur Passenger at Renigunta Station of Southern Railway on 28th May 1968

Home signal in their face.

Leverman Hari Gopa in Son Nagar East cabin said that he had just been about to deliver the token to train No. 2 when he had seen the goods train running past the Home signal at great speed, had shown a red light towards it and shouted, and the collision had occurred.

Switchman S.N. Mandal in Son Nagar West cabin testified that he had heard the noise of the collision at about 00:00 but had only been able to contact the East cabin some time later. The East cabin had not asked for a slot for No. 27 goods, of which he had known nothing.

Guard S.S. Sarkar in charge of train No. 27 testified that after leaving Gaya they had been stopped at Gaya West cabin, Rafiganj and Jakhim. Passing Anugraha Narayan at 23:48 he had exchanged *all right* signals with the cabins there. He had then sat down in his van and suddenly there had been a series of heavy jerks and his train had stopped. He had felt no previous brake application.

Driver S.K. Ojha of train No. 27 had disappeared after the accident, having feared for his safety. Later he reported at Gaya, was first taken to hospital, and then arrested and lodged in jail. Questioned by Additional Commissioner Arya Bhushan a week after the accident, he alleged that the Son Nagar Up Distant signal had been green and the Up Home yellow, then as he had got closer it had turned red. Previously, he said, had been disciplined for stopping at the Anugraha Narayan Down Starter when it had gone back in his face, so he had been reluctant to stop, assuming that the Son Nagar Home signal would be cleared. It had not, he had seen the headlight of train No. 2, had braked, and had seen the driver of the other train speeding up to get clear of the crossing, but the trains had collided. He criticised the switchman for not waiting until his train had stopped at the Home signal before letting train No. 2 across.

Assistant Driver S.K. Chakravarti agreed that the Son Nagar Up Distant signal had been green and the Up Home yellow, then the Home signal had gone back to red. He had called out this to his driver, who had replied that sometimes signals went to red temporarily and then back to green. Then the headlights of train No. 2 had appeared, they knew there would be a collision, and his driver had told him to save himself.

Additional Commissioner Bhushan put the time of the collision at 23:59, the speed of train No. 2 at 20 kilometres an hour and that of the goods train at about 30 kilometres an hour. Tests showed that had the Son Nagar Up Home been cleared it would have needed replacing at least 5½ minutes before the passenger train had reached the crossing, when the goods train would have been at least 2½ kilometres away and had ample time to stop. There was also no record of the West box having given a slot for the goods train.

It was clear, the additional commissioner continued, that the accident was wholly due to the driver of train No. 27 overrunning signals, while the assistant driver had failed to keep a good lookout and stop the train when his driver had failed to do so. Why the driver had failed to obey the signals was not clear. If the circumstances of the previous Anugraha Narayan incident were as reported by the driver, the fault lay with the switchman rather than with him. The matter should have been investigated.

The driver was correct in saying that the collision would have been avoided had the switchman given his train priority or else ensured that he had come to a stand before the passenger train was let across its path but no rules had been broken. Signals were meant to be obeyed and the flouting of them was bound to lead to accidents.

Driver Ojha had been involved previously in a head-on collision at Mughal Sarai, a collision with waggons and a derailment, all of which had been judged to be his fault.

The additional commissioner made three recommendations:

(1) Vigilance devices should be employed even on engines with an assistant driver, or dead man's handles as an inferior alternative.

(2) Quarters should be provided at Gaya for drivers' families and until they were drivers should be shifted to Mughal Sarai to be with their families (Driver Ojha had complained of serious family difficulties because he was away so much).

(3) Further trials of braking distances at high speeds should be undertaken.

The report noted that Driver Ojha and Assistant Driver Chakravarti (or Chakraborty, both names are given in the report) had been tried at Aurangabad Court and acquitted. [377]

A major catastrophe occurred in Hungary on the evening of Sunday 22nd December 1968. A passenger train from Budapest to Szolnok and a freight train coming from Szolnok collided head-on near Pusztaszentistván, between Mende and Sülysáp stations. Passenger train No.6616/a consisted of Class 424 steam locomotive No. 141 and eight coaches, while freight train No. 5565 consisted of Class M 62 diesel locomotive No. 056 and 53 waggons. Both trains were travelling at about 45 kilometres an hour when they met. As a result of the collision the leading coach of the passenger train was telescoped into the tender of the steam engine pulling it and its second and third coaches were also badly damaged. In the freight train the engine and 13 waggons were derailed and damaged. Forty-one passengers, the conductor and one of the two firemen of the passenger train were killed, while the driver and second fireman of the passenger train and the driver and conductor of the freight train were injured. In addition 15 passengers were seriously injured and 36 were slightly hurt. The track was torn up over a length of 150 metres. One line was blocked for 22 hours and the other for 27 hours. Long-distance passenger trains were diverted, local passenger trains were replaced by buses, and freight trains were delayed.

Passenger Andras Szabo, 16, who was injured, said: "Suddenly there was a monstrous crack. The lights went out, I felt a terrible pain and lost consciousness."

Many of the victims were children starting their Christmas holidays and their mothers, who were riding in the leading coaches, the ones that suffered the most.

An inquiry into the accident placed the principal blame for it on Signalman Fenrez Szantai at Mende. There had been single-line working in operation between the two stations but this had ceased and normal double-line working had resumed. Signalman Szantai had failed to notice that he had not reset the facing points leading to the opposite running line and had diverted the passenger train onto the wrong (left-hand) track, using an unauthorised "calling-on" signal. Driver Laszlo Papp of the passenger train was also found to have been negligent as he had disregarded written instructions received before starting the journey, had failed to stop to query the diversion when his train he had been

[377] Government of India Ministry of Tourism and Civil Aviation (Commission of Railway Safety): Report on the Collision of No. 2 GD Passenger Train and Electric Crack Goods Train No. 27 at Son Nagar on the Eastern Railway on 15th November 1968

switched onto the wrong track, and had taken the connection at 45 kilometres an hour, three times the speed that he should have done. [378] [379] [380] The signalman was arrested and the driver was liable for arrest when he had recovered from his injuries; his firemen and the passenger train conductor were found to have been at fault to a lesser degree.[381]

Following the disaster measures were taken to improve the training of signalmen and to modernise the signalling at the two stations, also not to undertake engineering work at times when there was heavy holiday traffic. [382]

A very brief report in a single Canadian newspaper spoke of a collision between two trains 20 kilometres from Cotonou, Dahomey, on the main line to Niger about Monday 10th February 1969. Ten people reportedly lost their lives and another 25 were seriously hurt. [383]

Accidents have occurred from time to time as a result of the drivers of passenger trains stopped at stations being advised that station work is complete and thereby setting off without thinking to check that the relevant signals have been cleared. A serious example of an accident of this type occurred at La Louviere, Belgium, on the morning of Tuesday 25th March 1969. Two passenger trains were involved; one was en route from Binche, near the French frontier, to Bruxelles, and the other was running from either Manage or Charleroi to Ghent. They collided at about 06:15.

The Brussels train had been stopped in La Louviere station and was given the right-away signal. It was just before dawn, there was a certain amount of fog and snow was falling intermittently. In these conditions of less than perfect visibility Motorman Paul Goelhen overlooked a stop signal 150 metres from the platform, his train ran through a set of points and then came into a head-on collision with the Charleroi train, which was running a few minutes late. Lineside factories and a curve meant that neither motorman saw the other train until the last moment and neither of them braked before the collision. [384] There was a speed limit of 40 kilometres an hour at the point of the accident and it was estimated that the Brussels train had been travelling at 36 kilometres an hour and the Charleroi train at 33 kilometres an hour when they collided. The leading car of the Brussels train was reportedly torn open for a length of ten metres and it was in this car that most of the casualties occurred. Two cars of the Charleroi train were derailed but remained upright.

There were an estimated 300 passengers in the Brussels train and 75 (some accounts say only 25) in the Charleroi train, mostly workmen on their way to jobs in local factories. Thirteen passengers were killed more or less outright and about 70 were injured, of whom 19 were badly enough hurt to be admitted to hospital and four of whom were reported the next day to be in a critical condition. Those hurt were taken to hospitals

---

[378] *The Holland Evening Sentinel*, Holland, Michigan, United States, 23rd December 1968
[379] *Edwardsville Intelligencer*, Edwardsville, Illinois, United States, 24th December 1968
[380] Wikipedia – Mende
[381] *The Times Recorder*, Zanesville, Ohio, United States, 24th December 1968
[382] *Railroad Wrecks* by Edgar A. Haine, pgs. 163-4
[383] *The Daily Colonist*, Victoria, British Columbia, Canada, 11th February 1969
[384] *De Volksrant*, Amsterdam, The Netherlands, 26th March 1969

in La Louviere, Bergen, and Hains-Saint-Paul. Many of the less badly injured had been thrown from their seats or cut by shards of glass. Motorman Mauroce Gaillex of the Charleroi train was killed outright and Motorman Goelhen was trapped in the wreckage critically injured, with both his legs shattered and head injuries. He lapsed in and out of consciousness as efforts were made to free him, was the last victim to be released and succumbed to his injuries five hours after the accident. [385]

Mr Jean Vermesch, who was probably the same person as the man identified in another newspaper as the owner of a lineside café who had been the first to alert the emergency services, lived only 20 metres from the point of the collision. He told a reporter that he had been awakened by the noise of the impact: "I thought that there had been an explosion at the steel factory opposite our house. Then I saw the two trains stacked on top of one another and people wandering around in a daze. Some fled along the line in a panic and when I got outside I wanted to flee as well." [386]

Red Cross ambulances were on the scene within five minutes and workmen from the adjoining Boël steelworks were also active in the initial rescue work. Oxy-acetylene torches had to be used to free a number of victims trapped in the wreckage and doctors had to perform at least two amputations. One of the rescuers told a reporter: "He (Motorman Goelhen) was the last to be taken out. It was as well that he was unconscious most of the time. The groans of those who were trapped lasted until noon. It was terrible experience to have people like that whom you could see and hear but you could not free. One locomotive was completely pushed into the first car of the other train. That's where most of the dead were."

The La Louviere stationmaster said that if the accident had involved the next trains that morning it would probably have been considerably worse, as those in the accident had each only consisted of four cars, whereas the next trains would have had seven.

Members of the Belgian judiciary and two government ministers visited the accident scene during the morning and the following morning Queen Fabiola visited the injured in hospital. A large number of people also went to the scene of the accident, some of them relatives or friends of passengers on one of the trains and others probably just curious onlookers. [387]

The lines were reopened to traffic the following morning. [388]

An unexplained error by the driver of a local passenger train from Łódź Kaliska to Kutno led to an accident at Jedliczach Lódzkich station, near Zgierzem, Poland, at 23:30 on Friday 31st October 1969. Instead of awaiting the arrival of a long-distance passenger train from Gdynia to Łódź the driver of the Kutno train set off prematurely and ran into the side of the Gdynia train as the latter was entering the station, derailing two coaches of it. Five passengers lost their lives in the accident; four of them were named as Stanisław Żak, 34, from Kościerzyny; Jan Rybiński, 38, from Smockza; Arkadiusz Chojnaki, 54, from Gdynia; and the 10-year-old daughter of Stanisławy Stróż of Chełmży. The identity

---

[385] *Nieuwsblad van het Noorden*, Groningen, The Netherlands, 25th March 1969
[386] *De Telegraf*, Amsterdam, The Netherlands, 26th March 1969
[387] *Algemeen Dagblad*, Rotterdam, The Netherlands, 26th March 1969
[388] *Limburgsch Dagblad*, Heerlen, The Netherlands, 26th March 1969

of the fifth victim, a woman aged between about 55 and 60, was not immediately known; she had not been carrying any identification on her. Twenty-six other people were injured, mostly not seriously. A fog which prevailed at the time of the accident might have contributed to it. [389] [390]

Sixteen people reportedly died and another 65 were injured when a Dakar – St Louis passenger train collided with a goods train near Lam Lam, Senegal, not far from Thies, on the evening of Wednesday 31st December 1969. [391] [392]

A passenger train and a light engine were in collision at Valašské Meziříčí, Czechoslovakia, near Horny Lidec in western Slovakia, on Monday 23rd February 1970. Eleven people died and 15 were injured, nine seriously. [393] [394] [395]

Early on Monday 26th October 1970 a passenger train and a goods train collided head-on at Khurja Junction, India, on the Northern Railway. Seven people were killed outright in the collision, one died en route to hospital and a ninth after admission. Eighty others were injured, 10 seriously. Those who died included the driver and first fireman of the passenger train.

Passenger train No. 1 AJD Up, consisting of a steam locomotive and 10 coaches, had left Khurja Junction station at 01:55, bound for Delhi, from Common Loop line No. 4 under clear signals. In the opposite direction train No. 456 Down Goods, consisting of a diesel locomotive, 60 waggons and a brake van, was approaching. The passenger train had entered the Down Main from Line No. 4 and was approaching the connection with the Up Main when it collided head-on with the goods train. Both engines were telescoped, the rear end of the first coach of the passenger train was crushed and the leading waggons of the goods train were wrecked.

Khurja Junction was a class B station on the double line between Sanwar and Sikandarpur and the junction for the single-line Meerut branch. East and West signal cabins were provided, their Home signals being controlled by slides from a panel in the stationmaster's office and slotted from the other cabin. Absolute block working was in force with Siemens General Electric double-line lock and block instruments and the lines through the station were track circuited.

Second Fireman Kashmiri Lal of No. 1 AJD told the inquiry into the accident that having seen that the departure signals for his train had been lowered he had been busy breaking up coal in the tender when the collision had occurred, shortly after they had started.

Guard K.M. Lal of the passenger train stated that he had exchanged *All Right* signals

---

389 *Dziennik Polski*, Kraków, Poland, 2nd and 3rd November 1969
390 *The Herald Journal*, Logan, Utah, United States, 2nd November 1969
391 *The Times*, London, England, 2nd January 1970
392 *Lubbock Avalanche-Journal*, Lubbock, Texas, United States, 2nd January 1970
393 Wikipedia – Liste von Eisenbahnunfällen in Tschechien
394 24th February 1970
395 *Lubbock Avalanche-Journal*, Lubbock, Texas, United States, 24th February 1970

with his driver, then had barely sat down when the trains crashed. He had been badly hurt.

Driver Harpassad of the goods train stated that neither light of the Khurja Junction Down Warner had been burning but the Down Outer Home had been lowered. He blamed the dazzling headlight of an engine on No. 4 line for his failure to see the Down Home and had not realised at first that he was being routed via the Down Main line. When he had realised he applied the automatic and independent brakes and the assistant driver had at the same time applied the emergency brake. He had begun to brake just after passing the Down Outer Home and did not know why he had not been able to stop in time.

Assistant Driver R.K. Purl testified that the green light of the Khurja Junction Warner had been burning, the red one had been unlit, while the engine headlight had shown its arm to be at danger. The Outer Home had been green, while the Home could not be seen because of the headlight of the passenger train, which at first had been directed towards the branch, then towards them. At this point he had warned his driver that there was a train on the Down Main and applied the brakes.

Guard V.S. Nim of the goods train testified that both the Down Warner and the Outer Home at Khurja Junction had been 'On' and when his train had not slowed down he had destroyed 10 to 15cm vacuum in his van. Before he could attempt to apply vacuum again the collision had occurred, throwing him around in his van and injuring him. Afterwards he first had gone forward and then back to protect the rear of his train. He had also remonstrated with his driver and told him that if he had been vigilant there would have been no accident.

Assistant Stationmaster D.P. Sharma stated that he had given permission to East cabin for the approach of train No. 1 AJD, which after changing engines had left at 01:55. Meanwhile at 01:35 he had given permission to the West cabin for train No. 456 but not for the signals to be lowered for it. The lights of all the Down signals were burning properly as could be seen from their backlights. The West cabin had informed him of the collision at 02:00. No defects of the signalling had been recently reported.

Stationmaster U.S. Srivastava detailed his checks of the signalling after the collision and confirmed that there had been no problems with it.

Switchman Kishan Lal in Khurja Junction West cabin testified to releasing the slot for the passenger train at 01:26, obtaining *Line Clear* for it from Sikandarpur East cabin at 01:48 and clearing his signals. With the permission of the assistant stationmaster he had given *Line Clear* to Sikandarpur East cabin for train No. 456 at 01:31 and had received the *Train Entering Section* signal for it at 01:55. As he had gone to the cabin window to give the *All Right* signal to the passenger train he had seen the headlight of the goods train approaching and immediately shown a red light towards it. Despite this the trains had collided at 01:57 or 01:58. There had been no problems with his signals, the weather had been clear and visibility good.

Switchman B. Lal in Khurja Junction East cabin confirmed that the headlight of the passenger train had been burning and that he had exchanged *All Right* signals with the engine crew and guard.

Assistant Stationmaster G.N. Sharma at Sikandarpur stated that the goods train had passed through at 01:57, travelling at 35 to 40 miles an hour, when he had exchanged *All Right* signals with the engine crew and guard.

Additional Commissioner G.S. Pandor put the time of the collision at 01:59. The passenger train, he noted, had only been travelling at 8 to 10 kilometres an hour at the time of the collision, the goods train, according to its guard, at 65 kilometres an hour. He was satisfied that there had been no failure of the signalling and that the signals had been clear for the passenger train to depart from line No. 4. The Down Outer Home of the West cabin could therefore have been off as alleged by the driver and assistant driver of the goods train.

The engine of the goods train was provided with a WABCO type vigilance device but it had been inoperative since July and the working conditions of diesel and electric locomotives were conducive to engine crews falling asleep. In the absence of a vigilance device *All Right* signals exchanged between the driver and the guard when passing through stations might help keep the engine crew awake but in this case no hand lamp had been provided for the driver and his torch had been defective. It was difficult to understand how the driver could have missed the signals unless he had dozed off.

The additional commissioner concluded that the accident had resulted from the gross negligence of Driver Harpassad in overrunning the Down signals at Khurja Junction and he was fully responsible for it. Assistant Driver Puri was responsible for failing to repeat the correct indications of the signals.

Additional Commissioner Pandor repeated an earlier recommendation that all diesel and electric locomotives should be fitted with vigilance devices and engines should not be dispatched from the works with the devices inoperative.

Measures should be taken to ensure that trains were correctly marshalled.

Mistakes in the recording of train numbers should be avoided.

The testing of the interlocking frames at Khurja Junction was overdue.

The Northern Railway should frame Subsidiary Rules making it obligatory for assistant drivers to remain in the engine cabs approaching stations and repeat signal aspects, also for drivers to exchange signals at stations and cabins.

The Sikandarpur East cabin switchman had only been able to record approximate times because his clock had been out of order. The administration should ensure that clocks in cabins and offices were maintained in good order. [396]

A mistake by a fahrdienstleiter (signalman) led to a double collision at Řikonín, Czechoslovakia, early on Friday 11th December 1970. A freight train was stopped at the home signal at Řikonín, on the Brno – Havlíčkův Brod line, but the fahrdienstleiter at the previous station at Níhov admitted a light engine, No. 489.0, into the occupied section. The rear collision that followed derailed the rear 12 waggons of the stationary freight train and one or more of these fouled the opposite running line. Two minutes later the derailed waggons were run into by the *Pannonia Express*, an international train between

[396] Government of India Ministry of Tourism and Civil Aviation (Commission of Railway Safety): Report on the Head-on Collision between No. 1 AJD Up Passenger Train and No. 456 Down Goods Train at Khurja Junction Station of Northern Railway on 26th October 1970.

Berlin and Belgrade. [397] Thirty-one or 32 people died and 17 or 18 others were injured. Several carriages of the express were reportedly derailed and telescoped, making the rescue operation difficult. Those who lost their lives were believed to have included several foreigners, probably Hungarians or Germans. [398]

Thirteen people lost their lives and 25 others were seriously injured in a collision between two trains near Atakpame in central Togo, on New Year's Day, 1st January 1971. [399]

Early on Saturday 1st May 1971 train EX-508, the *Italia* express, en route from Roma, Italy, to Wien, Austria, collided with a stationary freight train in Fürnitz station, Austria. [400] Driver Josef Feldrin had been travelling at 100 kilometres an hour as his train approached the station at 03:00 but seeing the obstruction he had made an emergency brake application and managed to reduce the speed of his train to 70 kilometres an hour at the moment of impact. He was seriously injured but survived. Eight of the 288 passengers on the train were killed and 23 were injured and were taken to hospital in Villach. Four of those who died were customs officials who had been riding in the leading coach, which was thrown over the locomotive; they were Carmelo Turna, 22, and Natalina Marta, 30, both Italian; plus Josef Santner, 64, and Josef Stabentheiner, 38, both of whom were Austrian. The other four who died had been passengers in a sleeping car which was derailed and ended up at 90 degrees to the track. They were named as Laszlo Deutsch, 72, a Hungarian-born American living in Wien; his wife Rosalia, 61; Deutsch's sister Ilona Hoffman, 63, from Wien, nationality unknown; and John Mitchell Kowall, 55, of New York City, United States.

An Austrian newsman, Walter Martinkovich, who was a passenger on the train, telephoned his wife after the accident and told her: "It was terrible. I escaped death by a miracle. It was like an air raid during the war." [401]

The accident was blamed on the failure of Fahrdienstleiter Franz Reichmann and Stellwerksleiter Guenther Zimmermann to remember that the freight train was occupying track No. 1 and accepting the express on that line. Both men were suspended pending the results of an investigation. [402] [403] [404]

An unusual accident occurred between Hotar and Magra Hat stations, India, on the Diamond Harbour branch of the Eastern Railway, early on Tuesday 6th July 1971. Down light engine No. 12253 collided head-on with Up light engine No. 12352 on the single

---

397 https://de.wikipedia.org/wiki/Liste_von_Eisenbahnunf%C3%A4llen_in_Tschechien#1939%E2%80%931945

398 *The Post Star*, Glens Falla, New York, United States, 12th December 1970

399 *Independent Press-Telegram*, Long Beach, California, United States, 3rd January 1971

400 *The Port Arthur News*, Port Arthur, Texas, United States, 1st May 1971

401 *Eau Claire Leader-Telegram*, Eau Claire, Wisconsin, United States, 1st May 1971

402 *The Daily Freeman*, Kingston, New York, United States, 1st May 1971

403 *The Mt Pleasant News*, Mt Pleasant, Iowa, United States, 1st May 1971

404 *Hamburger Aftenblatt*, Hamburg, West Germany, 6th May 1971

line and Up Diamond Harbour – Sealdah Push-Pull electric train No. SD 103 subsequently ran into the rear of the Up engine. Seventeen people were killed outright, one was fatally injured, 11 were seriously injured and the same number were slightly hurt. The collisions occurred at 03:58 and 04:51.

Assistant Stationmaster A.K. Guha at Hotar told the inquiry into the disaster that when he had come on duty at 20:00 he learnt that there had been a total failure of communications between his station and Magra Hat at 19:25. Since that time trains had been worked over the single line under the Rules and Regulations applying in such circumstances. Three Up and five Down trains had been worked over the single line after the failure before those involved in the accident. When engine No. 12253 had been at Sonarpur (on the Sealdah side of Hotar) at 02:15 Control had told him to let it go to Diamond Harbour but he had refused to accept it even to Hotar as *Line Clear* had already been given to Magra Hat for three Up trains (No. SD 101 Up, an Up Goods Special, and No. SD 103 Up) and had been requested for train No. SD 102 Down. Control had told him that the Up Goods Special could not come as its engine had failed, that engine No. 12253 would work that train, and had forced him to accept the engine to his station. After the engine had arrived at 03:07 Control had then demanded that he send it to Magra Hat after the arrival at Hotar of train No. SD 101 Up at 03:26, insisting that there was sufficient margin between train No. SD 101 Up and train No. SD 103 Up. When he had refused to comply with the order, saying that there might be an accident, Control had threatened him with serious disciplinary action. Feeling thus compelled, he had let engine No. 12253 down at 03:41. He pleaded that he had let the engine go, in defiance of the block working rules, not only because he had been threatened by Control but because he had been assured by Diamond Harbour on the P & T (Posts and Telegraphs) phone that the Up Goods Special had not started and that engine No. 12253 was urgently needed to work it.

Assistant Stationmaster D.K. Majumdar at Magra Hat stated that train No. SD 132 Down had arrived there at 00:58 with *Line Clear* for an Up Goods Special, No. SD 101 Up, and No. SD 103 Up plus a *Line Clear* enquiry for train No. SD 102 Down. He had despatched train No. SD 101 Up at 03:12, as the goods which should have left first was late. Engine No. 12352 had then been despatched at 03:48 and train No. SD 103 at 04:47. He had not cautioned the Up trains to proceed at 15 miles an hour on straight track, reduced to 5 miles an hour on curves and when visibility was reduced due to fog or rail or another cause. There had been about 300 passengers on train No. SD 103. He had not considered it necessary to advise Hotar through train No. SD 101 Up of the altered order of the Up trains as a memo had stated that the goods was only to come first if there was a sufficient time margin. He had only heard of the second collision at 05:25, when some passengers of train No. SD 103 had returned to Magra Hat.

Section Controller S.B. Bhattacharjee at Sealdah testified that he had been informed of the communications failure between Hotar and Magra Hat and that working under the relevant Rules and Regulations had been begun at 19:25. When engine No. 12253 was passing through Sonarpur the assistant stationmaster at Hotar had indicated that he could send it to Magra Hat after the arrival of train No. SD 101, which was the only Up train that he had understood Hotar had given *Line Clear* for. The section controller denied having been informed that Hotar had given *Line Clear* for three Up trains, denied insisting that engine No. 12253 must be sent to Magra Hat between trains No. SD 101 and No. SD 103, or having threatened to discipline the assistant stationmaster.

Assistant Booking Clerk B.C. Mukherjee at Bagha Jatin flag station testified to overhearing at 03:25 a conversation between the assistant stationmaster at Hotar and Control in which Control had asked the assistant stationmaster to let the Down light engine go after the arrival of train No. SD 101 Up, though the assistant stationmaster said that the conversation had occurred at 03:07.

Likewise Relieving Assistant Stationmaster P.K. Biswas at Baruipur (between Sonarpur and Hotar) stated that he had heard the same conversation between the assistant stationmaster at Hotar and Control between 02:30 and 03:00.

Assistant Booking Clerk P.K. Paul at Dhamua (between Magra Hat and Hotar) stated that his first knowledge of the accident had come from the injured driver of the Up light engine. Six or seven minutes later he had heard the sound of the second collision when he had sent a porter to protect the rear of the local train with fog signals and to inform Magra Hat while he had attended to the injured driver.

Assistant Stationmaster B.C. Chakravarty at Baruipur testified that Down light engine No. 12253 had arrived at his station at 02:15 and Control had told him to forward it to Hotar but the assistant stationmaster there had refused to give him *Line Clear* for the engine, saying that he had given *Line Clear* for three Up trains from Magra Hat. After informing Control of this he had overheard, at 02:25, Control telling Hotar that the Up Special Goods would not be coming as its engine had failed and that Hotar should let engine No. 12253 down as there would be ample time for it to reach Magra Hat between the two Up passenger trains. Hotar had then given him *Line Clear* at 02:40, and the engine had left Baruipur at 02:45.

Assistant Stationmaster A.K. Monda at Diamond Harbour said that he had had no conversation with Control over the P & T telephone.

Guard D.C. Dey train No. SD 103 Up stated that his train arrived at Magra Hat at 04:45 and left at 04:46 with the Starting Signal at yellow. He had received no caution order and had been unaware that there was a total failure of communications beyond Magra Hat. His train had proceeded at normal speed and with heavy rain visibility had been poor. He had been injured in the accident and some people had carried him to Dharma.

Driver Murath of Up light engine No. 12352 stated that he had been given a caution order at Magra Hat but had not been told why. Ten minutes after leaving there he had seen the other engine about 100 feet away and put his engine in reverse, then the collision had occurred. His engine had been moving at 10 or 12 kilometres an hour at the moment of impact. Its headlight and tail lamp had been burning and its brakes had been working properly. Losing consciousness in the collision, he only regained it after some local people had brought him to Dharma.

Railway Protection Force Sainik J.N. Sarkar testified that he had been a passenger on train No. SD 103 Up. Shortly after leaving Magra Hat he had seen people crossing the track, waving their arms and umbrellas to attract the attention of the motorman. Unfortunately, doubtless due to previous attacks on drivers by Naxalites, the motorman, instead of stopping his train, had speeded up.

Pointsman E. Gazi, on his way to take up duty at Chitpore, and riding in the rear coach of train No. SD 103, agreed that it had accelerated just before the collision.

Teacher Janab Mohamed Nural Islam spoke of his having been woken by a loud noise about 04:00. He had been unsure if it was an accident or robbers but had gone to the site with half a dozen others and they found that two engines had collided. He could not say if their headlights had been lit. Two unconscious crew members were carried to a nearby house. The headlights of a train coming from Magra Hat had then been seen and about 60 of them had run towards it shouting, raising their arms, and whistling. However, instead of stopping the train had appeared to accelerate. Other locals corroborated his evidence.

Additional Commissioner G.S. Pandor fixed the time of the first collision at 03:58 and of the second at 04:51. The speed of the Down light engine he calculated must have been about 15 kilometres an hour and of the Up engine about 26 kilometres an hour. Train No. SD 103 Up must have been going considerably faster, at least 75 or 80 kilometres an hour. He then reviewed the circumstances of the collisions as previously detailed.

The additional commissioner remarked that he did not think the section controller was as innocent as he pretended to be. He must have at least suggested to the assistant stationmaster at Hotar that the Down light engine could proceed to Magra Hat between the two Up passenger trains. The section controller had not acted correctly or wisely.

Regardless of the promotings or threats from the section controller the assistant stationmaster at Hotar had not been justified in dispatching the Down light engine in violation of the Rules. It was obligatory on him not to act upon any order from Control which violated the Rules. He was therefore primarily responsible for the accident.

Running tender first at night the light engines should not have been travelling at more than 8 kilometres an hour under the Rules. The drivers of both were therefore partly responsible for the initial collision.

If the visibility at Magra Hat had been poor due to heavy rain then train No. SD 103 Up should likewise not have exceeded 8 kilometres an hour and in any case not more than 25. Probably in panic from seeing what he believed were miscreants the motorman had instead accelerated. Given the extenuating circumstances he could not be held responsible for doing so.

After the initial total failure of communication on 5th July communication through the section should have been reopened with an Up passenger special from Magra Hat to Hotar, running empty (hence basically an empty coaching stock train) as per the instructions of the chief operating superintendent. However, when asked to leave the train at Magra Hat its passengers had refused to do so and threatened to assault the motorman. The train had also had to be driven at more or less normal speed because the motorman feared being assaulted if he lost time. Other trains had been driven faster than they should have in the same circumstances for the same reason, while at the end of April passengers had forced a motorman to pass a signal at danger and another motorman to drive almost 12 kilometres without a *Line Clear*. The additional commissioner commented: 'The prevalent tendency among Motormen, whether out of fear or under compulsion, to drive trains in violation of the relevant train passing Rules presents real hazards to the travelling public.'

The assistant stationmaster at Magra Hat was at fault for failing to issue a caution order to the guard of train No. SD 103 Up.

In the months of April to June 1971 there had been 21 total failures of communications between Magra Hat and Hotar. Only in a very limited number of cases had the train passing records relating to these been scrutinised by the transportation inspector as per the instructions.

In conclusion the additional commissioner decided that the cause of the disaster was the gross negligence of Assistant Stationmaster Guha at Hotar.

Regarding recommendations, the drivers of all trains should be given full protection to enable them to observe all Rules without fear. In addition where dense traffic coincided with frequent thefts of cables and overhead wires radio-patch arrangements should be provided to give an alternative means of communication between Control and all stations. Finally the Rules should be amended to allow the use of empty electric multiple units or push-pull rakes to open up communications in an emergency (as had, in fact, been done in the present case, though the Rules specified that a light engine should be used).

In response to these recommendations the railway board replied that the accident had occurred when the law and order situation in West Bengal, and especially Calcutta, had probably been at its worst. Measures were being taken by the Government to deal with the situation.

Limited resources prevented the provision of armed guards in the cabs of all trains.

The provision of microwave radio patch arrangements was very expensive.

Cable thefts could be repaired more quickly with the provision of armed guards to enable them to be carried out at night.

The suggestion of emptying a passenger train to enable it to examine the line was impracticable as the passengers would refuse to detrain, while the suggestion of using a loaded train was dangerous, as was shown by the accident. A light engine, push trolley, motor trolley or tower waggon were the only safe options, or perhaps a cycle or Jeep. [405]

One point not explained in the report was if the engine of the Up special goods had failed, why could not engine No. 12352 have been used to pull it, instead of being sent up light in its place? Perhaps the engine was needed urgently elsewhere or was not powerful enough to pull the goods train.

A train carrying a party of 250 (some sources say 400) schoolchildren and their teachers from Kragujevac, Serbia, and surrounding towns and villages, to the seaside resort of Gradac, ran into a freight train in the station at Semizovac, Yugoslavia, at 03:15 on Saturday 17th July 1971. The conductor on the diesel engine of the freight train saw the approaching passenger train and attempted to signal to its driver but without success, then together with his driver jumped from the cab of his locomotive to safety. The trains collided head-on and the leading coach of the passenger train was telescoped. The driver of the passenger train had run past a red signal and it was still moving at a reported 80 kilometres an hour at the moment of the collision. Eight girls and six boys aged between eight and 14 were killed, as was teacher Zlatmomir Ivanovic, aged about 32. Seven other children were hurt. Most of the children had been asleep at the time of the accident. For

---

[405] Government of India Ministry of Tourism and Civil Aviation (Commission of Railway Safety): Report on the Head-on Collisions, first between Up and Down Light Engines and then between SD 103 Up passenger train and the Up light engine between Hotar and Magra Hat Stations of Eastern Railway on 6th July 1971.

some of them it would have been their first trip to the seaside.

'Soldiers of an army unit at Semizovac were the first to rush to the scene of the crash. They said they saw a horrifying scene of the wrecked coaches with bodies and injured children screaming for help. Summer clothes fell on the rails and among smashed cars and children's rubber and plastic toys for playing in the sea were scattered around.' [406]

'People living near the railway line and soldiers from a nearby barracks rushed to the scene when they heard the crash. A local policeman said he was shaken from his sleep by the crash "like an earthquake," and saw a huge flash caused by the broken wires from the electrified line. He and his son raced to the scene in their pyjamas.' [407]

The driver of the passenger train and his assistant were arrested on the orders of an investigating judge. [408]

A collision occurred in the suburbs of São Paulo, Brazil, on the morning of Tuesday 30th November 1971 on the former Sorocaba line of the then newly-formed Ferrovia Paulista S/A (Fepasa). The accident happened near the Rua Joaquim Ferreira level crossing, near Água Branca station. Passenger train No. UM-28, with engine No. 2009 had left Mairinque at 05:15 for Júlio Prestes station, São Paulo. It consisted of an engine and eight coaches and was carrying over 1,000 passengers (some accounts said over 1,500) most of whom were workmen. Because of the overcrowding some of the passengers (called *pingentes* – literally pendants) were riding on the platforms and couplings of the coaches and it was they who were the main victims of the collision. The train was in a head-on collision with engine No 3614, which was coupled to a second locomotive and was performing a shunting operation prior to entering the Osasco yard, where it was to be attached to a train.

Crossing Keeper Alvaro Silva saw the two trains approaching one another on the same track and did what he could to save the situation, jumping onto the rails and waving a flag at the drivers. Driver Caetano Lemes Filho on train No. UM-28 and Driver Helio Bodo on the engines both applied their brakes and the speed of the trains was reduced, but the collision still occurred at a closing speed of 60 kilometres an hour at 07:15. The engines were not too badly damaged as a result of the collision but the leading coach of the passenger train was pressed against the rear of its engine, crushing the *pingentes* who had been riding on the platform between the two. Eight people were killed outright and over 100 were injured, 20 seriously, and one of whom subsequently died.

Train UM-28 was nicknamed *pau-de-arara* – literally macaw's perch and the name of a means of torture involving tying the wrists and ankles of the victim to a pole, causing severe joint and muscle pain, a reference no doubt to the discomforts endured by the train's daily passengers.

Passenger Dona Sebastiano Concieçãno, aged 44, who had been riding in the fourth carriage, told her story to a reporter between sobs, clearly still much affected by what had happened. "The journey was going normally. We should have stopped next at Barra Funda station, just after Agua Branca. Suddenly there was an infernal rumble and people

---

[406] *The Dominion Post*, Morgantown, West Virginia, United States, 18th July 1971
[407] *The Canberra Times*, Canberra, Australian Capital Territory, Australia, 19th July 1971
[408] *The Progress-Index*, Petersburg, Virginia, United States, 17th July 1971

were flying everywhere. I tried to get up and realized that my head was bleeding. The passengers seemed to be deranged and were screaming endlessly. Then I lost consciousness and when I came to I was in the casualty department."

The first ambulances arrived at the scene only five minutes after the collision and the injured were taken to the Emergency Room in Lapa and Sorocaba hospital. Firefighters, militiamen from the 27th Railway Detachment, members of the Military Police, doctors and nurses were also soon on the scene. Cordons were set up to keep sightseers at bay and by 10:00 the leading coach had been propped up by wooden beams supported by hydraulic jacks, prior to its being cut up by acetylene torches. Single-line working was apparently opened up early on over the other running line with trains slowly passing the accident site but serious delays to traffic could not be avoided. Relatives and friends in São Paulo awaiting passengers had an anxious time when the trains arrived very late, not knowing if the people they were waiting for had been killed or injured in the accident or merely held up.

Transport Minister Paulo Maluf gave a deadline of 72 hours for a committee of inquiry to determine the cause of the accident, which seems a remarkably short time. The minister expressed the opinion that the accident must have resulted from human error rather than a technical failure, as the line was operated by Centralised Traffic Control (CTC), "which practically eliminates the possibility of a signalling error."

The final report of the inquiry commission concluded that both drivers were to blame. Driver Filho of the passenger train had, it stated, been informed by the CTC (presumably meaning by signal indications) a few kilometres previously that he should stop his train and remain standing for two minutes before proceeding at not more than 10 kilometres an hour. Driver Bodo, meanwhile, should have taken the opportunity to read the notices issued by the CTC on the panel at Barra Funda station (presumably relating to not fouling the opposite running line until train No. UM-28 had passed). Both drivers were therefore to be summarily dismissed and placed at the disposal of justice.

At 12:46 the same day trains U-43 and U-46 collided at Júlio Prestes station itself, and the two drivers were seriously hurt. [409] [410] [411]

At least 76 people reportedly died and over 500 were injured when a crowded passenger train from the port of Khulna collided with a stationary goods train at Jessore, Bangladesh, on Sunday 4th June 1972. Due to an error by a signalman the passenger train, running at speed, was diverted onto the track occupied by the goods train. A telephone report from a deputy commissioner stated that ten coaches had been smashed to pieces and several others had been thrown from the track and had fallen into a large lineside ditch. Rescuers had already dragged 50 mutilated bodies from the wreckage. Rescue trains were rushed from Khulna and also from Calcutta in India. The injured were taken to three hospitals in Jessore and to Khulna. There were at least 50 seriously injured and it was feared that the death toll would rise as none of the hospitals had facilities for major surgery. Prime Minister Sheikh Mujibur Rahman ordered emergency relief measures and sent ministers and senior officials to the scene to open an inquiry into the cause of the

---

[409] *O Estado de S.P.*, São Paulo, Brazil, 1st and 2nd December 1971
[410] *Folha de São Paulo*, São Paulo, Brazil, 1st and 6th December 1971
[411] *Jornal do Brasil*, Rio de Janeiro, Brazil, 1st December 1971

disaster. [412] [413] A UN relief mission also flew to Jessore with medical aid. [414]

What at the time was the worst accident in Pakistan occurred at Liaquatpur station at 5 a.m. on Sunday 6th August 1972. No. 6 Down *Tezrao Express*, bound for Karachi, was apparently directed onto the wrong track by the pointsman and ran into the rear of No. 508 Down goods train, which was standing in the station. Having applied the brakes in emergency the driver and his fireman jumped from their engine and survived but the results were disastrous for the passengers on the express. The number of dead was early on put at 38, then climbed to 60 and by Wednesday 9th the death toll had risen to 90 and there were still believed to be more bodies in the wreckage. Among those who lost their lives was a bride, dressed in her red wedding suit. None of the other members of the wedding party could be traced, nor had the identity of the bride been ascertained several days after the disaster. There were about 250 others injured, who were taken to hospitals at Liaquatpur, Rahimyar Khan and Bahawalpur.

Mr Ghulam Mustafa Jatoi, the Central Minister for Communications and Political Affairs, ordered an immediate inquiry into the causes of the accident. [415] [416]

Ten people died and dozens were injured in a collision near Monapo, Mozambique, on the main line to the port of Nacala, on Sunday 29th October 1972. No further information on the accident seems to be available. [417]

'44 Die in I.C. Wreck'

'350 Hurt; Probe Cause of Disaster'

'Two trains ram in morning in morning rush'

The above were the headlines in a Chicago, Illinois, United States newspaper on Tuesday 31st October 1972, reporting a major disaster that had occurred in the city the day before, Monday 30th October 1972.

Below a large photograph of the aftermath of the disaster the newspaper gave an account of what had happened. 'Federal, state, and local investigations have been launched into yesterday's commuter train crash, the worst in Chicago history. Forty-four persons were killed and more than 300 were injured.

'The tragedy took place at 7.37 a.m. at the Illinois Central Railroad's 27th Street station. An old-style, six-car commuter train slammed into the rear of a modern, four-car bilevel commuter train, which had travelled past the station in error and then had backed up for a flag stop.

'The tremendous impact of the crash left a scene of bodies mangled within the train,

---

[412] *The Daily Gleaner*, Kingston, Jamaica, 5th June 1972

[413] *San Antonio Light*, San Antonio, Texas, United States, 5th June 1972

[414] *Anderson Daily Bulletin*, Anderson, Indiana, United States, 5th June 1972

[415] *Dawn*, Karachi, Pakistan, 7th and 9th August 1972

[416] *The Times*, London, England, 7th August 1972

[417] *The Canberra Times*, Canberra, Australian Capital Territory, Australia, 1st November 1972

passengers screaming and moaning, and bewildered and less seriously injured passengers staggering out of the coaches and wandering along the tracks.

'President Nixon sent John Volpe, secretary of transportation, to the crash scene, where he inspected the debris for 40 minutes and then interviewed the victims at Mercy Hospital. He then left Chicago. At Mercy Hospital Mrs Fannio Peterson, 53, of 8123 S. Brandon Ave., told Volpe the new cars "are built like pie crust."

'Volpe and Henry Wakeland, director of the department's Bureau of Transportation Safety, said the accident raised serious questions about the new lightweight steel and aluminium cars which commuter railroads having been buying for years.

'Investigations were began by the National Transportation Safety Board, the federal and state transportation departments, the railroad, the Illinois Commerce Commission, and the Cook County Coroner's office.

'About 240 firemen were at the scene quickly in response to an extra alarm, to begin the grim task of removing the dead and injured – a job not completed for six hours. The dead were taken to Cook County Morgue, where officials said there definitely were 44 bodies.

'The accident occurred less than 100 yards from the Michael Reese hospital complex, and two blocks from the hospital's emergency room. The hospital staff, including 650 doctors, was alerted and within a few minutes staff reached the scene, climbing into and onto the wreckage, giving advice to rescuers, and treating the injured.

'The crash was within a few blocks of the Mercy Hospital, which put into effect an emergency plan utilizing 50 doctors to treat victims.

'The platform of the 27th Street station was turned into an outdoor hospital where injured were given first-aid, their injuries evaluated, and priorities set up for transferring them to Michael Reese and six other hospitals. The injured were carried to hospital in dozens of ambulances, police squadrols and helicopters which were rushed to the scene.

'The two trains were filled with nearly a 1,000 passengers, as the second train, travelling at a speed estimated by railroad and other investigators at 30 to 40 miles an hour, plunged into the backing train. The lead car of the second train veered to the right after striking and virtually demolishing the rear car of the first train. The car jumped up over the base frame of the first train's rear car, splitting that coach in half.

'Most of the dead and those most seriously injured were in the rear coach of the front train. The other cars of the two trains remained on the tracks and appeared not to be substantially damaged; passengers in these coaches were thrown about by the heavy impact.

'Mayor Daley, an early visitor at the crash scene, ordered city flags flown at half-staff thru Saturday in memory of those who were killed.

'The cause of the tragedy had not been definitely ascertained. Robert W. Cavanagh, engineer of the rear train, was so seriously injured he could not be questioned. It was more than five hours before he was freed from the mass of twisted metal and rushed to a hospital.

'The 27th Street station was a flag stop for the front train, meaning it would only stop if someone wanted to get on or off. Jack Humbert, the railroad's vice president in charge of operations, said the overshooting of the station by the first train apparently led to the accident. It was explained by Humbert and others that the railroad uses block signals, with a green "go" light permitting speeds up to 65 miles an hour. A yellow "caution" light allows speeds up to 30 miles an hour, and a red light calls for the approaching train to stop. When the first train passed the station a block signal at 31st Street would have turned yellow, the signals at 35th and 39th would have remained green. By the time the train backed up, and the signal colors changed, they said, the second train had apparently travelled too far and was travelling too fast to stop in time. As far as could be determined, the officials said, the signal system and the two trains were functioning properly.

'Both trains originated at the railroad's South Chicago station at 91st Street. The first train of four cars, No. 416, was scheduled to leave at 7.6 a.m., and, after several stops, to arrive at the Randolph Street station at 7.40 a.m. This is the one which includes modern, air-conditioned, double deck cars, with a capacity of 196 passengers each.

'The second train of six, old-style, single-level cars, No. 720, was scheduled to leave South Chicago at 7.15 a.m., and was due downtown at 7.43 a.m., just three minutes after train No. 416 was due to arrive. This train's coaches have a capacity of 156 passengers each. This is an express train, with its last stop before arriving downtown at 71st Street and Stoney Island Avenue.

'H.G. Mullins, the I.C.'S superintendent of passenger services, said he was informed that both trains were on time. He said the engineer of train No. 416, identified as J.A. Watts, informed him he overshot the station by about 250 feet. He said the engineer and conductor, Ernest R. Cummings, who was at the rear, were in contact by radio. But the other train has no radio equipment, and thus there was no contact between the crews of the two trains. As to why he overshot the station, Mullins said, "Maybe he was going faster than he should have."

'Some passengers disputed Mullin's contention that train No. 416, the front train, was on time. Among these was Isaac Johnson, 22, of 337 South Shore Dr., who said the train was nine minutes late when he boarded it at the South Shore station, 71st Street and Yates Boulevard. Johnson said he believed the train carried more passengers than usual. After it passed the station, Johnson said, it seemed to him like it stood there for five minutes, before it began backing up to the station. As it moved into the station area, Johnson said, he saw the conductor, identified by the railroad as Cummings, jump out of the back coach, and after apparently looking at the other train approach, he stood there on the platform and put his hands over his eyes.

'Harold Melcher, 21, of 7411 S. Shore Drive, who was in the second car of the approaching rear train, said he heard the conductor shout, "We're going to crash. Everybody get down!" He said he dropped to the floor, as did many others, and he escaped without injury.

'Witnesses said the people inside tried to break the double plate glass windows of the modern coaches of the first train but failed. Firemen were apparently unaware that the way to break open these windows is to pry in from the top. Witnesses said some people were hurt as firemen smashed the windows with axes and crowbars.

'One of the first doctors on the scene was Dr. Edward Goldberg, staff surgeon at Michael Reese Hospital. He described the scene as, "grotesque beyond belief. Human limbs were hanging out the window," he said. "There was one 22-year-old fellow whose heart stopped, but he was revived. About the only thing doctors and nurses could do at the scene was to apply tourniquets and splints and give sedatives."

'Heavy equipment, including cranes, was rushed to the scene. While firemen broke in thru windows, others cut holes in the tops of the two most heavily damaged cars and climbed down. Doctors and nurses accompanied some of the firemen.

'Commuter services were restored partially in time for the afternoon rush hour, but in the morning rush hour power was cut off. The railroad has some 65,000 riders daily.

'The bodies were taken to the County Morgue. Not until mid-afternoon were persons admitted to attempt to identify bodies.

'The Illinois Central commuter service has been under fire from commuters for some months. The Illinois Commerce Commission recently has been conducting investigations into the corporate structure of Illinois Central Industries, of which the railroad is one of 50 subsidiaries, and the I.C.'s general quality of service and safety.

'Alan S. Boyd, I.C. president, rushed back to Chicago from a meeting in Arizona. He issued a statement which read in part, 'There are no words to describe our grief and sorrow for our patrons and employees who were killed in this tragic accident.'

'A railroad spokesman said the I.C. has had only one fatality since the suburban service was electrified 46 years ago.'

On another page the newspaper carried more stories of survivors of the accident. '"I looked to my right and the other train was inside our car. When I looked down, the floor was gone. It was just twisted steel with arms and legs sticking out."

'Cliff Retis, 23, of 8543 S. Muskegon Av., was sitting on the upper level of a double deck commuter car, when another commuter train crashed into it yesterday morning. Tho he lost his glasses in the wreck, the electrical engineer had little trouble seeing – or hearing – the horror around him. "I was thrown into the seat in front of me. Then windows started crashing down on top of me. The floor was gone and people were screaming for help."

'Retis was sitting on the left side of the commuter train that had stopped at the 27th Street Illinois Central station. The oncoming train had sheared away the right side of his car, peeling back the sides and top like a banana. "I just sat there until the firemen broke thru the windows on the lower level," Retis said. "Then I reached down with my foot and hit a fireman on the helmet, and they carried me down. I'm dammed lucky to be alive."

'Mrs Margie Andrade, 33, of 8422 S. Brandon Av., feels the same way. She was a passenger in the front car that rammed into the one carrying Retis. "The motorman ran out into the car and screamed, 'We're going to crash, we're going to crash!'" she said. "All of a sudden there was this huge crash and the wall on the front of the car just came down. The people who were in the first couple of seats were just buried. I just couldn't believe it, all those poor people – all bloody. Some of the passengers tried to dig out those trapped. It was terrible. Those that were pinned in just kept screaming."

'Xavier Rodriguez, 25, of 9042 S. Escanaha Av., riding in the rear train's third car, didn't think the accident had been serious until he left the train. "When we got out we didn't think it was that bad," he said. "Then we saw what had happened, there were people lying all over – trapped. Girls were just screaming hysterically."

'One car behind Rodriguez, Paul Martinez, 19, also thought at first that the accident wasn't so serious. Then he saw people running along the station platform. "I could see this terrible look on their faces. Then I knew something terrible had happened. People were trapped everywhere. There was lots of confusion. Some people were just walking around the platform badly in shock."

'Many of the passengers who were less seriously hurt tried to help treat some of the victims who had gone into shock. "We were trying to take care of the people," said Manfred Johnson, 26, of 7746 E. Yates Blvd. "Some were going into shock. One woman was dead." Johnson was riding in the next to last car of the lead train. "After the crash I could see six or seven people trapped in the stairway leading down from the upper level of the bilevel car. There was no way they could get down. The trapped people were those who were going to get off the train."

'Miss Ruby Smith, of 8458 S. Dante Av., was standing in the second car of the oncoming train when it collided. "It was a boom, baby, not a bang," she remembered. "I went flying forward like a bullet into the front car."

'Two cars behind her, Miss Rita Ortega, 20, of 8613 S. Houston Av., experienced the crash with a similar jolt. "There was no warning. It happened so fast. People were flying out of their seats. We didn't think it was really that bad until we saw what happened. There were girls that had their legs cut off and were just screaming. You knew a lot of them were dead. We saw people trapped and all cut up. It was really bad." A daily commuter on the Illinois Central, she says she's now looking for another way to commute. "The same train we were on today derailed Thursday," she said. "Then everyone was joking about it. They were saying 'what's going to happen next?'"

'The long string of Illinois Central double-deckers lurched forward and began rolling slowly out of the Randolph Street station into the gray Chicago dusk. Inside the new highliner cars passengers sat quietly gazing from windows or talking softly to those around them. The conversations weren't the usual ones you hear aboard southbound train No. 573. The men weren't talking about getting home for the Monday night pro football game; the women weren't discussing their day in the office or what they would cook for supper.

'It was 5 p.m. and just about 9½ hours earlier, two I.C. trains had crashed in Chicago's worst transit accident. So aboard No. 573 the talk was of death, of pain, of suffering. The talk was of the magnificent effort Chicago's police and firemen did in pulling the injured from the two trains; of the tireless work of doctors and nurses from nearby hospitals who saved countless lives at the scene. It was not idle talk. Many now headed south for their homes had been there.

'Bing Nisbura, 19, of 1527 E. 71ST Pl., sat alone. His eyes fixed on a window as the twilight forms of buildings and trees rushed by outside. His mind was recreating, reshaping, a nightmare. He had survived the crash. "I'll never forget the conductor's words," Nisbura, a student at the University of Illinois Chicago Circle Campus, said softly. "We were backing up and all of a sudden he began screaming, 'Stop, stop, stop, hold it, stop here!'

'Nisbura stopped talking for a moment. He had been in the double-decker car just in front of the one which was struck by the older northbound and had witnessed passengers flying through the air and the screaming. The agony developed in his mind quicker than he could the words to describe it. "The conductor's words had no more come out over the intercom system when they were punctuated by the crash and the crunching of steel and glass," Nisbura said. "As soon as it happened and I saw that I wasn't hurt. I helped get some of those who were injured off of the train." Nisbura then went on to class. "It never occurred to me to go home," he said, an incredulous tone in his voice.

'Lawrence Taylor, 29, of 2147 S. Colfax Av., had been aboard the older northbound train. He was thrown several feet thru the air when the trains crashed. "I was writing a letter to my grandmother when it happened," he said. "I just flew thru the air like a sack of apples when we hit and when I landed I still had the pain and notebook in my hand. I'm lucky to be alive but going to get a tongue lashing when I get home. After the accident I went on to work thinking how lucky I was and I forgot to call my wife and tell her I was all right till after lunch. She was worried sick."

'Suddenly the train slowed noticeably. Passengers rushed to the windows and stared silently out at the twisted and crumpled remains of several passenger cars covering track three at the 27th Street station. The slowly passing scene before them was illuminated by flashes of acetylene torches cutting thru the piles of wreckage. "My God," Nisbura whispered, almost inaudibly. "There's the car I was in. I'll never know how I got out."

'Then the southbound commuter train picked up speed and rushed thru the night toward the South Chicago station. Its passengers returned to their bright yellow and orange seats and their hushed conversations.' [418]

As usual the accident was the subject of a comprehensive investigation by the National Transportation Safety Board (NTSB). Its report stated that the collision occurred at 7.38 a.m., a minute later than stated in the newspaper account, and revealed that in fact it was a double collision, three trains being involved. According to the NTSB the final number of passengers killed in the accident was 45, while another 332 people were injured.

Illinois Central Gulf (ICG) train No. 416, the report stated, had consisted of four bilevel Highliner cars. It had started from South Chicago station on time at 7.06 a.m., and its brakes had operated correctly at the 12 stops between there and 53rd Street. At the latter station it had waited for train No. 718 to exchange passengers with it but the latter train had not arrived on time so train No. 416 had left at 7.31 p.m. on track 3. As the train passed 31st Street Conductor Cummings had announced the 27th Street stop over the intercom. Motorman James Watts stated that the train had been travelling at 62 miles an hour when he had applied the brakes for the stop there but the train had overshot the platform, finally stopping with the rear car 600 feet north of it, and crucially about 400 feet north of signal No. 3-3.10. Motorman Watts had then asked the conductor over the intercom for permission to back up, which was granted. The train had then begun to set back at a speed estimated at between 5 and 15 miles an hour. As the train had reached the platform and was just coming to a stand it was run into by train No. 720. Conductor Cummings had got out onto the platform immediately before the collision.

Train No. 720 consisted of six older cars in charge of Conductor G.F. Ryan. It had left South Chicago at 7.15 a.m., made seven stops between there and Stony Island Avenue, then had no more stops until Roosevelt Road (12th Street). At 76th Street the train had been routed onto track No. 3. As far as signal No. 3-3.60 at 31st Street the signals had all been green and Motorman Cavanagh put his speed at 55 to 60 miles an hour. Signal No. 3-3.60 had been yellow, which Motorman Cavanagh stated he had been expecting, as he

---

[418] *Chicago Tribune*, Chicago, Illinois, United States, 31st October 1972

had known he must be catching up train No. 416. He had begun to brake before he had seen the signal, which he estimated he had passed at 40 miles an hour. His view of signal No. 3-3.10 from 31st Street was obscured by the overhead walkway (footbridge) at 27th Street and it was the rear of train No. 416 which he had seen first, still backing along the 27th Street platform. Applying the brakes in emergency he had put his train in reverse, sounded the horn and evacuated his cab as he shouted a warning to the passengers.

In the collision the leading car of train No. 720 overrode the underframe of the rear car of train No. 416, sheared off its collision posts, veered slightly to the right and moved through the passenger section.

Meanwhile train No. 718, running slightly late, had left 53rd Street for Roosevelt Road and was approaching 27th Street at a speed estimated as 25 miles an hour by its engineer on track No. 4, gradually overtaking train No. 720. At the moment of the collision it had almost passed the latter train, but the wreckage of the collision sideswiped its rear car.

The line was equipped with automatic block signals with red, yellow, and green aspects. Some of the red signals did not indicate *Stop* but only *Proceed at restricted speed.* Movements were controlled by an operator at 67th Street station.

Train No. 416, which stopped at most stations, was always routed via track No. 3. Train No. 720 normally was routed to that track as well, but this depended on track occupancy. Eighteen trains passed through 27th Street between 7.30 a.m. and 8 a.m. on tracks Nos. 3 and 4 on weekdays. Of these only train No. 416 had what was according to the timetable a flag stop at 27th Street station, though in practice it stopped there every day to allow staff at the Michael Reese hospital to alight.

The NTSB noted that overruns of platforms in the suburban area were common, especially since the introduction of the Highliner cars, but rarely exceeded one car length. No one could recall a previous case of a train passing beyond the limits of a block, as train No. 416 had done. Nor could anyone recall any recent instances of the flagging of trains. Suburban trains did not commonly carry flagging equipment and Rule No. 99 (a) stated:

Within interlocking. Automatic Block System or Centralized Traffic Control limits, flag protection is not required against following movements on same track.

However, Rule No. 515 stated:

A train or engine having passed beyond the limits of a block must not back into that block except under protection as prescribed by Rule 99.

A relevant question put to crew members during rule testing, which required either a true or false answer, was:

A train or engine having passed beyond the limits of a block must not re-enter that block unless a member of the crew is sent far enough in advance of that movement to provide flag protection.

All the crew members of trains Nos. 416 and 720 had given the correct answer, which was, of course, *true*. Asked a variety of questions regarding the operating rules or circumstances of the accident the crew members gave answers which were inconsistent with the ICG interpretation in 18 out of 29 cases.

Calculations of the speed of trains Nos 416 and 720 indicated that if train No. 416 had been backing at 11 miles an hour and train No. 720 moving at 52 miles an hour, when train No. 416 first came into view, so had Motorman Cavanagh made an immediate emergency brake application he should have been able to avoid a collision. His train should also not have exceeded a speed 30 miles an hour after passing signal No. 3-3.60. His train was not, however, provided with a speedometer.

The day of the accident was the first day that Motorman Watts had worked train No. 416 and the NTSB concluded that he had probably forgotten about the flag stop at 27th Street station. Conductor Cummings was responsible for the safe reverse movement of the train and for protecting the movement by flagging, while Motorman Watts had not questioned the immediate authorisation by the conductor of the reverse movement.

In questioning none of the crew members of the two trains had recognised that a yellow aspect required a train to be prepared to stop at the next signal and two gave incorrect answers to the meanings of red signals with or without number plates (without a number plate – *Stop* – Stop; with a number plate – *Restricted proceed* – Proceed at reduced speed) This implied that the training, examination and enforcement of rules were inadequate and also that rules which were similar, such as those pertaining to red signals, were likely to be confusing.

The meaning of an application of other rules, such as Nos. 99(a) and 515, was unclear. A similar accident had occurred two years earlier at Riverdale, Illinois, but nothing had changed in the meantime. Violations of Rules Nos. 7, 35, 896 and 1003, relating to the carrying and use of flagging equipment, were disregarded by management, leading employees to the belief that the importance of flagging had diminished, while station overruns and short backing movements had also been accepted by management.

There had been a system breakdown – number of interrelated factors involving crew members, management, operating rules, and environmental factors had gone wrong to cause the accident:

Conductor Cummings had to stand in the centre vestibule of the rear car of train No. 416 to operate the intercom, at a point where he could not see behind the rear of his train.

The ends of the cars of train No. 416 were painted black and thus hard to distinguish, while its end marker lights were very small and incompatible with safety.

The walkway at 27th Street station and the positioning of signal No. 3-3.10 were inconsistent.

Motorman Cavanagh had not complied with the yellow indication of signal No. 3-3.60 which informed him that there a train ahead. He had not been especially concerned as there was a train ahead every day.

The signalling had worked as intended but had not prevented the accident. Trains normally ran with very close headways under yellow and red signal aspects. The NTSB commented, 'The system was safe as long as the rules were obeyed, {but} there was no margin for disrespect. There are signal systems available, e.g. cab signals, automatic train control, automatic train stop, that place less reliance on rule enforcement to ensure safety.'

The reversal of trains into previously vacated block sections was incompatible with the headway of only two to three minutes between following trains and the requirement to implement flag protection. The train should have continued to the next station.

Federal Railroad Administration regulations specifying the type of signalling to be used in particular circumstances were minimal.

Moving on to discuss the degree of damage caused in the collision the NTSB noted that the cars of both trains had heavy underframes and lighter superstructures, plus couplings designed to prevent the overriding (telescoping) of adjoining cars. These couplings had worked within the trains but the leading coupling of train No. 720 had been displaced downward in the collision, breaking the coupling carrier, and destroying any anti-climbing capacity between the dissimilar cars of the two trains. The underframe of the leading car of train No. 720 had then demolished the superstructure of the rear car of train No. 416, penetrating it to halfway and causing most of the deaths and injuries. Similar penetration had been observed in a collision at Darien, Connecticut, on 20th August 1969. Substantial research was needed to improve the resistance of lightweight cars.

The NTSB went on to list no less than 31 conclusions:

(1) The overrun of the 27th Street station platform by train 416 was caused by the failure of the engineer to initiate a brake application soon enough to stop the train at the platform from the speed at which the train was running.

(2) Train 416 passed beyond the limits of the block of Signal 3-3.60 and stopped 400 feet beyond Signal 3-3.10.

(3) Neither the engineer nor the conductor established whether train 416 had passed completely beyond the block of Signal 3-3.60 before they began to back up the train.

(4) The conductor of train 416 failed to provide the flag protection prescribed by Rule 515.

(5) The engineer of train 416 did not require that flagging be provided and thus did not fulfil the responsibilities assigned to him by Rules 106 and 1022.

(6) The collector of train 416 was available to perform flagging duties but did not do so, nor was he so instructed by the conductor or engineer.

(7) The conductor of train 416 located himself in the centre vestibule of the rear car to direct the backup movement, a position necessary for communication with the engineer. From this position, he was unable to see train 720 until too late because of the curve of the track.

(8) Signal 3-3.60 displayed a yellow aspect as train 720 approached it. The yellow aspect required that 720 reduce speed to 30 m.p.h. after passing the signal and approach Signal 3-3.10 prepared to stop.

(9) The rear of train 416 was camouflaged from the view of the engineer of train 720. The back end of the Highliner blended with the background, and the marker lights were not distinctive enough to be discerned by the engineer of 720.

(10) The speed of train 720 as it approached 27th Street exceeded the required 30 m.p.h. by at least 20 m.p.h., a speed difference large enough to be detected as in excess of 30 m.p.h. without a speedometer.

(11) If train 720 had been traveling at 30 m.p.h., it could have stopped short of impact if emergency braking had been initiated when the engineer first sighted train 416.

(12) If the engineer of train 720 had made an emergency brake application when train 416 first was clearly visible, the train could have stopped short of the impact point.

(13) The crewmembers' interpretations of the various signal aspects were not consistent with the interpretations of ICG representatives.

(14) Based upon the actions and knowledge displayed by the crewmembers of trains 416 and 720, the rule training, examination, and enforcement activities of the ICG were inadequate to insure rule compliance.

(15) A rule that imparts dual responsibilities to an employee invites partial employee understanding. The crewmembers of trains 416 and 720 understood the need for medium speed under authority of an approach signal but did not realize that the train must be prepared to stop at the next signal.

(16) Although in the 1970 written examinations, the crewmembers interpreted Rule 515 to require that a crewmember be sent far enough in advance of the movement to provide flag protection, evidence indicated confusion on the part of some employees regarding the relationship between Rule 515 and Rules 99, 99(a), and D-99 as well. It was not clear whether Rule 515 referred to Rule 99 only or to 99(a) and D-99 as well.

(17) Rules 7, 35, 896, and 1003, which require that proper flagging equipment be carried on trains, were not consistently enforced by management. This implied that flagging was not important in suburban service.

(18) The absence of a crewmember with the title "flagman" was not pertinent to the accident.

(19) The ICG did not comply with Illinois Commerce Commission General Order No. 196. It could not be determined whether this contributed to the accident.

(20) The design of the cars, which located the conductor's controls in the center vestibule of the Highliner car, was incompatible with the operating requirements that he govern the reverse movement and see that the track was clear.

(21) The walkway at 27th Street obstructed an approaching train's view of Signal 3-3.10 and diminished the effect of that signal.

(22) The automatic-block signal system did not adequately transmit sufficient information to identify fully the hazard to the engineer of train 720.

(23) The designs of the anti-climbing arrangements of the two trains prevented overriding between cars within the trains but did not prevent overriding between the two dissimilar cars which collided.

(24) Federal regulations did not require a design which would prevent overriding of colliding trains.

(25) The unsecured trucks of the first car of train 720 permitted the car to override the underframe of the Highliner car, which resulted in more penetration than would have occurred if the trucks had been secured.

(26) The Interstate Commerce Commission allowed the trucks of the "old" cars to remain unsecured in an order dated September 22, 1954.

(27) The collision posts of the Highliner car, even if constructed according to design, were not adequate to withstand the forces in the collision.

(28) Because of the improper welds, the collision posts of the Highliner car did not meet the shear requirements of the Federal regulations.

(29) It is probably technically feasible to design commuter cars to withstand crashes at moderate speeds without fatal injuries to the passengers.

(30) Although the Highliner cars were financed with Federal assistance, the design of the cars was not subjected to a review for compliance with Federal regulations by FRA or to a safety review by UMTA.

(31) FRA had no program to insure that design of new equipment complied with Federal regulations before the equipment was put in service. Instead, the FRA relied upon voluntary compliance.

The probable cause of the accident, the NTSB stated, 'was the reverse movement of train 416 without flag protection into a previously vacated signal block and the failure of the engineer of train 720, while operating faster than the prescribed speed, to perceive the train ahead in time to avoid the collision. Ambiguous rules which caused confusion among employees regarding the necessity to flag within automatic-block signal system limits and the reduced importance of flagging in suburban service implied by the management's failure to enforce Rules 7, 35, 99, 896, and 1003 also contributed to the accident.'

Contributing to the high incidence of fatality was the overriding of the underframe of the Highliner car by the older car, which allowed the older car to telescope the Highliner car.

Five recommendations were made:

(1) The Illinois Central Gulf Railroad ensure that its employees understand and comply with its operating rules. In order to do this, the ICG should improve its training program by developing:

(a) Books of standard interpretations of its rules in situations met both routinely and only occasionally to provide a basis for better use of the rule book in instruction; and

(b) A system of regularly testing the ability of employees to interpret actions required in specific operating situations.

(2) The Illinois Central Gulf Railroad review its organization systematically to ensure that safety is covered adequately in all interactions of equipment, personnel, rules, and procedures. For example, the Highliner design does not allow the conductor to use the intercom and emergency brake valve from the rear of the car, which is the logical location for supervising a reverse movement.

(3) The Federal Railroad Administration and the Urban Mass Transportation Administration cooperate in sponsoring an independent study to justify or disprove the need for a requirement that high-speed commuter train operations be governed by some form of automatic train-control system or some special procedures that will prevent a collision of two trains.

(4) The Federal Railroad Administration and the Urban Mass Transportation Administration initiate research to develop the technical approaches to crashworthiness in light-weight passenger cars for use in commuter or rail rapid-transit operations. These approaches should include crash testing as part of the design and development function for new equipment.

(5) The Federal Railroad Administration revise 49 CPR 230, (D) to the extent that data are now available to provide better protection for passengers in collisions. Further,

the FRA should enforce regulations on new cars before they are put into revenue service. [419]

Several months after the accident Motorman Watts died of a heart attack, perhaps brought about by the stress of it, and Conductor Cummings retired. [420]

It has not been possible to find the results of the inquest on the victims of the disaster.

A serious collision occurred in Turkey on the evening of Tuesday 31st October 1972 when a passenger train en route from Konya to Istanbul was in head-on collision with an engineering train, which due to a mistake had wrongly been let out of a siding onto the main line before the passenger train had gone by. The accident was stated to have happened 30 kilometres south of Eskisehir, indicating that it occurred at Alpu. Several coaches of the passenger train were stated to have become derailed and fallen down a cliff after the collision and the diesel tank of the locomotive of the passenger train was punctured, setting off a fierce fire that probably claimed some of the 190 passengers on the train. Civil and military rescue workers were also much hampered in their efforts until the fire had been extinguished. A relief train was sent from Eskisehir to convey the injured to hospital.

About 30 people lost their lives in the disaster and over 40 others were hurt, some seriously. A doctor at Eskisehir told reporters over the telephone that the number of dead could rise to 35 and that rescuers were searching the scorched cars for more victims.

Transport Minister Rifki Danisman went to the scene of the accident and an inquiry into how it had happened was opened. [421] [422] [423]

Sometimes the exact cause of an accident remains unclear even after a thorough investigation into it. Such was the case regarding a collision at Järna, Sweden, on Thursday 7th December 1972. Northbound local train No. 3828, bound for Södertälje Central and probably Stockholm was standing on a side track and the remotely controlled signals at each end of the station were clear for southbound express train X 37, bound for Göteborg. As the express was passing the local train moved out onto the main and a sidelong collision ensued. Four passengers and the driver of the local train died and ten or 22 (according to different accounts) people were injured.

Among the passengers on the train were some schoolgirls who were truant and on their way from Järna, a few kilometres into Södertälje They normally sat at the front of the train but when they boarded it they saw an elderly teacher sat there so they retreated to the rear. They survived; the schoolteacher was one of those who lost their lives.

The Royal train carrying King Fredrik and Queen Ingid of Denmark, on their way home, passed the scene of the accident some hours later, diverted onto a side track.

At the railway and police inquiries into the accident the train conductor, who had been

[419] National Transportation Safety Board report
[420] *Markham Star Tribune*, Markham, Illinois, United States, 22nd July 1973
[421] *The Salt Lake Tribune*, Salt Lake City, Utah, United States, 1st November 1972
[422] *The Lincoln Star*, Lincoln, Nebraska, United States, 3rd November 1972
[423] *The Modesto Bee*, Modesto, California, United States, 31st October 1972

stood alongside his driver, testified repeatedly that the starting signal for the local train had been cleared, though apparently he added that the signal had flickered between red and green. Some work on the track had been in progress at the time and it seems possible that the accident had resulted from a wrong-side signal failure. The Statens Järnvägar investigation was unable to reach a conclusion. [424] [425] [426]

An eastbound passenger train from Sofia to Tolbukin collided with a stationary freight train at the entrance signal at Venchan station, in the Varna district of Bulgaria on Sunday 10th December 1972. The leading sleeping car of the passenger train, in which there were 50 people, was wrecked and according to some reports caught fire. Twenty-six passengers lost their lives and ten more were injured. A government commission was set up to investigate the accident. The brief account of the accident from the Bulgarian news agency BTA did not clarify if it was a head-on, sidelong, or rear collision or who was likely to have been to blame for the accident. [427] [428]

Early on Friday 11th May 1973 a southbound passenger collided with a northbound freight train at a location stated to be 145 kilometres south of Taipei, Taiwan. Twelve people died and over 100 were reported injured. [429]

Direttissimo (express) train No. 2701 left Milano at 10:55 on Thursday 7th June 1973. It reached Roma Termini at 19:15 and left there again at 19:55, ten minutes late, for Napoli, Battipaglia, and Salerno, with Driver Diego Di Benedetto in charge of its locomotive. The train consisted of an engine, a luggage van, a postal van and 12 coaches and was carrying about 350 passengers. Ten minutes later, at 20:05, as it was travelling at 120 kilometres an hour and was within a few hundred metres of the station at Torricola, Italy, it met with disaster.

What exactly occurred is a little unclear from the newspaper reports but it would appear that at Torricola there were one or more sidings alongside the main line, on which some shunting operations were in progress, involving some ballast waggons. During these shunting manoeuvres there was either a collision between the waggons being shunted and another engine on the siding, or some uncoupled waggons ran away and collided with the buffer stops at the end of the siding. At any rate, as a result of this initial mishap one or more waggons fouled the main line as the direttissimo was passing, striking first the luggage and mail vans and then the second and third passenger coaches. The whole train was reportedly derailed, the mail van overturned and part of the train uprooted a concrete pole carrying the overhead line equipment.

Passenger Amalia Vernacchia, who had been riding in the third coach, the one in which most of the deaths and injuries occurred, was clearly still suffering from shock as

[424] https://web.archive.org/web/20140824145449/http://janne.goldnet.nu/trainwrecks/
[425] https://www.jvmv2.se/forum/index.php?mode=thread&id=105757
[426] https://www.tydal.nu/postvagnen/97689.htm
[427] *European Stars and Stripes*, Kaiserlautern, Germany, 12th December 1972
[428] *Progress-Bulletin*, Pomona, California, United States, 11th December 1972
[429] *Pacific Stars and Stripes*, Tokyo, Japan, 13th May 1973

she told her story to a reporter. She related, "There was a sudden violent braking. I was almost thrown to the floor. Within a very short time the train stopped. The windows were all shattered and smoke filled the compartment. You could hear the roar of metal being torn apart, the deafening thuds as the coaches were derailed, and the screams of the passengers. Then, when it all stopped, I saw men and women covered in blood running in every direction in terror."

Alessandro Vano, a 26-year-old Neapolitan, gave a similar account. "It was a terrible sight, a nightmare… the coaches windows were smashed, the walls of the coach were ripped open, metal sheets were crumpling at the sides… flying wreckage and smoke were everywhere, other passengers were screaming in fright. Then I must have passed out as I don't remember any more. I still don't know how I was able to escape."

The rescue operation quickly got under way, with numerous fire engines and Red Cross ambulances, police, carabinieri and airmen from a nearby Air Force barracks. Many of the injured were trapped in the wreckage of the two most badly damaged coaches and oxy-acetylene equipment was needed to free them. Ambulances took the injured to San Giovanni and Sant'Eugenio hospitals.

The accident was not a major disaster but five passengers lost their lives and another 32 were injured. The first of those who died to be identified was 46-year-old primary school teacher Mafalda Di Rosa, from Fregnano, who was killed instantly. The other four were later named as police officer Antonio Lamante, aged 21, of the Napoli IV Rapid Group, Cristina Grugni, aged 31, the mother of three children; Bruno Pararello, aged 28; and Vincenzo Pastore, aged 25. One of the injured, a Hungarian woman named Rita Loskaj, aged 21, remained in a coma at the San Giovanni hospital a couple of days after the accident.

Work on clearing the line went on through the night and single-line working was already in operation at 23:30 the same evening. Initially trains were diverted via Cassino. Deputy prosecutor Dr Callovini, with a body of experts; Dr Frazgranza of the railway police and Major Vitali of the carabinieri investigation unit went to the scene and a separate investigation into the accident was being undertaken by the Ministry of Transport. The deputy prosecutor interrogated Driver Domanico Fiorani, who had been operating the shunting engine; Assistant Driver Roberto Giardini and Shunter Stafano Rossi. Dr Callovini stated that he did not accept the version of the three railwaymen that the ballast waggons, after having been uncoupled from the engine, had run away along the siding into collision due to their weight and the slight slope of the siding. The driver and shunter were subsequently arrested on charges of multiple manslaughter and of causing a railway accident.

The blaming of the two railwaymen for the accident was denounced by the railway unions who blamed the accident on the very serious shortcomings and inadequacies of the railway network. In a press statement the unions commented, 'Within 70 days there have been two major railway disasters on the Roma – Napoli direttissimo high-speed line, supposedly the safest on the whole Ferrovie dello Stato (FS) network. The decision to blame an alleged error by a railwayman for the accident is a way of offered a scapegoat to public opinion so that it does not have to consider the fact that the real means of ensuring safety, by making accidents due to operator error impossible, does not exist.

'These facts once again confirm that the FS network is no longer adequate for the new traffic patterns and high train speeds, that there is a lack of political will to provide the indispensable improvements to signalling and rolling stock and that the arrest of railwaymen involved in accidents has the sole purpose of masking a reality which has not been addressed.'

The unions went on to condemn delays in railway investment and outlined the dangerous proximity of the sidings at Torricola to a line used by 400 high-speed trains daily. [430]

A passenger train collided head-on with a freight train standing at or near Sirajganj, [431]Bangladesh, either late on Sunday 16th or early on Monday 17th September 1973. One of the trains was reportedly standing at a platform and fire broke out after the collision. Newspaper reporters spoke of 100 fatalities and another 300 people injured. Railway officials would only confirm the death of the two train drivers. [432]

Carelessness on the part of a pointsman led to a collision at Marl-Sinsen, West Germany, on the Münster – Wanne Eickel – Essen line in the Ruhr, at about 18:10 on Friday 5th October 1973. The events that led to the accident began with the arrival from the south of a light engine, steam locomotive No. 044 405, which had to pick up a train of coal waggons from the Auguste Victoria colliery and return southwards with them. To do this it had to first cross over to the opposite running line at the South signal box before setting back into a siding. The engine was duly crossed over to the southbound running line by the pointsman and should then have been brought back into the siding. The pointsman thought that it had done so and reset the road for the main line. Unfortunately he reset the points prematurely, trapping the engine on the main line, though neither he nor the driver of the engine realised his mistake. The fahrdienstleiter in the north signal box then offered the pointsman in the south box passenger train No. D 632 from Flensburg to Düsseldorf and Köln, which as it was a Friday was carrying many Bundeswehr soldiers returning home from north Germany on weekend leave. The train was duly accepted and the signals were cleared for it.

In the twilight semi-darkness the driver of train No. D 632, travelling at 80 kilometres an hour, belatedly noticed the engine in his path, whistled and applied his brakes in emergency. There was no time for the pointsman to open his points and let the engine into the siding and the collision occurred almost at once. The force of the impact threw some of the wreckage onto the northbound main line where it was run into a few moments later by express freight train No. Dg 5535 from Wanne Eickel to Osnabrück, which was also travelling at 80 kilometres an hour. The wreckage of the three trains piled up in a heap and some of it fell down an embankment. Police in a patrol car witnessed the accident and so rescue work was immediately begun.

Seven people were killed outright and 62 were injured, one of whom, an elderly woman, died later in hospital. Three of those who died were railwaymen, two of whom, ironically, were drivers who should not have been on duty; one of them had swopped

[430] *L'Unica*, Roma, Italy, 8th, 9th and 10th June 1973

[431] *Winnipeg Free Press*, Winnipeg, Manitoba, Canada, 18th September 1973

[432] *The Daily Review*, Hayward, California, United States, 18th September 1973

shifts with a colleague and the other was standing in for a colleague who had reported sick. The driver and firemen of the steam locomotive (the passenger and freight trains were both hauled by electric locomotives) both jumped clear. In doing so the fireman broke his foot and the driver, suffering severe shock, did not return home until the next day, by which time it was supposed that he was dead and his body buried beneath his engine. [433]

As in the 1950 collision at Penmaenmawr in England, described in the previous volume of this series, there presumably was no track circuiting at Marl-Sinsen to assist the pointsman. A more modern box was opened there in 1975.

An initial newspaper report stated that at least 80 people had died and 100 had been injured in a collision between a passenger and a freight train in Bahia state, Brazil, at 20:30 on Monday 17th December 1973, due to a signalling error. Firemen, police, doctors, and members of all three armed forces were engaged in the rescue operation, searching the wreckage with the aid of torches. The accident was a head-on collision between passenger train S-40, consisting of locomotive and three coaches (or possibly as four-car multiple-unit set), and freight train CA-C2, consisting of an engine and seven waggons of mixed goods. The passenger train had left Calcado station at 20:30 for the suburb of Paripe and the freight train was on its way from Petrolina in Pernambuco state to Salvador.

The trains collided on a curve so that neither driver had sufficient warning to do much to prevent the collision. The freight train was clearly more massive than the passenger, the front two vehicles of which were described as having been entirely destroyed in the collision. The accident happened at a location called Plataforma, not far from Almeida Brandão.

A second report on the accident corrected the number of casualties to 17 dead and 67 injured; the driver of the passenger train was among the fatalities. The injured were taken to the casualty department at Getúlio Vargas Hospital. Two thousand people from the Platforma, Paripe and Periperi suburbs shuffled through the mortuary in Salvador from 06:00 to the evening, hoping not to find the bodies of relatives or friends. Five bodies, two unrecognisable, were not quickly identified.

Blame for the disaster was quickly placed on Telegraphist Benjamin Batista Santana at Almeida Brandão station, who had accepted the freight train on Track 1 instead of Track 2, despite the passenger train having already been despatched on Track 1 in the opposite direction.

Shortly after the accident Driver Eufrosino Manuel dos Santos of the freight train spoke to a reporter, stating that he believed the responsibility for the disaster rested with the telegraph operator and describing his dead colleague, Driver Lourival Nunes Duarte, as a genius, a good friend and a very professional driver, adding that it was a tragedy that Driver Duarte and the other 16 people had lost their lives in the horrible accident.

When it became clear to Telegraphist Santana that he was being regarded as solely responsible for the accident he disappeared. Neighbours said that he, his wife and their three children had left their home the morning after the disaster, taking a train to Queimadas, his home town. His superiors stated that he had been a good, reliable

[433] https://www.drehscheibe-online.de/foren/read.php?17,6590432,6590432

employee, who earnt 500.00 cruzeiros a month, and that they were very surprised at his disappearance. At the same time it was stated that Driver Manuel dos Santos had also disappeared, but this proved to be untrue, for the next day it was stated that in fact rather than having disappeared, his injuries in the accident had resulted in his being admitted to hospital. There he repeated that he had been in possession of an order authorising him to use Track 1.

The final newspaper report on the disaster stated that another passenger, Teresinha Conceiçao Pereira, had died of her injuries, increasing the number of dead to 18. Railway and police investigations into the accident were in progress and the tapes of the telegraph messages exchanged just before the collision had been seized. [434]

A mistake by a railwayman who operated the wrong set of points led to a major disaster at Zeytinlik, Turkey, on Monday 7th January 1974. The error, by a brakesman, led to a passenger train from Adana to Mersin and the *Eastern Express* from Ulukişla coming into head-on collision with one another. Passengers were reportedly thrown through windows as the trains met with a thunderous roar. Both trains were crowded with families returning home at the end of the Kurban Bayrami Islamic holiday, including many children.

Rescue work following the accident went on during the night with the help of floodlights provided by fire engines rushed to the scene from the town of Adana, 30 kilometres away, and from the American Air Force base at Incirlik, near Adana. After eight hours of work the bodies of 25 victims, many of them children, had been recovered from the wreckage together with 29 injured. It was unclear if there were any further bodies but the authorities promised to continue the search until they were sure that no one remained to be found. Government, police, and military officials went to the scene to take charge of the rescue operation. [435]

At least 40 people reportedly died and another 27 were injured early on Thursday 21st February 1974 in a collision between an express train and a freight train near Moradabad, Uttar Pradesh, India. [436]

A southbound Caminhos de Ferro de Moçambiqe passenger train from Bulawayo, Southern Rhodesia (now Zimbabwe) was in head-on collision with a northbound freight train from Lourenço Marques (now Maputo) Mozambique, at 07:15 on Wednesday 27th March 1974. The accident occurred at São Miguel siding, Mozambique, between Alvor and Vila da Manhica and about 70 kilometres from Lourenço Marques. The freight train included tank waggons loaded with petrol, which were damaged in the collision, and the oil was sprayed over the coaches of the passenger train and immediately caught alight. At least 40 people died and 27 were injured, while the Portuguese news agency ANI stated that there were over 60 dead in what it said was the worst accident in the history of the country.

According to witnesses the first- and second-class coaches of the passenger train were

---

[434] *Jornal do Brasil*, Rio de Janeiro, Brazil, 18th, 19th, 20th and 21st December 1973

[435] *The Weirton Daily Times*, Weirton, West Virginia, United States, 8th January 1974

[436] *Wisconsin State Journal*, Madison, Wisconsin, United States, 23rd February 1974

melted by the heat of the fire. Uninjured passengers managed to rescue some of the occupants from the burning coaches before the heat drove them back and ambulances, fire engines and helicopters were rushed from Lourenço Marques. Attempts by the firemen to reach the burning coaches were likewise frustrated by the intense heat. Late in the evening, after the fire had finally been extinguished, attempts were made to pry open the smouldering wreckage in the expectation of finding more victims. There were believed to have been 30 fatalities in one coach alone. Vilar Quelroz, the Mozambique secretary for communications, personally took charge of the rescue operations. The injured were taken to hospitals in Vila da Manhica and the most seriously injured were flown to Lourenço Marques by helicopter. Only two third-class coaches and the luggage van reportedly escaped the fire. 'Human error' was given as the cause of the disaster. [437] [438] [439] [440]

According to some reports 80 people died and many were injured in a head-on collision, possibly on the single line, between a passenger train and a freight train at a location given as 78 kilometres north of Cotonou, which would imply a point near Abomey, Dahomey (Dahomey was renamed Benin late 1975). Other reports spoke of 15 dead while one late report from the Dahomey embassy mentioned 200 fatalities. Fire broke out following the collision early on Wednesday 6th November 1974 and most of the victims were trapped in the wreckage. The cause of the accident was not known. [441] [442] [443]

Early on Monday 17th February 1975 a South African Railways express train from Cape Town to Johannesburg collided head-on with a goods train near Potchefstroom, South Africa. Three leading coaches of the express, two reserved for "non-whites" and the other for "whites," were wrecked. A shuttle service of ambulances took the victims to hospital in Potchefstroom and an urgent appeal was issued for blood donors. Fourteen passengers, all black, lost their lives, and about 53 were injured. A railway spokesman stated that sabotage had been ruled out as a cause of the accident, otherwise it was not immediately known how it had happened. [444] [445]

Various American newspapers carried a press report regarding a head-on collision between a northbound passenger and a southbound freight train at or near Magude, Mozambique, on the night of Friday 28th March 1975. It will be noted that this was clearly a different accident to that which had occurred at São Miguel siding, Mozambique, a year and a day earlier, though some sources appear to have confused the two events. It is not clear if this was a single-line collision or whether it occurred at the

[437] *Diário de Lisboa*, Lisboa, Portugal, 27th March 1974
[438] *The News*, Port Arthur, Texas, United States, 28th March 1974
[439] *The Cumberland News*, Cumberland, Maryland, United States, 28th March 1974
[440] *Chicago Tribune*, Chicago, Illinois, United States, 28th March 1974
[441] *Journal de Genève*, Genève, Switzerland, 8th November 1974
[442] *Raleigh Register*, Beckley, West Virginia, United States, 7th November 1974
[443] *Daily Colonist*, Victoria, British Columbia, Canada, 21st November 1974
[444] *The Times*, London, England, 18th February 1975
[445] *Times-News*, Twin Falls, Idaho, United States, 17th February 1975

station; the freight train was reportedly stationary when it was struck by the passenger train. Initial reports in the American press spoke of 50 or at least 70 dead and 200 injured. [446] Both locomotives and two coaches of the passenger train were said to have been completely destroyed. The injured were taken by road ambulances and helicopters to hospitals in Lourenco Marques (Maputo from 1976). [447] Perhaps because full information on the accident was still lacking an initial report on the disaster in a Portuguese newspaper spoke of only three dead in the accident with four seriously injured having been evacuated by helicopter. [448] However a follow-up report in the same newspaper two days later stated that there were at least 83 dead, with rescuers still searching for bodies in the wreckage, and another 75 injured. [449]

Lithuania, which at the time of the accident was the Lithuanian Soviet Socialist Republic and a constituent of the USSR, suffered a major disaster on Friday 4th April 1975 near Žasliai. On the afternoon of the day of the accident freight train No. 2719, consisting of a locomotive and 93 fuel tank waggons, was on its way from Paneriai, near Vilnius, to Palemonas, near Kaunas. Behind this relatively slowly moving freight train was passenger train No. 513 from Vilnius to Kaunas. The decision was therefore made to divert the freight train onto a side track at Žasliai and this was done, but through a signalling error the line was reported clear when the rear tank car was still foul of the main line and the passenger train was accepted.

At 17:35 the passenger train, consisting of an engine and four coaches and travelling at 70 kilometres an hour, collided with the obstruction in its path. The engine and leading two coaches were derailed and the third coach became embedded in the tank waggon and suffered particularly from the fire that broke out following the collision. As it was Friday afternoon the passenger train was carrying many Vilnius students returning home for the weekend. Following the accident passengers made desperate efforts to break coach windows to escape the fire, while locals lent what help they could and took the injured to Kaišiadorys. A pregnant woman whose hand was trapped beneath a coach was also freed by the locals while passenger Arvydas Garnys returned to the train to rescue three children, then went back again to get his friend and died in the fire. His body could not be found afterwards; he was posthumously awarded a medal for his bravery. As well as those who died from trauma, burns or smoke inhalation many of the survivors suffered severe burns when their polyester raincoats and nylon stockings melted onto their skins in the intense heat. Firemen from Kaišiadorys, Elektrėnai, Jonava, Kaunas and Kėdainiai all fought the fire, which was finally extinguished by military firemen 15 hours after the accident.

Information about the accident was not entirely supressed by the Soviet authorities but only the basic facts were initially stated, together with an incomplete list of dead and injured, given as 17 and 39 respectively. After the country had regained independence in 1990 the number of fatalities was corrected to 20 and it emerged that 80 injured had been treated at Kaišiadorys hospital alone, and others in hospitals at Kaunas, Vilnius and Elektrėnai as well as some in local clinics. As many of the bodies had been more or less consumed in the fire it was also believed that the real death toll

[446] *The Weirton Daily Times*, Weirton, West Virginia, United States, 29th March 1973
[447] *Pacific Stars and Stripes*, Tokyo, Japan, 31st March 1975
[448] *Diário de Lisboa*, Lisboa, Portugal, 29th March 1975
[449] Ibid, 31st March 1975

could be considerably higher.

Signalman Stasys Urbonavičius and Supervisor Motiejus Šiško were found to have caused the disaster and were sentenced to 13 and three years' imprisonment respectively and ordered to pay damages amounting to 253,406 roubles. Their colleagues petitioned for clemency for the accused on the grounds that the signalling equipment had not indicated that the freight train was still foul of the running line; presumably there was no track circuiting. The signalman was released after seven years and resumed railway service, with 20% of his wages confiscated for damages. [450]

Five people died and at least 34 were injured, half of them seriously, when express train No. E 3712 and local goods train No. 62 689 coming from Saarburg were in collision at Taben-Rodt station West Germany, on the Saarbrücken – Trier line. The accident occurred at 14:25 on Tuesday 22nd April 1975 and was blamed on 'incorrect train reporting procedure.' [451] [452]

In a collision reminiscent of that which had happened at Järna in December 1972 (see above) two express trains collided at Norrköping, Sweden, on Thursday 15th May 1975. Southbound express No. 1 from Stockholm to Malmö was just approaching the station when fast train No. 264 pulled out of the station prematurely and its engine collided with the rear coach of the arriving train. Travelling on the southbound express were 300 schoolchildren who had been picked up at various points along the line and were on a trip to the Kolmården Wildlife Park, near Norrköping. Many of them were standing in the aisles of the coaches preparing to alight when the accident happened. Eight people died in the accident, seven of whom were children aged between 10 and 12 from the same class of a Katrineholm school. Most of those who lost their lives had been thrown from the rear coach which then fell onto them. A number of other passengers were injured. The trains had both been moving at about 40 kilometres an hour.

Local doctor Kjell Kjellgren, who assisted at the scene, said: "It was the most dreadful thing I have ever seen. There were badly injured passengers on the ground, screaming and moaning, after being thrown from the train. I could hear the cries of children trapped inside the overturned coach."

Single-line working was in force north of the station at the time of the accident due to engineering work and the signals had been cleared for the southbound train when the northbound one set off. An investigation was of course launched into how the accident had happened but due to conflicting witness statements it was unable to draw any firm conclusions. [453] [454] [455] [456]

---

[450] Wikipedia – Žasliai railway disaster

[451] *Katastrophen der deutschen Bahnen Teil II* by H.J. Ritzau, pg. 154

[452] https://www-saarstrecke-de.translate.goog/DBgesch/Archiviertes/db_unfall1975/body_db_unfall1975.html?_x_tr_sch=http&_x_tr_sl=de&_x_tr_tl=en&_x_tr_hl=en&_x_tr_pto=sc

[453] *The Times*, London, England, 16th May 1975

[454] The *Galveston Daily News*, Galveston, Texas, United States, 16th May 1975

[455] *Altoona Mirror*, Altoona, Pennsylvania, United States, 15th May 1975

Having suffered a major disaster in 1961 (see Rear Collisions chapter) Hamburg, West Germany had another bad accident at 17:10 on Tuesday 22nd July 1975, this time near Hamburg Hausbruch station. In contrast to the earlier accident this was a head-on collision due to a mistake by a driver rather than an error by a signalman. One of the trains involved was local passenger train No. N 6858 from Hamburg Hbf to Buxtehude, the other was express freight train No. Sg 52323, which was on its way from the Hamburg Waltershof on the Hamburger Hafenbahn (harbour line) to the marshalling yard at Maschen. In order to reach its destination the freight train had to cross the Buxtehude line. The fahrdienstleiter decided to give the freight train priority, his decision doubtless influenced by the fact that the passenger train had to stop anyway at Hamburg Tempowerk halt, just short of the crossing of the two routes.

The semaphore signal protecting the crossing of the two routes was of course at danger against the passenger train and Indusi track magnets were provided at both the Vorsignal (distant signal) and Hauptsignal (stop signal). Unfortunately these did not prevent the accident. The Indusi at the Vorsignal did not come into action because the passenger train was already reducing speed to stop at Hamburg Tempowerk. The station work having been completed Motorman Ernst Schröter, 50, restarted his six-car electric train and in 800 metres had accelerated to 80 kilometres an hour, having completely forgotten during the brief stop at the halt that the Vorsignal had been against him. The 27-waggon diesel-hauled container train only had to cover a short length of track after coming off the branch before it gained its own track and had it been on time it would have done this before the passenger train arrived but by chance it was running three minutes late. The Indusi track magnet at the Hauptsignal applied the brakes of the passenger train in emergency at the same time as Motorman Schröter also applied them, but the speed of the train was too high for a collision to be averted. The passenger train was still travelling at about 40 kilometres an hour and the freight train at 30 kilometres an hour when the two met. The leading passenger car was crushed between the engine of the freight train and the five following passenger cars and was forced up into the air. Eleven passengers lost their lives and at least 65 were injured, another 60 were probably slightly hurt. Motorman Schröter threw himself to the floor of his cab and escaped with a few bruises, while the driver of the freight train jumped from his cab and also survived.

Deutsche Bundesbahn came in for much criticism after the accident, especially in view of the fact that almost exactly four years earlier an identical collision had almost occurred, the two trains involved stopping on that occasion within a few metres of one another. At that time drivers had complained of the poor visibility of the Hauptsignal against the background of a grey concrete motorway bridge, plus the lack of a repeater signal for the Vorsignal in advance of the halt, equipped with its own Indusi track magnet, which would have prevented the collision. All that had been done in the meantime, however, was to mount a white backboard behind the Hauptsignal to increase its visibility, a measure which was not very successful as dirt and exhaust fumes had soon turned the backboard grey.

A year and a half after the accident Driver Schröter appeared in court charged with 22 counts of manslaughter and 65 counts of causing injury through negligence. He was duly

[456] https://web.archive.org/web/20140824145449/http://janne.goldnet.nu/trainwrecks/

found guilty and received an eight-month suspended prison sentence.

The semaphore Hauptsignal was subsequently replaced by a colourlight signal in the same location and Tempowerk halt was closed. [457] [458] [459] [460]

At 21:35 on the evening of Sunday 14th December 1975 the *Sud Express* which had left Lisboa Santa Apollonia at 16:30 for Paris, France, collided head-on with special train No. 20010 in the station at Fornos de Algodres, Portugal. The trains were due to cross there and the special train was already stopped in the station on track No. 1 awaiting the arrival of the express when the latter train was diverted into its path. There were 150 passengers on the express and a large number on the special, which was bringing migrant workers back from Paris to spend their Christmases at home. The trains should have crossed one another at Mangualde but the special was running somewhat late, so the crossing of the two trains was postponed to Fornos de Algodres, about 25 kilometres further east. It was very foggy at the time of the accident and there was also deep snow on the ground. The engine of the special was described as having been destroyed in the collision and four coaches of that train were derailed. The engine of the express also suffered badly but only one coach of that train left the track. Firemen from Mangualde carried out the rescue operation; the injured were taken to hospitals at Veseu (15) and Coimbra (two). [461] There was reportedly a national appeal for doctors, nurses, and blood donors. [462]

The number of people who died in the collision is unclear. First reports in a Lisboa newspaper spoke of 10 or even 16 deaths, but the following day the number of dead was stated, in the same newspaper, to be only eight, though it is not clear if this figure included four bodies discovered in the snow some hours after the accident. [463] The accident occurred about 300 kilometres from Lisboa and the heavy snowfall made communication with the location difficult. The confirmed fatalities were passengers Joaquim Vitória Beja, Manuel Lourenço Jacinto, and Manuel Valente; Driver José Augusto Monteiro Lourenço of the special train and Conductor João Romano Proença, Driver Manuel de Costa Leitão and Assistant Driver António Barriga Branco of the express, plus an unidentified *Wagons-Lits* employee, apparently of Spanish nationality. [464]

The accident was blamed on either an error in the signalling of the trains or interference with the signals due to the severe weather. A rigorous inquiry was promised into its cause.

---

[457] https://www.spiegel.de/politik/volles-rohr-a-823db46e-0002-0001-0000-000041458152
[458] https://www.bahninfo-forum.de/read.php?5,348584
[459] *Katastrophen der deutschen Bahnen Teil II* by H.J. Ritzau, pgs. 154-5
[460] Wikipedia – Eisenbahnunfall von Hamburg-Hausbruch
[461] *Diário de Lisboa*, Lisboa, Portugal, 15th December 1975
[462] *The Charleston Gazette*, Charleston, West Virginia, United States, 15th December 1975
[463] *Indiana Evening Gazette*, Indiana, Pennsylvania, United States, 15th December 1975
[464] *Diário de Lisboa*, Lisboa, Portugal, 16th December 1975

CHAPTER 6

SINGLE LINE COLLISIONS

The operation of trains in opposite directions over a single line of rails carries the obvious danger of head-on collisions between them. By 1961 there were many hundreds of miles of single-track railways throughout the world, especially in places where the density of traffic did not justify the provision of a second line. In many countries these were worked by telephone, with or without lineside signals, decisions as to the meetings between trains being either made locally or, as in the United States, by a central dispatcher. In Great Britain by mid-1961 most lines were worked by the electric train tablet or its variants (electric train staff and key token), though these systems were not used extensively outside of the British Empire. The invention of a failsafe track circuit in 1872 by the American William Robinson[465] had led to the introduction in the United States of automatic block signals, by which train movements on a single line could be safeguarded by signals warning of the approach of an opposing train, though such signals were not installed everywhere due to cost. With the protection of trains running over a single line left solely to the exchange of telephone messages, rather than by the more secure electric train tablet or its later equivalents, and not supplemented by some form of automatic signalling, collisions due to human error in the transmission or interpretation of such messages could easily occur, or drivers could leave stations disregarding the fixed signals.

Collisions on what was normally double track where one line had been taken out of use due to accident or engineering work and trains were being worked in both directions over the remaining line are also included in this chapter.

The first accident we will consider in this chapter occurred near Eßlingen Hauptbahnhof (Hbf), West Germany on Tuesday 13th June 1961. To all intents and purposes this was a head-on collision during temporary single-line working, though at the point where it occurred, on the outskirts of Stuttgart, there was in fact quadruple track. The accident occurred between Eßlingen Hbf and Stuttgart-Obertürkheim stations, at a location where one pair of tracks was used exclusively for long-distance passengers and freight trains, the other pair of tracks being likewise reserved for suburban traffic. It was the suburban tracks that were involved in the accident. At the time of the accident the tracks and subgrade of the suburban lines needed renewal and it was decided to carry out

---

465 Wikipedia – Track Circuit

the work on one line at a time. The intensity of the train service was a problem: there were 230 trains daily on the long-distance lines and about 160 on the suburban lines, making the working of all the traffic over a single track difficult.

The Eßlingen Hbf – Stuttgart line was duly taken out of use for the track renewal, with single-line working over the line normally used by trains coming from Stuttgart towards Eßlingen Hbf. A crossover taking trains going to Stuttgart onto the line from there already existed at the Stuttgart end of the station platforms at Eßlingen Hbf, but in order to reduce delays a new crossover was laid immediately to the rear of the beginning of the track renewal area, with a separate block section being created between the normal starting signal, $F7^1$ at the end of the platform and the new starting signal, F7, 657 metres in advance of signal $F7^1$. Stuttgart-bound trains could thus travel rather more than half a kilometre further on their own track before entering the single line, saving perhaps two minutes.

The single-line working began at about 15:10 on the day of the accident and six trains were worked through the section in one or another direction safely. Train Nt 3885 from Stuttgart Hbf to Süßen then crossed train Nt 3900 at Stuttgart-Obertürkheim, left there at 16:49, called at the intermediate stop at Eßlingen Mettingen at 16:52 and proceeded towards Eßlingen Hbf. Meanwhile at Eßlingen Hbf, as soon as train Nt 3900 had cleared platform No. 7 at 16:41 train Nt 3902 to Stuttgart Hbf was brought to the platform in turn, arriving there at 16:47. It was normally booked to depart at 16:49 but in order to give train Nt 3885 time to clear the single line without train Nt 3902 having to be held for it at signal F7, the departure time of train Nt 3902 had been advanced to 16:52 for the duration of the single line working. Unfortunately neither the motorman of train Nt 3902, its conductor, nor its ticket collector was aware of this alteration, so when the train arrived at the platform at 16:47 they were under the impression that they only had two minutes to spare before their train was due to depart. During this time the motorman had to change from the rear to the front cab of the six-car multiple-unit train and in his rush he absentmindedly left his cap and briefcase in the rear cab.

Having seated himself at the front of the train the motorman peered out of his cab window and at 16:49 his conductor whistled twice and raised his arm. Starting signal $F7^1$ at the end of the platform was off and the train departed. The motorman apparently failed to see the new forsignal (distant signal) for signal F7 and accelerated away from the platform as per usual during normal working. As soon as his train had started the conductor had begun filling out the paperwork he had not had time to complete during the brief station stop, while the ticket collector began examining the tickets of the handful of passengers.

The signalmen in Stellwerk 1, situated close to the platforms, noticed train Nt 3902 starting away early and accelerating, although it had to stop 600 metres down the line. They warned their colleagues in Stellwerk 2, close to the temporary crossover. There was no time for the latter to run out and lay detonators but a signalman gave stop signs with his arms from an open window, signal B, which had been cleared for train Nt 3885, was thrown back to Danger, and a telephone call was made for the traction power to be turned off. Some platelayers also realised the danger and one of them removed his hat and waved it at the train as a warning. He testified later that the motorman had moved his hands as if to query what the platelayer meant. That the motorman had still been unaware of any danger was shown by the fact that on seeing another track worker at the new crossover (points No. 100) he had merely given a short warning blast to alert the platelayer to the approaching train. As his train was diverted through the crossover he had finally realised the danger and an emergency brake application reduced the speed of his train from 78 to 50 kilometres an hour at the moment of the collision.

Train Nt 3885 had just passed signal B when it returned to Danger, while an engineering train stabled on the dead track blocked the view of its motorman until the last moment when, according to a surviving passenger, he had applied his brakes in emergency and shouted a warning to the occupants of the front compartment. The trains collided at 16:54. The front cars of each train were thrown down an embankment to the bank of the River Neckar and both motormen and 33 passengers were killed. Another 92 people were injured, 36 seriously. Property damage was estimated at 337,795 Deutschemarks.

A German author has stated that the failure of the motorman of train Nt 3902 had left everyone puzzled. He was regarded as reliable, dedicated, and healthy, and apparently had no personal problems. He had not driven a train on the Stuttgart – Eßlingen Hbf line in the days preceding the disaster but was well familiar with the route. He had, however, complained of a headache during the morning and had indicated a desire to go off sick, though he had not done so. This appears to suggest to the present author that there might have been some medical reason for his mistake, such as a developing stroke. It is not clear if any post-mortem examination was carried out upon him and the motorman's age was not stated.

The investigation into the disaster decided that the fault for it did not rest, in any case, entirely with the deceased motorman. Questions were raised, for example, as to whether the conductor should not have positioned himself in the motorman's cab to assist in the observation of signals, what the maximum permissible speed should have been from the station platform to the temporary crossover, whether Indusi train protection should have been provided, whether the temporary crossover should have been installed at all, and so on.

Legal proceedings were begun against the surviving conductor of train Nt 3902 and other Deutsche Bundesbahn officials and the matter dragged on for almost seven years before the proceedings were finally abandoned. [466] [467]

Carelessness on the part of a stationmaster led to a single-line collision between Cali and Buenavista, Columbia, on the Occidente line of the Ferrocarriles Nacionales de Colombia on Friday 16th February 1962. Eight lives were lost in the accident, all of them railwaymen, and five others were injured. The trains involved in the accident were freight train No. 571, with Driver Saulo Salguera, coming from Buenavista, and what was described in a newspaper report as *carro-motor* No. 1837, probably what in the United States would be called a track motor car, which was basically a standard automobile fitted with flanged wheels to enable it to run on railway tracks. The *carro-motor* was carrying the crew of platelayers responsible for the section of line between kilometres 59 and 73 of the line. The accident occurred at 16:05, so the crew were probably on the way home at the end of their day's work. It would appear that the *carro-motor* was treated as

[466] *Katastrophen der deutschen Bahnen Teil I* by H.J. Ritzau, pgs. 67-84

[467] Wikipedia – Eisenbahnunfall von Esslingen

an ordinary train and was despatched from Puerto Dagua station towards the station at Loboguerro ahead of train No. 568, which was presumably a freight train. The circumstances of the accident are not entirely clear from the newspaper report but it seems that there was a second *carro-motor* already at Loboguerro which had arrived from Buenavista, while freight train No. 571 was also waiting there. In some manner the Loboguerro stationmaster mistook the second track motor car for that which was still on its way from Puerto Dagua, reported that the Puerto Dagua car had reached him, and allowed freight train No. 571 to proceed to the latter station. Two kilometres from Loboguerra the two trains met one another. The little *carro-motor* was no match for the diesel locomotive of the freight train and all the dead and injured were among the track crew. Foreman Florencio Rodriguez Hurtado, Nailer Simón Garcés, and labourers Julia Bonilla, Carlos Ortega, Guillermo Caicedo, Guillermo Villano Medina, and Leonilda Pretel Mesa were all killed. Claudio Melo, the driver of the track motor car; Guillermo Ledesma, Efrain Medino Rios, Octavio de Jesus, and Angel Maria Reinosa were all injured. For the last two named it had been their first day working on the railway. Several days later it was reported that an eighth worker, Luis Enrique Guerva, who was not initially listed amongst the injured, had also died.

Serious though the accident was, it would be followed by a considerably worse one of the same type only five days later and only 22 kilometres away on the same Columbia line.[468]

The second single-line collision in Columbia in less than a week occurred at 07:27 on Thursday 22nd February 1962 at kilometre 47 of the Cali – Buenavista line, between La Delfina and Cisneros stations. This time a passenger and a freight train were involved. Passenger train No. 171 had left Buenavista for Cali at 06:00. It consisted of diesel engine No. 328, a luggage van, four third-class coaches, a restaurant car and four first-class coaches, in charge of Conductor Ricardo Bolanos with Driver Roberto Arbeláez and Assistant Driver Leonel Montoya. Freight train No. 769 from Dagua consisted of diesel engine No. 307, seven waggons, a van, an empty tank waggon, two gondola waggons and a platform waggon, in charge of Conductor Mariano Espiniel with Driver Juan B. Gómez and Assistant Driver Fabio Benjumea (the names of the two conductors may be incorrect as the newspaper report is difficult to interpret). It was bound for Córdoba (a station on the line 20 kilometres from Buenavista and not the more obvious place of the same name in northern Columbia).

Among the passengers on train No. 171, riding in the first-class coaches, were girl students from the Liceo Fememino de Pacifico de Buenavista and 87 American sea cadets from the California Maritime Academy. The cadets had arrived in Buenavista on the 19th on the *Golden Bear* training ship, which was due to set sail again the following day. The captain of the ship, Carl Bowman, had gone to Cali by road the previous day to receive the cadets there.

Train No. 171 was running downhill at 60 kilometres an hour at the moment of the collision and the freight train was running at about the same speed. The topography of the ground at the scene meant that neither driver had much warning of the presence of the other train. One third-class passenger coach was completely destroyed in the collision,

[468] *La Tiempo*, Bogotá, Colombia, 17th February 1962

half of a second was likewise destroyed, a third was overturned and a fourth was damaged to a lesser extent. The restaurant car also suffered to some degree and five of the American sea cadets, who had been having breakfast in it at the time, were more or less badly hurt. The passengers in the first-class coaches at the rear of the train suffered from shock but were little hurt. An official announcement from the general manager of the Ferrocarriles Nacionales de Colombia spoke of 29 deaths and 46 injured, while a probably more up to date newspaper report mentioned 35 deaths and over 100 injured.

As soon as news of the disaster reached Cali the head of the Pacifico Division, engineer Francisco Jaramillo, arranged to set off for the scene with other officials using a railcar, while two ambulances were despatched by road with doctors, nurses, and medicines at 07:50. A relief train was likewise despatched from Buenavista to the scene at 08:50 with firemen, doctors, nurses, police, and oxy-acetylene equipment. The rescue work lasted until the early afternoon. The Cali ambulances met the train from there at Dagua and the medical staff and equipment were swopped onto the train to get to the accident site at kilometre 47. Rescuers placed the bodies of the dead along both sides of the track and covered them with the leaves of the breadfruit trees, which were plentiful at the scene, prior to their being taken away. The accident had occurred within the jurisdiction of the Buenavista municipality, so at the request of the mayor of that town, Helcias Martan Gongora, the injured and the bodies of the dead were mostly taken there, though the injured American cadets were flown by helicopter to the Clinica de Occidente in Cali, Colonel Calderón, of the Cali Air Base, having made any aircraft or helicopters that might be needed available to the railway company. Unfortunately there was some looting of money, watches, jewellery, and suchlike from the victims of the disaster, including items from some of the cadets, who had taken off their jackets to help in the rescue work and discovered later that dollars, timepieces, notebooks, and identification papers had been removed. It was stressed that anyone who had been found guilty of looting would be severely punished.

None of the railwaymen on the two trains lost their lives in the disaster, though most of them were more or less injured. A newspaper reporter was able to question both drivers, who were still trapped in the cabs of their engines two hours after the collision, awaiting release.

Driver Gómez, 41, of the freight train, was gently stroking the left side of his chest as he spoke to the reporter, having suffered a probable fractured rib. He related, "The accident was due to a very serious mistake made over the telephone by the official at La Delfino station, who gave me permission for my train to proceed thence from Cisternos. There are witnesses, including Miguel Arrallaneres Mosquera, that after I had set off the stationmaster at Cisternos a person (Mosquera?) waved a red flag to stop my train, but unfortunately my conductor failed to see it. Upon reaching kilometre 47, and entering an easy curve, I caught sight of the passenger train coming from Buenavista, running at its normal speed, when it was only about 70 metres away. It was a terrific shock. In that moment I knew that many people would likely die in a collision between our two trains and I had to do everything that I could to pull up. I applied the brakes in emergency – the air brakes – and put down sand. Then, clinging to the controls, I turned off the engine to prevent a fire and waited for the impact. It was brutal. The sound of shearing metal, breaking wood and human cries of agony and panic reached my ears. I did not lose consciousness. I saw one of the coaches leave the track and felt the earth tremble with the sudden braking of the 500 tons of waggons in my train. To be honest, I don't know how I survived, because the entire control panel of my engine was embedded in the control panel of the passenger locomotive. It was the first accident that I have been in during 19 years' service as a first-class driver on the Pacific line."

Driver Arbeláez, 48, of the passenger train, who had 30 years' service as a driver, passed his bandaged left hand over his face as he spoke to the reporter. "I am deeply upset by today's tragedy. I was on a normal journey, in charge of one of our most modern locomotives, and was travelling towards Cali with about 240 passengers. When my train reached kilometre 47 I suddenly saw a freight train advancing towards me. I did all that I could to pull up and shouted a warning to several cadets, who had been riding on the body of my locomotive to film a movie, to jump off. Several did so but two were trapped between the two engines, one of them suffering a torn arm. Immediately afterwards I heard a tremendous noise and then the crunching of metal, wood, and freight which had broken loose. For a time, I don't know how long, I sat paralysed in front of my controls. Then I realised that I had done all that I could to stop my train. Later I found that those who had suffered most had been the third-class passengers at the front of my train. I don't know who is to blame for the accident. As far as I was concerned I had a clear road and I was not expecting to find any obstacle on the way."

Also interviewed at the accident scene was Révérend Père Carlos Enrique Barrera, parish priest of the El Naranjo *corregimiento* (local district), who declared, "This has been the worst thing that I have witnessed in my life. Poor people. I was able to give absolution to 25 dying victims. Now all that remains to do is to provide timely help to the orphaned children, widows, and injured, to reduce their pain and distress."

An immediate investigation was launched into the causes of the accident and lawyer Enrique Camargo Maestro and his secretary, Julio Moreno, left at once for the scene to begin their inquiry, while the railway company initiated its own investigation in charge of José Eloy Diaz, chief of the transportation department, assisted by Jorge E. Quiroga, in charge of industrial safety. The initial finding was that the disaster was due to a failure, probably on the part of the stationmaster at La Delfina, to issue an order by telephone to hold freight train No. 769 at Cisternos for the passenger train. Quite how the mistake came to be made was not immediately apparent.

Whenever a bad railway accident has occurred newspapers love to be able to point to *yet another accident on the railway*, giving coverage to some minor mishap which would not normally have ever been thought worth mentioning. Thus the newspaper that had described the disaster was able to mention that the same day a waggon of a Sogamoso – Bogotá freight train had become derailed in the vicinity of El Mortiñal station. No one was hurt in this accident, the waggon was re-railed, and the train resumed its journey within half an hour. [469]

Two express trains met in a head-on collision in Spain on the early morning of Saturday 3rd November 1962. They were train No. 215/616, which had left Madrid at 19:20 the previous evening for Vigo, where it was due at 09:30, and its corresponding working, train No. 216/615, which had left Vigo at 19:15 and was due at Madrid Príncipe Pío at 10:00. The trains were due to cross one another at Abejera but the train from Madrid ran through the signals there and continued towards the next station at Sarracín de

[469] Ibid, 23rd February 1962

Aliste, despite the efforts of the Abejera station official to stop it. The official then telegraphed his colleague at Sarracín in the hope of getting the Vigo train stopped there, but was informed that it had already passed, when he knew that there was nothing more that he could do.

The trains collided at kilometre 55.55, just outside of Tunnel No. 3, at 02:43. The results of the accident were perhaps not as serious as they might have been but four people were killed outright and three were seriously hurt. All the dead and seriously hurt were railwaymen; none of the passengers on either train was more than slightly hurt. Driver Benigno Angado Antón and Assistant Driver Luciano Ruiz Méndez of the train from Madrid were both killed, as was Assistant Driver Manuel Méndez Alvarez of the train from Vigo. The body of a fourth victim could not immediately be identified. Seriously injured were Conductor Cesáreo González Maroto, Driver Andrés Pavón Carpena and Porter Antonio León Martin Gallego, all of whom had been on the train from Vigo. The engine crew of the train from Madrid had not reacted even when the train from Vigo had come into view and the driver of the latter had flashed the headlight of his engine to warn them. After the accident they were found both still sat in the seats in the engine cab and it was speculated that they must have fallen asleep before reaching Abejera, explaining why they had failed to stop there. Assistant Driver Alvarez had chosen to jump from the engine immediately before the collision but did so as the train was crossing a road underbridge and was killed in the fall. Driver Carpena chose instead to retreat into the interior of the locomotive and survived. [470] [471]

What is assumed to be a single-line collision occurred on the Rede Ferroviária Federal (ex-Central do Brasil) between Paraíba do Sul and Engenheiro Vieira Cortes, Brazil, on Monday 5th April 1965.

Shortly after the accident a Public Relations spokesman for the railway issued a statement regarding it: ‘At 12:55 today electric train No. F-54, en route from Barra do Pirai to Três Rios, after passing an automatic signal at kilometre 182 at Paraíba do Sul, collided with freight train No. CEC-51, coming from Três Rios towards Barra do Pirai. The violence of the collision caused the first car of the electric train and the engine of the freight train to telescope one another. Eight bodies have been removed from the wreckage of the second car of the electric train and resources are awaited to recover those beneath the leading car of the electric train and the freight locomotive, both of which were practically destroyed.

‘Motorman Aristides do Espirito Santo of the electric train is missing, while engineer Pedro Paulo Gonçalves de Andrade, from the 6th Inspectorate of Lines (IL/6) who was travelling in the electric train, was killed on the spot. It is possible that other bodies will be recovered from the wreckage, and this office will issue a detailed official note as soon as the necessary information becomes available.

‘A rescue team has been sent to the scene of the accident, consisting of doctors, nurses, engineers, operators, technicians from various specialities and staff from the public relations department.

---

470 https://www.trenzamora.es/Fotosvarias/abejera.html

471 *Correo de Zamora*, Zamora, Spain, 3rd and 4th November 1962

'As a result of the disaster certain trains have been cancelled: Service No. N-1, due to depart from Don Pedro II at 19:30 for Belo Horizonte; train D-3 from D Pedro II at 21:15 to the same destination and the railcar due to leave D. Pedro II at 16:50 for Juiz de Fora.' [472]

The accident turned out to be considerably more serious than the early announcement from the publicity office suggested. By 22:00 24 bodies had been recovered, including those of Driver Francisco José dos Santos of the freight train, Conductor José Guilherme Dias of the passenger train and Brakeman João Lima of the freight train. Fourteen other passengers were admitted to hospital with varying injuries. A panic occurred among the passengers on the electric train following the collision, with many reportedly jumping through windows and some, mostly children, being trampled in the confusion.

Upon becoming aware of the disaster Rio State Governor Paulo Torres ordered the dispatch of all possible assistance. Medical teams were sent to the scene from Paraíba do Sul, Casa de Saúde de Nossa Senhora de Conceição and Três Rios, while the Força Aérea Brasilaira (Brazilian Air Force) provided a helicopter which brought rhesus negative Group O blood supplies to the scene and ferried a woman passenger who had suffered multiple injuries to Rio de Janeiro.

Work on clearing the line was begun at 17:00 and traffic could be resumed just after 04:00 the following morning once the wreckage had been removed, the track itself not having been damaged. An investigation into the cause of the disaster was also opened, headed by General Technical Adviser to the Railway Julio Rouanet, assisted by Signalling Technician Aluísio de Faria and Assistant Traffic Manager Davi Schecheteter.

According to newspaper reports it was suspected that the accident had resulted from a wrong-side signal failure, which had caused the driver of the passenger train to enter the section still occupied by the opposing freight train. Possibly the light of the signal had just been extinguished; fuller details are not given in the newspaper reports. [473]

A tired railwayman, the theft of telephone cables and a failure to follow operating procedures were stated to have been the causes of a collision between two Philippine National Railways passenger trains on Luzon Island in the Philippines on the morning of Wednesday 16th June 1965. Northbound train No. 8 from Manila and southbound train No. 11, going to Manila, were involved. Relief Stationmaster Donaldo Deffino at San Fabian permitted train No. 8 to proceed towards the next station at Damortis before the arrival of train No. 11, because, he pleaded, he had been "sleepy." There was no communication between the two stations because the telephone wires had been stolen during the night, to be sold as scrap. Meanwhile train No. 11 had been despatched southwards from Damortis and the two collided at 8.15 a.m. Twelve passengers were killed outright and 32 were seriously injured, two of whom later died.

Conductor Z. Azana reported himself to the authorities as soon as he could after the accident, doubtless to show that his train had been in the right, as he was in possession of the train staff for the section. Meanwhile Relief Stationmaster Deffino had surrendered

---

[472] *Correio da Manhã*, Rio de Janeiro, Brazil, 6th April 1965
[473] *Jornal do Brasil*, Rio de Janeiro, Brazil, 6th and 7th April 1965

himself to the police at San Fabian, fearful of being lynched by angry passengers. [474] The crew of the northbound were presumably also at fault for leaving San Fabian without the train staff.

This was reportedly the worst accident in the history of the Philippine National Railways. [475]

Another single-line collision occurred in Spain on the afternoon of Monday 16th August 1965. The Almeira – Madrid mail train arrived at Gádor station on the Almeira – Linares line and after a brief stop was allowed to depart northwards by the assistant stationmaster, despite the fact that a southbound freight train, consisting of an engine and 28 waggons of ore, had not yet arrived. The two trains collided at 15:25 a kilometre and a half from Gádor and between there and Santa Fe, at a location known as Cuesta del Rayo. Ten people were killed outright and over 50 were hurt, of whom three later died of their injuries. Four of those who died were children. Thirty thousand people attended the funeral of the victims. A summer storm was raging at the time and possibly interfered with the communication between the stations. After the accident the official involved, accompanied by his family, surrendered himself at the police station and admitted his responsibility for it. [476] [477] [478]

What was effectively a third single-line collision in Spain occurred outside of the station at Villar-de-los-Álamos on the morning of Saturday 18th December 1965. One of the trains involved was No. 807 from Paris, France, to Lisboa, Portugal, the *Sudexpreso*, the reverse working of which was to be involved in a disaster at Fornos de Algodres, Portugal almost exactly ten years later (see previous chapter). Normally this train consisted of a single Class 242 steam locomotive, a van, two first-class coaches, a restaurant car and two sleeping cars, but on the day of the accident it had been strengthened to cope with the heavy pre-Christmas traffic. Its composition that day is not recorded but it was being hauled by two engines. It was in charge of Conductor Emilio Villaban Miranôn with Drivers Joaquin Passan and Alfonso Fernindez, Firemen Lorenzo Martin Guillén and a second unnamed fireman. Approaching Villar-de-los-Álamos from the opposite direction was omnibus and mail train No. 1802 from Fuentes de Oñoro to Medina del Campo, consisting of a Class 242T steam locomotive and five first-, second- and third-class coaches, in charge of an unnamed conductor with Driver Antonio Martiño Echevarris and Fireman José Antonio Simón Campos. This train had left Fuentes de Oñoro on time at 06:40 and was still running to time when it arrived at Villar-de-los-Álamos. The *Sudexpreso*, on the other hand, was running very late. It had left Irún, on the French border, at 21:40 the previous evening, should have departed from Salamanca at 05:01 and reached Fuentes de Oñoro, on the border with Portugal, at 05:01. Instead it did not leave Salamanca until 08:55, almost four hours late, and running fast reached Villar-

[474] *Pacific Stars and Stripes*, Tokyo, Japan, 18th and 22nd June 1965
[475] *San Antonio Express*, San Antonio, Texas, United States, 17th June 1965
[476] https://www.lavozdealmeria.com/noticia/3/provincia/45782/tres-tragedias-ferroviarias-en-la-historia-de-almeria
[477] https://www.diariodealmeria.es/almeria/comitiva-tren-fantasma_0_1268273231.html
[478] Federación Castellano Manchega de Amigos del Ferrocarril

de-los-Álamos at 09:11.

According to the safety regulations prevalent at the time the omnibus train had to be held outside of the station until the express had arrived. Ironically it was this stipulation which led to the disaster. Had the omnibus train been admitted to the loop line at the station no accident would have occurred, even had the express failed to stop at the station, as it could have passed through safely on the main line. Instead, the express ran through the station without stopping, to collide head-on with the stationary omnibus train waiting at the Fuentes de Oñoro end of it. Both engines, a van, two first-class carriages and a restaurant car of the express were derailed and one of the first-class carriages mounted the train engine. It was in this coach that most of the deaths and injuries occurred, many of the victims being scalded by escaping steam from both the damaged engine boiler and the steam heating pipes of the coach itself. In the omnibus train the leading wooden first-class coach, which was empty, took most of the force of the impact and was wrecked, but the passengers in the remainder of the train were thus spared and no one in this train was killed.

All three drivers, two firemen and possibly another railwayman riding on one of the engines lost their lives (the unnamed fireman survived) as did both conductors, a cleaner, two *Wagons-Lit* employees and a Civil Guard. Of the passengers 22 were killed and about 50 were injured.

A newspaper reporter spoke to a security guard and the night stationmaster. Security Guard Juan Comeron Frayle said that he had been one of the first people on the scene. When he realised what had happened he had stopped cars on the nearby highway to carry the injured to hospital. In one coach he had found a woman and two children, who were all dead. He had applied artificial respiration to victims who appeared to be still alive; if they were clearly dead he had moved on to others. There had been much panic and confusion among the passengers and it had been a struggle to obtain their cooperation.

Night Stationmaster Antonio León Cornejo said that he had just been coming off duty when the accident happened, having been relieved by Day Stationmaster Wenceslao Robledo Martin. Together with the signalman he had gone to meet the omnibus train and had had a lucky escape from being struck by a sleeping car and killed, though he was hit by some wreckage and injured. Asked if he believed that the drivers of the express had seen the closed disc he stated that they had apparently observed the signals at Quejigal, the previous station just up the line. Asked why they had not stopped at Villar-de-los-Álamos he said that he thought they had been travelling too fast to be able to do so. [479]

It had also been quite foggy at the time of the accident and it was suggested that the crew of the express should have been advised at Salamanca that they were to cross the omnibus train but then, of course, they should have obeyed the signals. [480] [481]

One of the Villar-de-los-Álamos stationmasters, it is not clear which one, and a second railwayman (the signalman?), were apparently charged with negligent homicide,

---

[479] http://vicenticoaa.blogspot.com/2018/06/accidente-ferroviario-de-villar-de-los.html

[480] https://www.salamanca24horas.com/cultura/fado-entre-encinas-cincuenta-cinco-anos-accidente-ferroviario-villar-alamos-1965-2020_2194682_102.html

[481] https://es-m-wikipedia-org.translate.goog/wiki/Accidente_ferroviario_de_Villar_de_los_%C3%81lamos?_x_tr_sl=es&_x_tr_tl=en&_x_tr_hl=en&_x_tr_pto=sc

it being alleged that the signals had been cleared for the express before the omnibus train had entered the station. The outcome of the court case is unknown. [482]

At least 16 of the victims had been poor Portuguese workers returning home for Christmas from jobs in France, Germany, and Switzerland. A special train was chartered in Portugal to take the survivors home and the accident was credited with improving relations between Spain and Portugal, both of which countries were living under dictatorships at the time. [483]

On the evening of Saturday 9th April 1966 railcar AT 535 left Chivasso station, Italy, at 22:02 for Asti. Eight minutes later it collided head-on with a freight train coming from Montigio near a bridge over the River Po at Abate. Five people died in the collision and two were injured, all the casualties occurring to the occupants of the railcar.

In charge of the railcar was Conductor Bruno Serra, aged 28, from Asti, and it was being driven by Motorman Bruno Montaldo, 34. The last train of the day, it was only carrying four passengers; a married couple, Brigadier Mario Mussino and his wife Elisa, both 35, who lived in Bolzano and were on their way to visit relatives; and two railway employees, Rinalda Perotta, 25, from Montiglio; and Paolo De Notaris. Rinalda had been in charge of Crossing No. 49 on the line for last four years and lived with her parents and sister. At 21:00 she had cycled into Chivasso to visit another relative there, leaving the crossing to be attended by one of her family, had left her bike in Chivasso and had to run to catch the train home. Paolo, 26, had worked on the railway at Cocconato for three years but ten days earlier he had been transferred to Saint Vincent. Every day since then he had commuted between Cocconato and Saint Vincent but as he told others, "This summer I'm getting married and finding a house in Saint Vincent." Alas it was never to be. There was one more 'passenger' on the railcar, ten-year-old Giani Montaldo, the son of the motorman. The boy, who reportedly did very well in school, had been promised a ride on the train for some time and after repeated reminders from his wife the motorman had finally agreed to let the boy accompany him that Saturday evening. They had come on the train from Asti to Chivasso, had had a stopover of several hours there, where they had eaten in the railway canteen, and were now returning to Asti. The motorman, his son, the conductor, and the two railway employees were all positioned towards or at the front of the railcar; the Mussinos were sat at its rear.

The freight train consisted of a steam locomotive and some empty waggons. It was in charge of Conductor Mario Illenghi, 46, from Asti, with Driver Elio Mossino, 31, Fireman Mario Duffini, 25, both from Chivasso, and Brakeman Sergio Goria, 34, from Asti.

The freight train had a mandatory stop at the little station of San Sebastiano Po, regardless of the indications of the signals. This Saturday evening the signal was at Danger anyway but Driver Mossino failed to see it, or to stop, and re-entered the single line at the Chivasso end of the station. He said later that as his train had left the tunnel approaching San Sebastiano Po station he had been distracted by a 'bonfire in the mountains' (literal translation of the Italian). The trains collided about six kilometres from Chivasso and a kilometre from San Sebastiano Po, just after the railcar had crossed the bridge over the

[482] Federación Castellano Manchega de Amigos del Ferrocarril
[483] https://www.elespanol.com/castilla-y-leon/region/salamanca/20221030/fatidico-accidente-villar-alamos/713678863_0.html

River Po and at a point where the line ran on a ten-metre-high embankment. The railcar and the locomotive of the freight train were both thrown down the embankment. Seeing what was about to happen Motorman Montaldo had jammed on his brakes and then thrown himself on top of his son in an attempt to protect the boy. It was no use. The motorman, his son, the conductor, and the two railway employees were all killed.

Motorists driving along the adjoining provincial road witnessed the collision and raised the alarm. Railwaymen, firemen, carabinieri, doctors and nurses went to the scene. The body of the conductor was the first to be discovered, lying on the embankment. Immediately afterwards Brigadier Mussino climbed out of the railcar through a window and then with a considerable effort extricated his wife. The two were taken to Chiasso hospital; they were not badly hurt. During the night the body of the Montaldo boy, then that of his father and finally that of the crossing keeper were removed from the wreckage; that of Paolo De Notaris took longer to get out.

Brigadier Mussino told a reporter later, "There was an immense noise as if a big bomb had exploded and the lights went out. Our train remained on the track for a few moments, then it began to roll down the embankment. There was broken glass and seats were flying. I tried to protect my wife. Eventually we stopped at the foot of the embankment and there were neither voices nor noises to be heard."

Unhurt, Driver Mossino fled the scene. He related, "Overcome by horror and panic, I ran through the fields without knowing where I was going. I found myself on the 'national' (the provincial road), flagged down a car and got a lift into Asti." There he had surrendered himself to the railway police, was arrested and taken to prison in Turin, charged with multiple manslaughter and causing injuries through negligence. He said, "It would have been better if I had also died in the collision."

Fireman Dubbini was not arrested. He explained that he had been shovelling coal in the tender as the freight train had passed through San Sebastiano Po and hence had not noticed the signals. He was in any case very inexperienced, having only finished railway school on 14th March. He had only been working as a fireman for a week and the Saturday trip had only been his second over the line.

The inquiry into the accident was told that the signal at San Sebastiano Po had been lit as usual by the railwayman on duty there, Porter Giovanni Malli. When Malli had seen that the train was not stopping he had tried at first to contact *casello* 46 (probably a crossing keeper's hut), 1,250 metres from San Sebastiano Po, via Asti station, only to find that the train had already passed there. Conductor Illenghi attempted to excuse his failing to notice that his train had run past the signal at San Sebastiano by saying that he had been studying his working timetables at that moment. [484]

A serious collision occurred in Sweden on the Norre Stambanen (northern main line) of the Statens Järnvägar on the afternoon of Sunday 24th April 1966. Train No. 5106, a northbound railbus travelling from Krylbo to Ånge had arrived at Holmsveden station. North of the station an engineering train consisting of a locomotive and a catenary maintenance vehicle (tower car) had completed its work and was returning southwards to the station on the single line, probably with the locomotive propelling the wooden

[484] *La Stampa*, Torino, Italy, 10th and 12th, and 14th April 1966

carriage forming the tower car.

The accident occurred because due to a misunderstanding the railbus set off at its normal departure time of 14:40 before the engineering train had returned to the station. The trains met on a slight curve about a kilometre and a half from the station, by which time the railbus had reached its full speed of 90 kilometres an hour and the engineering train was moving southwards at 40 kilometres an hour. The motorman of the railbus saw the other train when it was about 150 metres away, applied the brakes in emergency and he and the conductor ran back through the train, shouting to their passengers to take cover. In the collision the tower car mounted the steel railbus, the front half of which was crushed. Six women who had been sitting towards the front of the train were killed and another 12 of the 40 passengers on the train were slightly hurt. [485]

A disaster on the Rede Ferroviária Federal near Nova Iguaçu, Brazil, on the morning of Wednesday 16th November 1966 occurred during temporary single-line working. Relaying work had been in progress on line 1 near Nilópolis and all traffic was being worked over line 2. The accident was a head-on collision between special train SME-2, from Nova Iguaçu to the Don Pedro II terminus in Rio de Janeiro, and regular train UM-9, from Don Pedro II to Taireta. The latter train would normally have been running on line 1 but had been diverted onto line 2, possibly at Anchieta, because of the single-line working. Train SME-2 ran past a signal at Danger at Edson Passas, passed a second stop signal, ran through the points of the mains crossover which were set to divert train UM-9 back onto line 1 at Nilópolis, ran through Nilópolis station, where it was booked to stop, went past a third signal at Danger and shortly afterwards collided head-on with train UM-9 near Olinda station.

Four hours after the accident 150 rescuers from the fire service, military police and army were still working to free the injured and remove the bodies of the dead. One of the survivors was a 27-year-old newspaper seller, Erivala Tavares de Silva, who told a reporter that he had lost consciousness in the accident and had only come around, in considerable pain, when two firemen pulled him from beneath a metal beam. He had lost 72,000 cruzeiros (about £230 Sterling) worth of newspapers, plus some dictionaries.

Another survivor was Manoel Soares Elgo, a Spaniard, aged 52, who had been sat just behind the driver of train SME-2. He was badly hurt, with fractures of the skull, collarbone and shoulderblade, and was removed to the Pronto Socorro in Nilópolis. He had lived in Brazil for six years, was the father of two young children and said that his family had only heard of the accident at 17:00.

Soldier Ned Barcelos Pereira, of the Companhia Independente do Palácio Guanabara (military police), who suffered two broken ribs, spoke of shouting and waving his revolver to prevent passengers from jumping from the third coach of the still-moving SME-2 train. Instead most of them had moved forward to help the injured in the front coaches. Later five of his colleagues were able to frustrate a gang who were attempting to do some looting.

By the evening the number of dead in the accident was known to be 33, with about 75 injured, and the final number of dead was probably 35, including Motorman Aurino

[485] Wikipedia – Järnvägsolyckan I Holmsveden

Miguel da Silva of train SME-2 and Motorman Durval Ribeiro da Costa of train UM-9.

Later on the day of the disaster the railway company issued an official note: 'The Estrada de Ferro Central do Brasil regrets to state that at 05:11 today a special train from Nova Iguaçu (SME-2) ran past three signals at Danger, forced a set of points, and between Edson Passas and Olinda stations collided with the Paracambi train (UM-9), which was travelling in the opposite direction and was normally licensed. This regrettable occurrence resulted in 29 deaths and 67 cases of injury. All necessary measures were immediately taken by the Management of the Suburban lines, which had already placed all essential medical aid at the site within six hours. The Central do Brasil received support and assistance from the Fire Brigade, the Army, the Navy, the Air Force, the Guanabara Military Police and the Nilópolis local authorities. The injured were taken to the Carlos Chagas, Sousa Aguiar, Nilópolis and Casa de Saude Nossa Senhora de Loudes Hospitals; the Ferro Central do Brasil having them accompanied by doctors. The drivers of the two trains died on the spot. An inquiry commission has just been appointed to investigate the causes and responsibilities pertaining to the crash. The commission will be presided over by Central Assistant Engineer Wayr Augusto Ribeiro Beraldo.'

Both trains involved in the collision were electric ones and it emerged that when Signalman Nicolau at Nilópolis had reported to his Control that train SME-2 was overrunning signals an attempt had been made to stop the trains by cutting the electric current, but the measure had come too late to prevent the accident.

It was hoped to bring line 1 back into operation late on Friday 18th and line 2 perhaps the following day, following the removal of the wreckage.

Although an inquiry was launched into the cause of the disaster it seemed clear that the responsibility lay with 40-year-old Motorman da Silva. He had not, however, been in good health, suffering from hypotension, perhaps due to Addison's disease (adrenal insufficiency). In what seems a very insensitive act a newspaper reporter interviewed Senhora Alice da Silva, the motorman's widow, the same day that she had buried her husband, in their meia-agua (a low-budget house occupied by poor people). She told the reporter that she attributed the accident to her husband being overworked and added that he had not wanted to work the train that day because he had not been feeling well. He had complained of poor health for some time and had earlier requested four months' leave from the railway, presumably in the hope of getting his condition treated and perhaps cured, but had only been granted 15 days. He had been a good husband, who had never missed a day's work except through illness. She, herself, was an orphan and they had three children aged three, two and only eight days. Only one relative, she said, had come to her husband's funeral, her brother-in-law, José da Silva, and he had tried to console his sister-in-law, who was still physically weak after having given birth. Without her husband's income of 150,000 cruzeiros a month, Senhora de Silva said, she did not know how she was going to pay the monthly rent of 22,000 cruzeiros and support her children, unless she received support from the railway company. She feared that she might have to resort to begging on the street. Her fears in this direction were, apparently, to some degree at any rate, groundless, as legally she was entitled to compensation, probably of the order of 100,000 cruzeiros a month.

Professor Ribeiro da Graça, a cardiologist, stated as his opinion that given Motorman da Silva's known medical condition, his large family and his low wages, he should have been subject to two annual medical examinations, which should have included electrocardiograms. People suffering from hypotension could have a variety of symptoms such as dizziness, confusion, blurred vision and a general weakness. Furthermore a person suffering from severe hypotension was not really a suitable employee to be performing such an activity as driving a train. The possibility that the motorman had suffered a heart attack prior to the accident could also not be excluded, as a post mortem examination of his body was not possible due to its condition.

Suburban Regional Manager Leandro Gomez Coelho noted that neither train had been very full, train SME-2 in particular only having had a few passengers, and this had mitigated the consequences of the disaster. He announced that the railway company would, without hesitation, pay the funeral expenses of those who had lost their lives and the medical costs of the injured. With reference to the several accidents that had occurred on the former Central do Brasil lines he pointed out that even well-equipped railways in countries such as France, Japan and the United States suffered disasters. As to the cause of the present accident he did not believe that Motorman da Silva's medical condition alone could have caused it, pointing out that the driver's cab of train SME-2 had been provided with a 'dead man's handle,' which would have stopped the train even if the motorman had died in his cab. He did not know the cause of the disaster and an explanation would have to wait for the results of the inquiry.

Minister of Transport Marechal Juarez Távora was said to have been visibly irritated following an audience with President Castela Branco to discuss the accident. The minister said that he found it strange that the commission of inquiry investigating the disaster had not yet been able to explain how it had happened, given that it should have been absolutely impossible even with the collapse of the motorman or his having fallen asleep. There were other factors, such as the apparent poor condition of the trains, with deficiences in their brakes, faulty lighting, damaged walls, and broken windows, but none of these appeared to explain the disaster. Fully loaded each train had a capacity of 216 passengers.

Señor Hélio de Oliverira Melo, the president of the Rede Ferroviária Federal, after recommending that concrete measures be taken to protect the families of the victims of the disaster, suggested measures to prevent a recurrence of the accident, including a rigorous review of the signalling and safety devices and of the general working conditions of motormen and the avoidance of licensing trains to run in the opposite direction, for whatever reason (so presumably a ban on single-line working). Noting that most of the accidents on the system in the previous few years had been due to human error, he also requested that the social services department of the company carry out a survey of the living conditions of suburban train drivers to determine to what extent social and economic difficulties were a factor in causing them. [486]

The medical condition of train drivers has been a factor in a few train accidents through the years, if not that many. In an early example a minor collision in England in Victorian times was found to be due to an engine driver suffering from uncontrolled diabetes mellitus. In June 1964 the driver of a passenger train approaching the station at Bradford Exchange, England, suffered a severe heart attack approaching the terminus, lost control of his train and it collided violently with the buffer stops, killing him and two passengers (see below). The driver involved in the accident at Violet Town, Victoria,

[486] *Jornal do Brasil*, Rio de Janeiro, Brazil, 17th, 18th and 19th November 1966

Australia, in February 1969, knew that he had heart disease but continued to drive trains until an accident killed him and eight others (see below). Again in Australia the driver of a train at Waterfall, New South Wales, suffered a heart attack in January 2003 and his train ran away and was derailed, killing seven. Away from the railway the air crash at Staines, England, in June 1972, in which 118 people died, was attributed largely to the serious heart disease of the captain flying the aircraft.

A disastrous collision occurred in Spain on the morning of Sunday 18th December 1966, involving northbound omnibus passenger train No. 2004 from Teruel to Zaragoza and southbound freight train No. 6107, consisting of empty vans returning from Irún on the French border to Valencia. The passenger train consisted of a pair of motor units and a single coach and was due to leave Teruel at 07:05, arriving at Zaragoza Campo Sepulco at 10:45. The diesel-hauled freight train was the return working of a train carrying citrus fruits to France; it was running eight hours late.

The passenger train was timetabled to arrive at Villafranco del Campo station at 08:00 but on the morning of the accident it was running a few minutes late. The morning was cold and foggy. Anxious for the comfort of his passengers Conductor Teodoro Latorre advised them to move, if they wished, to the leading vehicle of the train, which had better heating, and most of them duly did so. It would be a fateful decision for them.

The crossing of the two trains was arranged to take place at Villafranco del Campo. The signal was cleared for the freight train to enter the station but the starting signal was of course at Danger, protecting the entrance to the single line. Somehow the driver of the freight train failed to observe the signal and proceeded out onto the single line, despite the frantic efforts of the station staff to stop him. Meanwhile the passenger train had already left the next station at Santa Eulalia del Campo, and disaster was inevitable. At about 08:08 the trains collided at kilometre 89.900, a little over two kilometres south of Villafranco del Campo, while both were travelling at about 80 kilometres an hour. The line was straight but the thick fog prevented the drivers from seeing one another until the very last moment. In the collision the heavier diesel locomotive overwhelmed the lighter passenger train, the front car of which was destroyed, while a fire that broke out immediately after the collision completed the destruction. Only two passengers in the front car survived together with five in the second. Conductor Latorre and his wife were among the fatalities; their young son had remained at Santa Eulalia. [487]

Assistant Driver Ramón Lamaspa Rubio from the second car was injured but managed to make his way back to Villafranca to raise the alarm. Help was immediately organised and special trains were sent from both Zaragoza and Teruel. The driver of the second car also survived, though injured, and speaking to a reporter from his hospital bed he said, "I had no idea at first what had happened. There was suddenly an emergency brake application from the leading car, followed at once by an almighty deafening crash. We were thrown over and lifted up and we crawled out of the coach. Beneath us we heard cries of despair. The front of our car had mounted the leading car and lay on top of it. Flames were already spreading beneath us and we could do nothing to save those trapped

---

[487] https://www.heraldo.es/noticias/comunicacion/2016/12/18/50-aniversario-tragedia-villafranca-del-campo-1148744-311.html

in the wreckage." [488]

The inquiry into the accident found no problem with the signalling, which had also been tested a month earlier, and determined that the disaster had been solely due to the freight train overrunning signals.

Either 29 or 30 people lost their lives in the accident including five railwaymen (presumably the driver and assistant driver of the freight train and of the leading car of the passenger train plus Conductor Latorre). Twelve other passengers were injured. [489 490]

Single-line collisions were hardly common in New Zealand but a minor one in terms of casualties occurred at 11.45 p.m. on Thursday 23rd March 1967. A freight train that had left Gisborne at 8.45 p.m. for Napier collided with a passenger railcar, which had left Wellington at 2.55 p.m. bound for Gisborne, the accident occurring near Whakaki. Driver D. Simkin and Guard W.T. Veale of the railcar jumped to safety at the last moment, having applied the brakes but without having had time to warn their passengers, 16 of whom were hurt. No one was killed. The railcar was pushed back about 60 feet, 18 of the 48 waggons of the freight train were derailed, and 100 feet of track was damaged. [491 492]

Driver Cyril Cordell Percy, 34, and Assistant Driver Eric Denis Grenhoefer, 23, subsequently appeared in court at Gisborne in November charged with injuring a named passenger and others in such a way that had death resulted, it would have been manslaughter. They faced an alternative charge of having had under their control an engine which without due precautions could endanger human life, and without lawful excuse having omitted their lawful duty to take such precautions and having thereby endangered the life of the named passenger and others.

Evidence was given that the risk of engine crews falling asleep at night whilst driving DA diesel locomotives was very high and this was apparently what had happened on the night in question. The engine was provided with a vigilance device that had to be cancelled by the driver at regular intervals but this had failed to prevent the collision. [493]

The outcome of the court case does not seem to have been reported.

147th Street, New York City, New York, United States, was the perhaps rather unlikely location of a single-line collision on the New York Central Railroad on Monday 22nd May 1967. The accident occurred on the 30th Street branch of the railroad, a double-track line that extended from Spuyten Duyvil for 11½ miles to East Bank Street. Track No. 1 was used by westbound (geographically northbound) trains and track No. 2 by eastbound (geographically southbound) trains. Trains moving with the current of traffic were controlled by automatic block signals, those operating against the current of traffic

---

[488] *Railway Accidents of Great Britain and Europe* by Ascanio Schneider and Armin Masé, pgs. 47-8

[489] Federación Castellano Manchega de Amigos del Ferrocarril

[490] Wikipedia – Accidente ferrroviaria de Villafranco del Campo

[491] *The Press*, Christchurch, New Zealand, 25th March 1967

[492] https://photonews.org.nz/gisborne/issue/GPN154_19670419/t1-body-d1.html

[493] *The Press*, Christchurch, New Zealand, 2nd November 1967

were operated by train orders issued by Dispatcher F.J. Carrigan [494] at Lexington Avenue (Grand Central station) to operators at DV (Spuyten Duyvil) and DO (72nd Street) towers. [495]

The trains involved in the accident were 1/NY-4, Extra No. 2020 east from Syracuse, New York, consisting of three diesel locomotives and 15 cars, in charge of an unnamed conductor with Engineer Howard Bruso and Fireman Franklin Jenkins; and ND-5, Extra No. 5305 west from 105th Street, consisting of four diesel locomotives and 60 cars, in charge of an unnamed conductor with Engineer Robert Orr with Fireman Charles Meuser Jr. [496]

At 8.45 a.m. on the morning of the accident the foreman of a track repair crew advised the operator at DO tower that he needed to conduct repairs on track No. 2 in the vicinity of 188th Street. The operator spoke to Dispatcher Carrigan at 9 a.m., the dispatcher approved the move and track repair cars left DO tower for 188th Street soon afterwards. At about 9.17 a.m. train B-7, the Manhattan switcher, left DO tower with four cars to conduct switching operations at 136th Street, its conductor having asked the yardmaster at DO tower if the train was to go there. Permission was not obtained from the dispatcher for the move, nor was he informed that train B-7 was proceeding from 105th Street via track No. 1 to 136th Street. Train B-7 arrived at 125th Street at about 9.30 a.m., crossed over to track No. 2 there, and its conductor advised the operator at DO that track No. 1 was clear. The operator did not inform the dispatcher.

At about 9.30 a.m. the DO operator cleared a dwarf signal for train ND-5 to enter track No. 1 at 105th Street from sidings there and told the yardmaster that he had done so.

Train 1/NY-4 had left De Witt yard, Syracuse, at 12.02 a.m. and entered the dispatcher's area at Croton, New York, at 8.46 a.m. At 9.16 a.m. it arrived at DV tower at Spuyten Duyvil and the dispatcher told the operator to hold it there. At about 9.17 a.m. the dispatcher asked the DO operator about train ND-5 and was told that it would not be ready to leave until 10 a.m. After several attempts, at 9.21 a.m. the dispatcher issued train order No. 301 to the operator at DO tower as a 31 order (one for which the train had to stop and the order be signed by the conductor and engineer) and to the operator, conductor, and engineer at DV tower as a 19 order (which could be hooped up to the train as it passed). Order No. 301 read:

Extra 2020 east has right over opposing trains om No. 1 track interlocking DV to West 105 Street, Signed LCL, FC

LCL and FC were the initials of the transportation superintendent and the dispatcher respectively. The DO operator read the order back to the dispatcher word by word, then the DV operator did the same and the dispatcher replied "Complete" to the two operators.

The 31 order required the DO operator to restrict the movement of train ND-5, despite which about eight minutes after receiving the order he had cleared dwarf signal 3-W for it to depart from 105th Street.

According to the rules the operator at DV should have delivered a Clearance Form A to the conductor and engineer of train 1/NY-4 together with train order No. 301, after

494 *The Times* Record, Troy, New York, United States, 15th July 1967

495 National Transportation Safety Board report

496 *The Kingston Daily Freeman*, City of Kingston, New York, United States, 23rd May 1967

having established a positive block on track No. 1 with the operator at DO tower. At about 9.23 a.m. the DV operator told the dispatcher, "One order, order 301, complete at 09.22," and the dispatcher had replied, "Okay." The DV operator did not mention the Clearance Form A. He had called his colleague at DO to ask if he could have a block on track No. 1, had received no reply and had then delivered the train order and the clearance form to an unidentified crew member of train 1/NY-4. The copies of the train order had not been signed by the DV operator and Engineer Brusco went to the operator to obtain his signature. At about 9.40 a.m. the DV operator asked his colleague at DO where train 1/NY-4 was to go and was told into the sidings at 72nd Street. The DV operator told the National Transportation Safety Board (NTSB) inquiry into the accident that having heard the dispatcher tell the DO operator to block track No. 1, the train order having been made complete, and the DO operator having given a route for train 1/NY-4, he had been convinced that he had been given a block on track No. 1 and could release train 1/NY-4. Train 1/NY-4 had left at 9.48 a.m. and he had reported its departure to the dispatcher and the DO operator.

A brake test having been conducted, train ND-5 entered track No. 1 at 105th Street at 9.50 a.m. as noted by the conductor. Seeing from his console that train ND-5 had entered track No. 1 the DO operator reported its departure to the dispatcher, whereupon the dispatcher and the DV operator quickly realised that the two trains were coming towards one another on track No. 1. The DV operator at once called DO, asked if train ND-5 was on its way, was told that it was, and replied that train 1/NY-4 was going down, at which the DO operator exclaimed, "Oh, my God, I forgot," and hung up the phone.

Various attempts were then made to stop one or both trains. The DV operator tried to contact train 1/NY-4 via the radio, the dispatcher tried to radio both trains, the DO operator replaced dwarf signal 3-W and when the conductor of train B-7 called DO a little later he was asked to flag down train ND-5. These attempts all failed. The radio calls were not answered, train ND-5 had already passed the dwarf signal and had also passed train B-7 before the message was received to stop it. [497]

The trains collided head-on on a curve at 9.55 a.m. in fine weather. Engineer Bruso, Fireman Jenkins, and Brakeman Jams Schulthies of train 1/NY-4 and Engineer Orr, Fireman Meuser Jr and Brakeman Herbert Fero of train ND-5 were all killed. The conductor and brakeman of train 1/NY-4 and a brakeman of train ND-5 were injured, according to the NTSB report. A newspaper gives the names of the injured railwaymen as Albert LeDuke and Harry Kelly (both seriously hurt) and Leo Balduc and Anthony Petronis (slightly hurt) but does not state which train which person was on, or who was the conductor. An uninjured crewman, James Sullivan, spoke of the brakes coming on in emergency just before the crash. He was thrown to the floor of the caboose, then got out. "I ran up to the head of the train and saw that the two engines had jack-knifed. I tried to help a guy on the ground but he was all burned and dead," he told a reporter. [498]

Another newspaper wrote, 'The diesel engines were travelling at about 30 mph when they collided near a freight yard. Witnesses said bodies were hurled from the wreckage, freight was spewed onto the tracks and fuel was sprayed over the area, touching off fires. Black smoke towered above the scene. It took firemen two hours to bring the blaze under

[497] National Transportation Safety Board report

[498] *The Kingston Daily Freeman*, City of Kingston, New York, United States, 23rd May 1967

control and three and a half hours to remove the bodies from the wreckage. Four other diesel engines and a dozen freight cars were derailed.

'A passerby, Totty Nieves, of Manhattan, said he heard an explosion and thought the train must be loaded with dynamite. "I raced to the tracks and tried to get as close as I could," he said. "I saw a body on the ground with all its clothes burned off. Then I heard a man screaming for help. Yesterday I saw a newborn baby and I thought to myself, 'That's the beginning of life.' Today I saw the end of life."' [499]

The NTSB listed a large number of causal factors involved in the accident, followed by 14 conclusions. Its conclusions were:

(1) The brake tests and hours of duty for personnel of trains NY-4 and ND-5 were in compliance with Federal Regulations.

(2) There were violations of Federal Regulations at DO station in that the prescribed Operators Daily Time Report and Station Record of Train Movements were not being used.

(3) Tests conducted after the accident indicated that the automatic block signal system on main Track No. 1 between 105th Street and DV was functioning normally prior to the accident on May 22nd, 1967.

(4) The dispatcher failed to assure positively that Track No. 1 had been blocked by DO before the train order was issued and failed to require the operators at DV and DO to use procedures specified in the Rules of the operating company. Specifically neither DV nor DO were required to state that "stop signals" were displayed for NY-4 and ND-5 after receipt of the train order.

(5) The dispatcher failed to address the train order to train ND-5 in addition to the operator, in violation of company Rules. Mitigating factors in this case were that the DO operator could restrict ND-5 by signal 3-W alone, and further that ND-5 was on a side track <u>between</u> DO and DV stations in a relatively inaccessible location insofar as delivery of train orders was concerned. However, as stated by the Rules Examiner, the order should have been addressed to ND-5. It is most unlikely that ND-5 would have departed until NY-4 had passed by, had the contents of the train order been known to the conductor and engineer of ND-5.

(6) The operator at DV failed to secure a positive block on main Track No. 1 prior to releasing NY-4.

(7) The operator at DO failed to block Track No. 1 as he had been instructed to do by the dispatcher before receiving the train order, and also failed to restrict ND-5 as required by the train order.

(8) The failure of the operator at DO to block Track No. 1 was preconditioned by his routine action in leaving the switch at West 105th Street open after the entry of ND-5's engine to the yard track and his action in giving train ND-5 a yellow signal to proceed prior to the time when ND-5 was ready to proceed. This action, which did not protect the main track and was in violation of the Rules, was caused by prior complaints of delay and inconvenience made to the operator by yardmasters.

---

[499] *The Citizen-Advertiser*, Auburn, New York, United States, 23rd May 1967

(9) The working conditions at the DO station were unsatisfactory. The lack of adequate space and excessive noise were not conducive to the level of concentration necessary for an operator to perform properly his duties.

(10) The assignment by yardmasters of operators at DO to throw switches in the area required the operators to leave their offices, in violation of a company rule which specifies that they be continuously on duty in their offices.

(11) (a) There was no evidence of fault on the part of personnel on train ND-5.

(b) The engineer of train NY-4 was in violation of the second part of Rule D 93a in that the speed of NY-4 within yard limits from West 153rd Street to the accident location at about West 147th Street was greatly in excess of a speed which would permit stopping short of an obstruction. However, the wording of the second part of Rule D 93a does not state the maximum speed permitted and there was no evidence of effort by the Company to enforce Rule D 93a by specifying a maximum speed. It is determined that a speed over ten miles an hour would not permit stopping short of an obstruction and Rule D 93a should have stated a maximum speed.

(12) There were numerous violations of company Rules in this case, and in the implementation of train orders by dispatchers and operators at DV and DO prior to this accident.

(13) The New York Central Railroad Rules are conflicting in a number of instances, are not clear in others and thereby lead to misinterpretation and misunderstanding that are deleterious to operating safety.

(14) The New York Central Railroad has not established or maintained adequate supervisory procedures to insure that its operating personnel comply with its Rules.

The NTSB determined that the probable cause of the accident was the failure of the operator at 72nd Street (DO) to restrict train ND-5. Contributory causes were the failure of the dispatcher to assure positively that Track No. 1 had been blocked by DO prior to the issuance of the train order, the failure of the operator at Spuyten Duyvil station (DV) to obtain a positive block on main Track No. 1 for train NY-4, and the failure of the New York Central Railroad company to establish explicit boundaries of authority, explicit operating rules, and procedures to insure that its personnel are instructed in and comply with such rules.

Turning to recommendations, the NTSB stated that, 'A wide variety of inadequacies in rules, operating practices and personnel training were revealed by the investigation of this accident on one of the largest railroads in the country. The investigation also identified ineffective organisation, inadequate use of modern communications devices (such as radiotelephones), and a lack of application of modern techniques such as the use of human factors engineering in designing operators' control stations.'

The NTSB recommended that the New York Central Railroad amend its operating Rules and require compliance with them by operating personnel and that the company reconcile its rules and procedures concerning their operators being on duty in their offices until relieved. [500]

---

[500] National Transportation Safety Board report

Switzerland suffered a serious single-line collision at 13:56 on the afternoon of Monday 24th June 1968. The Vallorbe – Lausanne – Brig main line on which the accident occurred was mostly double track at this time but there was a short stretch of six kilometres of single track between Sion/Sitten and St Léonard stations.

On the afternoon of the day of the accident passenger train No. 51070 left Sion/Sitten station at 13:52. It was an electrical multiple-unit formation consisting of a motor unit and five second-class coaches and was carrying a party of 270 farmers and their relatives from Lucerne canton, who had just completed a two-day excursion to visit agricultural establishments in Valois canton and were on their way home. The excursion was on the occasion of the 75th anniversary of the Reiden Agricultural Society, of which they were members.

In the opposite direction fast goods (mixed?) train No. 16845, consisting of an electric locomotive, two waggons and a coach, was approaching the other end of the single line at St Léonard, where it should have stopped to wait for the passenger train. The station was provided with remotely controlled colourlight signals. The signals stood at Danger and in addition a sort of automatic train control was provided at the St Léonard Distant and Starting signals protecting the single line. For some undetermined reason the driver of the goods train apparently forestalled the train control warnings at both signals, his train burst the points at the end of the loop, which were set for the oncoming passenger train, and re-entered the single line. The trains collided at a point almost exactly halfway between the two stations. According to a newspaper report the passenger train was travelling at 125 kilometres an hour, the goods train at a slightly lower speed. Another, perhaps more accurate report, states that both trains were travelling at 80 kilometres an hour.

As a result of the collision the front halves of the motor unit and the following car of the passenger train were destroyed; the second and third passenger cars were derailed but not badly damaged. The leading third of the goods engine was likewise destroyed but it was not derailed. Ten passengers, all members of the agricultural society, Motorman Edouard Mottaz of the passenger train and the driver of the goods engine were killed and perhaps 150 people were more or less injured.

A fleet of ten ambulances and private cars took the injured to Sion hospital. Leaking oil meant that the work of releasing the injured and cutting up the wreckage could only proceed very cautiously. It was 17:30 before the last of the injured could be released and the line was blocked until the evening of the following day, with trains being diverted via the Lötschnerg route and Berne. [501] [502]

The Chemins de fer de fédéraux suisses decided that it was impossible to determine how the accident had occurred. The signalling at St Léonard had been thoroughly tested after the accident and found to be functionally normally, with no signs of a technical failure or of malicious interference. There was no evidence that the driver of the goods train had suffered a heart attack or similar and in fact between St Léonard and the point of the collision he had been seen to wave to a group of track workers. He had also made an emergency brake application at the last moment. The tachograph of the goods engine unfortunately failed to resist the force of the impact and was not readable. The most

---

[501] *Railway Accidents of Great Britain and Europe* by Ascanio Schneider and Armin Masé, pgs. 108-112

[502] *Gazette de Lausanne*, Lausanne, Switzerland, 25th June 1968

likely supposition would seem to be that the goods train driver had let his mind wander and had absentmindedly and automatically twice cancelled the automatic train control warnings. [503]

A head-on collision between Medina del Campo and Gomeznarro stations, Spain, at 06:44 on Saturday 13th July 1968, apparently occurred during temporary single-line working due to engineering work. Express train No. 216 from La Coruña and Madrid via Santiago, Orense, and Zamora, due in Madrid at 10:05, collided with 'descending' freight train No. 6027. The express was running to time and due to a signalling error was allowed to leave Medina at 06:35, before the freight train had reached there from Gomeznarro. Nine people died and over 60 were hurt in the collision, which happened at the Casa del Francés level crossing. All of the fatalities occurred in a sleeping (couchette?) and second-class composite coach which as a result of the collision ended up in a vertical position and was then cut in two when it struck a post supporting the overhead line equipment. Both drivers were able to reduce the speeds of their trains before they met one another and the freight train had almost stopped. [504]

In what was presumably a single-line collision a diesel-hauled passenger train from Bogor to Jakarta collided head-on with another passenger train, bound for Bogor from Jakarta, on Friday 20th September 1968. Both trains were overcrowded, with passengers hanging onto the sides of the carriages, and initially it was reported that there had been at least 50 and perhaps 80 fatalities, local villagers describing the scene of the jungle collision as 'looking like a battlefield,' with overturned engines and carriages and screaming passengers trapped in the twisted steel of the wrecked carriages. A revised report stated that there had been 18 deaths with 38 people seriously injured. The collision occurred between Depok and Tjitajam stations, not far from the former place, and was stated to have resulted from 'a defect in the block signal system.' The stationmasters at the two places were detained for questioning. [505] [506] [507] [508]

A major disaster occurred in Romania at about 06:00 on the morning of Monday 7th October 1968. The trains involved were express No. 303 (or 340 according to another source) from Bucharest to Satu Mare and local train No. 3340 from Teiuş to Blaj. Previously the trains had normally crossed at Crăciunel station, just east of Bucurdea Grânoasă, but Clerk Teodor Petrişor at Bucurdea Grânoasă had been advised by Blaj that the crossing had been altered to his station. Unfortunately he was in a state of intoxication and having forgotten to tell Pointsman Alexandru Cristea of the change, he subsequently instructed the latter to clear the signals for the local train to proceed towards Crăciunel. He also neglected to obtain telephone clearance for the local train from Crăciunel. At about the same time the clerk at Crăciunel had cleared his signals for the express to

503 Ibid, 28th June 1968

504 Federación Castellano Manchega de Amigos del Ferrocarril

505 *The Sheboygan Press*, Sheboygan, Wisconsin, United States, 21st September 1968

506 *Albuquerque Journal*, Albuquerque, New Mexico, United States, 22nd September 1968

507 *The Bridgeport Telegram*, Bridgeport, Connecticut, United States, 23rd September 1968

508 *European Stars and Stripes*, Darmstadt, West Germany, 23rd September 1968

proceed towards Bucurdea Grânoasă, likewise without requesting or obtaining telephone clearance.

The trains met head-on at kilometre 368.862 in a dense fog. Steam locomotive 324.724 on the local train penetrated almost halfway through diesel locomotive 060-DA-328 on the express. The local train was commonly known as the School train because it carried many students going to Blaj, while it was also carrying lots of commuters. Twenty-two people died and another 72 were injured. Damage was estimated at 11,000,000 lei (about £2,000,000).

Clerk Petrişor and Pointsman Cristea later stood trial in the waiting room at Teiuş station for their responsibility for the disaster, in a trial reportedly attended by a thousand police officers. Initially the clerk was sentenced to death and to confiscation of his property, and the pointsman to either 25 years' imprisonment, or to 20 years' imprisonment and 10 years of 'civil degradation' (again according to different sources). The sentence on the clerk was later commuted to life imprisonment and he was released after 18 years. It was not stated if the sentence on the pointsman was likewise reduced. No action seems to have been taken against the clerk at Crăciunel, who would appear to also have been at fault, if, as stated, he had let the express proceed without obtaining clearance for it. [509] [510]

At the beginning of 1969 Australia's two largest cities, Melbourne, Victoria, and Sydney, New South Wales, were connected by an overnight express passenger train, *The Southern Aurora*, made up of first-class sleeping cars plus a dining car.

On the morning of Friday 7th February 1969 the southbound *Southern Aurora* departure from Sydney consisted of a locomotive and 14 carriages, in charge of Guard William Wyer with Driver Jack Bowden and Fireman Mervyn Coulthard. It was carrying 192 passengers and no less than 19 other crew, presumably mostly sleeping car attendants and dining car staff.

In the opposite direction a fast northbound freight train, consisting of a locomotive and 22 waggons, mostly loaded with new cars, had left Melbourne at 1.25 a.m. in charge of an unnamed guard with Driver Lawrence Rosevear and Fireman Arnfried Brendecke. The trains were timetabled to cross one another at Benalla at 6.44 a.m., the freight train being due there at 6.15 a.m. However both trains were running somewhat late and Controller Frank McDonnell at Melbourne decided to change the meet to Violet Town, Victoria, about 18 miles south of Benalla. The intention was to have what was termed a 'running cross,' whereby the freight train passed through the long loop at low speed without actually stopping, while the passenger train passed it on the main line.

The Southern Aurora duly passed through Benalla, travelling at 70 miles an hour and that speed was maintained onwards towards Violet Town. Approaching there Fireman Coulthard noted that the next automatic signal was yellow, and remarking to Driver Bowden that it looked as if they were going to have to stop at Violet Town, he got up from his seat and descended some steps to the nose of the engine to fill the electric kettle

---

[509] https://www.railnet.ro/viewtopic.php?f=119&t=4119

[510] https://ziarulunirea.ro/foto-video-catastrofa-feroviara-de-la-bucerdea-granoasa-22-de-morti-72-de-raniti-ceferistii-vinovati-au-fost-judecati-in-gara-teius-492821/

with water to make a cup of tea. As he was filling the kettle he heard the engine passing over the points at the start of the Violet Town loop, without having reduced speed at all. Returning to the cab he found his driver just sat there with the throttle wide open. The train was tearing along the track towards the far end of the loop, there was no train on the other line, and the signal protecting the end of the loop was at Danger. Shouting, 'Red light!' to his driver the fireman had at once retreated hurriedly to the relative safety of the engine room.

The freight train was moving at 50 miles an hour as it neared Violet Town, then its speed was reduced to 35 miles an hour preparatory to entering the loop. Driver Rosevear and his fireman watched the passenger train coming towards them, apparently without slowing up. At some point the driver realised that the passenger train was definitely not stopping. He slammed on his brakes, began flashing the headlight of his engine and moved to open the door of the cab. Fireman Brendecke took the hint and jumped from the engine. His driver did not have enough time to follow him and tried to retreat to the engine room; he was killed.

"When I hit the ground I rolled over and over," the fireman related later. "I looked up to see the *Aurora* climbing on top of our engine. Then there was an explosion. The locomotive landed about five feet from me. I dug my face into the dirt and put my arms over my head. A carriage appeared to fall right over me."

Driving along the parallel Hume Highway at that moment was commercial traveller Peter Fennessey. He related, "*The Southern Aurora* was about 300 yards to my right. It was doing just on 70 miles an hour. Then I saw the goods train near the crossing. I thought they would pass one another...I didn't realise they were on the same track. There was a hell of a smash...Everything seemed to shake. A goods waggon sailed high in the air through dust and smoke. Parts of both trains were in the air. They seemed to hang there. Then the carriages of the *Aurora* went crashing into the engine and into each other. They collapsed on each other like a telescope."

Both Fireman Brendecke, who suffered concussion and a dislocated shoulder, and Fennessey were active in the rescue work. Fennessey told a reporter, "People were screaming and there were faces at the window. People were trying to get out but many of the doors were jammed. Glass was breaking and falling as people hammered out windows with shoes. A couple of other motorists arrived and we began pulling people out. People were crying and moaning in the carriages. Many others were lying on the ground or wandering about, dazed. I got into one of the sleepers and found a baby crying...I don't know what happened to its parents. The goods engine was burning but the Aurora hadn't caught fire. A lot of liquid was flooding around though. It was dieselene. I said, 'God, the whole lot's going to go up.' There was a whoosh and flames were everywhere. We got out but there were still people trapped. One woman was beating at a window. We couldn't get to her. There was screaming all along the train. It was the most horrible thing I've ever seen."

A number of passengers also told their stories to reporters. 'Newlyweds Mr J. Culnane and his Chinese wife Tin, of South Yarra, were in car 9 when it was upended. "We were severely jolted then both of us were thrown to the floor," he said. "The fire was gutting cars 7 and 8 and was starting on ours. Smoke was filling up the carriage and people were coughing. There was a great urgency to get out but no one panicked. One of the children saw blood on his mother and started to scream. But we found we could not get out because we were too high from the ground. Then someone hacked through a window from the outside with an axe and a crowbar. About 30 of us were lowered gradually through the window."

'Mr J. Hawthorn, of Balwyn, said, "We didn't seem to be going as fast as they said – it was supposed to be 70 mph – and the next thing there was a series of jolts and the rattle of hardware, teacups, and things. Then the train stopped completely. I got out and saw this devastation down at the front. It was shocking. Smoke was pouring everywhere and carriages were piled on top of everything. Further down there seemed to be three carriages zigzagged one across another. There were a lot of elderly women in the carriage and I helped to get them out and save of their luggage."

'Mr and Mrs F. Mulready, of Belfield, Sydney, were breakfasting in the dining car when the collision took place. "Everything went bang," said Mr Mulready. "Food went all over the place and we were both hit severely in the chest by the dining table. Waiters were thrown all over the place. I thought we might get away through the kitchen…Then I saw it fall down. I looked out of the window and the kitchen was a mass of flame. We managed to get back to our carriage and found it being evacuated. People were falling down a lot but there was no panic. The passengers who were unhurt were wonderful. I can't say enough for them."

'Passenger Mrs Phyllis Cutts paid greatest tribute to the courage of Violet Town people. "I saw them go through flames with the greatest courage to bring out people," she told the Press. "Those who arrived with ladders and axes saved a lot of lives. There was little anyone could do at first until people arrived from Violet Town with crowbars and axes to smash open the sealed windows. It seemed ages until help came. Smoke was billowing in the air. There was no water but people were running away trying to fill up any receptacles they could find. You could not recognise the engines they were so badly damaged."

'But just as the passengers had their stories, so did the rescuers. Their favourite was six-year-old Heather Newell. Trapped for nearly four hours in the wreck of the train, Heather had fallen through broken flooring and suffered a broken leg, bad bruising, cuts on the face and head and superficial burns when the carriage she was in caught fire. Rescuers could hear her crying but they could not find her. Constable George Williams told *The Sun*, "There was fire all around but we could hear her crying somewhere in the wreckage. Firemen poured water into the carriage but we thought she must have been dead. When it cooled a little I climbed in and searched around but could see nothing. Suddenly I heard a voice say: 'Hey…mister.'"

'Former jockey Charles Shugg also crawled through the wreckage looking for her. "I saw two bodies," he said. "I was horrified but I could hear the kid crying and calling out and I concentrated on that. I shouted out, 'Just a minute mate, we'll get to you.' And when they pulled her out, this little darling, her face all black and bleeding and her hair still smouldering, said, 'I'm not a boy. I'm a girl.'"'

In total seven passengers and Drivers Rosevear and Bowden lost their lives in the accident and 117 people were more or less hurt. However the inquest that followed at Melbourne determined that only eight of the victims had actually died from injuries sustained in the accident. Medical evidence was to the effect that Driver Bowden had been dead from a heart attack for several minutes before the crash, explaining why he had run past two Caution and one Danger signals without making any attempt to apply the brakes. It emerged that Driver Bowden had been diagnosed with angina pectoris, a symptom of severe coronary artery disease, back in October 1967. He therefore was well aware that he was suffering from a condition that might make him likely to lose control of his engine without warning. Obviously he should have revealed this to the railway authorities, when he would have been relieved from driving duties on health grounds and given a desk job instead, but he was probably worried about losing out financially if he ceased driving and perhaps also had been concerned about loss of status. In June 1968, the day before he was due for a railway medical examination, Driver Bowden had again seen his doctor, who had challenged him as to whether he should continue driving trains, but the doctor was reassured by Driver Bowden that should anything happen to him the fireman would be there and would be fully qualified to take over. The doctor did not report the driver's condition to the railway management.

Coroner H.W. Pascoe commented at the inquest, "Although I have found the medical staff of the railways warrant criticism, the real blame is attached to Bowden for continuing a responsibility he was not physically well enough to perform. He was extremely negligent in persisting in driving what might well be called a prestige train, knowing that he was a candidate for death at any moment; and, in effect, admitting that if death did overtake him on the train, the safety of the passengers, crew and equipment would depend on the fireman pulling up the train."

The coroner continued, "And I have no hesitation in saying that the fireman and the guard were also negligent. Once the driver was dead, it only needed one of the surviving crew to be attentive to this duty and the accident would never have happened. I have no reason to doubt Coulthard's agreement with counsel that Bowden was a skilful driver with a reputation for being reliable and steady. But acknowledging this makes Coulthard's actions incomprehensible. When the steady and reliable driver was apparently acting in an unusual fashion by travelling at a speed which was prohibited under any circumstances (more than 70 mph), but in this case was aggravated by passing a caution automatic signal and had made no attempt to implement what the signal connoted, he left the cabin to fill the kettle. He did this at a time when other drivers (including Bowden on a prior trip) had slackened speed by at least 10 mph, and at a time when Coulthard knew there were other signals coming up and that it was his duty to observe."

Mr. Pascoe said he was satisfied that Bowden was dead or in a coma for at least six miles before the crash. He continued: "Coulthard was not the only person who considered that all he had to do was to be physically present on the train. The guard, William Frederick Wyer, was another. His log book was largely a piece of fiction and I am satisfied he was asleep shortly after clearing Albury."

Evidence was given that Wyer had recorded a six-minute delay in his log book at a point other than where it occurred – Wodonga – and he admitted that he saw the train go through the warning and red lights before he made any attempt to apply the train's brakes. When he finally did it was far too late – the collision was inevitable. "As both Coulthard and Wyer breached their duty they must share the responsibility for the loss of the trains and eight lives. The combination of a dead driver, an insensitive fireman and a somnolent guard caused this multiple tragedy. I have naturally given much thought as to whether I should commit Coulthard and Wyer for trial for manslaughter. But while I say they were extremely negligent, I am not prepared to attach to them the epithets of culpable, criminal, gross, wicked, clear or complete mentioned in the famous (and confusing) case of Andrews v. Director of Public Prosecutions and therefore content myself with a misadventure finding in the case of the deaths."[511]

Amongst recommendations made by the coroner was one suggested by a cardiologist, Dr Alan Goble, that drivers should be subject to annual medical examinations to the standard required by life assurance organisations, with electrocardiograms on alternate years. The coroner also suggested the provision of vigilance buttons in the guards' vans to ensure that guards did not fall asleep, of speedometers in the guards' vans and of modified vigilance controls on locomotives, which required the regular attention of both driver and fireman. [512]

The fireman and guard thus narrowly escaped having to stand trial on a manslaughter charge, but within a week Fireman Coulthard had been sacked while Guard Wyer was demoted to a station assistant, with a drop in his pay from $59 to $43 per week.

Then, in the New Year's honours list, nine men were honoured for their work in rescuing people from the crashed train. Dr. Robert Dunn, of Euroa, was awarded the OBE for risking his life by entering burning railway carriages to attend the injured and trapped. Eight others, including three of the *Aurora's* conductors, received British Empire Medals. [513]

A change of mind and a mix-up led to a single-line collision in Poland at 15:40 on the afternoon of Sunday 15th June 1969 between Kościerzyna and Skorzewo stations. Five passengers, one of them a six-year-old girl, and two railwaymen lost their lives in the accident, while another 18 people were injured.

One of the trains involved in the collision was a southbound 'aggregate' (ballast?) working, hauled by a large and powerful Ty 246 class locomotive, a so-called Truman engine, so named because they were provided by the United States after the Second World War, during the Truman presidency. Coming from, probably, Gdynia, this train had passed through the stations of Golubie and then Skorzewo under clear signals.

In the opposite direction northbound passenger train No. 913/929 from Bzdgoszcz to Gdynia had meanwhile arrived at Kościerzyna station. It was booked to stop here for 40 minutes to take water and to change engine crews, then to proceed to Skorzewo to cross the aforementioned southbound train there. Driver Jan Parzyka on the Ok 1 Class locomotive of the passenger train (a smaller and lighter engine than that of the freight train) had already been on duty for the regulation 12 hours, so he naturally wished to sign off and go home. Unfortunately no replacement driver was available. With the passenger train apparently not going any further very quickly, until a driver could be found for it, the decision was taken to advance the freight train to Kościerzyna instead, hence the clear signals for the latter at Skorzewo.

So far, all was in order, apart from the unavoidable delay to the passenger train, but after several minutes of negotiating with the officials Driver Parzyka was persuaded (browbeaten?) into taking the passenger train on to Gdynia, the signals were cleared for him and the passenger train left at 15:32. Only then, it seems, was a message belatedly

[511] *The Canberra Times*, Australian Capital Territory, Australia, 2nd July 1969
[512] *Australian Railway Disasters* by Kenn Pearce, pgs. 171-5
[513] http://messui.polygonal-moogle.com/incident/1969_violettown.pdf

sent to Skorzewo to hold the freight train there, when the signalwoman at Skorzewo replied that the freight was already on its way, too late to be stopped.

In the resulting collision the luggage van and leading coach of the passenger train were telescoped into the tender of its engine while the following coaches were overturned. Driver Parzyka and his fireman were crushed between the tender and the cab of their engine and killed. The heavier 'Truman' locomotive suffered less badly, though it and several waggons behind it were derailed.

Local villagers carried out the first rescue work, freeing victims from the wreckage despite the possible risk of one or both engine boilers exploding. Ambulances arrived at the inaccessible site later. The injured were taken to hospitals in Kościerzyna, Gdynia, Gdańsk and Starograd Gdański. The line was closed for 18 hours, the passenger engine was cut up on the spot and the freight engine was taken to Kościerzyna and later also scrapped.

Strangely an investigation into the collision seems to have put the blame for it on the driver of the freight train, it being alleged that he and perhaps his fireman had been drinking. From the facts as given the fault would seem rather to lie with those at Kościerzyna station. [514]

Under the system of working single lines in use in North America for many years the meeting places of trains were arranged by a central dispatcher who sent orders by telegraph or telephone to the operators at the various stations along the line. The most serious error a dispatcher under this system could make was to issue orders giving rights to two trains in opposite directions over the same stretch of single track at the same time, a so-called lap order. Robert B. Shaw states that not one such accident due to such a lap order was reported during the heyday of the train order system between 1901 and 1931. [515] Such accidents were therefore very rare, but one occurred near McManus, Louisiana, United States, on the Illinois Central Railroad, on Monday 7th July 1969.

The accident occurred on the line between Freight Yard, Vicksburg, Mississippi, and M.P. Junction, near North Baton Rouge, Louisiana. From Vicksburg southwards the intermediate points along the line were Cedars, Harriston, Roxie, Crosby, Gloster, Centerville, Norwood, McManus, and Slaughter.

Northbound second-class freight train No. 92 left North Baton Rouge at 9.05 a.m. on the day of the accident after receiving train order No. 105, which read in part:

Engine 8908 run extra Cedars to M.P. Junction has right over No. 92 Cedars to Slaughter.

This order restricted train No. 92 from proceeding beyond the north end of the loop at Slaughter until train Extra No. 8908 South had arrived there, had entered the siding there and had cleared the main track.

At 10 a.m. train No. 92 arrived at Slaughter, cars were set off and its crew received more train orders, among which was No. 110, which was addressed to it at Slaughter and

[514] https://www-trojmiasto-pl.translate.goog/wiadomosci/Katastrofa-kolejowa-pod-Koscierzyna-sprzed-prawie-50-lat-n98395.html?_x_tr_sl=pl&_x_tr_tl=en&_x_tr_hl=en&_x_tr_pto=sc

[515] *A History of Railroad Accidents, Safety Precautions and Operating Practices* by Robert B. Shaw, pg. 158

to Extra No. 8908 at Roxie. This order read in part:

Extra No. 8908 South meet No. 92 at Gloster. Complete 10.10 a.m.

The new order advanced train No. 92 from Slaughter to Gloster, 35½ miles north of Slaughter, and restricted train Extra No. 8908 South to the south end of the loop at Gloster. Train No. 92, consisting of four road-switcher units, 22 cars and a caboose, left Slaughter at 10.10 a.m., expecting to meet train Extra No. 8908 South at Gloster as per the train order. Travelling at 40 miles an hour it passed through McManus at 10.30 a.m. and about a mile further on its engineer saw train Extra No. 8908 coming towards him, applied the brakes in emergency and he and the head brakeman jumped from the cab of the engine.

Train Extra No. 8908 South left Freight Yard, Vicksburg, at 6.10 a.m. after receiving copies of order No. 105. Cars were set off at Harriston and it left there at 8.05 a.m. About 8.35 a.m. this train arrived at Roxie, where cars were set off and the crew ate a hurried breakfast. About 9 a.m., over an hour before the dispatcher issued train order No. 110, the train left Roxie, consisting of four road switcher units, 29 cars and a caboose. The train order signal had remained at clear all the time the train was at Roxie.

Train Extra No. 8908 South continued past Crosby, Gloster and Centreville, Mississippi and stopped for a maintenance of way force at Norwood, Louisiana, at 10.05 a.m. These places were 18¾, 26¾, 35¼ and 44½ miles respectively south of Roxie. The train order signals at these places, as at Harriston and Roxie, had all been clear. At 10.15 a.m., five minutes after the dispatcher had issued train order No. 110, the train left Norwood again, its crew knowing nothing of the new order. As the train was rounding a curve north of McManus just after 10.30 a.m. its head brakeman saw the headlights of train No. 92 and shouted a warning to his engineer. He remembered little more.

All eight diesel units, the first eight cars of train No. 92 and the first 11 cars of Train Extra No. 8908 South were derailed in the collision. Seven diesel units and eight cars were destroyed and the remaining unit and other cars were badly damaged by the impact and a post-accident fire. The conductor and fireman of train No. 92 and the engineer and fireman of train Extra No. 8908 South were all killed. The engineer, head brakeman, and flagman of train No. 92 and the head brakeman, conductor and flagman of train Extra No. 8908 South were injured.

At the time of the accident the railroad company had introduced Customer Service Agents who were in charge of several stations, driving between them in small vans. One such Customer Service Agent came on duty at 7.45 a.m. at Harriston. As he was leaving there in his van at 8 a.m. for Roxie, calling on several customers of the railroad on the way, he saw train Extra No. 8908 South arriving at Harriston.

The operator at Roxie had come on duty there at 8.45 a.m. and attempted to call the dispatcher but he was busy. Checking cars south of the station she saw train Extra No. 8908 South arrive and later depart southwards. At 9.35 a.m. she met the Customer Service Agent at Roxie but did not mention the train before she left the station shortly afterwards. About 10 a.m. the dispatcher contacted the Customer Service Agent at Roxie and asked him if train Extra No. 8908 was there. "No," the agent replied, "I saw it at Harriston at 8 a.m. and it probably hasn't had time to reach Roxie yet, though I'm not sure." Acting on this statement and without further ado the dispatcher had instructed the Customer Service Agent to put the train order signal at Roxie to Stop, the Customer

Service Agent having told the dispatcher that the operator was away checking waggons. The Customer Service Agent then accepted train order No. 110 at 10.10 a.m. for a train which had already left Roxie just over an hour earlier.

The Crosby and Gloster Customer Service Agent saw a southbound train (Extra No. 8908 South) as he was driving between the two stations at 9.25 a.m. but did not mention the fact to the dispatcher.

The dispatcher had come on duty at Chicago, Illinois, at 7 a.m. Just after 10 a.m., he stated, the Slaughter operator had advised him that train No. 92 had arrived there and was awaiting Extra No. 8908 South. He admitted at the Federal Railroad Administration inquiry that he had told the Slaughter operator to wait, had contacted the Customer Service Agent at Roxie and understanding from the latter that it was not likely that the southbound train had reached there yet, he had issued order No. 110 to the train there without being certain of its actual location, thus causing the accident. [516]

An overworked stationmaster was probably the principal cause of a major disaster in Spain on the evening of Sunday 9th August 1970. A large number of people from the city of Bilbao had spent that day on the beach at Plencia or in the nearby hills. In the early evening rain threatened and there was a rush of passengers to return home, perhaps somewhat earlier than they would otherwise have done, to avoid the expected downpour. The three-car electric passenger train of the Sociedad de Ferrocarrils y Tranvias de Bilbao which left Plencia just before 19:30 was thus carrying about 600 passengers. Meanwhile to cope with the large number of returning passengers a second empty three-car electric train was sent down to Plencia from Bilbao. The line being single track the empty train should have been held at Urdúliz station for the train coming from Plencia, but Stationmaster Juan Abando Amechuzarra at Urdúliz let the empty train proceed. He soon realised what he had done and attempted to stop both trains by having the electric current turned off but before this could be done they had collided head-on about two kilometres from Urdúliz and rather more than a kilometre from Plencia, meeting on a curve so that neither motorman could see the other train until the last moment. The collision speed was not especially high, only about 45 kilometres an hour, but the leading cars were embedded in one another and 33 people died and 165 were injured, 29 seriously.

After the accident both stationmasters were arrested, though it soon became apparent that the mistake had occurred at Urdúliz. It transpired that for the four days prior to the accident Stationmaster Amechuzarra had been working 16-hour shifts from 05:00 to 23:00 daily (this is of course 18 hours; presumably he must have had had a two-hour break at some point) because the other employee at the station had been on holiday. During these long hours he had had to issue tickets, attend to trains, operate the barriers of a level crossing, change points and so on. He spent several months in Basauri prison and was a broken man, unable to eat or sleep, and he died not long after the disaster. Prior to the accident he had had 29 years' service on the railway. [517] [518] [519]

---

[516] Federal Railroad Administration report

[517] Federación Castellano Manchega de Amigos del Ferrocarril

[518] Wikipedia – Accidente ferroviario de Urdúliz

[519] https://treneando.com/2020/08/09/50-anos-de-la-tragedia-de-urduliz/

A mistake by a signalman who cleared his signals for a passenger train, although a freight train was approaching from the opposite direction on the single line and was supposed to cross the passenger train there, led to a head-on collision near the station at Sasson, Iran, on New Year's Eve, Thursday 31st December 1970. The passenger train was carrying 300 railwaymen and miners who were on their way to Isfahan for the weekend holiday. [520]

The number of deaths in the accident was unclear. Government spokesmen would only confirm 15 fatalities and Transport Minister Hassan Shalchian merely spoke of "some casualties." The Kalyhan afternoon newspaper, on the other hand, reported that 50 bodies had been recovered from the wreckage by noon and that it was feared that 70 people had died. A correspondent for the newspaper said that 130 people were seriously injured, "some still crying under tons of steel," while the wreckage was described as "a heap of crumpled pieces of steel," with bodies scattered through the overturned coaches. The rescue operations were hindered by the isolated location of the accident but army and Red Lion and Crescent Society helicopters and rescue crews were rushed to the scene. The injured were taken to hospitals in Yezd (the starting point of the train) and Ardekan.

The Yezd to Isfahan railway, on which the accident happened, was a new line, built in the previous two years and still not officially open, constructed to convey raw materials to the country's first steel mill complex near Isfahan. [521]

Two Chemins de fer de fédéraux suisses passenger trains collided just after 20:00 on Monday 18th January 1971 on the railway line running along the right bank of Lake Zurich between Herrliberg-Feldmeilen and Meilen stations, Switzerland. The trains involved in the accident were No. 1077 from Zurich Stadelhofen to Rapperswil and No. 1076 from Rapperswil to Zurich Stadelhofen, both *Goldküstenexpress* (Gold Coast Express) services. [522] Initially it as stated thar six people had lost their lives in the accident, nine had been seriously injured and eight were slightly hurt. Two of the injured died in the following days. Material damage exceeded one million Swiss francs.

The events that led up to the accident began with a derailment at Meilen at about 19:30, which damaged a set of points and put the automatic block system out of action. A little later the agent at Herrliberg-Feldmeilen accepted train No. 1076 over the telephone from Meilen. There was then a discussion over the telephone, which led to a misunderstanding; the train numbers were not quoted or were given indistinctly, the agent at Meilen wrongly understood that train No. 1076 had already arrived at Herrliberg-Feldmeilen and freed the starting signal there for train No. 1077 to proceed to Meilen. Train No. 1076 was travelling at 10-15 kilometres an hour and train No. 1077 at 75-80 kilometres an hour when they collided. The 60-year-old agent at Herrliberg-Feldmeilen suffered a nervous breakdown after the accident. [523] [524]

A single-line collision in West Germany on Monday 29th March 1971 fortunately only

---

[520] *The Daily Herald*, Biloxi, Mississippi, United States, 31st December 1970

[521] *European Stars and Stripes*, Darmstadt, West Germany, 1st January 1971

[522] Wikipedia – Lake Zürich right-bank railway line

[523] *Gazette de Lausanne*, Lausanne, Switzerland, 20th and 26th January 1971

[524] *Journal de Genève*, Genève, Switzerland, 20th January and 10th February 1971

had minor consequences, thanks to the alertness and quick reactions of one of the drivers. It occurred between Wendlingen and Kirchheim-Ötlingen. Passenger train No. P 4801 and light engine No. Lz 75250 were timetabled to cross one another at Wendlingen but on this day the engine was running a little late, while the passenger train was on time, so to avoid delay to the passenger train the fahrdienstleiter at Wendlingen altered the crossing place of the two trains to Kirchheim. The passenger train arrived at Wendlingen at 05:41 and the engine had reached Kirchheim a minute earlier. Unfortunately, rather than waiting there for the passenger train, the driver of the light engine proceeded back out onto the single line against the Ausfahrsignal at Danger, running through a set of points in the wrong direction and over a level crossing, the barriers of which had not been lowered. Seeing the engine approaching the driver of the passenger train made an emergency brake application and was able to bring his train to a stand before the engine ran into it. Only material damage resulted.

It transpired that the driver of the light engine had been intoxicated and that he had already had two strong reprimands for careless driving. He pleaded that he had been suffering from nervous exhaustion due to family problems and could hardly sleep. He admitted that on the evening before the accident he had had no supper and had drunk a litre of cider while sat watching television. He was found after the accident to have a blood alcohol concentration of 0·8‰ (80 mg/100ml) (exactly equivalent to the legal blood alcohol limit for a car driver in the United Kingdom). Perhaps surprisingly he was not sacked by Deutsche Bundesbahn but was taken off driving duties and given a job in the workshops instead. [525]

Two Canadian Pacific Railway freight trains met in a head-on collision near Bowden, Alberta, Canada, on the evening of Tuesday 20th April 1971. The crew of the northbound train, Engineer A.E. Lewis, 48, and Fireman Colin Gradley, 42, jumped from the cab of their engine immediately before the collision but as a result of the collision a butane tank car, marshalled seventh from the engine, was ruptured by an adjoining flat car. Six thousand barrels worth of butane escaped and formed a sheet of flame which extended 500 feet from the car. The two railwaymen were burnt to death. There had been a crew of five on each train of whom four others suffered cuts and bruises. A box car loaded with beer kegs was placed ahead of the butane tank car and a series of relatively harmless explosions of the beer kegs followed. The northbound train of 56 cars had failed to wait at the siding a half mile short of the point of collision for the southbound 52-car train.

Four hours earlier the Dayliner passenger train from Edmonton to Calgary had collided with a sports car on a level crossing near Lacombe, Alberta, seriously injuring the car driver. [526]

The most serious accident in terms of casualties during the existence of the Deutsche Bundesbahn occurred on Thursday 27th May 1971 near Dahlerau, West Germany. Forty-six people died in the accident and 25 others were injured. Most of those who died (41) were senior pupils from Radevormwald Geschwister Scholl Hauptschule, the others who

[525] *Deutsche Eisenbahn-Katastrophen* by Hans-Joachim Ritzau, Jürgen Hörstel and Thomas Wolski, pg. 131

[526] *The Medicine Hat News*, Medicine Hat, Alberta, Canada, 21st April 1971

lost their lives being two teachers, the mother of one of the pupils and two railwaymen. Just one pupil escaped unhurt.

One of the trains involved in the accident was No. Eto 42 227, a special excursion train carrying 71 passengers, 63 of whom were pupils from classes 9b and 9c of the school, who had set off at 05:30 for a day trip to visit the sights of the city of Bremen, including its harbour and museum. The return trip had started promptly at 16:30 and the children had been on the final stretch of their journey, on a railbus and trailer from Wuppertal along the single-track Wuppertalbahn, when disaster struck. Some delay must have occurred during the return journey for the excursion train was half an hour late as it neared its destination; had it been on time it would have reached Radevormwald before local diesel-hauled freight train No. Ng 16856 had left there. Instead a crossing between the two trains had to be arranged and after a telephone conversation with his colleague at Beyenburg, the next station down the line towards Wuppertal, Fahrdienstleiter Gottfried Sengbart at Dahlerau had arranged that it would take place at his station.

At this time as an economy measure stations such as Dahlerau were not provided with stop signals to control the entry of trains onto the single line. Home signals were provided, equipped with telephones to the station office, which the driver could use if the signal was at Danger and remained so after he had given a whistle from his engine. When the signal was cleared arriving trains were permitted to run as far as a halt sign on the station platform, where they had to remain until cleared to depart by the fahrdienstleiter. Such stations were clearly marked by an 'H' in the working timetable.

What exactly happened on the day of the disaster was never really clarified. It was obvious that the freight train had entered the station slowly in accordance with the regulations, the Home signal having been cleared, and had stopped at the platform. It seems likely that Fahrdienstleiter Sengbart, anxious to be sure that the freight train remained where it was, given that the special train was already on its way from Beyenburg, had come out onto the platform with his Befelsstab – a signalling disc that had red and green discs with red and green lights inset for nighttime use. This was unnecessary, since the regulations dictated that the train must wait until the fahrdienstleiter had signalled for it to depart, and indeed it was in fact positively dangerous, as the switch changing the indication of the signalling disc from red to green could easily be inadvertently operated, or its light could be extinguished altogether, whereupon the train crew could take the raised baton as a proceed signal. This, the crew of the freight train alleged, was exactly what had happened. According to their accounts they had either had a green signal or had observed no light at all but had seen the signalling disc raised, whereupon train No. Ng 16856 had started away. The fahrdienstleiter related that when the freight train had begun moving he vainly attempted to run after it, then, when this failed, he had telephoned Beyenburg, to be told that the special was already well on its way. The crew of the freight train did not notice that they had run through the points at the end of the loop and after travelling 800 metres it met the excursion train. The diesel locomotive hauling the freight train was about five times heavier than the railbus, which was pushed back 100 metres and crushed.

Having failed to stop the freight train Fahrdienstleiter Sengbart did the only thing that there was left to do and alerted the emergency services, so that firemen, police, and other staff were quickly on the scene from Radevormwald, Wuppertal, Remscheid, and Solingen.

The poor fahrdienstleiter was questioned for seven days after the collision, until he collapsed from the strain. He insisted that he had exhibited a red light to the train. Not long afterwards he lost his life in a road traffic accident (which had not been his fault) and any further chance of elucidating what had happened died with him.

The Deutsche Reichsbahn subsequently banned the use of red signalling discs and amended the regulations so that when two trains were arranged to cross one another the first to arrive had to stop outside the station at the home signal until the second had come to a stand in the station. Later, Starting signals with Indusi train magnets were provided at Dahlerau and Beyenburg. Preparations were also pushed forward for the introduction of train radios. [527] [528] [529]

At least 40 people reportedly died at about 21:00 on the evening of Wednesday 4th August 1971 in a head-on collision between a six-car multiple-unit passenger train and a freight train between Kasapovac and Lipe, Yugoslavia. About 100 others were said to have been injured and four cars of the passenger train were wrecked. The southbound passenger train was en route from Belgrade to Pozarevac while the freight train was northbound. The accident occurred at a location that was inaccessible by road while the area was wet, muddy, and slippery after heavy rain. A special rescue train was sent from Belgrade that shuttled between the accident site and Vrčin station, where the casualties were transferred to ambulances. It was over five hours before the first injured reached hospital. About three hundred passengers had been returning on the train to their villages after visiting a fair in Belgrade.

The drivers of the two trains survived and both claimed that they had had clear signals.

Driver Svetomir Arsenijević of the passenger train stated, "We left Kasapovac and began to pick up speed, when we suddenly ran into the freight train on a curve. The crash was terrific. One of the cars fell into a ditch. We rescued many, but others we could not get to." [530]

Driver Miodrag Nestorović of the freight train testified, "I saw the lights of the passenger train coming around a curve but it was then too late to stop. There was a tremendous crash. Everything around me began to break to pieces. That's all that I can remember." [531]

The day after the accident it was announced that six railwaymen – four from Lipe station, plus Driver Nestorović and his assistant – had been arrested. It was alleged that the accident had occurred because the freight train had failed to wait at Lipe for the passenger train. [532]

In West Germany Deutsche Bundesbahn suffered another single-line collision on New

---

[527] *Katastrophen der deutschen Bahnen Teil II* by H.J. Ritzau, pgs. 146-7

[528] Wikipedia – Eisenbahnunfall von Radevormwald

[529] https://www.spiegel.de/politik/den-unfall-hat-er-allein-verschuldet-a-7dd35a62-0002-0001-0000-000043231198

[530] *The Gastonia Gazette*, Gastonia, North Carolina, United States, 5th August 1971

[531] *European Stars and Stripes*, Darmstadt, West Germany, 6th August 1971

[532] *The Times*, London, England, 6th August 1971

Year's Eve 1971, Friday 31st December, involving two diesel railbuses. Train No. Nto 3983 was on its way from Altenkirchen to Limburg while in the opposite direction train No. Nto 3972 was coming from Westerburg towards Altenkirchen. The two railbuses collided between Hachenburg and Unnau-Korb stations. Hachenburg station was equipped with fixed signals protecting the entrance to the single line but the smaller Unnau-Korb station had no fixed signals, the departure of trains being authorised, as at Dahlerau (see above) by the fahrdienstleiter exhibiting his Befehlsstab signalling disc to the driver. A further complication here was the fact that at times Unnau-Korb station was unstaffed, at which times it was presumably not used as a crossing place.

On the afternoon in question, when the railbus from Westerburg arrived at Unnau-Korb station, slightly behind time, its motorman, seeing no sign of the fahrdienstleiter on the platform, had wrongly assumed that the station must be unstaffed, and the station work having been completed he set off again towards Hachenburg. In fact the fahrdienstleiter was on duty and was in his office arranging with his colleague at Hachenburg for the two trains to cross at Unnau-Korb. When he heard the Westerburg train departing the fahrdienstleiter rushed out onto the platform and gave frantic stop signals to the receding train, which were probably seen by a passenger sat towards the rear of the railbus but were not understood as a warning of danger. Meanwhile a man on the platform, who had been waiting to meet a passenger off the Altenkitchen train, realised what had happened and jumping into his car attempted to intercept the Westerburg train at a level crossing up the line, only to find that the train had already passed that point. Having failed to stop the departing train the fahrdienstleiter had hurried back into his office and telephoned Hachenburg in an attempt to stop the Altenkitchen train, but again was too late.

The two trains collided at a curve in a cutting at kilometre 48.0, each travelling at some 60 kilometres an hour. Had the Westerburg train been 30 seconds later or the Hachenburg train been 30 seconds earlier they would have met on a long, straight stretch of line and the collision might have been avoided, or at any rate been less serious. Seven people were killed outright, probably including one or both motormen, and 34 were injured, some seriously. An eighth person died of their injuries in hospital.

After the accident radio communication was provided between trains and stations and almost all branch line stations were provided with Starting signals and Indusi track magnets.[533] [534]

The southbound *Acropolis Express* from München, West Germany, to Athens, Greece, which was much used by Greeks who had been working in West Germany, was in collision at about 16:45 on Sunday 16th January 1972 with northbound special passenger train No. 121, a military relief train bound from Athens to Thessaloniki. The trains met on the single line between Doxaras and Orfana stations, Greece, possibly during a sleet storm. Nineteen or 21 people died, according to different accounts, and about 50 were injured.

Stationmaster Nikolaos Gekas at Orfana was later found to have been responsible for

---

533 https://radundschiene.iphpbb3.com/forum/78852449nx65167/regional-historisch-f6/mein-artikel-zum-zugunglueck-an-silvester-71-t641.html

534 *Katastrophen der deutschen Bahnen Teil II* by H.J. Ritzau, pg. 149

the disaster and was sentenced to five years' imprisonment. [535] [536]

At 5.27 a.m. on the morning of Sunday 12th March 1972 Mr Lloyd Zeigler, whose house was only 50 feet from the Penn Central Railroad line in Herndon, Pennsylvania, United States, was awoken from his sleep. He said, "We heard a terrific crash that woke us up. I looked out of my back window and everything was in flames." A coal train and a freight train had collided head-on.

The population of Herndon turned out to assist after the accident, as a newspaper reported. 'All 600 residents of Herndon mobilised to aid firemen and work crews. They turned their two churches into canteens and rest centers. Police said the crash could be heard two miles away. But two elderly women whose small home bordered the track slept through the crash. Relatives, who said the women were hard of hearing, evacuated them from the house when the fire broke out. Coal from one of the derailed cars had spilled onto their back porch.' [537]

The line through Herndon was part of the 97½ mile long Penn Central route between Rockville, Sunbury, and Williamsport. This was single track but between the first two places there were three passing sidings (crossing loops) at Herndon (Boyles Siding), Millerburg and Ferry, each of which was about three miles long. These sidings were remotely controlled from Kase Tower, Sunbury, with position light signals.

On the morning of the accident train S-82 (Extra No. 7828 West) left Enola Yard and passed Rockville Tower at 4.10 a.m. It consisted of two diesel units, 70 loaded and 33 empty cars and a caboose, bound for Wilkes-Barre. The brakes of the train were tested before departure.

In the opposite direction train UY-328 (Extra No. 7095 East) was on its way from Clearfield to a power plant at York Haven. It consisted of two diesel units, 104 cars of coal and a caboose. A new crew for the train had reported at Newberry Yard, Williamsport, at 1.10 a.m. and were taken by taxi five miles to join the at Linden, 63 miles short of Herndon. En route to Linden the taxi stopped for 50 minutes at the crew's request at an inn. The owner of the inn told the National Transportation Safety Board (NTSB) inquiry later that the railwaymen had eaten sandwiches and two of the four had drunk soda and the other two a beer. The taxi driver said he had not noticed any smell of alcohol on any of the crew. The train left Williamsport at 3.05 a.m., crossed a westbound train at Muncy, 22 miles from Williamsport, and passed Molly, 45 miles from Williamsport and 4 miles short of Kase Tower, at 4.50 a.m. At this point the dispatcher instructed the Kase operator to route the train into the siding at Herndon at West Boyles (the west end of Boyles siding) to allow train Extra No. 7828 West to pass it on the main line. Kase Tower was passed at 5.02 a.m., the route having already been set for it to enter the siding at West Boyles. Neither train was informed of the meet; the eastbound train had no radio anyway. The westbound train did have a radio but it was not the practice to inform crews of meets, nor was it required by the rules.

Train Extra No. 7095 East entered the siding at West Boyles at 5.21 a.m. and was

[535] https://www.loquis.com/en/loquis/2176502/Doxaras+railway+station
[536] *The Times*, London, England, 17th January 1972
[537] *New Castle News*, New Castle, Pennsylvania, United States, 13th March 1972.

clear of the main line at 5.23 a.m. According to its conductor it then continued along the siding at a steady speed of 25 miles an hour (calculations made afterwards indicated that the speed had been closer to 40 miles an hour). Signal 26L CD, protecting the far end of the siding at East Boyles, was necessarily at Danger as train Extra No. 7828 West was still approaching the siding, but the eastbound train ran past the signal, ran through the points and back out onto the main line. The trains collided 520 feet further on, while each was travelling at about 30 miles an hour.

After the accident the conductors and flagmen of the two trains, who had been riding in the cabooses at their rear ends, went forward to the fronts of the trains searching for their engineers and front brakemen but could not find them, while soon the intensity of the fire that had broken out following the collision forced them to retreat. It transpired later that Engineer H.C. Billing and Head Brakeman Emil Serhan of train Extra No. 7828 West had both been killed, as had Engineer Robert Otto and Head Brakeman W.P. Kennedy of Extra No. 7095 East. Three of the four diesel units were destroyed and 29 cars of the eastbound train and 16 of the westbound one were derailed and more or less damaged.

The signalling was thoroughly tested after the accident and was found to be working correctly. The NTSB noted that according to the rules the conductor of a train was jointly responsible with the engineer for its safe operation. However his caboose could be over a mile behind the engine and he had no warning that the train was entering the siding before it did so. When the train in question had exited the siding again without having met another train the conductor had no way of knowing that the planned meet had not been altered, as sometimes happened. Had the train been equipped with a radio he could have been advised of the planned meet.

The NTSB noted that neither the conductor nor the flagman of the eastbound train had gone back to protect the obstructed main line against any following train; instead both had gone forward.

It was concluded that the engineer of the westbound train would not have had sufficient warning to appreciably reduce the speed of his train before the collision.

The fact that train Extra No. 7095 East had been operated normally between Williamsport and West Boyles but had then passed along the siding at excessive speed and had overrun the signal at its far end indicated that something must have happened to the engineer and brakeman at that point. Neither of the crew had previously complained of illness and in any case it was unlikely that both of them would have taken ill at the same moment. Smoke or fumes in the cab might have affected both crew members but there was no evidence of such a problem, while the possibility of the crew having been incapacitated by a lightning strike (there had been a thunderstorm about the time of the accident) was regarded as remote. What seemed much more likely was that the head brakeman had fallen asleep as the train had entered the siding and the engineer had then dozed off as well. In doing so the engineer might have still kept the dead man's handle depressed until the passing of Signal 26L CD at Danger had resulted in the handle initiating a full service brake application.

Had a continuous cab signal system been provided its warning whistle at Signal 26L CD would probably have awoken the engineer in time for him to take effective action to prevent the collision. An automatic train control system would have resulted in an automatic brake application to reduce the speed of the train if the engineer had failed to take action.

The dead man's control device on the locomotive could be easily defeated by placing a weight on the pedal or by depressing it with a stick or bar. The weight of an incapacitated or sleeping engineer would probably also prevent it from actuating. A vigilance device requiring repeated activation at intervals during a time limit of a few seconds was likely to be more effective.

Given that the conductor shared responsibility with the engineer for the safety of the train he should be provided with a two-way radio or intercom in the caboose. Alternatively he should ride on the locomotive with the engineer.

The NTSB listed its conclusions:

(1) The automatic brakes of both trains were tested properly and functioned properly en route.

(2) No defective conditions were found on the locomotive or cars of either train that could have contributed to the accident.

(3) The switches for Boyles siding were lined properly and the signals were displayed properly for train UY-328 to enter Boyles siding and permit train S-82 to pass on the main track.

(4) Train UY-328 was not equipped with radios, even though radios are part of the communication system for this section of the Penn Central.

(5) Train S-82 was operated from Enola yard to the collision point in accordance with the carrier's requirements.

(6) Train UY-328 was operated from Williamsport to a point approaching West Boyles in accordance with the carrier's requirements.

(7) Train UY-328 was operated over Boyles siding at a speed in excess of that permitted by the indication of the signal.

(8) At East Boyles, train UY-328 passed signal 26L CD, which was displaying a stop aspect ran through the siding switch, which was lined for the main track, and proceeded eastward on the main track.

(9) The brakes of UY-328's locomotive were applied as the train entered the main track.

(10) When train UY-328 entered the main track, train S-82 was too close for brake applications on either or both trains to prevent the collision.

(11) Because the locomotives' diesel-fuel tanks were ruptured, fuel spilled over the derailed equipment and caught fire.

(12) There is no backup warning system in the event that signal information is not received by the engine crew.

(13) According to the carrier's rules, the conductor and the engineer are equally responsible for the safety of the train. However, the conductor on the caboose was not in a position to receive the same information as the engineer on the operation of the train.

(14) The dead-man control device on the locomotive unit was inadequate.

The NTSB concluded that the accident had resulted from the engineer and brakeman of the eastbound train having fallen asleep. It made five recommendations.

(1) The Federal Railroad Administration (FRA), in cooperation with the Association of American Railroads, should develop a fail-safe device to stop a train in the event that the engineer becomes incapacitated by sickness or death, or falls asleep. Regulations should be promulgated to require installation, use, and maintenance of such a device.

(2) The FRA should include in its present investigation of the safety of locomotive-control compartments a study of environmental conditions that could distract crews from their duties or cause them to fall asleep at the controls. Regulations should be promulgated to correct any undesirable conditions disclosed.

(3) The ERA should promulgate regulations to require that a railroad equipped communication facilities, install radios in appropriate parts of trains and maintain them in operating condition, unless all personnel involved are notified to the contrary by appropriate railroad procedures, such as a train order or general order.

(4) The FRA, in the promulgation of regulations governing railroad operating rules, where responsibility for safe operation of the train is assigned jointly to the engineer and the conductor, should require that they be located and informed so that they can make quick, effective decisions.

(5) The Penn Central Transportation Company should establish a backup communications system in the event that signal information is not received by appropriate crewmembers.

Just over two months after the Herndon accident four railwaymen again died in a single-line collision in the United States, this time near Maquon, Illinois, on the evening of Wednesday 24th May 1972. The accident occurred on the Burlington Northern Railroad line between Peoria and Galesburg. Signalling on the line was by timetable, telephone, and train order, with the train orders being sent from a dispatcher's office at Hannibal, Missouri, 116 miles west of Galesburg and 167½ miles west of Peoria. At the time of the accident the only open station between Peoria and Galesburg was at Yates City, about 23 miles east of Galesburg.

One of the trains involved in the collision was Extra No. 2043 East. It consisted of two diesel units, 37 loaded cars, 29 empty cars and a caboose, in charge of an unnamed conductor with Engineer Frederick Mitchell and Brakeman Robert McGraw. This train left Galesburg at 6 p.m. with a clearance card and nine train orders, five of which concerned speed restrictions at different points, one bridgework west of Peoria, one a meet with train Extra No. 5720 West at Yates City, one of which authorised it to leave Galesburg after the arrival of train Extra No. 6050 West at Galesburg and the ninth, train order No. 263, which read:

Extra 2043 East take siding meet Extra 1577 West at Yates City. Extra 1577 West gets this order at Yates City.

Train Extra No. 6050 West had arrived at Galesburg just before 6 p.m.

Train Extra No. 1577 West consisted of three diesel units, ten loaded cars, 26 empty cars and a caboose. After a brake test it left Peoria at 12.50 p.m. with a clearance card and six train orders. Three of the orders related to speed restrictions, one to the bridgework between Peoria and Edwards, one established a meet with train Extra No. 5625 East at Oakhill Siding, 20 miles west of Peoria, and the sixth authorised it to leave Peoria after the arrival there of train Extra No. 5720 East, which it did.

The aforementioned bridgework delayed train Extra No. 1577 West at that point until 4.10 p.m., then it proceeded to Oakhill Siding where it met train Extra No. 5625 East at 5.10 p.m. as per the train order. The hours of the train crew expired at 5.30 p.m. and they were here relieved by a fresh crew, consisting of Conductor G.W. Weidenhammer with Engineer Glenn Way, Brakeman Ghlee Wilson, and Flagman R.W. Woodkirk. Meanwhile the Hannibal dispatcher had arranged a meet between Extra No. 5625 East and Extra No. 1577 West at Yates City and he issued train order No. 263, detailed above, to the conductor and engineer of the eastbound train at Galesburg and to the conductor and engineer of the westbound train at Yates City, the point at which the order applied. The dispatcher dictated the order to Galesburg and to Operator P.F. Naumann at Yates City at 3.21 p.m., the latter having first placed the westbound train order signal at *Stop*. The first trick operator at Yates City had hours of duty from 8 a.m. to 4 p.m. and the second trick operator's hours were from 9 p.m. and 5 a.m. Up until 6 p.m. the despatcher could call out the first trick operator if he were required and after 6 p.m. the second trick operator. Concerned about delivering the order to Extra No. 1577 West, Operator Naumann returned to the station at 5 p.m. and contacted the train by radio, to find it was still at Oakhill Siding. He did not mention train order No. 263 and was told by the dispatcher that the train would probably not reach Yates City before 6 p.m., when the second trick dispatcher would be called by the dispatcher to deliver the order. After confirming that the train order signal was at Stop Operator Naumann went home. The position of the train order signal was verified by independent witnesses at 5.20 p.m. and 7 p.m.

The dispatcher does not appear to have called out the Yates City second trick operator to deliver train order No. 263 and when Extra No. 1577 West arrived there at about 6.30 p.m. the station was unstaffed. Conductor Weidenhammer and Flagman Woodkirk told the National Transportation Safety Board (NTSB) inquiry later that they had both looked for the train order signal and it appeared to be green, that is Clear. They saw that Extra No. 2043 East was not in the siding but assumed that as there were no orders it must have been held at Galesburg for them. The train did not stop but continued towards there.

The trains collided just over 5 miles on the Galesburg side of Yates City and 2½ miles on the Peoria side of Maquon. Four diesel units were destroyed in the collision and the fifth was badly damaged. Fifteen cars of Extra No. 2043 East were derailed of which six were destroyed: nine cars of Extra No. 1577 West were derailed of which two were destroyed. Leaking diesel fuel from the engines caused a post-accident fire.

The NTSB report noted that post mortem blood tests by two separate laboratories had found blood alcohol levels of 0.095% and 0.078% respectively in the samples from Engineer Way of Extra No. 1577 West. Considerably raised levels of carbon monoxide were also found in the blood samples from Engineer Way and from Brakeman McGraw of Extra No. 2043 East but these were considered to have been a result of the post-accident fire, as there was no indication of excessive smoke or fumes having been produced in the cabs of the locomotives.

Checks of speed tapes of different trains showed that eight out of 13 of them had exceeded the speed limit on the curve on which the accident had occurred.

In the three months preceding the accident 83 train orders had been issued for meets at Yates City, 12 of them at Yates City for meets there, so-called restricting train orders. No special precautions had been taken other than keeping the train order signal at Stop.

Observation of the Yates City train order signal at the same time of day that Extra No. 1577 West had passed it and under the same weather conditions showed that a bright sun shining behind the signal when it was at Stop would make it appear to be showing a green light until the train was close to it, while its arm was partly obscured by a utility pole.

The Hannibal dispatcher, the NTSB noted, had apparently had no misgivings about issuing a restricting order to Extra No. 1577 West at Yates City because this had been done in the past with few problems. The dispatcher did not consider that any precautions other than maintaining the train order signal at Stop were necessary. Rule No. 78 required extra precautions to be taken by the dispatcher in cases of poor visibility or bad weather, which was unrealistic when the dispatcher was 147 miles away.

The misreading of the train order signal by the conductor and flagman might have been influenced by the failure the engineer to stop the train, leading them to believe that the signal must have been clear. Why the engineer and front brakeman had failed to see the signal could not be explained. The NTSB suggested that the fact that the engineer and perhaps the brakeman had been under the influence of alcohol night have been a factor (the brakeman's body had been too badly injured for a blood sample to be taken for alcohol measurement).

Rule No. 215, the NTSB pointed out, required that special precautions be taken when a restricting train order was issued but did not specify what they should be. Possibilities were:

(a) Notifying and obtaining permission from the chief train dispatcher before issuing such an order.

(b) Arranging for an operator to be on duty until the order had been delivered.

(c) Reviewing with the operator the special precautions to be taken. These could be the placing of torpedoes (detonators) on the track ahead of the train order signal, flagging the train with fusees, notifying the train crew by radio of the impending order, or displaying a special type of signal to indicate a restricting order.

The NTSB report listed 16 conclusions:

(1) Train order 263, which established the meet for Extra 1577 West and Extra 2043 East at Yates City, was worded and transmitted to the operators at Yates City and Galesburg as required by the rules, and was a restricting order for Extra 1577 West, in accordance with Rule 215

(2) It was possible to avoid the use of a restricting train order to establish the meet between the trains There was a pattern of noncompliance with this aspect of Rule 215 for some time preceding the accident.

(3) Rule 215 was ambiguous in that the rule did not make clear what special precautions were to be taken in the event of a restricting train order, or which employee was to take them.

(4) No special precautions were taken to protect Extra 1577 West other than those precautions required for any type of train order.

(5) The westbound train-order signal at Yates City was in the stop position when Extra 1577 West passed the station, but there was no operator on duty at the station.

(6) Extra 1577 West did not stop at Yates City as required by the indication of the train-order signal and did not receive train order 263.

(7) The design of the westbound signal at Yates City was such that at certain hours of the day the sun could shine through the green roundel of the signal and could produce a phantom clear aspect when the signal was viewed from certain positions. Furthermore, the semaphore arm was partially obscured at several locations.

(8) The phantom clear aspect and the partially obscured semaphore arm were seen by the flagman of Extra 1577 West when he checked the signal as the train approached Yates City.

(9) The same phantom clear aspect was seen by the engineer and front brakeman as Extra 1577 West approached Yates City. However, they had a clearer view as the train approached the station.

(10) The level of alcohol in the blood of the engineer on Extra 1577 West was sufficient to impair his mental and physical actions.

(11) The effect of the alcohol and the misleading phantom clear aspect of the westbound train-order signal at Yates City could have caused the engineer of Extra 1577 West to misinterpret the indication of the signal.

(12) Both Extra 1577 West and Extra 2043 East were exceeding the speed limit as they approached the point of the collision.

(13) Trains had been permitted to operate routinely at speeds in excess of the speed limit.

(14) Visibility between the trains was reduced to about 800 feet by the curve of the track.

(15) The fire which developed after the collision resulted from the ignition of diesel oil spilled from the ruptured tanks of the locomotive units.

(16) The relatively high carbon-monoxide content found in the blood of the engineer of Extra 1577 West and front brakeman of Extra 2043 East resulted from the fire which followed the collision.

(17) Radio communication by means of which the train order could have been transmitted to the crew of Extra 1577 West was available.

(18) The relief crew on Extra 1577 West was permitted to take over the operation of a train at an intermediate point and to proceed without notifying the train dispatcher.

The NTSB determined that the probable cause of the accident was the failure of the responsible employees at Yates City to deliver to the crew of train Extra 1577 West a restricting train order establishing a meet at that point, between Extra 1577 West and Extra 2043 East. Two factors had contributed to the failure. One was the failure of Extra 1577 West to stop at Yates City to obtain a copy of the order, in disregard of the stop signal, for undetermined reasons. The effects of alcohol on the engineer and a misleading phantom clear aspect of the westbound train order signal might have contributed to the train failing to stop. The other factor was the failure of the Yates City employees to take special precautions for the safety of train Extra 1577 West as per the operating rules.

The NTSB made several recommendations:

(1) The Burlington Northern change the design of their train-order signals so that there is no possibility that a phantom aspect would be displayed. (This was done soon after the accident.)

(2) The Burlington Northern review its operating rules and its interpretation of these rules to assure that train movements are adequately protected The management and employees of Burlington Northern should take any action necessary to insure that the operating rules are complied with.

(3) The Burlington Northern adopt a system of safety analysis of their operating rules and practices to disclose conditions of inadequate protection of train operations.

The Safety Board reiterates and emphasizes a recommendation made in the Board's report on the collision of two Penn Central freight trains at Herndon, Pa., on March 12th, 1972, The Board recommended that:

"The Federal Railroad Administration, in the promulgation of regulations covering the use of radio, require that where radio is part of an operating system, it be maintained in effective condition unless all personnel involved are notified by appropriate railroad procedure such as a train order or general order." [538]

The conclusions of the NTSB report were not well received by local railwaymen, especially the suggestion that Engineer Way and perhaps Brakeman Wilson had been under the influence of alcohol, which was strongly denied. [539]

A major catastrophe occurred in Spain on Friday 21st July 1972 between El Cuervo and Lebrija stations. The accident occurred at 07:36. Train No. 3201, a railbus which had left Cádiz at 06:00 and was due at Sevilla San Jerónimo at 08:54, should have waited at El Cuervo for express train No. 224 which had left Madrid at 21:30 and was due in Cádiz at 08:50. The railbus was either a four-car set or pair of two-car units and was carrying over 200 passengers. The express consisted of an engine, a van, a postal van, and a dozen carriages and was carrying 500 passengers.

The railbus left the station against the signals, entered the single line, and had covered 3.1 kilometres when it collided with the express head-on. Just before the collision the express was travelling at 90 kilometres an hour and the railbus at 80 kilometres an hour. Both drivers applied their brakes in emergency when they saw the other train but the collision was still a very violent one. The heavier express was more than a match for the more flimsy railbus, which was pushed back 300 metres. The leading three cars of the railbus disintegrated and the leading two metres of the remaining car suffered significant damage. No less than 79 people died and 121 were injured; significantly only two of the fatalities were in the express – two of the three postal workers in the mail van, which was the only vehicle of that train which was much damaged. Railbus Motorman Salvador Salcedo Fernández and Assistant Motorman Gabriel Hernández Castillo were amongst those who died. The conductor of the railbus chanced to have moved towards the rear of the train just before the collision and survived. Most of the passengers in the leading car of the railbus had been Marines returning home on leave from San Fernando. All or most of them lost their lives.

---

538 National Transport Safety Board report

539 *Galesburg Register-Mail*, Galesburg, Illinois, United States, 11th June 1973

Firemen from Cádiz, police from Jerez and Lebrija, Red Cross volunteers, the Civil Guard, soldiers, sailors from the Rota Naval Base, Marines from San Fernando, railwaymen, and locals Jerez and Lebrija all helped in the rescue work. Ambulances from Sevilla, Jerez, Cádiz. Utrera and the Rota Naval Base went to the scene while 15 Naval helicopters also took part in taking the injured to hospital.

With the death of the motorman and his assistant there could be no explanation forthcoming for why the railbus had not waited at El Cuervo. Possibly significant was that fact that it was running some minutes late, already due in the next station at Lebrija at 07:30. [540] [541] [542] [543]

On the foggy early morning of Monday 30th October 1972 a major disaster occurred just south of Schweinsburg-Culten station in East Germany, when train Ext 348, [544] the southbound *Karola* express from Leipzig to Karlove Vary, collided with train No. D 273, a northbound express train from Aue to Berlin. The single-track electric line was remotely controlled from Zwickau. The trains normally crossed one another at Werdau but because the *Karola* express was running rather more than ten minutes late the meet was altered to Schweinsburg-Culten, about 5 kilometres further north. Despite this the *Karola* express passed through Schweinsburg-Culten, travelling at about 100 kilometres an hour, without stopping, and the two trains collided about 700 metres south of the station just before 07:30. Twenty-five people, including the drivers, were killed outright and over 70 people were injured, a number seriously. Three of the injured passed away in hospital.

There had been about 1,000 passengers on train D 273, which consisted of two double-decker Class 175 articulated units and had left Aue on time at 06:05. Many of its passengers were deaf students, on their way to attend a special school for those with hearing difficulties in Leipzig. Unable to hear the noise of the collision they had reportedly wandered around the accident site in the fog afterwards and had only gradually become aware of what had happened. A fireman who had attended the accident recalled years later that in the early afternoon the firefighters had been ordered to leave by the Stasi secret police, who had begun an investigation into the accident, only to be called back in the evening when the fog which had cleared earlier had descended again and they (the firemen) were needed to illuminate the scene.

The accident was reported in contemporary newspapers but the results of the investigation into it were kept secret for a long time. It was clear that the collision had been due to the driver of the *Karola* express passing the Ausfahrsignal (Starting signal) at Schweinsburg-Culten at Danger and that he alone had been responsible for the collision. There were suggestions that he had been intoxicated, a stewardess on the train alleging that he had consumed 15 bottles of beer with her up until 01:30 on the morning of the accident, plus only getting three and a half hours' sleep between then and starting duty at

---

[540] http://www.lebrija.es/es/actualidad/noticias/Este-jueves-21-de-julio-se-cumplen-50-anos-del-tragico-accidente-ferroviario-entre-las-localidades-de-El-

[541] https://www.gentedelpuerto.com/2010/07/21/716-el-accidente-de-trenes-en-el-cuervo-21-de-julio-de-1972/

[542] https://www.republica.com/sociedad/el-cuervo-de-sevilla-una-catastrofe-ferroviaria-que-50-anos-sigue-sin-explicacion-20220721-19491469668/

[543] Federación Castellano Manchega de Amigos del Ferrocarril

[544] https://www.sachsenschiene.net/bahn/sys/unf.htm?134

05:00. There were actually two drivers on the train but the other had left the cab of the engine instead of being there to assist his colleague observing signals, while alcohol was said to have been detected on the bodies of both men, though possibly this might have been the result of post mortem microbial action. At any rate the eldest daughter of the driver directly responsible for the accident strongly denied that he would have been driving under the influence of alcohol. Possibly, with the trains normally crossing at Werdau and the fog, the driver had just missed seeing the signal and had assumed that it had been clear. Both drivers had exemplary records. [545] [546] [547]

Internally there was also criticism of the Deutsche Reichsbahn for not having yet redoubled the stretches of main line that had been singled by the Soviets in 1945, for the excessive hours train crews were expected to work, and for the fact that there had been no train stops at the Starting signals which would have automatically applied the train brakes were the signals overrun. [548]

A single-line collision between Karampada and Kiriburu stations, India, at 13:05 on Sunday 9th September 1973 fortunately did not result in any fatalities, though two people were seriously hurt and 55 suffered minor injuries. It happened on the South Eastern Railway and involved Divisional Material Train (DMT) No. 3, consisting of WGX Class steam locomotive No. 10466 and eight vehicles, and a light engine, WGX Class steam locomotive No. 9782. The line was worked with Neale's Block Token instruments, the irregular operation of which led to the collision.

Shunter (Passed Driver) P.V. Rao was the first person whose evidence appeared in the report of the subsequent inquiry. He stated that at about 09:30 on the day of the accident Stationmaster N.G. Choudhury at Kiriburu had asked him to proceed from Kiriburu to Karampada and back with his engine, so as to take staff to the market there and back. The stationmaster had operated the block instrument and withdrawn token No. 10 in his presence and the engine had left at 10:20, with the stationmaster, his daughter and ten other railwaymen riding on it! Before they had set off he had heard the stationmaster tell Token Porter Narayan Patro not to let the DMT train depart for Karampada until the engine had returned. The stationmaster had not given him the token but had, he had assumed, kept it in his possession. Karampada was reached at 10:40 and everyone had then set off for the market except the second fireman, who had been left in charge of the engine. They had set off back to Kiriburu at 12:45, running tender first. The stationmaster usually kept the token on these regular Sunday trips to the market, he said. The outward and return journeys were usually made with the same token because with all the Kiriburu staff other than the token porter having gone to Karampada there had been no one left at Kiriburu to release a token. After the collision the driver of the DMT train had shown him that he, the DMT driver, had token No. 10, and had said that the token porter had

---

[545] https://www.freiepresse.de/nachrichten/sachsen/zugunglueck-vor-40-jahren-dem-tod-ganz-nah-artikel8137969

[546] https://www.mdr.de/nachrichten/sachsen/chemnitz/zwickau/gedenken-zugunglueck-schweinsburg-culten-100.html

[547] Wikipedia – Eisenbahnunfall von Schweinsburg-Culten

[548] https://www-welt-de.translate.goog/geschichte/article241823159/Zugunglueck-bei-Neukirchen-Restalkohol-im-Koerper-des-Zugfuehrers.html?_x_tr_sl=de&_x_tr_tl=en&_x_tr_hl=en&_x_tr_pto=sc

told him that he could proceed, as it would be a while before the engine returned. The Down Outer signal at Karampada had not been lowered for the engine; he had been hand signalled in.

Shunter (Passed Driver) Premdhan said that the token porter had given the token and he had set off at 12:50, running at a restricted speed of about 15 kilometres an hour. Before he had left he had asked the token porter if the shunting engine, which had gone to Karampada, had returned, to be told that it had not, but that he was clear to go. As he was leaving Kiriburu he had had a green flag from someone, whom he had assumed was his guard; he had been unaware that the guard had gone to Karampada on the engine. In his experience the Up Outer signal at Kiriburu was not usually lowered, while various signals were not lit at night.

Guard G.N. Saha of the DMT train stated that in discussion with Stationmaster Choudhury it had been decided that the DMT train would not be ready to leave before 13:00, so he had gone to Karampada on the engine for his meal, after telling Fireman K.K. Rao and Material Train Supervisor Das where he was going (Supervisor Das denied having been told this). He had been partly knocked unconscious in the collision.

Assistant Stationmaster S.K. Saha at Karampada spoke of being asked for and giving *Permission to Approach* by Kiriburu, after the DMT train had reached there, for the shunting engine, which had duly arrived at Karampada at about 10:45. The token had not been handed over to him and the engine had set off back to Kiriburu at about 12:45. The Kiriburu stationmaster had informed him of the collision at about 13:30 and had told him that a token would be sent to him by messenger and that he should make entries in the train register indicating that *Line Clear* had been given for both the shunting engine and the DMT train. Substitute Token Porter Jugal Kishore Mahato had later arrived with token No. 12, which had been placed in the instrument, and he had made the false entries in the register, as instructed, but then had realised that these false entries would implicate not only the shunter driver but also himself, so he decided to tell the truth. He explained that his Up Outer signal was beyond a tunnel and was also up a hillside away from the track and difficult to access, so neither it nor the Down Outer signal was lit. The back seals of the block instruments on both the Kiriburu and Rangra sides of his station had been missing since he had started working there and he had been told to open the instrument if necessary, if it went out of order, though he had had no occasion to do so.

Block Maintainer Jotilal Mahato agreed that there were six block instruments without locks in his jurisdiction: the two at Karampada and one each Kiriburu, Rangra, Bimalagarh and Barsua. He denied having shown any assistant stationmaster how to remove a token by taking off the back cover of the instrument.

Assistant Block Signal Inspector P. Chinnoyya described sealing the backs of the block instruments at Karampada and Kiriburu after the accident. He explained that he had been repeatedly reminded by the block inspector about missing locks but new ones had not been received from the workshop.

Assistant Permanent Way Inspector S.R. Mahato described his part in the rescue work. He said that steps had been made up the side of the cutting to the Up Outer signal at Karampada six times, but monsoon rains had repeatedly washed them away. Pointsmen had also complained of visits from tigers, though he had never seen any, only snakes and the tracks of an elephant.

Token Porter Patro testified that the shunting engine had left at about 10:15, previous to which Stationmaster Choudhury had told him that when the engine reached Karampada he, the stationmaster, would advise the token porter over the telephone, and the token, which had been left in a drawer, could then be given to the driver of the DMT train. The latter train had duly left at about 12:30; he had not seen its guard. Asked whether Stationmaster Choudhury had spoken to him over the telephone from Karampada he had at first said that he had not, then that Assistant Stationmaster Saha had advised him that the shunting engine had arrived at Karampada, then that Stationmaster Choudhury had done so and told him to give the token to the DMT train.

Substitute Token Porter Mahato spoke of Stationmaster Choudhury giving him token No. 12 and telling him to give it to Assistant Stationmaster Saha. He said that he had not seen how the token was obtained.

Stationmaster Choudhury told the inquiry that after arranging with Shunter Rao for the latter to take the engine to Karampada and back he had obtained a token and placing it on a table, had asked Token Porter Patro to give it to Shunter Rao, then had gone for his bath. They had gone to Karampada and had been on their way back when Guard Saha suddenly shouted that the DMT train was coming, when he (the stationmaster) and others had jumped from the engine. He had then been much worried about his irregular departure being detected and so, after discussion with the Karampada assistant stationmaster, he had burgled a token removed from his instrument by removing its back cover and had it taken to Karampada, plus he had arranged for fictitious entries in the train registers in both the Kiriburu and Karampada train registers. He denied having told the token porter that he would telephone him from Karampada and stated that on the contrary he had told the porter to give the token to Shunter Rao. He also denied having previously kept the token himself on Sunday trips to Karampada with the shunting engine and alleged that the token porter was addicted to drink. He denied that the shunter driver had asked him for the token as the engine left Kiriburu, or that the fireman of the engine had similarly asked him at Karampada. He said that there had been no one to delegate responsibility to when he had left Kiriburu on the engine, as the post of assistant stationmaster was vacant. The non-recording of the movements of the Sunday engine trips in the train register had been the practice ever since he had been at Kiriburu. The Outer Home signal at Kiriburu was normally lit, except when no train was expected after the hours of darkness.

Additional Commissioner S.K. Mojumder fixed the time of the collision at 13:05.

As to the cause of the accident the additional commissioner stated that a line clear must have been obtained for the engine to leave Kiriburu as otherwise the interlocking would not have permitted the siding points to be set for the main line for the engine, as the key for the siding points lever could only be released when the block instrument was in the *Train Going to* position. The stationmaster had asked the token porter to give the token, No. 10, to the driver of the shunting engine, but this had not been done. The shunter driver's plea that he had asked the stationmaster for the token and the latter had indicated that he had it, was untenable. Witnesses such as Shunter Premdhan had stated that the driver was always in possession of the token, even on unrecorded journeys.

The statement of the token porter that the stationmaster had telephoned him from Karampada to give the token to the driver of the DMT train, and that having done so the stationmaster had then set off anyway from Karampada, with his own daughter on board the engine, was, said the additional commissioner, absurd. Had the stationmaster given such instructions to the token porter he would surely have waited at Karampada for the DMT train to arrive there. It seemed likely that the token porter, probably under the influence of drink, had given the token to the driver of the DMT train instead of to the driver of the shunting engine, as he had been instructed to do by the stationmaster.

The stationmaster had not verified with the driver of the shunting engine when he boarded the engine that the driver had the token and as the token pouch was not used, the missing token had not been noticed by anyone.

The lack of a back cover to the instrument at Kiriburu had prompted the stationmaster to attempt to cover up his absence from the station and his responsibility for the accident, by extracting another token from the instrument and having it taken to Karampada, plus making false entries in the train registers. It would then appear as if the engine had gone to Karampada with token No. 12, followed by the DMT train with token No. 10, and that the accident had been due to the engine returning from Karampada without a token, shifting the blame for the accident to Shunter Rao. This plan had worked perfectly, except that the stationmaster had forgotten the numbers of the tokens that had been issued (which had to be recorded in the train register).

The additional commissioner highlighted seven irregularities which had come to light:

(1) The shunting engine was making regular trips between Kiriburu and Karampada for the Sunday market and returning without any entries being made in the train registers. This practice had been going on for over five years.

(2) That six block instruments had been without locks since the overhaul of the signal workshops at Kharagpur in 1971 was, he stated, an extremely serious lapse.

(3) The Kiriburu stationmaster had left his station on Sundays without leaving anyone in charge, the post of assistant stationmaster having been vacant for eight months.

(4) The driver of the DMT train had left Kiriburu without exchanging signals with the guard, who, in fact, had not even been on the train.

(5) The guard of the DMT train had left it without informing his driver and there had been no detonators or First Aid kit in his brake van.

(6) The Outer Home signals at both Kiriburu and Karampada had both been left unlit at night and trains regularly passed them at Danger.

(7) Token pouches were rarely used at either station, except perhaps for through trains.

The additional commissioner concluded that the immediate cause of the collision was Shunter Rao having proceeded from Kiriburu without the token and later likewise having returned from Karampada without a token, while the token had instead been wrongly given by Token Porter Patro to the driver of the DMT train. Token pouches not being used and the stationmaster having left his post without competent relief had contributed to the accident.

Shunter Rao was held primarily responsible for the accident for leaving both Kiriburu and Karampada without the token.

Token Porter Patro was responsible for wrongly giving token No. 10 to the driver of the DMT train.

Stationmaster Choudhury was responsible for five rule violations:

(a) For adopting the irregular practice of allowing the shunting engine to proceed to Karampada and back with the same token.

(b) For not ensuring that the token had been given to the driver of the shunting engine, not using token pouches, and leaving his station without deputing a competent relief person.

(c) For not making entries in the train register for the shunting engine, then making false entries.

(d) For not obtaining a private number from Karampada for the shunting engine, without which the Line Clear was not valid.

(e) For unauthorisedly opening the block instrument and removing a token to conceal his responsibility for the accident.

Assistant Stationmaster Saha at Karampada was responsible for:

(a) Not recording the details in his train register of the Line Clear given for the shunting engine or its movements to him and back.

(b) Not asking for the token when the shunting engine reached him from Kiriburu.

(c) Making fictitious entries in his train register after the accident.

He had, however, come out with the truth at the inquiry.

Shunter Premdhan had left Kiriburu without a signal from his guard.

The additional commissioner made only two recommendations:

(1) That immediate and strict action should be taken to prevent unrecorded movements of engines or trains through block sections.

(2) That the signal engineering manual should be amended to provide for the locking and sealing of single and double line block instruments. [549]

The circumstances of a collision near Mustang, Oklahoma, United States, on the afternoon of Sunday 1st September 1974 were basically no different from those accidents which had occurred in that country a century earlier. Traffic on the St Louis-San Francisco Railway line involved, which ran between Oklahoma City, Oklahoma, and Quanah, Texas, was extremely light, consisting of only one freight train daily in each direction, together with a second freight train on weekdays that ran eastwards and westwards over the line on alternate days. The line was worked by timetable and train orders, plus special instructions. Westbound trains were superior to eastbound trains of the same class, which were required to reach meeting points five minutes before the scheduled departure times of the superior train or to send out flag protection. The trains could communicate with one another and with the dispatcher at Oklahoma City by radio but the reception of signals was patchy.

Westbound second-class freight train No. 3211 consisted on the day of the accident of

---

[549] Government of India Ministry of Tourism and Civil Aviation (Commission of Railway Safety): Report on the Head-on Collision between DMT-3 and Steam Light Engine between Karampada and Kiriburi Stations of South Eastern Railway at 13.05 hours on 9th September 1973

two diesel units, 23 loaded cars, 15 empty cars and a caboose, in charge of Conductor Edward Cook with Engineer Dave Baskett and Brakemen Gary Parks and A.E. Walker. The crew of the train checked their watches with the yard clock, the brakes of the train were tested and it left Oklahoma City at 1.18 p.m. according to the conductor, three minutes late.

Eastbound second-class freight train No. 3210 consisted at the time of the collision of two diesel units, 11 loaded cars, six empty cars and a caboose, in charge of Conductor Ed Burnley with Engineer Carl Martin and Brakemen Melvin Black and Walter Wendell. [550] The brakes of the train had likewise been tested before leaving Quanah, where the crew had compared their watches with the standard clock, from which their watches varied by one minute. They decided that their watches were correct. The train left Quanah at 8.30 a.m.

According to the timetable the trains were due to meet at Mustang at 1.52 p.m. During the previous three months they had met 20 times by timetable but only once at Mustang, on all the other occasions one or the other train had been late and so they had met elsewhere. Over the same period their meet had been arranged 72 times by train order, 10 times at Mustang. On the afternoon of 1st September when train No. 3210 had arrived at Cyril, about 118 miles from Quanah, Conductor Burney had asked the Oklahoma City dispatcher to issue a train order advancing their train to Wheatland, about 4½ miles closer to Oklahoma City than Mustang, to meet train No. 3211 there at 2.01 p.m., but the despatcher had told the conductor that the trains must meet as per the timetable at Mustang at 1.52 p.m. The train left Cyril at 12.20 p.m.

At about 1.08 p.m. Yardmaster Wayne Barber at Oklahoma City overheard the crew members of train No. 3210 speaking on the train radio and told some of the crew members of train No. 3211 that he had heard the crew of the opposing train attempting to contact the operator at Chickasha over the train radio. Chickasha was 27 miles on the Quanah side of Mustang.

About 1.35 p.m. Engineer Martin on train No. 3210 attempted to contact the crew of train No. 3211 over the radio to find out where they were. The reply he received was unintelligible but a little later he understood the crew of train No. 3211 to say that they had passed Mustang and were "starting down the hill." Realising the danger he had shouted an alarm over the radio. Train No. 3211 had then come into view, he had applied the brakes in emergency and his crew had jumped from the train, which by then had almost stopped.

Engineer Basket of train No. 3211 agreed that the radio conversations had been garbled. He told the subsequent National Transportation Safety Board (NTSB) inquiry that he had seen train No. 3210 about 400 feet away, when he had applied the brakes in emergency though they had not had much effect in the short distance remaining. He and the conductor had jumped from the rear cab of the engine and the brakeman from the front cab.

The trains collided about 1¾ miles west of Mustang. As a result of the impact the engines of train No. 3211 were destroyed and 23 cars were derailed. A tank waggon loaded with isobutane was punctured, causing a fire that also involved several carloads of jet fuel and set flames shooting hundreds of feet into the air. For fear of an explosion eight families living within a square mile of the accident had to be evacuated. Conductor

---

[550] *The Ada Evening News*, Ada, Oklahoma, United States, 2nd September 1974

Cook, Engineer Baskett and Brakeman Parks were seriously hurt and taken to the Baptist Medical Centre in Oklahoma City. Engineer Martin was treated for burns to his back and released. Brakeman A.E. Walker was initially missing, his body was found in the wreckage two days later.

In determining how the collision had occurred the NTSB noted that given that train No. 3211 had to wait, according to the timetable, at Mustang until 1.52 p.m. the exact time of the collision was the crucial factor. After the accident the rear brakeman of train No. 3210 had reported the collision to the Oklahoma City operator, who had timed the call at 1.50 p.m. The operator in turn had notified the police department, who timed his call at 1.54 p.m. A separate call to the Oklahoma Fire Department had been timed at 1.52 p.m. A Mustang police officer had noted the fire at a time he put as early as 1.45 p.m. while calls to the Mustang police department were timed at 1.49 p.m. Calls to other rescue units were timed between 1.54 p.m. and 2 p.m. The crew of train No. 3210 timed the collision at 1.44 p.m. Finally two bystanders who had rescued Brakeman Parks from train No. 3211 said that he had repeatedly protested to them, "I told them we were too early." The preponderance of the evidence, the NTSB concluded, was that train No. 3211 had passed Mustang ahead of schedule.

The NTSB decided that the crew of train No. 3211 must have been aware that Mustang was their timetabled meeting point and that they were leaving there early despite not having met the opposing train. The likely reason for their decision seemed to be the reported conversation mentioned by the yardmaster between the crew of train No. 3210 and the Chickasha operator. Thus Conductor Cook had testified, "(The yardmaster) said they are trying to come over on the radio. They are over the other side of Chickasha. Well, that's the only place you can get Oklahoma City is between Laverty and Chickasha. That's when they are up on the mountain," and, "When (the yardmaster) told me that he was over the other side of Chickasha, I didn't even dream he would come to Mustang. That was the last thing in my mind. He didn't have time to come to Mustang."

It seemed that the supposed location of train No. 3210 had been based on the fact that it was only between Laverton and Chickasha that radio messages would have been received in Oklahoma City. This would have been true on many occasions but on the day in question it had been heavily overcast and raining and radio reception had been extended.

At another point the conductor had stated, "Well... we had no meeting point... I was the superior train. I was the westward train. I'm the big dog. They are clearing me. They come as far as they can come to clear me and still take a siding. We don't have a meeting point any place."

The NTSB noted that meets at Mustang were rare and the crew of train No. 3211 thought that they knew where the opposing train was. This assumption was not justified and in any case under no circumstances could it justify passing the scheduled meeting point ahead of time.

The crew of train No. 3210 were not free of responsibility either. According to the rules their train should have been in clear at Mustang at 1.47 p.m. at the latest. If the collision had occurred between 1.44 p.m. and 1.46 p.m. it indicated that their train could not have covered the 1¾ miles between the point of the collision and Mustang, stopped to allow the points at the end of the loop to be operated, and moved in clear in the one to three minutes available. It could, however, have been in clear before the booked departure time of train No. 3211. Train No. 3210 had therefore been 'running on short time,' its crew stretching the rules, while illogically relying on the crew of train No. 3211 to adhere completely to them.

The NTSB listed its conclusions:

(1) The times indicated by the watches of the respective crewmembers of trains 3210 and 3211 probably agreed within prescribed limits even though the crew of 3210 did not set their watches to conform with the standard clock at Quanah, which evidently was in error.

(2) The crew register procedures at Oklahoma City did not assure compliance with rule G since beer was discovered in and around the locomotive of train 3211 after the collision.

(3) The braking systems of trains 3210 and 3211 were functioning normally before the collision.

(4) The dispatcher knew the location and work schedule of train 3210 about 1 hour before train 3211's departure, but he did not transmit this information to the crew of train 3211.

(5) The overcast weather conditions on the accident date allowed radio transmissions to be heard beyond their normal range.

(6) The engineer and conductor of train 3211 believed that train 3210 was farther from them than it actually was because of their past experience with sporadic radio transmissions from the Chickasha area.

(7) Both trains were moving about 40 mph as they approached the collision. The trains had operated at about the maximum authorised speeds since their respective departures from Oklahoma City and Cyril.

(8) The conductor of train 3211, after deciding to proceed, failed to heed the warning of the brakeman that they were passing Mustang ahead of schedule.

(9) The radio conversation between the crews of both trains alerted the crew of train 3210 of the impending collision and allowed them to brake their train and evacuate without serious injury; however, the radio transmission from the engineer of train 3210 was not understood by the crew of train 3211, and they did not take immediate action.

(10) The crew of train 3211 did not immediately recognise the approach of train 3210; however, if the emergency brakes had been applied on train 3211 when the other train was first visible the collision would still have occurred, but at a slower speed.

(11) The collision occurred at about 1:44 to 1:46 p.m., six to eight minutes before train 3211 was authorised to pass Mustang.

(12) The intended margin of safety of the rules was not produced by the crew of train 3210, since they would not have had sufficient time to clear train 3211 at Mustang by five minutes as required; however, they had time to clear train 3211 before that train's scheduled departure from Mustang.

(13) Although the procedures specified by rules 14(n), 34, S-90, 106, and 801 were designed to prevent a train such as 3211 from passing Mustang ahead of schedule, they were non-specific, outmoded, not followed, and therefore ineffective in preventing this collision.

(14) Although trains 3210 and 3211 were controlled by a timetable schedule, they seldom operated on schedule, which suggests that another method of operation may have been more appropriate for the circumstances.

(15) Before the accident, neither the railroad nor Federal authorities had prescribed the exact role intended for the use of radio in facilitating train movements.

The NTSB found that the probable cause of the accident was the failure of the crew of train No. 3211 to take preventive action after the train had passed Mustang ahead of schedule. Contributing to the passing of the meeting point ahead of schedule were the erroneous assumptions of the engineer and conductor based on radio transmissions from train No. 3210 that it was at a more distant point.

Three recommendations were made by the NTSB to the railway company:

(1) Enforce the exact requirements of the operating rules and establish a more effective system to insure that employees understand them. Particular emphasis should be given to the problems demonstrated in the application of rules G, 3, 3(b), 14(n), 34, S-87, S-89, S-90, 92, 106, and 801 in the circumstances of this accident.

(2) Promulgate operating rules that provide specific guidance for the use of radio in railroad operations.

(3) Revise its operating rules that pertain to train meeting procedures so that the rules are consistent with current operating conditions as to the meeting of schedules and so that they provide safeguards for the crews of both trains.

In addition two recommendations made to the Federal Railroad Administration in previous accident reports were repeated:

(1) "That the Federal Railroad Administration (FRA) issue as soon as possible regulations to provide for the use of radio in railroad operations. These regulations should include the traditional safeguards found in existing railroad operating rules where they apply to train movements. It is further recommended that, in drafting such regulations, consideration be given to the principles and procedures for radio used by military and civilian aviation authorities."

(2) "That the FRA require that railroads institute formal locomotive cab management procedures which will specify the duties of each crewmember to ensure the crew action required when the engineer does not function in a manner consistent with the safety of the train. These procedures should be integrated with the results of the ongoing FRA/industry locomotive-cab design project." [551]

On Saturday 22nd February 1975 Norway suffered what up to the present time (August 2023) has been its most serious railway accident during peacetime. Twenty-seven people died and another 25 were injured when two express trains collided just north of Tretten.

[551] National Transport Safety Board report

The events that led up to the disaster might be said to have begun with the late running of northbound express No. 351 from Oslo Øst to Åndalsnes. This train started about 10 minutes late from Oslo and further delays had occurred en route due to the large number of passengers disembarking and boarding, so that by the time it reached Hamar station it was 35 minutes late. Driver Bård Gundersen knew that his train would shortly have to cross southbound express No. 404 from Trondheim to Oslo, the *Dovreekspressen*, and thought that this would probably happen either at Losna or at Tretten. When the train arrived at Tretten the utkjørsignal (starting signal) at the far end of the crossing place was red. The train stopped as normal at the station, then restarted and was taken past the signal, which was still showing red, and out onto the single line. By chance a passenger on the train, Arvid Jensen, who had attended a course at the Norges Statsbaner signalling school and therefore was familiar with signals and their meaning, saw that the signal was red. He could have pulled the communication cord and stopped the train but he thought that the driver must have received authorisation to pass the signal at Danger and so did nothing. It was assumed that the low sun shining in the driver's face must have made him misread the signal as being green.

The trains collided at 13:45 about a kilometre north of Tretten. Both drivers applied their brakes in emergency when they saw the other train and jumped from their engines when the trains were about 200 metres apart. Both survived.

Controller Dagfinn Dyresen at Hamar saw from his panel that things did not seem to be right. He telephoned Tretten station to ask if train No. 351 was still there, to be told that it had left. He immediately cut off the traction current (the line was electrified) in an attempt to prevent the accident, but it was too late. [552]

The two leading carriages of train No. 404 suffered the most. Its leading luggage van was forced up into the air and coach No. 16 was crushed beneath it; it was here that most of the deaths and injuries occurred. Some however were luckier. *VG* newspaper reported on Monday 24th that a woman had managed to crawl out of one of the most damaged carriages. She had been travelling with two young children and in the accident had become separated from them. She could be seen desperately searching amongst the wreckage for her offspring. Her joy was indescribable when she found that they had been taken care of by strangers and were safe.

Among other passengers on the train were Jostein Nyføtt, his sister Mari Gullåsen and Mari's six-month-old son, who were on their way to visit their parents at Vinstra. Jostein was a student with the NRK (Norsk rikskringkasting – the state broadcasting company) in Oslo, and had a SLR camera with him, with which he took a number of photographs of the disaster. Mari said later that her son had been buried beneath a seat when the accident had happened, the coach had been filled with a clouds of dust and she feared at first that her little boy was dead. In fact he had been unharmed. She remarked years later that after the disaster the taxi fares of the three of them from Tretten to Vinstra had been paid by the railway company, but that beyond that the survivors had received no assistance; even as recently as 1975 there had been no post-accident counselling to help them get over the trauma of the experience.

Hjelpekorpsmann (an enlisted man with medical training) Rudof Hausstätter heard the noise of the crash from his house a couple of kilometres away. He jumped into his car

[552] https://lokalhistoriewiki.no/wiki/Tretten-ulykka

and drove to the scene. "I found a horrible sight when I got there," he said. "I spent a lot of time releasing a girl who was trapped. We had to remove a lot of wreckage and luggage to do so. While I was doing so I came across the hand of another woman. It was a lady who had been travelling with her daughters. They were all dead. However, we were able to save the girl."

Nurse Helene Aarli from nearby Øyer related that she had received a telephone call to come and assist. "I was probably the first to be told about the accident. The district doctor was away and so the call was put through to me. I first sent to the house of Ottar Hekneby, a retired district doctor at Tretten, and then the two of us went to the scene, where Dr Hekneby worked hard through the afternoon. I have many vivid memories of the disaster, for example I remember seeing a dead man sat by a window with a pipe in his hand."

Police Officer Ivar Ringen found himself in charge of the rescue operations. "I was a young and inexperienced officer at the time. I was ordered to go to Tretten because something serious had happened and there found that I was in charge. Many did an outstanding job that afternoon. At the time I had a 4-year-old daughter. One of the things that I had to do at the scene was to carry away the body of a young girl of about the same age. It was as much as I could cope with."

It was over half an hour before the first rescuers reached the scene. Eventually 20 ambulances were despatched as well as three military helicopters from Gardermoen and 300 people were involved. The rescue work went on through the afternoon and evening, and it was not until 20:00 that the last survivor was released. He was conscious and possibly was not that badly hurt, just trapped. A good many passengers had initially been imprisoned in the wreckage; some of them in much pain were given morphine injections. [553]

Driver Gundersen and Conductor Kjell Åge Kristiansen of train No. 351 were both subsequently charged with negligence resulting in the accident. However, when the case against the conductor proceeded in late June 1976 it resulted in his acquittal, whereby the case against the driver was dropped. [554]

Another disastrous single-line collision occurred in West Germany on Sunday 8th June 1975 near Warngau, on the line between Holzkirchen and Lenggries. One of the trains involved, No. E 3591, was a regular daily working, the other, No. E 3594, was a Sunday-only working which had only begun running with the introduction of the summer timetable and was only making its second trip of the year. A feature of the accident was that, with a view to permitting more flexibility in operating the trains should one or the other or both not be running to time, what was referred to as a so-called air crossing between them was set out in the working timetable, whereby they theoretically passed one another between stations, even though, given that there was only a single line, this was impossible.

The previous Sunday train No. E 3594 had been a few minutes late and the trains had

---

[553] https://www.nrk.no/innlandet/40-ar-siden-tretten-ulykka-_-se-hittil-upubliserte-bilder-1.12185370

[554] https://lokalhistoriewiki.no/wiki/Tretten-ulykka

crossed at Schaftlach, the next station to Warngau. On the day of the accident it had been on time. On duty at Schaftlach on 8th June was Fahrdienstleiter Ludwig Hiergeist; on duty at Warngau was Fahrdienstleiter Franz Habart, who had been away for three weeks and was only on duty that afternoon because a colleague was off sick. The operation of trains over the single track involved the exchange of telephone messages between the various fahrdienstleiters and it was the sloppy and careless manner in which these messages were exchanged that led to the disaster.

Train No. E 3592 left Schaftlach on the afternoon of the accident at 18:22 and arrived at Warngau at 18:28, where train No. E 3591 was waiting. Behind train No. E 3592 was train No. E 3594. Fahrdienstleiter Hiergeist at Schaftlach expected train No. E 3594 to follow train No. E 3592 through the section to Warngau, while his colleague Habart at Warngau wanted Hiergeist at Schaftlach to accept train No. E 3591 in the opposite direction first. The conversation that took place between them over the telephone was recorded and went as follows (the wording that should have been used according to the regulations follows in parentheses):

Warngau: 'Warngau signal box.' (Fahrdienstleiter Habart, Warngau here.)

Schaftlach: '3594 accepted.' (Train report Fahrdienstleiter Hiergeist, Schaftlach. Is train 3594 accepted?)

Warngau: 'From three…ah… 29.' (Train 3594 Yes.)

Schaftlach: 'From 29.' (Train 3594 from 29.)

Warngau: 'Yes.' (I repeat, Train 3594 from 29.)

… (Correct.)

At the end of the conversation each man was under the impression that the other man had accepted his train and each then let the train at his station proceed on the understanding that the other was being held for it. The fact that the train numbers were similar probably did not help.

The trains met one another at kilometre 44.312 at 18:31 or 18:32. Train No. E 3591 was running uphill and was able to reduce speed from 75 to 64 kilometres an hour at the last moment, while the speed of train No. E 3594 was slightly reduced from 92 to 88 kilometres an hour. In the collision the two diesel engines were crushed to two thirds of their length, the leading coach of train No. E 3594 was thrown off the track and the second coach of the train ended up on the roof of the engine. In train No. E 3591 the leading coach was crushed at each end and the second was telescoped over half its length. Thirty-eight passengers, both drivers and one of the conductors were killed or fatally injured. Another 122 people were hurt. The damage amounted to DM 4,000,000 and by the end of 1977 compensation had mounted to almost DM 3,500,000.

The collision happened close enough to Warngau for the sound of it to be heard there by Fahrdienstleiter Habart, who had thereby telephoned his colleague at Schaftlach and their mistake had been discovered, too late to prevent the disaster.

After the accident Deutsche Bundesbahn received severe criticism in the Press for the way in which the line and others were worked and in particular the absence of Streckenblock, which would have prevented the accident. As it was, safety on such single lines relied entirely on the correct train-reporting procedures being followed. A railway official admitted that there had been nine previous single-line collisions in the previous ten years in the area, all of which, due to the vigilance of the train drivers or to luck, had been minor.

The two fahrdienstleiters subsequently found themselves in court and were convicted of manslaughter and causing injury through negligence and of dangerous train operation. Hiergeist was given an eight-month suspended sentence and fined DM 3,000, to be paid to the Red Cross, while Habart received a suspended sentence of 12 months and a fine of DM 5,000. Habart received the heavier sentence on the grounds that he had interrupted the report being made by his colleague, that he had not during the afternoon taken the trouble to read the local instructions for Warngau, and that according to those instructions the trains should normally have crossed at his station.

That the two fahrdienstleiters could have found themselves in court, given the casual manner in which they had done their duty, is hardly surprising. What was much more surprising is that Amtmann Stephan Dorfner, who had compiled the working timetable and who had deliberately introduced air crossings to increase the flexibility of the train operation, also found himself in court, and furthermore was convicted and given an eight-month suspended prison sentence. To the author this seems very unjust, as whatever the timetable might say, it was patently obvious that two trains could not pass one another on a single line, except at a crossing place, and thus air crossings should hardly have been a feature of the timetable likely to cause an accident. Such crossings had, in any case, also been a feature of the timetables in the Hamburg, München, Stuttgart and Karlsruhe districts for the previous 20 years. An unsuccessful attempt was also made by the prosecution to indict more senior railway officials for responsibility for the accident.

After the accident all air crossings were removed from the working timetables and Streckenblock was later introduced on the line. [555] [556] [557] [558]

---

[555] *Katastrophen der deutschen Bahnen Teil I* by H.J. Ritzau, pgs. 229-244

[556] *Eisenbahnunfälle in Europa* by Erich Preuß, pgs. 135-6

[557] Wikipedia – Eisenbahnunfall von Warngau

[558] https://www.spiegel.de/politik/vorschrift-verkuerzt-a-ad1ded69-0002-0001-0000-000041251910

CHAPTER 7

RUNAWAY TRAINS AND VEHICLES

A variety of situations can lead to a train or part of a train running away out of control. If there are railway staff in attendance the train or engine can only run away out of control if there is a failure of the brakes or if the brakes are incapable of holding it on a steep gradient. In the case of unattended vehicles, the action of gravity alone is sufficient to precipitate a runaway. In contrast to the situation on a road, a train or vehicles can sometimes run for many miles without mishap, an accident occurring only if and when the runaway meets with another train, comes to the end of the line, or gets to a curve too sharp to negotiate at the speed at which it is travelling.

Thailand's *Northern Express* was derailed somewhere in the vicinity of Mae Tha in Lampang province on the night of Saturday 8th July 1961. [559] The train was on its regular southbound run from Chiang Mai, the northern terminus of the line, to the Thai capital, Bangkok, when it left the rails on a curve as it was speeding down an incline, possibly out of control. Two diesel units and the first six of the eight coaches of the train plunged from a 15-foot-high embankment into the thick jungle that lined the single track. Twenty-four people lost their lives and over 60 were seriously injured. There had been over 200 passengers on the train. The driver, a brakeman and three other railwaymen were amongst those killed. The passengers who died included an American Presbyterian missionary named William James Sinclair Thompson, originally from Albion, Iowa, [560] but who had lived in Thailand for some years; two holidaying Malayan Railway employees, Miss Tan Cheng Ee and Miss Violet Nunis; [561] and several unnamed American technicians who had been working on a mining project in the north of Thailand. [562]

A major disaster occurred near Vrbno pod Pradědem, Czechoslovakia, on the line from Milotice nad Opavou on Sunday 27th August 1961. Some waggons loaded with logs ran away and collided with a diesel railcar, which then caught fire. Eighteen people died and another 17 were injured. After the accident it was ordered that all 'track barriers'

[559] *Pharos Tribune*, Logansport, Indiana, United States, 10th July 1961
[560] *The Cedar Rapids Gazette*, Cedar Rapids, Iowa, United States, 10th July 1961
[561] *The Straits Times*, Singapore, 11th July 1961
[562] *Corpus Christi Times*, Corpus Christi, Texas, United States, 10th July 1961

(scotches?) must be kept closed. [563]

On the evening of Saturday 16th September 1961 Red Nacional de los Ferrocarriles Españoles freight train No. CR 9 was proceeding southwards from Mérida, Spain, with waggons of iron ore from the Burguillos and Fregenal mines. The train was assisted up an incline towards Zafra by a banking engine at its rear. This engine was not coupled to the train and when the summit was reached the banker was detached and ran back down the incline to Los Santos de Maimona station, where it was shunted into a siding. Unknown to anyone the rear four waggons of the freight train, weighing with their load of ore 252 tonnes, had become detached from the train, presumably at some point near the summit, and they began to follow the banking engine back down the incline, moving slowly at first but then gathering speed down the 17‰ (about 1 in 60) gradient. By the time the runaway waggons had reached Los Santos station the points had been reset for the main line after the banking engine had been disposed of and nothing could be done to arrest their progress northwards.

Behind the freight train was omnibus train No. 1728 from Mérida to Sevilla, consisting of a locomotive, front and rear brake vans, a mail van, a second-class coach and two third-class coaches, with 64 passengers, many of them railwaymen and their families. At 22:25 a message was sent from Los Santos to the next station northwards at Villafranca de los Barros to stop the omnibus train, but it had just left. Some Civil Guard motorcyclists reportedly also witnessed the runaway from an adjoining highway and they too hurried to Villafranca but were likewise too late.

The runaway waggons reportedly had reached a speed of 100 kilometres an hour when the collision occurred at about 22:45 at kilometre 47.8. In the collision the second-class coach overrode one of the third-class ones and it was in the latter that most of the casualties occurred. In all 14 passengers, a brakesman(?) and the driver of the omnibus train were killed more or less outright. The fireman was very seriously injured and died in hospital the next day. About 20 other passengers were injured and were taken to the Villafranca Red Cross centre and to hospitals in Mérida and Badajoz in private cars. The Los Santos stationmaster, Carmelo López, was in an agony of despair for several hours knowing that his own daughter, Ana Mari, was a passenger on the train. She worked in the city hall at Almendralejo (just north of Villafranca) and during the afternoons at an electrical appliance store there and was a daily passenger on the train. Eventually she arrived at Los Santos by car, unharmed. Much less fortunate was the operator of the telephone switchboard in Los Santos, who because of her job quickly learnt of the accident. Her sister, Purificación Alvares, who was in charge of the Los Santos telephone exchange, was another passenger on the train and lost her life. Other victims included another woman from Los Santos, two women from Zafra, four students from the Colegio La Granada in Llerena who had been returning from an examination in Mérida, and a doctor from Peñarroya. [564] [565]

---

563 https://de.wikipedia.org/wiki/Liste_von_Eisenbahnunf%C3%A4llen_in_Tschechien#1939%E2%80%931945

564 https://lossantosdemaimona.hoy.es/anos-septiembre-1961-20201015120936-nt.html?ref=https%3A%2F%2Fwww.google.com%2F

An express from Bandung to Jakarta or Jakarta to Bandung, according to different reports, became derailed on Thursday 11th April 1963 near Tagekapu, West Java, Indonesia, the diesel engine, and a goods waggon leaving the rails and plunging into a valley before demolishing a house that had been vacated only the previous day. Because the train was grossly overcrowded passengers were riding both on the locomotive and on the goods waggon. Initial reports spoke of 28 dead and 35 injured but a search for bodies was hampered by the darkness. A further report stated that there were 37 known dead and 126 injured. The driver was amongst those who lost their lives. [566] [567] [568] [569] [570]

Initially the cause of the derailment was not known but a couple of days later the Land Communications Ministry announced that the train's brakes had failed, implying that the train had run away. [571]

On the evening of Wednesday 18th September 1963 two railcars came into a violent collision at the Noir bridge over the Foron River, between Chêne-Bourg station in Switzerland and Annemasse station in France. The collision occurred in Switzerland but within a few metres of the French border. One railcar had left Eaux-Vives station, Genève, at 19:13, bound for Annemasse, and was carrying passengers, mostly French workers from Haute-Savoie province. The other railcar was empty and being shunted at Annemasse when it ran away. The whole of the front of the railcar from Eaux-Vives was demolished and the empty railcar also suffered some damage. About 30 passengers were injured, seven seriously. Most of the injured were taken to Annemasse Hospital, though three were taken to the Policlinique in Genève. [572] The motorman of the Eaux-Vives railcar saved his life when he leapt from his cab upon seeing the other train coming towards him along the single track. He suffered a head injury. The motorman of the runaway train, in his position at the Annemasse end of it, escaped unhurt.

It was estimated that each railcar had been travelling at at least 50 kilometres an hour, the runaway perhaps at rather more than that due to the steep descent from Annemasse.

A rescue operation was quickly organised by the firemen and police of Gaillard commune, in which the accident had occurred, assisted by police from Annemasse and Chêne-Bourg and firemen from Genève. The release of the injured was made more difficult by the fact that the wreckage was entangled with the superstructure of the Noir bridge. During the evening a breakdown crane arrived from Annemasse to remove the damaged trains, so that the line could be reopened the following morning. Floodlights were set up to enable the clearance operation to be continued through the night.

An inquiry into the accident was opened by M. Guéraud, the Société nationale des

---

[565] Federación Castellano Manchega de Amigos del Ferrocarril

[566] *The Holland Evening Sentinel*, Holland, Michigan, United States, 12th April 1963

[567] *Mason City Globe Gazette*, Mason City, Iowa, United States, 12th April 1963

[568] *The Press-Courier*, Oxnard, California, United States, 12th April 1963

[569] *Press-Telegram*, Long Beach, California, United States, 12th April 1963

[570] *The Times*, London, England, 15th April 1963

[571] *The Corpus Christi Caller-Times*, Corpus Christi, Texas, United States, 14th April 1963

[572] *Gazette de Lausanne*, Lausanne, Switzerland, 19th September 1963

chemins de fer inspector in Genève. [573]

A runaway train accident with some similarities to the 1969 Violet Town disaster (see Single Line Collisions chapter) occurred at Bradford Exchange station, England, on Wednesday 3rd June 1964.

The train involved in the accident was the 16:25 passenger train from Manchester Victoria to Leeds, composed of a two-car diesel multiple unit. From Bowling Junction, about a mile out, the line to the terminus at Bradford Exchange fell very steeply at a continuous 1 in 50.

Prior to its forming the Leeds train the empty multiple-unit set had been worked from the nearby Newton Heath depot to Manchester Victoria station by Driver S. Booth, who told the Ministry of Transport inquiry into the accident that everything about the train had been in order and its brakes had worked correctly each time he had used them, with no sign of sticking.

When the train passed Bowling Junction the signals there had been clear for it, as they were at the next box at Coal Shoots, which was switched out. At Mill Lane Junction, the following signal box, however, about half a mile from the terminus, the signals were at Danger as Signalman T.P. Carroll was still awaiting the *Train out of Section* signal from Bradford Exchange box. He said that the train had been travelling at 50 miles an hour when he had first seen it, rather than the usual 30 miles an hour, and he had quickly realised that it was not stopping. There had been no sign of braking as the train passed his cabin and he had immediately sent the *Train Running Away on Right Line* signal to Bradford Exchange box, following it up with a telephone call.

Signalman C. Searle, in charge of the west side of Bradford Exchange box, said that when he had received the emergency bell code and phone call the road had been set for No. 4 platform. All four platforms had been occupied at the time: No. 1 by parcels vans and restaurant cars, Nos. 2 and 3 by parcels vans being loaded, No. 4 by a parcels train and No. 5 by a loaded passenger train, so even had he had time there would have been no better route through which to divert the train. The horn of the train had been "popping" as it had come into view and its driver appeared to be sat normally at the controls. There had been no sign of braking.

Signalman H. Scott, in charge of the east side of Bradford Exchange box, said that he had not been busy when the running away signal had been received from Mill Lane and he had gone to the end of the box to view the incoming train. He agreed that its horn had been sounded and added that he had seen the driver sitting upright with his eyes open and his face apparently impassive. The train had appeared to be running freely.

Driver G.H. Gott of the engine of the stationary 17:47 Bradford to London King's Cross parcels train spoke of hearing the horn of the runaway train, seeing it coming in fast and taking cover, shouting to his fireman to do likewise.

Passed Fireman G.A. Cliffe confirmed his driver's evidence. [574]

The train was still travelling at about 50 miles an hour when it collided with the

---

[573] *Journal de Genève*, Genève, Switzerland, 19th September 1963
[574] Ministry of Transport report

engine of the parcels train. As a result of the collision the cab and leading compartment of the multiple unit were telescoped by the parcels train engine. The driver of the runaway and Permanent Way Inspector Aubrey Mortimer, who had been travelling as a passenger, were killed outright and 18 passengers were injured, one of whom later died in the Bradford Royal Infirmary, where six of the passengers had been detained. [575]

Passenger Guard Joseph Hansell was in charge of the empty stock working from Newton Heath and the Leeds train it formed. He told the inquiry that his driver, John Arthur Smith, had appeared to be quite normal when he had spoken to him at Manchester. The early part of the journey had been normal. His first idea that something was amiss had been the sounding of the train horn from a point between Bowling Junction and Mill Lane Junction onwards. He estimated the speed of the train past Mill Lane Junction as 30 miles an hour, whilst agreeing that it could have been higher. He said that he had begun to be alarmed at the speed and looked forward towards the driver but could not see him because passengers were blocking his view. He insisted that the vacuum gauge had been at zero at that point though there had been no sign of the brakes having been applied. Guard Hansell admitted that according to Rule 148(a) he should have been keeping a careful watch as the train was approaching a terminus. He said that although he had realised that the train was going too fast before he had seen the vacuum gauge at zero, he had not thought to apply the brakes. Approaching the terminus he described the passengers as having looked "panicky" and related that they had been moving back along the train, when he had sat in a seat and braced himself for the coming shock, without doing anything to help or advise the passengers or making a belated attempt to apply the brakes. His only excuse for not having applied the brakes at any point was that there had been "no vacuum." The inspecting officer, Colonel J.R.H. Robertson, commented that he did not find the guard's evidence convincing.

Passenger D.N. Hudson, a traction control designer with the English Electric Company, whom the inspecting officer described as a very good witness, had been sat at the front of the train with a very good view of Driver Smith. He considered that the train had run a little fast earlier in the journey and testified that he had become concerned about the speed after passing Low Moor station and going through Bowling Tunnel to Bowling Junction. At the latter point he described the driver having moved the brake handle from left to right five or six times, though the brakes had not been applied since, he thought, leaving Halifax. The driver had then begun sounding the horn continuously. Approaching Mill Lane Junction, Mr Hudson said that he had realised that the train was definitely out of control and he had begun to move back through the coach, shouting to other passengers to do the same. Driver Smith, Mr Hudson added, had been sitting a little stiffly and not relaxed, as if he was concentrating on the line ahead. No movement had been made by the driver towards the emergency or hand brakes, nor had he risen from his seat.

Passenger H. Semper, an assistant lecturer at Halifax Technical College, stated that he had become alarmed when the brakes had not been applied after passing Mill Lane Junction, where the train was usually diverted through a crossover, then had been reassured when he had seen that the points were set for straight ahead. Later he had realised that the train was running at speed towards an occupied platform and had started towards the rear.

---

[575] *The Times*, London, England, 4th June 1964

Passenger D. Brown, a gardener, who was a regular passenger, spoke of the horn being sounded continuously after Mill Lane Junction while the driver's right arm was making slow push-pull movements. After watching the driver for a short time and realising that the train was not stopping he too had hurried towards the rear.

Passenger J.T.M. Mackenzie, Acting Goods Agent at Halifax, testified that he put the speed of the train at 40-50 miles an hour at Mill Lane Junction and that he had seen that the train was running past the home signal there at *Danger*. Driver Smith appeared to be moving the brake handle to the right three or four times but the brakes had not come on. The horn had also been sounding from Mill Lane Junction. He had watched the driver until the train had reached the start of platform 4, when he (Mackenzie) had dropped to the floor. At that point the driver had still been sat upright and facing forward.

Passenger G. Strong, Goods Agent at Sowerby Bridge, recalled the last brake application as having been at Wyke Tunnel (between Halifax and Low Moor) and he thought that the train had been accelerated in the vicinity of Bowling Tunnel.

Passenger H. Aspinall, Chief Goods Foreman at Halifax, thought that the train had been accelerating near Bradford.

Struck by the similarity of the accident to a runaway buffer stop collision at Royton, England, in 1961, where the driver had mishandled the braking and then lost his head, but had jumped from the runaway train and survived, the inspecting officer asked the passengers who had given evidence of Driver Smith's action to watch a reconstruction of the earlier accident. The witnesses agreed that Driver Smith's action had been similar to those of the Royton driver, but faster.

Driver Smith was 63 years old. He had passed medical examinations in 1960, 1961 and 1962. He had been off sick with sciatica in August and September 1962 and had suffered chest pain, diagnosed as having been due to a spasm of the lower oesophagus, in January and February 1964. An electrocardiogram done at that time had shown no evidence of a coronary thrombosis or of a myocardial infarction. A post mortem examination showed a heart of normal size with a degree of coronary atheroma usual in a man of his age, plus an old thrombus in his right coronary artery that had become recanalised. His death was attributed to the major trauma inflicted on his head and body in the collision. The pathologist added, however, that should death occur within a few minutes of a heart attack no changes in the heart muscle would be visible upon examination.

Discussing the causes of the accident, Colonel Robertson outlined three possibilities:

(a) The brakes of the train had failed.

(b) Driver Smith had mismanaged the brakes and lost control of the train.

(c) Due to a heart attack Driver Smith had lost control of the train.

Dealing with the first possibility the inspecting officer noted that it could only have been caused by a sudden blockage between the driver's brake valve and the leading brake cylinder, or to the connection to the next brake cylinder if the leading cylinder was out of action. No evidence of any such blockage had been found despite a careful search. If such a failure had occurred the driver would have realised it at once, released the dead man's handle, operated the emergency brake valve, belled the guard for assistance or even applied the hand brake. He would also not just have remained sat in his seat, with an impassive face, and made no attempt to warn the passengers of the impending accident by a shout or gesture. The possibility of a sudden brake failure could therefore be ruled out.

The possibility that Driver Smith had badly mismanaged the braking met with the same objections as that of a brake failure, namely that he would not in such circumstances just have remained sat impassively in his seat and not made any attempt to save himself, as the Royton driver had done, or to warn the passengers.

The post mortem had revealed that Driver Smith had suffered a previous coronary thrombosis. Such a person was liable to a severe attack of angina pectoris, causing agonising pain, and it was possible that Driver Smith had suffered such an attack as his train was approaching Bradford Exchange.

Concluding his report, Colonel Robertson concluded that the only possible explanation for the accident was that the driver had indeed suffered a severe attack of angina. Possibly he had passed through Bowling Junction slightly too fast, had been taken by surprise by the Mill Lane Junction signals at *Danger* and had feared, prematurely, that his train was out of control. The sudden stress of this might have precipitated the angina attack. Alternatively the attack might have occurred slightly earlier in the journey and caused the initial speed mismanagement, though this would not explain the severity of the angina so well. In his agony the driver might have held down the safety valve and made some of the right movements, such as sounding the horn, whilst forgetting others with his mind being almost wholly concentrated on his pain.

It went without saying, the inspecting officer continued, that no man who had had a coronary thrombosis should be driving a train. As an examination of the driver by a specialist when he had complained of chest pain had attributed the pain to an oesophageal spasm and not to heart disease no one could be blamed for allowing him to continue driving trains (cf the Violet Town accident). The inspecting officer did not apparently feel disposed to blame the specialist for his wrong diagnosis.

With such a steep approach to Bradford Exchange, Guard Hansell should have been particularly careful to keep a good lookout. He had not done so and had only become aware of the emergency, by his own admission, when he had seen passengers hurrying to the rear. Had he obeyed Rule 148 he would have seen that the train was overrunning the Mill Lane Junction Home signal and would have applied the brakes in emergency at once. If the brakes had been working properly, as seemed likely, he would then have prevented the collision or at least greatly mitigated its effects. Even when he had seen the passengers rushing back he had not applied the brakes. Guard Hansell must therefore bear some responsibility for the severity of the accident.

The inspecting officer decided that the accident had resulted from a combination of a number of unfavourable factors that were unlikely to recur and made no recommendations at the end of his report. [576]

Fourteen people died and another 18 were injured when an overloaded freight train from Kladno, carrying goods from a local steelworks, collided with a stationary

[576] Ministry of Transport report

passenger train at Zákolany, Czechoslovakia, on Saturday 25th September 1964. [577] [578]

At least 33 people lost their lives and a number of others were injured when a passenger train crowded with holidaymakers collided with a freight train near Amatilán, Guatemala, on Sunday 24th January 1965. The brakes of the passenger train had reportedly failed. [579]

Early on Friday 26th February 1965 a passenger train bound for Khartoum (?) and consisting of an engine and four coaches suffered a brake failure, ran away, and collided head-on with a goods train bound for Port Sudan (?) near Haiya, Sudan, about 100 kilometres from the port (the destinations of the two trains are unclear, each being reported as heading for Port Sudan and for Khartoum in different press reports). The goods train was carrying oil seed and oil cake. The force of the impact hurled its waggons on top of the passenger coaches and early reports stated that over 100 people were trapped in the wreckage, which caught fire after the collision. [580] 'Groans and screams heard through the night by rescue teams as they groped through the wreckage were not heard by dawn,' a correspondent wrote, implying that many of those trapped had become unconscious or had died before they could be released. The same correspondent stated that at the time he wrote on Saturday, at least 46 bodies had been recovered. Twenty-nine survivors were injured. [581]

Three railwaymen had a lucky escape from death when a freight train ran way between Katoomba and Wentworth Falls, New South Wales, Australia, on Friday 16th July 1965. The train, consisting of an electric locomotive and 44 waggons, was in charge of Guard Gordon Pereia with Driver Stephen Kipp and Fireman David Wolfenden. It had left Katoomba at 4.30 a.m. and was approaching Leura when its brakes failed. The train tore through Leura as Driver Kipp sounded blasts on his horn to warn lineside residents. The speedometer on the engine was reportedly indicating a speed of 95 miles an hour as the train continued down a steep incline. When the train reached a curve at Wentworth Falls the engine was travelling too fast to negotiate it, it broke free from the train behind and leaving the rails overturned at the end of the station platform. Behind the engine the waggons of the train piled up in a 50ft high pile of wreckage, while houses near the line were bombarded with ballast and pieces of broken sleepers. The line was blocked for 12 hours.

Driver Kipp was pulled unconscious from the engine cab and was confined in Katoomba District Hospital overnight with concussion, head cuts and slight back injuries. He reportedly refused to discuss the accident.

Guard Pereia wandered away from the wreckage and ended up at a nearby house, dazed and bleeding.

577 https://de.wikipedia.org/wiki/Liste_von_Eisenbahnunf%C3%A4llen_in_Tschechien
578 https://www.waymarking.com/waymarks/WMXT0V_Train_Station_Zakolany_Czech_Republic
579 *Ironwood Daily Globe*, Ironwood, Michigan, United States, 26th January 1965
580 *The Times*, London, England, 27th February 1965
581 *Wisconsin State Journal*, Madison, Wisconsin, United States, 28th February 1965

Fireman Wolfenden was found standing beside the engine, with numerous cuts and suffering from shock. He and the guard were also taken to hospital, treated, and released. [582]

A passenger train and a freight train, which was also apparently carrying about 50 passengers, collided at 08:25 on Wednesday 16th February 1966 between Kaštel Stari and Sadina, Yugoslavia, not far from Split. It was the freight train that ran away; according to one account 19 waggons went running away down an incline during shunting operations. [583] Eighteen waggons were overturned in the ensuing collision, as were four coaches of the passenger train. Some of the occupants of the passenger train, seeing what was about happen, jumped from the train windows, and thereby escaped but others who did so were crushed by the derailed vehicles. Rescue operations, using cutting equipment to free passengers from the wreckage, were hampered by a fire that broke out after the collision, fed by saltpetre, sulphur and coal being carried on the goods train. Twenty-nine passengers were initially reported as having been killed and at least 30 as injured, 20 of them seriously and five critically. Almost all the victims were workmen. Hundreds of people responded to a radio appeal for blood donations and attended clinics in Split and other local hospitals. The line was of course blocked and passengers travelling between Zagreb and Belgrade were taken by bus between Split and Labin. The Ministry of Transport appointed a commission of inquiry to investigate the disaster. [584]

The final death toll was 31. In April a tribunal at Split found four railwaymen responsible for the disaster; they were sentenced to terms of between two and 15 years' imprisonment. [585]

A runaway accident in West Germany on Thursday 17th November 1966 had a somewhat unusual cause. When a three-car diesel multiple unit arrived at Kelkheim-Hornau station on the Frankfurt (Main) – Königstein line its driver alighted from his cab to relieve himself. What he failed to do before he left his cab was apply the brakes, while the line fell steeply towards Frankfurt, the gradient between Kelkheim-Hornau and the next station at Kelkheim being 1.77% (1 in 56). While the train was unmanned it began to run away down the incline, through Kelkheim, Kelkheim Münster, Liederbach and Liederbach Süd stations and halts, reportedly reached a speed of 100 or even 130 kilometres an hour, according to different reports. At Kelkheim Münster an unsuccessful attempt was made to derail the train. There was only one passenger on it, a 22-year-old man, who for some reason does not appear to have pulled the communication cord, possibly he was asleep.

The runaway ended its journey at Unterliedbach when it collided with steam-hauled passenger train No. P 2177. The driver and fireman of the steam locomotive had time to jump clear but six passengers in train No. P 2177 and the sole passenger in the runaway train were killed and at least 66 people were injured. The rear car of the runaway was demolished in the collision and the three leading coaches of train No. P 2177 telescoped

---

582 *The Canberra Times*, Canberra, Australian Capital Territory, Australia, 17th July 1965
583 *The Bridgeport Telegram*, Bridgeport, Connecticut, United States, 17th February 1966
584 *La Stampa*, Torino, Italy, 17th February 1966
585 *Journal de Genève*, Genève, Switzerland, 7th April 1966

one another. [586] [587]

A runaway accident in Canada on the evening of Friday 3rd February 1967 was much less serious than it might have been. It began with a collision between a two-car passenger train of the Canadian National Railway and a breakdown lorry on a level crossing near Windsor, Quebec. The crew of the passenger train alighted after the collision to inspect the damage to their train but failed to properly secure it before they did so, and it began to run away down a long falling gradient, covering about 7 miles before it finally came to a stand near Sherbrooke. The train crew ran to a nearby yard to secure an engine and pursued the runaway but the train had stopped before they reached it. The lorry driver was injured and was taken to hospital; none of the 20 train passengers was hurt. [588]

A major catastrophe occurred on the Southern Railway at Kuppam station, India, at about 21:50 on Sunday 21st May 1967. Train No. 26, the Up Bangalore City – Cochin Harbour Terminus Island Express, ran into the platform line at high speed and became derailed in a sand hump. Fifty-eight passengers and five railwaymen lost their lives and 73 passengers and seven railwaymen were injured, 20 passengers and three railwaymen 'grievously.' The train consisted of 4-6-2 steam locomotive No. WP 7075 and 11 coaches, in charge of Guard C.R. Ekambaram, with Driver G. Jeremiah, First Fireman R. Bhaskaran and Second Fireman B. Parthasarathy.

Guard Ekambaram stated at the inquiry into the accident that his train had started on time. The vacuum gauge in his van had indicated 40cm, though he had recorded it as 48 on the brake certificate. He thought that the brake blocks had been binding at Krishnarajapuram, and at Whitefield the second fireman had been checking for vacuum leaks. Approaching Tyakal the train had stopped suddenly near the Home signal and the vacuum had dropped to zero; then the train had restarted. At Bangarapet the brakes had been examined. Approaching Kuppam his driver had whistled for brakes and he (the guard) had applied the hand as well as the vacuum brake, then he had felt a sudden shock.

Driver Jeremiah said that leaving Krishnarajapuram station he had felt the train dragging heavily and at Whitfield he had sent the second fireman to check for leaks. The fireman had not found any but said that the guard had spoken of the brakes binding on the rear two coaches. At Devangonthi he and the second fireman had found a few pistons jammed. Approaching Tyakal the vacuum had suddenly dropped, the train had stopped, and he discovered a crack in the vacuum pipe near the 'T' joint above the footboard, which he sealed by applying grease. At Tyakal station he had asked for a fitter to be ready to attend to the vacuum pipe at Bangarapet. At the latter place he had also requested that another engine be available at Kuppam just in case it was needed. Approaching Kuppam, down a series of 1 in 100 falling gradients, he had found that the train was not slowing in response to the vacuum brake, had applied the steam brake, reversed the engine, asked the fireman to put on the hand brake, and whistled for the

[586] https://bahnblogstelle.com/16375/17-november-1966-50-jahre-nach-dem-zugunglueck-von-liederbach/

[587] https://www.gralsmacht.com/wp-content/uploads/2018/05/2731-artikel.pdf

[588] *The Lethbridge Herald*, Lethbridge, Alberta, Canada, 4th February 1967

guard to apply his brake. He had tried to recreate vacuum but the train pipe had separated at the joint. He had kept whistling to warn the people in the station and his train had run into the sand hump. There had been, he said, 50cm of vacuum on leaving Bangalore City and between 45 and 50 on the journey. He blamed the failure of the vacuum brake on the vacuum cylinders being inoperative on some coaches, some cylinders having excessive piston travel and the release valves of some cylinders having been left in the released position by the examiners at Bangarapet.

First Fireman Bhaskaran corroborated his driver's evidence.

Second Fireman Parthasarathy stated that he had checked for leaks at Whitefield and Devangonthi. He had not released the brakes at any station.

Carriage and Waggon Fitter T. Rathinam testified that he had released the binding brakes at Bangarapet. The pistons had dropped freely. He had not kept the release valves pressed down by any irregular means.

Power Controller A. Aiyasamy told of Driver Jeremiah reporting by telephone from Tyakal that the train vacuum pipe had broken and he had secured it with a wire. The driver had requested that an engine fitter be available at Bangarapet and that engines be available for an engine change or assistance at Bangarapet and Kuppam, though he had not spoken positively of trouble with brake power.

Power Controller A.S. Krishnamurty, who had relieved his colleague at 21:00, said that he had been advised of the train stop at Tyakal and the driver had told him over the telephone at 21:08 that he might need a replacement engine at Kuppam.

Section Controller S.A. Hameed gave evidence of having been advised by the assistant stationmaster at Tyakal that the train had stopped there.

Section Controller V.N. Marghabandhu described being informed by the assistant stationmaster at Kuppam that the train was entering the station at high speed and then being told that an accident had occurred.

Fitter Chargeman-in-Charge M.S. Gopal at Bangalore described attending to the head air valve joint on the morning of the accident.

Assistant Stationmaster D. Padmanaban at Tyakal said upon the pointsman at the west cabin telling him that the train had stopped at the Home signal, he had gone to investigate and twice asked the driver what was wrong but received no answer. He had heard the driver asking Control for the banking engine to be available at Kuppam and for the train not to be stopped at the Bangarapet Outer Home signal. The train had left at 20:42.

Assistant Stationmaster R.S. Gnanavadivelu at Bangarapet described receiving messages from the guard of the train asking for a train examiner and from the driver for a fitter chargemen. The engine fitter had not turned up but the carriage and waggon fitter had attended to the brakes.

Assistant Stationmaster S. Subramanian at Kuppam detailed the train entering his station at a terrific speed, its driver whistling for the guard's brakes. It had passed the Starting signal at Danger and crashed into the sand hump.

Fitter Chargeman C. Sundaravadivelu described fixing a leak in the vacuum pipe in the days before the disaster.

Train Examiner V.C. Kurien at Cochin Harbour Terminus stated that he had found the vacuum cylinders on three coaches, two of which had the stock of the train involved in the disaster, dummied (inoperative) the day before but had not had time to deal with them.

Senior Train Examiner P.S. Sripathy at Bangalore City stated that he had dealt with the faults on the two coaches of the express before it had left. He could not explain why the same cylinders had again become inoperative on the journey to Kuppam.

Continuing his report into the disaster, Additional Commissioner V. Venkataramayya gave a detailed account of his observations after the accident and described the events leading up to it, as per the testimonies of the various witnesses. He concluded that the main vacuum pipe must have been leaking after leaving Krishnarajapuram, causing the brakes to bind. After the stop at Tyakal, when the fireman must have released all the brakes by pulling the cylinder release valves, the driver must have continued with only 15 or 20cm of vacuum in the train pipe, most of the release valves would not have returned to the closed position and there would have been no vacuum in the cylinders. By driving cautiously down the falling gradients the driver had been able to stop at Bangarapet using just the steam brake, and this must have given him more confidence to continue to Kuppam, where he could have changed the defective engine with one that had banked a Down train to there, without delaying any other service.

The accident had therefore been caused by the inability of Driver Jeremiah to control his train, because he had started from Tyakal with insufficient vacuum, in contravention of Rules 3 (b), 5 (c) and 6 (g), and he was mainly responsible for it.

Guard Ekambaram was partly responsible for the accident. He should have detained his train at Tyakal upon finding that it had insufficient vacuum.

Senior Train Examiner Sripathy had failed to rectify the defective brake cylinders or give a correct brake power certificate but this had only indirectly contributed to the accident.

The additional commissioner made three recommendations:

(1) The location of the train vacuum pipe should be altered to prevent it being struck by the reversing lever (the cause of the breakage at Tyakal) and a thicker pipe should be used.

(2) The anti-telescopic feature of the coaches of the express, which had not been very effective, needed further investigation.

(3) The connecting pipes between the vacuum and steam pipes has been removed in the engine sheds at the request of drivers. This should not be permitted.

Driver Jeremiah and Guard Ekambaram had subsequently stood trial at Kuppam in connection with the disaster and were both acquitted, the report noted. [589]

A major disaster in Spain on Friday 15th March 1968 involved an air-conditioned diesel railcar set, a Tren Español Rapido, or TER. TER No. 251 from Madrid Norte, the

---

[589] Government of India Ministry of Tourism and Civil Aviation (Commission of Railway Safety): Report on Accident to No. 26 Up Bangalore City – Cochin Harbour Terminus Express at Kuppam Station of Southern Railway on 21st May 1967

train involved in the accident, consisted of two two-car units, No. 9732, bound for Coruña, in front of No. 9736, bound for Vigo. The motor car units were at the front and rear of the train with the two trailer units in the middle. The TER had left Madrid at 12:35 with about 200 passengers and a crew of 12, including the driver, assistant driver, conductor and two sets of catering staff.

Between Robledo de Chavela and Santa Maria de la Alameda stations the northbound line was closed for engineering work, with single-line working over the southbound line.

Approaching from the opposite direction, from either Ávila or Venta de Baños, according to different accounts, was a series 305 shunting engine, No. 10510, described in newspaper reports as a tractor. After the engine had passed Las Navas del Marqués station its driver realised that its brakes had failed. It should have stopped at Santa Maria to give priority to the TER over the single line but was unable to do so and its driver sounded the alarm on his horn as the engine reached the station. The Santa Maria stationmaster then attempted to derail the engine as a last resort, by shifting the points at the exit from the station, but the engine just ran through them. The engine and the TER collided at a point between Santa Maria and Robledo stations at 13:56. Having just finished their meals, the TER passengers were being served coffee at the moment of the accident. The runaway engine had reportedly reached a speed of 100 kilometres an hour at the point of collision, while the TER was travelling at about the same speed. In the collision the leading car of the TER climbed up over the runaway engine and leaking fuel from its tank started a fire, which spread to the second unit of the train. The driver, assistant driver and conductor of the passenger train, the driver of the shunting engine and probably 24 passengers lost their lives, including two who succumbed to their injuries on the way to hospital. According to some accounts four railwaymen riding on the runaway engine died; the accounts agree that the total number of fatalities was 28. At least 50 passengers were injured, about 20 or more seriously. Helicopters and road ambulances took the more badly hurt to El Escorial. The two rear units of the TER suffered very little damage and they returned to El Escorial with the slightly injured victims and presumably also those who had escaped unhurt. After the accident unit No. 9732, which was all but destroyed by the impact and fire, was scrapped on the spot but the engine, which was relatively little damaged, was repaired and returned to service.[590] [591] [592]

The Mount Washington Cog Railway in the United States was a cog (rack-and-pinion) line that climbed three miles from Base Station (elevation 2,700 ft) to the summit of Mount Washington (elevation 6,288 ft). It had a maximum gradient of 37% (about 1 in 3). Opened in 1869 the railway could boast in August 1967 that it had never had an accident in which a passenger was killed, which was strictly true, though a railway photographer had lost his life in a runaway accident in 1929. Its good safety record was to end on Sunday 17th September 1967.

The train involved in the accident was the 5.10 p.m. departure from the summit, the last working of the day, known as the Sunset. It consisted of locomotive No. 3 Base Station, and a single new, aluminium car, named Chumley 11. The locomotive was at the

---

[590] https://redondelanexpress.blogspot.com/2013/08/hace-45-anos-se-produce-el-primer.html
[591] Federación Castellano Manchega de Amigos del Ferrocarril
[592] *Railway Disasters of the World* by Peter Semmens, pg. 178

lower end of the car in both directions, pushing the car up the incline while the car, which was not coupled to the locomotive, rested against it on the descent. Riding on the engine were Engineer Gordon Chase, 56; Trainee Engineer Charles Kenison, 18; Fireman Guilliaem (Rusty) Aertsen, aged about 25; and off-duty Brakeman Peter Carter, 20. Peter's brother Nathan, 23, was the actual brakeman of the train and Peter's girlfriend Julia Swenson was a passenger in the car. The car had 56 seats but was carrying 81 passengers and two railwaymen that evening.

The train had descended a mile when the accident occurred at Skyline Siding, New Hampshire (elevation 5,600 ft). As their engine reached the siding the engineer, trainee engineer and fireman all stated later that they had failed to see that anything was wrong, but in the words of Peter Carter, he "felt the locomotive bounce into the air and return again to the trestle." The engineer and his trainee had applied the brakes and the locomotive slowed but then it seemed to have been bumped by the car and was thrown from the trestle on which the railway ran.

Brakeman Nathan Carter thought that the speed of the train had been normal when the accident occurred. He saw the engine "pop up and down," pull away and flip off the trestle. He applied the brake and hoped that the car would stop; instead it picked up speed and reached between 20 and 40 miles an hour when it, too, left the trestle. Eight passengers lost their lives and 70 people were injured; Brakeman Carter suffered a dislocated shoulder.

The investigation into the accident found that the engine had struck a flip rail (a short steel hinged rail that supported cars as they moved on or off the track at the points) lying across the cog rack at an angle of nine degrees. The engine striking the flip rail had been lifted up into the air and off the cog rack and running rails, rendering the cogwheel and brakes useless. The engine had then run out of control for 68 ft before leaving the trestle. The passenger car had then run away and covered 516 ft before it too left the trestle, many of its passengers jumping from it on the way. The car had fallen a couple of feet onto the rocks, slid along the ground about 60 ft, then hit a large, jagged rock. Freeing himself from the engine Peter Carter ran down to the wreckage of the car to find his girlfriend shaken up but not badly hurt. She had given up her original seat at the front the car to a couple, who had been both killed. Carter and his girlfriend then worked to help the injured.

After the alarm had been raised and a rescue operation had begun, four trains were used to bring the passengers down to Base Station, the first bringing the uninjured passengers, the next two the injured and the last the bodies of the dead. The injured were taken to Littleton Hospital.

There were suggestions that the initial accident had been due to hikers having left the flip rail in the wrong position (the same suggestion had been made to explain a minor accident in 1946). There was no evidence that this had been the case and in any case, it was the duty of the engine crew to check that the line was safe, while given the presumed low speed of the descending train it should easily have been stopped short of any obstruction had it been noticed in time. Five reasons were given for the failure of the engine crew to see the obstruction:

(1) Restriction of the view ahead by the tender of the steam engine, which was working tender first.

(2) At the time of the accident (5.30 p.m.) the sun had been low in the sky, possibly also interfering with the view of the engineer.

(3) The engine crew might simply not have been paying attention and had distracted one another.

(4) The engine, being driven by the trainee engineer, might have been going too fast.

(5) The possible incompetence of Engineer Chase, who had previously been fired because of his lax attitude to safety.

The accident was investigated by the New Hampshire Public Utilities Commission (PUC) whose preliminary report listed five findings:

(a) It is our conclusion that the cause of the accident was human error and not due to any structural or mechanical deficiency of the railroad.

(b) It has been established beyond doubt that the switch at the Skyline Station was not in the proper position at the time of the mishap. It is our conclusion that it was the duty of the crew of the involved train to notice this fact and to stop the train prior to its entering the switch.

(c) While the train crew responsible for changing the switch insists that the switch was properly closed by them, and while conjecture has been raised that possibly a hiker changed the position of the switch afterward, we can come to no positive conclusion as to the exact facts of the matter in this preliminary investigation.

(d) It is our conclusion that there was no relationship between the number of people on the train and the cause of the accident.

(e) It is our conclusion that the braking capacity of both the engine and the passenger car were more than adequate to stop the train, had the cog not been disengaged from the rack.

The PUC stated four requirements for the future operation of the line:

(1) Trains ascending the mountain must stop before passing over any switch, following which the brakeman shall examine the switch, and the train will proceed upon his motion only after he is satisfied that it is properly set.

(2) All descending trains shall stop before passing over any switch, and both engineer and fireman shall make a physical inspection, acknowledging to each other that it is safe for passage and that passage through the same be made at very slow speed, not to exceed 1½ miles per hour.

(3) The brakeman, while ascending the mountain, shall be stationed in the forward end of the car in a position to observe and ascertain that the Cog rack, rail, and structure is in normally safe condition without obstruction.

(4) The descending trains, while under way, shall have a qualified engineer and fireman in their proper positions in the locomotive and a qualified brakeman at the brake control in the car with no other duties to perform.

The railway reopened already in October and remains very much in use in 2023. Skyline Switch was later removed. [593]

---

[593] https://digitalcommons.dartmouth.edu/cgi/viewcontent.cgi?article=1181&context=appalachia

A news report from Belgrade radio detailed a major disaster in Yugoslavia on the night of Wednesday 30th July 1969. Shunting operations were in progress at Gostivar station when in some manner the shunters lost control of three 40-tonne tank waggons, which escaped onto the main line and began to run away down an incline in the direction of Tetovo, 30 kilometres way. Meanwhile a two-car diesel multiple-unit passenger train had left Tetovo after having discharged about half its occupants. The collision that ensued was of such violence that the leading unit of the passenger train, which was almost 11 metres long, was reduced to a length of only two metres. Twenty-nine people were known to have died and another 17 were injured, many seriously. Only 18 of the fatalities had been identified at the time of the report. Several railwaymen at Gostivar were taken into custody pending possible negligence charges. [594]

A serious runaway accident occurred in Sweden on Monday 10th November 1969. Passenger train No. 4565, consisting of an engine and six coaches, was stopped at the *infartssignal* (Home signal) at Upplands Väsby, which was at Danger. Its driver applied the engine brake and alighted to use the signal post telephone. For some reason the driver was out of his cab for an appreciable time, during which the train began to roll back down the hill. In making an attempt to regain the cab the driver reportedly tripped and fell onto the track. Apparently having run back some distance, the train then reversed direction and began to roll forwards, past the *infartssignal,* in the process clearing the previous block signal for the following long-distance train, No. XI 2048. The train then ran back again and collided with train No. XI 2048 at Skavstaby, between Upplands Väsby and Roserberg. Four passengers died in the accident. [595] [596]

On Friday 27th March 1970 a freight train left Pau in France for Canfranc in Spain, over the secondary line through the Pyrenees between these two places. It consisted of two Société nationale des chemins de fer français (SNCF) Class BB electric locomotives and nine waggons of corn. Lescun-Cette-Eygun station was passed, after which the train had to surmount a 35mm/m (about 1 in 29) incline. It was a cold morning and the Urdos electric substation was out of action, leading to a fall in the voltage between the Bedous and Forges d'Abel substations, on the steepest part of the line. At 06:45, as the train exited Sens Tunnel, it slipped to a standstill on the frost-covered rails. Having applied the rheostatic brakes the two drivers alighted from their locomotives and began to place pieces of ballast on the rails in an effort to gain some adhesion. Unfortunately the Forges d'Abel substation then tripped out, the brakes of the engines released and the train began to run back down the bank. The drivers could do nothing about it. Gathering speed the train passed through Lescun-Cette-Eygun station at over 100 kilometres an hour, over an automatic level crossing before the barriers had had time to descend – fortunately there were no vehicles using it at the time – and got as far as the Estanguet bridge over the

[594] *The Times*, London, England, 1st August 1969

[595] https://www.goldnet.nu/trainwrecks/index.html

[596] https://www-jvmv2-se.translate.goog/forum/index.php?mode=thread&id=112214&_x_tr_sl=sv&_x_tr_tl=en&_x_tr_hl=en&_x_tr_pto=sc

Gave d'Aspe River at Accous, France. This was situated on a sharp curve and a waggon caught the metal superstructure of the bridge, causing it to collapse, together with the train. No one was hurt and a bus service was provided between Bedous and Canfranc for train passengers.

The bridge could easily have been rebuilt but the SNCF, facing a serious financial deficit at the time, decided that it could not afford to replace it, and the line was closed instead. [597]

A tragedy occurred in Chile at 13:30 on Thursday 11th March 1971. A special train carrying 350 passengers, many of them children, was organised by the Valparaiso Central Unica de Trabajadores (CUT) (the Central Workers Union) to take them from La Serena to Vicuña, where they were to visit the Gabriela Mistral Museum. The trip to Vicuña and the museum visit were uneventful. After lunch the engine of the train was turned on a turntable, the passengers reboarded the train for the return journey and the crew went into the stationmaster's office to eat and to work out the details of the return trip. While they were away the train began to move off. The passengers, unaware that no one was on the engine, naturally thought nothing of this. When the train crew realised what was happening they attempted to reboard the train but failed. Quickly gathering speed on the steep gradient the train was reportedly travelling at 120 kilometres an hour when it reached a sharp bend in the line near the Elqui River, at a point near Huayihuaica station, about eight kilometres from Vicuña, and here it was derailed and fell into a ravine. The engine and five coaches left the track and were wrecked. The accident occurred near the old town of Gualliguaica.

Twelve passengers were killed and 53 were injured, 11 very seriously, according to one source. Another source stated that there were 13 dead and 60 hurt. Initially it had been reported that there had been 25 deaths.

There were allegations that a six-year-old boy had climbed onto the engine and had accidentally released its brakes while the train crew were having their lunch. [598] However an investigation determined that it had not been the practice to apply the hand brakes to secure the train, even when it was left standing for a long period, and it seems possible that the Westinghouse brakes had simply leaked off during the long interval. [599]

The driver, assistant driver and the Vicuña stationmaster were arrested. Initially the Huayihuaica stationmaster and train conductor had been detained. [600]

Eighteen children lost their lives in an accident near Namwon station, South Korea, on Wednesday 13th October 1971. Fifteen of the children, 14 girls and a boy, were among a party of 232 elementary school pupils on an excursion from Namwon to the port of Kunsan. All were sixth-graders (aged 11 or 12). The other three children were high school boys bound for a school at Chonju, about 50 kilometres away. Forty-nine others

---

[597] http://transpyreneen.free.fr/histoire/histoire6.htm
[598] *The Times*, London, England, 13th March 1971
[599] Wikipedia – Gualliguaica rail accident of 1971
[600] *La Vanguardia*, Barcelona, Spain, 13th March 1971

were injured, 16 seriously. [601]

The passenger train had gone about a kilometre and a half from Namwon when it stalled on an incline. Behind it was a train of oil tankers, the engine of which was detached to push the passenger train up the bank. Meanwhile the passenger train began to run back (some accounts state that its driver had deliberately set back, perhaps to get a run at the incline) and train and assisting engine collided. [602] The rear coach of the passenger train was derailed in the collision and then telescoped by the penultimate coach. [603]

Every 4th October there are festivities at Real de Catorce, Mexico, in San Luis Potosi state, to commemorate the feast of St Francis of Assisi, who died on 3rd October 1226. In 1972 large numbers of pilgrims came to Real de Catorce, not only from San Luis Potosi state but also from Coahulla, Nuevo León and Tamaulipas states. It was a train carrying returning pilgrims to Saltillo, the capital of Coahulla state, which was involved in an accident that was possibly the worst in Mexican history, when it was derailed at Puente Moreno, only seven kilometres short of its destination. The Ferrocarriles Nacionales de México train had left Real de Catorce with 1,564 fare-paying passengers, not including babies and small children who did not require tickets. Officially it was stated that 234 people had died with an unknown number of injured. Unofficially it was estimated that there had been over a thousand deaths, with the bodies of many of the victims having been entirely consumed in the post-derailment fire.

The train had two engines, Nos. 8405 and 8408, and 22 coaches. Driver Melchor Sánchez Echeverría had reported problems with his engine at La Encantada station. As the train reached a curve and the beginning of an incline its brakes failed, the train gained speed and it became derailed at 23:25. A tachograph showed a speed at derailment of 120 kilometres an hour; the maximum permitted speed at that point was 60 kilometres an hour. Engine No. 8408 fell down an embankment, the two leading coaches were telescoped into engine No. 8405 and other coaches piled up in a heap behind. The middle and rear coach were derailed but not badly damaged. Fire then broke out in the wreckage and soon the heat made it impossible to approach it.

The magnitude of the disaster and the number of injured overwhelmed the hospitals and clinics of Saltillo and some of the casualties were transferred to Monterrey and Torreón. There were insufficient ambulances to take all the injured to hospital so lorries, vans and private cars had to be brought into use. At the hospitals a lack of beds meant that some patients had to be treated on the floors, while the pressure of work meant that some wounds had to be sutured without anaesthesia.

An amphitheatre (?) was filled with the bodies, parts of bodies and charred corpses of victims of the disaster. Some were later identified by grieving relatives, while many could not be, and in some cases entire families had died, leaving no one left alive to recognise them. The unidentified bodies were subsequently buried in a common grave while the official figure of the number of dead was based solely on the number of identified bodies.

---

[601] *The Straits Times*, Singapore, 14th October 1971

[602] *The Cedar Rapids Gazette*, Cedar Rapids, Iowa, United States, 13th October 1971

[603] *Appeal Democrat*, Marysville, California, United States, 13th October 1971

María de la Cruz Vázquez, a surviving passenger, recalled later, "Thank God I'm alive. I was barely ten years old when the accident happened. Fourteen of my relatives died. I was trapped in the twisted metalwork for three days and gangrene forced them to amputate one of my legs. My brothers died. I knew nothing after the accident, not even if it was day or night."

Maria Ines Ramos, another passenger, said she had been riding in the second coach and been awakened by the crash. The coach had immediately burst into flames. Then she had lost consciousness and an unknown man had pulled her from the wreckage. She suffered concussion, a broken arm, and extensive bruises.

A newsman who flew over the accident site said, "It looked like at least half the cars were off the track and none of them seemed to be seated normally. Three or four of the cars were upside down, some were piled on top of one another. About a dozen ambulances were standing by with other emergency vehicles." [604]

There was controversy as to the causes of the catastrophe. Víctor Manuel Villaseñor, the director of the Ferrocarriles Nacionales de México, travelled to Saltillo to open an investigation into the derailment. Initially it was officially stated that the accident had resulted from the train crew having been drunk, in addition to which it was alleged that they had had women on board (the engines?) 'for fun'. The day after the accident the train crew were arrested and taken to the state penitentiary in Saltillo. When blood tests on the train crew proved negative for alcohol Señor Villaseñor released a statement, published in a newspaper on Saturday 7th to this effect. Railway company officials then made an attempt was to cover this up, but Luis Morales Benavidez, the director of the Hospital Ferrocarrilero in Saltillo, refused to sign a statement that the train crew had been drinking on the grounds that the laboratory tests had detected no alcohol. For his impudence he was dismissed; later he was reinstated. Statements that the train crew had had women with them also proved untrue. Undeterred by these setbacks the directors of the railway accused the trainmen of disruption to communications, manslaughter, causing injuries, and damage to property due to gross negligence. As a result Drivers Mechor Echeverría and Jesús Rocha Serna, Firemen Ignacio Gonzáles and Garcia Carrizales, and Guard Juan Picón Alvarado were sentenced to terms of imprisonment ranging between eight and 18 years. Guard Vincente Martinez Torres was likewise found guilty of gross negligence but he had died in the accident, while Driver Serna apparently fled after the accident, [605] [606] so presumably was sentenced *in absentia*. (The names of the railwaymen vary according to different accounts.) On 11th October the accused railwaymen were transferred to a prison at Piedras Negras after an angry crowd had besieged the penitentiary at Saltillo, threatening to lynch them. Several years later the railwaymen were all released on the grounds that their guilt had not been proven.

As to the probable real cause of the catastrophe it appears that, not for the first time in such circumstances, the train had been grossly overloaded, with an excessive number of tickets having been sold for it at Real de Catorce. As a result the railway company had been obliged to add extra coaches to it, and the only rolling stock available had been old coaches, in poor condition, from Aguascalientes, which had been waiting to be taken to

[604] *The Arizona Republic*, Phoenix, Arizona, United States, 7th October 1972

[605] https://periodicolavoz.com.mx/coahuila/monclova/la-tragedia-del-puente-moreno/97336

[606] https://www.capitalcoahuila.com.mx/especial/puente-moreno-a-46-anos-de-la-tragedia

Mexico City to be scrapped. Even then, the coaches had been overcrowded, each having a seating capacity of 88 but carrying an average about 120 passengers. Señor Villaseñor put the financial cost to the railway, in terms of the loss of two locomotives and 11 'recently acquired' coaches alone, not including medical expenses or compensation to victims, at 7,000,000 pesos [607] but there were allegations that this figure was exaggerated, and a former railwaymen, Enrique Navarro, stated that all the rolling stock of the train had been 'museum pieces.' [608]

Another cause of the runaway was stated to have been the presumably accidental closing of the brake pipe somewhere along the train, preventing its brakes from working. [609] [610]

A major disaster on the Eastern Railway in India on Friday 23rd August 1974 began with the derailment of Up goods train No. EC 1211 at 06:30 at Km 412/1-5 between Dilwa and Nathganj block huts on the Grand Chord line. Both lines were blocked. The Gomoh Accident Relief Train (ART) proceeded to the site via the Down line with a 120 tonne diesel crane leading towards Gurpa. During the rerailing operation the crane was detached from its match truck while four of the derailed waggons fouling the Down line were rerailed and coupled to the match truck. The engine that had brought the ART to the site had meanwhile been detached and had its fire dropped. At the completion of the rerailing at 18:10 the crane set off towards Gurpa, drawn by engine No. 12230, which had come from there. About 40 minutes later the ART engine, coupled to another engine, bumped against the remainder of the ART train, upon which it, the match truck and the four waggons coupled to the match truck, 12 vehicles in all, began to run away down the 1 in 85 gradient at that point. After travelling about 12 kilometres the runaways caught up and collided with the crane, between Yadugram block hut and Gurpa. In the collision the four rerailed waggons were smashed and the match truck and other vehicles of the ART were more or less damaged, as was the crane. The crane and engine No. 12230 were pushed down the line almost to Gurpa. Fourteen railwaymen riding on the platform of the crane and the second fireman of the engine pulling it were killed in the collision, two people were seriously injured and five were slightly hurt.

Transportation Inspector S.D. Singh told the inquiry into the accident that the ART had arrived at the derailment site at 12:40. Wooden wedges had been placed under the Gurpa end wheels of each vehicle and hand brakes had been applied in the guard's van, then the crane had been detached. Four derailed waggons had been placed on the Down line by the crane, their wheels wedged and brakes applied, while a fifth waggon was thrown clear of the line. At 16:15 engine No. 12230 had arrived from Gurpa to take the crane to there, so that it could return on the Up road to clear that line. Nathganj block hut had been passed at 18:18, Baskatwa block hut at 18:42 and Yadugram block hut at 18:52. After the crane had passed Gurpa Up Distant there had been a severe bump at its rear, whereupon the crane had been pushed at high speed to the Gurpa Down Loop. Originally it had been planned to take the whole ART train to Gujhandi but the jib of the crane could not be lowered far enough

[607] https://eldiariodecoahuila.com.mx/2020/10/05/trenazo-la-tragedia-que-enluto-a-miles/
[608] https://notassinpauta.com/2020/10/05/una-tragedia-en-tren/
[609] https://www.elsiglodetorreon.com.mx/noticia/2008/el-trenazo-el-desastre-ferrocarrilero.html?from=old
[610] Wikipedia – Accident ferroviario de Saltillo de 1972

onto the adjoining rerailed waggon to clear tunnels on the route.

Loco Foreman M.S.W. Hasan testified that he had arrived at the derailment scene with the Gaya ART from Gurpa at 10:00, materials had been unloaded and the train had returned to there. He had come back with engine No. 12230 at 16:15. Meanwhile the Gomoh train had arrived at 13:00. At 17:15 it had been proposed to take the entire Gomoh ART train and rerailed waggons to Gujhandi (in the opposite direction), subsequently the plan had been changed to take the crane to Gurpa with engine No. 12230 and the rerailed waggons plus the remainder of the ART train to Gujhandi. Engine No. 12230 and the crane had almost reached the Gurpa Down Advanced Starter when there had been a severe jerk, followed by a number of minor jerks, and the speed of the engine increased from 8 to about 35 kilometres an hour, before it was stopped by the steam brake. Someone had shouted that men had fallen from the crane while the second fireman and two others had been missing from the engine.

Driver N.C. Roy of the Gaya ART engine (No. 12230) said that he had proceeded towards Gurpa with the crane at about 12 to 13 kilometres an hour, having been told by the guard not to exceed 15 kilometres an hour. Near Gurpa a severe bump had thrown Second Fireman R.K. Sinha from the engine, which had then attained a speed of 30 to 35 kilometres an hour before it could be stopped.

Guard B.B. Chakravarty of Divisional Material Train (DMT) No. 2 stated that he had reached the derailment site at 18:40. There Traction Driving Inspector A. Maitra had told him to attach his engine to the dead engine (that which had brought the Gomoh ART there) as per the orders of the Divisional Safety Officer. Meanwhile Carriage Foreman S.K. Patra had told him not to delay matters as train No. 9 Up was waiting at Gujhandi. He had duly coupled the DMT and dead engines together, then had gone to check the load of the ART. Having checked three coaches he had heard the vacuum trouble whistle sounded and had begun to return to his engine, when he had found the ART rolling down and heard Patra remark that the driver did not know how to shunt and had dashed against the load.

Carriage Foreman Patra more or less corroborated the evidence of Guard Chakravarty.

Traction Driving Inspector Maitra denied telling the guard to couple the engines together.

Driver Thanu Mahato, in charge of Class WG engine No. 8719 attached to the DMT, said that he had attached his engine to the dead engine after checking that its hand brakes were applied. The vacuum brake hoses had been attached by Second Fireman Gaffur Ansari of the dead engine. The vacuum not increasing, he had sounded the vacuum trouble whistle. After four or five minutes he had heard a shout that the ART was moving. He originally stated that he had moved the two engines towards the ART but had not bumped it, then at the inquiry admitted that he had not moved the engines an inch.

Fireman D.H. Khan of engine No. 8719 corroborated his driver's statement.

Driver Kamal Ram of Class WG engine No. 10506 of the Gomoh ART stated that after arriving at the derailment site the steam pressure of his engine had fallen and Assistant Mechanical Engineer (AME) Mohammad Murtaza had told him to uncouple the engine from the ART, move towards Dilwa and drop the fire, which he had done under protest. The AME denied giving him any such order and said that the pressure had been increasing at 15:00. Driver Ram recalled hearing the vacuum trouble whistle being blown and a shout that the ART was running away. He denied that his engine had bumped the ART.

Loco Foreman M.K. Bose of Gomoh explained that he had wanted to move the crane up the bank clear of the waggons to swivel its jib but the low steam pressure of engine No. 10506 had prevented it doing so. Lack of space had required the rerailed waggons being placed on the Down line and the jib of the crane not being able to be lowered enough onto the derailed waggon to clear the tunnels, it had been decided to send the crane alone to Gupta, pulled by engine No. 12230. At 18:50 he had heard a heavy bump and shouts that the ART was rolling down. Driver Mahato had denied bumping the ART. He had not deputed anyone to check that the wheel wedges holding the ART vehicles did not fall out as they were usually driven in hard, nor anyone to man the hand brake in the van, while suitable chains to secure the wheels of the ART vehicles were not available.

Fitter-in-Charge M. Satyanarayana of Gomoh stated that 24 wedges had been put under the wheels of the ART and the rerailed waggons; there had been 90 wedges available. He had been in the staff van when he felt a severe bump and the runaway had begun. He and other staff had tightened the handbrake in the van without the speed diminishing, though the brake blocks of the coach had become red hot. He estimated the runaway had attained a speed of about 100 kilometres an hour.

Train Examiner C.K. Jha of Gujhandi spoke of seeing Guard Chakravarty signalling to the engines with a torch, hearing the vacuum trouble whistle, seeing the guard running towards the engines and the engines bumping the ART. The ART had first moved slightly, then had begun rolling, and he had shouted to the men on it to jump off.

Several other officials gave evidence.

Additional Commissioner S.K. Mojumder fixed the time of the collision from the time that the traction power had been cut by damage to the overhead line equipment as 18:58.

According to the driver of engine No. 12230 pulling the crane he had not exceeded 12 or 13 kilometres an hour, but about 12 kilometres had been covered in the 48 minutes between 18:10 and 18:58, including a 10-minute stoppage to attend to the brakes, giving an average speed of 19 kilometres an hour, and a speed of about 20 kilometres an hour at the moment of the collision. The runaway had taken eight minutes to cover the 12 kilometres, giving an average speed of 90 kilometres an hour and a probable speed at impact of 100 kilometres an hour.

There was overwhelming evidence that the runaway had begun after the ART had been bumped by the coupled engines. It appeared that the guard had signalled the two engines forward to couple onto the ART and it was at that point that the vacuum trouble had occurred. The two coupled engines and a rest van attached to them had weighed 363 tonnes. Moving about 33 feet down a 1 in 85 gradient they would have been able to cause a bump of sufficient magnitude to overcome the resistance of the wooden wedges of the ART train weighing only 285.5 tonnes.

The additional commissioner noted that:

(a) No one had been deputed to keep a watch on the wedges and tighten the hand brakes of the vehicles of the ART having them.

(b) Very little reliance could be placed on the pinning down of the hand levers of ordinary waggons and safety chains should have been used.

(c) The guard of the ART had been deputed to the different job of taking the crane to Gurpa, without anyone else being left in charge of the ART.

(d) The driver of engine No. 10506 had acted most indiscreetly and irresponsibly in detaching his engine from the ART without permission or even warning the staff on the ART.

The additional commissioner concluded that the collision had resulted from the running away of the ART due to its having been bumped by the coupled engines. The absence of safety chains had contributed to the accident.

Driver Ram was responsible for detaching engine No. 10506 from the ART and dropping its fire without permission from the guard of the ART or his superior officers, in violation of General Rule No. 129, also for releasing the hand brake on the engine.

Driver Mahato was responsible for attempting to attach the coupled engines to the ART on a steep falling gradient without taking adequate precautions, in violation of General Rule No. 147 (2).

Guard Chakravarty was likewise responsible for trying to attach the engines without any orders from superior officers or taking adequate precautions, in violation of General Rule No 151.

Additional Commissioner Mojumder made seven remarks and recommendations:

(1) Suitably designed steel skids and safety should be included in the standard equipment of ARTs and made obligatory on sections with heavy gradients, as wooden wedges were not very dependable.

(2) The initial derailment had resulted from the displacement of imported steel plate rolls which had not been properly secured on the waggons of a goods train. A large number of such derailments had occurred and the matter of the proper loading of such rolls should be taken up with the Calcutta port authorities.

(3) The failure of engine No. 10506, the absence of a device to keep the steam brake handle of engine No 8719 applied, and the repeated repairs needed to the steam brake of engine No. 12230 en route to Gurpa indicated extremely poor maintenance of steam locomotives, which must be improved.

(4) The foreman of the ART should have known that the 120 tonne diesel crane could not be turned through 180 degrees with waggons on the adjoining line and should have turned the jib previously. ART crane crews should undergo regular drill at suitable intervals so that they did not forget vital points.

(5) In order to overcome difficulties caused by a guard's work to rule on the Eastern Railway General Rules 93 and 112 and Supplementary Rule No. 102 had been dispensed with on 11th March, meaning that trains could be dispatched from any station:

(a) Without a guard in daylight hours (GR 93)

(b) Without a guard in the brake van (GR 93)

(c) Without a brake van at their rear (GR 93)

(d) Without standard screw couplings (GR 112)

(e) Without a rear tail lamp, tail board or red flag, the number of the rear vehicle being transmitted to the station ahead (GR 146)

(f) Without the guards having their standard equipment. (SR 102)

The additional commissioner was not impressed, considering that the 'dispensation had been issued in a rather off-handed manner' without at the time issuing the special instructions required by the rules, or excluding sections with heavy gradients where the dispensation involved risk to the public. The chief operating superintendent had withdrawn the dispensation on 5th June but trains without brake vans were still running on 4th July and the feeling that guards were redundant and could be dispensed with when required might explain why the ART had been left without a guard on a steeply falling gradient.

Whenever dispensation of any rule was unavoidable special measures should be issued simultaneously and the dispensation should not apply to sections where there was special risk.

(6) The travelling of staff on crane platforms should be strictly prohibited.

(7) Transportation Inspector Singh was to be commended for his detailed record of the course of the runaway, which had been extremely useful to the inquiry. [611]

Gormanston station, Eire, was the scene of a serious accident involving three trains on the morning of Monday 21st October 1974. The 06:50 Up passenger train from Dundalk to Dublin and Bray was standing at the Up platform of the station and the 07:00 Dublin Connolly to Skerries Down empty coaching stock had just arrived at the Down platform, when the latter train was run into from behind by the 07:00 Dublin Pearse to Howth Junction Down empty coaching stock train. As a result one of the vehicles of the Connolly train was derailed and struck the side of the Dundalk train. Two passengers were killed and 26 passengers and the guards of each train were injured. The accident was the subject of a Department of Transport and Power inquiry, conducted by Mr J.V. Feehan.

The Dundalk train consisted of engine No. B 219, two heating vans, six passenger coaches and a guard's van and was carrying between 75 and 80 passengers.

The Connolly train was a railcar set with a motor coach at each end and two intermediate coaches.

The Pearse train was a push/pull set consisting of a control car, two passenger coaches, a brake van and engine No. B 202, with the engine at the rear at the time of the accident.

Driver W. Scanlon told the inquiry into the accident that he had been driving the Connolly train. His first indication that anything was wrong had been seeing the signalman at Rusk waving his hands as he had passed through the station. Arriving at Skerries, his destination, he had slowed to five miles an hour, when the signalman had shouted to him to keep going as there was a runaway train behind. He had been travelling at maximum speed at Gormanston, where the runaway had caught his train up halfway along the platform.

Guard J. Monahan of the Connolly train said that he had known nothing of the runaway until they reached Skerries, when he had retreated to the driver's cab for safety.

---

[611] Government of India Ministry of Tourism and Civil Aviation (Commission of Railway Safety): Report on the rolling down of a part of the Accident Relief Train along with rerailed waggons and its collision with the 120 tonnes diesel crane at KM 424/4-6 between Yadugram block hut and Gurpa on the Sown Grand Chord line of Eastern Railway at about 19:00 hours on 23rd August 1974

Guard P. Salmon of the Pearse train stated that he had checked his train before starting, though he had not advised Driver P. Rogers, whom he had not worked a push-pull train with before, of its load and make-up, as per the rules. He was not sure if he had read the vacuum gauge before starting. He had advised his driver to start from Pearse station, the train had been stopped by signals at Connolly, had restarted there over the crossover to the Northern line, then had suddenly come to a stand. His driver had then come back through the train and told him that the brakes had come on in emergency. They had both alighted and gone to the rear of the train, where a hissing sound told him that a brake pipe had become disconnected. He had found the problem, reconnected the hose, shouted to his driver that he had done so, and reboarded the train. The train had been stopped five or six minutes. He had assumed that his driver had also reboarded the train and when the vacuum pressure had risen to 17 inches of mercury he had rung the bell twice and the train had restarted.

Guard Salmon said that he knew his train was timetabled to stop at Howth Junction but when it had not done so he assumed that there had been an alteration in its schedule. He had been convinced that its driver was on board. He knew that he was in charge of the train and how to operate the brake in the guard's van but thought he must have lost concentration at some point after Howth Junction. He had not heard any detonators exploding, nor had he realised that his train was running at high speed.

Driver Rogers stated that he had come on duty 20 minutes late on the morning of the accident and had not booked on. He had not supervised the coupling up of the vehicles or realised that the shunter had not coupled up a push-pull train before. He had found the safety control equipment handle in the control cab in the "isolate" position and had placed it in the "run" position. It was not unusual to find matters such, and if a sealing wire were available he would tie the handle in the "run" position, otherwise he would ask an inspector to reseal the handle. No inspector had been available that morning. He had never reported that the safety equipment handle was unsealed and was unaware that in such a case there should have been a second man in the cab. He thought that there had been 25 inches of vacuum upon starting from Pearse.

When the train had stopped out of course, Driver Rogers continued, he had gone back with the guard and left the guard to examine the vacuum brakes, while he went to the locomotive at the rear of the train. He had had no reason to move the control handle in the control cab from the "run" position after leaving Pearse. He had seen the guard reconnecting the vacuum pipe and found nothing wrong with the locomotive, which had been running at idling speed. Seeing the guard reboard the train he had alighted from the locomotive and had been on his way along the track forward to the control cab at the front of the train when it had begun to move off. After unsuccessfully trying to reboard the train he had gone to Driver A. O'Reilly's engine nearby and given a series of "crow" whistles. Then he had gone to the East Wall signal cabin, told the signalman that it was a runaway train and telephoned the Howth Junction signalman and told him the same thing, and for the Howth Junction signalman to let the train down the main line.

Driver E. O'Grady of the Dundalk train said that he had left Drogheda at 07:22, arriving at Gormanston at 07:36, and the accident had occurred three or four minutes later. He had been unaware of the runaway until he had seen the two trains approaching, one close behind the other.

Guard M. Lee of the Dundalk train said he had likewise been unaware of the runaway train until he had heard the noise of its approach, then witnessed the accident.

Driver O'Reilly testified that he had seen the push/pull train move off from East Wall, when Driver Rogers had shortly come to his engine and had begun sounding its horn, saying that the train was running away. He had reassured his colleague that the train would not get far because the "deadman" would apply the brakes but Driver Rogers had indicated that the "deadman" was either isolated or disconnected and would not stop the train.

Signalman P. Garrahan in East Wall cabin related that the push/pull train had left at rather high speed. About five minutes later Driver Rogers had come into his cabin, looking agitated and excited, to say that the train was running away. He (the signalman) had telephoned Howth Junction and told his colleague there to keep the runaway on the main line and Driver Rogers had then taken the phone and remarked to Howth Junction that his guard would realise that something was wrong when the train did not stop there. Later the Howth Junction signalman had informed him that the runaway had passed through there, moving at 60 or 70 miles an hour.

Controller P. McKay stated that he had been advised by the Howth Junction signalman at 07:27 that the Pearse train was running away behind the Connolly train, with seven minutes between them as they had passed him, the runaway travelling at 70 miles an hour. Following a conversation with the Drogheda signalman it had been decided to divert the runaway into the goods yard there, this being the first point at which there were any manned facing points at this time of the morning. He had not at that time been aware of the Up Dundalk train but did not consider that he would have acted any differently in any case.

Relief Signalman L. Cunningham at Howth Junction said that the Connolly train had passed him at about 07:20. When the East Wall signalman had advised him of the driverless runaway his points had been set for the Howth branch and an Up train had been approaching from Howth. He had reset the points for the main line. His signals had been at Danger but the runaway had passed him 'like an express.' There had been no time to lay detonators. He had seen no one on the runaway.

Relief Signalman W. Martin at Rush stated that he had had time to place detonators on the line in front of the runaway, which had exploded them. It had passed his box 'faster than an express train.'

Signalman H. McKenna at Skerries explained that it would have taken ten minutes to cross the Connolly train over to the Up line, so knowing that runaway was close behind he had told the driver of the Connolly train, before it had stopped, to keep going as fast as he could, then had replaced his signals to Danger and put down detonators. Only two or three minutes later the runaway had passed him, going very fast, and exploding the detonators.

Signalman J. Collier at Gormanston said that the Dundalk train had arrived there at 07:34 and was due to leave at 07:37. At 07:31 the Skerries signalman had advised him of the runaway and to keep the line to Drogheda clear for it. At 07:36 Skerries had sent the *Train Entering Section* signal for the Connolly train and the same signal had followed for the runaway only 16 or 20 seconds later. He had seen the two trains crossing the Delvin viaduct and the runaway had collided with the Connolly train on the Dublin side of his cabin at 07:40. He had held the Dundalk train at his station upon being advised of the runaway. There had been no time to warn its crew. He had been the only railwayman on duty at the station at that hour.

Signalman P. Matthews at Drogheda stated that the Dundalk train had left him at 07:22. He had been advised of the runaway at 07:30. The boxes at Laytown and Mosney, between him and Gormanston, had both been switched out (as was Balbriggan cabin on the Skerries side of Gormanston), so the Dundalk train could not be stopped before it reached Gormanston.

Acting Foreman P. Dunphy at Pearse said that neither he, nor he thought, any of the shunters he was in charge of had been instructed in the shunting of push/pull trains. Prior to the accident he had not been aware that it was not possible to release the brakes on a push/pull set with the engine at the control car end without isolating the safety control equipment. He had not seen the instruction from the assistant stationmaster dated 23rd August regarding the shunting of push/pull trains.

Shunter L. Millar at Pearse admitted that he had isolated the safety control equipment on the evening of Saturday 19th October, when he had had to shunt the push/pull train with the pilot engine at the control cab end, something which was only necessary on Saturday nights when the train engine was taken away for servicing. He knew that it was not possible to create a vacuum with the engine at the control car end without isolating the safety control equipment, though the brakes could be released by pulling the carriage vacuum strings. Shunters considered shunting with just the engine brakes to be unsafe. He, too, denied having seen the assistant stationmaster's instructions of 23rd August.

Shunter N. Doran said it was he who had coupled up the push/pull train on the morning of the accident, something he had not done before. He had heard the sound of a slight leak from the brake pipe but the sound had stopped before he could locate its source so he had taken no further action. He had not seen the assistant stationmaster's instructions.

District Manager V. Ryan stated that Acting Running Superintendent Hand had told him 6 to 8 weeks before 21st August and again on 20th/21st August that the seals on the safety control isolating equipment handle at Pearse were being broken. He had then written to Assistant Stationmaster J.J. Brady at Pearse outlining the correct procedures for releasing the brakes of push/pull trains during shunting operations. He had received no reply to his letter.

Assistant Stationmaster Brady said that upon receiving the letter from District Manager Ryan he had issued an instruction to his inspector that the brakes of push/pull trains should be released by pulling the carriage vacuum strings when necessary during shunting operations, with copies to the shunters. He had assumed that these instructions had been implemented and had only discovered after the accident that some staff had not received them.

Mr Feehan came to eight conclusions regarding the accident:

(1) The accident had resulted from the 07:00 train from Pearse overtaking and colliding with the 07:00 train from Connolly and a derailed vehicle of the striking the side of the Dundalk train. The Pearse train had run away because Driver Rogers had failed to secure it properly before leaving the control cab and Guard Salmon had failed to take any action when it had run through Howth Junction without stopping.

(2) Despite the failures of the driver and guard the train would not have run away if the safety control isolating equipment handle in the control cab had been properly sealed, or if unsealed had been left in the "run" position.

(3) The couplings on the high vacuum hose between the locomotive and the adjoining coach had separated as the train was crossing to the Northern line at Connolly, due to excessive speed, not being properly connected, or another cause and the brakes had then come on automatically.

(4) The wire seal on the safety control isolating equipment handle had already been broken when Driver Rogers had taken over the train, having been broken deliberately by the shunter on the Saturday night.

(5) Some Córas Iompair Éireann officials had been aware that the seals of the safety control isolating equipment handles of push/pull trains were being broken without authorisation to facilitate shunting moves at least three months before the accident. Despite the issuing of instructions as to the correct procedure to be followed the breaking of seals had continued.

(6) The training and supervision of the Pearse shunting staff, arrangements for the distribution of the weekly circular at Connolly shed, and the supervision of drivers signing on, were all unsatisfactory.

(7) Neither the mechanical condition of any of the trains nor the medical condition of the staff of the Pearse train appeared to have contributed to the accident.

(8) That the drivers and guards of suburban trains were expected to use both Córas Iompair Éireann (CIE) and Great Northern Railway (Ireland) rule books, the two not being identical, was not satisfactory and might explain a slackness in the observation of the rules.

Mr Feehan concluded his report with six remarks and recommendations:

(a) Improved methods and procedures should be the training and supervision of the Pearse shunters, the supervision of drivers signing on at Connolly shed, and the circulation of the Weekly Circular to the Connolly shed staff.

(b) The CIE Signalling Regulations and Appendix to the Working Timetable should apply over the whole CIE system while the Driver's Instruction Manual should be updated to include information on push/pull trains and the safety control equipment.

(c) The apparent conflict between CIE Rule 149 (4) (b) and the instructions to drivers for working push/pull trains, regarding who was responsible for checking couplings between vehicles, should be removed.

(d) CIE Weekly Circular No. 156 of 1st March 1975 stated that the CIE Rule Book was to apply throughout the system and Rule No. 50 had been amended to include the signal which a driver much exchange with a guard when a train was stopped in exceptional circumstances.

(e) The layout of the Weekly Circular should be improved to draw particular attention to changes in the Rules, Regulations and Operating Instructions with such notices being repeated at regular intervals until such amendments were accommodated in the Rule Book, Regulations, Operating Instructions etc.

Consideration should be given to fixing securing clips on the coupling joining the sections of high vacuum hose between the locomotive and coaches of push/pull trains. [612]

---

[612] Department of Transport and Power report

CHAPTER 8

BRIDGE FAILURES

Some of the most serious railway accidents have resulted from the failure of bridges. Bridges may fail for a variety of reasons. The inherent design of the structure may be faulty, the materials used in its construction may be poor and inferior, the materials may deteriorate over time, or its foundations may be washed by a stream or river that has become exceptionally swollen following abnormally heavy rainfall. With improvements in bridge engineering failures due to faulty design had greatly diminished by the middle of the twentieth century, but accidents, including major disasters, continued to occur from causes such as bridge foundations being washed away. A new danger, mainly though not entirely affecting road underbridges, would be so-called 'bridge bashing' by vehicles that were too high to pass beneath the structure.

Thirty-six cars of a Southern Pacific Railroad freight train from St Louis, Missouri, to Los Angeles, California, fell 60 ft into the Colorado River near West Point, Texas, United States, on Friday 26th May 1961, when two spans of a bridge collapsed. Seventeen of the cars were loaded with miscellaneous merchandise and the other 19 were empty refrigerator cars. No one was hurt. [613] The engine and 26 cars reached the south bank of the river safely and the rear 31 cars and the caboose stopped short of the bridge. One hundred and fifty men went to the scene with equipment from as far away as Houston (over 100 miles away) but the line was expected to be blocked for some time. A load of steel being carried on one of the cars apparently shifted and struck the bridge superstructure, causing its failure. [614]

At about 5 a.m. on Friday 7th December 1962 a section of a trestle across Echeconnee Creek, Georgia, United States, 12 miles south of Macon, collapsed under a Central of Georgia Railway freight train and 24 cars fell into the stream. No one was hurt. It was expected that it would take two weeks to clear the wreckage and repair the trestle. [615]

613 *Ogden Standard Examiner*, Ogden, Utah, United States, 26th May 1961
614 *Corsicana Daily Sun*, Corsicana, Texas, United States, 26th May 1961
615 *The High Point Enterprise*, High Point, North Carolina, United States, 8th December 1962

In contrast to the previous two accidents three railwaymen were injured and a fourth was missing following the collapse of a trestle at Eagle Rock, North Carolina, United States, beneath a Norfolk Southern Railway at 7 a.m. on Tuesday 2nd July 1963. The engine and three cars of the train, which was en route from nearby Raleigh, North Carolina, to Norfolk, Virginia, fell into Marks Creek. Fire had weakened the trestle, apparently after it had been struck by lightning. In mid-morning the fire was still burning and firemen stood by to prevent it from spreading to three waggons of gas (petrol), the first of which was marshalled 14th from the front of the train. [616]

Two or four people died and 50 people were injured when one coach of an express train was derailed as it crossed a rain weakened bridge between Adana and Elâziğ, Turkey, on Boxing Day, Sunday 26th December 1965. [617] [618]

At 12.40 p.m. on Thursday 27th June 1968 a 111-car Green Bay and Western Railroad freight train en route from Winona, Minnesota, to Wisconsin Rapids, Wisconsin, was wrecked at Merrillan, Wisconsin, United States. The train was in charge of Conductor Edwin Losinski with Engineer Rudy Pflanze, Fireman Roy Rickhoff, and Brakemen William Habeck and Leonard Sullivan. It was crossing a 75 ft long steel bridge with a height of about 40 ft above Halls Creek. [619]

Engineer Pflanze related afterwards, "I could feel the bridge give way as the first diesel passed over it." Fortunately the momentum of the train carried the leading two diesel units off the bridge and onto solid ground beyond. The rear diesel unit and following cars fell with the bridge, either into the creek or onto its banks, and in all three diesel units and 11 cars were derailed. No one was hurt but several hundred feet of track were ripped up or otherwise damaged. The Chicago St Paul Minneapolis and Omaha Railway (CMO), which crossed the Green Bay and Western on the level at Merrillan, was blocked, and Highway 12 through the village was also blocked by derailed cars.

The Green Bay agent at Winona, Vernon Zimmerman, said that his company's trains were being temporarily diverted over the track of the Chicago Milwaukee St Paul and Pacific Railroad via New Lisbon, Wisconsin. He added that no final decision had yet been taken but it was possible that half a mile of temporary track might be laid to connect with CMO tracks near Merrillan, thence reaching the Green Bay tracks via a Y. This would prevent Green Bay trains having to make a long diversion until the bridge had been repaired.

The cause of the bridge failure was not stated. [620]

According to an official report 27 people died and over 140 were injured when a bridge across the flooded Baoule River in Mali collapsed beneath a passenger train on Thursday 17th September 1970. Nine of the 14 coaches of the Chemin de fer Dakar-Niger

---

[616] *The Daily Independent*, Kannapolis, North Carolina, United States, 2nd July 1963
[617] *Pacific Stars and Stripes*, Tokyo, Japan, 28th December 1965
[618] *Nevada State Journal*, Reno, Nevada, United States, 27th December 1965
[619] *The Daily Telegram*, Eau Claire, Wisconsin, United States, 28th June 1968
[620] *Winona Daily News*, Winona, Minnesota, United States, 28th June 1968

train fell into the water. Unofficial sources stated that the death toll might be even higher. [621] [622]

The *Hellas Express,* an international train from Köln and Munich, West Germany, to Athens, Greece, met with an accident just on the Athens side of Leptokarya, Greece, a few kilometres south of Katerini, at 06:45 on Tuesday 7th March 1972. Following prolonged heavy rain a bridge over the Ziliana stream collapsed and the engine, two coaches and a restaurant car fell 10 ft into what was described as a fast-flowing river. Some reports speak of a dry river bed, presumably indicating that the stream was dry part of the year. Three other coaches were derailed, one of which was left suspended over the gap. One or two of the approximately 280 passengers on the train was killed and 13 or 15 others were injured. [623] With the coaches that had fallen from the bridge being partially submerged in the water, the rescue operation was a difficult one. [624]

In an accident similar to that West Point, Texas, back in 1961 (see above) the boom of a crane on a Union Pacific Railroad work train, swinging loose, struck the superstructure of the Orodell Bridge over the Grande Ronde River, a mile north west of La Grande, Oregon, [625] United States, on the night of Thursday 1st August 1974, causing the steel bridge to collapse at one end. The crane car and 14 other cars were derailed, four or five cars fell into the river and 150 feet of track were torn up. None of the train crew nor of 140 members of a track-laying crew riding in the bunk cars of the train was hurt. The train was part of a large track laying operation. It had completed work in the Blue Mountains, between La Grande and Pendleton, Oregon, and was on its way to a new job east of Baker, Oregon. [626]

After special permission had been obtained from the Oregon Environmental Quality Commission culverts were placed in the shallow river so a temporary span could be placed across the river, allowing traffic to resume on Sunday 4th August. [627] The line was the only link between Portland, Oregon, and Boise, Idaho and 16 trains had been held up while the line was blocked. [628]

A Skopje – Belgrade passenger train, which was also described as a night express from Bitola, [629] was derailed near Vranje, Yugoslavia, just before midnight on Thursday 15th May 1975. The engine and six leading coaches of the train were derailed but made it across a steel bridge over the Korbevačka stream, but part of the bridge then collapsed and the seventh coach fell into the water. At least 13 people died and about 169 were

---

[621] *The Daily Colonist*, Victoria, British Columbia, Canada, 20th September 1970
[622] *The Daily Gleaner*, Kingston, Jamaica, 20th September 1970
[623] *The Times*, London, England, 8th March 1972
[624] *The Daily Report*, Ontario-Upland, California, United States, 7th March 1972
[625] *Appeal Democrat*, Marysville-Yuba City, California, United States, 3rd August 1974
[626] *Idaho State Journal*, Pocatello, Idaho, United States, 2nd August 1974
[627] *The Register*, Orange County, California, United States, 5th August 1974
[628] *Walla Walla Union-Bulletin*, Walla Walla, Washington, United States, 2nd August 1974
[629] *The New York Times*, New York, New York, United States, 17th May 1975

injured, 38 seriously. [630] Most of the fatalities had been riding in the unlucky seventh coach. Civilian rescuers and soldiers worked through the night to remove the injured and dead from the muddy waterlogged terrain, while villagers living downstream were asked to look out for the bodies of passengers who might have jumped or been thrown out of the train and carried away by the torrent. The injured were taken to hospital in Vranje; most were Yugoslavs but one Briton was also among the injured. The main line was expected to be out of use for three days while the bridge and 100 metres of track were repaired. It was presumed that floodwaters had undermined the approach embankments, causing the bridge to fail. [631]

"One minute the bridge was there, the next it wasn't," said Engineer John Ruscio, after an accident at Big Creek Bridge, near Lavender, Washington, United States, just after midnight on Tuesday 2nd December 1975. The Chicago Milwaukee St Paul and Pacific Railroad Tacoma to Othello freight train was carrying radio parts, truck trailers, chains and rubber and its engineer said that it had been moving at 40 miles an hour through a heavy rainstorm at the time of the accident. The leading diesel unit, occupied by the engineer and Head Brakeman Bob Starkovick, made it across the 50 ft wooden span, supported on pilings, but the following two units and a flat car went down with the bridge and ten boxcars went down the approach embankment. Ruscio, Starkovick and the three railwaymen riding in the caboose all escaped unhurt. [632]

Heavy rains since Sunday had swollen all the rivers and streams in the area. More than eight inches had fallen in that period at Stampede Pass, in the Cascade Mountains. [633]

A problem that has become widespread on railways worldwide in modern times is so-called 'bridge bashing,' where an overbridge, usually across a road though occasionally across a waterway, is struck by a vehicle, in the case of a road bridge often a double-decker bus or a similar tall vehicle, which is too high to pass beneath the structure. Such accidents usually just result in delays to trains while the bridge is examined for damage but occasionally the results are more serious. An example of a serious train accident due to 'bridge bashing' occurred near Gorey, Eire, on Córas Iompair Éireann (CIE), on New Year's Eve, Wednesday 31st December 1975. The 08:05 passenger train from Rosslare Harbour to Dublin, consisting of an engine and eight vehicles, in charge of Guard T. Burlong with Driver Joseph O'Neill, was derailed as it was crossing Bridge No. 292 (known as the Cain or Clogh Bridge). The train was carrying 94 passengers. As a result of the derailment five passengers were killed, and four railwaymen and 39 passengers were injured. The passengers who lost their lives were William Hayes, 80, from Ardcavan; Philomena Moss, 50, of Kilmore; Anthony O'Brien, 42, from Wexford; Richard O'Neill, 24, from Dublin (an off-duty CIE employee); and Peter Roche, 16, from Wexford. The engine went down the embankment, the first three vehicles of the train were wrecked, the fourth and fifth were badly damaged and the sixth was partly derailed.

---

[630] *The Modesto Bee* and News Herald, Modesto, California, 16th May 1975
[631] *The Bakersfield Californian*, Bakersfield, California, United States, 17th May 1975
[632] *The Herald*, Provo, Utah, United States, 2nd December 1975
[633] *Redlands Daily Facts*, Redlands, California, United States, 2nd December 1975

The track was badly damaged. Had not the leading two coaches been empty the accident would doubtless have been worse.

The bridge had been constructed by the Dublin Wicklow and Wexford Railway in 1863 and had two wrought iron girder spans over a public road. According to the initial plans the bridge had a clearance over the road of 15 ft but at the time of the accident this had been reduced to 13 ft 10 in. There was no warning sign as to the restricted clearance on the road approaches to the bridge, nor was the CIE empowered to erect one, though it could place signs on the bridges themselves, and had been doing so at different locations since the previous year, beginning with bridges crossing major roads.

About two minutes before the train arrived, the bridge had been struck by an excavator being carried on a low-loader articulated lorry, which had been travelling from east to west along the road. As a result of the collision the wrought iron girder on the east side of the track was fractured and that on the west side was carried 42 ft along the road. The rails were then left unsupported and out of alignment.

Roy Hill, 28, was driving a tractor on his parents' farm and witnessed the accident. He said later, "Suddenly the carriages started to pile up on top of each other. It was like looking at an old Western film or something. For a second or two I couldn't believe my eyes." Roy's brother Christopher tried to stop the train. "He ran over the little field and climbed onto the railway track and waved his arms wildly, to indicate to the driver," Roy added.

Driver O'Neill recalled, "A man hopped out of a ditch to try and stop me but it was too late. I was only 200 yards from the bridge and that was the first indication I got."

The accident was investigated by the Department of Transport and Power.

Driver J. Lawler told the inquiry into the accident that he had worked the 05:25 newspaper train from Dublin on the day in question, passing over the bridge four hours before the accident, when all had been normal. He had been a passenger on the 08:05 train but could remember nothing of the accident.

Driver O'Neill told the inquiry he had been travelling at about 53 miles an hour as his train approached Cain Bridge. He had whistled for an accommodation crossing, then about 400 yards from the bridge he had seen a man beside the track waving his arms and apparently trying to stop the train. He had sounded his hooter and applied the brakes, then had seen the top of a yellow machine, which he had thought might be trying to come up the embankment to cross the line. Then he had seen that two rails and parts of the bridge were displaced and knew that his train would be derailed. The speed of his train, he estimated, had fallen to between 20 and 30 miles an hour when it reached the bridge. There had been nothing wrong with the bridge the day before. His brakes had been working normally.

Guard Burlong, Dining Car Attendant T. Boyce and Travelling Ticket Collector T. O'Donoghue could not really add anything to the evidence of the driver.

Platelayer J. Doyle said that a neighbour had informed him of the accident. He had previously seen low-loader lorries travelling on the road under the bridge but was not aware of any previous instances of the bridge having been struck.

Ganger M. Morris stated that he had inspected the track the day before the accident and it had been satisfactory. He likewise had never heard of the bridge having been struck before.

Permanent Way Inspector D. Donovan said that the bridge was inspected annually and had been cleaned and repainted in 1968. The previous year he had discovered a mark on one of the bridge girders suggesting that it had been struck by a high road vehicle but the mark had been so slight that he had not bothered to report it. The bridge had last been inspected in February.

Christopher Hill told the inquiry of hearing a bang, having been told by a man whom he believed to have been the driver of the excavator that the bridge was down, and of his attempt to warn the train driver.

Garda Sergeant French stated that the excavator driver had told him that his vehicle had struck the bridge. Fresh yellow paint marks on the bridge girders matched the paint of the excavator. He was not aware of any previous report of the bridge having been struck.

Area Manager J. Leonard described meeting the driver of the lorry involved a week after the accident, when the latter had told him that having hit the bridge he had shouted to the excavator driver to contact Gorey and Ferns (the stations on either side). The driver had added that the bridge was 'rotten through' and he was unfortunate to have been the first to damage it.

Various experts gave evidence on the condition of the bridge and agreed that although there was a small amount of corrosion, there had been nothing in its condition which would have made it unsafe for rail traffic prior to it having been struck by the excavator.

Concluding his report the inspecting officer, Mr J.V. Feehan, stated that the bridge had been structurally adequate for the loads imposed upon it by the rail traffic and the derailment had resulted from the track having been left unsupported after the bridge had been damaged in a road accident, shortly before the arrival of the train. Nothing in the condition of the train had contributed to the accident and the train crew were not responsible for it.

The inspecting officer made three remarks and recommendations:

(1) There had been no signs or notices on the bridge itself or on its approaches to warn drivers of the low clearance. In recent years the CIE and local authorities had begun to erect such warning signs, concentrating first on primary and secondary roads, later including certain tertiary roads. In view of the increasing number of high vehicles using public roads and the concomitant risk to rail traffic adequate signs should, he suggested, be provided on the approaches to all railway overbridges on public roads where the clearance height was 15 ft. 6 in. or less.

(2) Although the bridge had been adequate for the rail traffic using it, there had been some corrosion in the lattice girder bottom plates, drainage holes in the bottom plates had been blocked, and a mark on a girder indicated a previous impact from a road vehicle. Bridges should be examined frequently for signs of corrosion and blocked drainage holes and appropriate action should be taken. Bridges over public roads should also be frequently examined for signs of road vehicle impacts and all such impacts should be reported.

The action of Mr Hill in trying to warn the train had been most commendable. [634] [635] [636]

[634] Report of Inquiry into the Railway Accident that occurred near Gorey, Co. Wexford, on 31st December 1975

[635] https://www.irishtimes.com/news/ireland/irish-news/remembering-the-new-year-s-eve-train-crash-of-1975-1.2480479

[636] https://www.independent.ie/regionals/wexford/news/remembering-the-gorey-rail-crash-of-december-31-1975-in-which-five-passengers-lost-their-lives/34320740.html

CHAPTER 9

OPENING BRIDGE ACCIDENTS

Accidents involving opening bridges appear to have occurred in only a handful of countries, with the majority in the United States. It seems that most countries managed to a large degree to avoid building such bridges, which from a railway point of view are a constant source of expense and worry. Robert Shaw states that the early railroads on the eastern seaboard of the United States were forced by opponents of the new form of transport to build such bridges over even insignificant waterways where they were never really needed. [637] Only two accidents involving opening bridges appear to have occurred between 1961 and 1975.

A fatal accident occurred at Leith Docks, Scotland, on British Railways, Scottish Region, on Monday 20th July 1964. A diesel shunting engine was pulling a train of eight waggons loaded with wood pulp between the Albert and Imperial Docks, which were separated by a swing bridge. Mr Alexander McLean witnessed the accident. He related, "The bridge was open for a small boat coming through. Amid the jangling of the waggons I heard the brakes being applied but it was too late." The shunting engine and leading five waggons plunged into 22 ft of water. The other three waggons remained on the quayside, the front one hanging precariously over the edge of the water.

When Edinburgh Fire Brigade arrived with their tender they found that their lifting apparatus was inadequate. Instead Leith Dock Commission divers were called to attempt to reach the driver of the shunting engine, George Fyfe, 56, who was pinned in his engine beneath the waggons and drowned. The engine and waggons were later retrieved from the water. How the accident had happened was not clear. [638]

Late in the evening of Sunday 16th September 1973 the two diesel units and leading 16 cars of a Texas and Pacific Railway freight train of 100 cars plunged off a drawbridge at Morley, Louisiana, United States, south west of Baton Rouge, into a canal, the Gulf Intercoastal Waterway, and were submerged. The train was travelling at 35 miles an hour at the time. After the accident, as at Leith Docks, divers were searching for the two

[637] *A History of Railroad Accidents, Safety Precautions and Operating Practices* by Robert B. Shaw, pg. 187

[638] *Liverpool Echo and Evening Express*, Liverpool, England, 20th and 21st July 1964

railwaymen who had been at the front of the train, who were missing, presumed drowned. The conductor and others riding in the caboose said that they had felt the brakes coming on in emergency and then the train had stopped suddenly. When they had made their way to the front of the train they found out what had happened when they saw that the engines and leading cars were in the canal. Traffic on the canal, mainly barges, was interrupted.

The bridge tender explained to police that having opened the bridge for a barge a power failure had occurred, which had prevented him from closing it again. An auxiliary gasoline engine was provided to be used in such a situation but the tender said that he had been unable to start it. Why the engineer of the train had failed to realise that the bridge was open was unknown. What sort of signal protection was provided at the bridge was not stated in the newspaper reports. [639] [640]

[639] *The Ruston Daily Leader*, Ruston, Louisiana, United States, 17th September 1973

[640] *Hattiesburg American*, Hattiesburg, Louisiana, United States, 17th September 1973

CHAPTER 10

TRAIN ACCIDENTS INVOLVING BRIDGES

Bridges could feature in train accidents in one of two ways. Either the bridge itself could collapse for a number of reasons such as faulty design, faulty construction or because its foundations had been damaged; or a train could fall from a bridge, the structure itself remaining intact. Bridge failures have been dealt with in a previous chapter; this chapter will concern itself solely with accidents where the bridge itself was not at fault. It may seem perverse to list such accidents separately instead of in the chapters dealing with primary causes of the accidents, the procedure adopted elsewhere in this book, but the fortuitous presence of a bridge has such a potential for transforming what might have been only a minor accident into a major disaster that it has been decided to treat such accidents separately, in the same way as accidents occurring in tunnels have been given a chapter of their own.

A broken rail derailed 17 cars of a Baltimore and Ohio Railroad freight train on the evening of Monday 27th February 1961. The accident happened at Wells Creek, Pennsylvania, United States on the Rockwood to Johnstown line. One of the cars, breaking loose from the train, caused the brakes to apply automatically, but another 16 cars behind piled up in a heap on a bridge over Wells Creek and seven of them plunged into the creek, while others were thrown about the right of way on either side of the stream, scattering the ore they had been carrying. One end of the bridge was badly damaged and several hundred yards of track were torn up. No one was hurt.

A local newspaper added, 'The main part of the wreck was on the Somerset {west} side of the bridge. Workers crossing the stream had to climb over the tangled cars and walk across one of the beams supporting the bridge. Linemen were at work installing telephones to co-ordinate work at both ends. Poles carrying telephone and power lines were broken and knocked out of line by the crash. There were no reports of service being disrupted.'

It was hoped to have the bridge repaired and the line reopened to traffic by the end of the following day. [641]

---

[641] *Somerset Daily American*, Somerset, Pennsylvania, United States, 28th February 1961

The worst peacetime train accident in Italy occurred at about 07:50 on the morning of Saturday 23rd December 1961 at the Fiumarella Viaduct, about two kilometres from Catanzaro. The accident occurred on the 950 mm gauge Cosenza-Catanzaro line of the Societa Italiana per le Strade Ferrate del Mediterraneo. [642]

Passenger train AT 7 consisted of a Breda diesel railcar pulling a trailer, in charge of Conductor Luigi Aristodemo with Motorman Ciro Miceli. It had left Soveria Mannelli at 06:43 for Catanzaro and was carrying a large number of passengers, many of them students on their way to Catanzaro for the last day at school before the Christmas holidays. Other passengers were workmen, farmers, and women going into Catanzaro to do some pre-Christmas shopping. There were about 80 passengers in the railcar and 101 in the trailer.

The viaduct was situated on a curve and immediately followed a tunnel. As the train left the tunnel and came onto the viaduct it became derailed, the coupling between the railcar and the trailer snapped, the trailer knocked down about ten metres of the iron railing bordering the viaduct, then fell 54 metres from the middle of the structure to the stream below. The railcar was derailed but remained on the viaduct.

As soon as the railcar had stopped Motorman Miceli made his way along the track to Catanzaro station to raise the alarm, while Luigi Posca, the son of Crossing Keeper Domenico Posca, who had been riding on the railcar, alighted from the train to make his way to his father's house to do likewise. The site of the accident was inaccessible by road but as soon as possible a railcar set off for the viaduct from Catanzaro station, carrying doctors, nurses, soldiers, railwaymen, firemen and police, who made their way in the rain down the muddy slope of the ravine to the wreckage of the trailer. Blood transfusions were given at the scene and the badly injured were carried up the slope to the railcar to be taken to Catanzaro station, from where ambulances took them to the hospital. After all the injured had been removed firemen used cutting equipment to free the bodies of the dead, which Red Cross nurses carried up the slope to be laid along the side of the track. Initially the number of dead was put at 40, then with the discovery of more bodies it rose to 45, then 50, then 60, and by 21:00 it had risen to 69. About 28 others were injured, ten seriously, and two of whom later died. News of the disaster quickly spread and large numbers of people who had relatives or friends on the train rushed to Catanzaro, where they besieged the station and the civil hospital, while others made their way to the viaduct.

Various survivors were interviewed by reporters. Raffaele Sacco, 30, from San Pietro Apostolo, said, "I was in the trailer which fell from the viaduct. Suddenly I felt a jolt and I remember nothing more until I came to some little time later, when I felt a very severe pain in my shoulder."

Salvatore Mussari, 25, a tractor driver from Isola Capo Rizzuta, had been on his way home from Soverina Mannelli for the Christmas holidays. He related, "The coach suddenly skidded as it reached the bridge, then it ended up on the ground and there were the most desperate cries from people within it."

Students Domenica and Stella Bonacci, one from Adani and the other from Decollatura, were doing a third master's course, and sat at the same desk at school. "What we have seen and heard is too much to relate," they said. "We have neither the strength nor the courage to do so. Where are our classmates? What has happened to

[642] https://www.internationalsteam.co.uk/trains/italy12.htm

them? Were they saved? Where are we?" At this point, the reporter wrote, the two girls had burst into tears.

Student Angelo Bruscino, from Adami, who attended the Catanzaro Technical Institute, was only slightly hurt. He said, "I was riding on the train, as I do every morning, on my way to school in Catanzaro, the last day before the start of the Christmas holidays tomorrow. I was just chatting to some friends about how I was going to spend the holiday, when there was a violent jolt as if an earthquake had happened. The coach was in the air and off the line. The next thing was we found ourselves in a coach which had crashed to the bottom of the ravine. I came to myself in a few minutes. The scene in front of me was indescribable: twisted sheets of metal which prevented us – the handful of survivors – from getting out the inanimate bodies from the wreckage, the screams of many hapless passengers, the gasps of the dying. The first rescuers were there in a few minutes and with difficulty I was carried up to the bridge, from where a railcar took me and the other injured to Catanzaro station and from there to the hospital."

Employee Franco Catanzriti, 30, on his regular daily journey from his home in Decollatura to his job in Catanzaro, was a passenger in the railcar. He told the reporter, "As soon as the train left the tunnel I was preparing my bag and getting my coat, as it was less than two kilometres to Catanzaro. I felt a jolt to the left followed by a jolt to the right. Then there was a sudden brake application and I didn't know what was happening. It was only when the train had stopped and I alighted that I saw what was wrong: the trailer had detached itself from the train and gone off the viaduct, while the tracks leading up the viaduct were all twisted. Due to the rain and the hazardous terrain only a few of the passengers from the front of the train were able to get down to the wrecked trailer in the valley below. Firemen with tools soon appeared, with railwaymen, carabinieri and doctors." [643]

On his way to his home Luizotto, as the Crossing Keeper Posca's boy was commonly known, met tenant farmer Antonio De Frazio, who understood from the garbled phrases of the breathless lad that something very serious had happened on the railway. Antonio called fellow tenant farmer Francesco Critelli and the two made their way to the scene. They were the first two rescuers to get there; it was almost another hour before more help arrived. The scene was so horrific and there were so many who needed help that the two men did not know where to start. Antonio said later, "Our families have been on this farm for 40 years. Nothing like this has ever happened before."

Shortly after the disaster five railwaymen were taken into custody for interrogation. Two of them were Motorman Miceli and Conductor Aristodemo; the others were Depot Manager Raffaele Morrone, Chief Inspector Luigi Marazzo, and Deputy Chief Inspector Luigi Calvelli. Two investigations into the accident were begun. One was an administrative one conducted by Engineer Michele Costa, deputy director of civil traffic and transport run under concession, together with Professor Ernesto Stagni, transport professor at the University of Bologna, and Engineer Mario Fanelli, formerly general manager of the Fellovie della Stato. The second inquiry was the judicial one, carried out under the supervision of Luigi Ammirati, chief prosecutor of the Republic by Engineer Saverio Rafaelli, Catanzaro provincial traffic inspector, and a team of technicians. [644]

---

[643] *La Stampa*, Torino, Italy, 24th December 1961
[644] Ibid, 26th December 1961

A couple of days after the accident a national newspaper was already able to write that the accident seemed very probably due to an excess of centrifugal force on the curved viaduct, because the train had been going too fast. Another likely factor in the accident, the newspaper pointed out, was that because of the imminent Christmas holiday the train had been very crowded, with about 100 people in a single narrow-gauge coach.

The same edition of the newspaper noted that 22 of those who had lost their lives had been from Decollatura, 11 from Cicala, and 8 from San Pietro Apostolo. Others had been from Gimigliano, Catanzaro, Caropoli, Motta Santa Lucia, Sorbo San Basile, Soveria Mannelli, Conflenti, Serrastretta, and Adami.

The disaster had resulted in many tragedies, with children losing one or both parents, parents losing children who could have provided for them in their old age, and so on. Some of the sad cases were described in the newspaper.

Vittorio and Maria Pettinato from San Pietro Apostelo both died in the accident, leaving three children. Santo, the eldest, aged 14, was already working as a labourer. Their second child, Maria, was 12, and their youngest, Rosetta, was not even three. There would have been a maternal uncle, Giuseppe Sirianni, but he, too, had lost his life in the disaster, leaving a wife and three children. 'Who will now look after the Pettinato orphans and the Sirianni children?' the newspaper asked and spoke of little Rosetta left with her brothers in a cold and dark house, without a spoonful of soup or a hand to wash her face or change her clothes.

Railway lengthman Carmine Bonaccio and his son Alberico likewise lost their lives in the accident, leaving his wife Maria and three other children. Asked what his most heartfelt New Year's wish was, Maria replied that it would be to be granted the post of a crossing keeper at Adami, so that she could earn some money to care for her family.

Salvatore Audino, from Decollatura, was away working in an equestrian circus in München, West Germany, and was performing dressed as a clown when he heard news of the death of his student son Ilario in the accident.

Artisan Domenico Salvino had returned from his job as a bricklayer in Torino to his home in San Bernardo to find his house empty and shut up. He discovered that his wife Marietta and their other children had gone to Catanzaro to attend the funeral of their 17-year-old daughter Giuliana, another victim of the disaster. His mother recalled, "The girl always said: 'Pappa, let me study, and then I'll take care of you and Mother.'"

Depot Manager Morrone, Chief Inspector Marazzo, and Deputy Chief Inspector Calvelli were soon released from custody, it having quickly become clear that there had been no problem with the train itself and that the accident had resulted from excessive speed. The speedometer of the railcar was found to be working perfectly. It must have therefore indicated to Motorman Miceli the speed at which he had been travelling, while it was clear that the speed of the train upon leaving the tunnel had been well in excess of the 35 kilometres an hour laid down in the working timetable as the maximum permitted at that point. Due to the excessive speed the train had left the rails, when the shock of the derailment had caused the drawbar between the railcar and the trailer to break, whereupon the trailer had plunged from the viaduct into the ravine.

Conductor Aristodemo was also detained in custody. Although not mentioned in the newspaper reports one might suppose that he would have been held partly responsible for the accident, because he had not taken any action to apply the brakes when he found that the train was going too fast at that point, though, as shown in other accidents, guards and conductors were always very reluctant to take control of trains out of the hands of their drivers. The other responsibility of the conductor concerned the overcrowding of the trailer, which theoretically had a maximum capacity of 80 passengers (45 seated and 35 standing). Here, though, the newspaper pointed out, the responsibility did not rest entirely with the conductor, for the local management of the railway had been well aware that trains on some routes were overcrowded at peak traffic periods but had done nothing about it.

The rolling stock of the railway was also criticised, it being alleged that some of it dated back to 1895, the coaches resembling those seen in films of the American Wild West, though the railcar and trailer had been relatively new, dating from 1954. Also the track had been laid to cope with trains weighing six or seven tons and travelling at 25 kilometres an hour, but now had to cope with 30 ton trains moving at up to 40 kilometres an hour. [645]

An unfortunate aftermath of the accident, and a measure of the outrage it had caused, was an attack on Tuesday 27th on four stations of the railway by locals armed with axes, pitchforks, shovels, bludgeons, shotguns, torches, and cans of pitch, oil, and petrol. The attacks took place while the local carabinieri were attending the funeral of 30 of the victims of the disaster at Decollatura. There were only two railwaymen on duty at Soveria Mannelli station when the mob appeared: Stationmaster Angelo Timpano and Deputy Stationmaster Giovanni Nicoletti. Also upstairs in the station buildings were the stationmaster's wife Carla and their three children: Franco, Maria Teresa, and Ettore. Naturally, the stationmaster feared for the safety of his family and he pleaded with the attackers to have mercy upon them, whereupon a man came out of the crowd and assured him that no one would be harmed. No one was, but the crowd proceeded to wreak havoc on the station fittings and equipment, smashing and damaging everything they could find. A pile of wrecked items was then made, petrol was poured onto it and it was set alight, making a giant bonfire. With this the work of the mob was not still finished. Trains from Catanzaro and Cosenza were due to cross at the station at 12:57 and the mob apparently planned to ambush and set fire to them (the tank of each railcar could hold 370 litres of diesel, which could have helped to incinerate the whole train). The planned ambushes were thwarted because Deputy Stationmaster Nicoletti managed to find a working telephone down the line in time to warn the train coming from Catanzaro, while labourer Felice Chiodi, dodging behind hedges out of sight of the crowd, was able to make his way along the line in the opposite direction and stop the train coming from Cosenza by waving a red handkerchief. [646]

The accident led to a heated debate in parliament. In late 1963 the concession to the railway company was revoked and a government commission was appointed to manage it. Traffic over the line was disrupted for a long time and the portion between Soveria Mannelli and Catanzaro was temporarily taken out of use. [647]

Conductor Aristodemo was apparently later released without charge but over three years after the disaster, on Saturday 2nd April 1966, Motorman Miceli was convicted of

---

[645] Ibid, 27th December 1961

[646] Ibid, 28th December 1961

[647] Wikipedia – Incidente ferroviario della Fiumarella

manslaughter and was sentenced to 10 years' imprisonment. [648]

A shifting load on a freight train again caused an accident late on Saturday 28th April 1962. This time seven to ten cars of a 142-car Louisville and Nashville Railroad train in charge of Conductor B.R. Porterfield with Engineer E.P. Estes fell from a single-track span bridge into the Tensaw River, 27 ft deep at that point. The accident happened a mile south of Hurricane Alabama, United States, on the main line between Montgomery, Alabama; Mobile, Alabama; and New Orleans, Louisiana. No one was hurt but Assistant Superintendent C.B. Matthews of the Mobile division of the railroad estimated the damage at more than $1,000,000. He explained that the bridge itself had been extensively damaged and half a mile of track had been torn up. It was thought that it might take as long as a month to repair the damage and reopen the line. About 100 undamaged cars were removed on Sunday. Meanwhile freight trains were being diverted over Southern Railway tracks to Selma and thence over Western Railway of Alabama to Montgomery. Passengers were being taken by bus between Mobile and Pensacola, Florida, and vice versa.

The accident was caused by the shifting of a load of large concrete culverts being carried on a flat car. These caught the side of the bridge, throwing the following cars off the track. The cars that fell into the river had been mostly carrying grain, wheat, and corn. Elsewhere on the train were loads of limestone and miscellaneous products such as paper. [649] [650]

Train No. 53, an express that had left Paris Lyon at 12:30 on Monday 23rd July 1962 for Marseilles met with disaster near Velars-sur Ouche, France, about 12 kilometres short of Dijon. The express consisted of a Société nationale des chemins de fer français Class 2D2 electric locomotive and 19 coaches. It was travelling at almost 140 kilometres an hour, had just emerged from the long Blaisy tunnel, and was passing through a cutting, on a curve leading to the Combe-de-Fins viaduct, when first the second vehicle and then the 12 following coaches became derailed, at 15:19. The presence of the viaduct made the accident much more serious than it otherwise would have been for the seventh vehicle, a coach that was reportedly crowded with passengers, broke through the parapet of the structure, and fell 50 metres to the wooded valley below. The remainder of the train continued for 500 metres, ploughing up the track before coming to a stand. The passengers in this part of the train were only slightly hurt but of those in the coach that fell from the viaduct 39 were killed and others badly hurt. A total of 47 other passengers were injured.

A passenger who had been riding in the sixth coach said, "I had been reading peacefully in my first-class compartment when I felt the train swaying and heard splintering noises underneath the coach. I knew we had left the rails. At that moment the windows fell in, then the speed slowed down and the train stopped."

The alarm was quickly raised by the train conductor, who had escaped injury, and the ORSEC emergency plan was activated. Firemen, police, ambulances and almost every doctor from Dijon went to the scene, as did a large number of onlookers. Much effort was

---

[648] *Florence Morning News*, Florence, South Carolina, United States, 4th April 1966.
[649] *Sarasota Herald Tribune*, Sarasota, Florida, United States, 30th April 1962
[650] *The Daily Herald*, Biloxi, Mississippi, United States, 30th April 1962

required to extricate the dead and injured from the crushed remains of the seventh coach and two 50 tonne cranes were later brought into use. Somehow, one young man was taken from the wreckage almost unhurt but he was the only occupant of the vehicle to have such good fortune. The victims included a number of children, some of the compartments of the coach having been booked for a school party. Other victims were mostly Parisians bound for holidays on the Riviera but also included British and American foreigners.

The line was blocked for a considerable time following the accident and trains between Paris and the South of France, Switzerland and Italy had to be diverted to routes from Paris Est. [651] [652] [653]

The mechanism of the derailment was not immediately clear; initial theories included excessive speed, track defects and even sabotage by the OAS (the Organisation armée secrete – a far right paramilitary group opposed to Algerian independence). The tachograph from the engine showed that the train had been travelling at about 130 kilometres an hour at the time of the derailment while the permitted line speed was 180 kilometres an hour, so excessive speed could be ruled out as a factor in the derailment. There was no real evidence that it had resulted from a terrorist attack either. It had, however, been a very hot day, and in addition the driver of the express spoke of having seen a distortion of the track as his train left the curve approaching the viaduct, leading to the conclusion that warping of the track due to the rails being unable to expand sufficiently was the cause of the disaster. [654]

Early on Tuesday 23rd March 1965 a Chesapeake and Ohio Railway train of three diesel units and rather more than 80 cars, mostly empties, was bound for Kentucky coal mines from Peru, Indiana. It was in charge of Conductor Scott Newby with Engineer James Benham, Fireman C.E. Goodwin, and Brakeman Chester Lee Wilburn and R.E. Arthur. Brakeman Wilburn, from Flatwoods, Kentucky but temporarily resident in Peru, was aged 20 or 21 and had only been doing the job for two months. This was to be his last trip and in fact the last day of his life.

At 2.40 a.m. the train reached Shortcreek Trestle, 75 ft above a ravine and about halfway between Richmond and Boston, Indiana, United States. Engineer Benham said afterwards that as his train had got onto the 1,000 ft long trestle it had suddenly reared up, separated from the remainder of the train, and he had then been able to continue to the end of the trestle. Behind, the second and third units plunged from the trestle to the creek below and then caught fire. The third unit also exploded. Brakeman Wilburn, who had been riding on the second engine, was crushed to death. Behind the third unit 38 or 42 of the cars were also derailed but they did not fall from the trestle, while the rear 43 cars, including the only two loaded ones, and the caboose remained on the track. None of the other railwaymen was hurt. The north west end of the trestle was damaged by the fire.

It was hoped to reopen the line by the end of the following day, meanwhile trains were

---

[651] *The Times*, London, England, 23rd and 25th July 1962
[652] *Gazette de Lausanne*, Lausanne, Switzerland, 24th July 1962
[653] *Journal de Genève*, Genève, Switzerland, 24th July 1962
[654] https://archive.wikiwix.com/cache/index2.php?url=http%3A%2F%2Fwww.gazette-cotedor.fr%2F2010%2F01%2F28%2Fhorreur-sur-le-viaduc#federation=archive.wikiwi

diverted over the Pennsylvania Railroad. It was believed that the derailment must have begun with the second engine but its cause was unclear. [655] [656]

An overheated axle journal ('hot-box') caused the derailment of a 72-car Chicago Burlington and Quincy Railroad freight train at Bellflower, Missouri, United States, on Tuesday 9th April 1968. The dragging effect of the derailed car caused the train to separate and the front 25 cars remained on the track. Sparks from the derailed car set fire to the 65 ft high wooden trestle that the train was passing over at the time. The Bellflower Fire Department proceeded to the scene but backed off for fears that a car in the train loaded with ammonia under pressure might explode. Another car in the train was carrying potentially dangerous titanium tetrachloride. The front cars were drawn away by the engine of the train and another engine drew away the cars at the rear, but eight cars in the middle of the train were burned and by the end of the afternoon the fire had caused part of the of the trestle to collapse. It was feared that the remainder might also collapse.

The accident fortunately occurred in an unpopulated area and no one on the train was hurt. Only two trains daily used the secondary line and these were diverted while the line was out of use. [657]

A 97-car Canadian National Railway freight train was derailed at about 2.45 a.m. on Tuesday 23rd April 1968 as it was crossing Birdtail Creek Trestle, Manitoba, Canada. The latter structure was 290 ft long with a 79 ft central steel span and wooden approach spans; it had a maximum height of 80 ft. The four diesel units and 24 leading cars of the train fell from the trestle into the creek or onto the approach embankment and fire broke out in the wreckage. Fireman Alfred Varga jumped as the train left the rails but was fatally injured and Engineer Robert Emerson and Trainman Herbert Degerstedt were trapped in the units, which were buried underneath the derailed cars, and also lost their lives. The two railwaymen in the caboose at the rear of the train escaped unhurt. The accident happened in hilly and inaccessible terrain on the Birdtail Indian Reserve, about a mile east of the point where Birdtail Creek joined the Assiniboine River.

A newspaper reported that, 'The fire was still burning Tuesday afternoon and clouds of black smoke could be seen as far as 10 miles away...Diesel fuel was keeping the fire alive and mixed cargo in the freight cars was also burning. The freight included a shipment of trucks and household goods such as rugs. Slight explosions were heard at regular intervals and were keeping a crowd of curious onlookers at a distance. Railroad emergency crews had their own pumpers on the scene and were trying to douse the blaze with the muddy waters from the Birdtail Creek. Men from the air force rescue squad at the Rivers base were flown to the scene at dawn in a helicopter and clad in asbestos suits were attempting to reach the diesels at the bottom of the burning pile. Rescue officials on the scene said little could be done until the fire was put out. Meanwhile, with the railroad bridge blocked by the stalled freight cars the only means of transportation across the creek was an old beaver dam a few yards away.'

---

[655] *The Logansport Press*, Logansport, Indiana, United States, 24th March 1965
[656] *Kokomo Morning Times*, Kokomo, Indiana, United States, 24th March 1965
[657] *The Joplin Globe*, Joplin, Missouri, United States, 10th April 1968

The cause of the derailment was unknown, though there was speculation that a protruding part on one of the diesel units might have caught the guard rail of the bridge. [658]

With the main line between Portage la Prairie, Manitoba, and Melville, Saskatchewan, blocked for an indeterminate length of time, trains had to be diverted via Dauphin, Manitoba, with considerable delays. On the day of the accident, the eastbound *Super-Continental* passenger train was expected to be 5½ hours late into Winnipeg and the east bound *Panorama* 5 hours late. [659] Passengers for local destinations were being conveyed by bus.

Fifteen people reportedly died and several were injured when an express passenger train fell from a bridge into the Inmadat River at Langsa in northern Sumatra, Indonesia, on Thursday 9th May 1968. [660] [661] [662]

A train carrying hundreds of holidaymakers back to Karachi, Pakistan, late on the evening of Sunday 29th March 1970, plunged from a trestle 25 miles from its destination in the darkness. The engine and four carriages were derailed on the Karachi side of the small trestle and two carriages reportedly jack-knifed into a dry gully, while three carriages overturned. Rescuers spoke of 20 dead and about 100 injured; railway officials would only admit to five dead and 58 injured. [663] Rescuers worked through the night removing dead and injured by the light of car headlights. [664]

A train carrying 400 passengers and on its way from Santiago de Cuba to Havana was derailed as it crossing a bridge near Vicentica, Cuba, just outside of Santa Clara in Las Villas province, on Friday 16th March 1973, according to an announcement from Havana Radio. One coach of the train went off the bridge and fell into a river bed. Twenty-four people lost their lives and 38 were injured. Only 14 people, one of them the conductor, had been identified at the time of the report. [665]

A 131-car Kansas City Southern Railway freight train was passing through Nederland, Texas, United States, just after 8 a.m. on Saturday 12th January 1974 when a rail apparently broke about the 35th car in the train as it was crossing a small wooden bridge over a drainage canal. The weight of the cars falling on the bridge weakened it and the ninth derailed car fell into the canal. Twenty cars in all were derailed. The derailed vehicles blocked Nederland Avenue, Boston Avenue and Helena Avenue. Wheat and petroleum products were being carried on the train and a substantial quantity of wheat was spilt along the right of way. It became a play area for local children for a while at the

---

[658] *Winnipeg Free Press*, Winnipeg, Manitoba, Canada, 23rd April 1968
[659] *The Brandon Sun*, Brandon, Manitoba, Canada, 23rd April 1968
[660] *Het Vrije Volk*, Amsterdam, The Netherlands, 10th May 1968
[661] *The Daily Banner*, Greencastle, Indiana, United States, 10th May 1968
[662] *Norwalk Reflector*, Norwalk, Ohio, United States, 11th May 1968
[663] *The Arkansas City Daily Traveler*, Arkansas City, Kansas, United States, 30th March 1970
[664] *Gazette Telegraph*, Colorado Springs, Colorado, United States, 30th March 1970
[665] *Sunday Journal and Star*, Lincoln, Nebraska, United States, 18th March 1973

same time as some adults helped themselves to the grain and carted it off.

Don Beagle, the owner of a welding shop adjoining the railway, witnessed the derailment. "'I was standing there by the heater looking out of the window when I heard the brakes. The train couldn't have been making over 20 miles an hour. The car seemed to jump slightly and leave the track and the rest just followed," Beagle said. Beagle discounted the theory that a small trestle had been the cause of the accident. "It stood the weight of several cars after the brakes were applied. Then a wheat car was thrown some 15 feet into the air crushing the trestle when it came down," he said.

The blocked roads were cleared at 5 p.m. and the line was cleared at about midnight. [666]

A serious railway accident occurred in Belgium on Thursday 15th August 1974. This was Assumption Day and a public holiday in the country. A passenger train from Charleroi via Brussels to Antwerp had not travelled very far when it was derailed at 19:15 shortly after leaving Luttre station and as it was crossing a ten-metre-high bridge over the Brussels – Charleroi canal. The second and third coaches left the rails but remained on the track, the fourth coach fell from the bridge and ended up upside down on the canal towpath and the parallel road, while the fifth coach became enmeshed in the metal superstructure of the bridge, where it was crushed by the remained three coaches of the eight-coach train. A fire also broke out in one of the front coaches of the train but it was extinguished before it could cause much damage.

The aftermath of the accident was described as horrific. Oxy-acetylene cutting equipment was required to free victims trapped the twisted metal, surgeons performed on the spot amputations and the condition of some of the injured was so serious that rescuers were reluctant to move them. The rescue operation continued through the night with the help of floodlights and the injured were taken to hospitals at Nivelles, Jumet, Gosselies and Charleroi. Most of the 100 passengers on the train were returning to Brussels from Charleroi. Fifteen people, one of them a two-year-old child, lost their lives in the disaster and according to one report 84 were injured, 41 badly enough to need to be taken to hospital. The train driver was slightly hurt. All the casualties were believed to be Belgians.

The line was of course blocked, though it was hoped to be able to reopen it on the morning of Saturday 17th. Meanwhile passengers were conveyed by bus between Luttre and Nivelles stations.

Excessive speed was ruled out as a cause of the derailment, as the train was shown to have been travelling at less than the 120 kilometres an hour permitted at that point. With the temperature on the day of the accident having been as high as 36 degrees, heat distortion of the track was considered to have been the most likely culprit. [667] [668] [669]

[666] *The News*, Port Arthur, Texas, United States, 13th January 1974
[667] *Tubantia*, Enschede, The Netherlands, 16th August 1974
[668] *Het Parool*, Amsterdam, The Netherlands, 16th August 1974
[669] *The Times*, London, England, 17th August 1974

CHAPTER 11

TUNNEL ACCIDENTS

In his classic work on British railway accidents, *Red for Danger*, L.T.C. Rolt remarks on the acute trepidation felt by Victorian travellers when they were passing through a tunnel, and their oft repeated pleas to the railway authorities that all tunnels should be artificially illuminated, pleas which, probably for practical reasons if nothing else, fell upon deaf ears. He goes on to point out that these fears were by no means unjustified, especially in the period before the introduction of the absolute block system, when trains were merely spaced by the time interval system. If a train should chance to break down in the open there was a good chance that the crew of a following train might be able to see it in time to pull up, even if there was insufficient time for the guard of the first train to go back and protect its rear. In the smoke- and steam-filled darkness of a tunnel there was no such hope. Furthermore, should a serious collision occur in a tunnel the confined space would be likely to result in the wreckage being hideously compressed, and should a fire break out in a long tunnel provided with ventilating shafts, the latter could act as chimneys. In addition the rescue work would be hampered by the darkness and the difficulty of working in a confined space where the use of breakdown cranes was impossible, and the only resort was to drag the wreckage bodily out of the tunnel. Operation of the railways a century later was of course much different from that in Victorian times but the dangers of tunnels remained, as shown by continuing bad accidents and major disasters within them, several really bad tunnel accidents occurring between 1961 and 1975.

The *Tirreno* rapido (express), an ETR 200 Class electric multiple unit, which left Torino for Roma at 16:34 on Thursday 30th March 1961, met with an accident in the approximately 1,400-metre-long single-track Marmi e Salici Tunnel, Italy, between Framura and Bonassola stations. A fire broke out on the train and five passengers lost their lives and about 60 others were injured, 15 of whom were taken to hospital. All were expected to be well enough to be released from hospital by the following day, after having been interrogated by the judicial authorities inquiring into the accident.

The Ferrovie dello Stato train had arrived at Genova Piazza Principe station late and left there at 18:40, now 24 minutes behind time. Between Genova and La Spezia the line ran along the coast of the Ligurian Sea through a whole series of tunnels. At 19:46 the train was two-thirds of the way through Marmi e Salici Tunnel when there was a jolt and tongues of flame appeared below the fourth coach of the train, the leading second-class coach of the seven-car train. A woman passenger, seeing the flames just below her, operated the emergency alarm, as did a sailor and another man. The train then ground to a halt, still 200 metres from the end of the tunnel.

Ascertaining the reason for the stopping of the train Driver Burchi, Assistant Driver Riccardo Scarpellini and Conductors Blondi and D'Andrea rushed to tackle the blaze with fire extinguishers. (In passing it may be added that the train seems to have been somewhat overstaffed as there was apparently also a second driver, Annibale Cannalire, as well as an electrical technician named Stafando Montando.) The railwaymen were able to douse the flames but it was replaced by thick black smoke from beneath the car, which the wind was driving towards the rear of the train. The staff decided that any danger was now over and they could take the train forward. This proved not to be so easily done, as the sudden stoppage had brought down the pantograph and the train could not be restarted. While efforts were being made to correct the problem, the passengers were moved by staff from the fourth car, most of them going to the rear, to the other second-class cars, while a few went forward into the first-class section.

The train remained stationary in the tunnel. At first the passengers stayed calm and waiters continued to serve meals as if all was normal, but as the time went on smoke began to enter the carriages, despite the doors and windows being kept closed, and the passengers began to get increasingly anxious and asked the waiters how much longer it was going to be before the train started moving. Then the passengers felt themselves unable to breathe in the increasingly noxious atmosphere until finally those in the second-class carriages forced open the coach doors and descended into the narrow gap between the train and the tunnel wall. They then made their way forward along the side of the train and through the tunnel to the open air. The 70 passengers in the first-class coaches (there were 405 passengers on the train in all) were the last to evacuate the train. Perhaps due to the air-conditioning system keeping the first-class cars initially free of smoke, none of them apparently became aware of the cries from the rear of the train. It was only when the air conditioning failed, smoke began entering their coaches and then grew worse, that they became alarmed. One passenger, saying that he had an overwhelming feeling of being trapped, tried to break open a window, without success. Finally the doors were forced open and as more smoke poured into the coaches the passengers made their escape, helped by the railwaymen. To make matters worse the wind was now blowing in the opposite direction, towards the front of the train. By now the danger had become obvious to all. When it appeared that everyone was off the train Technician Morando, a handkerchief over his mouth and his eyes bloodshot, made his way through the whole train to check that no one was left on it, calling out loudly to anyone who might be there. No one answered. Unfortunately in the blinding smoke he failed to see five passengers in one of the first-class coaches.

Several passengers gave detailed accounts of their experiences. Dino Grossi, 30, from Genova, was on his way to La Spezia to join the *William A.M. Burden* boat, bound for Buenos Aires, Argentina. He related, "I was in the last car. I heard a screech of brakes and the *rapido* stopped. No one imagined that there was a fire. We thought there must have been a signal at danger; there was a lot of work going on on the lines. The passengers remained calm and someone began eating his dinner. After about 20 minutes we realised that smoke was coming into the coach and noticed that there was a funny smell. We began to get concerned. A child coughed and moaned. The smoke began to rise to the ceiling and I suggested people open windows. They were blocked and couldn't be lowered. We began to panic. 'We have to get out,' a woman screamed. The doors were closed. Four or five of us tried to open them. We eventually did so and got down onto the track in the tunnel. We assumed that the passengers in the other carriages were already safe; we couldn't see them, because there was no light. We squeezed between the train and the tunnel wall, a space of less than half a metre, and walked on, until we were in the fresh air. We had left our suitcases behind."

Eraldo Baratto from Vercelli, 37, said, "I was in the fifth car. I waited 50 minutes after the train stopped. Around me the passengers were getting restless. The smoke and the unpleasant smell were getting worse. I got up and heard a railwayman muttering to a colleague, 'What are we going to do? Shall I let them down? We have to warn them…' With that I realised that the situation had become serious. I got off, as did dozens of others. Now everyone was screaming. They were afraid, they looked for each other in the dark. Shapes in front of me slipped on the sleepers, lost their luggage, got up with difficulty. They coughed and pressed handkerchiefs to their mouths. We felt suffocated. Groping along, almost in anger, we jumped at each other, we made ourselves fall. I slammed my face against the tunnel wall, afraid of being swept away by the tide of people. The cries were impressive, women wept and fainted. Luckily there wasn't far to go. If the tunnel had been 100 metres longer I don't know how many of us would have survived."

The Hon. Randolfo Pacciardi, a Centre-Left politician, was riding in the second car of the train. He said, "Near me were a general and a German lady. We waited patiently in the dark, for the train to move. Someone lost their temper. We suggested passengers sit quietly in their seats and if I have one regret, it is that at my suggestion the evacuation of the train was delayed. Then there was a panic. The restaurant staff began shouting, 'Get off the train, otherwise you're lost.' I found myself among people who were trampling over one another, with the flames grazing us. The air was unbreathable; my lungs ached. After going a few metres I felt that I could not escape any more. A railwayman cried, 'Come on, we're safe, don't stop!' I heard an exclamation, a woman screamed, 'I'm dying, help me!' but I couldn't stop." The politician suffered mild carbon monoxide poisoning; he was treated at Levanto hospital and discharged after a couple of hours.

Help was slow in arriving because news of the accident was slow reaching anyone. The accident also happened at an inaccessible location, unreachable by road. Finally at 21:30 Stationmaster Mario Podestà at Chiavari raised the alarm, because a train due from La Spezia had not arrived, and the Chiavari firefighters were called at 21:35. The alarm was raised somewhat later at La Spezia. A steam engine was then sent from the latter station and was able to pull the leading three cars out of the tunnel; the brake blocks of the fourth car had been welded to the wheels by the heat and it could not be shifted. When the engine and cars reached Bonassola station fire broke out again in the second carriage. There was no water available at the station but some fishermen who had been washing their boats connected a hose to a fountain in the village square and the small quantity of water that they could thus obtain proved sufficient to extinguish the blaze. The coaches were then examined and the lifeless bodies of the five passengers were discovered. A second steam locomotive and two coaches later came from La Spezia with firefighters.

The passengers who had lost their lives were three men, a woman, and a boy. One man was identified as Dr Carlo Marcotulli, 33, an Agip (Italian oil company) official, who had been returning to Roma with his family for the Easter holidays. The second man was Dr Francesco Paolo Nascuti, President of Section II of the Genova Criminal Court, who had been on his way to visit relatives in Chieti province. The third man was Ettore Bacigalupo, 60, deputy director of the autonomous institute for popular housing in Genova. The woman was his wife, Lina Galloreto, 55, and the boy was their 11-year-old grandson, Ettore Perrero, whom they were taking to visit his parents in Roma. All had died of carbon monoxide poisoning. It was believed that having initially remained calmly sat in their seats as requested by the railway staff, by the time the other passengers had begun to evacuate the train they had already become unconscious and had possibly already died from the carbon monoxide, thus they had not responded.

Other passengers had lucky escapes. Professor Eugezznio Guglielminetti, a set designer of the Teatro Stabile in Torino, was on his way to Roma to meet actress Lilla Brignone, accompanied by Dr Robert Marchietti, director of the Alfieri Studies Library in Asti. Dr Marchietti stayed too long on the train and lost consciousness, but was rescued by his friend, the professor, and a fireman. The professor was understandably critical that the passengers had not been warned of the potential danger during the period of almost an hour that they had been stopped in the tunnel.

A woman from Genova, having struggled to the end of the tunnel and recovered somewhat, was crying and calling out the name of her little daughter, from whom she had become separated in the confusion and whom she feared had perished in the fire. Later, to the mother's intense relief, the child was found lost and stumbling along the line amongst another group of passengers.

The fire was blamed on a short circuit in the electric motors of the train. It eventually spread to all the cars of the train and 200 metres of the overhead line equipment was also destroyed.

After the accident some of the railwaymen blamed it on the woman passenger who had panicked and activated the alarm, stopping the train, and asked the police to try and identify her. A newspaper protested that this was most unreasonable and what else was the passenger meant to do with a fire around her feet? It was implied that on an Italian train as soon as the emergency alarm was operated the train came to a stand, whereas in Great Britain if the communication cord was pulled the driver was able to continue to run the train until it reached an opportune point at which to halt. If this was the case, then it seems strange that the train did not continue the short distance until it was out of the tunnel. [670] [671]

On Sunday 13th January 1963 the 7.31 p.m. passenger train from York to Liverpool caught fire in the 657-yard-long Sowerby Bridge Tunnel, England, on British Railways, North Eastern Region, two minutes after it had stopped at the station of the same name. It was the rear coach of the six-car diesel multiple unit that caught fire. Firemen from Sowerby Bridge, Halifax and Elland attended and were able to extinguish the fire before it spread to the remainder of the train. A senior fireman said, "We were lucky the train was not entirely in the tunnel or we should have been considerably hampered and the fire may have been much more serious than it was." Five fire engines in all took part. Despite the efforts of the firemen the rear car was burnt out.

---

[670] *La Stampa*, Torino, Italy, 31st March and 1st and 2nd April 1961
[671] *L'Unita*, Roma, Italy, 1st April 1961

The 70 passengers were forced to alight from their train in the pitch-dark tunnel and walk back three-quarters to Sowerby Bridge station in temperatures which were well below freezing, waiting at the station on a relief train until the tunnel could be reopened after almost three hours. The guard of the train was slightly hurt but none of the passengers was injured. Some, however, missed the night sailing of their boat to Ireland. A passenger said, "There was no panic. When we saw smoke filling the coach we just grabbed our luggage and walked up the train. Then, when it stopped, we filed out and walked past the coach, which was well alight and filling the tunnel mouth with smoke and flames." [672]

The annual report of the railway inspectorate outlined the cause of the fire: 'The ducting of an oil-fired heater which was made of an inflammable glass fibre material caught fire. A technical investigation, much of it carried out in conjunction with the Fire Research Station, disclosed a number of unsatisfactory features in the design and fabrication of the heater concerned and of its ducting and in the way in which heaters of that type were maintained. These features have now been much improved.' [673]

Two suburban trains were involved in a serious collision in a tunnel at Lyon, France, at 14:20 on the afternoon of Monday 26th August 1963. An electric omnibus train from Charbonnières-les-Bains proceeding through the 1,421-metre-long Loyasse Tunnel to the Lyon Saint-Paul terminus at 10 to 15 kilometres an hour was run into from behind by Micheline railcar No. 27100 from Villefranche-sur-Saône, towing three trailers and travelling at 80 kilometres an hour. The omnibus train was propelled forward 50 metres by the impact but not greatly damaged, while most of its not very many passengers were sat towards its front end and escaped with slight injuries. The Micheline railcar and its leading trailer were, however, telescoped into one another and it was here that most of the casualties occurred. Its driver and seven passengers were killed outright, while 16 more of its 50-odd passengers were injured, three of them seriously.

The dazed conductor of the omnibus train struggled several hundred metres through the tunnel to the terminus to report the accident. Railwaymen from St Paul at once went to assist the injured passengers and the local prefect launched the ORSEC emergency plan. Rescuers accessed the tunnel from both ends but the lights of the two trains had been extinguished in the collision and the darkness made the rescue work difficult. The uninjured and slightly injured passengers made their way out of the tunnel on foot as best they could. While the line was blocked the Société nationale des chemins de fer français (SNCF) organised replacement buses for the passengers between St Paul and Gorge-de-Loup stations.

The investigation into the accident noted that the driver of the Micheline train might possibly have seen the three red lights at the rear of the omnibus train rather earlier but his vision had probably been affected to some degree by the bright sunlight outside, then the sudden transition to the darkness of the tunnel. Blame for the accident lay in any case not with him but with one or both of the signalmen at Lyon Saint-Paul and Lyon Gorge-de-Loup, who had got confused between the two trains and had not clearly identified

672 *The Times*, London, England, 14th January 1963

673 Railway Accidents: Report to the Minister of Transport on the Accidents that occurred on the Railways of Great Britain during the year 1963.

which train it was that they were referring to in a telephone conversation, so that the Micheline train received clear signals at Gorge-de-Loup while the omnibus train was still in the section. This was evident from an examination of the tachograph of the Micheline railcar, which showed that its driver had received clear signals at the entrance to the tunnel. The investigation board recommended that the signalling on the line be modernised, in view of its considerable traffic. [674] [675] [676]

On the early evening of Friday 15th April 1966 what could have been a very serious accident occurred in the 7,971-metre-long Borgallo Tunnel in Italy. The double-track Ferrovie dello Stato tunnel lay between Grondola Guinadi and Borgo val di Toro stations and carried the La Spezia – Parma main line beneath Monte Cisa. At about 17:30 northbound freight train No. 6540 entered the tunnel at the Grondola Guinadi end. For a reason that was not immediately determined a wheel of the rear waggon of the train failed, the train was derailed and fouled the opposite running line. Shortly afterwards *diretto* passenger train No. A261 from Palma to La Spezia collided with the derailed vehicle.

A workman from La Spezia who was a passenger related, "I was sat quietly reading a newspaper when I got up to take out a packet of cigarettes from my bag. As I doing so I heard a crash like a bomb going off and was covered by bits of glass. I didn't know what was going to happen next but feared the worst as the coach was shuddering terribly and I thought that the train was about to leave the track. Finally things settled down. To be honest I was very frightened."

Many windows of the coaches of the passenger train were shattered, 30 passengers were slightly hurt and something of a panic was caused amongst them, especially those towards the front of the train. Emergency vehicles were got ready and ambulances were sent but all the passengers were treated at the scene and none needed to go to hospital. [677] The passengers were forwarded to Pontremoli station by bus, from where they continued by train to La Spezia. Late at night the line was still blocked and trains were diverted.

The passengers on a train from San Baudillo de Llobregat that arrived at the Plaza de España underground terminus of the Ferrocarrils de Catalunya in Barcelona, Spain, at 09:45 on Thursday 8th September 1966, found themselves confronted with smoke and flames coming from another train stabled in a siding. More or less trapped on the platform, a panic arose among them as they fought to escape from the station. Three people died, presumably from smoke inhalation, and 115 were injured. [678] [679]

About 1,200 passengers were riding on two four-car Massachusetts Bay Transportation Authority (MBTA) passenger trains that were in collision near Maverick

---

[674] http://www.maligne-ter.com/ouest-
[675] *Le Monde*, Paris, France, 28th August 1963
[676] *The Times*, London, England, 27th August 1963
[677] *La Stampa*, Torino, Italy, 16th April 1966
[678] https://www.licexballet.com/wp-content/uploads/2019/03/1966-09-09-Festivales-de-Espan%CC%83a-Presentacion-en-el-Liceo-del-Ballet-Moisseiev-LVG001-cu.pdf
[679] Federación Castellano Manchega de Amigos del Ferrocarril

station, Massachusetts, United States, at the height of the morning rush hour on Monday 8th January 1968, in the underwater tunnel beneath Boston Harbour connecting the main part of the city and East Boston. The front train was stopped at signals waiting to enter Maverick station when a second train ran into its rear. Many standing passengers were knocked to the floor by the impact and then trampled on by other passengers. Police, firemen, uninjured passengers, and newsmen took the injured to ambulances waiting at Airport station, though the rescue work was hampered by deep snow and near-zero temperatures. The electric current was quickly shut off after the collision, allowing passengers to walk about 100 yards along the tunnel to its east end. At least 15 passengers were treated at Massachusetts General Hospital and about the same number at the City hospital, both in Boston. Only one passenger was detained with knee injuries.

Passenger Miss Jennie Cohen of Revere, Massachusetts, a bookkeeper, who was an uninjured passenger in the colliding train, said it had, "Stopped, started, then stopped and started again. The driver seemed to be blowing his horn as a warning…Then suddenly we stopped again as though we were hitting something."

Nancy Yakuvian, 21, of Chelsea, Massachusetts, riding in the front train, said, "We were stalled nearly 20 minutes between the Airport and Maverick stations. I heard three whistles and then we were stuck from behind with a terrific impact."

Henry Goode, 25, of Winthorp, Massachusetts, another passenger in the front train, said that the "trip from Orient Heights (three stops before Maverick) had been a constant stop and go affair. Everyone was thrown to the floor when we were hit from the rear."

MBTA General Manager Leo Cusick stated that the motorman of the second train had done all that he could to stop his train. The signals had been working correctly but the cold and snow had prevented the brakes of the train from gripping the rails, so it had slid along at five or six miles an hour until it struck the train in front. An investigation into the accident would be held. He added that there had been four pregnant women passengers; he did not know if they were among the injured.

'Uninjured passengers,' newspapers reported, 'took refuge from the biting cold in stores along nearby Bennington Street and clogged telephone lines to assure relatives they were all right. The MBTA rushed several buses to the area to pick up Boston-bound passengers.' [680]

Perhaps the most feared type of railway accident that can occur is a single-line collision in a tunnel, with the wreckage of two trains being compressed together in the narrow confines of the bore. Such accidents have always been rare but one happened in Sicily, Italy, early on Sunday 15th June 1969, on the electrified line between Messina and Palermo.

Accelerato (slow) train No. 2910 had left Palermo Centrale at 22:35 on the evening of Saturday 14th June and was due at Messina at 04:34 on Sunday 15th. It consisted of a locomotive, a mail van, a first-class coach, three second-class coaches and seven boxcars, in charge of Conductor Biagio Bonifacio, 30, with Driver Filadelfia Di Leo, 54, and Assistant Driver Pasquale Pugliatti, 56. There were three first-class passengers and 127 second-class passengers.

[680] *The Bridgeport Post*, Bridgeport, Connecticut, Unites States, 8th January 1968

In the opposite direction freight train No. 7441 was en route from Messina to Palermo. It consisted of a locomotive and 39 waggons, some loaded with tar, in charge of an unnamed conductor with Driver Antonio Saglimbeni, 43, and Assistant Driver Antonino Barbera, 33. (The various newspaper reports are not entirely clear. They mention another railwayman, Antonio Nigrelli, who apparently was not on duty at the time of the collision but was travelling to take up duty at Messina. It is also possible that Conductor Boniacio was in charge of the freight train rather than the passenger train.)

The trains collided at 03:20 on Sunday morning in the 150-metre-long Sant Antonio Tunnel, between Barcellona and Castroreale Terme stations, while both were reportedly travelling at 90 kilometres an hour.

Though he was seriously hurt Assistant Driver Barbera somehow survived the collision and he was reportedly one of the first to raise the alarm. He was transported to Messina for urgent surgery, having suffered a head injury that it was feared might cost him his sight. Meanwhile some of the passengers from the rear of the train had made their way out of the Palermo end of the tunnel in the darkness and had also raised the alarm. About 60 other passengers found themselves trapped in their carriages and they were not released until Sunday night, after what must have been a long ordeal. Several hundred firemen, railwaymen, police, doctors, nurses, and volunteers took part in the rescue work but the task was extremely difficult in the confined space, and oxy-acetylene equipment could not be used to cut through the twisted metal wreckage because of the possibility of starting a fire, though in fact a fire broke out anyway later on (see below). Special ventilating equipment was installed to try and disperse the fumes from the damaged tar waggons. Early on it was known that both drivers, Assistant Driver Pugliatti and Postal Worker Francesco Di Salva had been killed, together with a passenger later identified as financier Franco Cardile. Trapped in the wreckage for many hours were Conductor Bonifacio and a 17-year-old athlete named Salvatore Santamaria. The young man lingered in agony, trapped by his legs, pleading for the rescuers to hurry and release him. Doctors feared that any attempt to amputate his legs would result in a fatal collapse but were able to administer pain killers and sedatives into one of the victim's hands to somewhat ease his suffering. After 31 hours both the young athlete and the conductor, who had long been unconscious, succumbed, before they could be released. Later the body of another passenger, who had not been on the list of missing persons, 26-year-old student Claudio Fiscauli, from Bologna but living in Palermo, was also discovered. Meanwhile the discovery of a wallet belonging to a 64-year-old French tourist from the Gironde, Roger Pierre Tachoires, led to fears that he was a ninth fatality. Eventually it was found that this was not the case; happily the tourist had escaped, losing his wallet in the confusion. About a dozen people in all were hurt, including seven firemen affected by the fumes.

Three days after the accident, on the afternoon of Wednesday 18th, while the rescue operation was still in progress, fire broke out in the wreckage, consuming the bodies of Driver Di Leo, Assistant Driver Pugliatti and Conductor Bonifacio. It was finally brought more or less under control after about eight hours by firemen clad in asbestos suits.

A mistake on the part of the Castroreale Terme stationmaster, Angelo, or Giuseppe Venuto, apparently led to the disaster, it being surmised that he had permitted the passenger train to proceed towards Barcellona while the freight train was already in the section from there. An order was issued for his arrest by the Public Prosecutor, Dr Rocco Scisca. Presumably realising what he had done, the stationmaster disappeared shortly after the accident and several days later had still not been traced. Engineer Bianchi, from Roma, interviewed various railwaymen as part of the Ferrovie dello Stato investigation and the public prosecutor interrogated the Barcellona stationmaster and other officials.

Newspaper editorials condemned the fact that the line was only single track – as if every railway, however little traffic it carried, should be double track – the lack of radio-telephones on all trains, the inadequacy of the rescue operations, and the fact that the railways of southern Italy allegedly received less financial support than their counterparts in the northern part of the country. [681] [682] [683]

Shortly after passing Iselle di Trasquera station and entering the south portal of the Simplon Tunnel (probably the 19,803 metre Simplon I Tunnel) at about 10:30 on Saturday 8th November 1969 some of the passengers on the northbound *Trans-Europa-Express* thought that they smelt burning and began to feel uneasy. A kilometre further on flames appeared, the communication cord was pulled and the train stopped. Passengers from the affected coach were moved to the first one not involved in the fire and the engine continued through the tunnel with the front of the train to Brig station at the Swiss end. A dozen or so passengers from the rear of the train fled back through the tunnel to Iselle station. No one was injured either from burns or from smoke inhalation. The tunnel was partly reopened a little after 13:00. It was believed that an overheating of the electric motors of the train might have caused the fire. [684]

Only four days after 45 high school boys from Seoul had been killed in a level crossing collision, ten schoolboys from Inchang High School in Seoul were killed in another train accident, this time a head-on collision between a passenger train and a freight train in a single-line tunnel on Saturday 17th October 1970. Two teachers and two civilian passengers were also killed and 55 students were injured.

The first six coaches of the passenger train, which had stated from Chonggyangri station in Seoul at 07:30, [685] were carrying 700 boys and girls from three Seoul high schools to the Kyongju resort, 450 kilometres south of Seoul, for a four-day holiday. Those who died had been riding in the leading coach of the train, which police said had jumped onto the engine of the freight train in the collision and had been crushed between the top of the engine and the roof of the tunnel. The rescue operation was in charge of the transportation and education ministers. As usual in such situations it was hampered by the darkness in the tunnel and the very confined space that was available to work in. The rescuers included about 300 South Korean soldiers and police, among them 50 Army medics from 1st Korean Army headquarters in Wonju. The injured were taken to hospital in Wonju.

A breakdown in the centralised traffic control system was stated to have been the cause of the disaster. [686] The trains collided in the 150-metre-long Bongsandong Tunnel at 11:45.

---

[681] *L'Unita*, Roma, Italy, 17th, 18th and 19th June 1969
[682] *Stampa Sera*, Torino, Italy, 16th June 1969
[683] *La Stampa*, Torino, Italy, 19th June 1969
[684] *Gazette de Lausanne*, Lausanne, Switzerland, 10th November 1969
[685] *The Bakersfield Californian*, Bakersfield, California, United States, 17th October 1970
[686] *Pacific Stars and Stripes*, Tokyo, Japan, 19th October 1970

One of the surviving passengers, 17-year-old Kim Yun-han, said, "I was in the first coach sitting just behind the teacher and was about to move to the second coach when I heard a terrible crash and the coach jumped upwards. Every light went out and I could hear screams and feel my friends wriggling around in the darkness." Kim escaped unhurt; his friend who had been sat next to him was killed. [687]

The breaking down of a diesel locomotive was the precipitating event in a disastrous train fire in Vranduk Tunnel, Yugoslavia, on Sunday 14th February 1971. Local train No. 820 from Žepče to Zenica consisted of the diesel electric locomotive Končar and nine coaches, with about 200 passengers, mostly mine workers and workers at the RMK steel mill at Zenica, on their way to work, together with a few students.

The single-line tunnel was about 1,500 metres long and the train stopped only about 200 metres from its Zenica end. What exactly happened next is unclear but it would seem that while efforts were being made to restart the engine an electrical spark set fire to diesel fuel that had spilt from the tank of the engine and the blaze quickly spread to the passenger coaches. When the train at first stopped, its passengers were not unduly concerned as it was often stopped at that point by signals. Then, however, the conductor informed them that the train was on fire and they must abandon it and make their way back through the tunnel towards Nemilia. Going forward, was of course not an option, even though it was a much shorter distance to the tunnel exit, as that would mean going through the fire. As the tunnel was on a rising gradient towards Zenica there would have been a possibility of detaching the carriages from the engine and allowing them to run back through the tunnel but apart from the difficulty of carrying out such an operation and of controlling the runaway coaches, there would have been the risk of running down passengers who were already making their way back through the tunnel on foot. It was a long, hard journey to the Nemilia end of the tunnel for the frightened passengers, groping their way through the darkness and choking smoke. Some of them would never make it.

One of the survivors, Mr Hasan Mujkanović, described his escape to a reporter. "When the train entered the tunnel I caught the smell of oil. This is unusual. When the train was in the middle of the tunnel I felt braking, and then the train stopped. I was in the last carriage. We started to open windows, but quickly closed them again because smoke and fumes poured in...Someone proposed that we abandon the train, which we did. It was pitch dark in the tunnel and difficult to find our way, but we kept close to the walls of the tunnel and made our way out. I heard some screams for help, but no-one could help anyone. We were all trying to save ourselves. When we came out we were all exhausted and sick. About 50 workers came out in the first group." [688]

The alarm was reportedly raised by the first passengers to escape from the tunnel and police, ambulances, fire engines and volunteers from the steel mill rushed to the scene. The accident had occurred at about 06:00 and the rescue operation lasted until about noon. A swift response from the rescuers was credited with saving many lives. The injured were taken to hospital in Zenica. After the fire had finally been extinguished and the heat had subsided the whole tunnel was searched, when rescuers found that the rails had been distorted by the heat and came across places where passengers had vainly

[687] *The Daily Report*, Ontario/Upland, California, United States, 18th October 1970
[688] *The Times*, London, England, 15th February 1971

excavated holes in the ballast in an attempt to escape the fire and fumes. There and elsewhere were discovered the blackened bodies of victims, many clutching handkerchiefs to their mouths to try to keep out the smoke. In all 34 passengers lost their lives, most, if not all, dying from carbon monoxide poisoning, and another 113 were injured, 60 seriously. As the accident happened on a Sunday the number of passengers on the train was considerably less than it would have been on a weekday. [689]

At the start of the blaze the driver and his assistant had attempted to fight the fire with an extinguisher. Both survived the accident and were afterwards arrested by the police, though the accident would hardly seem to have been their fault. The locomotive involved had apparently failed five times previously in the two months before the disaster. [690]

A destructive accident that was also fatal to two railwaymen occurred at 04:22 on Saturday 20th March 1971 in the 226-metre double-track Crozet Tunnel, France, between St Martin-Sail-les-Bains and Roanne stations, in the Loire department. Two freight trains were involved in the accident, which was a derailment followed by a collision. Travelling in the southbound direction and running two minutes early at the time of the accident, was a train of 55 waggons of mixed goods. Proceeding in the opposite direction was a train of 24 tank waggons of hydrocarbons: fuel oil, heating oil, petrol, and premium (high octane) petrol; this train was 22 minutes behind time. Both trains were being pulled by Class BB diesel locomotives. The locomotives were of a lightweight design and had plastic front grilles which provided little resistance to impact and were inflammable.

The cause of the derailment of the northbound train, which was the initiating event of the accident, could not be determined with certainty afterwards. A broken rail was found during the clearance work but it was not clear if this had caused the derailment or was a consequence of it. As the trains were passing one another in the tunnel the 18th waggon of the northbound train, carrying heating oil, was derailed, fouled the opposite running line, and was struck by the engine of the southbound train. As a result this tank waggon was pushed back against the following waggons of the northbound train, while the leading hopper waggon of the southbound train, loaded with 39 tons of silica, penetrated seven metres into the rear cab of the engine of the southbound train before rebounding into the tank waggons of the northbound one. Behind the hopper waggon in the southbound train were ten light empty tank waggons, which due to the sudden stoppage were compressed by the heavier waggons behind, bending the front of the train 'like an accordion.' To complete the destruction fire broke out immediately after the collision and quickly involved all of the derailed waggons and the engine of the southbound train.

By chance a volunteer fireman lived in one of the houses on the hill above the tunnel. Awoken by the noise of the collision and several muffled explosions he realised that something must be wrong, quickly dressed. and went to the north end of the tunnel, where he found the two railwaymen from the northbound train. He was joined by a second volunteer fireman from Le Crozet. Despite the smoke and heat the two firemen entered the tunnel and were able to uncouple the tank waggons behind the 13, so that these would be drawn out of the tunnel. They were later decorated for their bravery.

With no engine available on the southbound line it was not possible to do anything

[689] Wikipedia – Željeznička nesreća u Vranduku 1971

[690] *The Times*, London, England, 15th February 1971

about the waggons of that train and soon the increasing intensity of the fire, fuelled by about 600,000 litres of oil, made it impossible to enter the tunnel again. The temperature of the fire was estimated to have reached 2,000 degrees Centigrade, sufficient to melt the glass of bottles of mineral water being carried by the southbound train and the granite lining of the tunnel. It was not until about 17:00, 13 hours after the accident, that the last loaded two tank waggons of the northbound train, which had not caught fire, could be extricated from the south end of the tunnel and the rear 34 waggons of the southbound train, mostly badly burnt, could be likewise pulled from the north end of the tunnel.

Over Saturday night and Sunday morning the roof of the 113-year-old tunnel collapsed at three separate points, while a further problem was caused by leaking unburnt fuel polluting the local water supply.

Further collapses of the tunnel hampered the clearance operations and it was eventually decided that there was no alternative but to convert it into an open cutting, an operation that also involved the demolition of the four houses that had stood on the hill above it. It was not until Saturday 17th April, four weeks after the accident, that the engine of the southbound train could be reached. The front of the engine was completely crushed in and it appeared that Driver Gaston Mathieu and Assistant Driver Charles Calvez must have been killed instantly. All that remained of them was a few charred bones and the driver's watch, stopped at the time of the accident. The line was not finally reopened until Thursday 24th June.

After the accident it was suggested that the timetables should be amended such that trains carrying dangerous substances should not pass other trains in tunnels, a proposal that was not really very practical. Another, probably more sensible measure, was the reinforcement of the driving cabs of locomotives so that their occupants had more protection in the event of a collision or derailment. [691]

The Simplon tunnels were the scene of another accident early on Thursday 22nd July 1971, more serious than the 1969 fire (see above). At 04:28 a train left Domodossola in Italy for Brig and Visp in Switzerland. It was carrying 60 passengers, mostly Italian workmen, and consisted of a Swiss Class Rbe 4/4 electric railcar coupled to a driving trailer. When the train reached the middle of the tunnel it was diverted through one of the facing crossovers situated there onto the single line of the second tunnel and it was at this crossover that the accident occurred. As the train was negotiating the crossover the driving trailer left the rails and crashed against the wall of the tunnel. The lights of the train were extinguished and some of the coach doors were jammed.

The alarm was at once raised and two special trains with doctors and nurses were sent from Brig, the seriously injured being in due course taken to the hospital there, where nine were admitted. The less seriously hurt were also taken to Brig for treatment, then a special train took them home to Italy. In all five passengers were killed outright, four were very badly hurt and about 20 were less seriously injured. The damage caused was estimated at several hundred thousand Swiss francs. The workmen were all employed at the Chipis and Steg factories in Switzerland.

---

[691] http://www.tunnels-ferroviaires.org/tu42/42078.1.pdf

At a press conference it was stated that rumours that the train had been travelling at 100 kilometres an hour at the time of the derailment were probably greatly exaggerated. There was a 50 kilometres an hour speed limit through the crossovers. An inquiry into the accident was immediately opened at Brig. The likely causes for the derailment were assumed to be excessive speed or a defect in the points.

Traffic through the tunnel resumed at about 09:00. Inevitably there were delays to trains, including some international expresses between Paris and Milano. Chemins de fer de fédéraux suisses officials and federal councilor Bonvin went to the scene at the end of the morning. [692]

The inquiry into the derailment concluded, a few days later, that the speed of train had indeed been excessive, at 80, 90 or even 100 kilometres an hour through the crossover where speed was limited to 55 kilometres an hour. Motorman Dreyer then stated, "It's wrong, I was not running at 80 or 90 kilometres an hour. It was 120 kilometres an hour...I was entitled to. When you have to reduce speed to 55 kilometres an hour for the crossover there are two lights displayed in the tunnel. There was only one light, I am certain. It was the automatic device which failed," he added. "As soon as I saw that I was being routed through the crossover I applied the brakes in emergency, but the train takes 800 metres to stop." [693]

The author is not familiar with the signalling in the Simplon tunnel but that a signal indicating that a driver must reduce speed should have two lights and that one indicating he could proceed at full speed should have only one light seems unlikely, as the failure of one light would then create a dangerous situation. Possibly the newspaper account has its facts wrong.

A fire that occurred on a train in an Italian tunnel just after midnight on Friday 30th July 1971 had an unusual cause. Train No. 13644 from Roma Ostiense consisted of an engine and 14 carriages at the time of the accident, six carriages originating in Roma, three in Milano and five in Genova. It was a train carrying pilgrims to Lourdes in France, organised by the Marian Priestly League, an organisation that provided support for members of the Roman Catholic clergy and their families who were elderly, sick or disabled. The departure from Genova was delayed about 10 minutes due to the difficulties getting the disabled and sick passengers on board but it left there at 22:45 with 450 people. The train was due to arrive at Finale Ligura Marina at 23:30 on Thursday and when it did not appear the alarm was raised.

It transpired that the train had come to a stand in a tunnel between Varigotti and Finale Ligure Marina because an explosion and fire had occurred in one of its carriages, due to an accident with an oxygen cylinder belonging to one of the passengers. The fire spread to two and three other carriages, which were detached and taken out of the tunnel after their passengers had been moved to an unaffected part of the train. The fire brigade and carabinieri attended and the adjoining Via Aurelia coastal road was closed while the rescue operation was in progress. A number of passengers suffered from the effects of smoke inhalation and were taken to local hospitals; none was apparently

[692] *Gazette de Lausanne*, Lausanne, Switzerland, 23rd July 1971
[693] Ibid, 27th July 1971

seriously hurt. [694] [695]

A major disaster occurred in Japan at about 16:00 on Monday 25th October 1971 on the Kinki Nippon Railway. A four-car express train from Osaka to Nagoya with 96 passengers collided head-on with a seven-car Nagoya – Osaka – Kyoto express carrying 396 passengers. Both trains were electric multiple units. They met in the 356-metre-long Sotani Tunnel, between Higashi Aoyama and Sakakibara Onsenguchi stations. The train bound for Nagoya should have stopped at Higashi Aoyama until the other train had cleared the single-line section. Instead it continued, travelling past Higashi Aoyama at an estimated 90 kilometres an hour instead of the normal 65 at that point. Upon seeing this the signalman there immediately sent a message to Sakakibara Onsenguchi to stop the Kyoto-bound train but it had already entered the section from the other end.

Seeing the other train approaching the conductor of the Nagoya-bound train shouted to his passengers to get on the floor and by doing so probably saved some of their lives. He was found dead in the wreckage, still holding his microphone. [696] Also killed were the conductor of the other train, Motormen Hiroyuki Okumara, 28; and Katsumi Nishiyama, 27; and possibly 21 passengers. About 230 passengers were injured, some 40 of them seriously. All four cars of the Nagoya-bound train were derailed, three of them overturning, while 'cars on the other train were crumpled and wrecked and sections on all the cars thrust out in jagged chunks to block rescuers.' There was also a delay in the rescue operation as there was no road access to the tunnel, while downed overhead line equipment reportedly blinded rescuers and wreck-related power failures knocked out telephones. [697] Several hundred people took part in the rescue work and hotels and inns in the area rushed buses and other vehicles to the scene to carry the injured. [698]

As to the cause of the disaster Fumio Ohtsuki, the vice president of the company, announced that it had been due to an 'operational error' on the part of Assistant Motorman Masahiro Miyata of the Nagoya-bound train. The vice president explained that the assistant motorman had closed the valves of the train's air brake system to check that they were working properly before the train reached the tunnel but had then failed to reopen them as the train descended a steep gradient, so that it had run away. [699]

From time to time accidents have occurred when the roofs of tunnels have collapsed and trains have subsequently run into the mass of rubble that ended up on the line but up until Friday 16th June 1972 there had never been any accidents approaching the magnitude of the one which occurred on the evening of that day, in the 1,390-metre double-track Vierzy Tunnel in France, just north east of Vierzy station, on the line between La Plaine and Hirson. This catastrophe claimed 108 lives and resulted in 111 injured.

At 20:21 on the day of the accident a light engine passed through the tunnel without incident. Shortly after the passage of the engine the roof of the tunnel collapsed and 700

---

[694] *L'Unita*, Roma, Italy, 30th July 1971

[695] *The Times*, London, England, 31st July 1971

[696] *Gazette Telegraph*, Colorado Springs, Colorado, United States, 27th October 1971

[697] *Pacific Stars and Stripes*, Tokyo, Japan, 27th October 1971

[698] *Muscatine Journal*, Muscatine, Iowa, United States, 25th October 1971

[699] *Pacific Stars and Stripes*, Tokyo, Japan, 28th October 1971

cubic metres of cement, rock and sand piled up in a mass on the track. At 20:54 a Paris Nord to Laon passenger train No. 2841, consisting of a railcar, entered the tunnel, travelling at 108 kilometres an hour. The train was crowded with students and workers on their way home for the weekend and soldiers going on leave. A few moments later the railcar hit the mass of debris and stopped abruptly, its second car became separated from the leading one and came to rest alongside of it and the third car telescoped the first two. Then at 20:56 a passenger train No. 7844, bound for Paris, another railcar, entered the tunnel from the opposite end and collided with the mass of debris in turn. Its leading car was forced up towards the roof of the tunnel and was crushed by the second.

The conductor of train No. 7844 raised the alarm from a lineside telephone at 21:11 and the conductor of train No. 2841 did likewise about the same time, having made his way back to Vierzy station. Eight rescuers initially went to the scene and a major alert was issued at 22:00, when the Aisne prefect launched the ORSEC emergency plan.

At the height of the rescue operation about 500 rescuers laboured in the darkness and heat of the tunnel. The rescue work had to be done with the workers well aware of the possibility of further debris falling onto them and oxy-acetylene equipment could not be employed. Dr René Noto commented that "The SNCF staff told us to hurry up. If more of the tunnel collapses the victim dies and so do we." As might be expected some of the rescuers were considerably affected by the stress of the situation, as the doctor noted, "At the time, we thought that the firefighters were made of brass. We found that it wasn't true. They too experienced fear and anxiety."

Jean François Devars du Mayne, at the time a junior house officer, spoke many years later of his experiences. He described the rescue of the last survivor, 22-year-old Bernard Debraye. "We were called back after 36 hours. When we got there we were told that the search was over." A few moments later though signs of life were heard deep in the wreckage. "Neither the nurse nor the doctor at the scene felt that they had the strength to return," du Mayne continued. "It was difficult to go back when we were so affected by the magnitude of the disaster. They accompanied me. I had to climb up a mound of earth and scrap metal. They had made a hole 50 centimetres in diameter, in which I could make out an arm, a hand. They hung me down by my feet and I shook the hand of the trapped victim…That was all that we could see of Bernard…He had a pulse and poor oxygen saturation. I had tried to comfort him with my voice by telling him that I was going to perfuse him and hanging by my feet I was able to insert a venous catheter into him. He was unresponsive and I decided that although he still had a pulse he must have lost consciousness." It was still going to be another eight hours before the victim could be released. "I stayed like that," the house officer went on, "until 17:00, when they took me aside and told me that they intended to pull him out by attaching ropes under his shoulders. I didn't see him again. I knew that he had had both legs amputated but had survived." Debraye was taken by helicopter to Val de Grâce. Fifty years later the house officer, now a retired Professor of Medicine, and the victim, now a grandfather, met one another once again at a survivors' reunion. [700]

Another survivor had been freed after many hours but subsequently died. The hunt for

---

[700] https://france3-regions.francetvinfo.fr/hauts-de-france/aisne/temoignage-il-y-a-50-ans-la-catastrophe-ferroviaire-de-vierzy-faisait-108-morts-il-nous-arrive-encore-d-en-faire-des-cauchemars-2562116.html

further survivors ended the following day. The last of the wreckage was removed on Friday 23rd June.

M. Jean Chamant, the Minister of Transport, immediately established a committee of inquiry to investigate the causes of the disaster and to determine what measures needed to be taken to prevent a repetition of the accident. The committee was chaired by General Engineer Rerolle.

The inquiry determined that the tunnel had been constructed using inferior materials in 1862 through ground that was also not very geologically stabile. It had suffered damage in 1870 during the Franco-Prussian War and a section of it had been dynamited during the First World War in 1914 to hinder the German advance. In 1915 the tunnel had been repaired but the bell formed in the 1914 dynamiting had not been filled in, instead had been integrated into the vault of the tunnel, by reinforcing it with masonry. The tunnel had been damaged again during the Second World War. At the beginning of 1972 cracks had been noted in the vault and bricks were falling from the roof. The Société nationale des chemins de fer français (SNCF) began repair work, involving removal of part of the masonry and the temporary use of props prior to the spraying of concrete. This operation was still in progress in mid-June and the inquiry decided that it was the repair work that had been the direct cause of the tunnel's collapse. M. Yves Guéna, the new Minister of Transport, stated in May 1973, "The commission of inquiry has concluded that the ultimate cause of the break in the tunnel vault appears to have been the stripping of the bricks over a great length of the bore during the repair work." [701]

Meanwhile in late July 1972 the SNCF restarted the repair work on the tunnel. The rolls of bricks were demolished, the vault was cleaned, reinforcements were inserted and finally shotcrete (concrete or mortar pneumatically sprayed through a hose at high velocity) was applied. This new lining of the tunnel considerably reduced its internal dimensions and as a result the line through it had to be reduced to single track. Other tunnels, such as Vauxaillon on the same line and Coudray on the Paris – Beauvais line were similarly dealt with. [702]

In February 1977 the Amiens Court of Appeal sentenced the SNCF structures controller to eight months' suspended imprisonment for involuntary manslaughter and the causing of injuries.

A more recent study, however, suggests that in fact it was not a failure of the tunnel vault that had led to the collapse, but that the ground above the tunnel had been weakened by the 1914 explosion, which had caused a series of cracks that would have been undectable by the technology of 1972, and that the vibrations of passsing trains had eventually precipitated the collapse. [703]

The accident remains the third worst in French railway history after Modane (1917) and Pomponne, near Lagny (1933).

Japan suffered its second railway disaster in 1972 involving a tunnel on Monday 6th

[701] https://france3-regions.francetvinfo.fr/hauts-de-france/aisne/catastrophe-ferroviaire-de-vierzy-quelles-sont-les-lecons-tirees-du-drame-qui-a-fait-108-morts-en-1972-2563428.html

[702] Wikipedia – Effondrement du tunnel de Vierzy.

[703] http://www.inventaires-ferroviaires.fr/hd02/02799.r.pdf

November. This time it was a fire, which occurred in the 13,870 metre Hokuriku Tunnel, at the time the sixth longest tunnel in the world. The double-track tunnel dated from 1957 when the line had been electrified and the new line through the tunnel replaced the original steeply graded single track.

The train that caught fire was the 15-coach *Kitaguni* late-night working from Osaka to Aomori, carrying 760 passengers. At about 01:10 passengers reported to the conductor that a fire had broken out in the dining car, the 11th vehicle of the train. In accordance with the regulations of the time the conductor immediately stopped the train, now 5,300 metres from the Tsuruga entrance. Two railwaymen first tried to tackle the fire with extinguishers and when this did little to subdue it, the decision was made at about 01:24 to divide the train between the 11th and 12th vehicles and evacuate the passengers. This was done and at 01:29 help was requested from Minami-Imajō and Tsuruga stations, at the further and nearer ends of the tunnel respectively. Donning gas masks, the crew then separated the train again between the ninth and tenth coaches but an attempt to proceed out of the tunnel with its front vehicles was frustrated after a short distance had been covered when at 01:52 the current failed, which also meant that the train was plunged into darkness. A train arrived on the opposite running line from Tateyama and picked up about 100 passengers and took them to Tsuruga. A relief train entered the tunnel from Tsuruga at 02:43 and although the thick smoke prevented its crew going all the way to the burning train they were able to pick up many passengers who had attempted to escape the tunnel on foot. Another relief train entered the tunnel at 06:43, probably from the Minami-Imajō end, but the last passengers and crew were not evacuated until 14:00; about 290 had eventually made their out of the tunnel on foot. [704] A survivor who took two hours to walk out of the tunnel said, "I heard somebody shout, 'This way, this way,' as if he was trying to calm down panicking passengers. Many bruised themselves as they groped and fell repeatedly on the slippery concrete sleepers." [705] A lot of the passengers had been returning from weekend holidays, most of them had been asleep when the fire broke out and had been awoken by the conductor shouting that there was a fire.

In total 29 passengers and one railwayman lost their lives in the accident; they had all died by poisoning from carbon monoxide and other noxious gases. The body of the last victim was only discovered in the tunnel a week after the accident. Over 700 passengers were more or less injured due to smoke inhalation.

Two days after the fire Prime Minister Kakuei Tanaka told his cabinet ministers that they must take "absolutely secure measures" to prevent any recurrence of what he termed an "unbelievable" train fire. Meanwhile Japanese National Railways (JNR) President Satoshi Isozaki conceded that the fire had been 100% the fault of the railway and added, "I must admit something must have been lacking in our effort." The JNR had already paid 100,000 (about $330) to each of the victims as solatium (damages for emotional rather than physical or financial harm) and said that it would negotiate compensation with the families of the victims. [706]

Initial theories as to the origin of the fire were that it had been caused by a stove that had not been turned off properly after the dining car had stopped serving at midnight, or

---

[704] https://www.shippai.org/fkd/en/cfen/CA1000605.html

[705] *The Lima News*, Lima, Ohio, United States, 6th November 1972

[706] *Pacific Stars and Stripes*, Tokyo, Japan, 9th November 1972

by a discarded cigarette end, but a police report on Wednesday 17th attributed it to the burning out of a copper plate in an electric heater, beneath a sofa in the smoking car adjoining the dining car, which had caused a short circuit.[707] The rapid spread of the fire was blamed on the flammable material used in the construction of the railcar and it was noted that the tunnel itself had not been designed with fire safety measures in mind, thus:

(1) The long tunnel was not provided with any ventilation other than that provided by the passing trains themselves.

(2) Radio communication was not possible in the tunnel, though lineside telephones were provided at 300-metre intervals.

(3) No diesel-hauled vehicle was provided that could negotiate the tunnel when the current had been switched off.

The local fire department had pointed out these deficiencies to the JNR, which had simply replied that it would review the fire safety measures.

The JNR undertook a number of safety measures in response to the accident:

(1) All dining cars of the OSHI-17 type, such as that which had been involved in the fire, were withdrawn, with the result that for a time express trains ran without dining cars.

(2) New railcars were to be built from incombustible materials and fire-retardant structures and be provided with fire extinguishers and emergency lighting.

(3) The regulations were amended such that should a train catch fire in a tunnel it was not to be halted until it had been brought out of the tunnel.

(4) Long tunnels were to be equipped with powered rescue vehicles, smoke control equipment and evacuation corridors. [708]

The 790 foot Quitaque Tunnel, at the time of the accident the only remaining tunnel in use in the state of Texas, United States, was partly demolished on the morning of Saturday 4th August 1973 when it was damaged by the derailment of a 98-car Burlington Northern Railroad freight train from Lubbock, Texas, to Seattle, Washington, carrying cotton, milo (sorghum) and wheat. Engineer J.D. Logan said later, "We were going through the tunnel at 5 a.m. when one of the big grain hoppers climbed the tracks and the walls of the tunnel." He added that the following cars had run alongside the tunnel walls until it had just fallen in, and continued, "At first there was a little tug, and then another small tug, and then the tunnel fell in." He had not heard the tunnel collapse because by then his engine had been beyond its far end. [709]

Burlington spokesman C. J. May thought it likely that the derailed vehicles had first hit the guard rails in the tunnel, which had then sent them into the beams supporting the structure. No one was hurt in the derailment, the cause of which was not immediately apparent. Thirteen cars in the middle of train left the track. Most were empty, though one was loaded with wheat and one with cotton.

---

[707] Ibid, 17th November 1972

[708] https://www.shippai.org/fkd/en/cfen/CA1000605.html

[709] *Amarillo Sunday News Globe*, Amarillo, Texas, United States, 5th November 1973

Following the derailment the line was closed for 18 months, with trains being diverted via Amarillo and over the Atchison Topeka and Santa Fe to Plainview. Eventually the line reopened, the tunnel having been shortened in length to 582 feet. [710]

As the *Italicus Express* from Roma, Italy, to München, Germany was passing through the 18.507 kilometre Apennine Base Tunnel, Italy, and was close to its far end, a bomb exploded in its fifth coach at 01:23 on Sunday 4th August 1974. The effect of the explosion was somewhat reduced by the fact that the momentum of the train carried it out of the tunnel but even so 12 passengers were killed and another 48 were wounded. Many of the passengers were Italians going to work in Germany, while the dead included one Japanese and the injured Dutch and American tourists. There were thought to have been about 1,000 passengers on the train.

The bombing was the work of extreme right-wing terrorists. The day after the attack a statement was issued by the group Ordine Nero (Black Order): 'We took revenge for Giancarlo Esposti. We wanted to show the nation that we can place a bomb anywhere and however we please. Let us see in autumn, we will drown democracy under a mountain of corpses.' (Esposti was a terrorist who had been killed by the police at the end of May, two days after another bombing.) The statement continued by claiming that an attempt to defeat it by banning its predecessor, Ordine Nuovo (New Order) had failed, and that the Nazi banner had not died in Berlin in 1945 but continued to live for a great, Fascist and Nazi Italy. It accused President Leone, Signor Taviani (the Minister of the Interior), Signor Longo (the head of the Communist Party) and Signor Santillo (the head of the anti-terrorist squad) of being responsible for those who had died in the bombing, because by dissolving right-wing organisations they were placing Italy under Marxism. [711] [712]

'At least 29 people were killed and more than 70 injured when a rush-hour Underground train crashed into a blind tunnel at Moorgate station, in the City of London. It was feared early today that many more bodies were still to be found. Rescuers worked all day in heat, foul air, and cramped conditions to free the injured from the telescoped wreckage of the first three coaches. An operating theatre was set up on a platform but some people died before doctors could reach them. A policewoman, aged 19, was trapped for 12 hours, and the last survivor, a man aged 26, was brought out at 10 p.m. The driver was among the dead. A preliminary investigation indicated that the brakes had not been applied.'

Thus read the first report in a national newspaper on a disastrous and mysterious accident in England on Friday 28th February 1975.

What had happened was detailed in longer accounts in the same newspaper. 'More than 200 rescuers struggled all morning to reach the first and second coaches and it was not until after 1 p.m. that they were able to cut their way in through the side and roof of the first. The train, the 8.37 from Drayton Park to Moorgate, was full of rush-hour office workers. It came in on platform nine, and several passengers said later that it shuddered and appeared to accelerate through the station.

---

[710] Wikipedia – Quitaque Railway Tunnel

[711] Wikipedia – Italicus Express bombing

[712] *The Times*, London, England, 5th, 6th, and 8th August 1974

'There is an overrun tunnel at the end of the platform, a terminal line, but the train careered through tons of sand laid in the tunnel to help as a brake, demolished buffers and smashed into a brick wall, which a London Transport official said was "as solid as a granite cliff." The last remaining carriages remained on the line, but the third from the front blocked the mouth of the tunnel. Inside the front two carriages were telescoped together.

'The crash threw the station into darkness and sent up clouds of dust. It was also thought that a small fire was started. All roads to the station were sealed, as a large fleet of ambulances waited for the dead and injured who were brought up every few minutes. Often 20 to 30 minutes would pass between the casualties being brought up.

'Rescue workers strove late into the evening to make contact with survivors in the front coach, although it was established early that the driver, Mr Leslie Newson, aged 55, of New Cross, London, had been killed. By mid-morning conditions underground had become so bad that the rescue force was reduced from about 200 to about thirty, while oxygen and fans were taken down. One of the many soot-covered, sweating rescue workers who returned to the surface said it was "hell down there, just like a furnace."

'One of the first policemen on the scene was Police Constable Graham Bridges, who was on motorcycle duty outside the station. "It was the most awful mess," he said. "I could see that the front coaches were telescoped together. People were screaming and shouting, there were a lot of injured people around and the dust was terrible."

'Mr Barry Gilbert, aged 23, from Chingford, Essex, told of what happened on the train as he left hospital after receiving stitches for a large cut on his head. He said, "I was standing in the fourth carriage. As we came into Moorgate I noticed we were not stopping and I saw the platform going past fast. Normally you can hear the train slowing down." Mr Gilbert, with soot still covering his neck and shirt, said: "There was a small screech and then a large one. I fell flat on my face and cut myself. I could hear a woman screaming. It was pitch black and the air was filled with soot. Then people were picking themselves up and asking where they were and saying, 'Don't panic.' I tried to make for the back of the train and emergency lights came on. It took me a quarter of an hour to get out through a window and firemen started to come in and open a door." Mr Gilbert made his way back to the platform, where he saw the train had overrun the buffers. Injured people were lying round him.

'A woman who would not give her name, said as she left the hospital: "As we came into Moorgate, instead of slowing down there was this acceleration. Then there was a terrible crash."

'One of the survivors from the second coach from the front, Mr John Ryder, aged 38, said: "After we stopped there was a long silence and then people started moaning but there was nothing we could do to help them. One woman was lying across me and I was holding her head in my hand." Mr Ryder, who was with Miss Katharine Hobenden, said: "We were sitting near the middle of the carriage, close to the central doors and the door arches strengthened the roof and formed a sort of tent which probably saved us. On either side the roof had been squashed down to the floor."

'Rescuers who came to the surface after many hours underground painted a grim picture of the scene and conditions. The first coach was crushed to about two thirds of its normal length, the second was forced above it, and the third was flattened into the back of the second. A fireman said there was a great deal of noise at first. "Then everything started to go quiet. There were no screams from the tunnel...you could tell when somebody was alive; when you touched them they screamed." Many were "pretty badly mangled. It appeared that some had been forced through the floors."

'Many different people were involved in the rescue. One team of three surgeons from St Bartholomew's Hospital spent eight hours at the crash, giving emergency first aid and administering pain-killing drugs. The Salvation Army were also involved, with the Fire Brigade and the police. Many of the victims brought out were given drip-feeds as they were carried to ambulances. Conditions were so cramped that special bamboo folding stretchers had to be used. One of the casualties received by London Hospital was dead on arrival and another died later. At least 11 were seriously injured, with broken limbs and bones; four people had crushed chests. Twelve have been discharged after treatment for cuts, bruises, and shock.

'Mr Edward Markwardt, aged 20, a student, of Ecclesbourne Gardens, Palmers Green, who was treated for cuts and shock, said the train was going very fast when it crashed. "The lights went and I found myself on the floor. I climbed out of the shattered window and on to the top of the carriage. I walked along the top of the carriages until I got to the platform."

'Mr Frank Gaunt, aged 45, of Buttersbridge House, West Hythe, Kent, said the second coach "lifted up and the roof crunched down." He found himself sitting on a girl, while other people were trapped. He climbed out through a trap door in the floor but could not help others because his shoulder had been hurt.

'Policemen toured neighbouring streets with loudhailers to appeal for blood donors after the crash. Within a few minutes a queue of more than 100 formed at the National Blood Transfusion Service's centre in London Well, a hundred yards from Moorgate station. People continued to arrive throughout the morning, while workers in office near St Bartholomew's Hospital hurried there to offer blood. Students from the City of London Polytechnic withdrew from the big protest rally at Hyde Park to give blood.''

'At 9 a.m. yesterday the City of London police got in touch with the casualty department at St Bartholomew's hospital with the first news of the Underground accident. By mid-morning the first minor casualties were leaving the hospital, while doctors worked in the tunnel to free other injured and dead.

'By late afternoon the hospital, which became the main treatment centre, had dealt with 40 patients, admitting 18 of them. The London Hospital dealt with 32 and Guy's took four. During the day the hospital said it had never received a warning from the London Ambulance Service that it would be dealing with a major emergency. In London that means at least 50 casualties. Mr Dennis Jones, administrator for the group of hospitals including St Bartholomew's, said: "We were not told it was a major disaster but we decided to invoke our disaster plan." The absence of a warning seems to have been caused by the fact that the first injuries were not serious and the situation was confused by the nature of the crash. A London Ambulance brigade official denied that there had been any breakdown in communications. He said St Bartholomew's was told it was a major accident by the "red alert" procedure. "We alerted them using a special telephone line, a matter of minutes after we were informed of the accident. They were told at 9 a.m. They could not have been told any quicker."

'Dr Anthony Madden was sent from St Bartholomew's with a medical student to see what was happening at Moorgate and by 9.15 he was following firemen as they made a path for him through the wreckage. He was joined by a full medical team from the hospital, of two doctors and two nurses. Dr Madden said: "After about a quarter of a hour I was able to find out what was going on and by 9.45 I had reached the second carriage. There were six or seven trapped, caught by their legs in the wreckage. A number of the dead had been crushed by one carriage falling on another. Most of the people were extraordinarily quiet." He was joined by Dr Phillip Finch, an anaesthetist, who went up into the front carriage.

'The rescuers were working in a very limited space, with less than 4ft to stand in at the front of the train. The doctors worked on their knees or were sitting while the firemen used compressed air to cut into the carriages. Oxy-acetylene cutters could not be used because of the lack of air and space.

'A London Hospital team ran a casualty clearing station on the platform, and later relieved doctors in the tunnel. After a couple of hours the air became very hot: one doctor estimated it was 120 deg. F. The white coats of the doctors were covered in soot and sweat. Dr Finch said: "It was chaos. We had to step on one dead body to reach someone."

'At one stage the rescuers burst into relieved cheers as they released a victim, but in another spot they found a young girl with a wedding ring on her finger who had been dead for hours, and another girl trapped by her thighs.

'A London Underground filled with rush-hour commuters overran the platform at Moorgate yesterday morning and ploughed into a blind tunnel killing at least 29 people and injuring more than 70. It was the worst accident in the 100 years of the London Tube system.

'Early today engineering teams were trying to separate the telescoped coaches. In view of the number of people usually commuting at the time of the crash, police, London Transport officials and medical teams feared that the final death toll would be higher than the 29 originally estimated. Sixteen bodies have been brought to the surface. The train's third coach was being winched back from the rest of the wreckage at a rate of an inch every 10 minutes in the faint hope of there being more survivors.

'Workers at the scene said that many more people than the 70 injured and 29 feared dead would have been travelling in the first three coaches of a rush-hour Tube train. London Transport estimates that about 430 people could travel in three coaches of such a train. Rescuers toiled all day in the heat and dust to free the bodies of dead, dying, and injured people from the tangled mass of metal. Twelve hours after the accident a young policewoman trapped in the front coach of the train was freed after her foot was amputated. Then, shortly after 10 p.m., a man, aged 26, believed to be the last remaining survivor, was extricated. He had a fractured leg, but an amputation to free him, which had earlier been feared, proved to be unnecessary. Eye witnesses said rescue workers and doctors had to crawl through a hole 3ft by 2ft bored into a second coach, which overrode the first in which the young man, Mr Geoffrey Benton, a stock jobber, and Policewoman Margaret Liles, aged 19, were trapped.

'At 8.30 p.m. six pints of blood were taken to the station and a fresh team of surgeons and nurses arrived to relieve the four doctors and two nurses who had been working since the morning in an emergency operating theatre set up on a platform. Earlier doctors had managed to reach the trapped couple, who were conscious though under heavy sedation. Their condition deteriorated markedly during the day. The doctors and nurses had to work in extremely cramped conditions. Eye witnesses said the heat was tremendous and the smell of stale air almost unbearable. One said: "It is just a mass of human endeavour down there." A doctor was barely able to describe the day's experience. He said: "It is so ghastly I just can't explain. This is much worse than the Old Bailey bomb explosion. There have been terrible injuries and people very badly mutilated."

'Disaster struck at 8.46 a.m. when the 8.36 six-coach Northern Line train drew into Moorgate, its terminus. Instead of slowing as it came along the platform, it appeared to gather speed and smashed through buffers and a barrier of sand blocking a blind tunnel. The second coach telescoped into the first, and the third one rode up on top of the second, embedding its front in the roof of the tunnel. The last three coaches blackened and with broken windows lay in sequence along the platform.

'The rescuers had to cut through tons of twisted metal as frightened and injured people cried for help. At one stage the temperature in the tunnel rose alarmingly and there was a lack of oxygen. The team had to crawl through gaps 18in high to reach the two front carriages. Most of the victims were in these coaches, which had buckled into a tangled mass in the tunnel. A senior police said: "It's like working in a sardine tin: it is just a lump of consolidated metal."

'As London Transport officials started an immediate investigation, Mr Graeme Bruce, the operating manager, said that preliminary inquiries showed the brakes of the train had not been applied. That is bound to lead to an examination of Tube train safety devices, hitherto regarded as more than adequate. In addition to the "dead man's handle" (designed to stop the train in the event of the driver's collapse or death) there are two emergency braking systems.

'London Transport confirmed last night that the driver, Mr Leslie Newson, had died in the crash. His body had not been recovered by late last night.

'Efforts last night centred on the struggle to free Miss Liles and Mr Benton. Miss Lile's mother and Mr Benton's wife, Valerie, waited for news with other relatives of the victims for many hours. Miss Liles was trapped in her seat by the twisted wreckage, with only her head and shoulders visible, rescuers said. She was in good spirits. Her courage was noticeable on the microphone link when she was asked whether she would like one of her girl friends in the force to be at the scene. She replied: "No, I'd rather have one of the boys." Police Constable Robert Fox, based at the Wood Street station, said he was able to get within two feet of Miss Liles, who is attached to Bishopsgate police station. He added: "She is in very good spirits and she is a very courageous girl." Although she was trapped she did not seem to be in any pain.

'Mr Gerard Kemp of *The Daily Telegraph,* representing Fleet Street newspapers, was allowed on Moorgate station platform two hours after the disaster. He said: "Platform one looked like something out of a wartime film. The platform was choked with helmeted firemen, police, uniformed ambulance men, doctors, nurses, and welfare workers. The 'up' escalator was still working, bringing up blackened and tired rescuers. Some were assisting survivors, still able to walk, others were struggling to get stretcher cases up to the clear air and waiting ambulances."

'Dr Drummond Dymond, a registrar in cardiography, said he found people with blackened features and could smell burning from electricity. "There was one young man who had stopped breathing. I could hear his heart. After five minutes of resuscitation he began breathing again and the firemen passed him back. A couple of girls whom we got out of the first carriage and who were not so badly hurt were laughing in a hysterical and shocked way. It was really controlled hysteria. I saw about six dead in the first carriage, including a young blonde wearing a wedding ring. It was just a horrible mess of limbs and mangled iron. One of the great problems was the intense heat down there. I reckon it must have been 120 degrees. It was like opening the door of an oven." He paid special tribute to the work of ambulance men at the scene. "As they came out," he said, "they were weary, haggard and filthy."

'Mr Michael Robbins, managing director of London Transport railways, said yesterday: "It is a very damaging, very upsetting, and very grievous thing." London Transport undertook to meet in full all proper claims for transportation.' [713]

On Monday 3rd March the same newspaper was able to able to report that the death toll in the accident remained unclear and the rescue work was still in progress: 'A death roll of at least 42 is expected in the Moorgate Underground disaster. Twenty-six bodies have been recovered and identified, and last night the police revised their estimate of the remaining dead in the front coach from at least 12 to at least 16. The final toll is unlikely to known until Wednesday.

'There was a rush of 20 to 30 genuine new telephone inquiries about missing persons between 7 p.m. and 10 p.m. yesterday. By 11 p.m. all but four had been accounted for after inquiries in north London. But the 50ft long coach had been telescoped into less than half its length, and, so far, it has not been possible to tell exactly how many bodies remain in the wreckage. Heavy engineering equipment was taken late last night to reinforce what police described as "a mining operation."

'Only yesterday, two days after the accident, rescue workers finally got through to part of the main section of the first coach. Such is the state of the wreckage that they thought they had reached the coach on Friday. In fact, they had reached the small back section of the front coach, where they rescued Margaret Liles, the young policewoman. The men hacking at the wall of twisted metal are London Transport breakdown engineers, who are working with four men at the front of the wreckage, with another four behind cutting up the section as they are passed back. Progress has been painfully slow. When the team reached the rear of the first coach, they could not start cutting into it for fear that the massive front wheels of the second coach perched over it would crash on top of them. The end of the tunnel is virtually packed solid with twisted metal.' [714]

The final toll was 43 dead: 42 passengers and Motorman Newson. Another 74 passengers required hospital treatment for injuries. The body of the motorman was not removed until 22:00 on Tuesday 4th March. It was only at 07:00 on Thursday 6th March that the last of the wreckage was removed and the station was not reopened until Monday 10th March.

The Department of the Environment report on the disaster noted that platforms 9 and 10 at Moorgate station served the terminus of the London Transport Highbury branch. From the previous station at Old Street the line fell, mostly at 1 in 150, for 642 yards,

---

713 Ibid, 1st March 1975
714 Ibid, 3rd March 1975

then it was level for 233 yards to the south end of No. 9 platform, before falling at 1 in 80 for 67 feet to a single central hydraulic buffer stop about 12 feet from the end wall of the tunnel. A sand drag was also provided.

The terminus was controlled by an 11-lever signal box which worked standard London Transport two-aspect colourlight signals, with Red/Green stop signals preceded by Yellow/Green repeater signals. In addition there was a fixed red light on a post at the beginning of the sand drag.

The train consisted of two three-car units of steel 1938 Tube stock. It was equipped with a combination of Westinghouse and electropneumatic brakes. The dead man's handle required 15 lbs force to depress it and then a force of 3½ lbs to keep it depressed.

Motorman Newson had worked for London Transport since March 1969, initially as a guard and from January 1974 as a guard/motorman, transferring from Barking to Drayton Park in January 1975. On the day of the accident he had been making his fourth trip that morning along the short branch with train No. 272, which was in charge of Guard Robert Harris.

A large number of witnesses gave evidence at the inquiry into the disaster.

Acting Leading Car Examiner Frederick Wager and Car Examiner J. James confirmed that they had examined train No. 272 on the night of 27th/28th February and its brakes had been in order.

Relief Clark C. Brown stated that he had been acting station manager at Drayton Park. He said he had seen Motorman Newson, who had appeared perfectly normal.

Guard Harris had been late reporting for duty at Drayton Park that morning so Motorman B. Rosario had worked the first service to Moorgate at 06:40 with Motorman Newson, Rosario driving the train to Moorgate with Newson acting as guard. Guard Harris had been at Moorgate when they had arrived there and had taken over from Rosario. Rosario agreed with Brown that Newson had appeared perfectly normal and fit for duty. He (Rosario) stated that approaching Moorgate it was his normal practice to apply the brakes to reduce the speed of his train first to 15 miles an hour and then to 10 miles an hour at the platform.

Motorman H. Lang said he had worked train No. 871 in Moorgate at 08:33. He did not know Newson well but thought that he had looked his normal cheerful self that morning. His description of his driving technique approaching Moorgate varied somewhat from that of Motorman Rosario.

Motorman C.A. Gladding of train No. 273 confirmed that the red marker light at Moorgate had been alight just before the accident. He said that earlier in the day he had asked Newson for some sugar to put in his cup of tea and the latter had replied, "Go easy with it, I shall want another cup when I come off duty."

Motorman J.P. Bray testified that he had considered Newson a cautious driver, who approached stations more slowly than his colleagues. He had had a cup of tea with Newson that morning and the latter had appeared perfectly normal.

Guard J.T. Catney, who knew Newson well, was another railwayman who considered that Newson had been quite normal that morning. Newson, the guard stressed, had been very cautious driving into stations.

Motorman Raymond Deadman stated that he had trained Newson in January 1975 and had considered him to be cautious but competent. Newson had stated that he had been satisfied with the training he had received.

Station Foreman G. Fox at Drayton Park described Newson as a very conscientious driver. He, too, thought Newson had appeared quite normal that morning.

Guard Harris stated that although he worked regularly with Newson he did not know him very well, though he had seemed a normal stable person with a good sense of humour. The motorman's technique had been to shut off power early and to coast for some distance, then to make a series of gentle brake applications to reduce speed to 10 miles an hour at the crossover outside the station and to 5 miles an hour at Moorgate signal box. He agreed that he had been late for work on the day of the accident and had taken over the first train from Dayton Park when it had arrived at Moorgate. Between then and the accident he had made three round trips with Newson, whose driving had been very good, with use of the Westinghouse brake at Highbury on northbound trips and at Old Street on southbound trips. He had chatted a little to Newson about camping. With regard to the final trip the guard admitted that after Old Street station he had gone to look for a newspaper and had been occupied reading the advertisements in one that he found, when he suddenly realised that his train was entering Moorgate station at 35 miles an hour. Before he had time to do anything he had been thrown to the floor by the impact of the collision.

Moorgate Relief Signalman Walter Wade stated that Guard Harris had appeared at his cabin when he (Wade) had come on duty at 06:40 and he had allowed the guard to report for duty to Dayton Park from his cabin. He remembered the guard and Newson chatting about camping. The motorman had appeared quite cheerful and normal. The signalman described the accident train as having arrived at the station "like an express." The train had continued along the platform at 35 to 40 miles an hour, still under power, and it had been obvious that it could not stop. There had been a terrific crash and what looked like black smoke. He had at once telephoned the line controller, telling him that a major disaster had occurred and to call out all the emergency services.

Railman T. Andrews at Moorgate spoke of seeing the train coming in very fast and it crashing before he had time to telephone Control. He thought that the train had been under power and had heard no sign of braking. He had not seen the motorman.

Guard B.A. Friar had been on Moorgate platform and said he had been alerted by the noise of the incoming train. He had glimpsed Motorman Newson, whom he described as having been sitting up in a normal position, his hat on his head and looking straight ahead. He had not been able to see if the motorman's eyes were open or what his hands had been doing. He had witnessed the collision and put the train's speed at 30 to 35 miles an hour.

Assistant Chief Operating Manager C.H. Cope spoke of finding a rule book, working timetable and a notebook with information on how to deal with defects in trains in Newson's bag after the accident, indicating that he was a conscientious employee.

Chief Medical Officer P.A.B. Raffle stated that the motorman had passed the usual medical examination. Newson's injuries were consistent with his having been sat at the controls with his left hand on the brake handle. There was no indication of having attempted to shield his face.

Passenger I.A. Butt had been on platform 9 at Moorgate. He recalled seeing Motorman Newson sitting erect with his arms forward as if on the controls.

Passenger Miss H.M. Stewart had been stood halfway along the Moorgate platform. She said that she had realised as soon as she had seen the train that it was travelling much faster than usual, at 30 or 40 miles an hour, with no sign of braking. The motorman she recalled as having been sat upright, looking forward, apparently quite aware of the situation but looking very tense.

Another passenger on the platform, Mr A.R. Board, put the speed of the train at 40 miles an hour. Motorman Newson, he recalled, had been sitting completely upright with his hands in the normal operating position. He appeared as if he were paralysed or frozen in position. Mr Board explained that he had especially noticed the motorman's eyes, which appeared to be staring straight ahead and larger than life. The motorman's mouth had been closed and his face was a normal colour.

Regular passenger C.B. McGovern said that the train had appeared to slow down slightly at the crossover and then accelerate and he had braced himself for the impact.

Passenger M.F. Taylor agreed that the train had appeared to accelerate at the point where it would normally have begun to slow down. He put the speed at that point at 35 miles an hour and had noticed no sign of braking.

Passenger R.C. Stewart agreed that the train had continued into Moorgate under power. What he at first thought had been a final deceleration he later believed must have been due to the train entering the sand drag.

Passenger J.M. Ball placed the speed of the train at the crossover at 50 miles an hour and said that he had been waiting for a brake application that had never come.

Passenger H.H. Holderness said that when the train had not reduced speed at the usual spot he had not at first been alarmed but thought that the motorman was just leaving the braking a little late, until the train had reached the platform, still not slowing up. He estimated its speed there at 44 miles an hour.

Passenger E. Markwardt, presumably the student mentioned in the newspaper report, said that he had not realised anything was wrong until the train had been entering the station. He had noticed no brake application.

Finally passenger P. Patterson described an emergency brake application lasting a second or a second and a half as the train had entered the station, after which it had ceased.

Depot Engineer S.J. McManus, Assistant Divisional Manager T.B. Savin, Rolling Stock Engineer G.H. Hafter and Mechanical Engineer T.J. Lowe described the examinations of the train they had undertaken after the accident.

The results of the post mortem on Motorman Newson were described in detail. There was no evidence of heart or brain disease, no evidence of pre-accident ingestion of drugs or alcohol and no evidence of alcoholic liver disease or of any electrical injury. Death had been due to multiple blunt trauma. A small amount of alcohol was detected in his remains but there were conflicting views between different experts on whether this had been present before death or was the result of post mortem bacterial action.

Mrs Newson testified at the inquest on the victims that although her husband enjoyed the occasional glass of brown ale he rarely touched spirits. He did not take sedatives, tranquilisers or sleeping pills and had not suffered from depression. He had been planning to purchase a car for his daughter. None of the railwaymen who worked with Motorman Newson could remember him using alcohol on duty.

Dr Raffle, the medical officer, did not consider that the 80 milligrams/millilitre of alcohol detected in the body of Motorman Newson would in any case have been sufficient to affect his driving ability. The doctor theorised that one of two rare medical conditions could explain the accident. The first was 'akinesis with mutism,' a condition due to a tiny blood clot in the mid-brain, which would cause a person to freeze without his muscle tone being affected. The second was 'transient global amnesia,' due to a spasm of the blood vessels in the brain, though this condition would not have prevented the motorman raising his hands to protect his face, which he had not done.

The inspecting officer, Lt Col I.K.A. McNaughton, attempted to determine from all the evidence how the disaster had happened. From witnesses' accounts the train had run normally as far as Old Street. After leaving there the motorman had likely centralised the controller, as was common practice among motormen to reduce the effort needed to hold down the dead man's handle. What was most notable was the complete inaction of the motorman during the last half minute of the journey. Had he merely allowed his mind to wander the noise as his train passed through the crossover outside of the station combined with the viewing of the red marker light at the end of the line should have been enough to bring him to his senses. Even at the late stage merely releasing the dead man's handle would have prevented the accident. But he had done nothing, despite being apparently alert due to the statements of several witnesses.

The inspecting officer concluded that there had been no fault with the train. Had the motorman collapsed, fallen asleep or otherwise lost consciousness the dead man's handle would have stopped the train. All the eyewitnesses also agreed that the motorman had been sitting upright. It must be concluded therefore that the motorman had been conscious but had made no attempt to stop the train.

The approach to Moorgate was distinctive and the possibility that the motorman had misjudged or mismanaged his approach could be ruled out. It was possible that Newson had consumed some alcohol but there was no evidence that he had been drunk or incapable of driving his train. In any case the alcohol detected in his body might have been produced by post mortem changes.

Had Newson been daydreaming he would surely have come to his senses before his train struck the buffer, at least to the extent of trying to protect his face.

Suicide was another possibility but there was no evidence to support it.

'Akinesis with mutism' or 'global transient amnesia,' as suggested by Dr Raffle, were other possibilities, but there was no evidence for them either.

Not surprisingly Guard Harris came in for severe criticism from the inspecting officer, who conceded that the guard was young and inexperienced but stated that he had also been 'idle and undisciplined,' in coming to work late and failing to pay attention to the running of his train as it had been approaching the terminus. However, Lt Col McNaughton conceded that the guard could not have been expected to realise that anything was wrong until the rear of his train, where he was, had reached the crossover, by which time the front of the train would have been halfway along the platform. Even then it would have been hard for the young and inexperienced guard to take over control of his train from a driver in whom he had confidence. The behaviour of the guard had not, therefore, contributed to the accident.

Motorman Newson was thus solely responsible for the disaster, but why he had failed to act could not be determined.

Minor buffer stop collisions were not uncommon on London Transport. After a fatal one at Tooting Broadway in 1971 on a reversing siding in a tunnel, time-released train stops had been installed at such sidings. It had not been thought necessary, though, to install such protection at termini. The last buffer stop collision caused by a driver being taken ill had happened at Edgware in 1946, where the motorman had suffered a heart attack. There had been no previous accident similar to that at Moorgate.

There remained the possibility that Motorman Newson had committed suicide, and the only protection against a motorman deliberately driving into the buffers was the fully automatic control of trains at termini, but the enormous cost of installing such protection at all termini could not be justified in view of the very small likelihood of such an event.

Although it was not practical or economic to provide automatic train control at all termini with conventional signalling the inspecting officer recommended that it be introduced as a normal requirement when the tube lines were resignalled. Meanwhile the installation of a series of train stops at termini was recommended, or alternatively the installation of a device to prevent sufficient current to be drawn for an incoming train to re-accelerate.

The arrangement of the signalling, as at Moorgate, where a green aspect was provided at the last stop signal before the end of the line, was also criticised. London Transport had resolved that in future the last signal would not be cleared until the train speed had been properly reduced, while large illuminated speed-restriction boards were being erected at all termini and dead-end sidings. On the Highbury branch new three-aspect signalling was being provided with train stops at the signals and three additional train stops at the terminus, which should be adequate to prevent both a repetition of the Moorgate disaster and relatively common low-speed buffer stop collisions. A traction current limiting device to prevent a train re-accelerating was also being considered.

Finally the inspecting officer was critical of 'the level of observance of routine instructions and safety precautions' on the Highbury branch as evidenced, for example, by various staff coming on duty late and by motormen failing to test hand sets.

The inquest on the victims of the disaster returned a verdict of *Accidental death.*[715]

A freight train and a train of oil tank waggons collided in a tunnel in southern Iran on Sunday 21st September 1975. In what was very likely a head-on collision on a single line five railwaymen were killed and several hundred barrels of crude oil 'went up in flames.' The location of the accident was given as near Andimeshk, a place 725 kilometres south west of Tehran. An investigation into the accident was begun.[716]

---

715 Department of the Environment report

716 *Winnipeg Free Press*, Winnipeg, Manitoba, Canada, 22nd September 1975

An accident occurred in Spain on the Barcelona Metro at 20:45 on Tuesday 30th October 1975. Train No. 8 had left Virrey Amat station for Vilapicina but for undetermined reasons it then set back through the tunnel towards Virrey Amat and was struck by the following train, No. 9. Motorman Enrique Cuartero of train No. 9 was killed and 84 passengers were injured and taken to San Paulo Hospital. Four of those injured died in the following days. The accident led to the hospital's catastrophe plan being put into operation, the first time it had been used in the eight years of its existence.

The cause of the accident was a mystery. The trains were provided with four different brake systems – three automatic and one manual. The driver of train No. 8 stated later that he had operated the manual brake but it had failed to hold the train. The automatic brakes all apparently failed to work. The train had run back down a gradient of 42‰ (1 in 24).[717] [718]

An engine and nine coaches of train No. 617, the *Rias Baixas* express, which had left Madrid at 21:10 and was due in Vigo at 08:45 and Pontevedra at 09:09, were derailed at 05:05 on Saturday 8th November 1975 in Tunnel No. 66, between Laza and Villar de Barrio stations, Spain. Five lives were lost and seven people were injured. A broken rail had caused the derailment; the low temperature and wheel slipping were possible factors in the fracture of the rail.[719]

[717] *La Vanguardia*, Barcelona, Spain, 31st October and 1st and 26th November 1975
[718] Federación Castellano Manchega de Amigos del Ferrocarril
[719] Ibid.

CHAPTER 12

LEVEL CROSSING COLLISIONS

Wherever roads and railways intersect they must do so either by means of a bridge or on the level. Bridges have the advantage of allowing road and rail traffic to operate entirely without interference from one another, but they are much more expensive to provide, especially in flat countryside, due to the need to provide extensive earthworks for approach embankments or cuttings as well as the cost of providing the span itself. Where underbridges are provided there is often also the need to build retaining walls and there can be problems with drainage.[720] The provision of level crossings, especially if, as in many parts of the world, they are unstaffed, is the much cheaper alternative, but where they exist they introduce an additional source of danger to both road and rail users. Robert B. Shaw in his outstanding book on American railway accidents *A History of Railroad Accidents, Safety Precautions and Operating Practices* writes that he has excluded ordinary grade (i.e. level) crossing collisions from consideration on the grounds that they are not proper train accidents, inasmuch as trains usually suffer much less harm from such accidents than the road vehicle involved. This series of volumes, on the other hand, will attempt to include, amongst others, all such accidents that resulted in five or more fatalities, even if, as was often the case, none of those who died were train passengers or railwaymen. As in previous volumes, accidents involving school buses have also been included.

The First World War provided a great stimulus for the development of the internal combustion engine and the number of motor vehicles in use, and consequently of level crossing collisions involving them, greatly increased following the war. The surge in ownership of motor vehicles continued in many richer countries after the Second World War and the removal of the most dangerous level crossings, or at least the provision of more sophisticated crossing protection, did not keep pace, so there was a substantial increase in serious level crossing accidents. This chapter is therefore a long one.

The American occupation of Japan had ended in 1952 but under a security treaty large numbers of American military were allowed to remain in the country indefinitely. Thus it was that a United States sergeant, his wife and two children were killed in a level crossing collision in Tokyo on New Year's Day, Sunday 1st January 1961. The victims were Sgt.

---

[720] *Backtrack* Volume 24 pg. 88.

James Schmidt, 29; his wife Toshiko, 28; their children Thomas, 8, and Susan, 5; and the sergeant's father-in-law, Kenzo Ota, 62. Another son, James Schmidt, 6, survived the accident and was taken to Tachikawa Air Force Hospital. The accident happened at a crossing variously described as near Ikebukuro station and near Ozumi station. The 1958 Oldsmobile car was proceeding along M Avenue at 19:00 when it reached the crossing, was struck by a Seibu Railway train and carried more than 100 metres. After the accident the car was found to be in reverse gear, from which it was theorised that having ignored the flashing lights and warning bell and entered the crossing, the sergeant had seen the train close at hand, had panicked and attempted to reverse his car.

On Tuesday morning James Schmidt was recovering in hospital but no one dared to tell him that his mother and father, his siblings and his grandfather were all dead. Family services officer Capt. Eugene Strathman said: "We are trying to contact someone in Schmidt's outfit who was friendly with them and who might even know young Jack. We haven't even been able to find out if he has grandparents in the U.S. yet. We'd like to have some friendly family take Jack in their home for a little while until we can make arrangements to either send him home with an Air Force family or an escort officer or to have one of his grandparents come for him. Frankly we don't know what to do just yet." [721]

Japan suffered another serious level crossing accident at 16:16 on Friday 13th January 1961, this time involving deaths and injuries to the passengers of two electric trains. Fumio Sato, 25, was in charge of an empty gravel truck, although he had no driving licence. He drove onto an unprotected level crossing at Akibamachi, Totsuka Ward, Yokohama, without stopping and his lorry was struck by a northbound train and thrown into the path of a southbound one. The northbound train was derailed in the collision and it also struck the southbound one. Coaches at the front of each train were ripped open by the impacts and train passengers were torn from their seats, while the gravel truck was wrecked and became a mass of flames. Sato was thrown clear of his lorry but badly hurt. Five passengers were killed more or less outright and 98 were injured, 17 seriously.

"There was a terrific noise, like thunder," a passenger said later. "The shock was tremendous. It threw passengers down. Someone was calling for help. A woman was crying. Someone else said, 'Let's run.'"

A newspaper reporter chanced to be a passenger on the northbound train and was thus on hand to give a detailed account of what happened. He wrote:

'It looked as though a giant can opener had ripped into the sides of the trains, gouging out huge chunks of metal, shattering windows, and wrenching passenger seats from the floor. The "can opener" was a 2½ ton dump truck. In attempting to negotiate a double track railroad crossing near Totsuka City, about six miles from mid-town Yokohama, the driver had pitted the lumbering vehicle against two crack electric passenger trains bearing down on him from opposite directions. What happened next was attested to by a tangle of twisted steel.

'The truck became a bouncing ball. The crowded trains – one northbound for Yokohama, the other southbound for Yokosuka – in effect became paddles. Slammed into by the Yokohama train, the truck flew apart, and its bed or chassis was sent hurtling

[721] *Pacific Stars and Stripes*, Tokyo, Japan, 2nd and 3rd January 1961

into the side of the lead car of the southbound train, which screeched to a halt, its sides torn in gaping holes, three of its passengers dead – one of them decapitated. And the truck driver still lived, but he was reported near death late Friday evening.

'The southbound train, its driver dazed and bleeding at the controls, hurtled about half a mile down the track, where it derailed, its wheels fouled by bits of the twisted trucks wreckage. In the lead car two passengers were dying, at least 40 others were injured, many seriously.

'When I arrived on the scene searchlights spotlighted wrecking crews straining to clear the tracks of the crumpled remains of the truck. The truck's bed lay sprawled across the tracks, apparently fused to the rails by a combination of great speed and heat. Behind it, not more than two feet, stood the lead car of the southbound train – a limp, metal monster disembowelled, nothing intact except the driver's seat, spared by inches from the ripping action of the truck.

'Further down the track, and down an embankment, the truck's seat rocked gently in a flooded rice paddy. Two wheels lay lodged in a culvert beneath the tracks.

'The centre cars of the southbound train were undamaged. I reached the other end of the tragedy – the lead car of the northbound train – I found an identical scene. The gaping holes were the same. It could have been the other car. I looked back down along the track. Hundreds of spectators had gathered, police sirens wailed, and children squealed with excitement.'

The lines were reopened after about five hours.

A sixth passenger died later and the lorry driver became the seventh fatality when he succumbed to his injuries the following Tuesday. [722]

In an accident reminiscent of the Rosedale disaster of March 1960 described in the previous volume of this series on Tuesday 17th January 1961 the *City of New Orleans* crack passenger train of the Illinois Central Railroad, en route from Chicago, Illinois, to New Orleans, Louisiana, collided with a Standard Oil gasoline (petrol) tanker on a level crossing at Magnolia, Mississippi, United States on an unguarded level crossing. Fire broke out immediately after the collision and involved the train, the tanker, some cars on a siding and an adjoining warehouse. This time no passenger lives were lost but the crash killed Engineer Jimmy Livingston, Fireman John Collins, the lorry driver Morris Piggott, Bob Gwin, Charles Hughes and Hancey Johns. The latter three men had been unloading grain from three freight cars in the station yard.

'R.J. Williams, a Magnolia painter, was in a store along the main street and heard the *City of New Orleans* and its whistle "blowing like hell."

'"The train smashed into the truck," he said, "and shot flames as high as the trees and spread the length of the train. I couldn't see anything but fire, screaming and blazes." He said, "It looked like hell. I knew somebody was burning up and I couldn't do anything about it. I could feel the flames."

'Mrs Wilma Grey, who worked in a drug store near the crossing, related, "Fire went everywhere. The front sections of the train were burned and killed several people. I don't

[722] Ibid, 14th and 18th January 1961

guess I ever heard so loud an explosion."

'John Emmerich, editor of the *Enterprise Journal* in nearby McComb, said most of the bodies were horribly burned.

'Dr Claude Pettey, a dentist whose office is near the train yards, rushed to the scene when he heard the crash and saw Percy Nelson, a white train yard laborer, staggering down the tracks with his clothes aflame. Pettey said he pulled the burning clothes from Nelson, who appeared calm and kept asking "about the other boys."

'Fire units from nearby Mississippi communities – McComb, Osyka, and others – rushed to the scene, along with the volunteer fire department from Kentwood, La., just across the state line in southeast Louisiana.

'Hospital corridors in Magnolia were jammed with high school students, offering to help the injured and offering blood to those who needed transfusions.'

The express stopped 300 yards down the line due to the action of the dead man's handle in applying the brakes.

Both the freight cars and the warehouse were badly burned. Six people were injured, none of whom were train passengers.

It was theorised that driver Piggott had neither seen nor heard the train. The sound of its horn was presumably drowned out by the noise of the lorry's engine and he had been fatally unobservant. [723]

There was only a 'red and white sign' at the crossing. [724]

On Saturday 11th March 1961 a military truck carrying Chinese Nationalist Army recruits was struck by diesel-hauled express passenger train No. 3001 on a level crossing between Linfengyin and Longtien, Taiwan, not far from Tainan. Initial reports spoke of 22 dead and 19 injured, all on the lorry, and a couple of days later the number of dead had risen to 28. An army officer who asked not to be named stated later that the driver of the truck had been speeding, and that this was a common practice among military drivers, who mistakenly believed that by doing so they could save fuel and sell the excess on the black market. [725] [726]

A school bus stalled on a level crossing at Estaire, Ontario, Canada, on Wednesday 15th February 1961 and driver Percy Hone was unable to restart it. He then shepherded 13 elementary school pupils off the vehicle and they and its driver stood and watched as it was demolished by a Canadian Pacific Railway train, which carried it 200 feet down the track. No one was hurt. [727]

A picnic outing for three families ended in tragedy on Sunday 19th March 1961. The

[723] *Hattiesburg American*, Hattiesburg, Mississippi, United States, 17th January 1961
[724] *San Mateo Times*, San Mateo, California, United States, 17th January 1961
[725] Wikipedia – List of rail accidents 1960-1969
[726] *Pacific Stars and Stripes*, Tokyo, Japan, 12th, 13th and 14th March 1961
[727] *The Daily News*, Saint Johns, Newfoundland, Canada, 18th February 1961

Herron, Fuller and Lennox families were on their way from Norwalk and Downey to a place in the mountains called Lytle Creek, a journey that involved their crossing the Atchison Topeka and Santa Fe Railway main line at Cucamonga, California, United States. Initially just the Herrons were in the leading vehicle with their two children and the Fullers and Lennoxes were following in a second car, but the Fuller and Lennox children wanted to be with the Herron children and so just before the accident they transferred to the Herron station waggon.

As the front vehicle reached the crossing with the Santa Fe line Mr John Herron, 29, apparently failed to notice, until the last moment, an approaching eastbound six-car Santa Fe freight train travelling at 55 miles an hour. He then jammed on his brakes and swerved, but the train struck his station waggon squarely. He, his wife Audrey, 30; their children Kathleen, 5, and Bill, 4; Cim Fuller, 5; and Edward Lennox, 3, were all killed. Mr and Mrs Fuller and Mr and Mrs Lennox could only watch helplessly as the 1957 sedan was smashed and carried some distance down the line.

Engineer James Case said that he had only seen the car at the last moment. He said, "I blew the horn but it was too late." The engine of the train was damaged in the collision and it had to go into the yard at nearby San Bernardino for repairs.

There were no gates at the crossing but it was equipped with warning lights, which were found to be working properly after the accident. [728] [729]

On Wednesday 12th April 1961 a Missouri Pacific Railroad passenger train from Houston, Texas, to New Orleans, Louisiana, collided with a car carrying eight people on a level crossing with a country road about two miles from Port Barrie, Louisiana, United States. Lois Bernard, 30, who had been driving the car; her children Charles, 5, David, 3, and Mark, 2; her father and mother, Mr and Mrs Walter Andrus; and Joanne Andrus, 3, were all killed. Tena Andrus, 5, sister of Joanne, was critically hurt and taken to Opelousas Hospital, where there were hopes for her recovery. The bodies of one of the women and two of the children had to be recovered from the Little Bayou Teche, which the railway passed over by a trestle close to the crossing.

The engineer of the train said that he had whistled for the crossing, " Two or three times, but I guess they didn't hear me." His train, slightly behind time, had been travelling at 60 miles an hour.

The party had been visiting Port Barrie and were returning to Opelousas.

There were no barriers, warning lights or bells at the crossing. [730]

A Missouri Pacific Railroad train was involved in another bad level crossing accident early on Thursday 1st June 1961, this time on the outskirts of Angleton, Texas, United States. William Rose Sr., 36, and Manuel Oliver, 68, were both employees of the Fritz Toomer Packing Co. in Freeport, Texas, and were on the first day of a fortnight's vacation. They were on a trip to Thunderbolt, Georgia and had been only 15 minutes into

---

[728] *Pasadena Independent*, Pasadena, California, United States, 20th March 1961

[729] *Oakland Tribune*, Oakland, California, United States, 20th March 1961

[730] *The Victoria Advocate*, Victoria, Texas, United States, 13th April 1961

their journey when the accident happened. Also in the car were William's wife Gypsie and the Rose children; Eunice, 15; Ellen, 13; Linda, 8; and William Jr., 6.

The train was engaged in shunting operations and was only moving at 10 miles an hour when the car ran into it with such force that a gondola car was derailed. Engineer O.A. Whittington said things happened so fast that, "I didn't even have time to put on the brakes. I don't think he ever saw the train" Brakeman A. Nelston, standing on another car, was thrown 10ft into the air but escaped with bruises. The automobile though was wrecked.

L.O. Hadley, a wrecker driver who was soon on the scene, commented, "It was the most terrible thing I have ever seen. The automobile rammed against the freight car so hard it was literally welded to the train. The bumper was pushed against the front seat and the whole front end of the car was in the seat. The steering wheel was bent like a pretzel and crushed against the roof of the car. Every person in the car rode it the whole way – nobody was thrown out. The man, woman and little boy were crushed together in the seat, which was thrown back against the back seat. The three girls and the older man was lying in one big mess."

All the occupants of the car were killed except Linda Rose, who was critically injured and taken to the John Sealy Hospital in Galveston with a fractured skull and broken legs.

Coffee and picnic items were scattered about the scene.

The accident was hard to explain. Highway 288, on which it occurred, was a dead straight four-lane road. The weather was clear and warning lights and bells at the crossing were operating correctly. There were no skid marks to indicate that the brakes of the car had been applied before the collision. Possibly the fact that it happened at 3.15 a.m. was relevant. Why the party chose to set off at such an early hour, instead of getting a good night's sleep first, is not known. [731] [732]

A car carrying a family out on an afternoon shopping trip was in collision with a slow-moving Chicago Milwaukee St Paul and Pacific Railroad freight train, making its thrice weekly journey between New Glarus and Brodhead, on the afternoon of Saturday 24th June 1961. According to witnesses the car, a 1955 Buick, rounded a gentle curve on Highway 59 coming out of Albany, Wisconsin, United States, just as the train reached the crossing. Skid marks 104 feet long indicated that the car driver had been taken by surprise by the obstruction in front of him. The car collided with the train, spun sideways and ended up wedged against the concrete abutment containing the crossing controls. It then caught fire and it was only the quick action of the railwaymen on the train in pulling them from the vehicle that prevented its occupants from being burnt to death. The driver, John Smith, 28; his wife Marilyn, 25; and their daughter Tammy Marie, 3, were killed outright. Another daughter, Beckie Linn, 2, died 20 minutes after admission to St Clare Hospital in nearby Monroe and son Roger, 8, passed away in the same hospital at 11.58 p.m. that evening. Another son, Randy Lee, 7, survived with a broken leg and was placed under sedation, unaware for a while that he was now an orphan.

There were a warning bell and standard crossbucks at the crossing but no lights and no

[731] *The Childress Index*, Childress, Texas, United States, 1st June 1961

[732] *The Denton Record Chronicle*, Denton, Texas, United States, 1st June 1961

barriers. The train had also been whistling as it approached the intersection. A high knoll and foliage partially obscured the view of trains however, and the crossing was located just past the end of the speed limit through the town, at a point where car drivers tended to accelerate. [733]

The train, in charge of Conductor V.A. Marsh with Engineer Oren Inman, was travelling at 25 miles an hour at the time of the accident. Head Brakeman Amil Tekale said that he just had time to shout to the engineer when he had seen the car, then the crash had occurred.

The Smiths had been visiting Mrs Smith's parents, Mr and Mrs Herbert Kubly, on a farm southwest of Albany before they went shopping there. They left the grocery store in Albany at 4 p.m. and the accident happened a few minutes later.

There were fears that Mr Smith had not been insured and donations were being sought to pay for the funerals of the victims. [734]

Formosa (now Taiwan) suffered a disastrous level crossing collision on Sunday 9th July 1961. A diesel-hauled express train collided with a bus on an open crossing at Minghsiung, near Chiayi, in the south of the country, cutting it in two and throwing its parts into a canal. Forty-nine people were killed more or less outright and 27 others were seriously injured. Two coaches of the southbound train were derailed. Those who died included the bus driver, the female conductor of the bus and, according to some reports, also the train driver, two coaches of the train being derailed. Three of the injured died later. The bus had stopped at the crossing for a northbound freight train and had then entered it and been struck by the express. A witness said that "The bus was hit in the middle and crushed like a matchbox. Passengers were thrown into the air."

Police blamed the bus driver, Lin Ching-Long, for the accident, saying that according to the regulations he should have waited for the conductor to proceed over the crossing on foot to check that it was clear before proceeding. The bus owners distributed the equivalent of $50 to the families of each of the dead. [735] [736] [737] [738] [739]

The level crossing tragedies in the United States continued as well. On Thursday 17th August 1961 Mrs Betty Jane Olszewski, 33, a traffic supervisor in the switchboard of the General Telegraph Co., had taken a day off work to take her children swimming at Bass Lake, Michigan. They were on their way home again to Alma, 15 miles away, and were nearly back when the station waggon that she was driving was hit by an eastbound 41-car Chesapeake and Ohio Railway freight train, from Grand Rapids to Saginaw, in charge of Conductor Joseph Larabell with Engineer John Meusser, 60.

"I can't understand how she didn't see the train," said Donald Brink, 17, who was a

---

[733] *Janesville Daily Gazette*, Janesville, Wisconsin, United States, 26th June 1961
[734] *Monroe Evening Times*, Monroe, Wisconsin, United States, 26th June 1961
[735] *The Modesto Bee*, Modesto, California, United States, 10th July 1961
[736] *The Waco News-Tribune*, Waco, Texas, United States, 12th July 1961
[737] Ibid, 11th July 1961
[738] *Wisconsin State Journal*, Madison, Wisconsin, United States, 11th July 1961
[739] *Indiana Evening Gazette*, Indiana, Pennsylvania, United States, 10th July 1961

witness to the accident.

Engineer Meusser said his train had been travelling at about 50 miles an hour at the time. "I saw her coming but by the time I realized she wasn't going to stop, it was too late," he said later.

The train dragged the car for over a quarter of a mile and it was "welded" to the front of the leading engine. Mrs Olszewski and her children Christine, 12; Michael, 9; Andrew, 7; and Mary Jane, 4, were all killed. Acetylene torches had to be used to free the bodies from the wreckage. Mr Olszewski, 38, a lineman for the Consumer Power Co., was somewhere out in the country on a job at the time. He collapsed when told that his whole family had been wiped out and had to be placed under sedation.

The crossing, with Leuce Road, was protected only by crossbucks but there was clear view up and down the line. [740] [741]

The day after the Michigan accident, Friday 18th August 1961, a similar one happened in Canada, near Dugald, Manitoba. Mr Antoine Walleyn, 31, a lorry driver, had come home early from work and had taken his six children from their home at St Boniface, Winnipeg, to visit his sister, Mrs Prosper Willems, who lived on a farm near Dugald, 18 miles east of Winnipeg. The family were returning home at the time of the accident.

The train that struck the car was Canadian National Railway eastbound extra No. 4, consisting of a locomotive and 10 vehicles, the first section of the eastbound *Continental* express from Winnipeg. This portion of the train was composed solely of express and mail cars, the passengers following in the second section, when the train was run in two parts. It was travelling at between 50 and 60 miles an hour when it struck the car, a 1958 Chevrolet sedan, on a crossing with a secondary road a few hundred yards off Highway 15. As at Alma the crossing was unprotected except by crossbucks and it was just after dusk at the time of the collision, a little before 9 p.m. Both the road and the railway were dead straight at the crossing and the flat prairie country meant that there was good visibility. Engineer Francis English said that he had whistled as usual for the crossing.

The car was carried almost a mile down the track to a second level crossing and Mr Walleyn, his daughters Karen Camellia, 9; Lynne, 8; Maureen, 7; Patricia Ann, 6; and Deborah Kim, 2; and son Gerald Michael, 4, were all killed instantly. Rescuers searched along the track for the bodies and only found that of the last child when they were told of the size of the family by a relative. It took a good time to complete the search and remove the wreckage from the line and the second portion of the *Continental* was delayed for three hours, while the westbound *Super Continental*, bound for Vancouver, British Columbia, lost two hours. The diesel engine of the train was somewhat damaged. After repairs on the spot it was able to limp the short distance to Dugald station, where it was replaced by another locomotive.

The only survivor of the accident was the mother, Mrs Germaine Elodie Walleyn, who had stayed at home. As might be expected she suffered severe shock upon learning what had happened and had to be sedated.

---

[740] *Marshall Evening Chronicle*, Marshall, Michigan, United States, 18th August 1961
[741] *Oshkosh Daily Northwestern*, Oshkosh, Wisconsin, United States, 18th August 1961

There could easily have been another victim, for Timmy Moore, the 5-year-old son of a neighbour and a friend of the Walleyn children, had wanted to accompany them on their visit, and had been very disappointed and had a tantrum when told by his mother that he could not do so, because his aunt was coming to take him out. [742] [743] [744]

An inquest upon the victims of the accident was opened in October but as there was no evidence as to what had happened, there being no witnesses to the collision, it had to be abandoned. A police investigation had already been abandoned for the same reason. The family of Mrs Walleyn had also asked that there be no inquest on compassionate grounds. [745]

In the third major level crossing accident in three days five members of an American family were killed in a collision near Menomonie, Wisconsin, on Saturday 19th August 1961. Mr Ernest Yarnell, 39, a tire builder at a rubber plant, from Osseo, Wisconsin, was returning home from a day's outing to a summer cottage of one of his sisters, where family members had helped drill a well. With him in a pickup truck were his wife Fern, 36, and three sons, William Robert, 9; Jeffrey Allen, 7; and Ronald Howard, 4.

Once again the crossing where the accident occurred, two miles east of Menomonie on County Trunk Highway 2, was protected only by crossbucks. The pickup truck ran into the side of a nine-car Chicago and North Western Railway passenger train, the *Twin Cities 400*, about 25 feet behind the engine, and was thrown 78 feet. Head Brakeman H.L. Whitney said that the westbound train had been 35 minutes late and travelling at 79 miles an hour. It had been due in Menomonie at 6.02 p.m. and its engineer was E. Johnson. Damage to the train was limited to its air hoses.

Farmer Herb Harrison, who lived near the crossing, had been eating his supper when he heard the noise of the crash and it was he who alerted the emergency services. Rescuers found Mr Yarnell's body 100 yards from the crossing; the bodies of the other victims were near the wreckage of the pickup truck.

Five Yarnell daughters, aged between 2 and 17, had remained at the family home. After the accident they went to their paternal grandparents. [746]

On Sunday 20th August 1961 it was the turn of Australia. Late in the day the *Brisbane Express*, en route to Sydney, collided with a car containing seven people on a level crossing near Kyogle, New South Wales. Factory worker Neville Moylen, 42, the driver of the car, who lived in Kyogle; his wife Maureen, 34; his son Stephen, 14; his daughters Louise Frances, 9; and Janine, 3; plus Anthony Norman McLennan, 20, from Casino, were all killed. The one survivor was junior railway porter Denis Ritchie, 18, from Casino, who was also Australian light welterweight boxing champion. He suffered shock, cuts, and bruises.

The small sedan was coming from Casino when it was struck by the second portion of

---

[742] *The Winnipeg Tribune*, Winnipeg, Manitoba, Canada, 19th August 1961
[743] *Winnipeg Free Press*, Winnipeg, Manitoba, Canada, 19th August 1961
[744] *The Medicine Hat News*, Medicine Hat, Alberta, Canada, 19th August 1961
[745] *The Winnipeg Tribune*, Winnipeg, Manitoba, Canada, 27th October 1961
[746] *The Daily Telegram*, Eau Claire, Wisconsin, United States, 21st August 1961

the express and carried 150 yards down the line. The train had just stopped at Kyogle station and was travelling at about 30 miles an hour.

Fettler Thomas Dennis Finnigan, who lived near the crossing, ran to the scene, and helped the train crew remove the victims from the car. The guard of the train was unaware what had happened and asked Finnigan why the train had stopped. It took railwaymen, assisted by passengers, half an hour to release all the victims from the wreckage. The diesel engine pulling the train was damaged and disabled.

This time the accident was not the fault of the road user but was blamed on 18-year-old crossing keeper Reginald Bagshaw, who suffered shock and was taken to hospital along with Ritchie. Later he was released from hospital and was arrested, charged with manslaughter.

Visibility at the crossing was restricted by a sharp curve and a cutting. [747]

On 5th October Kyogle Assistant Stationmaster Vincent Cyril McRae told a court that the express had left his station on the day of the accident at 2.25 p.m., three minutes early, and that if it had left at the proper time the accident might have been averted. Mr Teece, the Stipendiary Magistrate, then dismissed the charge. [748]

The failure of a crossing keeper to lower the barrier was likewise blamed for a level crossing collision near Taipei, Formosa (later Taiwan), on Saturday 30th September 1961. As a result of his error a speeding train collided with a bus, killing 10 bus passengers, and injuring more than 20. [749]

In typical American fashion it was not uncommon by the 1960s for more senior high school pupils to possess their own car and use it to commute to and from school. Thus it was that on the afternoon of Monday 2nd October 1961 five students of Franklin Central High School in Indianapolis, Indiana, United States, all close friends, were on their way home the two or three miles to the Acton area in a car driven by 16-year-old Stephen Pratt. Also in the car were Stephen's 15-year-old sister Nadine, Robert Altmyer, aged 17, who was not always there but was that evening because there was no football practice; Robert's sister Laurel, aged 14; and Lenny "Junior" Taylor, also 14.

The pupils' 1949 model car had only gone a short distance from the school when it reached a crossing with the New York Central Railroad at about 2.50 p.m. Behind them in another car was one of the teachers at the school, Mrs Florabel Willsey. She saw the students' car pull up at the crossing and stop for an approaching train. Then the car moved forward again, onto the crossing and into the path of the *Cincinnati Special* express. Eleven minutes late, the train was travelling at between 70 and 80 miles an hour. The car was hit broadside and cut in two. One part of the wreckage, 'crumpled like a piece of waste paper,' was found 25 yards from the crossing and the other part, mostly the roof, ended up the same distance further on. Both sections of the vehicle were thrown into a lineside ditch. All five of the pupils were killed.

---

[747] *The Canberra Times*, Canberra, Australian Capital Territory, Australia, 21st August 1961
[748] Ibid, 6th October 1961
[749] *Anderson Sunday Herald*, Anderson, Indiana, United States, 1st October 1961

Mrs Willsey thought that the windows of the students' car must have been closed and that they had failed to hear the whistle of the approaching train. She said that Pratt had been "driving very carefully," and, "He stopped and waited but apparently didn't see the train. I heard the train whistle and was dumbfounded to see them go across...You don't know what to expect."

Engineer Verna Zander said, "They stopped, then they just drove up on to the tracks. I don't think they were trying to beat me. I don't know what they were doing. I was within 100 feet of the crossing when they drove up on the tracks. I was running a little late but would have been on time by the time we hit Shelbyville."

Another, perhaps less likely explanation for the accident, would seem to be that the driver had seen the train but had supposed that it was a slow-moving freight and thought that he could beat it over the crossing.

Mrs Altmyer reportedly lamented afterwards, "I told them, 'I wish you'd take the school bus today,' but my son told me that the bus just got them there in time for school, and he had some history homework to do before classes began." [750] [751]

Five railwaymen and the lorry driver responsible for the accident lost their lives in a level crossing accident three miles east of Miles City, Montana, United States, at dawn on Wednesday 25th October 1961. Engineer Charles Guidice related how as he approached the unguarded crossing he had seen two gravel trucks moving towards the tracks. He had whistled, despite which the trucks did not stop. The first just made it across, the second did not, the leading diesel unit smashing into its cab. A sheet of flame, presumably from the fuel tank of the lorry, then enveloped the three diesel units and burned for an hour and a half before fire crews could extinguish it. Behind the engines eight of the 22 cars of the train were derailed and two were telescoped, trapping the five railwaymen in the adjoining lavatory and smoking rooms.

The railwaymen who died were Conductor Howard Henry and Porters W.G. Davis, K.P. Richie, E.A. Sims, and Charles Thomas. It took six hours to remove their bodies. The Porters' Union in Chicago said that they could not recall an accident in which so many porters had died. The lorry driver, Louis Evering, had been at the end of his night shift, so tiredness was perhaps a factor in the accident.

Eight of the train passengers were taken to hospital for X-rays and observation but were found not to need much treatment. The accident had occurred near the end of the Berlin Crisis between the United States and the Soviet Union, and the Chicago Milwaukee St Paul and Pacific Railroad train was carrying 385 men of the 32nd Red Arrow division of the Wisconsin National Guard, who were en route to Ft Lewis, Washington, for active service. It had been travelling at 65 miles an hour. [752]

After the accident the men were taken to the National Guard Armoury in Miles City. By just after midnight a half mile "shoo-fly" had been constructed around the wreckage and the troops proceeded on another train. [753] [754]

---

[750] *The Terre Haute Tribune*, Terre Haute, Indiana, United States, 3rd October 1961
[751] *Garland Daily News*, Garland, Texas, United States, 3rd October 1961
[752] *Scottsdale Daily Progress*, Scottsdale, Arizona, United States, 26th October 1961
[753] *Kenosha Evening News*, Kenosha, Wisconsin, United States, 26th October 1961

At the inquest on the victims in November there was some dispute as to whether or not the train had whistled. The jury returned a verdict that the deaths had "not been caused by the criminal negligence of any person now living." [755]

Nine or ten factory workers were killed on Sunday 19th November 1961 when a train collided with the bus in which they were riding on a level crossing at Jaratjt, Iran, about 40 kilometres north-west of Tehran. About a dozen other workmen were injured in the accident. [756] [757]

Two people died in the first of two school bus collisions on level crossings in the United States in just four days in December 1961. Just after 3.30 p.m. on Monday 11th a school bus left Coolidge, Arizona, with 50 students on board. At 4.40 p.m. seven pupils were dropped off and the bus was nearing its final destination, on the far side of a level crossing with the Southern Pacific Company line at La Palma, Arizona. The only remaining occupants of the bus were its driver, Ted Andrews, 27, and Agnes or Ignacia Garcia, 13, a sixth grade pupil. The bus reached the unguarded crossing at 4.45 p.m., at the same time as the *Imperial* passenger train, on its way from Tucson to Phoenix, three hours behind time and running at 70 miles an hour.

Fireman Truman Ellard said later, "I saw the bus and we were under full whistle. It appeared the driver tried to stop and then figured he couldn't and gunned us in an attempt to beat us across the tracks." State law required school buses to stop dead at all level crossings before proceeding. The train stopped down the track with the front of the engine afire. Engineer Arl Booher confirmed that he had whistled for the crossing and that the bell at the front of the locomotive had been ringing. Wreckage of the bus was scattered over three quarters of a mile and the bodies of its driver and sole passenger were found 100 yards from the crossing. With no one knowing how many had been on the bus every available ambulance in the area was sent to the scene but in the end only one was needed to take away the bodies.

Driver Andrews had been operating the bus since August. The vehicle had been inspected a couple of months previously by the State Highway Patrol and again only a week previous to the accident by an insurance company. There were only crossbucks at the crossing. [758] [759]

At 8.05 a.m. on the morning of Thursday 14th December 1961 a disastrous school bus accident happened on a level crossing near Greeley, Colorado, United States. Between 7.30 a.m. and 8 a.m. the Greeley School District (No. 6) bus had picked up 37 children to take them to Delta School, east of Greeley: Arlington Elementary School, Meeker Junior High School, and Greeley High School. En route to the schools the bus had to negotiate a

---

[754] *Idaho Free Press*, Nampa, Idaho, United States, 26th October 1961

[755] *The Billings Gazette*, Billings, Montana, United States, 9th November 1961

[756] *Sarasota Herald Tribune*, Sarasota, Florida, United States, 20th November 1961

[757] *Albuquerque Journal*, Albuquerque, New Mexico, United States, 20th November 1961

[758] *Casa Grande Dispatch*, Casa Grande, Arizona, United States, 13th December 1961

[759] *Tucson Daily Citizen*, Tucson, Arizona, United States, 12th December 1961

level crossing with the Union Pacific Railroad main line. Yet again the crossing was provided only with crossbucks and a sign. The line was crossed on the diagonal and there was good visibility in both directions. The weather was clear, with temperatures below freezing. There had been a recent three-inch snowfall but the gravel road at the crossing was clear.[760] [761]

Witness accounts differed as to whether the bus driver, 23-year-old Duane Harms, had or had not stopped his bus before entering the crossing. At any rate he drove onto the tracks into the path of train No. 111, the 16-car *City of Denver* streamliner, en route from Chicago to Denver and running 90 minutes late due to delays with Christmas baggage. Running at 79 miles an hour the streamliner caught the rear of the bus and cut it in two. The rear wheels and four back seats were carried 455 feet down the line, the main part of the bus was carried 192 feet.

Harms could not remember afterwards if he had stopped at the crossing or not but said, "I think I did (stop). Well, I'm quite sure, because I usually do."

Bus passenger Jerry Hembry, 16, agreed, saying, "The bus stopped and then started up again," while Mrs Henty Kauffman, who lived near the crossing and whose two children had been waiting to board the bus, also thought that it had stopped.

On the other hand Engineer Herbert Sommers stated, "He stepped on the gas and drove right in front of the train. Sommers said he had begun sounding his horn 500 or 600 feet before the crossing and had applied the brakes in emergency 50 or 75 feet before the crossing. The train finally pulled up reportedly about a mile beyond the crossing but probably at considerably less distance." Sommers added, "You could look down in the bus and see the windows were slightly frosted," implying that this might have prevented the bus driver from seeing the train. Harms agreed that the windows had been slightly frosted or steamed up but said that the windscreen defroster had been working.

Fireman Melvin Swanson said, "I thought he (the bus driver) was going to stop. Then he drove right in front of the train." Possibly the bus had stopped just momentarily.

Colorado State law required a school bus driver to stop not more than 50 feet or less than 10 feet from the nearest rail of a crossing and look and listen for a train before proceeding.[762]

Twenty of the children on the bus were killed and the other 17 were injured, of whom all but four were admitted to hospital. Driver Harms escaped with slight injuries and shock.

A newspaper reporter described the aftermath of the accident:

'The scene at the crossing after the crash was one of tragedy and hysteria. Mothers and fathers of schoolchildren all over the area rushed to the crossing. A number were accompanied by children too young to attend school. One woman uncovered the body of a dead child and screamed, "My baby, my baby." Then she found the body of a second child and sobbed tragically.

'Men at the scene broke into tears at the sight of mangled bodies of schoolchildren.

'Women who had no children aboard the bus went to the scene to try to comfort

---

[760] *The Greeley Daily Tribune*, Greeley, Colorado, United States, 14th December 1961

[761] *Colorado Springs Gazette Telegram*, Colorado, United States, 14th December 1961

[762] *The News*, Sarasota, Florida, United States, 15th December 1961

parents who lost youngsters in the accident. A doctor was called to give treatment to parents suffering from shock.

'Officers told the Greely Tribune they decide to give the bus driver shelter in jail for his own protection. Later the State Highway Patrol refused to discuss this angle, but said the driver was not being held officially and no charges had been filed.' [763]

As might be expected the local newspaper had extensive coverage of the disaster. In one report, published the following day, it described the experiences of people on the train:

'Among the 200 passengers on a Union Pacific streamliner that collided with a school bus Thursday was L.J. DiCampli, Boise, Idaho, missile technician. Twenty children died and 17 were injured. "We saw papers and chunks of yellow stuff flying through the air, said DiCampli, who was eating breakfast in the diner of the 16-car train. My first reaction was, 'What are they dumping trash out here for?' Then we heard the train had hit a school bus." When the train stopped, DiCampli said, "Most of us went back to look. I wish I hadn't. It was quiet, too quiet," he added. "It was hard to tell the living from the dead, except that the train crew and ambulance men were using blankets to cover the dead."

'DiCampli said he was going "to have a mass said for those kids as soon as we get to Boise. I'm not even going to go home first."

'"We didn't even feel any impact at all," said a dining car steward, Nels Harding, of Portland, Ore. "The train just slowed down. We had been going about 79 miles an hour, which is the prescribed speed for the stretch. It was a miracle the engineer was able to stop as quickly as he did without derailing the train."'

In another piece the same day the newspaper described the tragedy of some of those who had lost their children in the accident:

'Grief is a sombre tenant in 17 farm homes in the Auburn district southeast of Greeley. Brightly decorated Christmas trees stand in most of the houses. But it will be a sorrowful yuletide for parents and brothers and sisters of 20 youngsters – killed in a grinding smash-up of school bus and train Thursday.

'Ten-year-old Linda Alles was one of the 14 little girls who died. "She was always on the go," said her father, Reuben Alles, as he sat drooped-shouldered in the kitchen. "She was going to decorate the Christmas tree tonight." Alles said Linda, "always sat in the back of the bus." It was the rear section that caught the full impact of the speeding train.

'Two miles south of the Alles house is the house of Mr and Mrs Joe Brantner. Two of their eight youngsters – Kathy, 9; and Mark, 6 – were killed. Their father was one of the first to reach the nightmarish scene of death. "I found Kathy right away," Brantner said, "I knew she was dead." He and his wife told how they numbly picked up small forms, some screaming in pain, others limp, to their station waggon for a fast trip to the Greeley hospital. "Kathy just loved to be with me," Brantner said, "Mark, too. When I'd been irrigating they always wanted to come too."

'Anger was mixed with the grief for the mother of 11-year-old Cynthia Dorn, who also perished. "The folks around here wanted to keep Auburn school open this year, but they wouldn't let us," she said. "It's only a quarter mile down the road. Cindy used to

[763] *The Chillicothe Constitution-Tribune*, Chillicothe, Missouri, United States, 14th December 1961

walk to it. If we still had our school, Cindy'd be alive today.'" [764]

The school bus resumed its journeys the following week but took a different route, avoiding the crossing altogether.

Driver Harms was subsequently charged with manslaughter and stood trial in March 1962. He gave evidence in his own defence. "I don't remember for sure stopping. But I remember I opened the door. I noticed my visibility was poor, so I looked through the rear-view mirror over my head in the bus and was able to see better through it." He said that using the mirror he could see to the east as far as the foot of a block signal post 728 feet east of the crossing.

He went on to explain that his usual practice was to stop at the crossing and opening the door, listen for an approaching train. "Then I turn around in my seat and raise up to look over the children's heads out of the windows to see if a train is coming, and, as a secondary precaution, I look through the mirror."

Harms said that the bus windows had been steamed up on the morning of the accident. "I guess about two inches at the top of the windows were clear." He said that having checked that there was no train coming, "I closed the door. I remember that. As for us putting the bus in gear or going forward, I don't remember." The next thing that he remembered was lying on the ground after the accident.

Harms admitted that at another crossing three-quarters of a mile further east he had occasionally just reduced speed to about one mile an hour, shifted to second gear and then crossed without actually stopping. [765]

After a four-day trial and having discussed their verdict for 8½ hours, over the night of Friday to Saturday, 23rd to 24th March, at 7.05 a.m. on Saturday morning the jury returned with a verdict of "Not Guilty." Harms, his wife and his parents were obviously pleased; he told reporters though that he was too tired to comment. He was discharged by the judge and the four of them left the court. Judge Carpenter had suggested to the jury that they retire to bed at midnight but they had replied that they wished to continue their deliberations. [766]

Five people lost their lives in a level crossing accident at Placedo, Texas, United States, at 4 p.m. on New Year's Day, Monday 1st January 1962. They were Onesimo Vargas Jr., 28; his wife Mary Lou, about 24; their children Evelyn, 4; and Patricia Ann, 2; and a nephew, Jesse Escamilla Jr., 8. The accident happened at a crossing with US Highway 59 when their car was struck by the *Eagle* streamliner of the Missouri Pacific Railroad, en route from Houston to Kingsville. No one on the train was hurt. The crossing was provided with warning signals, which were apparently working normally, and skid marks indicated that Vargas had attempted to stop his car. The train was travelling at an estimated 75 miles an hour; it was little damaged. Coming from their home at Port Lavaca, the family had been going to visit Vargas' parents at Yorktown. [767] [768]

---

764 *The Greeley Daily Tribune*, Greeley, Colorado, United States, 15th December 1961
765 Ibid, 23rd March 1962
766 Ibid, 24th March 1962
767 *The Cuero Record*, Cuero, Texas, United States, 2nd January 1962
768 *Corpus Christi Times*, Corpus Christi, Texas, United States, 2nd January 1962

Thick fog was blamed for a level crossing collision that killed 11 farm labourers on the evening of Wednesday 31st January 1962. The men had been employed in cotton fields on the Azhderian Ranch, west of Mendota, California, United States, and at the end of their day's work were returning to a labour camp owned by Rose Lee Foster, who also owned the bus. Nine of the men were killed outright, one died en route to Fresno hospital and the 11th shortly after arrival there. Roosevelt Crutchfield, the driver of the bus was among those who died. Three brothers – Edward, Ray, and Willie Varella – escaped because they had got off the bus to visit a grocery store two blocks short of the crossing.

Conductor Ernest Burnal and Engineer Orville McRay of the eastbound 22-car Southern Pacific Company freight train that struck the bus said it had been travelling at 60 miles an hour at the time. The engineer added that he had whistled a mile before the crossing and that he had not seen the bus because of the fog.

The bus was a 1949 panel truck converted into a labour bus, then to a lunch waggon, then back to a labour bus. [769]

Mendota City councillor Nolan Lankford stated that there were only standard crossbucks at the crossing, explaining that "the small farming communities can't afford an automatic crossing signal." [770]

What was described as a fast-moving Canadian National Railway 10-car freight train collided with a large yellow school bus on a level crossing with the Trans-Canada Highway on Saturday 24th March 1962. The accident occurred at a location between Moose Jaw and Archydal, Saskatchewan, Canada. There were only two people on the bus at the time, Andrew Kulnych, 18, and Ernest Ronald Howes, 17, both of Southey, Sask. It was not known which of them had been driving. The body of the bus was found 90 feet from the crossing on the east side of the track, its motor was 150 feet away on the west side of the track. The crossing was protected only by flashing lights. [771]

An Ann Arbor Railroad freight train collided with a car on the rural Ruess Road Crossing, four miles north-west of Owosso, Michigan, United States, at 8.30 a.m. on Friday 30th March 1962. Engineer Adlbert White said his train had been travelling at 45 miles an hour and the car had been going at about 40. "The whistle was blowing but the car kept coming. It didn't slow down or pay any attention to the train. I applied the brakes just before the train hit the car," he said. An independent witness verified that the train whistle had been sounded.

The car was being driven by Fred Crawford, 41, of Durand, and was carrying Mrs Barbara Kunzi, 45; Mrs Donna Cutter, 25; and Mrs Cutter's children, Ronnie, 8; Tommy, 6; Billie, 5; Kennie, 4; and Charles, 18 months. The group were bound for the circuit court at Corunna with plans to drop off the children at a nearby school.

The train carried the car 1,000 yards down the line and its occupants were thrown out

[769] *The Modesto Bee*, Modesto, California, United States, 1st February 1962
[770] *Scottsdale Daily Progress*, Scottsdale, Arizona, United States, 1st February 1962
[771] *The Medicine Hat News*, Medicine Hat, Alberta, Canada, 26th March 1962

of it. The three adults, Tommy and Kennie were killed. The other three children were injured and taken to the Memorial Hospital in Owosso. Charles, who suffered serious head injuries, was later transferred to Hurley Hospital in Flint, where he underwent a six and a half hour emergency operation. Ronnie had a bruised forehead and a possible knee fracture and Billie suffered two broken legs.

Adrian Cutter, the children's father, worked at an Owosso furniture factory with Charles Kunzi, 27, son of Mrs Kunzi.

The crossing was protected only by crossbucks. [772] [773] [774]

A car carrying two families, the Gagnons from Thetford Mines and the Fortiers from Robertsonville, both in Quebec, Canada, was on their way to visit relatives when it was struck by a Canadian National Railway freight train on a level crossing near Drummondville, Quebec, on Sunday 1st April 1962. The car was cut in two and Michael Fortier, 24, the driver; his mother, Mrs J.H. Fortier, 56; his brother Claude, 20; his sister Louise, 30; Mr and Mrs Bertrand Gagnon, aged 33 and 27 respectively; and their children Richard, 8, and Andrea, 6, were all killed.

The police theorised that the driver's vision might have been obscured by falling snow. [775]

A lorry carrying 17 relatives from the village of Mishmar Heshiva to a memorial service at a cemetery was struck by an express from Jerusalem to Tel Aviv on a level crossing near the latter place at Beit Dagon, Israel, on Sunday 8th April 1962. The train driver said that he saw the lorry entering the crossing but was unable to stop in time. At least 11 people were killed and six were seriously injured. Bodies and parts of the lorry were scattered for a considerable distance. [776] [777] [778]

Seven waggons broke away during shunting operations in India on Monday 23rd April 1962, ran down an incline, smashed through some level crossing gates and collided with a bus at Dhanbad, in eastern Bihar state. Sixteen people died and another 30 were injured. [779]

A bus carrying Muslim pilgrims returning from the shrine at Mecca was struck by a train in the Ma'an area of southern Jordan on Wednesday 30th May 1962. Eight of the pilgrims were killed and 33 were injured. Seven of the dead were Turks, one of them a woman, and 19 of the injured were also from Turkey. The other victims were

772 *Ironwood Daily Globe*, Ironwood, Michigan, United States, 31st March 1962
773 *Traverse City Record-Eagle*, Traverse City, Michigan, United States, 30th March 1962
774 *Marshall Evening Chronicle*, Marshall, Michigan, United States, 30th March 1962
775 *Winnipeg Free Press*, Winnipeg, Manitoba, Canada, 2nd April 1962
776 *The Times*, London, England, 9th April 1962
777 *San Mateo Times*, San Mateo, California, United States, 9th April 1962
778 *The Australian Jewish News*, Melbourne, Australia, 13th April 1962
779 *The Times*, London, England, 25th April 1962

Jordanians. [780] [781]

Another bad level crossing accident happened in India on Monday 11th June 1962. A bus negotiating an unmanned level crossing between Phulera and Sambhar Lake at 11.25 a.m. was struck by a passenger train from Rewari to Marwar. Eighteen bus passengers were killed outright, one died on the way to hospital and six after admission. Another 22 were injured, six seriously. No on one the train was hurt.

Upon news of the accident being received medical aid was at once sent from the civil hospital at Jaipur, 60 kilometres away, and from the salt department hospital at Sambhar Lake (Sambhar is India's largest inland salt lake and there was a salt extraction industry there). Relief trains with doctors and railway officials were also sent from Jaipur and Phulera. [782]

The accident was mentioned in Parliament the following day when Railway Minister Singh pointed out that the railway could hardly be blamed for the accident, which had happened in broad daylight, and he stressed that the drivers of road vehicles had a duty to see whether the track was clear before trying to cross it. [783]

At 08:20 on the morning of Monday 11th June 1962 Argentina suffered what is possibly in terms of fatalities the most serious level crossing ever involving a school bus. The accident happened near Presidente Illia station in the Villa Soldati suburb of Buenos Aires. The bus was carrying 120 children, all boys, from the working-class neighbourhood of Villa Soldati, 40 of whom were killed in the accident, together with a woman teacher and the bus driver, a 59-year-old Spanish man, named Ricardo Costoya. About 80 children were injured. The children were aged between 13 and three and were being taken to school or, the younger ones, to the Antonoi Zacanini recreation centre in Nicolás Avellaneda Park, as both parents of most families had to work to earn sufficient money to support themselves and their children.

The bus was a Dodge model, registration number 01971. It was proceeding along the Avenida Lacarra towards the crossing on a foggy morning, added to which smoke from burning rubbish was drifting across the road, so visibility was down to 20 metres. Crossing Keeper Alfonso Jesús de Aragón had lowered the crossing barriers for a diesel-hauled commuter train but had then raised them for a lorry, and the bus following it, to pass when the bus driver (lorry driver?) had flashed his headlights. Possibly the crossing keeper had failed to notice the ringing of a warning bell advising him of the approach of the train. Driver Clodomiro Vilches and Fireman Juan Carlos Miranda saw the bus enter the crossing and applied the brakes but it was too late, and the Ferrocarriles Argentinos train was only pulled up some several hundred metres beyond the crossing.

The bus was carried 150 metres down the track and all but destroyed. The first casualties were taken to the Piñero hospital in a lorry belonging to the owner of a nearby grocery store. Later, numerous ambulances, police and fire engines went to the scene and

---

[780] Ibid, 31st May 1962

[781] *Pacific Stars and Stripes*, Tokyo, Japan, 1st June 1962

[782] *The Indian Express*, Chittoor, India, 12th June 1962

[783] Ibid, 13th June 1962

the firemen used special tools to remove the dead and injured from the wreckage of the bus. In all the injured were taken to four different hospitals and 1,500 people offered blood for transfusion within two hours.

A mother who had heard the noise of the crash said that she dared not go to the scene. Four of her 14 children had been on the bus. One was killed and others were not expected to live.

One of the survivors, a little girl, said that she had seen the outline of the train emerging from out of the fog and had jumped through a window. A six-year-old girl said, "We were laughing and watching the fog. There was a noise and my ears hurt and everyone seemed to scream."

The crossing keeper suffered a nervous breakdown. After the accident an infuriated crowd had attempted to lynch him and he had locked himself in his cabin until police arrived and rescued him. He was arrested, as were initially the driver and fireman, though the two enginemen were later released. It is not known what happened to the crossing keeper. [784] [785] [786] [787]

Eight labourers were killed and six others were seriously hurt when the *Kamrup Express* collided with a lorry carrying labourers near Tilrah station in Bihar province, India, on the North Eastern Railway, on Thursday 21st June 1962. [788] [789]

A Tehran – Meshed freight train collided with a crowded bus on a level crossing near Garmsār, Iran, at dawn on Monday 9th July 1962. Forty-five people were killed outright and about 30 others were injured. Rescue and first-aid teams rushed to the scene and the injured were taken to hospitals in Tehran, 95 kilometres to the north west, and other local towns. The line was blocked for 12 hours. At the time this was the worst railway accident in the country's history. [790] [791] [792]

By the following day the number of dead had risen to 54 and more deaths were expected. [793]

The Rev Jack Shipman, 43, a Baptist minister, was on the way to conduct the Sunday service at the church at Melrose, Kansas, where he was pastor, at 9.35 a.m. on the morning of Sunday 15th July 1962. With him in the car were his 35-year-old wife Avis, their children Jacki Mae, 15; Ross, 13; and Rebecca Ann, 8; as well as Darrell Evans, 13,

---

[784] http://www.trascarton.com.ar/ediciones-impresas/2012/150-junio29/848-un-suceso-que-enluto-a-familias-obreras.html

[785] Wikipedia – Accidente Ferroviario de Villa Soldati de 1962

[786] *The Lethbridge Herald*, Lethbridge, Alberta, Canada, 12th June 1962

[787] *The Daily News*, St Johns, Newfoundland, Canada, 12th June 1962

[788] *The Indian Express*, Madras, India, 22nd June 1962

[789] *The Times*, London, England, 23rd June 1962

[790] Ibid, 10th July 1962

[791] *The Winnipeg Tribune*, Winnipeg, Manitoba, Canada, 10th July 1962

[792] *European Stars and Stripes*, Darmstadt, West Germany, 10th July 1962

[793] *The Times*, London, England, 11th July 1962

son of the Rev. and Mrs Alfred Evans of Pilcher., Oklahoma. All the occupants of the car except Jacki Mae were killed when the almost new Chevrolet, bought only a month earlier, was struck by a northbound St Louis-San Francisco Railway train heading for Fort Scott, Kansas. Jacki was taken to the Baxter Springs Memorial Hospital with two broken legs, a broken pelvis and hand and head injuries.

The accident happened at a level crossing at Quipaw, Oklahoma, United States, only a short distance from the Shipmans' home. The morning was described as misty and visibility at the crossing was obscured by trees and a bush.

Ther train consisted of just an engine and a caboose. Fireman Robert Warford, who presumably was driving the engine, said that he applied the brakes in emergency when he had seen the car but the distance had been too short to be able to stop. [794]

A packed bus owned by the Korea Tourist Board of Seoul and carrying passengers from Taegu to Pohang was struck by a train travelling between the same two places in the opposite direction, at 07:35 on Wednesday 1st August 1962, on a level crossing at Yongkang, near Kyongju, South Korea. The bus had failed to give way to the train at the crossing. Ten people on the bus died and 43 were injured. Local police took the injured to hospital and were assisted by the 18th ROK Army Hospital Unit from Kwangju. [795]

Seven teenagers from the Olton-Springlake area, bound for a Sunday School picnic, lost their lives on the night of Thursday 9th August 1962 in a level crossing accident near Shallowater, Texas, United States. Their car was struck by a Panhandle and Sante Fe Railway passenger train from Clovis, New Mexico, to Houston, Texas.

Engineer Thomas Norman Bickers told a reporter afterwards, "I saw the car stop on the tracks about 100 yards from it. A boy jumped out of the car on the right side, turned round and then ran back to help a woman out of the car." By then the train had been very close and the teenager, Jimmy Lawson, 18, had to jump away to save himself before he could rescue anybody. The other occupants of the car had all either been frozen with fear or too slow to get out of the car and were killed.

Those who died were Everett Baldwin, 20; Carolyn Carson, 16; Barbara Fancher, 16; James Loveless, 17; Sharon Miller, 16; Bill Shirley, 18; and Diane Smith, 18.

Mr A.J. Blair, a Lubbock salesman, took Lawson, who was suffering from shock, in his car to the Methodist Hospital in Lubbock, where he was given a sedative. "He didn't talk much," Blair said, "but he told me that he was on a Sunday School picnic and that his girlfriend was in the car." Lawson told Blair he understood that the car had tried to stop but had come to a stand on the crossing. In another part of the same newspaper report it was stated that asked by a highway patrolman Lawson had replied that he did not know why the car had stopped. Perhaps, seeing the train close at hand, its inexperienced driver had panicked and stalled it. The report did not indicate who had been driving the car, presumably not Lawson. At the hospital Blair attempted to contact a relative of Lawson's and was finally able to reach an uncle.

---

[794] *Miami Daily News Record*, Miami, Oklahoma, United States, 16th July 1962
[795] *Pacific Stars and Stripes*, Tokyo, Japan, 3rd August 1962

The accident happened shortly before 7 p.m. on road F.M. 401, 20 yards from its junction with U.S. Highway 84. The party had been en route to MacKenzie State Park for a picnic of the intermediate department of the Olton First Baptist Church.

The engine of the train, which was moving at 79 miles an hour, carried the car just over a mile down the track before it could pull up. All its occupants were thrown out over a distance of 105 yards except Barbara Fancher, whose body remained in the car.

'One by one,' a Lubbock newspaper wrote, 'the parents began to arrive, first coming to the hospital, then to the Rix funeral home, where the victims were taken.

'Mrs Cleo Smith was brought to Lubbock by a neighbour. She clutched Jimmy, and the boy finally let his emotions go. Together they seemed to sense that Mrs Smith's daughter, Diane, was dead. "I can't stand it, I can't stand it any more," Mrs Smith cried. "Not my Diane. I just lost my husband less than a month ago. How can God do this to me. I've loved him so."

'Cleo Smith, widely known area farmer, died in Methodist Hospital July 19 after suffering a heart attack at Buffalo Lakes.

'Soon the pastor of the Baptist church which sponsored the outing arrived. He read over the list of those killed. Then he bowed his head, as if in prayer.' [796]

Seven people died at Temple City, California, United States, at 10.40 p.m. on Monday 20th August 1962, when their car was struck on Encinita Avenue Crossing by a Southern Pacific Company passenger train, the *Sunset Limited* streamliner, en route from Los Angeles, California, to New Orleans, Louisiana.

Engineer Earl Knorr said his train had been travelling at between 40 and 45 miles an hour when it hit the car. "I didn't see it coming and I didn't know I'd hit anything until I saw the sparks," he related. By then, he added, he had travelled another 100 yards and an emergency brake application stopped his train 700 yards further on. "The wig wags were going and my lights were on and I used the air horn," the engineer said. Conductor James Christiansen concurred.

As a result of the collision the 1958 hardtop sedan was carried nine blocks (about half a mile) and having been lodged beneath the train was ground almost flat. A witness described the car as, "just a piece of compressed metal. There was glass all over and blood splattered inside. You could hardly tell it was a car." None of the three diesel units or 20 coaches of the train were derailed and no one on it was hurt. The train was running 55 minutes late, having had to await a connection at Los Angeles with the *San Joaquin Daylight* train from Oakland.

Those who lost their lives, all but one from La Puente, California, were Joe Donnell, 22; his wife Dixie Lee, 19; their children Kathleen, 2, and Kendall, 1; Wanda Boatwright, 18, of Rio, California, Mrs Donnell's sister, whom she had been visiting; Madonna Lydia Finley, 18; and Stanley Earl Peacock, 19, her fiancée. The latter two had been due to marry the following month. The party had been returning from a visit to Mrs Donnell's parents, Mr and Mrs Clifford Adams of Temple City.

The car belonged to Mike Bollinger of Timberlake, South Dakota, who had himself

[796] *Lubbock Avalanche-Journal*, Lubbock, Texas, United States, 10th August 1962

been visiting the Donnells, and whose car had been parked in front of the Donnell vehicle so with his consent they had used his instead. [797]

The State Public Utilities Commission (PUC) had recommended improvements to the crossing, as well as those at Lower Azusa and Temple City Boulevard, just ten days earlier, after a collision with a pickup truck on the Encinita Avenue Crossing on 30th March. Flashing red lights and crossing gates (barriers) would be provided at all three crossings, with the cost, estimated at about $21,000, being shared equally between the railway company on the one hand and the cities of Temple City and Rosemead on the other. After the second accident it was stated that the improvements to the crossing were being expedited.

At the time of the accident protection for southbound vehicles (the direction in which the Donnell car had been travelling) consisted of a small round yellow reflector 500 feet from the crossing, a warning sign painted on the street 300 feet from the track and the wigwag signal and bell at the crossing itself, which for a train travelling at 50 miles an hour began operating 35 seconds before the train reached the crossing. Sergeant Mae Wrona, of the sheriff's homicide squad, stated that, "I don't think the present warning system is adequate."

Local resident Herbert Longley, who lived near the crossing, said, "I don't blame the railroad. People are careless. I sit here (in front of his home) most of the day. I see women with children in the car drive over when the wigwag is in operation and all they do is laugh."

Dean Babcock, the only witness to the crash, stated, "I saw the (Donnell) car coming 30 to 40 miles an hour. The wigwag was working and I heard the train blowing its whistle. The train was going about 60 miles per hour." [798]

A Southern Pacific Railroad of Mexico train bound for Nogales collided with a passenger bus from Mexico City heading for Guadalajara on a level crossing at a location 35 kilometres north of Guadalajara, a place variously described as being at El Arenal and near Amatitlán, Mexico. The bus stalled on in the middle of the crossing late on Saturday night, 22nd September 1962, or possibly early on Sunday 23rd. Six bus passengers were killed and 21 were critically hurt. [799] Among the passengers who lost their lives was Mrs Lois Hayes, 43, an American missionary from Lincoln, Nebraska. [800]

Passenger train No. P 2013 collided with a tractor and trailer on an unprotected level crossing between Bestwig and Frankenberg, West Germany, on Wednesday 26th September 1962, the force of the collision being sufficient to derail the engine of the train. Five people died and one was injured. [801]

[797] *Star News*, Pasadena, California, United States, 21st August 1962
[798] Ibid, 22nd August 1962
[799] *Corpus Christi Times*, Corpus Christi, Texas, United States, 24th September 1962
[800] *The Lincoln Star*, Lincoln, Nebraska, United States, 25th September 1962
[801] *Katastrophen der deutschen Bahnen Teil II* by H.J. Ritzau, pg. 251

On Saturday 13th October 1962 a railcar collided with a bus carrying pilgrims who had visited the Sanctuary of Fátima on the Volta de Ferradura level crossing, on the Lousã line, not far from Coimbra, Portugal. Six bus passengers lost their lives and 16 were badly hurt, though only one, Conceição Garcia de Carvalho, 50, whose husband was away in Mozambique, was thought to be critically hurt. Those who died were Maria Guilhermina, 74; Rita Marquesdos Santos Dinis, 65; and Augusta de Jesus Rosa, 42, all from Espariz; and João Alves Correia, 66; his wife Maria de Conceição Brito, 69; and Deolinda Gil do Carmo, 23, all from Pinheiro de Coja.

The crossing was provided with gates and controlled by a woman crossing keeper, Armandina Ramos de Carvalho, though at the time of the accident her husband, Candido de Sousa Martins Jacob, had been in charge of the crossing. He maintained afterwards that the crossing had been open for road traffic because he had not received the usual telephone call from Ceira station, advising him that the train had left there.

Señor Manuel Graça Santos, head of the Coimbra Municipal Fire Department, chanced to be following the bus on a motorcycle and was thus the first rescuer on the scene. He said that he had not heard the crossing keeper's telephone ringing. He despatched the injured to the Hospitais de Universidade in Coimbra in private cars, one of which on the way apparently ran down a pedestrian, builder Armando Vilea, 56, who suffered a broken leg and also ended up being taken to hospital.

The police were attempting to determine who was responsible for the accident and the Caminhos de Ferro Portugueses launched its own investigation. [802]

Five people lost their lives in a horrific level crossing accident in the United States on Sunday 14th October 1962. A 93-car Southern Pacific Company freight train collided with a car on Airport Way Crossing, California, 10 miles south of Stockton.

Labour contractor Samuel DeLeon Sr., 36, his wife Beatrice, 31; and their children Edward, 10; Maryann, 9; Timothy, 4; and Theresa, 3, were on their way to visit Mrs DeLeon's sister at Manteca, California. Engineer Mario Rinaldo, 51, said he saw the car approaching the line; his engine horn and bell were sounding. The automobile appeared to slow down, only to then roll onto the track. His train was travelling at 45 miles an hour. The train struck the vehicle, which exploded into flames, burning the husband and the four children to death.

Ray McCurdy, 20, from Bakersfield, who had been riding the train, presumably as a trespasser, saved the life of Mrs DeLeon by pulling her from the remains of the car. She was taken to San Joaquin County General Hospital, where she was said to be poorly. Two other DeLeon children, Samuel Jr., and Mario, escaped the accident because they had been left at home.

It was supposed that Mr DeLeon's view of the train might have been obscured by a fence as he had rounded a curve approaching the crossing. [803]

Five members of a family were killed and three others were critically hurt when their

---

[802] *Diário de Lisboa*, Lisboa, Portugal, 14th October 1962

[803] *The Modesto Bee*, Modesto, California, United States, 15th October 1962

car was struck by a Canadian Pacific Railway freight train on a crossing near Lac Mégantic, Quebec, Canada, on Sunday 21st October 1962. Those who lost their lives were Mrs Adrien Isabelle, 51; her sons Julien, 21, and Jacques, 7; and her daughters Noelle, 19, and Yolande, 13; all from nearby Nantes. Mr Isabelle, the driver of the car, and sons Renald and Rolande were injured and taken to hospital in Lac Mégantic. The crossing, with an unpaved road, presumably lacked barriers, and it was foggy at the time of the accident. [804]

Fourteen of 32 children riding in a lorry lost their lives when it was struck by an express train on a level crossing in southern Kazakhstan, USSR, on or about Tuesday 4th December 1962. The lorry driver was arrested and charged with murder. [805]

Nine people died when a Moscow – Berlin express collided with a Przedsiębiorstwo Komunicaji Samochodowej (PKS) bus on a level crossing with Siennicka Street in Mińsku Mazowieckim, Poland, about 30 kilometres east of Warsaw, at 19:02 on Wednesday 5th December 1962. One of those who lost their lives was Crossing Keeper Piotr Kowalski, killed as he tried to stop the bus entering the crossing. Although the newspaper report does not specifically say so it seems likely that the crossing keeper had lowered the barriers for the train and that the bus had rammed them.

Seven of the bus passengers who died were Zenon Gańko, from Wielgolasu; Stanislaw Zwiersz, from the Radzynski area; Jan Banaszek from Generalowa; Stanislaw Jastrzębski from Kęt; Feliks Ośkow from Iwowa; Feliks Szczęsny from Siodlowa and Czeslaw Galadyk from Iwowa. The remaining bus passenger could not immediately be identified. Another 12 bus passengers and the bus driver, Roman Nalążek, were seriously injured, while seven were slightly hurt. They were taken to hospitals in Mińsku Mazowieckim and Warsaw.

A commission of the Dyrekcja Okręgowych Kolei Państwowych (DOKP – the regional state railway directorate) and the Ministry of Communications launched an investigation into the accident, assisted by the public prosecutor. [806] [807]

Eight young Canadians lost their lives in a level crossing collision at about 11.30 p.m. on Saturday 5th January 1963. The accident happened near Anse-à-Gascon, Quebec, in the Gaspé Peninsula. The victims had been returning from a bowling party, when their car was struck by a diesel-hauled Canadian National Railway freight train and carried 2,700 feet along the track. Rescuers needed three hours' work with an oxy-acetylene torch to free the bodies from the wreckage.

Those who lost their lives were Lucian Roy, 19; his brother Jean Marine, 17; his sisters Marie Claire, 21, and Andrea, 16; Giles Soussy, 19; his brother Jean Claude, 18;

---

[804] *The Daily News*, St John's, Newfoundland, Canada, 23rd October 1962
[805] *The Winnipeg Tribune*, Winnipeg, Manitoba, Canada, 7th December 1962
[806] *Dziennik Polski*, Kraków, Poland, 7th December 1962
[807] *The Times*, London, England, 7th December 1962

Hubert Langlois, 18; and Gerard Normandeau, 21. [808]

Engineer John Prentice said that the occupants of the car had apparently seen the train and tried to stop. "I saw by the tail lights of the car that the driver had applied the brakes at least twice." [809]

Almost exactly a month after the December accident another bad level crossing collision occurred in Poland, on Monday 7th January 1963. Again a Przedsiębiorstwo Komunicaji Samochodowej (PKS) bus was involved, this time one was struck by two steam locomotives, possibly a pair of light engines, on a crossing with Szubinska Street in Bydgoszcz, south of Gdansk, in the north of the country. Again the crossing was protected by barriers, but for some reason these had not been lowered. Nine bus passengers were killed and 46 were injured, 11 critically.

Eight of the passengers who lost their lives were Zygmunt Koziol of Wyrzysk; Mieczyslaw Cyganek of Bydgoszcz; Józef Berait, from Kowalewo, Szubin area; Kazimierz Czaczyk, from Szubin; Józef Reszka from Żnin; Bronislaw Mankowski from Kruszyn; Remgiusz Frytz from Szubin and Anna Bucek from Bydgoszcz. One of those who lost their lives could not be at once identified.

Investigations into the accident were begun by the Gdansk Dyrekcja Okręgowych Kolei Państwowych (DOKP – the regional state railway directorate), the PKS and the public prosecutor. Advances in compensation were quickly paid out to the families of the victims by the Państwowy Zakład Ubezpieczeń (PKU – the state insurance company).

By the following day some of the less badly hurt had been released from hospital but 24 were still under treatment, two of whom were in a critical condition. [810]

On the night of Tuesday 8th January 1963 a Rede Ferroviária Federal train collided with a lorry loaded with workers in the town of Arcoverde, in the drought-stricken area of north eastern Brazil. Sixteen workers died and another 20 were injured, despite which the accident seems to have received little attention in the Brazilian press. [811]

A school bus with 76 pupils on board, making its regular run between the area west of Jackson, Michigan, United States, and the Western High School there, stalled on Glasgow Road Crossing, three miles from the town, on Friday 25th January 1963. Fortunately there was no train in the immediate vicinity and the bus driver, William Trumble, was able to calmy disembark the students and march them a block away in accordance with a drill they had gone through at school. He was not, apparently, able to restart the vehicle, and ten minutes later it was struck by an eastbound New York Central Railroad passenger train, which he had not had time to flag. The impact sheared off the front end of the bus but no one was hurt. [812]

---

[808] *Las Vegas Sun*, Las Vegas, Nevada, United States, 7th January 1963
[809] *The Brandon Sun*, Brandon, Manitoba, Canada, 7th January 1963
[810] *Dziennik Polski*, Kraków, Poland, 8th January 1963
[811] *Press-Telegram*, Long Beach, California, United States, 9th January 1963
[812] *The News-Palladium*, Benton Harbor, Michigan, United States, 25th January 1963

Mrs Helena Sherwood, 35, of Warren, Wisconsin; her children David, 13; Sharon, 11; and Patricia, 9; and her nieces Carol Woods, 16, of Warren; and Edith Paddock, 16, of Wisconsin Rapids, were all killed when the car they were riding in was struck by the Chicago and North Western Railway *Minnesota 400* passenger train on a level crossing at Tunnel City, Wisconsin, United States. The accident happened at 4.45 p.m. on Sunday 10th March 1963. Mrs Sherwood had waited for a southbound Chicago Milwaukee St Paul and Pacific Railroad passenger train to pass at the crossing with that line, and had then pulled onto the track into the path of the northbound North Western train from Chicago to Mankato, Minnesota, at the crossing with that line 150 yards further on. There were warning lights at the first crossing but there was no protection at the second.

Mrs Allen Lind, who lived near the crossing, saw the accident happen from a window of her house. She said, "I heard the train. The car went onto the tracks and I wanted to scream. I saw bodies flying." Too upset to go to the scene Mrs Lind called her brother, Mr Robert Betthauser, who lived next door. He found that three bodies had been thrown from the car onto his sister's front yard. The other three remained in the vehicle, which was carried almost 2,000 feet to the mouth of the tunnel from which the place takes its name. All six had been killed instantly.

Mrs Sherwood, who had been separated from her husband, was returning home after visiting her mother, Mrs Bernice Paddock, in Tomah hospital, and had stopped to call on a relative in Tunnel City.

The weather had been overcast and the roads were somewhat slippery but it was not snowing.

Engineer A.G. Harnish said his train had been travelling at 50 miles an hour. No one on it was hurt. [813] [814]

Five people lost their lives at 1.25 a.m. on Saturday 13th April 1963 when the 1956 Ford station waggon in which they were riding was struck by a St Louis-San Francisco Railway freight train on a level crossing in Lebanon, Missouri, United States, and demolished. They were Thomas Floyd Barr, 22, from St Louis, Missouri; his unnamed infant daughter; Olin Zane Long, of Oklahoma City, Oklahoma; Billy Thornton Fitzerald, 29, of Tulsa, Oklahoma; and John Neely, 33, of St Louis, Missouri. Wanda June Barr, 20, wife of Thomas, survived the accident and was said to be in fair condition in hospital at Springfield, Missouri. The vehicle was carried a short distance down the line. Mrs Barr was a deaf mute and it was theorised that others in the car might also have been. [815]

The only witness to the accident was Fireman D.L. Samuel, who had shouted to Engineer C.C. Blakeslee to stop when he had seen the car. [816]

Two families who were apparently on their way home from Easter church services

[813] *Oshkosh Daily Northwestern*, Oshkosh, Wisconsin, United States, 11th March 1963
[814] *The La Crosse Tribune*, La Crosse, Wisconsin, United States, 11th March 1963
[815] *Arkansas City Daily Traveler*, Arkansas City, Kansas, United States, 13th April 1963
[816] *Mt Vernon Registre-News*, Mt Vernon, Illinois, United States, 13th April 1963

were killed when a Southern Pacific Company passenger train smashed into the rear of their car on a level crossing at Boutte, Louisiana, United States, about 20 miles south west of New Orleans, on Sunday 14th April 1963.

The victims were Clarence Theriot, 34, of Baton Rouge, Louisiana; his wife Anna, 31; their sons Charles, 9, and Alvin, 7; Leonard Theriot and his wife Mary, both 28; and their year-old daughter Pamela. The men were brothers and the wives were sisters.

The engineer said that his train had been travelling at 50 mile an hour and he had whistled repeatedly. [817] The car was knocked almost 150 feet and torn in half. [818]

On about Wednesday 24th April 1963 seven people died and 15 were injured, four seriously, when a Colombo-bound train ran into a bus about ten miles north of Colombo, Ceylon (Sri Lanka from 1972). All those killed were on the bus. [819]

A party of six young women from Youngstown, Ohio, United States, all close friends, left there on the morning of Saturday 15th June 1963 to visit a friend's lakeside cottage at Geneva-on-the-Lake, on the shore of Lake Erie, to go swimming. Originally they had planned to stay the night there but later they changed their minds and decided to return home late Saturday for Father's Day. It was on their return journey that tragedy struck at nearby Geneva, when they drove onto a level crossing into the path of an 89-car New York Chicago and St Louis Railroad freight train. Just previously they had stopped at the adjoining level crossing with the New York Central Railroad line, where no trains had been about.

Fireman Kyles Andes, who had been driving the engine of the train, said afterwards that things had happened so quickly that he had not seen the car before he hit it. His train had been travelling at 59 miles an hour and the flashing lights at the crossing had been working. Two young men who had been following the car on motor scooters confirmed that the flashing lights at the crossing had been working, adding that they had been amazed to see the car pull into the path of the train.

The car was converted into a pile of twisted metal and part of it was carried a mile down the track at the front of the engine. Jennie Angelo, 21; Barbara Cowan, 19; Dariene Muzenic, 21; Juliane Palko, 20; Sylvia Sanker, 18; and Sondra Van Sack, 18; were all killed. It was not clear who had been driving the car. [820]

A family either on an outing to see relatives, or, according to another account, driving home from a picnic, was involved in a level crossing collision that resulted in the deaths of seven of the nine persons in the four-door sedan, at about 5.50 p.m. on the night of Sunday 11th August 1963. The accident happened on an unguarded crossing five miles from Saginaw, Michigan, United States, when the car was struck by a 54-car Chesapeake and Ohio Railway freight train.

---

[817] *Ogden Standard-Examiner*, Ogden, Utah, United States, 15th April 1963
[818] *The Mexia Daily News*, Mexia, Texas, United States, 15th April 1963
[819] *Indian Express*, Chittoor, India, 25th April 1963
[820] *The Salem News*, Salem, Ohio, United States, 17th June 1963

Ellis Gary Jr., 22, of Saginaw; his wife Judy, 20; their daughter Dawn Marie, 4; their son Ellis Walter, 3 weeks; Mrs Audry Glynn, Mrs Gary's aunt, 28, from Auburn, Michigan; the Glynns' daughter, 8, and son Douglas, 6, were all killed, Dawn Marie dying after arrival at St Luke's hospital, Saginaw. Also taken to the hospital were Mrs Glynn's husband, Lester, 29, who was in a critical condition, and the Glynns' son Michael, 2, who was said to be in a serious condition.

Engineer William Mills said that he had not seen the car before the impact with it, when he thought that his train was derailed and had stopped in, according to the conductor, about 27 waggon lengths. The car was thrown into a ditch and wreckage was scattered over a distance of half a mile. One of the train crew commented, "You can't describe it, I tell you. I have children of my own, it was horrible."

Visibility at the crossing was restricted by a clump of bushes. There were skid marks about 100 feet long, indicating that the driver of the car must have seen the train at the last moment and tried to stop. [821] [822] [823]

A family of five lost their lives in a collision on Corrigan Road Crossing, near Noble Park station, Victoria, Australia, at 9.02 p.m. on Sunday 25th August 1963. They were schoolteacher Bernard Carrigg, 42, from South Oakleigh; his wife Eileen, 37; and their sons Peter, 8; Michael, 6; and Gerard, 3. A daughter, Jane, 10, was trapped in the wreckage for 40 minutes and was taken to the Alfred Hospital in Melbourne. Initially she was reported to be in a critical condition but she was somewhat better by the following day. The train was only delayed 50 minutes. At the time of the accident the crossing was completely unprotected but approval had been given for the installation of flashing warning lights. [824]

Presumably due to a mistake on the part of the crossing keeper the barriers of a level crossing at Großen Linden, West Germany, on the line between Frankfurt Main and Gießen, were not lowered on Saturday 31st August 1963 for approaching railbus 51 730. As a result a car entered the crossing, was struck by the train, and five people lost their lives. [825]

Mr Richard Alfred Kuykendall Sr., 27; his wife Mabel Shearer, 25; and their children Richard Jr., 4; Elizabeth, 4; and Cynthia Lynn, 2; from Brooksville, Florida, were killed on Saturday 31st August 1963, when their 10-year-old car was struck by an Atlantic Coast Line Railroad train on a level crossing at Blanton, Florida, United States. The car was obviously being driven too fast and its driver had been caught by surprise to find a train crossing his path. This was shown by skid marks 138 feet long, at the end of which the car struck the locomotive of the train. The automobile was carried 63 feet with parts of it

---

[821] *South Haven Daily Tribune*, South Haven, Michigan, United States, 12th August 1963
[822] *The Ludington Daily News*, Ludington, Michigan, United States, 12th August 1963
[823] *The Cedar Rapids Gazette*, Cedar Rapids, Iowa, United States, 12th August 1963
[824] *The Canberra Times*, Canberra, Australian Capital Territory, Australia, 26th and 27th August 1963
[825] *Katastrophen der deutschen Bahnen Teil II* by H.J. Ritzau, pg. 251

carried three-quarters of a mile and its occupants were all thrown out. [826]

Jerry Wheeler, aged about 23, a student at the Oklahoma Christian College in Oklahoma City who had been substituting the pastor of Harrah Church of Christ; Mrs Phyllis Watson, of Edmond, Oklahoma, wife of the director of development of the college; Mrs Watson's daughter Angie, 3; college students Karen Heirick and Janice Novak of Wichita, Kansas; and Sharon Stewart, of Albuquerque, New Mexico, were all killed when a train struck their car on a level crossing two miles from Harrah, Oklahoma, United States, on the night of Wednesday 11th September 1963. The car was hit by a 63-car Chicago Rock Island and Pacific Railroad freight train during a rainstorm, while a lack of skid marks indicated that the car driver had entirely failed to notice the train. [827] [828]

In an especially serious accident of its kind, a makeshift bus carrying a party of Mexican labourers (bracero workers) from the celery fields where they had been working back to the Earl Myers labour camp was struck by a Southern Pacific Company freight train at about 4.23 p.m. on Tuesday 17th September 1963 a mile south of Chualar, California, United States. Thirty-two of the labourers were killed or fatally injured and another 25 were hurt. It was described as the worst motor vehicle accident in California history.

Engineer Robert Cripe said that his train of 71 sugar beet gondola cars and a caboose had been running at 60 miles an hour. He had seen the bus stopped at the crossing and whistled but the bus had then moved onto the crossing anyway. It took him half a mile to bring his train to a stop, with the remains of the bus adhering to the front of the locomotive.

Francisco Gonzales "Pancho" Espinosa, the driver of the bus, declared that he had stopped at the crossing and looked to the right without seeing or hearing the train until the front wheels of his vehicle were on the track. A highway patrol spokesman stated that Espinosa would be held on an open charge of felony manslaughter with charges likely to follow. He was the only person of the 58 people on the bus to escape injury.

Tony Vasquez had been working in an adjoining lettuce field and saw the converted truck ripped to pieces. It was he who had alerted the California Highway Patrol, before going back to see what he could do to help. He said, "Bodies just flew over the place. Two of those men died in my arms."

Coroner Christopher Hill Jr. went to the scene and described what he saw. "One body was hooked under the engine. Shoes, hats, and cutting knives were all around. Everywhere you could hear the injured moaning." He added that identification of the dead would not be attempted until the next day. "These workers are supposed to carry passports, but so far we have found only one."

The workers had been sat on four board benches running lengthwise along the flatbed truck. Twenty-two died at the scene, others in the 15 ambulances taking them to three Salinas hospitals or after arriving there. The staff at the hospitals were almost

---

[826] *Bradenton-Manatee County Call*, Bradenton, Florida, United States, 1st September 1963

[827] *The Greeley Daily Tribune*, Greeley, Colorado, United States, 12th September 1963

[828] *The Sheboygan Press*, Sheboygan, Wisconsin, United States, 12th September 1963

overwhelmed by the number of casualties, eight of whom were initially stated to be in a critical condition. [829]

The crossing, with a lane known as Thomas Ranch Road, was completely unprotected, though there was a good view along the line in each direction.

Driver Espinosa duly stood trial on 32 charges of misdemeanour manslaughter at Monterey Superior Court in December 1963. Defence counsel pleaded that Espinosa's peripheral vision had been poor and that his view along the track had been interrupted by a line of utility poles, plus the weather had been overcast. Prosecutor Ralph Drummond countered that Espinosa had been negligent and should have looked more carefully for approaching trains. After deliberating for two hours the jury returned a verdict of *Not Guilty*. Had the driver been found guilty he would have faced a maximum penalty of 32 years' imprisonment. The original charge had been one of felony manslaughter but this was thrown out by the grand jury. Driver Espinosa initially remained employed by David L. Myers Co. who would not say if he would still be driving a bus,[830] but in any case, fearing retribution from the families of the victims, he shortly returned to Mexico, where according, to an unconfirmed report, he was later slain in revenge for having caused the disaster.

The Interstate Commerce Commission investigated the accident. Its report concluded that driver Espinosa had had "an unobstructed view of the railroad through a considerable distance" and that "it is evident on this record that, as the bus approached the crossing, the driver could have readily observed the approaching northbound train if he had looked southward along the track." [831]

A further inquiry into the tragedy was made by Ernesto Galarza on behalf of the US House of Representatives. This disputed the allegation at his trial that utility poles had obscured his line of vision but found that a foreman sitting to his right in the bus had done so, also that Espinosa had diabetic retinopathy. As Espinosa could not speak English there was a question as to how he had obtained a chauffeur's licence and the inquiry was critical of the growers' disregard for the safety of its workers, as exemplified by the illegal conversion of a flatbed truck into a bus and the employment of poorly trained drivers. It was recommended that there be regulation to improve the transport of migrant workers.

In 1967 a federal court awarded $1,500,000 to 24 injured workers and 164 heirs, the sum to be paid by the Southern Pacific Company, the Growers' Farm Labor Association, Hardens Farms and Myers Corporation. A small sum had already been paid by the State Compensation Fund with more to follow. [832]

An accident involving two trains that occurred at Fukuoka, Japan, on Kyushu Island, on Friday 20th September 1963 began with the stalling of a lorry loaded with gravel on an unprotected level crossing. The lorry was in the process of being towed away by a second lorry when it was struck by a four-car train bound for Moji. The impact derailed the train,

[829] *The Bakersfield Californian*, Bakersfield, California, United States, 18th September 1963
[830] *The Humboldt Times*, Eureka, California, United States, 13th December 1963
[831] *The Fresno Bee*, Fresno, California, United States, 25th January 1964
[832] Wikipedia – Chualar Bus Crash

which fouled the opposite running line and was struck in turn by a diesel railcar travelling in the other direction. Eight people died and 100 were injured, 18 seriously. The line was expected to be blocked until late the following day. [833]

A freight train collided with a bus in Mexico City, Mexico, on Sunday 20th October 1963. Four adults and an 11-year-old girl lost their lives and another 22 passengers were injured. Driver Juan Flores said that his train had only been travelling at 10 miles an hour at the time. Two buses had tried to cross the line in front of him, the first had made it across and the second was hit. The bus driver involved had fled and was being sought by the police. [834]

On, probably, Wednesday 4th December 1963 a freight train collided with a school bus on a level crossing at Greenville, North Carolina, United States. The bus was dragged 250 feet down the track and the bus driver and two black pupils were hurt. The children were taken to hospital. [835]

At 10.30 a.m. on Saturday 7th December 1963 an Atlantic Coast Line Railroad freight train collided with a station waggon on a level crossing at Midland City, Alabama, United States. The accident was witnessed by Rufus Riley from his home 400 yards away; he said that he had looked up when he heard the train whistle twice. "I saw the station wagon about to cross the tracks," he related. "It rolled onto the track and apparently stalled. The train hit it…When we got there, there was a deadly quiet. There was no noise. There just wasn't anything happening. It was so quiet. There was no moaning."

Mr and Mrs Henry Langford, their children Donald, 18 (who died in hospital); Ray Allen, 10; and Jeannie, 6; their daughter-in-law Mrs Kenneth Langford and her children Shirley Ann, 7 (who also died in hospital); Henry Allen, 3; Vernelle and Judy Carol (ages unknown); and another daughter-in-law, Mrs Thomas Langford, were all killed, nine of them more or less outright.

Mrs Thomas Langford's daughter Margaret Ann, aged 12 months, was the sole survivor of the occupants of the vehicle. She was taken to Dothan hospital with serious head injuries. [836] [837]

In addition to having lost his wife and various relatives Mr Thomas Langford, a seasonal farm worker, was faced with a number of bills from the Southeast Alabama General Hospital and for funeral expenses and insurance cover possessed by some of the victims had expired. He and his daughter, who recovered from her injuries, moved in with his sister, while various churches and townspeople combined to raise just over $1,000 to cover his bills, and Mayor Daniels of Midland City gave little Margaret Ann a

[833] *Pacific Stars and Stripes*, Tokyo, Japan, 22nd September 1963
[834] *San Antonio Express*, San Antonio., Texas, United States, 22nd October 1963
[835] *Press Telegram*, Long Beach, California, United States, 5th December 1963
[836] *The Decatur Daily*, Decatur, Alabama, United States, 8th December 1963
[837] *The Humboldt Times*, Eureka, California, United States, 8th December 1963

huge bright red Santa Claus doll as a Christmas present. [838]

Sixteen bus passengers and the train driver died when a train collided with a small bus on a level crossing at Yilmaz, Turkey, near Manisa, in the west of the country, on the night of Sunday 8th December 1963. [839] Two bus passengers survived the accident, which occurred on the line between Ankara and Izmir. [840]

In the second school bus accident in North Carolina, United States, in slightly over a week, a Norfolk Southern Railway freight train collided with a bus bringing 45 pupils home from P.S. Jones and Washington elementary schools, a mile from Washington, on Wednesday 11th December 1963. This time one black student, Stanley Jones, 14, was killed. Thirty of the students were taken to the Beaufort County Hospital in Washington, all but three of whom were released after treatment.

The bus was struck at its left rear end and almost torn in half. Patrolman J.C. Lee commented that, "it was sheer luck that more weren't killed."

Bus driver Melvin Bailey, who himself was only 17, alleged that he had stopped the bus at the crossing but had not been able to see the train, nor had he been able to hear it "because the kids were making so much noise." Having driven onto the line, he said he had then seen the train and tried to get the bus across the track in front of it, but the rear of his bus had been hit.

Engineer James Bell, on the other hand, denied that the bus had stopped at all. He said that his train had been travelling at 28 miles an hour and that "If the bus had stopped I would have seen it. He didn't stop."

There was no protection at the rural crossing. [841] [842]

A serious level crossing accident happened in Colombia at 08:00 on Saturday 21st December 1963. Passenger train No. 162 from Santa Maria to Fundación, consisting of engine No. 53 with Driver Ermenegilo Castillo, and 14 carriages, had not travelled very far when it collided with a bus owned by the La Veloz company, on Gaira level crossing. The bus, registration number KA-6901, was en route from Barranquilla to Santa Maria and was being driven by Efren Villada, aged 43, from Pereira. The accident happened at the level crossing with the Ciénga – Santa Maria highway at kilometre 12 of the Magdalena Division of the railway.

As a result of the collision ten bus passengers were killed outright and about 35 were injured, of whom four had died in hospital by the following day.

Some of those who lost their lives were named as Carlos Osorno, a railway conductor who was travelling as a passenger on the bus; Ramón Avendaño, Ezequiel Polo,

---

[838] *The Chronicle*, Pascagoula and Moss Point, Mississippi, United States, 17th December 1963

[839] *The Canberra Times*, Canberra, Australian Capital Territory, Australia, 8th December 1963

[840] *Oakland Tribune*, Oakland, California, United States, 9th December 1963

[841] *Statesville Record and Landmark*, Statesville, North Carolina, United States, 12th December 1963

[842] *The Robesonian*, Lumberton, North Carolina, United States, 12th December 1963

Florentino Patiño, José Maria Marín, Guillermo Campo, Amelia Suárez, and Roberto Coliante. Three bodies were not immediately identified and the names of three bus passengers who had died in hospital were presumably not included.

A newspaper correspondent interviewed a newly married couple, Jaime and Etelvina Salazar, who after being wed in Bogotá, had honeymooned on the Atlantic Coast and chanced to be passengers on the bus. Both said that they had been busy talking and had not realised that there was any danger before the accident happened. Jaimie was apparently the more badly hurt of the two and was placed in a separate room on his own. The first thing that he asked the correspondent was how his wife was, plus that his relatives in Bogotá be informed. Etelvina likewise was desperate to know the condition of her husband and asked that her family be informed of the accident. She had suffered bruises all over her body and a cut in her forehead.

It was pointed out that there was good visibility at the crossing and there were clear signs to warn of the danger. The accident was blamed on the bus driver, who had reportedly seen the train approaching but decided that he had time to cross anyway. He was also the only person to escape unscathed from the accident, presumably because the train hit the rear of the bus. [843]

A train collided with a packed bus at Ghamra, on the outskirts of Cairo, United Arab Republic, early on New Year's Day, Wednesday 1st January 1964. (Egypt continued to be officially known as the United Arab Republic between the dissolution of its union with Syria in 1961 and 1971.) [844] Initially stated as 25 and then 32, by the following day the number of dead bus passengers had risen to 34, with over 50 more injured. [845] Rescuers reportedly had to lift the coaches of the train to remove victims from the wreckage. [846]

A freight train collided with a bus on a level crossing near Cuernavaca, Mexico, on Thursday 30th January 1964. Ten bus passengers lost their lives and 32 were injured. [847]

On Friday 31st January 1964 a major level crossing accident occurred at La Picada, Argentina, about 25 kilometres from Paraná. A two-coach Fiat diesel railcar on the General Urquizia division of the Ferrocarriles Argentinos, in charge of Conductor Ignacio Olivera with Motorman Guillermo Camejo, collided with Villa Federal company bus No. 336, driven by Lucindo Vázquez, who was accompanied by his three-year-old son. There were 90 passengers on the train and 25 passengers on the bus, which was en route from Paraná to Curuzú Cuatiá and Paso de los Libres. There were no gates or barriers at the crossing in question and visibility was restricted by lineside vegetation.

The bus should have passed the level crossing on national route 126 well before the train was due to arrive at it but mechanical problems had led to the departure of the bus from Paraná being delayed by 40 minutes. According to witnesses the train was close at

[843] *El Tiempo*, Bogotá, Colombia, 22nd December 1963
[844] *Charleston Daily Mail*, Charleston, West Virginia, United States, 1st January 1964
[845] *The Lowell Sun*, Lowell, Massachusetts, United States, 2nd January 1964
[846] *The Times*, London, England, 2nd January 1964
[847] *The Bakersfield Californian*, Bakersfield, California, United States, 31st January 1964

hand when the bus reached the railway, nevertheless its driver attempted to cross the line, probably because he was anxious to make up lost time, having driven fast from its starting point. The violent collision cut the bus in two and dragged it several metres. Of the bus occupants 22, including its driver, were killed, according to one report. The remaining bus passengers were injured and taken to the San Martin hospital in Paraná. Those who lost their lives were said to have comprised 16 adults, two teenagers and four children. Other reports spoke of only 17 fatalities, possibly some of the badly mutilated corpses were ascribed to two people but had in fact only belonged to one. Whether the bus driver's young son survived the accident was not stated. No one on the train was hurt. The accident happened at 11:20 and firemen worked at the accident scene all afternoon, first releasing the injured and the bodies of the dead, then removing the wreckage.

Two potential bus passengers were lucky. Señor José Luna normally caught the bus every Friday to pay a visit to his brother in Curuzú Cuatiá but on the day in question, unusually for him, he had dozed off, and so missed it. Twelve-year-old Dionisie Sambiana had also planned to catch the bus but in the end he decided to catch the next one instead. As he was boarding it, he heard of the accident that he had narrowly avoided being a victim of.[848] [849]

A 68-year-old man from Stanley, Wisconsin, United States, and four of his grandchildren were killed in a level crossing collision two miles from Stanley at about 10.15 a.m. on Saturday 1st February 1964. The victims were Mr Henry Steinmetz Sr., 68, and Bruce, Roxanne, Douglas, and Gregory Smasal, aged 9, 8,7, and 6 respectively, the children of Mr and Mrs Duane Smasal. Mr Steinmetz was returning the children to their farm home a mile from the accident scene after they had a catechism class at St Anne's Roman Catholic Church in Stanley.

The Minneapolis St Paul and Sault Ste Marie Railroad freight train that struck the car was a special working; train No. 43 from Stevens Point to Chippewa Falls, in charge of Conductor V. Diesche with Engineer Harry Brown and Fireman Le Roy King. The engineer said that he had whistled for the crossing and Steinmetz appeared to slow down but had then continued over the crossing in front of his engine. The children were thrown from the car, which was dragged 97 feet before falling 15 feet into a ditch. Steinmetz was trapped in the vehicle; all five were killed instantly.

The crossing was unprotected except for a 'black and white wooden crossing marker.' The gravel road climbed gently to the crossing from either side. There was good visibility in each direction and no previous fatal accidents had been recorded there. Given that the train was a special and the line was apparently not very busy perhaps Mr Steinmetz was not looking out for a train, as he was not expecting to encounter one at that hour.

The car was a 1960 Rambler.

The engine suffered an estimated $300 damage. The train was stopped about a third of a mile down the line. It continued its journey shortly before noon, in time to clear the line for the regular eastbound freight train.

Mrs Betty Smasal was due home from Marshfield hospital that afternoon, having

---

848 https://archivoferroviario.com.ar/choque-de-colectivo-a-cm-fiat-en-la-picada/

849 *Clarín*, Buenos Aires, Argentina, 1st February 1964

undergone an operation there the previous Monday. Mr and Mrs Steinmetz were staying at the farm while she was away to give Mr Smasal assistance with his large family (there were three other Smasal children; James, 5, Lori, 2, and Mary Ann, 10 months).

Mr Smasal came to the crossing to identify the bodies. His wife was not immediately told of the accident. [850] [851]

The Right Reverend Richard Emery, 53, of Fargo, North Dakota, the Episcopal Bishop of North Dakota, was one of five fatalities in a level crossing accident at Grand Forks, North Dakota, United States, which occurred at about 1 p.m. on Sunday 23rd February 1964. The Rev. Edwin Bigelow, 29, chaplain to Episcopalian students at the University of North Dakota in Grand Forks; his wife Phyllis, 34; their daughter Pamela, 9; and Sharrell Simons, 19, a university freshman and Sunday School teacher, from Mandan, North Dakota, were the others who died. Surviving the accident, though badly hurt, were Christopher Bigelow, 5; and Kimberly and Bethanie Schick, aged 7 and 6 respectively, from Grand Forks. All the dead and injured were thrown from the vehicle and it was not realised that there was anyone alive until a train passenger, newspaperman Lauren Clayman, heard a cry from little Christopher. Bethanie was rushed to hospital in Fargo for neurosurgery. Also taken to hospital and sedated, clearly suffering badly from shock, were Engineer William Norman and Fireman Thomas McMahon of the Northern Pacific Railroad (NP) passenger train from Winnipeg, Manitoba, Canada, to Minneapolis, Minnesota. Another Bigelow son, Lindsay, 10, was not on the trip.

The party had been returning from morning services at the Washville Mission near Oslo, Minnesota, 20 miles further north.

Yet again the accident happened at an unprotected crossing. Another passenger on the train, Joe Christensen, an NP policeman, noted that the engineer had whistled for the crossing with a dirt road, while skid marks, only about a foot long, showed that the driver of the car had seen the train at the very last moment. Dr G.G. Thorgrimsen, the county coroner, noted that it had been very cold at the time of the collision and the windows of the station waggon had been tightly closed, adding, "Maybe they couldn't hear the train whistle." Clayman speculated that the car driver's view might have been obscured by a grain elevator and another large lineside building. [852]

Late on Friday 6th March 1964 Michael Hamel, 39, a Canadian National Railway carman, left Winnipeg with his wife and their daughters, Michelle, 14; Dixie Doris, 11; and Kymie, 9; to visit Mrs Hamel's parents at Redditt, Ontario, Canada, and participate in the annual winter carnival there.

At 2.45 p.m. on Saturday afternoon the family were in a car driven by Mrs Hamel's father, Mr Frank Jakubec, 75, when it reached the east level crossing on the outskirts of Redditt. At the same time a slowly moving diesel-hauled Canadian National freight train was also approaching the crossing. The train hit the car broadside and despite its low speed the car was carried several hundred feet down the line. Mr Hamel, the three

---

[850] *Eau Claire Sunday Leader*, Eau Claire, Wisconsin, United States, 2nd February 1964
[851] *La Crosse Sunday Tribune*, La Crosse, Wisconsin, United States, 2nd February 1964
[852] *Winona Daily News*, Winona, Minnesota, United States, 24th February 1964

children and Mr Jakubec were all killed and Mrs Hamel was injured and taken to St Joseph's hospital in Kenora. Four of the occupants of the car were killed more or less outright. Dixie Doris was critically hurt and in an effort to save her life was flown by Ontario Central Airlines to Winnipeg but she passed away in the General Hospital there at 7.45 p.m. on Sunday.

Until recently the Hamels had lived in Redditt, where Mr Jakubec had been a Canadian National agent. [853]

At the inquest on the victims of the accident the jury returned a verdict of 'Accidental death' and recommended that speed of trains over the crossing should be reduced until more adequate protection was provided. At the time of the accident the crossing was only provided with crossbucks. [854]

A fast-moving Union Pacific Railroad passenger train collided with a station waggon on Sambo Road crossing, on the Fort Hall Indian reservation a mile and a half south of Blackfoot, Idaho, United States, just after 1 a.m. on Sunday 8th March 1964. Engineer B. Patterson said that his train had been travelling at 75 miles an hour at the time. "I saw the car approaching, and it was going slow, about 15 miles an hour," he stated. "It seemed to be stopping, but it then pulled onto the tracks."

Melvin Wagon, 24, and Wesley Stevens, 25, both of Fort Washakie, Wyoming; Clifford Dixey, 36; Mrs Patricia Nagashoah Honena, 30; and Leland Pinto Edmo, 50, all of Fort Hall, Idaho, were all killed. Their bodies and the wreckage of the 1961 vehicle were strewn along the track for 250 feet. Mrs Sybil N. Weiser, 28, of Blackfoot, a sister of Mrs Honena, was thrown from the station waggon and critically hurt. She was taken to hospital in Blackfoot, then transferred to Idaho Falls. [855]

Sgt. Harold Ackers, of Tempe, Arizona, testified at an inquest that he had been driving west along Baseline Road, three miles from Mesa, Arizona, United States, on Monday 11th May 1964, and had seen that the warning lights were flashing at the level crossing with the Southern Pacific Company line. About 300 or 400 yards from the crossing he had also heard the whistle of an oncoming train. He continued, "I saw an oncoming car and the more it came toward me the more concerned I became. It failed to slow down until it was about 75 to 100 feet from the crossing and then it began to swerve from side to side." Then the train had hit it and, "The car literally exploded and was cut in two."

Engineer William Witt, in charge of the engine of the 14-car express passenger train that hit the station waggon, stated that he had not seen it until it was too late to apply the brakes.

Killed in the collision were Mrs Lola King, 26, of Phoenix; her children Victoria, 4; and Michele, 2; and Mrs Ina Lowe of Winkelman and her children Clifford, 10; Michael, 8; Linda, 6; Karen, 5; Mark, 3; and Gregory, 1. The two women were sisters. Mrs Lowe had been heavily pregnant with her seventh child.

---

[853] *The Winnipeg Tribune*, Winnipeg, Manitoba, Canada, 9th March 1964
[854] *Winnipeg Free Press*, Winnipeg, Manitoba, Canada, 23rd March 1964
[855] *Times News*, Twin Falls, Idaho, United States, 9th March 1964

Deputy Sheriff Jack Gilchrist said the flashing lights had been working 20 minutes before the crash. Sgt. Ackers and Donald Smith, whose lorry had been following Ackers' car, and who had both frequently travelled along the road, agreed that trees, a barn, and brush restricted visibility for the drivers of cars travelling east, as the station waggon had been.

John (Bud) King and Harold Lowe, both 30, the husbands of the victims, had to give evidence of identification at the inquest, after which Lowe broke down and cried uncontrollably. [856]

Lowe suffered from an unspecified lung complaint and his only means of income had been derived from covered waggon lamps he and his family made and sold. He had no money to pay for the funeral and the local Veterans of Foreign Wars (VFW) organisation asked for contributions to help him. Lowe wished the VFW to supply him with the names of the donors so he could repay them at some point when he could afford to do so. [857]

On the night of Wednesday 17th June 1964 an Illinois Central Railroad freight train collided with an automobile on a rural level crossing north west of Freeport, Illinois, United States, killing its six occupants. The victims were farm labourer Ray Douglas, 31, of Freeport; his wife Judy, 20; their son Mervin, 4; their daughters Peggy, 2; and Donna, 10 months; and Mrs Douglas's sister Dorothy Cantrill, 16, also of Freeport. Another Douglas child, Ray III, two weeks old, was in the care of friends at the time.

Engineer Harry Bierce said that his 13-car train from Madison, Wisconsin to Freeport had been travelling at 25 miles an hour when it had hit the car broadside and carried it 400 feet along the line. It was reportedly the only train of the day over the line.

The crossing was unprotected by either barriers or lights. [858]

A single-coach train collided with a bus at Sunchon, South Korea, 300 kilometres south of Seoul, on Saturday 1st August 1964 and dragged it 100 metres down the line before the train could stop. Five people were killed outright, four were fatally injured and 12 were seriously hurt. All the casualties were bus passengers. [859]

Colombia suffered another serious level crossing collision on Tuesday 11th August 1964. A lorry driven by Señor Miguel Cadena entered a level crossing in the path of railcar No. 1689, coming from Medellin at considerable speed. The accident happened at Palestina Crossing, near Pailitas. The lorry was demolished and its driver, his three young sons, Armando, Miguel, and Alberto, and two other passengers on the lorry, Gumersindo Montaño and Rito Granados, were all killed. On the train Drivers Hernando Orozco and Jesús Madrigal, plus an unnamed cook, were seriously hurt. [860] [861]

---

[856] *The Arizona Republic*, Phoenix, Arizona, United States, 14th May 1964
[857] *The Yuma Daily Sun*, Yuma, Arizona, United States, 15th May 1964
[858] *La Crosse Tribune*, La Crosse, Wisconsin, United States, 18th June 1964
[859] *Pacific Stars and Stripes*, Tokyo, Japan, 3rd August 1964
[860] *El Tiempo*, Bogotá, Colombia, 12th August 1964
[861] *San Antonio Express*, San Antonio, Texas, United States, 13th August 1964

Train No. 232, a Canadian Pacific Railway Ottawa – Montreal passenger train, commonly known as the *Businessmen's Special*, had covered about 17 miles on the morning of Friday 21st August 1964, having left Ottawa at 8 a.m., and was passing through the farming community of Leonard, Ontario, Canada, at 8.20 a.m. when its day coach was run into by a gravel truck. Derailed, the day coach careered into a concrete pump house and ended up in a grove of poplars. The two following parlor cars were derailed but remained upright and a fire broke out in the dining car. The diesel locomotive, baggage car and leading passenger car were not derailed.

Heward Grafftey, Conservative Member of Parliament (MP) for Brome-Missisquoi, was in the dining car at the time. He said, "It was the most dreadful thing you could imagine." He had tried to help the injured but there was little that he could do. "About all I could do was to console a little boy who watched as his mother and father died." He added that he had heard some priests praying as they lay dying or critically injured, and that Jean-Luc Pepin, Liberal MP for Drummond Arthabaska, and Secretary of State Ernest Steele had also been on the train and were unhurt.

Doug Innes, 17-year-old son of Commons Committee E.W. Innes, said he believed that he was one of only two or three survivors of the 12 passengers who had been in his coach, saying that, "Some people were thrown out of windows and the train rolled on them." He thought those who died included priests and nuns and several elderly women. He had saved himself "by cuffing my arms around my head and neck, then getting down in the seat and holding on." [862]

Another newspaper wrote, 'Residents of Leonard arrived at the scene to find the dead and injured scattered for about 100 feet along the track. The truck was ripped into three sections and the driver's body was found 50 feet from the crossing.

'Mrs T.S. Woods, an elderly woman who lives near the track, said the crash woke her about 8.30 a.m. She rushed downstairs to find people crawling, staggering for help. Her back porch was stained with blood.

'Her daughter-in-law, Mrs F.J. Woods, said she had just driven her husband to work and returned to see the wreckage, including one coach on her mother-in-law's lawn. She said: "It was just awful. I saw some people whose bodies were in pieces. A lot of people were crying but there was no panic."

'In the coach were 52-year-old Reverend Gaston Lassonde, who with his brother, Reverend Andre Lassonde, were travelling to Montreal for their father's 80th birthday. Both are Roman Catholic priests from the Redemptorist Father's Seminary at Aylmer, Quebec. "I was sitting on the on the window side of the seat," Father Andre said. There was a loud crash and I was thrown up against the baggage rack. My brother was pushed out the window. All the windows in our car were smashed and most of the people fell out on the ground. The gravel truck seemed to hit our coach from the opposite side. I'm still spitting gravel out of my mouth." Father Gaston Lassonde had his left foot severed as he was thrown through the window.' The priest's head was also cut by some licence plates being carried by Doug Innes, who had been on his way to Montreal to pick up a new car, and who had dropped the plates when he dived to the floor.

---

[862] *The Brandon Sun*, Brandon, Manitoba, Canada, 21st August 1964

'Another passenger in the coach, 40-year-old Jack Hardman, of Burnaby, B.C., ripped the cord from a package of paintings he was carrying and used it as a tourniquet on Father Lassonde's leg.

'Bodies were left at the scene, covered with blankets, until the injured were taken to hospital.' [863]

Ten ambulances took the injured to hospital in Ottawa after Leonard's general store had briefly been used as an emergency hospital.

The driver of the lorry was identified as Basil Ozopyk, 43, of Ottawa.

In all seven train passengers died in the accident in addition to driver Ozopyk and 23 passengers were injured, some seriously.

The only protection at the crossing was a pair of wooden crossbucks.

An inquest on the victims of the accident was opened at the end of October.

Engineer John Arden Simpson told the court that his train had been travelling at 70 miles an hour. Conductor Edward McMullin stated that the train had departed Ottawa at 8 a.m., five minutes late, and that it was due to run non-stop to Montreal at a maximum speed of 75 miles an hour. Engineer Simpson, however, said that the maximum permitted speed was 85 miles an hour.

Leonard farmer Jack Neely spoke of following Ozopyk's lorry along County Road No. 4. "I was travelling 35-40 miles an hour and the truck was pulling away from me. I saw the truck veer suddenly to the right; it didn't stop. I saw no brake lights." He put the speed of the gravel truck at 50 miles an hour.

The owner of the lorry, Joe Ambrozas of Ramsayville, said that Ozopyk had driven the lorry for two successive summers, that the lorry had been in good working condition, and that Ozopyk was a very good driver who had never had an accident. There was a statement from another person, however, not presented at the inquest, that Ozopyk had boasted from time to time of beating the morning train over the crossing.

Police Constable R.E. Collins told the inquest that the view westward from the crossing was blocked almost entirely to within 50 feet of the track by buildings and hedges.

Allan Dobson and Peter Hopkins, students working at Rathwell's gravel pit near Leonard, the destination of the gravel truck, agreed that the vehicle had been noisy, perhaps due to a faulty muffler.

At the close of the inquest the jury, not surprisingly, took less than an hour to find that the deaths had resulted from Ozopyk failing to use proper care and caution when approaching the level crossing. They recommended the installation of suitable automatic signals at the crossing. [864]

Flashing lights were installed at the crossing in January 1965. [865]

Sixteen or possibly 20 people died when a freight train collided with a bus on

---

[863] *The Citizen*, Prince George, British Columbia, Canada, 21st August 1964

[864] https://churcher.crcml.org/circle/Wreck%20Details/1964Leonard.html

[865] *The Winnipeg Tribune*, Winnipeg, Montana, Canada, 13th January 1965

Thursday 24th September 1964 near Byblos, Lebanon. [866] [867]

Six men died when a car was driven into the side of a train at San Jose, California, United States, early on Saturday 31st October 1964. They were named as Fielding Lundy Jr., 34, the driver; Joseph Duarte, 34; Robert Enos, 28; Lester Glenyer, 34; Donald Louisano, 33; and James MacMillan, 31. All were from San Jose.

The Southern Pacific Company engine, which was struck by the automobile, was only pulling one boxcar, and had been moving at no more than about 10 miles an hour. The car, on the other hand, must have been moving at considerable speed as a portion of its bonnet was embedded in the side of the engine by the force of the impact. The victims were described as being entangled with the wreckage "like spaghetti."

There were no indications in the way of skid marks that the car driver had braked before the collisions, nor was there any suggestion that the men had been drinking. [868]

The failure of farm worker Eulogio Villalba Bejarano, 31, to see a Southern Pacific Company freight train early enough resulted in the death of him, his wife Toresa, 28, and their children Eduardo, 6; Eulogio Jr., 4; and Anna Belle, 1. At the last moment he apparently did see the train and attempted to swerve his four-door sedan, which was northbound on the O.T. Smith cut-off road at Tornillo, Texas, United States, at 5.45 p.m. on Friday 6th November 1964.

The 133-car train was travelling at an estimated 55 miles an hour. Engineer Jimmy Mayfield said, "I saw the car approaching the crossing. I blew my whistle as we do for all crossings but either he didn't see me or thought he could beat me across." The cowcatcher of the engine caught the left front of the car, which spun around and was struck a second time. The occupants of the car, who were killed outright, were thrown out of it and their bodies strewn along 400 feet of track; the car was wrecked. There were no warning devices at the crossing other than a pair of signs.

Three other children of the family, who had not been in the car, were taken care of by Mrs Villalba's sister, Mrs Delfina Castillo, of Acala, who said that the family had been on their way home to Acala when the accident had happened. Other relatives lived in Aldama, Chihuahua, Mexico. [869]

On the afternoon of Wednesday 9th December 1964 No. 236 Down passenger train collided with a private bus on an unmanned level crossing between Dichwa Daboli and Sidhwalia stations, Uttar Pradesh, India, on the Chapra – Thawe branch of the North Eastern Railway. The exact location was a point 69.8 kilometres Gorakhpur. As a result of the accident the bus was wrecked and it was initially reported that 22 bus passengers, the bus driver, and a cleaner on the bus had been killed and another 70 bus passengers injured. The engine of the train was derailed and the driver and fireman had also been

[866] *The Salt Lake Tribune*, Salt Lake City, Utah, United States, 25th September 1964

[867] *The Times*, London, England, 25th September 1964

[868] *The Press-Courier*, Oxnard, California, United States, 31st October 1964

[869] *El Paso Herald-Post*, El Paso, Texas, United States, 7th November 1964

slightly hurt but all the train passengers escaped. [870]

By the following day five of the injured bus passengers had succumbed and the magnitude of the disaster led to it being discussed in parliament. Mr S.K. Patil, the Railway Minister, stated that it was beyond the power of the railway to prevent the accident. This led to a call from the Opposition for an adjournment motion to discuss the accident, which was rejected by Speaker Hukam Singh, and a call for the railway minister to resign, the government being blamed for failing to ensure the safety of the public by allowing unmanned crossings. It was pointed out that the bus was dangerously overloaded as it had been carrying 96 passengers.

Mr Patil assured members that a committee had already been set up to investigate the accident.

Mr Nath Pai (Shiv Sena Party) argued that the railways had failed to implement all of the far-reaching recommendations of the Kunzru Committee, to which Mr Patil replied that the committee had only called for the manning of 1,200 more level crossings, thus bringing the number of manned crossings to almost 18,000 out of a total of 33,000, and that the crossing where the accident had occurred was not one of them. He pointed out that there was a good view along the line from the crossing in both directions and that warning boards were provided on the road and whistle boards on the railway. To man every level crossing would cost the Government 30 crores (300,000,000 rupees) annually. Mr Nath Pai retorted that the lives of Indian citizens were more valuable than the cost but the Speaker supported the Government, saying that no country in the world had undertaken to man every one of its level crossings.

Mr Karni Singh (Independent) suggested that bells could be installed at crossings to warn road users, as was the practice in Japan. Mr Patil replied that this suggestion would be looked into and added that 1,200 more crossings would soon be manned. [871]

A seven-car eastbound Chicago Rock Island and Pacific Railroad passenger train travelling at 50 miles an hour collided with a car travelling at about the same speed on a presumably unprotected level crossing near Merritt, Oklahoma, United States, about seven miles south west of Elk City on Wednesday 30th December 1964. Engineer John Bernell said that the car had made no attempt to stop. Alcy David Tippens, 55; his wife Edith, 52; their daughter Nancy Sue, 15; their nephew R.L. Tippens Jr., 23; and their niece Stella Louise Tippens, 15, were all killed. The car was dragged 1,513 feet in the accident, which happened only a mile from the Tippens' home. [872]

A train collided with a bus on an open level crossing at Puntalara, Argentina, south of Buenos Aires, on Sunday 10th January 1965. Ten people died and 35 were injured. The bus driver apparently 'failed to see the signals,' implying that there were warning lights at the crossing. [873] [874]

---

[870] *The Indian Express*, Madras, India, 10th December 1964
[871] Ibid, 11th December 1964
[872] *The Daily Ardmoreite*, Ardmore, Oklahoma, United States, 31st December 1964
[873] *The Salt Lake Tribune*, Salt Lake City, Utah, United States, 12th January 1965
[874] https://tapas.clarin.com/tapa.html#19650111

Argentina suffered another bad level crossing accident on Sunday 17th January 1965 when a train collided with a car on an open level crossing at General Rodriguez in Buenos Aires province. Six people lost their lives in this accident. [875]

Panhandle and Santa Fe Railway passenger train No. 54, consisting of a diesel engine, a passenger car and a mail car, en route from Lubbock to Amarillo collided with a car at Kaffir, Texas, United States, six miles north of Tulia, just before 5 p.m. on Saturday 23rd January 1965. James Murry Mason, 66, of rural Swisher county; his wife Ella, 63; their daughter Mrs Leonard Green, of Tulia; and her children Terry Lynn Green, 3; Kerry Von Green, 2; and Teresa Gayle Green, 9 months, were all killed. Apart from Mrs Mason, whose body was discovered 250 feet from the crossing, the victims were pinned in the wreckage.

The accident happened on Farm to Market Road 214, only 50 yards west of US Highway 87, which the car had just turned off of. It was not clear who had been driving the car. [876]

A 92-car Chesapeake and Ohio Railway freight train collided with a school bus carrying about 65 children, mostly agd 13 or 14, on a level crossing at Midland, Michigan, United States, on the morning of Monday 1st March 1965. The children were bound for the Midland Central Intermediate school and St Brigid's Catholic School. Seven children were badly enough hurt to be taken to hospital, though none were critically injured.

Driver Rodney Hickerson said that when he had seen the train he jammed on his brakes but the bus had skidded onto the crossing. The train hit the right rear of the bus and pushed it into a ditch. Hickerson himself was unhurt and he helped pupils escape via an emergency door. The crossing was an unprotected one, lacking barriers, warning lights or bells. Engineer George O'Leary said that he had not been able to stop his train in time when he saw the bus entering the crossing. [877] [878]

Local residents blamed the lack of barriers and warning lights at a level crossing for a collision between a train and a bus in Brazil, on the evening of Sunday 11th April 1965. Rede Ferroviária Federal (RFF) passenger train No. PPG-14 was due to leave Guapimirim station at 18:35 and to call briefly at Parada Modelo at 18:40. At 18:42, moving at 30 kilometres an hour, it collided with a bus of the Rio-Roma tourist company, registration number 80-13-76, on the level crossing with the Rio de Janeiro – Nova Friburgo main road, 100 metres from Parada Modelo station. The bus was dragged 20 metres and the impact was sufficient to derail the engine and four of the 13 coaches of the train, causing a panic amongst its passengers, who attempted to jump through doors and windows. Ten people died in the accident and about 50 were injured. The newspaper

[875] https://tapas.clarin.com/tapa.html#19650118
[876] *The Abilene Reporter-News*, Abilene, Texas, United States, 24th January 1965
[877] *Marshall Evening Chronicle*, Marshall, Michigan, United States, 1st March 1965
[878] *The Hillsdale Daily News*, Hillsdale, Michigan, United States, 1st March 1965

report does not specify if the dead and injured were on the train or the bus, though it seems likely that most of the casualties at any rate were on the bus.

Parada Modelo station was likely unstaffed for the first call for assistance was sent from Guapimirim station at 19:00. Ambulances and private vehicles took the injured to hospitals in Magé, Cachodeias de Macacu, São Gonçalo, Teresópolis and the António Pedro Hospital in Niterói.

There were three versions of how the accident had occurred. One was that the bus driver had failed to observe the headlight of the locomotive due to undergrowth alongside the track and a large tree adjoining the line. A second version was that the bus passengers had shouted to the driver that a train was approaching but he had either failed to hear or had ignored them. Others again claimed that the bus had stalled on the track.

After the accident it was pointed out that residents had previously petitioned the railway company for more protection at the crossing, which was on a busy road with traffic to and from Nova Friburgo, Teresópolis, Cantagalo and Cordeiro, but nothing had been done.

The accident brought down the telegraph wires, interrupting communication between Guapimirim and Parada Modelo stations. Later the disabled train was towed back to Guapimirim. The nearest hospital in Magé was hard pressed to cope with all the injured it received and the following day some of those less critically hurt, but who still required hospital treatment, were transferred to Nilopolis and to the António Pedro Hospital in Niterói.

An inquiry into the accident was opened by the Magé police authorities and Colonel Paulo Leal, Superintendent Director of the RFF, announced that a special commission of inquiry had been opened to investigate the accident. Meanwhile the bus driver, José Raposeiras Blanco, who was injured in the collision but did his best to help his passengers, and the train driver, José Alves de Sousa, had already been interviewed by the police.

One of the owners of the bus company, Señor Alfonso Maldase, appealed to the railway company for compensation, stating that the damage to the bus, which had only recently been purchased and was not insured, had amounted to 35,000,000 cruzeiro reals.

Señor Luis Botelho, Secretary for Health and Assistance visited the injured in the different hospitals and announced to the 3rd Infantry Regiment and the Military Police that there was a need for blood donors, whereby 30 soldiers immediately presented themselves at the António Pedro Hospital to give blood.

The newspaper report on the accident mentioned that one of the victims who had lost her life had been 14-year-old Sonia Maria de Freitas, described as one of the best and sweetest students at the Lars de Jesus orphanage in Nova Iguaçu, who had been travelling to Nova Friburgo on the bus. [879]

A desperate effort to save their car resulted in the death of five members of a family in a level crossing accident three miles from Lincoln, Alabama, United States, on the night of Monday 24th May 1965. The Groce family were within a mile of their home when their car stalled on Embry Crossing. Clindon Groce, 33; his wife Sarah, 28; and some of the

[879] *Jornal do Brasil*, Rio de Janeiro, Brazil, 13th April 1965

children had got out of the car and were attempting to push it clear when a 16-car Southern Railway passenger train appeared, travelling at 40 to 50 miles an hour. The parents were apparently trying to get the remaining children out of the vehicle when the train hit the car, killing them; their son Hanit, 3; their daughter Rheatta, 18 months; and their niece Dollie Mae Cunningham, 2. Four other children: Katherine, 10; Robert, 8; Albert, 7; and Terry, 6, had fled upon seeing the train close at hand and escaped. The engineer said that he made an emergency brake application when he had seen the stalled car but had been unable to stop. The automobile was carried 1,300 feet from the dirt road crossing and the bodies of the victims had to be pried from the wreckage. There was presumably no protection at the crossing, though this does not appear to have been a significant factor in the accident. [880] [881]

Six people travelling in a Biscúter (a very small and basic car produced in Spain as a result of post Second World War material shortages and economic difficulties) [882] lost their lives in a collision with a train on a level crossing at Oreitia, Spain, ten kilometres from Vitoria-Gasteiz, on Monday 28th June 1965. The car was dragged 600 metres along the line. Those who lost their lives were the driver, Pedro Salgado García, 34, a nurse; his wife Gloria Valmaseda, 33; their daughters Aránzazu, Gloria and Yolanda; and the driver's mother, María Ramos García, the widow of Salcedo, 58. [883]

Only two days after that at Oreitia Spain suffered a worse level crossing accident, this time involving a bus and due to a mistake by a crossing keeper. At 21:00 on Wednesday 30th June 1965 a bus left Seville for Morión, Barriada de San Juan, Algámitas and El Saucejo with about 40 passengers. At 21:45 it reached the level crossing at El Arahal. The lifting barriers at the crossing were raised and the bus entered, only to be struck by a railcar coming from Granada to Seville and dragged along the track about 50 metres, embedded in the front of the railcar, which was also badly damaged at its front end and derailed. Twelve people were killed outright and 30 injured, five of whom died later.

Passing motorists were the first to give help and they took some of the injured into El Arahal town, a short distance away, and raised the alarm. Doctors and ambulances went to the site from neighbouring towns and the Civil Guard later cordoned off the scene.

Living in a house adjoining the level crossing was 9-year-old Antonio Conde, the son of a railwayman, who along with his sisters was the only witnesses to the accident. Years later he told a reporter of his experiences that day: "In 1965 it was completely dark, because the clocks had not changed. I was having supper with my sisters, my parents had gone into town to run some errands and we had been left in the charge of my eldest sister, Araceli, who was 14 at the time. Suddenly we heard the noise of the Granada train and realised that the crossing barriers had not been lowered. Araceli and my little sister, Ana, 6, went to the front door, while with my 11-year-old sister Rosario I looked out of the window. We saw the train arrive, it seemed to be slowing down, while at the same time the bus came around the curve and entered the crossing. It was terrifying." Antonio said

[880] *The Daily Independent*, Kannopolis, North Carolina, United States, 25th May 1965
[881] *Lebanon Daily News*, Lebanon, Pennsylvania, United States, 25th May 1965
[882] http://dictionary.sensagent.com/Biscuter/en-en/
[883] *ABC*, Madrid, Spain, 29th June 1965

he thought that the railcar had been braking, "because, otherwise, with the force that the train had hit the bus, it would have destroyed it and no one would have escaped alive. The motorman was doubtless reducing speed because he saw that the crossing barriers had not been lowered, but he could not avoid the collision. The bus was carried along the line and thrown against a shallow cutting, while the railcar, which we call the baby pig because of its shape, was derailed.

"First," Antonio continued, "there was a heart-breaking silence. Then I heard screams and moans which I still remember. I will never forget them. My older sister went closer, before any help had arrived, but she became frightened and turned away." Antonio had left the house by now but his sister shepherded him back indoors to protect him from the horrors of the scene. Then the crossing keeper, who he believed was called Miguel Andrade, also came into the house, and looked so scared that they had hugged him. "My sister Araceli took him by the shoulders and asked him what had happened. In a state of shock he just kept repeating, 'I don't know, I don't know.' The crossing keeper asked for water and then left."

Soon after the children's parents, Rafael Conde and Ana López, had returned home. His mother, obviously having heard that something had happened on the railway, had feared that one of her children had been caught by the train. Her offspring all hugged her. Meanwhile Rafael had taken charge of the situation and when the crossing keeper, who had been hiding in a ditch, reappeared, Antonio's father advised him not to go to his gate box until the Civil Guard arrived.

The following day reporters arrived and wished to interview the children but their mother, afraid, forbade them to say anything.

Crossing Keeper Andrade was dismissed by Red Nacional de los Ferrocarriles Españoles, arrested, and subsequently stood trial, Antonio's parents, and his aunt Dolores (Andrede's wife?) who was also a crossing keeper, testifying in court on his behalf. He was sentenced to several years' imprisonment, dying, Antonio believed, after about a year from cancer. No explanation was given for the crossing keeper's mistake. [884] [885]

Two brothers, their wives and two children were killed in an Independence Day tragedy near Sallisaw, Oklahoma, United States. On Sunday 4th July 1965 Leonard Hudson, 21, of Wichita, Kansas, was on a visit to his brother James, 31, who lived at Muldrow, Oklahoma. The two families then embarked on a picnic outing which took them along the Old U.S. 64 road and across the track of the Kansas City Southern Railway. It was at the crossing with this railroad that their car was struck by a freight train, demolished, and carried three quarters of a mile down the track. Leonard Hudson, the driver of the car; his wife Judy, 20; James Hudson, his wife Mavis Dean, 28, and two of James and Mavis' children, Pamela, 9; and Terry, 3; were all killed. James and Mavis' daughter Sheila, 2, was critically injured and was taken to hospital at Fort Smith, Arkansas. [886]

---

[884] http://www.memoriavisualdearahal.com/el-accidente-de-trafico-mas-tragico/

[885] http://unpaseoporarahal.com/accidenteferroviario.html

[886] *The Salina Journal*, Salina, Kansas, United States, 5th July 1965

A 1964 pickup truck carrying ten people was struck by an Atlantic Coast Line Railroad freight train consisting of an engine, two cars and a caboose, on a level crossing near Maxton, North Carolina, United States, on Saturday 10th July 1965, killing nine of its occupants. The pickup truck had been en route to a baseball game at Bowmore, 11 miles from Maxton. Two of those who died were members of the Red Hill Blue Devils, a Lumbee Indian sandlot (amateur) baseball team, and the others who died had been supporters of the team. All came from Maxton. The other members of the team had travelled in two other cars. The players who died were Cecil Locklear, 21, and Eddie Jacobs, also aged 21. The spectators were Marlen Blue, 12; his cousin Roy, 9; Jesse Clark, 46; his wife Elizabeth, 42; Howard Locklear, 34; his wife Viola, 28; and Hethrow Lowrey, about 42.

The owner and driver of the pickup, Angush Blue, 31, and the team manager, critically hurt with head and chest injuries, was taken to Laurinburg Hospital. Robert Collins, 29, who should also have been a passenger in the vehicle and was due to umpire the game, had declined to ride in the pickup because he said that it was overloaded. Having stayed behind he heard the crash, hurried to the crossing, and later helped to identify the victims. Those who died had only got into the pickup a mile short of the level crossing at a filling station.

A highway patrolman said that high banks probably obscured the pickup driver's vision, as might have two women passengers, each holding a child, who had been sat on the front seat. The vehicle was knocked about 50 feet and the bodies of the victims were strewn along the track between 50 and 300 feet from the crossing. The other two cars had already gone over the crossing and continued some way, before they became aware that the pickup truck was not following them. [887]

Engineer Hugh Gross Merriman Jr. said that he had only seen the pickup truck a moment before his engine hit it, saying, "I'm just heartsick about this thing." According to him the pickup had been travelling at about 50 miles an hour and made no attempt to stop. His train had been travelling at about 40 miles an hour and he had whistled for the crossing. His train was one of four regular freights daily over the crossing. He added, "I don't understand it. They were all residents of the area."

There were only wooden crossbucks at the crossing. [888] [889]

Late on the night of Thursday 26th August 1965 a car carrying a family of six from Fort Worth, Texas, was struck by a St Louis Southwestern Railway (Cotton Belt) train travelling at 60 miles an hour on a level crossing with the Arkansas 27 highway at Lewisville, Arkansas, United States. Mr D.A. Harrison, 36; his wife Lequita, 26; his son David, 15; their daughter Davette, 6 months; and Mrs Harrison's children by a previous marriage, Rodney and Reba Hogue, aged 8 and 4 respectively, were all killed. The car was dragged 200 feet and its occupants were all thrown out.

The view of the driver was obscured by a shunting engine backing up on another track.

The engineer of the colliding train was A.A. Adams.

---

887 *The High Point Enterprise*, High Point, North Carolina, United States, 11th July 1956

888 *The Gastonia Gazette*, Gastonia, North Carolina, United States, 12th July 1965

889 *The Daily Independent*, Kannopolis, North Carolina, United States, 11th July 1956

The family had been visiting a sister of Mr Harrison, Mrs Barrett Bright, who lived in Lewisville. [890]

In another collision involving a family of six, a car driven by Enrique Rodriguez, 35, from Corpus Christi, Texas, was struck by a Missouri Pacific Railroad on a level crossing with FM 88 road near Delta Lake, Texas, United States, at 4.55 p.m. on Saturday 4th September 1965. Mr Rodriguez, his daughter Ninta, 3, and son Ricardo, 4 months, died at the scene. His son Enrique Jr., 5, passed away in Raymondville Memorial Hospital at 7.45 p.m. and his daughter Christina, 2, had succumbed there by the following day. Mrs Manuelite Rodriguez, the wife and mother, was seriously hurt.

The train, driven by Engineer C.T. Kelley, consisted of a switch (shunting) engine, one car and a caboose and was on its daily run between Raymondville and Edinburg.

What protection was provided at the crossing was not stated in the newspaper reports of the accident but almost certainly there was none other than a pair of standard crossbucks. [891] [892]

The bad level crossing accidents in the United States continued with one at Stuart, Iowa, on Wednesday 15th September 1965. A car carrying Mr Harold Jones of Stuart, 51; his wife Vada, 50; Mrs Dorothy Rohwer, 74, of Menlo, Iowa; her son Donald Woodward, 57, of Menlo; and Raymond Green of Greenfield, Iowa, aged 47, was struck by a Chicago Rock Island and Pacific Railroad freight train. 'The battered and twisted auto was carried three-fourths of a mile down the track on the engine's buffer.' One of the victims, Mrs Jones, who apparently had been the driver, was hurled from the vehicle. The others remained crushed in the car. All were killed.

The crossing where the 73-car freight train hit the car was about a block from the depot (station). In this case there was some protection provided at the crossing in the form of a warning bell but the car driver apparently did not hear it, for according to eyewitnesses she drove onto the track directly into the path of the train. [893] Alcohol was a possible factor in this case as the victims had come from a tavern; an unbroken half-gallon glass container of wine was found on the front seat of the car after the accident.

The car had turned south off of Highway 90, from where it was less than half a block to the crossing.

Engineer L.J. Rollins said he had been travelling at 60 miles an hour, and added, "The car came from the east and turned right in front of us." [894]

Conductor Bryce Morehouse described his Council Bluffs to Des Moines train as a "red ball express", which was not scheduled to stop at Stuart. It was delayed over two hours by the accident. [895]

---

[890] *Northwest Arkansas Times*, Fayetteville, Arkansas, United States, 27th August 1965
[891] *The Brownsville Herald*, Brownsville, Texas, United States, 5th September 1965
[892] *The Odessa American*, Odessa, Texas, United States, 5th September 1965
[893] *The Washington Evening Journal*, Washington, Iowa, United States, 16th September 1965
[894] *The Daily Gate City*, Keokuk, Iowa, United States, 16th September 1965
[895] *The Telegraph-Herald*, Dubuque, Iowa, United States, 16th September 1965

According to an announcement from Havana radio 14 people died in a collision between a diesel and a bus at an unstated location in Cuba on Tuesday 21st September 1965. No further details were given. [896]

At least 30 more people died when a Veracruz to Mexico City passenger train collided with a bus on a level crossing near the ancient pyramids at Teotihuacan, 40 kilometres north east of Mexico City on the evening of Monday 1st November 1965. All the casualties were on the bus, which the train cut in two, dragging its rear portion about 85 metres down the track. The diesel engine, leading van and two box cars of the train were derailed but remained upright; the five passenger coaches stayed on the track. Twenty-six people died at the scene and four more after being brought to Mexico City. The bus driver was among those who died. Another 29 were injured, some of whom the hospital authorities feared might not survive.

The driver of the train testified, "He was trying to beat me to the crossing. I was only going about 20 kilometres an hour but I couldn't stop."

One person had died in a previous level crossing involving a bus in the Mexico City area on Tuesday 26th October. [897] [898]

Something of a freak accident occurred on Tuesday 11th January 1966 at La Salada, Argentina, in the suburbs of Buenos Aires. A fast passenger train collided with a bus on a level crossing but the principal casualties were neither on the bus nor the train, but were pedestrians, 10 or 11 of whom were killed and about 27 injured. On the far side of the level crossing was a railway bridge across the Matanza River, to one side of which was a pedestrian footbridge. Shortly before noon the train struck the bus in its middle as it was trying to cross the line and pushed it 500 metres down the line and onto the bridge, the steel railings of which were ripped off. The train came to a stand towards the far side of the bridge with the overturned bus in front of its engine. Some of the pedestrians, who had been using the footbridge at the time, were crushed by the bus, while others jumped into the muddy river to escape. The bus driver and the only two passengers on it were injured. [899] [900]

Nine of ten occupants of a speeding station waggon lost their lives when it was struck by an express train on a level crossing near Tainan, southern Formosa (now Taiwan) on Tuesday 25th January 1966. The remaining occupant was critically hurt. Police said that the driver of the vehicle, who was amongst those killed, had ignored warning lights and bells at the crossing. [901]

---

[896] *Star-News*, Pasadena, California, United States, 22nd September 1965
[897] *The Daily Independent*, Kannopolis, North Carolina, United States, 2nd November 1965
[898] *Corsicana Daily Sun*, Corsicana, Texas, United States, 2nd November 1965
[899] *Clarín*, Buenos Aires, Argentina, 12th January 1966
[900] *Independent*, Long Beach, California, United States, 12th January 1966
[901] *The Courier Express*, DuBois, Pennsylvania, United States, 27th January 1966

On Sunday 30th January 1966 Edward Kratochvil and his family missed a church breakfast at St Joseph's Catholic Church on the outskirts of Agnew, Nebraska, United States. Mr Kratochvil explained that he wanted to "hurry up and get a stove fixed." The parents and their six children jumped into their car and set off home to Raymond, travelling east on Nebraska 79. On the edge of Agnew their car reached a level crossing with the Union Pacific Railroad at the same time as a southbound freight train. Engineer Fred Aden said that his train had been travelling at about 42 miles an hour. The car had slowed down, "but just didn't stop." The collision, at about 10.15 a.m., knocked the car about 40 feet along the line, where it landed upside down in a ditch.

Lloyd Kratochvil, 14, and his brother Richard, 12, were killed more or less outright and their sister Patty, 19, died at the scene. Edward Kratochvil died at 1.30 p.m. in the St Elizabeth Hospital in Lincoln, Nebraska, and his son Gary, 18, died at 5 p.m. in the same hospital. Mrs Josephine Kratochvil, 44, and children Bernadette, 22, and Jerome, 20, were said to be in a critical condition the following day.

The priest at the church the family had just left was called to the accident, as a newspaper related: 'Fr. Richard Brombach, pastor of St Joseph's church, which the Kratochvils attended, with tears in his eyes and a very shaky voice, announced to his parishioners attending the noon mass Sunday that he had "just given the last sacraments of the Catholic Church to the eight members of the Edward Kratochvil family."

'Then the priest told his parishioners about the accident he had been called to just a short while before. Some of the family had received communion only a while before the accident, Fr. Brombach said.

'Fr. Brombach described the accident as "the worst human carnage he had ever seen.

'"I had to climb over the train to get there because the train was blocking the road," he said. "I started giving absolution to those members of the family who were alive. Bernadette was the only one able to talk. She was next to Patty telling Patty to 'take it easy, take it easy.' I gave Patty absolution and then she died. I started giving conditional absolution to those who were already dead. Those bodies were so badly mangled," the priest said.'

It was impossible to tell if any of the people in the car had seen the train. The only protection at the crossing was a pair of crossbucks. Lawrence Masek, the owner of an Agnes television shop, told reporters that he and some other citizens had contacted the railway authorities about a year previously to "see if they would put a signal light at that intersection. They said they would check into it." [902] [903]

Jerome Kratochvil succumbed to his injuries a week after the accident. By then Mrs Kratochvil and Bernadette were somewhat better. [904]

A dilapidated old green bus carrying 34 Puerto Ricans from the strawberry fields where they had been working back to their migrant labour camp was struck by a train on an unguarded Eureka Drive Crossing near Sterling, Florida, United States, 15 miles south of Miami, at about 5 p.m. on 1st February 1966. Nineteen of those on the bus were killed

---

[902] *The Lincoln Star*, Lincoln, Nebraska, United States, 31st January 1966
[903] *Lincoln Evening Journal*, Nebraska, United States, 31st January 1966
[904] Ibid, 7th February 1966

and the other 15 were badly hurt.

The Seaboard Air Line Railroad train, on its way to pick up a load of produce, consisted of only an engine, in charge of Engineer B.E. Lee, and a caboose, and was travelling at about 30 miles an hour. Conductor Edward Shaw related, "The bus just ran in front of us. Then I heard the racket. I looked back but couldn't see anything but dust. Then it settled and I saw bodies lying at the side of the road. The brakeman yelled, 'Good Lord, we've killed a bunch of people.' They were all hollering and crying."

The newspaper report continued, 'The bus was hit squarely in the middle and wrapped around the front of the engine. Battered straw and felt hats were scattered along the road bed. Inside the bus were shoes, shreds of clothing, a battered tin cup and a paper bag that carried lunch to the field.' [905]

Another report stated, 'The dead were left while the injured were taken through a snarl of traffic caused by sightseers and homeward bound motorists to South Miami and Jackson Memorial Hospitals.

'Luis Santiago, 21 years old and powerfully built, talked loudly and rapidly in Spanish as he lay in the corridor of Jackson Memorial's emergency room waiting for hurried doctors to work on his shattered left arm.

'Alejandro Cato, 54, mumbled through cut lips of the horror of the crash. "There was no warning, no whistle, nothing," he said. He told how he saw the driver, Manuel Padilla Marando, 49, then he saw him no more. Padilla Marando was critically injured.

'Another of the injured, 18 year old Manuel Acosta, said that he had heard a warning. "I heard the horn of the bus. I heard the train horn. But the train couldn't stop and the bus couldn't stop."' [906]

'Friends of the Puerto Ricans heard the news at the labor camp where the bus was bound following a day in a pole-bean field. "Did these men have families?" a man asked one of the few who spoke English. "They were Puerto Ricans," was the reply. "In Puerto Rico everybody has a family."' [907]

A helicopter from one of the hospitals brought doctors to the site and returned to the hospital with the most seriously injured patients. The bodies of the dead were covered with coloured blankets and laid beside the track pending their removal to the mortuary. Three hours after the accident and with all of the injured removed work was in progress under floodlights and using oxy-acetylene equipment to cut the remains of the bus, which had been pushed 430 feet down the line, from the front of the locomotive.

Visibility at the crossing was restricted by trees along both sides of the road and at the time of the accident it was provided only with a pair of crossbucks. [908] Preparations were in hand, however, to install warning lights following another accident that had occurred a few weeks earlier, when a 'tomato bus (had) knocked a train off the track.' [909]

'Rickety farm labor buses are a familiar sight in south Florida's winter vegetable

[905] *Reno Evening Gazette*, Reno, Nevada, United States, 2nd February 1966
[906] *The Tipton Daily Tribune*, Tipton, Indiana, United States, 2nd February 1966
[907] *Reno Evening Gazette*, Reno, Nevada, United States, 2nd February 1966
[908] *The Arizona Republic*, Phoenix, Arizona, United States, 2nd February 1966
[909] *Washington C.H. Record-Herald*, Washington C.H., Ohio, United States, 2nd February 1966

gardens. They carry the workers – usually including husbands, wives, children, and grandparents – to the fields from their flimsy labor camp homes,' a footnote in a newspaper informed its readers. [910]

On the morning of Friday 4th February 1966 a postal bus was making its daily run and leaving the town of Heimemkirch, West Germany. It was carrying 38 people, most of its passengers being schoolchildren bound for Lindenberg. As it reached a level crossing with the Buchloe – Lindau railway the barriers were raised and it began to cross the line, only to be struck by local goods train No. 8418. The train was travelling at about 60 kilometres an hour and the bus was carried 200 metres. Five or seven people lost their lives according to various sources and about 26 were injured.

The accident was due to a mistake by the crossing keeper. A passenger train had passed in the opposite direction immediately before the goods train and the crossing keeper had then raised the barriers, having apparently forgotten about the goods train, the approach of which he had been notified of, and which was running 20 minutes late. This was not the first time he had made a similar mistake and it was concluded that he was not suitable for the job. No further action appears to have been taken against him. [911] [912]

Two Church of Christ ministers were among five people who lost their lives in a level crossing collision about three miles west of Seguin, Texas, United States, on Wednesday 9th February 1966. Those who lost their lives were the Rev. Robert Breitkreutz, 38, minister of Cibolo United Church of Christ, the driver of the car; his wife Elizabeth, 37; their children Mark, 4, and Amy Elizabeth, 3; and the Rev Raymond Milton Maggart, 41, pastor of the Bethany United Church of Christ in San Antonio, Texas.

According to Engineer C.J. Logan of the Southern Pacific Company freight train he had seen the car moving towards the crossing at the same time as his engine approached it. The car driver had apparently failed to notice the train until the last moment, when he had jammed on his brakes but the car had skidded on the gravel into the path of the train. Four of the occupants of the car were thrown from it and killed more or less outright. The Rev. Breitkreutz, the only person in the car who had been wearing a seat belt, survived long enough to be taken to hospital.

The crossing was unprotected but the weather had been clear. [913] [914] [915] [916]

Driver inexperience was a likely factor in a tragic accident on Saturday 12th February 1966 on Fairview Avenue Crossing, Barberton, Ohio, United States. Mrs Ella Louise Barnes had returning home from a shopping trip with her four young daughters, Susan, 11; Nancy, 9; Margaret, 5; and Janet, 4; when their car was struck by an Erie-

---

910 *The Humboldt Times*, Eureka, California, United States, 2nd February 1966
911 Wikipedia – Busunglück von Heimenkirch
912 *Katastrophen der deutschen Bahnen Teil II* by H.J. Ritzau, pg. 251
913 *The San Antonio Light*, San Antonio, Texas, United States, 10th February 1966
914 *The Salem News*, Salem, Ohio, United States, 10th February 1966
915 *Corpus Christi Times*, Corpus Christi, Texas, United States, 10th February 1966
916 *Lubbock Avalanche-Journal*, Lubbock, Texas, United States, 10th February 1966

Lackawanna Railroad freight train moving at 50 miles an hour and carried over half a mile. Engineer T.T. Kelly said, "I sounded the horn as I rounded the curve. I saw the car about 100 yards ahead." He had immediately applied his brakes in emergency but the distance was, of course, far too little to be able to pull up in. Mrs Barnes and all but her youngest daughter were killed more or less outright. Janet was taken to Barberton Citizen's Hospital where she died the next day. The father, John Barnes, who had not been in the car, had to be placed under heavy sedation.

Mrs Barnes had obtained her driving licence only a fortnight earlier. [917] [918] [919]

Two level crossing collisions involving buses occurred in Mexico on Sunday 13th February 1966.

Near Torrecon, in the north of the country, a local bus was struck by a train on the outskirts of town and all 25 occupants of the bus were injured, some critically.

The more serious accident, early on Sunday, happened at Texoco, and involved a school bus, though all of the victims were adults. Alberto Gutierrez Lozano had only been hired 10 days earlier to drive the school bus. On Saturday night he had taken the bus without authorisation to take a party of friends on a drinking spree. About dawn on Sunday they were returning from an all-night party when they reached an unprotected level crossing. According to Driver Vicente Rodriguez, Lozano had ignored his repeated whistling and drove the bus slowly onto the crossing directly in front of his engine. The bus was demolished and it took rescuers five hours to free the bodies of the nine dead. By Sunday evening only two bodies had been identified, one of which was that of Lozano. The passenger train did not escape unscathed either. Nine of its coaches were derailed and one passenger was seriously injured and 14 were slightly hurt. A spokesman for Ferrocarriles Nacionales de México estimated the damage to the train and track at $16,000. [920] [921] [922]

A level crossing accident in Brazil on Thursday 3rd March 1966 had unusual features. The crossing at Tatuapé, in the suburbs of São Paulo, like many on the Central do Brasil line in question, was worked at this time by a crossing keeper and protected by barriers. Treadles or similar on the track alerted the crossing keepers to the approach of trains in time for them to lower the barriers before the train reached the crossing.

At about 13:30 on the day in question a bus of the Auto Ônibus Vila Carrão company, registration 12.146, on its way from Jardim Nove de Julho to Parque D. Pedro II, reached the crossing in Tatuapé with the Rua Viela. The barriers were raised and the bus entered the crossing, only to be struck by passenger train No. UPE-6, which had left Roosevelt station at 13:18 for Mogi das Cruzes. Seeing the train approaching Crossing Keeper Jovelin Pereira da Costa attempted to stop the bus but was unable to do so. Both the bus

[917] *The Evening Independent*, Massillon, Ohio, United States, 14th February 1966
[918] *The Middletown Journal*, Middletown, Ohio, United States, 13th February 1966
[919] *The Republican-Courier*, Findlay, Ohio, United States, 14th February 1966
[920] *The Register*, Santa Ana, California, United States, 14th February 1966
[921] *San Antonio Express*, San Antonio, Texas, United States, 14th February 1966
[922] *Clovis News-Journal*, Clovis, New Mexico, United States, 15th February 1966

driver, Antonio Rodrigues de Matos, and the train driver, Ivadir Santana, saw the danger and applied their brakes but the collision was still a violent one. The bus was carried about 10 metres, thrown against a pole, and almost destroyed; the train on the other hand suffered only a few broken windows. Of the 34 passengers on the bus 10 were killed outright, as was the bus driver, and 22 were injured. The bus conductor, Ianao Satori, and a child aged about 18 months were the only ones on the bus to escape unscathed. The crossing keeper rescued the child from the wreckage and handed them to an unknown person. Only after several days was the child identified. Three of the injured died after arrival at the Hospital das Clinicas and two more passed away the following morning, increasing the number of dead to 16.

It transpired that the accident was not, as might at first appear, the fault of the crossing keeper. For a reason that was not stated the train was travelling on the wrong running line and hence did not operate the treadle to warn the crossing keeper of its approach, nor, due to an oversight on someone's part, had the crossing keeper been advised of the approach of the train by either Quarta Parada or Sebastião Gualberto stations. [923] [924]

Two schoolchildren were hurt, though only one of them was detained in hospital, when a Baltimore and Ohio Railroad freight train collided with a school bus on Bell's Mills Crossing, Pennsylvania, United States, two miles from Punxsutawney, at 3.55 p.m. on Thursday 10th March 1966. The bus, being driven by Allen Miller, 45, was travelling along Legislative Route 33036, as the road was known, following a small lorry. The vehicles stopped at the crossing, which presumably was an unprotected one, and waited for a northbound freight train to pass. They then started over the crossing, only for both drivers to observe that a southbound coal train was approaching on the double track. The lorry was able to make it over the crossing but the train, which was being driven by Engineer I.J. Jacobson, caught the rear of the school bus, which was pushed into the rear of the lorry by the impact. There had been about 15 pupils on the bus. [925] [926]

Denmark suffered a serious level crossing accident on the evening of Sunday 1st May 1966. The driver of a fast train to Herning was advised by the signal protecting Crossing No. 137 at Hammerum that the barriers had been lowered, but when his train neared the crossing he found that in fact this was not the case. He applied his brakes and sounded his horn in an attempt to warn road users but a Danske Statsbaner (DSB) service bus entered the crossing and was struck by the train at 19:23. The bus was severely damaged, cut in two and its rear portion caught fire. The train, which had been travelling at 90 kilometres an hour, was partly derailed. Ten bus passengers were killed outright and about 60 people were injured, 11 seriously. Two more of the victims died later. At least nine of those who lost their lives were from nearby Ikast, including six schoolchildren who had been returning from a concert featuring Denmark's top singer, Ulla Pia, in Herning. [927] [928] [929]

---

[923] *O Estrada de São Paulo*, São Paulo, Brazil, 4th and 6th March 1966
[924] *Jornal do Brasil*, Rio de Janeiro, Brazil, 4th March 1966
[925] *Indiana Evening Gazette*, Pennsylvania, United States, 11th March 1966
[926] *The Titusville Herald*, Titusville, Pennsylvania, United States, 11th March 1966
[927] *Redlands Daily Facts*, Redlands, California, United States, 2nd May 1966
[928] https://www.herningfolkeblad.dk/artikel/4ab1940b-d4ee-4aaa-b4d5-e609f8e3bf35/

[930] [931]

A mistake by a crossing keeper who failed to inform his colleague about a delayed engine led to a fatal level crossing accident at Deštnice, Czechoslovakia, on the evening of Monday 2nd May 1966. Unaware that the engine had yet to pass, the crossing keeper who had taken over failed to lower the barriers of the crossing on the Praha – Chomutov line and the engine collided with a lorry carrying a party of 48 engineering students to a state farm. Ten students lost their lives and 37 were injured. [932] [933] [934]

Eight people, five of them members of the same family, lost their lives in an accident at Jerico Park Road Crossing, Bowie, Maryland, United States, shortly after noon on Sunday 15th May 1966. The car in which they were riding was struck by the *Midday Congressional* passenger train of the Pennsylvania Railroad. Fireman Brown DuVall was the only witness to the accident. He said that the train had been travelling at 78 miles an hour and that the car had appeared to maintain its speed as it approached the crossing, indicating that its driver had been completely unaware of the train. The train's brakes had been applied but the train could not stop in time. The main part of the car was carried 120 feet and parts of it as much as 1,500 feet. After the accident the northbound train eventually proceeded to Baltimore.

Those who died were Raleigh Howard Jr., 30; his wife Mary, 34; their 11-year-old twins Rosalind and Roxanne; their daughter Sharon, 9; Mr and Mrs Benjamin Manson, 50; and James Kent, 7, the son of a neighbour, all from suburban Baltimore.

The crossing was provided with crossbucks marked 'Railroad Crossing' but no barriers or other warning devices. [935] [936]

According to first reports 25 Africans lost their lives and another 28 were injured, three critically, when a train collided with a bus on a level crossing at Potgietersrus (now Mokopane), South Africa, on the night of Monday 11th July 1966. Fourteen other people on the bus escaped unhurt. The bus was carried at least 200 metres down the track. The main line to Rhodesia was blocked for seven hours. [937] [938]

A follow-up report stated that only 14 of the 67 bus passengers had been killed, while 39 were injured, 19 of them seriously. [939]

---

[929] https://jyllands-posten.dk/jpaarhus/ECE4454633/En-venlig-mand-kom-ufrivilligt-i-rampelyset/
[930] https://amtsavisen.dk/danmark/liste-de-stoerste-togulykker-i-danmark-2022-11-29
[931] https://www.farallon.dk/dokuwiki/doku.php?id=jernbaner:skolebus_kollideret_med_tog_-_hammerum
[932] Wikipedia – Liste von Eisenbahnunfällen in Tschechien
[933] *The Press-Courier*, Oxnard, California, United States, 3rd May 1966
[934] *Wellington Daily News*, Wellington, Kansas, United States, 3rd May 1966
[935] *The Frederick Post*, Frederick, Maryland, United States, 15th May 1966
[936] *Cumberland Evening Times*, Cumberland, Maryland, United States, 15th May 1966
[937] *The Sheboygan Press*, Sheboygan, Wisconsin, United States, 12th July 1966
[938] *Diario de Burgos*, Burgos, Spain, 13th Jul 1966
[939] *The Daily Gleaner*, Kingston, Jamaica, 13th July 1966

Thirty-eight people reportedly died and 20 were injured when an express train collided with a bus on an unmanned level crossing at Lodhran, Pakistan, on Thursday 21st July 1966. Most (all?) of the casualties were on the bus. [940]

Early on Sunday 28th August 1966 a diesel-hauled passenger train collided with a bus on a level crossing near Denizil, Turkey, somersaulting it several times and pushing it more than 130 metres along the track. The bus had been on its regular run between Konya and Izmir with about 50 passengers. The bus driver and 15 passengers were killed instantly and nine more passengers died either on the way to hospital or after admission. Twenty-three other bus passengers were reportedly injured, six seriously. Police stopped vehicles to take the injured to hospital in Denizil. No one on the train was apparently hurt, though its engine was derailed. [941]

A complex accident happened at Andancette, France, early on Monday 5th September 1966, involving a lorry and two trains. At 03:10 a lorry travelling along departmental road No. 132 arrived at level crossing PN 349 in Andancette and its driver, finding the barriers raised, entered the crossing, only for his vehicle to be struck by a crowded passenger train from Paris to Briançon. The engine of the passenger train was derailed and fell from a small bridge into the Bancel stream, while the four leading coaches also left the rails. Shortly afterwards a fast freight train arrived on the opposite running line and collided with the derailed carriages, though its speed had been somewhat reduced by a loss of electric power, because the initial collision had brought down the overhead line equipment.

Three train passengers lost their lives in the accident: Jean-Laurent Fond, 20; Auguste Maistre, 19; and Gérard Paludo, also 19. All were soldiers returning from leave. The two men on the lorry, Messieurs Benado and Maure, were also killed and 15 passengers were injured, six seriously. Drivers Catalon and Chevalier on the locomotive were likewise hurt.

The crossing was manned throughout the 24 hours by three keepers who each worked eight-hour shifts. Crossing Keeper Monty, 47, was on duty at the time of the accident. After the collision he was interrogated and a blood sample was drawn, presumably to determine if he had been drinking. At the end of the morning he was arrested and taken to prison in Valence. It was initially unclear whether the crossing barriers had already been raised when the lorry arrived or they had been raised at the lorry driver's request, but in any case a mistake on the part of the crossing keeper appeared to be the cause of the accident. [942]

Five schoolboys and the school janitor who was taking them home were killed when the pickup truck in which they were riding was struck by an Illinois Central Railroad freight train at a level crossing situated a mere 75 feet from their secondary school at Eldena, Illinois, United States, on Tuesday 20th September 1966. William Bothe, 13; Martin A. Cruze, 13; Robert M. Enlow, 14; Reid Miller, 12; and James Shoemaker, 12,

---

[940] *The Kingston Gleaner*, Kingston, Jamaica, 22nd July 1966

[941] *Pacific Stars and Stripes*, Tokyo, Japan, 30th August 1966

[942] *Le Monde*, Paris, France, 6th September 1966

had been helping the janitor, Everitt Mullins, 73, load a drag grader onto a pickup truck to return it on the opposite side of the tracks after using it to smooth an athletic field. One of them had then sat in the cab with the janitor and the others were riding in the back of the truck holding the drag grader in place. [943]

A witness to the accident, Mrs Mabel Beire, told police: "The train was about 75 feet away from the truck. I don't understand how it happened. The truck was moving so slowly. I don't know why it was moving so slowly." [944]

A southbound limited express on the Tobu – Isesaki line collided with a bus bound for the station at Koshigaya, Japan, near the latter place on Thursday 22nd September 1966. Six people died and more than 40 were injured. [945]

A major disaster involving a school bus occurred at Dorion, Quebec, Canada, at about 7.40 p.m. on the night of Friday 7th October 1966. Nineteen people were killed instantly and another died later, while all of the remaining 41 people on the bus were injured. Apart from the bus driver, Marcel Fleury, all were students aged between 12 and 19 from a student centre at Cite des Jeunes who had been on their way to a dance at Hudson, Quebec. The bus stopped at the barriers of St Charles Street, Dorion, for an eastbound train. The barriers were then raised, according to one account by a 14-year-old boy, [946] and the bus proceeded onto the crossing, to be struck by a westbound Canadian National Railway (CNR) freight train from Montreal to Toronto, consisting of several locomotives and about 100 cars. Part of the bus ended up in a ditch and the remainder was dragged half a mile.

Driver Paul Zowchuk on the leading engine stated, "As we approached the crossing the fireman yelled 'bus.' I pulled the emergency cord until we came to a complete halt. The frame of the bus was on fire and it was wrapped around the engine. I guess the impact caused sparks that hit the gas tank. I thought we all were going to burn like rats."

George Shaar, who owned a store near the crossing, said that the noise of the impact, "felt like an explosion." He saw that the bus had been cut in half and there was fire underneath the train. "Kids were shouting and screaming and crying," he added.

Dorion Police Chief Jean Lapointe said, "This is the worst tragedy that I have seen in all my years on the force. We don't know what caused this horrible thing. All we know is that there are many versions."

'A passing police car sounded the first alarm,' a newspaper reported. 'Ambulances came roaring into Dorion from as far away as Montreal. Most of the injured were taken to Lakeshore Hospital but four went to the Montreal Neurological Institute for treatment of brain injuries. Their condition was described as critical. "I just want to be sick," said one police officer who helped take six bodies away from the scene.' [947]

---

[943] *Lubbock Avalanche-Journal*, Lubbock, Texas, United States, 21st September 1966
[944] *The Post-Standard*, Syracuse, New York, United States, 21st September 1966
[945] Wikipedia – List of level crossing crashes
[946] *The Brandon Sun*, Brandon, Manitoba, Canada, 18th October 1966
[947] *Winnipeg Free Press*, Winnipeg, Manitoba, Canada, 8th October 1966

Another newspaper wrote, 'Debris was spread over a wide area after the crash. The ambulances, sent to the scene from Montreal, took the victims to Lakeshore and other hospitals. Numerous cars crowded the highways in the area, where routes 2 and 17 divide and take separate courses westward in the direction of Ottawa. Many who rushed to the scene were relatives of the teenagers, who watched and wept as the searchers combed the debris and nearby areas.

'Searchlights played on the scene of the accident which occurred shortly before 8 p.m. EDT. Police were issuing calls for car owners who had arrived on the scene early to report whether they had taken any of the victims to hospital on their own. Police said the search for possible victims would go on far into the night. There were flares and flashlights being used as the searching through the sometimes charred wreckage of the bus went on.

'Before police, firemen, and ambulances arrived, private motorists were being flagged down and asked to help with the rescue efforts.' [948]

The dance was called off after the accident and the band who would have played at it was sent home.

Investigations into the disaster were begun at once by the railway company and the board of transport commissioners.

On Monday a CNR spokesman in Montreal stated that the signals and barriers at the crossing had been working properly, stating, "We are getting close to the end of an exhaustive investigation and have not found any abnormalities in the function of the signals or crossing protective devices." [949]

On Tuesday the accident was discussed in Parliament. Transport Minister Pickersgill told the Commons that the government would make every effort to determine the real cause of the disaster and to take steps to prevent a repetition, then went on to express his sympathy for the relatives of those killed and for those injured.

Opposition leader Diefenbaker joined in expressing sympathy and stated that the continuing "maximum slaughter" of Canadians on the highways was a challenge to the federal government and the provinces to take remedial action and that something should be done to guard against a repetition of the accident.

New Democratic Party leader Douglas suggested that perhaps the inspection of safety measures at level crossings had become too routine, with the railways not checking them carefully or frequently enough.

Creditiste leader Caouette said that highway traffic in the Dorion area had been dense for years and that the level crossing should have been eliminated long ago.

Social Credit leader Thompson said that the accident underlined the need for further precautions and the elimination of dangerous level crossings. [950]

Guy Boucher, 15, of Dorion, became the last student to succumb eight days after the accident. [951]

---

[948] *Victoria Daily Colonist*, Victoria, British Columbia, Canada, 8th October 1966
[949] Ibid, 12th October 1966
[950] *The Lethbridge Herald*, Lethbridge, Alberta, Canada, 12th October 1966
[951] *The Brandon Sun*, Brandon, Manitoba, Canada, 18th October 1966

The inquest on the victims of the accident opened in January 1967. On its opening day evidence was given of the barriers being tampered with while the bus had been stopped at them. 'Mrs Frances Hardy, 34, a Toronto housewife, driving a car immediately behind the bus, said she saw three boys tampering with the barrier seconds before it went up...Mrs Hardy first testified that she saw a young man lifting the barrier. She later modified this to say only that the barrier had been raised and that at least one youth had his hands on it. "I was conscious of the barrier going up," she said. "I was conscious of arms holding up the barrier." She said she could describe the boys although she had seen the face of one of them.' [952]

On the second day of the inquest Robert Brosseau, 19, of Dorion, gave evidence. 'Mr Brosseau said he was in his car which was stopped on the north side of the St Charles St. because the north side barrier had just come down. An eastbound train passed. "As the last car moved I saw a large vehicle on the south side of the crossing and between the vehicle and the south barrier were three shadowy figures. The figures walked toward the barrier and either one or two of them had their hands underneath the barrier. As I watched the barrier was rising and at that time it was level with their chins. I then turned and saw another train, this time westbound. When I turned to look at the bus the south barrier then was vertical. I then looked between the bus and the train until the moment of the collision. I did not see the shadowy figures again. I saw no warning signals at the crossing and heard no signal bell or whistle."' [953]

On the third day of the inquest Lewis Hawkins, chief signal officer of the board of transport commissioners, described the crossing as "fail safe" while conceding that a boy of 13 or 14 could lift a barrier to an angle of 45 degrees, sufficient for a bus to pass beneath. He added that all crossings were subject to regular maintenance. [954]

At the close of the inquest the five-man jury returned a verdict of 'Accidental Death.' Its foreman, speaking in French, stated, "We unanimously find that the collision was not criminal in origin but accidental. We find that the level crossing is extremely dangerous and to avoid such accidents in future we shall make certain recommendations. In the meantime we recommend that the security measures now in operation at the level crossing be maintained."

Since the accident the speed of trains over the crossing had been reduced to 25 miles an hour. [955]

In February 1967 approval was given in principle to a plan to replace the Dorion level crossing by an overpass (bridge). However, as usual, 'the big problem is how the proposed $2,300,000 four-lane overpass would be paid for.' [956]

A federal inquiry by the board of transport commissioners finally reported its findings at the beginning of November 1967. A newspaper wrote, 'The board took testimony from a score of witnesses who thought they had seen students near the gates at the time but was unable to establish blame. The judgement said that Daniel Cuerrier and Guy Vendette, who had admitted being near the gate, but denied that they had touched it, had

[952] Ibid, 10th January 1967
[953] *Victoria Daily Colonist*, Victoria, British Columbia, Canada, 11th January 1967
[954] *The Brandon Sun*, Brandon, Manitoba, Canada, 12th January 1967
[955] *Winnipeg Free Press*, Winnipeg, Manitoba, Canada, 13th January 1967
[956] Ibid, 24th February 1967

acted unusually after the crash. It said:

'"They had gone to the crossing with the intention to cross it and go to a doctor's office. The gate went up, the bus moved onto the tracks, but they did not make a move to go where they intended to go.

'"At the time of the collision, or a moment or two before or after it happened, they ran into a dark lot on the east side of the street. They did not then go to where the bus was lying wrecked or where the occupants of the bus were lying injured or killed.

'"We scarcely credit it as believable that it was in order to escape flying debris that they ran into the dark lot. On the witness stand they created an impression on us that they were clever young men, clever and evasive rather than frank and willing to tell the full truth as they knew it."

'The board found that the gate was working properly before and after the accident.

'"In our opinion there was a preponderance of credible evidence that the gate was raised by arms and hands of one or two persons who were at it. Having regard to the consequences, it is not surprising that no one has admitted to raising the gate."' [957]

The crossing was replaced by a bridge in November 1972. [958]

Five teenagers died and two were critically hurt when the stolen car in which they were riding was struck by an Atchison Topeka and Santa Fe Railway freight train on Saturday 15th October 1966. The accident happened on an unprotected crossing near Del Rey, California, United States. The party in the car had left a post-football-game party and were headed into town to eat. It appeared that the car driver, Billy Overby, 16, from Modesto, one of those killed, had not seen the train until the last moment when the car, travelling at an estimated 60 miles an hour, had gone into a 'wild, broadside skid' and collided with the side of the train.

Engineer Norman Kjar said his seven-car southbound train had been travelling at 40 miles an hour. The car driver had attempted to stop when he heard the train whistle. [959]

The others who lost their lives were Shirley Garabedian, 16; Gilbert Juvinall, 18; Eugene Singleton, 20; and Levonne Sullivan, 16. Gloria Chatten, 16, and Pedro Sesate, 20, were both critically injured.

The car had been reported stolen in Fresno earlier in the day, a fact only one or two of its occupants were apparently aware of, it was not clear which. [960]

Six steelworkers lost their lives when the bus in which they were riding was struck by a freight train on a level crossing at Leoben, Austria, on Tuesday 25th October 1966. Twelve or 29 others were injured according to different reports. A crossing keeper had failed to lower the crossing barriers; he was arrested by the police. [961] [962]

---

957 *Victoria Daily Colonist*, Victoria, British Columbia, Canada, 1st November 1967
958 Wikipedia – Dorion level crossing accident
959 *Las Vegas Sun*, Las Vegas, Nevada, California, United States, 16th October 1966
960 *The Modesto Bee*, Modesto, California, United States, 16th October 1966
961 *El Paso Herald Post*, El Paso, Texas, United States, 25th October 1966

A Ferrocarriles Nacionales de México diesel locomotive bound for Toluca collided with a crowded bus of the Mexico – Huixquilucan line heading for Toluca on a level crossing near Mexico City, Mexico, on Friday 4th November 1966. Five people died and 30 were injured. [963]

Only apparently very briefly reported in the Press was a level crossing accident on the night of Saturday 5th November 1966 on the western edge of Kansas City, Kansas, United States. A car being driven by Mrs Patricia Ross, which was bringing six children home from a skating rink, was struck by a Union Pacific Railroad train. Mrs Ross, her daughter Patty Ann, 5, and Tommy Collins, 10, were killed outright. David Ross, 9, died early the next day and Mike Hutson, 9, died on Monday. Shana Ross, 7, and Peggy Sue Collins, 15, were injured. [964]

A school bus carrying 33 passengers was struck by an Illinois Central Railroad passenger and mail train on an unprotected level crossing with Louisiana 151 road at Calhoun, Louisiana, United States, at 5.30 p.m. on the evening of Tuesday 13th December 1966. The bus was taking basketball players from Eastville High School at Farmingdale to play in a pair of games at Jasper High School in Chatham.

Mack Cleveland, 17, who had been stood in the entrance well of the bus, directing the driver, was killed instantly, and four students and two teachers/coaches were injured, one of them Edward Moore, 38, who in his third role was the bus driver. The Shreveport to Vicksburg and Meridian train had been travelling at 55 miles an hour; it carried the bus 144 feet down the line, shearing off its front end. It was said that the accident would have been more serious had the train hit the centre or rear of the bus. The injured were taken to Glenwood Hospital in West Monroe. [965]

Driver Moore said that he had not seen or heard the train, also that he had not been able to stop the bus because of a problem with its brakes. He was charged with negligent homicide and negligently causing injury. [966] [967]

The *Red River* passenger train of the Great Northern Railway, which had left St Paul, Minnesota for Fargo, North Dakota, at 5.45 p.m. on Wednesday 21st December 1966, collided with a car travelling along Stearns county road No. 2 on a level crossing at St Joseph, Minnesota, United States, at 7.30 p.m. Train No. 11, which consisted of two diesel locomotives and six coaches, carried the car 75 feet down the track, demolishing it and killing its five occupants. The victims were Mr and Mrs James Roeser, both aged about 23; their four-month-old daughter Michel Mary; James' uncle Raymond Roeser, 50, who lived with the couple; and their nephew Michael Roeser, 12, all from

---

[962] *Ogden Standard-Examiner*, Ogden, Utah, United States, 26th October 1966

[963] *Albuquerque Journal*, Albuquerque, New Mexico, United States, 5th November 1966

[964] *Daily Tribune*, Great Bend, Kansas, United States, 7th November 1966

[965] *The Biloxi-Gulfport Daily Herald*, Biloxi, Mississippi, United States, 14th December 1966

[966] *The Ruston Daily Leader*, Ruston, Louisiana, United States, 14th December 1966

[967] *Dixon Evening Telegraph*, Dixon, Illinois, United States, 14th December 1966

Paynesville. The crossing was an unprotected one. No one on the train was hurt but it was delayed for an hour and a half. [968] [969]

Five people died on Boxing Day, Monday 26th December 1966, when the car in which they were riding was struck by a Ferrocarriles de Cataluña train on a level crossing with Gasómetro street in Tarrasa, Spain. The car was pushed 200 metres along the line. Those who lost their lives were José Albert Roma, 42, the driver of the car; Luis Albert Bagur, 71; José Vilalta, 80; Santiago André Casals, 20; and Benjamin Albert, 14, the driver's son. All were neighbours from Castellar del Vallès. [970]

The disaster that occurred at 12.10 a.m. on Wednesday 28th December 1966 at Everett, Massachusetts, United States, was another of those rare level crossing collisions that result in significant fatalities to train passengers. A Boston and Maine Corporation diesel railcar collided with an Oxbow Transport Corporation road tanker loaded with 8,300 gallons of No. 2 fuel oil. As a result of the collision the railcar was slightly derailed and the contents of the road tanker were splashed over the front of it, and immediately caught fire. Eleven passengers, the motorman and the conductor died in the accident and other passengers were injured. The deaths and injuries were caused by smoke inhalation and burns rather than the initial collision.

Train No. 563 had left Boston North station at 12.01 a.m. in charge of Conductor Edwin Hunt with Motorman Thomas Bagley [971] and 28 passengers. At 12.08 a.m. it had reduced speed to about 20 miles an hour to cross a drawbridge, then increased speed to about 55 or 60 miles an hour until, about 900 feet short of the crossing, Motorman Bagley apparently had seen the tanker obstructing the crossing, applied his brakes and attempted to warn the train crew and passengers of the impending collision. [972] According to a surviving passenger, Marvin Marnik, 18, of Lynn, Massachusetts, the motorman had shouted, "Lean forward, we're going to hit!" [973]

The lorry driver, Raymond Bouley, had begun work at Lexington, Massachusetts, at 6 p.m. the previous evening and had made two deliveries before leaving Lexington on his third trip at 11.57 p.m. On the approach to the crossing he had stopped twice before entering it because of vehicles in front of him. He had then started over the crossing, the poor state of which had caused his tanker to bounce badly, and he thought that he might have applied the brakes to steady it. Without warning the brakes of his trailer had then come on fully. Because of this, the tanker had stalled on the crossing and he saw that the brake pressure had fallen to zero. Efforts to restore the brake pressure or restart the lorry in low gear failed. Realising his dangerous position, he had then asked a car driver to contact the police and to get them to warn the railway, and had also asked the motorist if the latter had any flares or similar. The car driver, who used the crossing regularly, warned the lorry driver that a passenger train usually passed at about that time. They had

[968] *The Brainerd Daily Despatch*, Brainerd, Minnesota, United States, 22nd December 1966
[969] *Winona Daily News*, Winona, Minnesota, United States, 22nd December 1966
[970] *ABC*, Madrid, Spain, 27th December 1966
[971] *Somerset Daily American*, Somerset, Pennsylvania, United States, 29th December 1966
[972] National Transportation Safety Board report
[973] *The Record Herald*, Waynesboro, Pennsylvania, United States, 28th December 1966

then become aware of the approach of the train and both had run up the line towards it, signalling as best they could to its motorman.

The railcar stopped 158 feet from the crossing with its leading wheels derailed. The impact of the collision ruptured the fuel tanker and a portion of its cargo splashed over the front part of the train and immediately burst into flame. As the flames and smoke spread six passengers and the flagman were able to escape via the rear of the train but a jam of passengers up against an inward opening door trapped those remaining, while the lights of the train went out, leaving it in darkness.

The noise of the impact, a low-order explosion and the flare-up of fire attracted the attention of two police officers nearby, who were quickly on the scene and realised that passengers were trapped on the burning train. Unable to open the train door due to the crush of passengers inside they and the lorry driver had then tried to break the coach windows but had not had the tools to do so. After five minutes Everett Fire Department firemen arrived and after failing to get in via a door had succeeded in smashing the side windows with axes and bars. [974]

'Chelsea Fire Chief Herbert Fothergill said most of the victims were trapped in the train for 20 minutes until the fire was put out. He said firemen used their bare hands to extricate the passengers. A fireman said one door was blocked "because there was a body lying on the other side."

'Policeman Ralph DeVito said he and officer George Stewart "tried breaking the windows with our night sticks but that wouldn't work. We couldn't open the door because there was this guy trying to get out the door with his head stuck through the window. We pushed him away but then there was so much smoke we couldn't see."'

The injured and dead were taken to Whidden hospital in Everett, Chelsea Memorial and Naval hospitals and Massachusetts General Hospital. The injured included two rescuers, Fireman Nunzio Iozza and a passer-by, Navy corpsman John Mann, both of whom suffered from smoke inhalation. [975]

The disaster was the subject of a comprehensive report concluded by the National Transportation Safety Board (NTSB), though it was begun by the Interstate Commerce Commission, which the NTSB had superseded in April 1967. The report's conclusions were:

(1) The collision at 12:10 a.m. on December 28, 1966, at the Second Street railroad-highway crossing in Everett, Massachusetts, between the single-car passenger train 563 operated by the Boston and Maine Corporation and the motor tank truck operated by the Oxbow Transport Corporation, resulted in the death of 11 of a total of 28 passengers and two of the three train crew members, and other injuries and extensive damage to both the train car and the motortruck.

(2) The collision ruptured the 8,500-gallon tank of the truck resulting in a portion of the oil cargo being thrown against and into the forward section of the train. Low-order explosions and the rapid spread of fire from the flammable fuel oil contributed to the fatalities which followed the collision.

---

[974] National Transportation Safety Board report

[975] *Indiana Evening Gazette*, Indiana, Pennsylvania, United States, 28th December 1966

(3) The train was being operated in conformity with all applicable rules and regulations as it approached the crossing.

(4) The driver of the motortruck exercised all due caution in stopping before entering the crossing; and at the time he commenced the crossing of the railroad tracks, the warning signals were not indicating the approach of a train.

(5) There was a loss of air pressure in the brake system of the motortruck from an undetermined cause, and this loss led to an automatic application of the brakes on the tractor and semitrailer which stopped the motortruck directly across the Boston and Maine tracks at the Second Street crossing.

(6) The brake system on the motortruck was installed in compliance with Massachusetts State law and the regulations of the Interstate Commerce Commission.

(7) The arrangement of the brake system on the motortruck was such that with loss of air pressure, the driver in the cab of the tractor was unable to release the brakes and move the vehicle from its precarious position before the collision.

(8) It could not be determined at which point the train's brakes were applied in emergency before the collision, but based upon tests made after the collision, it is reasonable to presume that the application of brakes was made at some point between 1,200 feet and 800 feet from the collision location. From tests made after the collision, it is evident that, at the maximum sighting distance of 1,200 feet from the crossing, the train traveling at a speed of approximately 60 miles per hour could not stop before reaching the crossing location, even with emergency application of brakes.

(9) There was no system in existence at the crossing by which either the driver could warn the engineer on the train, or the engineer could be advised of the presence of the truck stalled across the tracks at the crossing.

(10) The automatic railroad crossing-warning signals were functioning properly but were not pertinent in this case since the truck had stalled on the tracks before the signals were actuated by the presence of the approaching train.

(11) There were no emergency exits in the railway car. The emergency equipment that was in the car that could have been used to break windows was located so that when the only door available for escape was opened to lead to the rear vestibule, the emergency equipment was hidden behind the door. There was no emergency lighting system on the train car. When the lights of the train went out, confusion and panic were increased during the attempt of the passengers to escape from the burning and smoke-filled car. The normal passenger-exit doors proved inadequate as a means of escape under conditions of darkness and panic.

The NTSB made a number of recommendations:

(1) The Secretary of Transportation, with representation from the National Transportation Safety Board, should continue to study and review the railroad-highway grade crossing problem in all of its aspects. Reference is made to the short range and long range programs announced by the Secretary of Transportation in his news release of August 8, 1967.

(2) The Secretary of Transportation should seek legislation to authorize the Federal Railroad Administrator to prescribe regulations requiring:

(a) Emergency means of escape from railroad passenger cars.

(b) Emergency lighting for railroad passenger cars.

(3) The Board notes that on passenger-carrying railroad cars, the exit doors normally open inward. The Board recommends that the Federal Railroad Administrator initiate studies and action that will insure that, in emergency, passengers can reliably escape from regular exits of passenger-carrying railroad cars.

(4) The Department of Transportation should include in its grade-crossing protection study and action program the problem of motor vehicles stalling on railroad tracks and methods of warning approaching trains to prevent a collision. The study should include all technical methods which could have a bearing on the problem such as means of warning the crew of the oncoming train, special means to be carried by vehicles loaded with hazardous cargo for activating the railroad signal system, and methods of reducing the emergency stopping distance of existing and future rail equipment.

(5) The Federal Highway Administrator should take under immediate consideration the revision of existing regulations under his jurisdiction in order to:

(a) Require an emergency means within the cab of a motor-truck by which brakes, which have been applied automatically to the tractor and/or trailer because of the loss of air pressure in the braking systems can again be released.

(b) Require motor vehicles of unusual size and those carrying flammable, toxic, or other hazardous cargo to use grade crossings offering minimum risk of vehicle stalling or stopping on the crossing. Criteria for designation of such crossings might include approaches free from steep grades and curves, freedom from nearby traffic lights on highway and other sources of traffic congestion, smoothness of crossing pavement, sight distance along the railroad track, and positive grade separations such as overpasses.

Where grade crossings are designated, consideration should be given to developing means of displaying a stop signal to rail traffic during the time truck {sic – traffic is presumably the intended word} is actually crossing the tracks.

(c) Require emergency flares of high brilliance from a self-contained power source to be carried on all motortrucks subject to Bureau of Motor Carrier Safety regulations, in order to provide visual warning in an emergency. Emergency flares should not be of a type that might represent a fire hazard.

(d) Require all drivers of motor vehicles subject to Bureau of Motor Carrier Safety regulations to demonstrate knowledge of and use of emergency signals and emergency procedures.

(6) The Federal Highway Administrator should study the feasibility of fire resistance regulations for tank trucks carrying flammable fluids to prevent low-order explosions and rapid propagation of flame from such tanks when they are ruptured. Such techniques as lining tanks with soft material or filling tanks with special reticulated foam are known to be technically effective in preventing such rapid flame spread. These methods would also be important in preventing fires following highway accidents and their feasibility as to future cost and weight should be evaluated. [976]

[976] National Transportation Safety Board report

Five teenagers from Three Rivers, Michigan, lost their lives late on New Year's Eve, Saturday 31st December 1966, in a level crossing accident about 25 miles further north, near Kalamazoo, Michigan, United States. It was believed that their car had been struck by both the eastbound and westbound *Twilight Limited Flyer* passenger trains of the New York Central Railroad. Both trains were travelling at about 80 miles an hour at that point and the wreckage of the vehicle was reportedly carried seven miles, fused to the front of the engine of the westbound train. Those who died were David Hagerman, David Hanchon, and Dennis Lockwood, all 16; Teryil Smith, 17; and Gay Warner, 16. [977]

Seventeen young people returning from a dance lost their lives late on Sunday 22nd January 1967 when passenger train collided with bus they were riding at Arroyo Seco, Santa Fe province, Argentina, south of Rosario. Another 16 of those of the 60 on the bus were injured. [978] [979]

What was described as a 'station wagon school bus,' being driven by Thelma Ellis, 33, of Havelock, and carrying five children aged between six and nine who were returning home from school, was in collision with a Canadian Pacific Railway freight train near Havelock, Ontario, Canada on the afternoon of Wednesday 8th February 1967. The front and rear of the vehicle were torn off and wreckage was strewn 100 feet along the right of way. David Cole, 7, of Havelock, was the only casualty, suffering a broken collarbone. [980]

A husband, his wife and seven of their 12 children died in a level crossing collision in the United States on Wednesday 22nd February 1967. Reflecting, presumably, its magnitude, the accident was the subject of a full National Transportation Safety Board (NTSB) investigation. It happened on Fruitridge Road Crossing, Sacramento, California. Those who died were Herbert Hill Sr., 42, a retired Air Force sergeant; his wife Annie, 38; and their children Douglas and Herbert Jr., both 10; Caroline, 8; Edward, 7; Timothy, 6; Michael, 4; and Deborah, 3. The family came from Broderick, eight miles from the accident scene. [981]

The Hills' 1961 Ford station waggon was struck by freight train Extra No. 6335 West, which had left Fresno Yard at 9.02 a.m. Engineer Calvin Cearley told the NTSB investigators that he had sounded the horn of his diesel locomotive as he had passed the whistle board 1,370 feet short of the crossing. About 300 feet from the crossing he had made a service brake application because of a restrictive signal indication a quarter mile beyond the crossing. Neither he nor Brakeman Edwin Mitchell, who was also in the engine cab, had seen the station waggon before their engine struck it at 12.05 p.m., when he had applied the brakes in emergency, stopping rather more than three-quarters of a mile beyond the crossing. No one on the train was hurt, it was not derailed and there was only slight damage to the front of the engine. The station waggon was demolished.

---

977 *The Salem News*, Salem, Ohio, United States, 2nd January 1967

978 *Clarin*, Buenos Aires, Argentina, 24th January 1967

979 *The Press Courier*, Oxnard, California, United States, 23rd January 1967

980 *The Brandon Sun*, Brandon, Manitoba, Canada, 9th February 1967

981 *The Humboldt Times*, Eureka, California, United States, 23rd February 1967

The station waggon had been proceeding eastwards along Fruitridge Road, which was a four-lane highway with the eastbound and westbound lanes separated by a traffic island, indicated by double yellow lines painted on the road. According to witnesses the station waggon had passed a pickup truck and a car stopped on the curb (inside) lane, had slowed momentarily and then accelerated as it entered the crossing.

At a distance of 318 feet from the crossing were advance railroad warning signs, consisting of intersecting diagonal stripes and the letters "RR" in black on a yellow background. Close to the crossing automatic warning signals of the flashing light type were cantilevered over the road with two pairs of red lamps, supplemented by a bell and crossbucks marked "RAILROAD CROSSING." There were no gates (barriers). The lights began to flash and the warning bell would ring when a train reached a point 2,530 feet from the crossing, 24 seconds before a train travelling at 70 miles an hour reached it and 43 seconds before a train travelling at 40 miles an hour. The crossing equipment was inspected weekly and was found to be working correctly after the accident. The view of a northbound train for the driver of an eastbound vehicle was limited to a distance of 33 feet from the crossing and the view of a road vehicle for an engineer was likewise restricted. (The train was geographically northbound, though its railroad designation was westbound.)

It was fine and sunny at the time of the accident.

Fruitridge Road was a very busy road, with an average volume of 11,000 vehicles daily. An average of 29 trains used the crossing daily, all but two of which were freight trains. Passenger trains could travel at 70 miles an hour. A 100-car freight train travelling at 35 miles an hour would cause a delay of about two and half minutes to a motorist from the time the warning signals were activated, while shunting movements at an adjoining siding could block the crossing for as much as five minutes or more.

Discussing the circumstances of the accident the NTSB noted that there was no evidence of any problem with the station waggon, the weather had been excellent and the train crew were experienced and familiar with the crossing. Even if the warning bell had not been audible in the station waggon due to the noise of a radio and closed car windows Mr Hill must have been aware of the other vehicles stopped at the crossing as he had overtaken them. The warning horn of the locomotive might not have been audible.

A post mortem had shown that Mr Hill had had a perforated eardrum and severe cardiovascular disease. It was possible, therefore, though not likely, that he had suffered a heart attack as he approached the crossing.

The train had been travelling at 60 miles an hour, within the speed limit.

Discounting the possibility of a sudden medical condition, the question was why Mr Hill had ignored the flashing warning lights and proceeded onto the crossing. It appeared that he was familiar with the crossing and with train movements there. During shunting operations, which occurred six days a week, the flashing lights and bells would operate for long periods without any trains passing over the crossing. Observation showed that motorists frequently passed over the crossing while these shunting movements were in progress despite the operating warning lights and bell, although legally they should have stopped. Delays at the crossing could be five minutes or more and Mr Hill might not have been aware of the one fast passenger train daily in each direction and supposed that it was a shunting move or at least a slow-moving freight train.

It seemed likely that Mr Hill had seen the lights beginning to flash as he approached the crossing and with his previous knowledge of a possible long delay and the likelihood that it was a slow freight train had elected to cross. Instead it had been a fast freight train. The NTSB concluded that:

(1) The 1961 Ford station wagon driven by Mr. Herbert Hill moved onto the main track immediately in front of the approaching train. This action violated Section 22451, Subsection A, of the California Motor Vehicle Code requiring a motorist to stop before crossing a railway grade crossing when a signal device gives warning of an approaching train.

(2) The Southern Pacific freight train was being operated in conformity with all applicable rules and regulations.

(3) The automatic highway-railroad grade crossing warning signals were functioning properly and were indicating the approach of the train.

(4) The use of a train horn and bell and/or bells at the crossing to warn motorists is not systematically effective because driver practice does not assure that a driver can hear the warnings. On the surface, existing regulations appear to create an audible warning system, but that system is spotty and defective.

(5) The automatic highway-railroad grade crossing warning signals met all applicable standards but failed to prevent the fatal accident.

(6) Wide variation in warning times, false warnings and unclear and misleading warnings caused by switching operations existed from time to time at this crossing. Taken with the general industrial and railroad yard atmosphere, the result was a potential "booby-trap" for motorists.

(7) Motorists frequently proceeded, without stopping, across the grade crossing against properly operating warning devices.

(8) There are many grade crossings where the railroad operating conditions cause a wide variation in signal warning times, false warnings, and unclear and misleading warnings. This results in motorists becoming excessively familiar with low-risk conditions that may change quickly, creating a "booby-trap" situation. Familiarity breeds contempt for such warning devices and may encourage the motorists to make unsafe decisions to cross in the path of oncoming trains.

(9) Based on the circumstances of the accident and the past record where crossing gates supplement flashing lights, there is a possibility that this accident would not have occurred if the grade crossing had also been protected by crossing gates and a predictor.

(10) Substantial work to upgrade crossing protection is being performed in California, using State and local funds. Though it would have been desirable to place automatic gates at this crossing earlier, the large amounts of Federal funds available could not have been used under existing laws because the grade crossing involved was not on the Federal Highway System.

The NTSB made six recommendations:

(1) The Board recommends that the Department of Transportation include in its current and future study and action programs and expedite review of the significant problem of "booby-trap" crossings which may cause grade crossing accidents.

(2) The same recommendation is made to the Association of American Railroads, the Railway Labor Executive's Association, the American Trucking Association, Inc., the National Safety Council and all other associations and public-interest groups. Even awareness and publicizing of the critical "booby-trap" phenomenon would pay dividends by reducing such accidents.

(3) The Board recommends that agencies for law enforcement at grade crossings consider adequate enforcement at grade crossings to be as important as enforcement at signals governing highway intersections; and that special enforcement attention be paid to potential "booby-trap" crossings.

(4) The Board recommends that the Federal Highway Administration study the problem of questionable audibility of external sound signals within motor vehicles, both in relation to grade crossing signals and train horns, and to all other audible warnings which can assist a driver. Such a study should be executed in cooperation with the Federal Railroad Administration and should be aimed toward creating a unified system of warnings and reliable reception which could be made effective through Federal regulations and State laws.

(5) The Board recommends that the Department of Transportation proceed apace to prepare broadly acceptable grade crossing hazard ratings or other objective criteria of grade crossing protection needs so as to formalize comparisons of the grade crossing values on a broader basis than local judgments and local surveys. The existence of such broadly recognized criteria would speed the consideration of optimum use of funds on a state-wide or nationwide basis.

(6) The Board recommends that the Congress and the Department of Transportation consider the minimal use of Federal funds for grade crossing safety protection by the States. Investigation should determine the degree to which these funds are not being employed because they are limited to Federal Highway System; and also the degree to which these funds are not being employed because States prefer to use the funds for general highway expansion or improvements. Since these funds are intended to reduce grade crossing fatalities as a safety benefit to all motorists, it would therefore be appropriate to consider whether legislation should extend the use of these funds beyond the Federal Highway System.

The City of Sacramento had already agreed the previous November to help fund the installation of gates at the crossing and three weeks after the accident the Southern Pacific Company agreed to share the cost 50/50 with the city. Gates were duly installed in June 1967, together with a predictor which meant that the gates lowered at the same interval of time before the arrival of a train regardless of its speed, thus minimising delays to road users.

Another school bus was involved in a level crossing accident at Lutzville, Pennsylvania, United States, at 7.35 a.m. on Friday 24th February 1967. The bus driver, Mrs Bernard Diehl, 55, had just picked up three students on a dirt road and was backing the bus to turn around when its rear end projected over the Pennsylvania Railroad line and was struck by the engine of a freight train.

Engineer Matthew Calvin said that he had seen the bus on the crossing with Legislative Road 05036 as he had rounded a curve 150 yards away, travelling at 25 miles an hour. His engine had picked up four freight cars at Mt Dallas and was heading towards Bedford. The engineer said that he had sounded his horn, though Mrs Diehl denied having heard it. Pupils Robert Stiffler, 12, whom the engineer had seen running towards the front of the bus, and his brother Thomas, 10, both of whom were slightly hurt, were the only casualties in the accident. Six other pupils on the bus escaped injury.

Mrs Diehl stated that in many years of driving the school bus this was the first time that she had ever seen a train. She added that some years ago local residents had sought permission for a bus turn around but nothing had been done. [982]

Three days after that, at Sacramento a Southern Pacific Company freight train was again involved in a bad level crossing accident in California, United States. This time it was at Midway Road Crossing, [983] two miles from Dixon in Solano county (the location was also known as Nelson's Corners). Shortly after 11 a.m. on Saturday 25th February 1967 a new pickup truck into which were crowded two men and three boys was struck by an eastbound freight train consisting of two (?) locomotives and 95 cars. The train was en route from Schellville and Roseville with Engineer M. Stranahan, Brakeman C.F. Chandler and an unnamed conductor. Its leading engine was struck by the pickup truck about 14 feet from its front end.

Erven Hilbert Sr., 35; his brother Clovis, 34; Erven's son Erven Jr., 10; and Clovis' sons Mitchell, 9; and John, 6, were all killed. They were all from Antioch, California. [984]

There was good visibility at the crossing in both directions but no protection other than the standard crossbucks. The pickup truck apparently initially paused at the crossing, then darted across in front of the train. It was carried 109 feet down the line. [985]

At 19:25 on Saturday 1st April 1967 an express commuter train on the Nankai Electric Railway collided with a gravel truck that had stalled on an unmanned level crossing between Ozaki and Tarui stations, Japan. The crossing adjoined a bridge over the Onosato River and two of the five cars of the train fell about six metres into the river, which was only a few centimetres deep, while a third car dangled from the bridge. The train was packed with 300 homeward-bound office workers and excursion parties and five passengers were killed and 205 were injured, six seriously.

Motorman Yoshinori Sugimoto, 26, who was slightly hurt, stated afterwards, "I was travelling at full speed and couldn't bring the train to a stop in time. Then there was a violent shock and after that I don't know what happened. I must have passed out."

The lorry driver, Shinichiro Sato, 21, and his mate, Shuji Shirokawe, 19, were unhurt. Both were arrested, charged with "negligence of duty." [986]

The recklessness of a teenage car driver cost five lives at about 11.30 p.m. on

---

[982] *Bedford Gazette*, Bedford, Pennsylvania, United States, 25th February 1967
[983] *The Humboldt Times*, Eureka, California, United States, 26th February 1967
[984] *The Modesto Bee*, Modesto, California, United States, 26th February 1967
[985] *The Times*, San Mateo, California, United States, 27th February 1967
[986] *Pacific Stars and Stripes*, Tokyo, Japan, 3rd April 1967

Saturday 8th April 1967 at Hicksville, New York, United States. The half barriers at the level crossing with the Long Island Railroad were lowered, the train whistle was blowing, the warning lights were flashing and a gong was clanging. Ignoring these warnings the 1957 car was driven around the crossing barrier and into the path of the train, to be carried 200 feet along the line and cut in two. Jeffrey Bloom, 17; Thomas Coffee, 16; Francis Dove, 18; and Carl Goodfield, 17; all of Hicksville, and Gregory Marrone, 19, of Westbury, New York, owner of the car and the presumed driver, were all killed. [987]

A tractor pulling a trailer loaded with agricultural workers was struck by a passenger train on a level crossing at Futog, Yugoslavia, not far from Novi Sad, on the night of Saturday 13th May 1967. Nine of the workers were killed and another 13 were seriously injured. The train driver said that he had whistled in an attempt to warn the tractor driver but it had entered the crossing anyway. [988]

Six teenagers, four boys and two girls, lost their lives just before midnight on Friday 17th June 1967 when the car in which they were riding was struck by a Missouri Pacific Railroad freight train near Newark, Arkansas, United States at an unprotected crossing with State Highway 122. They were named as Kenneth Hackworth Jr., 15, thought to have been the car driver; Meredith Ann Koontz, 15; Jack Lindsay, 14; Paul Milligan, 14; David McLain, 14; and Terry Self, 15, all from Newark.

Deputy Sheriff Bill Harris saw the train and car approaching the crossing and attempted, unsuccessfully, to warn the car driver by waving his flashlight, then witnessed the accident. He said that "it sounded like throwing a rock through a plate glass window." The 108-car freight train carried the car over half a mile before it could stop. The three occupants of the front seat were trapped inside it while those in the back were thrown out; all were found to be dead at the scene. The teens had been apparently riding around in several cars during the evening and were on their way home at the time of the accident.

Engineer Wayne Roberts was in charge of the locomotive of the train. [989]

Five residents of Port Arthur, Texas, United States, lost their lives when a Kansas City Southern Railway passenger train collided with the 1961 model sedan in which they were riding on Savannah Avenue Crossing, just outside of the city limits, at about 6 p.m. on Saturday 1st July 1967. The train, driven by Engineer Henry Adams, carried the car 102 feet down the track. Olevia Brooks, Walter Deal, the probable driver of the car; Betty May Wallace, Vera Wallace, and Kenneth Wayne Parker, aged four were all killed. The ages of the other victims were not known. A sixth occupant of the car, Betty Jean Parker, 24, was critically hurt. All the occupants other than Walter Deal were believed related. [990]

Six people lost their lives and 38 were injured in an accident that happened at 00:30

---

[987] *The Bridgeport Post*, Bridgeport, Connecticut, United States, 10th April 1967

[988] *Florence Morning News*, Florence, South Carolina, United States, 15th May 1967

[989] *The Brownsville Herald*, Herald, Texas, United States, 18th June 1967

[990] *The Port Arthur News*, Port Arthur, Texas, United States, 2nd July 1967

on Monday 3rd July 1967 at Villabilla, Spain, seven kilometres from Burgos, at km. 363.068 of the Madrid – Hendaye line and at a level crossing with the Burgos – León road. Forty-two people who lived along the road between Burgos and Astudillo, and who used the buses of the Nemesio Pérez company, had gone into Burgos on Sunday morning to the fair of San Pedro and San Pablo. They were on their return journeys just after midnight on Monday morning when the accident happened.

When the bus reached the level crossing the barriers were raised and it passed onto the crossing, only to be hit by a freight train en route from Madrid to Burgos. The bus, registration BU-12182, was struck on its left side and pushed 20 metres, while a large hole was made in its side. The rear of the bus was thrown against an earth bank before rebounding onto the road, transformed into little more than a pile of scrap metal. Driver Julio Pérez Rodriguez stopped the train and he and other members of the train crew went to assist the victims of the accident, helped by local railwaymen and passing motorists. The driver then went to Villabilla station to report the accident.

The alarm was raised and doctors and ambulances went to the scene. Dr Carazo Carnicero, the head of the local health department, organised the evacuation of the injured in collaboration with the Civil Guard.

It was difficult to understand how the accident could have happened. The device that should have announced the approach of a train to the crossing keeper when it reached a point 2,200 metres from the crossing was apparently working correctly. After the accident Crossing Keeper Andrés Dorado Villa, crying continuously, was held under guard in his gate-box by two members of the Civil Guard. [991]

On Thursday 6th July 1967 a bus left the city of Korat, Thailand, for a village six miles distant. It was crammed with villagers returning from the morning market, with 86 passengers inside and another ten riding on the roof. Before it had gone far it reached a level crossing near Korat station at the same time as an express. According to the bus driver a large advertising billboard had obscured his view of the track; the police thought that he had been going too fast or possibly the brakes of the bus had failed.

Seeing the train at the last moment the bus driver had wrenched open a door and jumped to safety, leaving the bus and its screaming passengers, many of them women and children, to roll onto the crossing in front of the train, which caught its rear end. Forty-three bus passengers were killed and the same number were badly hurt. The driver and the ten passengers who had been riding on the roof escaped.

The bus driver, who was also the owner of the vehicle, attempted to flee after the accident but was soon apprehended and charged with negligent homicide. [992] [993]

The most serious railway accident in East German history occurred on Thursday 6th July 1967 at Langenweddingen, near Madgeberg. Just after 08:00 on the morning of that day accelerated passenger train No. BP 852 from Salzwedel left Madgeburg for Thale in the Harz Mountains. Its 500 passengers included a large party of schoolchildren bound for a

---

[991] *ABC*, Madrid, Spain, 4th July 1967
[992] *Raleigh Register*, Beckley, West Virginia, United States, 7th July 1967
[993] *European Stars and Stripes*, Darmstadt, West Germany, 8th July 1967

holiday camp. The train consisted of a steam locomotive (No. 22 022), a luggage van, four double-decker coaches, another luggage van and four more double-decker coaches. The party of 50 schoolchildren was in the leading coach, second behind the engine.

At Langenweddingen the single-track line was crossed on the level by Bundesstraße (federal highway) 81, the crossing being protected by four lifting barriers, two on either side of the line, worked by a crossing keeper. Several overhead utility lines which ran alongside the road also crossed the railway at this point, including a Deutsche Post telephone cable. The crossing barriers had been run into and damaged by a bus in January 1966 (1967?) and had had to be replaced. Since that time a problem had arisen with the telephone cable fouling one of the raised crossing barriers and at the beginning of February 1967 Deutsche Post had applied for the cables to be taken underground over the crossing. Meanwhile as a temporary measure for, it was hoped, only three months, the 3mm diameter lower cables were replaced by cables of only 0.5 mm diameter without supporting tension wires. There was no Deutsche Reichsbahn regulation requiring the crossing to be closed before a train had been accepted or before the protecting signals had been cleared. Dense vegetation at the crossing also meant that trains could not be seen by road users until they were almost on the line and road vehicles could likewise not be seen by train drivers.

At 07:54 the fahrdienstleiter (signalman) in signal box Lof, which adjoined the level crossing, accepted a local goods train from the opposite direction to run into track 3 at the station, while the barriers were still up. At 07:58 he was offered and accepted the passenger train, which was travelling at about 85 kilometres an hour as it neared the crossing. Also approaching the crossing at the same time from one side along Bundesstraße 81 was a road tanker loaded with 15,000 litres of petrol, bound for a rubber plant at Ballenstedt, while coming from the opposite direction was a lorry converted into a bus, which was carrying 34 passengers. The Langenweddingen Schrankenwärter (crossing keeper) then lowered, as he thought, the crossing barriers and the Lof fahrdienstleiter cleared his signals for the train. However, as had happened before, the hot weather caused the telephone cables to sag and foul one of the crossing barriers, which thus failed to descend properly. The crossing keeper, seeing what was happening, raised and lowered the barriers several times in an effort to free them, without success. Meanwhile the fahrdienstleiter waved his flag at the approaching bus from his signal box window and was able to stop it. What he failed to do was to throw his signals to danger against the advancing passenger train, nor did he signal to the road tanker, which he could not see coming from the other side. The driver of the tanker, seeing the crossing barrier raised, naturally assumed that it was safe for him to cross and began to do so.

The driver of the train standing in Langenweddingen station also saw the impending danger and attempted to warn his counterpart in train BP 852 by giving an emergency signal, but the driver of the latter train did not hear it above the noise of his engine, and it was only when he was about 200 metres from the crossing that he finally caught sight of the tanker and applied his brakes in emergency. Striking the tanker, it was thrown against the side of the train, breaking many carriage windows. The contents of the tanker were splashed onto the leading two double-decker coaches of the train and the station area, there was an immediate explosion and the train was enveloped in flame. Temperatures reached 800 and possibly 1,000 degrees inside the coaches. Seventy-seven people died at the scene and another 17 succumbed later from their injuries. All but one of those who lost their lives were train passengers, including 44 of the children. The other fatality was the lorry driver, who died from injuries sustained when he was thrown from his vehicle. Another 37 passengers and the engine driver were seriously hurt; the fireman was less badly injured. The station clock had stopped at 08:06, though the accident appears to have happened a couple of minutes earlier, so perhaps the clock was a little fast, or possibly it was only after the heat became more intense that it was enough to stop the clock working.

There were several individual cases of great bravery. Teacher Werner Moritz rescued 12 children from the train, sustaining fatal burns in the process from which he later died in hospital. Seventeen-year-old student Hans-Günter Bodewell jumped from the train onto the station platform, then returned to the train and dragged a woman and a child from the coach, tearing off his shirt to smother the flames enveloping the toddler. The intense heat prevented him from rescuing anyone else. He survived.

Langenweddingen volunteer fire brigade were quickly on the scene and were joined by their Madgeburg colleagues at 08:32. No fire hydrants were available and water had to be taken from a nearby pond. The heat was so great that the first water applied to the wreckage evaporated with almost explosive force, and it was some time before firemen wearing protective gear could enter the coaches to search for bodies. The fire was extinguished by 10:00.

Subsequently the crossing keeper was charged with not having taken any other measures to protect the crossing despite the problem with the barrier sticking having been present for some time, while the fahrdienstleiter was likewise charged with not having done anything to remedy the problem. Both were sentenced to five years' imprisonment. The accident had been investigated by the State Security Police (Stasi) who at the time classified the results of their findings as a state secret. There was a feeling that the crossing keeper and the fahrdienstleiter were scapegoats, who took the complete blame for the disaster to conceal failings among more senior officials, who had likewise allowed the known problem with cables fouling the crossing barrier to drag on without doing anything decisive to fix it.

A number of measures were taken after the accident to prevent a recurrence of it. New regulations for the transport of dangerous goods were promulgated and came into force on 1st March 1968. The minimum time that crossing barriers must be lowered before a train reached a crossing was increased. Trains were prohibited from passing over crossings unless the barriers had been lowered. Crossing keepers were required to confirm to fahrdienstleiters that the barriers had been fully lowered before the protecting signals were cleared. Buses and lorries carrying dangerous goods were ordered to stop before negotiating a crossing and road vehicles were not to cross unless the barriers were completely vertical. Any instance of a crossing barrier remaining raised upon the approach of a train was to be reported and punished, even if no accident resulted. Finally a programme was begun to install automatic crossings and to interlock crossings with signals, so the signals could not be cleared unless the crossing was closed. [994] [995] [996] [997] [998]

[994] Wikipedia – Eisenbahnunfall von Langenweddingen

[995] https://www.mdr.de/geschichte/ddr/wirtschaft/verkehr/zugunglueck-langenweddingen-100.html

[996] https://www.welt.de/geschichte/article166266049/Es-war-ein-Wunder-dass-hier-noch-einer-herauskam.html

Just over a fortnight after the Villabilla accident in Spain a similar collision occurred at Astorga, though this time it seems to have received little attention in the Press. At about 03:00 on Wednesday 19th July 1967 a 'Seiscientos' (a SETA 600 car, a small inexpensive vehicle produced in Spain between 1957 and 1973) [999] registration B546.526, on its way from Léon to Astorga, was struck by a train and its seven occupants were killed. Again the crossing was manned by a crossing keeper and protected by barriers, which were raised, luring the car and its human cargo to destruction. [1000]

A Pennsylvania-Reading Seashore Lines freight train collided with a school bus on a level crossing at Somerdale, New Jersey, United States, on Thursday 21st September 1967. Southbound freight train Extra No. 6020 South, consisting of a diesel locomotive, six cars and a caboose, had left West Collingwood at 2.35 p.m. At about 2.55 p.m. it was approaching the level crossing concerned, with Somerdale Road, at a speed of 20 to 25 miles an hour. The whistle had been sounded for the crossing and the locomotive bell was ringing. As the train neared the crossing the front brakeman saw that the school bus was also approaching the crossing and not reducing speed. He shouted a warning to the engineer, who could not see the bus from his position on the engine, and the brakes were applied in emergency, but the bus was struck. As a result of the impact the bus was spun around and its driver and three children were thrown out. The bus driver, Hazel Cahilley, 43, and one pupil, Barry Shanks, 9, were killed. [1001] Three other children were injured.

The bus had started from its driver's home, picking up intellectually disabled (referred to as retarded in contemporary reports) children from different schools in the area to take them home. It had called at Osage elementary school at 2.50 p.m. and had four passengers as it reached the crossing. Three vehicles: another school bus, a lorry, and an automobile, were already stopped at or near the crossing waiting for the train and their drivers witnessed the accident. They testified that they had heard the horn of the train and the ringing of the crossing's warning bell, and had seen the red lights at the crossing flashing. The school bus that was hit had approached at 30 to 40 miles an hour and its driver had made no attempt to brake or given any indication that she was aware of the train.

No skid marks were found after the accident but the brakes of the bus were found to have been applied, implying that the driver might have seen the train at the last moment.

The Federal Railroad Administration (FRA) report on the accident was unable to determine why the bus driver had failed to either see or hear any indication of the approaching train. It noted that the bus driver had a valid licence, had passed a physical examination and a State test for school bus drivers, and had a clear driving record. She had been driving buses for two years.

The bus was a 1966 General Motors Handi-bus with a seating capacity of nine passengers. It did not meet the construction requirements for school buses prescribed by

---

[997] https://www.mz.de/mitteldeutschland/sachsen-anhalt/ungluck-in-langenweddingen-1967-ein-uberlebender-berichtet-vom-grossen-ddr-bahnungluck-1328370

[998] *Katastrophen der deutschen Bahnen Teil I* by H.J. Ritzau, pgs. 296-7

[999] Wikipedia – SEAT 600

[1000] *ABC*, Madrid, Spain, 20th July 1967

[1001] *The Evening Sun*, Hanover, Pennsylvania, United States, 22nd September 1967 the New Jersey State Department of Education and bore no markings describing it as a school bus.

Under New Jersey state law the drivers of school buses were required to stop, look, and listen at all level crossings and check for the approach of trains before driving onto the line. However, according to the New Jersey State Department of Education and the New Jersey Police because the bus involved in the accident did not meet the specifications for a school bus and was not carrying 12 or more passengers it was not classified as a school bus, and hence was exempt from the requirements to stop, look and listen at crossings. It was stated by the authorities that this was the first collision between a small bus and a train in thirty years.

The FRA concluded that:

(1) The train was being operated in accordance with applicable rules and regulations.

(2) The automatic railroad-crossing warning signals were functioning properly and were indicating the approach of the train.

(3) The school bus was moving in a school zone at a speed somewhat in excess of the limit for vehicles moving in such zones. The excessive speed, however, does not appear to be a significant factor in the accident.

(4) The driver of the school bus was unable to see the train approaching, and apparently neither saw nor heard any of the signals that were giving a warning of the approach of the train to the crossing.

(5) The school bus entered the crossing without stopping, as required by State law when a clearly visible mechanical or electrical device gives warning of the immediate approach of a train, causing the accident. The reason for the failure of the driver to stop the bus short of the crossing could not be determined. [1002]

Less than a fortnight after the Somerdale accident, on Monday 2nd October 1967 a school bus was again involved in a level crossing collision with a train in the United States. This time the location was County Road No. 29 Crossing, near Waterloo, Nebraska. At 8.10 a.m. that day a bus carrying nine children on their way to school was struck by a Union Pacific Railroad train, killing four of the children and injuring the other five. This time the accident was investigated by the National Transportation Safety Board (NTSB).

Train Extra No. 708 West consisted of two diesel electric locomotives coupled in multiple operation, 96 empty box cars and a caboose, in charge of an unnamed conductor with Engineer Glenn Peterson and Brakeman Weldon Nybbelin on the leading engine.

The school bus was owned and operated by Waterloo Public School and was being driver by Thomas McMahon, 28, who was also a teacher at the school.

County Road No. 29 was described in the NTSB as a two-lane gravel-surfaced highway. The crossing was an unprotected one. The first warning of the crossing should have been a railroad approach sign, 380 feet from it, but this had been knocked over and was lying on the ground, invisible to drivers. Thirty feet from the crossing was a stop sign, closely followed by a standard crossbuck bearing the words "RAILROAD

[1002] Federal Railroad Administration report

CROSSING" with a separate sign on its post reading "2 TRACKS." Visibility was restricted by trees until a vehicle was 320 feet from the crossing and then by a row of telegraph poles until 90 feet from the crossing. The stop sign and crossbuck could have blocked the view of a train had the bus been stopped at a certain point.

The weather on the day of the accident had been clear with no haze. At 8.10 a.m. the sun had been eight degrees above the horizon.

Driver McMahon told the NTSB inquiry that as he had approached the crossing he had not been able to see very well due to the glare of the sun, and he had had the windscreen visor down. He had not been wearing sunglasses. He stated that he had stopped the bus at the stop sign, opened the boarding door and listened for a train but had not heard one. His view of the track eastwards had been obscured by the stop sign. He had then closed the bus door and proceeded over the crossing in low gear, remaining unaware of the train until after the collision.

Engineer Peterson testified that he had sounded his horn for the crossing, the locomotive bell had been ringing and its headlight had been lit. About 1,400 feet from the crossing the school bus had come into view. It was slowing down and had then stopped short of the crossing. The engineer said he did not see the bus door opened and as his engine was close to the crossing the bus had started up again and entered the crossing, whereupon the engineer had applied the brakes in emergency and the collision occurred.

Several of the surviving schoolchildren were interviewed. They agreed that the bus had either stopped or nearly so and none of them had seen the train until the last moment.

The driver of a cattle lorry on a road parallel to the track witnessed the accident. His attention had been drawn by the sound of the train horn and he confirmed that the locomotive headlight had been lit. He saw dust due to sand being put down by the engine and "fire" from the engine wheels. He did not see the bus stop but witnessed the collision. Although he was 500 feet further from the train than the bus driver he said he had been able to see the train and its headlight clearly.

The train was travelling at 56 miles an hour and the bus at seven to nine miles an hour at the moment of the collision. The right rear of the bus was struck and at least four seats were ripped out and thrown 50 feet. It was in this part of the bus that the fatalities occurred, while the front of the bus was little damaged. [1003]

A newspaper wrote, 'Some of the wreckage was strewn as far as 100 feet away. Among the pitiful debris were school books and papers on which homework had been laboriously done.

'Classes were called off immediately by Supt. Milton Parsons, who was on the verge of tears.

'The entire rural community {of Waterloo} of some 520 persons about 20 miles west of Omaha was stunned and the mournful wail of the big diesel engines which pull freights through the town added to the pall. Flags at the school and the small city park were lowered to half-staff.

'Parsons said, "The entire community is shocked. But let us thank God the bus wasn't filled to capacity." The superintendent said the bus run was just beginning at the time of

[1003] National Transportation Safety Board report

the crash and that it could have carried as many as 50 children.' [1004]

Another newspaper stated, 'One of the first to arrive at the scene was a Roman Catholic priest. He was administering last rites to the fatally injured when the first newsmen arrived.

'Most of the injured were taken to Children's Memorial Hospital in Omaha. A spokesman at the hospital said seven of the nine admitted were not seriously hurt and would not be kept overnight.' [1005]

The NTSB report noted that the school bus had been a standard type, a 1964 International Schoolmaster model, with a tinted transparent plastic sun visor, the vehicle had been painted yellow with high gloss paint that might cause some reflection of the sun's rays from its hood into the driver's eyes.

Tests conducted with similar locomotives and buses indicated that the train horn could be heard in the bus when the train was six seconds from the crossing with the bus doors closed and 12 seconds from the crossing when the door was open.

The bus driver held a valid school bus driver's permit. He had passed a physical examination in August 1967 when he had 20/20 vision and was not colourblind. He had been driving the bus over the same route since the middle of the 1966/67 school year.

A traffic count showed 147 motor vehicles using the crossing in 24 hours. An average of slightly less than 34 trains had used the crossing daily over the previous month.

There had been two previous accidents at the crossing since the beginning of 1963, the first of which was attributed to a car driver being blinded by the sun and hence failing to see the train.

The NTSB commented, 'This is a very complex accident, and many significant factors contributed in a causal way to the occurrence of collision and to fatal injury when the collision occurred.' The main causal factors were:

(a) The bus route chosen by the local authorities, which involved passing over two unprotected level crossings. An alternative route avoid both crossings would have increased the journey length from 3 to 10½ miles. The standards of the Nebraska State Board of Education for school bus operation did not consider route selection or possible hazards on a route.

(b) The bus driver had not seen the train. The brief stop that he had made had given him little time to search for a train. In addition his vision had been compromised by the glare of the sun, the obstructions caused by the stop sign and crossbuck, and the upward inclination of the road, which had caused the sun to shine directly into the driver's eyes. Reflection of the sun from the hood of the bus and dust on the windscreen were other possible factors.

(c) The bus driver had not heard the train. The train had been too far away for its horn to be heard when the bus was stopped and when the horn was heard by the children, if not by the driver, the train had been too close anyway. The unreliability of train horns to warn the occupants of road vehicles had been noted in the report on the Sacramento accident (see above).

[1004] *The Joplin Globe*, Joplin, Missouri, United States, 3rd October 1967
[1005] *The Washington Evening Journal*, Washington, Iowa, United States, 2nd October 1967

(d) The speed of the train. Approaching the crossing the train had been moving at 56 miles an hour, slightly in excess of a railroad order imposing a maximum speed of 50 miles an hour, but this order was for traffic purposes and not with regard to safety considerations.

(e) The engineer had not made an emergency brake application when the bus had restated until he had been sure it was not going to stop short of the line. Whether it would have made any difference had be braked a few seconds earlier was unclear. The decision as to when to brake was a difficult one in such circumstances.

(f) Irregularities on the locomotive such as handrails and footboards had snagged the bus and made the collision more severe. Possibly future locomotives might be designed to reduce such collision damage.

(g) School bus bodies might be designed to increase their resistance to collisions.

The NTSB report continued with its conclusions:

'The school bus driver, after stopping at the signs, failed to ascertain that there was a train approaching and drove the bus onto the tracks in front of the oncoming train. The engineer did not make an emergency brake application after the bus started beyond the stop sign until he realized that the bus was not going to stop again. It has not been possible to determine completely why the bus driver proceeded across the tracks after stopping, or why he continued in the face of the normal warnings. Some conclusions can be drawn, however.

'It is concluded that, as the bus driver reached the crossing, he encountered a difficult problem in maintaining adequate vision due to the direct rays of the sun and lesser indirect reflections from the vehicle hood. If during that time, the sun's rays did shine directly into his eyes, his ability to see objects would be impaired for some undetermined period of time even after he looked away from the sun. He was not wearing dark glasses. Whether or not the driver succeeded in protecting his eyes against the direct rays of the sun, some part of his attention would necessarily have been occupied by this task, tending to distract him from other driving tasks. The time of stopping was momentary.

'The audibility of the train horn within the bus was very low and inadequate to provide warning when the bus engine was accelerated in low gear and the door was closed. Nebraska law does not require that the door be opened or maintained open for any period of time. In the circumstances of the accident, moreover, the horn sound could not be heard even if the door was open at the time the bus was stopped at the crossing, due to the distance of the train. Thus, warning by train horn is an unreliable system as a backup for visual sighting, especially when there is no legal requirement that the door be opened. Due to conflicting evidence, no conclusion has been made as to whether the door was opened.

'Because it is not known definitely at what point the emergency application of the train brakes was begun, it is not possible to conclude whether the normal 8-second time lag for full effective brake application over the length of the train or the low rate of deceleration achieved by existing brakes advanced the time of the train's arrival at the crossing sufficiently to allow the collision. It is concluded, however, that the stopping rate of the train under emergency brake application did not exceed 37% of the theoretical maximum braking rate achievable on normal rails. The theoretical maximum is that of the highest adhesion obtainable without inducing wheel sliding. This problem is a shortcoming in the state of the art of railroad braking.

'The protuberances on the front of the locomotive are, at times, needed for functional purposes in switching cars, but serve no purpose when the train is engaged only in over-the-road operations. In this case, their existence contributed to the severity of the accident.

'The rear structure of the bus disintegrated due to a combination of forces caused by protuberances on the locomotive and low structural strength of the bus. The structural strength of the bus would have been greater if the structural elements and the skin had been more effectively joined. The blow received by the bus is considered not beyond the magnitude of impacts which might be delivered by collision with a truck moving at highway speeds. Although gross injuries might be expected under the degree of disintegration which occurred, no final conclusions can be drawn as to relationship between bus disintegration and the occurrence of fatalities, because the patterns of fatal injury are unknown. It can be concluded, however, that similar disintegration in future collisions to school buses would tend to make ineffective future standards if they are based only upon safety of school bus interiors.

'Had the crossing been protected by a flashing signal light, or railroad gates, this accident would probably have been avoided. The low density of vehicular traffic at this crossing might not be considered sufficient to justify the installation of flashing signal lights or automatic railroad gates by most authorities. However, the Voorhees Report indicates that on an average, railroad grade crossings of all types experience one accident involving a vehicle and a train each 54 years. The previous record of two accidents in 5 years was, therefore, far above average and might have led to the provision of protective devices or closing of the crossing, had the unusual record been analyzed. The crossing now has a record of three accidents in 5 years.'

Finally the NTSB listed 11 recommendations:

(1) The Safety Board recommends that the States not now having a requirement that the door of a school bus be opened for a sufficiently long period while stopped to allow the whistle or horn of a train to be heard at unprotected railroad grade crossings, consider establishing such requirement.

(2) The Safety Board again recommends to the Federal Highway Administration and the Federal Railroad Administration that they study the questionable audibility of external sound signals within motor vehicles and to work toward creating a unified system of warnings and reliable reception which could be made effective through Federal regulations or State laws. A similar recommendation was made on January 15, 1968, in the Board's report of a grade-crossing accident which occurred near Sacramento, California.

(3) The Safety Board recommends that the Federal Highway Administration consider the need for requirements for structural strength of school bus bodies in connection with its study of desirable standards for protection of school bus occupants. In particular, the Board recommends that program A.1.1.4 of the National Highway Safety Bureau, titled "Design, Fabrication, and Test of a Safe School Bus Interior," be expanded in scope to include consideration of structural integrity and intrusion into the school bus interior.

(4) The Safety Board recommends that criteria for school bus operating safety of the Federal Highway Administration, States, and local school governing bodies include school bus routing to avoid grade crossings whenever possible.

(5) The Safety Board recommends that when it is absolutely necessary that school buses operate over unprotected grade crossings, provisions should be made for a

responsible individual, other than the driver, to alight and determine that no train is approaching, and to signal the driver to proceed over the crossing.

(6) The Safety Board recommends that the Federal Highway Administration establish a vehicle safety standard to protect driver vision against external sun glare.

(7) The Safety Board recommends that the Federal Railroad Administration, under its authority to regulate railroad brakes, study the existing state of the art of railroad passenger and freight brake systems, and issue descriptive reports of the capabilities of such systems to efficiently stop trains in an emergency.

(8) The Safety Board recommends that the National Education Association, the National Professional Driver Education Association, and the Supervisors of Driver Training of the Boards of Education of the several States, review their driver training programs to insure that, in the normal course of driver instruction for all types of drivers, specific attention be paid to the visual and perceptual task of searching a location of possible hazard to ascertain that the hazard is not present.

(9) The Safety Board recommends that the American Association of Motor Vehicle Administrators place greater emphasis on proper grade crossing procedures during the examination process for drivers' licenses applicants.

(10) The Safety Board recommends that the Federal Railroad Administration study the visual effectiveness of the white, fixed, sealed-beam headlights on locomotives in contrast with the bright daylight as compared with an oscillating or gyrating headlight unit, a flashing strobe light or other high intensity-type light, possibly of a more contrasting color.

(11) The Safety Board recommends that the Office of High Speed Ground Transportation, the Association of American Railroads, railroads operating in the Northeast Corridor, and States having safety regulatory authority over railroads, consider the implications of this accident analysis for logical and necessary train operating speed reductions under restricted visibility wherever tracks cross unprotected grade crossings. The time needed by motor vehicles to cross tracks requires that drivers be able to detect the train at a considerable distance in order to be certain of crossing safely. This distance, as illustrated by this case and others, is already beyond the range of typical present-day train horns when the actual conditions under which the horn is to be heard are considered. Conditions which limit audibility at a distance, including enclosed vehicle passenger compartments, local vehicle noises, and restricted use of horns or bells, are now found so frequently as to be a normally anticipated situation. As train speeds rise, persons crossing a grade crossing must rely increasingly on ability to see approaching trains in order to determine that it is safe to proceed. [1006]

On Wednesday 4th October 1967 a Canadian National Railway from Toronto, Canada, to Port Huron, Michigan, United States, was running about five minutes late, as it often was due to the volume of passengers and luggage to be taken on at Kitchener and the need to pass another train at Breslau. Whether or not its late running had any significance in relation to the accident is unknown but it collided with a station waggon on a level crossing at Stratford, Ontario, Canada, killing its six occupants. The victims were Mrs

[1006] National Transportation Safety Board report

Carl Seltzer, the driver; her son Karl, 16; Terry Swartzentruber, 19; Robert Sparling, 16; and Robert Schlitt, 16, all of Tavistock, Ontario; and Ronald Zehr, 16, of Shakespeare, Ontario. The teenagers were all students at the Waterloo-Oxford District High School in New Hamburg, Ontario. Mrs Seltzer had picked the students up from a football practice at Baden to take them home.

The station waggon was carried half a mile along the track. The little-used crossing was protected only by crossbucks, while trees restricted the view of westbound trains, but it was not considered dangerous. A motorist had been killed there by a train about 30 years previously. No one on the train was hurt. [1007]

There was a lucky escape on Friday 20th October 1967 when a Seaboard Coast Line Railroad freight train struck the rear of a school bus on a dirt road crossing near Schofield, South Carolina, United States, about five miles north of Ulmer. The train caught the rear of the bus, shearing off a foot or so of it, but as chance would have it all the children were sat towards the front of the vehicle and no one was hurt. The accident happened at about 2 p.m., as the dozen children were on their way home from school. [1008]

If the Schofield accident was a lucky escape, the fact that the bus driver and her five passengers escaped with their lives in the collision of a train with a Madison Community school bus at Wever, Iowa, United States, on Monday 30th October 1967 was described as a near miracle, though all six were injured. Mrs Wyona Lee Tuttle, the bus driver, had picked up about 40 pupils from the junior and senior high schools at Fort Madison just after 3 p.m. and at about 3.39 p.m. had dropped off all but five of her passengers and was heading west when her bus reached the Chicago Burlington and Quincy Railroad (CB&Q) crossing near Wever. At the same time northbound 23-car CB&Q train No. 79 was also approaching the crossing. Engineer Braxton Gilbert stated afterwards that he had been travelling at 30 miles an hour and had seen the bus stop short of the crossing, then it had pulled forward into his path. He had sounded his whistle but the warning had come too late. The train struck the bus broadside and carried it about 620 feet down the line.

Taken to the Sacred Heart Hospital in Fort Madison and admitted were Mrs Tuttle, 47; and pupils Linda Chesnut, 15; Theresa Chesnut, 13, sister of Theresa; and Lynn Chesnut, 13, cousin of Linda and Theresa. Treated and discharged were Stephen Beebe, 15; and Connie Gintz, 13.

Stephen Beebe related later, "I must have been thrown through the window. I was sitting two seats back of the front door, reading a book – I didn't hear anything." He found himself lying in a ditch about 80 feet from the crossing. "I was dazed and tried to walk around to see if I was hurt," he said. Then he saw Lynn Chesunt also lying in the ditch and tried to help her to her feet but she was unable to stand because her leg had been broken. "She wasn't unconscious, but she was all bloody, and had a bad gash over her eye," he added. Working in a field 400 feet away was Stephen's father, Phillip. "I was looking at some hogs when I glanced up and saw the stopped train, and the wreckage in front of the diesel. I couldn't find Steve and ran back down the tracks in an effort to find

[1007] *The Medicine Hat News*, Medicine Hat, Alberta, Canada, 5th October 1967
[1008] *Florence Morning News*, Florence, South Carolina, United States, 21st October 1967

him. I found him near the crossing trying to comfort the girl," Mr Beebe said.

Driving northwards at the time along US 61, which ran parallel with the railway, was Walter Gintz, who told a reporter, "I saw the train hit the bus, but I got the impression it was a cement truck that was hit." He had turned his car around and gone back towards Wever to report the accident at a service station. Leaving there he was told that it was the school bus that was involved. "My first thought was that Connie was on that bus," he said. He had raced back to the crossing to find Mr Beebe tending the injured.

The crossing was protected only by crossbucks, without barriers or warning lights. It was raining quite hard at the time of the collision and visibility was further restricted by tank cars stabled on a siding south of the crossing and east of the running line.

Mrs Tuttle had been driving the bus for six years. She had been making her first trip of the afternoon and would have returned to Fort Madison to pick up the elementary school pupils, whose classes finished an hour later.

Fortunately all the pupils on the bus when it was struck had been sitting on its right-hand side, opposite the point of impact. Mrs Tuttle was trapped behind the steering wheel and it took several minutes to free her.

There had been a previous collision with a cement lorry on the same crossing on 2nd October. [1009] [1010]

An announcement from Cuban radio stated that a train from Santa Clara to La Habana had collided with a bus on a level crossing at Guanabacoa, Cuba, on the outskirts of La Habana on about Tuesday 28th November 1967. Eleven people died and 12 were injured. [1011]

Six children squeezed into Mrs Vonnie Culver's pickup truck for the ride home from school in Foyil, Oklahoma, United States, on the afternoon of Thursday 21st December 1967. Five of them were all set for their parts in a school nativity play the next day. The pickup truck had only travelled a quarter mile when it was hit by a St Louis-San Francisco Railway freight train, killing five of the children, critically injuring Mrs Culver and injuring the sixth child. Three of Mrs Culver's own children lost their lives: Harold, 10; Darrell, 9; and Ricky, 8. Two children of her neighbours: Linda Dotson, 14; and Dianne Snider, 9, also died. Another of Mrs Culver's children, Donald, 8, was injured.

'A service station attendant nearby heard the train sound its whistle as it neared this Rogers County community of 150. Then he heard the sound of metal hitting metal – hard. He looked up and saw the train taking the pickup down the tracks. It was carried 1,600 feet.

'Highway Patrol trooper Don Blizzard said Mrs Culver apparently pulled onto the railroad tracks at a city street crossing before she saw the train. "There were no skid marks – she didn't slide up to the tracks – it appeared she pulled up on the tracks, saw the train, and hit her brakes. They were astride of the tracks when it hit," he said. Blizzard described the accident as, "The worst I've ever investigated."

[1009] *Evening Democrat*, Fort Madison, Iowa, United States, 31st October 1967
[1010] *The Burlington Hawk-Eye*, Burlington, Iowa, United States, 31st October 1967
[1011] *The Victoria Advocate*, Victoria, Texas, United States, 29th November 1967

'Game Ranger Ray Tanner was riding with Blizzard in his patrol car when the call came in. "It looke terrible," he said. "It was just full of children. They were just crammed in there. There was one little boy. We knew he was alive because he was crying all the time." That was eight-year-old Donald.

'I finally felt life in one of them – the little fellow," said Tanner. I went in through the windshield and tried to give him mouth-to-mouth resuscitation. I got hold of him and he was warm and I went to working on him. I finally got life out of him but he died." That was Ricky Culver, 6.

'When the school principal heard about the accident he cancelled school for the day – and the Christmas pageant. It was a nativity scene. Darrell was to have played a shepherd. Little Ricky was to have sung in the first-grade chorus. Dianne was to have been an angel.' [1012]

On 4th January 1968 an English newspaper noted that the residents of Waterloo, Lancashire, were petitioning the Minister of Transport to defer for 12 months the installation of an automatic half barrier crossing at nearby Brook Road Crossing after two men had died two days earlier on an automatic half barrier crossing in Kent.

The newspaper wrote, 'Controversy has dragged on for years: are unmanned automatic barriers of the Continental type as safe as the older, manned gates? The British Railways Board says "yes;" time and again people living near crossings where an automatic barrier has been proposed have said "no."

'The board sees considerable advantages in the automatic crossing, apart from a saving of money in that they are unmanned. Most manned crossings need to be closed for three or four minutes each time a train passes. The unmanned barrier lowers automatically in time to prevent road vehicles being caught on the line by even the fastest train, bur remains closed for between something under a minute in the case of a fast train and 90 seconds for a slow freight.' [1013]

Only two days later, on Saturday 6th January 1968, eight train passengers and three railwaymen lost their lives in a collision on an automatic half barrier crossing at Hixon, England, on British Railways, London Midland Region. The 11:30 express from Manchester Piccadilly to London Euston collided on the crossing with a road transporter carrying a 120 ton electrical transformer and was totally derailed. Forty-five more of the approximately 300 passengers on the train were injured, six seriously, and the electric engine and five carriages were destroyed. In view of the concerns raised by the accident Barbara Castle, the Minister of Transport, promised a full public inquiry. Because the Railway Inspectorate was involved in the introduction of new half barrier crossings one was held under Section 7 of the 1871 Railways Act, chaired by a barrister, Mr Brian Gibbens, Q.C. It was the first inquiry to be held under the act since the 1879 Tay Bridge disaster.

The transformer was to be moved from the English Electric works at Stafford to a

---

[1012] *Miami News Record*, Miami, Oklahoma, United States, 22nd December 1967
[1013] *The Times*, London, England, 4th January 1968

storage depot on the disused Royal Air Force airfield at Hixon, situated on Station Road. The journey was authorised by a special order issued by the Ministry of Transport, which had worked out a somewhat circuitous route for the transporter, via the M6 motorway and the A34 and the A51 main roads. No mention was made in the description of the route of the automatic half barrier crossing.

To carry the transformer a massive transporter vehicle owned by Robert Wynn and Sons Limited was used, 148 feet long with a 32-wheel trailer, impelled by a tractor unit at each end and manned by a crew of five. The laden weight of the transporter was 162 tons. A police escort was provided by two constables in a patrol car.

The crossing was situated on a long straight stretch of road with good visibility in each direction. On the approach to the crossing from the A51 main road a traffic sign 173 yards from the crossing indicated to motorists that an automatic crossing lay ahead. Then came a sign warning of the danger of electrocution from the overhead line equipment of the electrified line and indicating a headroom of 16 feet 6 inches. Finally there were two notices at the crossing itself. One read:

"IN EMERGENCY or before crossing with exceptional or heavy loads or cattle PHONE SIGNALMAN."

The other notice read:

"Another train is coming if lights continue to flash."

A census taken in February showed an average of 870 motor vehicles and 26 trains over the crossing in 24 hours.

The operation of the barriers was set in motion by a track circuit and two treadles (thus requiring a failure of all three for the barriers to fail to descend). The treadles were placed 1,000 yards from the crossing and for a train travelling at the maximum speed of 85 miles an hour there was an interval of 24 seconds between the activation of the crossing and the train reaching it. For the first eight seconds of this period the alarm bells would ring and the warning lights flash, over the next eight seconds the barriers would descend and eight seconds after they were fully lowered the train would reach the crossing (the bells would continue to ring and the warning lights to flash throughout the sequence). The "brisk operation" of the crossing was intended both to discourage drivers from going around the barriers and to reduce delays to road traffic. The report noted that an Up train such as the Manchester express, travelling down the 1 in 600 falling gradient to the crossing at maximum speed and weighing 400 tons would take 1,520 yards to stop. The engine and 12 coaches of the express involved in the accident weighed 491 tons and so would have taken longer still to stop.

The transporter set off from Stafford at about 9.30 a.m. At about 12.20 p.m. it turned off of the A51 into Station Road. Short of the crossing it stopped briefly while the police went ahead to check where the entrance was and this having been confirmed the journey was resumed. From the main road the transporter had followed the police patrol car at a distance of 50 to 75 yards, moving at four miles an hour, but on reaching the crossing speed was reduced to two miles an hour, so that three of the transporter crew could check the clearance between the top of the transformer and the overhead line equipment and between the bottom of the transporter and the road. When the transporter arrived at the crossing the half barriers were raised, the warning lights were out and the bell was silent.

The main bulk of the transformer was astride the tracks when, at 12.26 p.m., the Manchester express reached the treadle, the alarm bell began to ring and the warning lights to flash, then, eight seconds later, the half barriers descended upon the top of the transformer. Looking around to his left Mr Bryn Groves on the leading tractor saw the express coming and, realising that it was not going to be able to stop, shouted a warning to his crew. He also tried to accelerate the transporter, as did Mr A.L. Ilsley in the rear tractor, even though this meant potentially putting himself in the path of the train. Their actions lessened the extent of the disaster to some extent as instead of striking the main part of the transformer squarely the train struck it towards its rear end, close to the connection to the rear tractor. The bravery of the two men, and especially of Mr Ilsley, was praised by the inquiry. As a result of the impact the transformer was knocked off the trailer and to the left of the Up line, while the engine and leading coaches of the train were demolished. Driver Stanley Turner, Secondman James Toghill and Driver J. Atkinson, who was also in the engine cab, were killed instantly, as were eight passengers in the following coaches. The casualties would have been higher had not there been few passengers in the first-class coaches at the front of the train and but for the strong construction of the rolling stock.

After the accident Guard W.C. Final ran back a mile along the line and put down detonators to protect the rear of his train. Having reported the accident to Crewe from a lineside telephone he then hurried southwards, going through fields to bypass the wreckage, and put down protection a mile south of the crossing. He told the inquiry, "I never stopped running." He, too, was praised for his actions at the inquiry.

Fireman Leslie Hockenhull, travelling as a passenger on the train, attempted to use the telephone at the crossing to call the emergency services and finding it inoperative got a lift in a van to a house on the A51 from where he was able to call them.

Signalman M.J. Woodcock in Meaford box, 12 miles north of Hixon, told the inquiry that when the track circuits around the crossing had indicated in his box had failed to clear he had telephoned his colleagues at Colwich, south of the crossing, to find what was wrong. Unable to contact them he had rung the panel box at Stoke-on-Trent to the north instead and had replaced his signals to danger.

Signalman B.S. Register in Colwich box said that eight track circuits on his panel had lit up, there had been indications of a power failure and a failure of the half barriers. He had then replaced his signals to danger and telephoned Control at Stoke-on-Trent and the linesman's office at Stafford. Technician E. Bickley had then gone from Stafford to the crossing, was able to place more signals at danger and had informed Colwich of what had happened.

Thanks to the police escort notice of the accident was immediately sent by radio telephone to police headquarters at Stone and the first ambulance arrived at the scene less than 20 minutes after the crash at 12.45 p.m. Had the police not been present, the inquiry noted, the delay in calling the emergency services might have been considerably longer.

Mr Gibbens stated: 'The immediate cause of the accident is plain. The level crossing was thirty feet long from the nearest half-barrier to the furthest rail and no vehicle of the length of the transporter could traverse it within the 24 seconds' warning period before the arrival of an express train unless it moved at more than six miles per hour: but this transporter was going at only two miles per hour. Neither the crew of the transporter nor the police escort knew the time sequence of operation of automatic crossings, and so did not realise that they would have such short warning of the onset of a train. Consequently, no one paused to consider whether a train might be imminent. Nor had any of them observed the Emergency Notice or become aware of the provision of a telephone in the half-barrier apparatus, so no one telephoned the signalman to enquire whether it was safe to cross.

'The use of the telephone by the driver of any large slowly moving vehicle to ascertain from the signalman whether he will have time to cross safely is vital to the safe use of automatic crossings. If that had been done on this occasion the disaster would not have happened.'

Mr Gibbens then went on in his report on the disaster to examine in detail the part played by the various individuals and bodies in causing the disaster.

Whatever the responsibility of the Ministry of Transport in designating the route and of the police in agreeing to escort it, the immediate responsibility for the safe movement of the transporter, Mr Gibben noted, lay with its five man transporter crew under the leadership of Mr Groves. Two questions arose:

(i) Why did not the leading driver, Mr. Groves, telephone the signalman at Colwich before going on to the crossing, as required by the Emergency Notice? And

(ii) Why was the equipage going so slowly over the crossing?

Mr Groves was an experienced and skilled driver. He had driven the load carefully from Stafford and had abided by three "cautions" set out in the Ministry of Transport route, two relating to crossing over bridges and one to a low bridge under a railway on the A51. Although three was no mention of the crossing in the route plan he had been aware of it, having been advised of it by Mr Preston of the English Electric Company.

Mr Groves' actions and conduct on reaching the crossing were, Mr Gibbens commented, remarkable. He said that he had not stopped short of the crossing to telephone the signalman because he had a police escort and he thought that somehow an engine driver would receive notice if the crossing were obstructed. It had not occurred to him that if the train were too close its driver could do nothing to stop anyway. He had not thought that the police had any control over the rail traffic but had assumed that as the police patrol car had passed over the crossing in front of him it was all right to follow. He was vaguely familiar with automatic crossings but had not been aware of the short interval between the activation of the warning system and the arrival of the train at the crossing, and he had assumed that there must be some mechanism for stopping the train should the crossing be obstructed.

As regards checking the headroom under the overhead wires Mr Groves added that the transporter might have had to have stopped on the crossing for the headroom to be adjusted, a process which would take several minutes.

The low speed over the crossing was, Mr Groves said, because the road over it was little wider than the transporter, though he normally reduced speed to 3 miles an hour over level crossings anyway.

Mr Gibbens stated that he was very troubled by Mr Groves' responsibility for entering the crossing with such an abnormally large vehicle in the path of an express train and his general attitude to the unfamiliar type of level crossing which he knew was unattended and operated by approaching trains. It could not be accepted that he really thought that the police escort could ensure his safety, as he would have known as an experienced driver that the police were principally concerned with traffic control. Had he considered the matter he would have realised that his assumption that the barriers would not operate with a vehicle on the crossing must be wrong, or else a queue of road traffic would completely disrupt the train service.

None of the other members of the transporter crew knew that their route took them over an automatic crossing, none had seen any of the crossing notices, knew anything about such crossings or had received any instruction about them from their employers.

The next thing to be considered was the responsibility of Robert Wynn and Sons Limited. Their supervision of drivers was minimal. Drivers recruited were tested for their driving skills, had a probation period with an experienced driver, then were more or less left to themselves, working up from smaller to larger vehicles.

At the opening of the inquiry the counsel representing the company had stated that they had no knowledge of such crossings, but this was not true. More than a year earlier, on 8th November 1966, a low-loader lorry belonging to the company carrying a crane had grounded on an automatic crossing at Leominster. A disaster had only been averted by the lorry driver, Mr Horton, courageously remaining in his cab and by violent acceleration and letting out of the clutch making the front of the lorry rear up in the air and drag the trailer off of the crossing, just as the express reached the crossing.

Mr Horton had duly reported the matter to his employers, who in turn had written to the Chief Civil Engineer of British Railways, Western Region, detailing what had happened. The company thus were clearly aware of the problem back then.

The company had received what Mr Gibbens characterised as an arrogant reply, lacking in insight, from the British Railways, Western Region, Assistant General Manager, Mr Lattimer. In part this read, 'In brief, road vehicles must not become immobile on these crossings. If they do, they not only become a hazard to themselves, but they have become a hazard to trains and whoever is travelling in them.' Unfortunately the letter only concerned itself with vehicles which had stopped on the crossing and not with the possibility of a very long, very slow-moving vehicle, as at Hixon.

The haulage company were apparently unaware of the installation of an automatic crossing at Pontsarn, only 20 miles from their headquarters in Newport.

Edward Clark, the manager of their Manchester depot, had discussed the difficulties of getting to the English Electric site at Hixon with Mr Preston of the company, but this was before the introduction of the automatic crossing and he said that even had he become aware of the installation he would have left the matter to the individual drivers.

The company was clearly aware, from the letter they had received from British Railways after the Leominster incident, of the very short interval of time left once the crossing controls had been activated, but somehow had not linked this to the problem of slow-moving abnormal loads. 'In this state of ignorance, and failure to draw the glaring inferences from the events which had taken place in their own business, the management failed to give any instructions or warning to any of their drivers as to the hazards of automatic crossings, even about the hazard (stalling) which they did appreciate.'

Since the accident the company had begun issuing a special note to drivers:

"DRIVER TO NOTE. BEFORE CROSSING ANY AUTOMATIC CONTROLLED RAILWAY CROSSING, HE MUST ENSURE THAT THE POLICE ESCORT OR HIMSELF CONTACT THE SIGNALBOX TO OBTAIN PERMISSION TO CROSS."

Mr J.H. Preston, Chief of Heavy Transport with the English Electric Company at Stafford, told the inquiry that he was familiar with Hixon Crossing but he had never seen it in operation, he had been unaware of its "brisk operation," and basically had never given any thought to the question at all. Mr Gibben considered it 'surprising and disappointing that he {Mr Preston} never paused to consider (as he had done when the line was electrified) how this new apparatus on his Company's doorstep would affect the movement of transporters. If he had done so, he would probably have observed the Emergency Notice and would probably have told Mr. Groves about the necessity to telephone before crossing.'

The role of the police was next considered. It was only on the morning of the accident that they had been advised that the exceptional load was ready to leave Stafford. Neither P.C. Anthony Nicholls nor P.C. Ephraim Prince, the two officers who had escorted the transporter, had ever been to Hixon before. During the journey they had controlled traffic and cleared the way ahead in the usual manner. Both men were vaguely familiar with automatic crossings but neither had realised the very short warning time. P.C. Nicholls commented afterwards, "I was absolutely astounded by the speed of the whole process." Neither of them knew there was a telephone at the crossing, neither had seen the emergency notice or was aware that it was a crossing with main line expresses. They said that they had been concentrating on traffic control and had not realised how slowly the transporter would move. Had they been aware of all the facts they would have insisted that the haulage crew telephone the signalman.

A note regarding automatic crossings had been sent by the Home Office to the Chief Constable of Staffordshire back in November 1963, and a further note in August 1966, but little attention had been paid to them. The Chief Constable, Mr A.M. Rees, and the Assistant Chief Constable, Mr S.E. Bailey, claimed to have no more knowledge of automatic crossings than the two constables, "which was very little," and blamed this on the Home Office not having picked out the points relevant to police work from the long document. There were, however, already six other automatic crossings in the Staffordshire area, including the first one installed in Great Britain, at Spath Crossing.

Every time a new automatic crossing was installed British Railways sent leaflets regarding it to the police but at police stations they were only displayed for members of the public to take and were not specially studied by the police officers, who regarded them as "mere flotsam."

Since the action the Chief Constable had sent a memorandum to all police stations regarding the action to be taken the escorting of abnormal loads over automatic crossings.

As regards the role of British Railways it was clear that they were aware of the problem of very slow-moving abnormal loads and automatic crossings, as evidenced by the notice instructing the drivers of such vehicles to telephone the signalman, but Mr Gibbens remarked that 'somehow it did not dawn on them that there was danger of a serious accident every time such a vehicle crossed the railway.' Other than the warning notice and the telephone British Railways had taken no other steps to address the danger of slow-moving vehicles on automatic crossings and at the time of the disaster there were still 18 automatic crossings that lacked telephones anyway. Mr Gibbens was of the opinion that, 'British Railways did not think to give the telephone procedure the importance it deserved, probably because they believed that the entire responsibility for safe use of the crossings had been passed to the road user who, being provided with a telephone, should be left to use it.' No telephone calls regarding abnormal loads had been made to Colwich or Meaford signal boxes on any occasion prior to the disaster.

No one in British Railways had thought to send a publicity leaflet regarding automatic crossings to heavy haulage companies such as Robert Wynn and Co. or Pickford's. The leaflets also stated that the drivers of heavy vehicles or exceptional loads *should* phone the signalman, rather than that they *must* phone.

The Ministry of Transport was next considered. Mr Gibbens noted that at the meetings prior to the opening of the crossing, conducted by Major P.M. Olver of the Railway Inspectorate, no mention had been made of abnormal loads using the crossing and Major Olver had not made any inquiry regarding local traffic. He was defended by his colleagues in the inspectorate, Colonel McMullen, and Colonel Reed, for not having done so, but criticised by Mr Gibbens.

No caution note was included in the Ministry of Transport route for the transporter regarding the crossing because British Railways had not indicated that one was needed.

Prior to the accident the attention of the Ministry of Transport had not been specially directed to the special hazard of slow-moving vehicles on automatic crossings.

After the accident the special orders issued for the movement of abnormal loads required their drivers to ring the signalman, circulars were issued to all the haulage companies stressing the importance of their drivers using such telephones and all the Chief Constables were likewise informed by the Home Office.

Before the opening of the first automatic crossing in Great Britain working parties had studied the practices in France and the Netherlands. Telephones were not provided at any level crossings in the Netherlands, at least at first, nor at all crossings in France, hence they were not provided initially at all British crossings either. Neither France nor the Netherlands seemed to have many abnormal or slow-moving vehicles, so this possible hazard had received little attention. When Colonels McMullen and Reed had later considered slow-moving vehicles their minds had been fixed on agricultural machinery such as combine harvesters. Mr Gibbens commented, 'It is difficult to find an explanation for the failure of these competent and experienced officers to appreciate that there were other vehicles, besides the agricultural, which deserved consideration in this respect.'

The Bridges Engineering Design Standards Division of the Ministry of Transport was well aware of the movement of exceptional loads but with regard to the non-mentioning of Hixon crossing in the route devised by the Ministry Mr A.D. Holland 'declared that when the question was considered by the Route Section as to what problems were posed by automatic level-crossings it was decided that they were a self-evident hazard, marked by signs and signals, and that, therefore, no different attitude should be adopted towards them from the attitude in regard to ordinary traffic lights at crossroads. But he agreed that the analogy was not an exact one, and that, perhaps, the problem ought to have been studied in greater depth.'

Finally the inquiry considered the publicity given to the introduction of automatic half barrier crossings, which was shared between British Railways (locally) and the Ministry of Transport (nationally). Mr Gibbens considered that 'though they seem very good and convey a simple message adequately, they do not state explicitly how soon the train may arrive at the crossing after the lights begin to flash. I think a more emphatic definition of the imminent arrival of the train is necessary if, as the Ministry maintain, "the formidable onset of the train so soon after the falling of the barriers reinforces the traffic signals."'

For the benefit of the people most likely to be affected by the new crossings, namely long-distance lorry drivers and the drivers of heavy slow-moving vehicles, the Ministry of Transport made amendments to the Highway Code, though these did not mention the existence of telephones. In addition the Ministry distributed to the police, the Road Haulage Association, the Magistrates Association and so on a document headed the 'Requirements of the Ministry of Transport with regard to Automatically Operated Half Barriers at Public Level Crossings.' These documents, Mr Gibbens remarked, 'were inept and wholly ineffective. They contained a great deal of technical detail which could be of no interest whatsoever except to the railway technicians who were concerned with the installation or maintenance of the crossings or those manufacturers who might hope to tender for the supply of the equipment.'

Beginning his conclusions, Mr Gibbens wrote that, 'The real cause of the disaster was ignorance, born of lack of imagination and foresight at the sources where one would expect to find them. It is an odious task to criticise anyone unfavourably for having failed to foresee a danger, when many intelligent minds and experienced and talented people have conscientiously considered the same problem before the danger manifested itself yet failed to appreciate it.'

Mr Groves and the police officers were, Mr Gibbens considered, mainly the victims of shortcomings at more responsible levels. 'The essential fact which Mr. Groves and the police officers ought to have known was the need for the driver of an abnormal load always to use the telephone before attempting to cross the line. Without a telephone at Hixon, there would have been only the alternatives of asking the police to radio to police headquarters for advice, or of taking the vehicle across in the hope that British Railways had attended to the safety measures.

'Mr. Groves did not know, and never had the means of knowing, that he might have only 24 seconds to get his enormous equipage over the railway; but he knew that half-barrier crossings were automatically operated by approaching trains, and he probably had read that they worked more rapidly than the traditional kind. His task required a great deal more thought and perception than driving a car. I have come to the conclusion that Mr. Groves did not think what he was doing when he negotiated the crossing. If he had thought about the matter when he realised that he was approaching a new automatic level-crossing, instead of basking in the shadow of the police escort, he might have recalled that its operation was rapid and he would at least have paused to seek advice before launching his huge vehicle across the railway. A driver of such experience as he had of heavy haulage ought not to have risked, as he did, becoming immobilised across the railway for an indefinite time. Moreover, he ought to have instructed his statutory attendant, Mr. Parsons, to look out for signs and notices, and either he himself or Mr. Parsons ought to have observed and obeyed the Emergency Notice. On the other hand, he had not the foreknowledge that there might be a notice, other than a traffic sign, which governed his behaviour at the crossing.

'It was a lack of awareness which led Mr. Groves not to concern himself about the level-crossing problem except as regards clearances. However, I have formed the opinion that he is the kind of man who, though a most experienced driver, requires to he told things specifically and who does not draw more than the superficial inferences from what he sees. In this instance, he had not the advantage, which he ought to have been given, of guidance from his employers as to the inherent risks and the proper procedure to be adopted.

'Police constables Prince and Nicholls, likewise, were almost totally ignorant of the important but latent features of the working of automatic crossings. I do not think they can be blamed for failing to observe the Emergency Notice at the crossing for it was not a traffic sign and was not readily observed from a car moving at a normal speed. (They did notice the regular traffic signs.) Even if they had noticed its presence, read rapidly when passing it would have seemed to apply only to the case of an emergency. They had forgotten the paragraph in the Highway Code referring to automatic crossings, perhaps because its bare statement that the half-barriers would descend "just before a train reaches the crossing" (without indicating how very quickly that would happen) had not made the necessary impact on their understanding and there was no mention there of the telephone procedure. Nevertheless, if they had thought intelligently about the significance of an unmanned level-crossing, they might well have become as apprehensive of the safety of the venture as Mr. Groves ought to have been, though his was the prime responsibility. (It is a strange fact that neither the police officers nor Mr. Groves seem to have felt that sense of danger that so many people experience when traversing an automatic half-barrier crossing.)

'The principal responsibility for Mr. Groves' failure to use the telephone to obtain permission to cross the railway immediately before the accident lies upon his employers, Robert Wynn and Sons Limited. The directors of that Company were aware of the introduction of automatic crossings into Britain, and ought to have been inquisitive to discover how the new automation affected their business. It is true that others, such as Pickford's Limited, appear to have been no more inquisitive or perceptive than they, but Robert Wynn and Sons had an advantage that no others, in the trade or even in the Ministry, had: in November 1966 the near-catastrophe at the level-crossing at Leominster had brought home to them the exceedingly short time which their large vehicles would have for traversing an automatic crossing and yet, though this had occurred in their own business, they gave no real thought to the problem because they were annoyed by the tone of the letter received from British Railways. They gave no warning or information at all to their drivers, even of the foreseen risk of stalling, and continued to rely on their principle that their experienced drivers should be left to deal with all hazards of the road themselves, even though this was an invisible hazard which had taken the directors (themselves with long experience of the highways) by surprise. They ought to have made immediate enquiries, discovered the object of the telephone, and put a caution on all their special order and general order movement routes. In my opinion, the default of the directors of Robert Wynn and Sons Limited was the principal factor contributing to the disaster.'

The Chief Constable of Staffordshire did not escape criticism, Mr Gibbens noting that, 'according to the evidence, police officers in Staffordshire knew less about the brisk operation of these crossings than the local schoolchildren.' Although by the end of July 1967 there were seven automatic crossings the police had not appreciated the hazards that they presented. Information as to the crossing had, however, not been well presented to them.

British Railways were aware of the dangers of very slow exceptional loads but considered that the warning notice at the crossing and the provision of a telephone were sufficient to deal with the hazard. Their main failures were not impressing upon Robert Wynn and Company after the Leominster incident the imperative need for the drivers of heavy vehicles to telephone the signalman and their failure to send full information on automatic crossings to the haulage companies and to English Electric.

The Ministry of Transport accepted their collective responsibility for the disaster. It seemed astonishing that the problem of slow-moving vehicles had not occurred to anyone at the Ministry. There were however 13,000,000 registered road vehicles in Great Britain but only a few abnormal loads, and most of these were not as slow as the transporter. The Ministry had relied too much on statistics that indicated vehicles were unlikely to stall on crossings and, probably more importantly, had been keen to keep the waiting time for motorists to a minimum to prevent drivers from zig-zagging around the barriers, without considering the time needed for a slow-moving vehicle to clear the crossing.

The necessity of using the telephone at the crossing had not been stressed.

It was not clear what purpose was intended by sending the Requirements for Crossings document to lay bodies.

The Ministry, more than British Railways, should have sent information directly to the road haulage companies.

The wording of the notice at the crossing: "IN EMERGENCY or before crossing with exceptional or heavy loads or cattle PHONE SIGNALMAN," was wholly inadequate and required revision.

Mr Gibbens concluded, 'When the Ministry and British Railways decided that the responsibility for safety at level-crossings should be transferred from the Railways to the public on the road, it behoved them to make sure that the public knew of that change and knew what they had to do. There can be no doubt that, in collaboration with the Railways, Mr. Scott-Malden and the officers of the Railway Inspectorate gave the most conscientious and careful thought to all questions of safety at these new level-crossings and it is, therefore, tragic that the proper way of dealing with and eliminating the comparatively infrequent hazard of the slow-moving abnormal vehicle escaped them. Such oversights may be inexplicable, but they are not unknown. Therein lay the origin of an accident which was both foreseeable and avoidable.'

The report continued with general considerations regarding automatic half barrier crossings, a general review of their safety and hazards, possible methods of improving their safety and publicity regarding such crossings, site inspections and recommendations. This part of the report occupied another 45 pages (the whole report is 170 pages long) and is therefore too long to be reproduced here, instead we will content ourselves with the summary of the findings of the report and its recommendations.

Findings:

(1) The immediate cause of the accident at the automatic crossing at Hixon on 6th January 1968, was that the driver of a huge transporter vehicle, 148 feet long, carrying a load of 120 tons, failed to comply with a notice erected near the stop line on the approach road directing him to telephone the signalman before attempting to

cross, and drove the vehicle across the railway at two miles per hour when the arrival of an express train at 75 miles per hour was imminent.

(2) The driver of the transporter, Mr. B. H. Groves, did not know of the existence of such a notice nor of the need to telephone for permission to cross the line, but he or his statutory attendant, Mr. R. C. Parsons, ought to have seen it and complied with it.

(3) Mr. Groves knew the crossing was automatically operated by trains but he did not pause to think before venturing upon the crossing about the possibility of the arrival of a train. The responsibility for his neglect and ignorance belongs principally to his employers, Robert Wynn and Sons Limited. Those employers were gravely at fault for failing to discover the proper procedure for their heavy vehicles when using automatic crossings and to instruct their drivers accordingly. That failure by the directors of Robert Wynn and Sons Limited was the principal factor contributing to the disaster.

(4) The two police constables escorting the transporter, P.C. Prince, and P.C. Nicholls, were almost wholly ignorant of automatic crossings and the way they worked. They had not observed the Emergency Notice and were not responsible for Mr. Grove's failure to comply with the telephone procedure, but they were at fault for failing to consider intelligently whether such a vehicle could safely cross the railway, and to make inquiries.

(5) The Chief Constable and senior officers of the Staffordshire Police failed to instruct their constables about the working of automatic crossings, and especially of the need for drivers of heavy vehicles to use the telephone procedure, because they themselves were ignorant of those matters. They had failed to consider the possible hazards presented by automatic crossings already working in North Staffordshire and had not read with care the explanation contained in the Requirements and Explanatory Note which were sent to them. They were less to be expected readily to relate those hazards to slow abnormal vehicles than Robert Wynn and Sons Limited or British Railways or the Ministry of Transport but nonetheless were at fault in failing to do so.

(6) Since early 1964 officials of British Railways knew of the hazard of large vehicles which would be unable to clear an automatic crossing within 24 seconds but assumed that the Emergency Notice, which had been drafted and approved by the Ministry, was sufficient warning to drivers of such vehicles.

(7) The principal faults of British Railways contributing to the accident were:

(a) Their failure when replying to a letter of the 19th November 1966, relating to an alarming incident at an automatic crossing at Leominster, to inform Robert Wynn and Sons Limited of the imperative necessity for drivers of heavy transport to comply with the telephone procedure, and

(b) Having undertaken to conduct the necessary local publicity campaigns, they failed to send any information direct to heavy haulage contractors.

(8) The Ministry of Transport failed to consider the individual types of traffic that must be catered for, and so failed to consider the need for special precautions for the abnormal slow-moving vehicle.

(9) In July 1966 the Ministry took a policy decision which was wrong, and which directly contributed to the accident at Hixon, namely that cautions relating to automatic crossings should not he inserted in Special Order routes.

(10) The Ministry of Transport's attempt to convey information about the working of automatic crossings to a number of lay bodies, including the police, by circulating the technical Requirements, though accompanied by an Explanatory Note, was inept, and should have been recognised as such. The Explanatory Note, which was expected to be understood by lay addressees, made no mention of the obligation on drivers of abnormal heavy vehicles to telephone the signalman.

(11) The Ministry is responsible for the fact that the Chief Constable of Staffordshire was not properly briefed by the Home Office about the significant matters which would concern constables escorting heavy loads.

(12) The Ministry drafted the Emergency Notice in such a form that it was likely not to attract the attention of drivers who ought to have complied with the telephone procedure.

(13) The origin of the accident was in the failure of officers of both the Ministry and British Railways in collaboration to appreciate the measures necessary to deal with a hazard of which they were aware.

Recommendations:

(1) Automatic half-barriers are a valuable answer to the needs of modern transport and can be made acceptably safe without the introduction of obstruction detection devices, linked with either full or partial signal protection. These devices, even if technically feasible, would be costly; and full protection would involve the loss of most of the benefits for road-traffic achieved by automatic half-barriers.

(2) The conversion of crossings to automatic working should be pressed forward at greater speed, concentrating on particular areas, so that the public become familiar with the new equipment.

(3) Automatic half-barriers should not be introduced at crossings with three or more railway tracks until further experience of automatic operation has been gained.

(4) Certain basic changes in the operation of all automatic crossings are necessary and should be put into effect as soon as possible:-

(i) Extension from 24 to 32 seconds of time before arrival of fastest train on the crossing;

(ii) Alteration of equipment so that if a second train would reach normal strike-in point within 20 seconds after the first train strikes out, the half-barriers will remain down and the tights continue to flash;

(iii) "Second train coming" notice to be illuminated from within.

(5) Zig-zagging to be discouraged by:-

(i) Installation of dual carriageways, where road is wide enough or where road traffic is sufficiently heavy to justify the expense;

(ii) Extended use of double white lines.

(6) Flashing red lights to be retained, with addition of preliminary amber warning light.

(7) Extension of use of yellow box markings.

(8) More stringent standards of road profile, to diminish risk of grounding.

(9) Extra precautions for slow-moving vehicles:-

(i) Special Order and V.R.1 movement orders to include a caution or direction that the driver must telephone the signalman for permission to cross and where there is a prescribed route the location of every automatic crossing to be indicated on it;

(ii) Mandatory traffic signs at crossing, requiring drivers of large vehicles (over 55 feet long, or 9 feet 6 inches wide, or 32 tons gross weight) or vehicles unable to cross at over 10 miles per hour, to telephone the signalman.

(10) Revised paragraph about automatic crossings for Highway Code.

(11) More severe penalties for drivers who zig-zag or fail to use the telephone procedure.

(12) Need for more intensive local publicity.

(13) Reconsideration of design of automatic crossings and associated road signs.

(14) Need for experiments with early warning road signal.

(15) Suggestions for improved procedure in relation to site meetings and inspections. [1014]

(16) The installation of new half barrier crossings was halted after the disaster until the publication of the findings of the inquiry. [1015]

The crossing was replaced by a bridge in 2002 at the cost of £2,000,000. [1016]

At 17:00 om the evening of Monday 29th January 1968 a collision occurred between a taxi, registration 51-88-30, and train U-93, of the Estrada de Ferro Santos-Jundiaí on a level crossing with Avenida Santa Marina in Água Branca, Brazil, a suburb of São Paulo. The taxi driver, Ivo Rodas, and four passengers in the taxi were killed, the taxi being badly mangled and engulfed in flames after the collision. Two other people, either in the taxi or on the train, were badly hurt. The train was being driven by José de Azevedo. A newspaper report of the accident includes its supposed cause but the text is unfortunately unreadable. [1017]

Indonesia suffered a major level crossing disaster on Wednesday 6th March 1968 near Tebing Tinggi in north Sumatra. A passenger train collided with a bus, killing 26 people on the crowded vehicle, and injuring another 35, four seriously. It was initially unclear if there were any casualties aboard the train. The accident was reportedly due to the brakes of the bus failing to work. [1018] [1019]

[1014] Report of the Public Inquiry into the Accident at Hixon Level Crossing on 6th January 1968
[1015] *The Times*, London, England, 10th January 1968
[1016] Wikipedia – Hixon rail crash.
[1017] *Folha de S Paulo*, São Paulo, Brazil, 30th January 1968
[1018] *The Times*, London, England, 7th March 1968
[1019] *Farmington Daily Times*, Farmington, New Mexico, United States, 6th March 1968

Ten people died and another 43 were injured in India at about 03:00 on Saturday 9th March 1968 when train No. 44 on the South Central Railway, the *Circar Express*, running between Kakinada and Madras (renamed Chennai 1998) collided with a tourist bus on a level crossing at Appayya Gate, between Venkatachallum and Manoboholu stations.

There were four tracks at this point and the crossing was provided with gates or barriers worked by a crossing keeper. According to one account, one of the gates was closed and the crossing keeper was proceeding over the tracks to close the other when the bus entered the crossing. 'Words were exchanged' between the crossing keeper and the bus driver, perhaps the crossing keeper telling the bus driver to go back and the bus driver refusing to do so, when the train arrived and hit the vehicle.

Seven people, including a child, were killed on the spot. Lorries took 24 of the injured to Gudur hospital and 22 to Nellore hospital. Three of the injured died on the way.

Three hours after the accident a medical relief van reached the scene from Bitregunta. The Chief Operating Superintendent and Chief Medical Officer of the South Central Railway also went to the scene by car from Secunderabad while the police opened an investigation. [1020]

Two more of the injured bus passengers had died by the following day. The railway ordered assistance of 400 rupees to be paid to the seriously injured and 200 rupees to the others.

The bus driver surrendered to the police and the crossing keeper was taken into custody. [1021]

The following day Railway Minister C.M. Poonacha stated that the level crossing had been "open to {road} traffic at the time" of the accident, that the signal had been on, and that the train had not been "supposed to pass at that time," implying a mistake by the train driver. [1022]

A mother and five of her children were killed when an eastbound Penn Central Railroad passenger train from Chicago to Cleveland and New York struck the car which she was driving on a level crossing with County Road D between Archbold and Pettisville, Ohio, United States, on Monday 1st April 1968. The victims were Mrs Barbara Anne Foxe, 29, of Holgate, and her children Scott, 8; Shane, 5; Shannon, 3; Tommy, 32; and Earl Jr., 7 months. Another child, Andrew, 8, was at school at the time.

Mr Earl Foxe, the husband and father, was working nearby with a telephone company crew. They all witnessed the accident and rushed to the crossing. It was believed that Mrs Foxe had been on her way to meet her husband at the time.

Engineer R.G. Bowman said that his train had been travelling at about 80 miles an hour and that the car had started across the tracks without stopping. The train caught the rear of the car, carried it about a mile down the track and the car had then burst into flame. No one on the train was hurt. [1023] [1024]

[1020] *The Sunday Standard*, Madras, India, 10th March 1968
[1021] *The Indian Express*, Madras, India, 11th March 1968
[1022] Ibid, 12th March 1968
[1023] *East Liverpool Review*, East Liverpool, Ohio, United States, 2nd April 1968

While the inquiry into the Hixon disaster was still in progress another serious collision occurred in England on an automatic half barrier crossing, though under different circumstances to that at Hixon. The location of this accident, which occurred at about 11:13 on Tuesday 16th April 1968, was Trent Road Crossing, Beckingham, on British Railways, Eastern Region. A car was involved this time, an old Austin A40 saloon, and its driver and four passengers died in the accident.

The car driver, Mr John Hilton, 50, a licensee and chef from Blyth, Nottinghamshire, was taking his children, Sheena, 13; Irvine, 9; and Stewart, 8; and his mother-in-law, Mrs Maria Moody, on a day trip to Cleethorpes. [1025] When he arrived at the crossing, on the A631 main road, the warning lights were flashing and he stopped his car short of the crossing barrier, his being the leading vehicle in a short queue. A Down goods train passed, the barriers rose and Mr Hilton moved forward onto the crossing. Almost at once the crossing alarms began to sound again for an Up express, the 10:00 from York to Yarmouth Vauxhall, 1E63, and it was this point that the car stalled on the crossing. Mr Hilton apparently first tried to restart it, then to push it off the crossing, and was still trying to do so when it was struck by the train.

The accident was the subject of a Ministry of Transport inquiry, conducted by Colonel J.R.H. Robertson.

The report on the accident showed that the crossing was a busy one. A census taken after the accident, over a 12-hour period between 07:00 and 19:00, showed rail traffic of six passenger trains and 36 goods trains and light engines, while traffic on the main road from Rotherham to Louth comprised no less than 7,597 cars, lorries and buses, 174 motorcycles, 58 bicycles and 17 pedestrians.

Several of the motorists who had witnessed the accident gave evidence at the inquiry.

Mr R. Charman, a grocery manager of Misterton, said that his car had been stopped on the Gainsborough (Up) side of the crossing behind a lorry when the lights began to flash for the goods train. That train passed, the lorry began to move slowly forward, the warning lights started again and the barrier descended, just missing the back of the lorry. As the lorry cleared the crossing he had seen for the first time the A40 car stopped on the crossing. Its driver had scrambled out and tried to push it clear, then the express had arrived and swept it away.

Mr N. Eggett, departmental manager of the Becker Electric Company, Rotherham, said that he had been following the A40 from the Beckenham (Down) side. The A40 was being driven carefully and well and had stopped without hesitation at the flashing warning lights of the crossing. The barriers had gone up as soon as the goods train had passed and the A40 had moved forward, apparently normally. He had then turned his attention to the lorry starting off from the other other side of the line, had prepared to follow the A40 over the crossing and at that point the warning lights had begun to flash again and he had remarked to his passenger that the lorry had better be quick. He had then seen that the A40 was stationary on the line and its driver had got out and had given it a "tremendous heave – he really put his weight in it." The car had not moved. Its driver

---

[1024] *The Times* Recorder, Zanesville, Ohio, United States, 2nd April 1968
[1025] *The Times*, London, England, 17th April 1968

had then looked up at the approaching express, given his car another heave and the train had struck it. Mr Eggett said that he had then driven over the crossing and reported the accident to the Beckingham signalman, who had quickly answered the phone.

Mr R. Adams, a Beckingham builder, was in a minivan behind Mr Charman. His impression was that the A40 had entered the crossing as the barriers had begun to descend for the second train. He thought that the car had faltered in its progress and stalled as its driver had seen the Up side barrier come down and that the driver had first tried to restart it, then jumped out and had attempted to push it clear, and that he had moved it about a foot.

Mr G.D. Pinchbeck, who owned the restaurant where Mr Hilton worked, said that Mr Hilton had bought the A40 about three months earlier. Two days before the accident Mr Hilton had been tinkering with the car and had asked for Mr Pinchbeck's help, saying that he did not know much about cars and needed to go to Scunthorpe. He (Mr Pinchbeck) had examined thar car and had told Mr Hilton that the starter motor was definitely defective and would be most unlikely to restart the car, except by chance, while repairing it would be expensive. Mr Hilton had replied that the car would not be worth the expense of the repair, whereby he had assumed that Mr Hilton would hire a car for the journey instead.

Mr E. Mound, a handyman at the restaurant, said that he had helped Mr Hilton start the A40 on the morning of the accident. Attempts to start the car with the starter motor and starting handle having failed, he had been able to start it by depressing the accelerator while Mr Hilton turned over the engine by hand. In Mr Mound's opinion the car could not have been quickly restarted once it had stalled, even with the starting handle.

Police Constable W. Webster, who was a motor vehicle examiner, said he had examined the wreckage of the A40. He described the car as not having been in good condition or well maintained. It had several defects and in his opinion there were several reasons why it might have stalled. Once it had done so the chances of restarting it would have been remote.

Relief Signalman C.F. Pickering at Beckingham stated that he had accepted both the Down Empties and the Up express at 11:08, had closed his own level crossing gates and cleared his signals for the two trains, which had passed one another near his box. From his indicator (he monitored the Trent Road barriers) he saw the barriers descend for the goods train and rise again after it had passed, then a lorry crossed and a car moved onto the crossing. His view was then hidden by the express. He had heard the express whistle once, then it was whistling again continuously.

Driver A. Grundy, of the Down Empties, 7B26, stated that he had been travelling at about 30 miles an hour. He had only heard of the accident when he reached Park Drain, four signal boxes down the line.

Goods Guard Muhammed Azad Khan agreed as to the speed of his train. He said that when he had heard the express whistling repeatedly he had gone out onto the veranda of his brake van and had seen the car move onto the crossing but had not seen it stop before it was hidden by the express.

Driver G. McMahon of the express stated that it had been travelling at 58 miles an hour past Beckingham. His secondman had remarked that the lorry "was leaving it a bit late" using the crossing. The A40 he had not seen at first because it had been obscured by the goods train. He thought that the car had first stopped across the Down line, restarted, then had stopped again across the Up line. Its driver had then got out and looked towards his train. By then he had been applying the brakes in emergency while sounding his horn continuously. He thought that the brakes had begun to take effect before the collision.

Passed Fireman J.G. Walker was secondman to Driver McMahon. He said he had not been watching the speedometer but put the speed of his train at 55 to 60 miles an hour. He also was of the impression that the car had stopped twice and that its driver had been pushing it when it was struck.

Colonel Robertson concluded that Mr Hilton had driven onto the crossing slowly as soon as the barriers had lifted after the goods train and before he would have been able to see the warning lights begin to flash again for the express. He had finally stalled the car when he had seen the Up side crossing barrier begin to descend. He had then been unable to restart his car using the starting motor and had no time to push the car clear. He had no chance to save his passengers and it was to his credit that he had not abandoned them to save himself.

The cause of the accident was that an old car in poor condition was driven onto the level crossing under "critical second train situation" conditions, was stalled across the line in the path of the closely approaching second train, and that the car's starting arrangements were defective, so that, once stalled, it could not be started again. The barrier equipment had worked correctly and the driver of the express was in no way to blame. [1026]

An inquest on the victims returned a verdict of Accidental Death. [1027]

Early on the morning of Thursday 16th May 1968 a private bus on its way from Ashta to Shajapur carrying 70 passengers had almost made it across an unmanned level crossing near the station at Brecha, India, when it was struck by a speeding express with disastrous results. According to unofficial reports seven men, seven women and six children were killed outright and 43 were injured, 15 grievously. Five more victims were said to have died on the way to hospital and another five after arrival at Shajapur hospital. Officially it was stated that there had been 27 deaths. All the victims had been on the bus; the bus conductor was among the dead and its driver was missing. Some of the bus passengers had been on their way to a wedding. Three other passengers were two police constables and a murder suspect. The accused person was killed, the two constables escaped with minor injuries.

Rail traffic was not disturbed and the train involved in the collision took the injured to Shajapur and Ujjain. 'First aid was mobilised through doctors from Ercha even as medical aid was being rushed from Ujjain, about 70 kms away, from Shajapur and from Bhopal.' Railway medical and district officers also rushed to the site.

Mr G.H. Khosla, general manager of the Western Railway, cut short an inspection tour of the Bulsar – Surat section to go direct to Ujjain and the Chief Medical Officer of the railway flew from Bombay to Indore en route to Ujjain. [1028]

[1026] Ministry of Transport report
[1027] *The Times*, London, England, 25th April 1968
[1028] *The Indian Express*, Madras, India, 17th May 1968

Another disastrous level crossing collision involving a bus carrying migrant farm workers occurred in the United States at 6.20 p.m. on Tuesday 23rd July 1968. The location was Malaga Road Crossing, Newtonville, New Jersey. The 1953 Mack old school bus had been bringing 32 workers back to Chester, Pennsylvania, from a day spent blueberry picking at Crescent Farms, May's Landing, New Jersey, and was being driven by Love Clark, aged 33. The southbound train was a Pennsylvania-Reading Seashore Lines passenger train from Philadelphia to locations on the New Jersey Atlantic shore which was carrying about 150 passengers. The two-car diesel train struck the rear of the bus as it was crossing the single track and carried it 200 feet. Nine bus passengers were killed and the remainder were injured, four seriously. Five of the dead were children aged between 11 and 16 who had been sat at the rear of the bus. The injured were taken to Kessler Memorial Hospital in Hammerton, the Newcomb hospital in Vineland and the Atlantric City Hospital. [1029]

The crossing was unprotected except by warning signs, though there was a good view along the line and the weather was clear. The bus had a capacity of 33 passengers and so was not overloaded.

A newspaper quoted State Trooper Andrew Robertson, who had arrived at the scene five minutes after the accident. "Seven bodies were strewn along the railroad bed, some of them hidden in the bushes. Two of the bodies, including the Tweedy girl, were imbedded on the front of the train."

The newspaper continued, 'Mrs George Costen, said today her sons were picking berries to earn spening money, "and to help get their school clothes. They didn't go every day, but the majority of the time. There are 15 children in my family, so they helped with the clothing. Eugene was making his little money, and spending it as fast as he got it – you know how kids spend money," she said.

'Mrs Costen was not notified in the accident. "I heard the news on television and when they were not home by 11 o'clock I got worried," she said. The mother learned of the deaths after calling Kessler hospital, she said.

'Trooper Robertson said Clark, who was treated for minor injuries, was carrying a Delaware {county, Pennsylvania} driving licence which had expired in March of this year.

'The bus was owned by Charles Bradley, of 460 Bradley St., Chester, police said. Bradley said today he usually drives the bus and transports and transports the farm labor to various locations in New Jersey, but missed the trip Tuesday because he was ill. Asked if he was aware that Clark's licence had expired, he said, "No, I wasn't. He showed me a chauffeur's licence about a month ago, but I didn't know it was expired."

'According to Robertson, there were four other children injured on the bus. Bradley said he was "surprised" there were so many youngsters on the bus because, "I usually don't carry children that much."

'Robertson said Charles Heck, senior inspector for the New Jersey Department of Labor's Bureau of Migrant Labor, would investigate the accident.

'(A *Daily Times* reporter recently posed as a migrant laborer at a New Jersey

---

[1029] *The Titusville Herald*, Titusville, Pennsylvania, United States, 24th July 1968 blueberry farm and found violations of wage and child labor laws. The reporter found children not even 10 years old picking berries, although state law says they must be 12, and workers receiving only half the minimum wage. Some workers were forced to sit on crates in the bus which took the workers to New Jersey).

'"The driver (Clark) told me he stopped, and I was told by someone else he didn't come to a complete stop," said Robertson.

'Police could give no estimate on the speed of the train, but said the crossing is located in a 70 mile an hour zone. Police were to receive statements today from the train's engineer and Clark.' [1030]

Engineer David Morris stated the next day that when he had seen the bus on the track he attempted to stop but the distance had been too short. [1031]

Bus driver Clark and owner Bradley were both arrested and charged with various offences related to the accident before being released on bail. [1032]

Investigations into the accident were begun by the New Jersey State Police, the state Public Utilities Commission, the migrant labor division, and the division of motor vehicles. [1033]

In late August Charles Bradley was fined $300 on four motor vehicle charges [1034] and the following day another $225 on two further charges. [1035] At this date driver Clark was still awaiting trial on a charge of causing 'death by auto.' It has not been possible to trace any report of his trial.

An accident at Scott's Crossing, near Corry, Pennsylvania, United States, on Saturday 3rd August 1968 resulted from the fatal decision of a car driver to enter the crossing before his exit was clear. On the far side of the crossing, the road that passed over the railway line made a T junction with a main road, highway US 6, and there was only room for one vehicle between the line and the main road. When a car driven by Mr Walter Sturzman, 31, of Corry, entered the crossing there was already a car waiting at the stop sign on the far side of it to enter the main road. Before the car ahead could do so the warning lights of the crossing began to flash and shortly afterwards an Erie-Lackawanna Railroad freight train arrived. Engineer George Jeshner, seeing the car, began to sound his whistle and then, when the car did not move, applied his brakes, but could not, of course, stop his train short of the crossing. Sturzman, unable to go forward, attempted to move his car parallel to the line and stalled the vehicle in the process.

Mrs Sturzman, 27; the Sturzmans' children Brian, 5; Brenda, 3; and James, nine months; and Mrs Sturzman's mother, Mrs Norman Gleason, 60, were all killed. Mr Sturzman was taken to Corry Memorial Hospital with severe head injuries. Mrs Gleason had had time to get out of the car and was attempting to get the children out when the

---

1030 *Delaware County Daily Times*, Upper Darby, Pennsylvania, United States, 24th July 1968
1031 Ibid, 25th July 1968
1032 Ibid, 27th July 1968
1033 Ibid, 31st July 1968
1034 Ibid, 22nd August 1968
1035 Ibid, 23rd August 1968

train struck it. [1036]

On Wednesday 7th August 1968 a bus sponsored by the Young Men's Christian Association, which was returning from a sightseeing tour of the Panama Canal and was carrying 46 people – American soldiers and their Panamanian dates, was struck by a Ferrocarril de Panamá engine on a level crossing at Diablo Heights, Panama. Initial reports spoke of seven dead and 39 injured, all bus passengers. [1037] Pentagon sources in Washington D.C. would only admit to one death and many injured, [1038] but two days later it was stated that eight persons had died [1039] and a week after the accident it was announced that Airman Ronald Richards of Mount Vernon, Ohio, had died of his injuries, making a ninth fatality. [1040]

A woman and four of her children were killed in an accident on an ungated level crossing near Yutan, Nebraska, United States, on the night of Wednesday 4th September 1968. The car driven by Mrs Cecil Jordan, 33, was struck by a southbound Chicago Burlington and Quincy Railroad freight train consisting of an engine and 13 cars. Engineer R.H. Burns stated that his train had been travelling at 41 miles an hour at the time. He had whistled for the crossing and the crossing signals had been working.

Mrs Jordan had been driving westbound on Highway 92-30A and it was thought that she had been blinded by the sun and failed to see either the crossing lights or the train in time to stop. According to a witness the brake lights of her car had gone on at the last moment and it had swerved before colliding with the second car of the train. The train had continued for 1,300 feet before it could pull up.

As well as Mrs Jordan her children Michael Ray, 10; Debra, 7; Julie Annette, 5; and Marty, 2, lost their lives. Taken to St Elizabeth Hospital in Lincoln, Nebraska, were Randy Lee, 9, who was critically hurt; Lisa Ann, 3, who was seriously injured; and Kimberly, 13 months, who was only slightly hurt. The family had apparently been on the way to Wahoo, Nebraska, where they had lived until recently. [1041]

Five Michigan teenagers and young men on their way home from a football game at Vicksburg High School were killed early on Saturday 5th October 1968 when the station waggon in which they were riding collided with a Grand Trunk Western Railroad freight train on a crossing near Schoolcraft, Michigan, United States. The driver of the vehicle apparently had attempted to stop but skidded into the path of the train, which was travelling at 60 miles an hour and the engineer of which was, of course, also trying to pull up.

"It didn't carry the car down the track. It more or less picked it up and threw it down," a witness said.

[1036] *The Titusville Herald*, Titusville, Pennsylvania, United States, 5th August 1968
[1037] *The Daily Telegram*, Eau Claire, Wisconsin, United States, 8th August 1968
[1038] *The Lincoln Star*, Lincoln, Nebraska, United States, 9th August 1968
[1039] *Washington C.H. Record-Herald*, Washington C.H., Ohio, United States, 10th August 1968
[1040] *Xenia Daily Gazette*, Xenia, Ohio, United States, 15th August 1968
[1041] *Lincoln Evening Journal*, Lincoln, Nebraska, United States, 5th September 1968

The victims were later named as Evan Erkkila, 16, of Scotts; Edward Turk, 20; Owen Green, 19; and Patrick Pinkerton, 17, all of Vicksburg; and Keith Thomas, 18, of Leonidas. [1042] [1043] [1044]

On about Wednesday 16th October 1968 a Philippine National Railways multiple-unit passenger train collided with a small car, specifically a Volkswagen Beetle, on an open level crossing near the station at Calamba, Laguna Province, the Philippines. The Volkswagen had overtaken a jeepney that was already stopped at the crossing for the train. Eight people in the car were killed. [1045]

Only a week after the Calamba accident, on Tuesday 22nd October 1968, a speeding Jeep serving as a school bus was struck by a train on a level crossing at Iloilo, the Philippines. Two students were killed and 12 injured in this second accident. [1046]

Two high school students, Shelley Riel, 18, a Grade 12 student, and Shirley Cunningham, 13, a Grade 9 student, lost their lives when their school bus was struck by a Canadian National Railway freight train on a level crossing on Lily Lake Road, five miles from Peterborough, Ontario, Canada, at 8.45 a.m. on the morning of Thursday 21st November 1968. Another student was critically hurt, three were said to be in a fair condition and others suffered broken bones, cuts, and bruises. They were taken to St Joseph's General Hospital in Peterborough, where the two who died had also been taken, only to succumb shortly after admission. There had been about 39 people on the bus, which was bound for Crestwood Secondary School. The train, travelling at about 35 miles an hour, had been on its way from Peterborough to Lindsay, Ontario. [1047]

Bus driver Clem Riley was shaken but unhurt. Afterwards, he said that he had picked up two children about 100 yards short of the crossing and then started up the hill towards it, adding, "At this time of the year you have to take a run at it because the road is so slippery. If there was a whistle I didn't hear it because the kids were making so much noise. I didn't see the train until it was almost on top of me."

Student Larry Norrad, 17, said, "There was just this bump and then everyone was screaming."

'Student Peter Cleary,' a newspaper reported, 'shouted a warning as the train neared, fell out of the back emergency door, and then returned to help others out.

'Al Porteous, city editor of the *Peterborough Examiner*, arrived at the scene after the collision to find the bus draped around the front of the train. The diesel locomotive stopped above a steep embankment about half a mile from the crossing. "On one side the embankment drops about 75 feet to a swamp," he said. "It was a godsend the bus wasn't

---

[1042] *The Herald-Press*, St Joseph, Michigan, United States, 5th October 1968
[1043] *The Holland Evening Sentinel*, Holland, Michigan, United States, 5th October 1968
[1044] *Marshall Evening Chronicle*, Marshall, Michigan, United States, 8th October 1968
[1045] https://www.facebook.com/PTERCOfficial/photos/a.2604928076422903/1983750001874050/
[1046] *Pacific Stars and Stripes*, Tokyo, Japan, 24th October 1968
[1047] *The Brandon Sun*, Brandon, Manitoba, Canada, 21st November 1968

flipped over the embankment."

'Police used axes to rescue a half dozen students trapped by twisted seats in the crushed left side of the bus.

'Train crew members were uninjured.' [1048]

Three couples, two from Wisconsin and one from Minnesota, were killed when the car in which they were riding was struck by a fast-moving Northern Pacific Railroad freight train on a level crossing with County Road 9 at Harris, Minnesota, United States, at 11.50 p.m. on Saturday 30th November 1968. Those who died were Leo Parmeter, 23, and his wife Patricia, 20, of St Croix Falls, Wisconsin; Carl Hacken, 24; and Dorothy Hockenson, 17, his fiancée (in some reports referred to as his wife), of Eureka Center, Wisconsin; and Wayne Parmeter (Leo's brother or cousin according to different reports), 27; and his wife Lois, 23, of Taylor's Falls, Minnesota, across the St Croix River from St Croix Falls. The car was crushed and carried 3,000 feet from the unprotected crossing. [1049] [1050] [1051] [1052]

The *North Coast Daylight Express* collided with a school bus taking children home from Kempsey High School and Kempsey Convent School on Middleton Street Crossing, South Kempsey, New South Wales, Australia, at about 4 p.m. on Monday 9th December 1968. Pupils Margaret Lynne Barnett, 12, of Belgrave Falls; Rosemary Dawn Lawrence, 14, of Kalateena; Sharon Leone Perrott, 14, of Kempsey; and Rhonda Joy Searle, 13, of Sherwood, were all killed more or less outright and 15 children, the 36-year-old bus driver and a 60-year-old male passenger were admitted to hospital. Graham Burley, 15, died of his injuries on Tuesday 17th and Patricia Ramsey, 16, became the sixth fatality on Thursday 19th, after she had been flown to Sydney for specialist treatment.

The accident was the result of a mistake by Crossing Keeper Desmond Stewart, who was subsequently arrested on six charges of manslaughter and eight of causing grievous bodily harm. Meanwhile a deputation from Kempsey Municipal Council, led by the Mayor, Alderman Melville, urged Mr Morris, the Minister for Transport, to replace the crossing with an overbridge, a request that was apparently agreed to. [1053]

At his trial in April 1969 it emerged that Crossing Keeper Stewart had had no warning of approaching trains other than the times set down in the working timetable and by looking out and listening for them. On the day of the accident he had failed to notice the approaching express and hence to close the gates against the school bus. Driver Grantley Small told the court that an embankment had obscured his view of the bus but when his fireman had shouted upon seeing the bus from his side of the engine, he had applied the brakes in emergency, while the fireman had sounded the horn.

[1048] *Daily Colonist*, Victoria, British Columbia, Canada, 22nd November 1968

[1049] *The Brainerd Daily Dispatch*, Brainerd, Minnesota, United States, 2nd December 1968

[1050] *Wisconsin State Journal*, Madison, Wisconsin, United States, 2nd December 1968

[1051] *Waukesha Freeman*, Waukesha, Wisconsin, United States, 2nd December 1968

[1052] *Janesville Daily Gazette*, Janesville, Wisconsin, United States, 2nd December 1968

[1053] *The Canberra Times*, Canberra, Australian Capital Territory, Australia, 10th, 14th, 18th and 19th December 1968

At the conclusion of the court hearing Stipendiary Magistrate J. Featherstone ruled that although Stewart was in some degree responsible for the accident his actions did not amount to criminal negligence and the case was dismissed. [1054]

Nine people died and 23 were injured when a train collided with a bus on a level crossing at a location given as 75 miles north of Bangkok, possibly in the vicinity of Lop Buri, Thailand, on the Northern line on Sunday 29th December 1968. The bus driver and six of its passengers were killed outright. Eight cars of the train were derailed. [1055] [1056]

Eleven farm labourers were killed and 12 more were injured when the lorry in which they were riding was struck by a fast-moving freight train on a level crossing in Afyon, Turkey, 200 miles south-east of Istanbul, on New Year's Day, Wednesday 1st January 1969. [1057]

On the night of Wednesday 8th January 1969 a train coming from Chechon and travelling at about 30 miles an hour collided with a crowded bus on a level crossing in the eastern part of Seoul, South Korea. The bus caught fire following the collision and two of its passengers were trapped inside and burnt to death. In all, 18 bus passengers had died by the next day and 71 were injured, ten of them critically. The train dragged the bus 70 metres. Initially the cause of the accident was variously stated as the crossing gate having been lowered too late and the bus having failed to stop. By the following day it was stated that the brakes of the bus had failed.

While the investigation into the Seoul accident was still in progress a second bus collided with an Iri to Taejon train on about 10th January on a crossing near Kanggyong, 160 kilometres south of Seoul, injuring 20 bus passengers.

Three people had died in an accident at the Seoul crossing in 1966 and a train had collided with a taxi on the same crossing in December 1968. [1058]

A dump truck carrying workers from the Federal Electricity Commission on their way to a Volkswagen car plant was struck by an engine at Puebla, Mexico, on Thursday 30th January 1969, killing 15 of them and injuring ten more. Following the collision the engine was derailed and struck a bridge. [1059]

Twenty-nine people died and 11 were injured when a diesel railcar collided with a bus on a level crossing near Raiwind Junction, Pakistan, about 50 kilometres from Lahore, on

[1054] Ibid, 16th April 1969

[1055] *Traverse City Record-Eagle*, Traverse City, Michigan, United States, 31st December 1968

[1056] *Eastern Sun*, Singapore, 31st December 1968

[1057] *Chicago Tribune*, Chicago, Illinois, United States, 2nd January 1969

[1058] *Pacific Stars and Stripes*, Tokyo, Japan, 10th and 12th January 1969

[1059] *The Daily Courier*, Connellsville, Pennsylvania, United States, 1st February 1969

Monday 3rd February 1969. [1060]

Bad weather, with a considerable snowfall, contributed to a level crossing involving two trains and a van at Hogebrug, the Netherlands, between Gouda and Oudewater, at 10:45 on Sunday 9th February 1969. Two families were on their way to the Roman Catholic church in nearby Oudewater. Driving the van was 39-year-old casual labourer C. Schoonderwroed, who was accompanied by his wife, also 39, their children Mary, 14; Nelleke, 11; and Adje, 7; and their neighbours, A. van der Marel, 40; and his wife, who was 35.

Normally the two families, who lived less than 100 metres from the crossing, went to church on a Saturday evening but because of the bad weather they had delayed their trip until the Sunday morning. The crossing was protected by flashing lights and a warning bell but had no barriers. The occupants of the van knew the times that all the regular trains passed over the crossing, however because of the severe weather conditions the trains were running late. As the van arrived at the crossing none of its occupants apparently heard the warning bell, which was shown to have been working properly after the accident, while the flashing lights were obscured by a blanket of snow. The van was driven onto the crossing directly in front of an express coming from Gouda that was travelling at 130 kilometres an hour. Thrown onto the opposite running line by the impact, the van was then struck in turn by a train coming from Utrecht. The driver of the latter said afterwards, "It was like an explosion. I think my train jumped half a metre up into the air. It was a miracle that we were not thrown off the line. If we had been the trains would have collided with one another."

Six of the occupants of the van were killed outright and Mevr Schoonderwroed succumbed to her injuries a few hours later. No one on the trains was hurt. A police officer spoke of the scene of the accident as being like a battlefield and said that he had never seen such a sight before. Wreckage of the van was scattered over hundreds of metres.

The line was blocked until 13:30, when single-line working was begun. The two trains involved were pulled backwards clear of the crossing and normal operation resumed at 15:15. Damage to the trains was estimated at several tens of thousands of guilders. [1061]

A 77-car Chicago Milwaukee St Paul and Pacific Railroad freight train collided with a school bus on a level crossing half a mile east of Chillicothe, Missouri, United States on the morning of Monday 31st March 1969. Fortunately the bus was only beginning its run to pick up pupils and only its driver, Hubert Stewart, 47, as on board at the time. The bus was carried 60 feet and demolished; Stewart suffered a broken leg and other injuries and was admitted to hospital.

Engineer John Houchard said that his train had been travelling at about 40 miles an hour at the time of the collision. The front of the locomotive was somewhat damaged. [1062]

At 07:32 on Monday 21st April 1969 a train on the Lausanne – Lyss line collided with

[1060] *The Indian Express*, Madras, India, 4th February 1969

[1061] *Algemeen Dagblad*, Rotterdam, The Netherlands, 10th February 1969

[1062] *Maryville Daily Forum*, Maryville, Missouri, United States, 31st March 1969 a lorry on a level crossing at Chiètres, Switzerland. The lorry driver was quite badly hurt and there was significant damage caused but this was only a prelude to a more serious level crossing accident on the same line only two hours later.

At 09:30 near Galmiz, Switzerland, a train collided on a level crossing with a lorry owned by the Fasel Frères company of Vuissens, which was loaded with wood. The force of the impact was sufficient to throw the engine from the track and Driver Hans Zehnder, 46; Driver Jean-Pierre Bieri, 43, who was piloting the engine; and Assistant Driver Juerg Kuhn, 44, were all killed, as were the two men on the lorry, Messrs Paul Fasel, 53; and Bernard Nicolet, 38. Both the engine and the lorry were destroyed and the line was blocked, passengers having to be ferried by bus between the stations on either side of the accident. [1063]

Tragedy struck a large wedding party on their way to Mathura, India, in the early hours of Monday 12th May 1969 when the bus in which they were riding was struck by train No. 356, a Bareilly – Agra passenger train, on a crossing 14 kilometres from Aligarth. The bus caught fire after the collision and 15 members of the party were killed more or less outright and six more succumbed after being taken to Aligargh civil hospital. Another 35 were injured, 16 of them seriously, including the bridegroom.

General Manager G.M. Beri of the Northern Railway stated that the bus had probably hit one of the side ports of the crossing while the crossing keeper was closing the gates. Before the bus could be freed the train had arrived at the crossing.

The railway authorities announced *ex gratia* payments to the families of each of the dead of 400 rupees of 200 rupees to those seriously hurt, and of 50 rupees to those slightly hurt, while an investigation was opened into the disaster. [1064]

At 08:27 on the morning of Tuesday 3rd June 1969 railbus No. Et 2827 travelling from Anrath to Krelfeld collided with a Krefelder Verkehrs AG bus on a level crossing with Bellenweg at Forsthaus Haltepunkt, West Germany, just west of Krefeld. Bus driver Hans de Wever was fatally injured and the six bus passengers, all women, killed outright, while ten train passengers were slightly injured by their train's sudden stop. One axle of the railbus, which was travelling at 110 kilometres an hour at the moment of the collision, was derailed. The bus was carried over 100 metres and destroyed and the signal box adjoining the crossing was struck and badly damaged. The signalman was another casualty, seriously injured and suffering from shock, while a cyclist who had passed over the crossing in the opposite direction was slightly hurt when he was struck by a piece of wreckage. The train driver, seeing the bus on the crossing, had applied his brakes in emergency and then fled into the passenger compartment. He escaped unhurt. There could have been more passengers on the bus but heavy rain had made a number of forestry workers postpone their start.

After the accident, the cyclist testified that the barriers had been down when he arrived at the crossing, a freight train had passed, the barriers had then been raised, he had started across and narrowly escaped being killed by the passenger train. The accident

---

[1063] *Gazette de Lausanne*, Lausanne, Switzerland, 22nd April 1969
[1064] *The Indian Express*, Madras, India, 14th May 1969

was clearly due, therefore, to the premature lifting of the barriers by the signalman. A new timetable had come into operation the previous day and the railbus no longer stopped at Fortshaus, as it had before. It was theorised that the signalman might have forgotten that the train no longer stopped there and thought that he had time to let traffic across between the two trains. However, he had cleared his signals for the train and it seems more likely that because he had been speaking on the telephone to a colleague at the time he had allowed himself to become distracted and automatically raised the barriers once the freight train had cleared the crossing. He appeared in court in February 1970 and in view of his previous unblemished service, the mental strain the accident had caused him and the fact that he had suffered a serious back injury in the accident, he escaped with a year's suspended sentence. [1065]

Late on the afternoon of Friday 20th June 1969 Engineer William Thomson was approaching a dirt road crossing near Vienna, Georgia, United States with a Southern Railway local freight train from Valdosta to Macon, consisting of two locomotives and eight cars. About a fifth of a mile from the crossing he saw a white automobile, a 1961 Ford sedan, coming along the road to the crossing, and when he had covered about half that distance he saw the car stop on the crossing. He at once applied his brakes in emergency but his light train was travelling at 40 miles an hour and he could not stop, in fact the car was dragged three-quarters of a mile before he could finally pull up.

There had been 12 people in the car, and ten of them, one adult and nine children, were killed. Six of those who died were all the children of Mrs Frances Lewis of Vienna, who was not a passenger in the car – Thelma, 12; Holsey, 9; Eric, 8; Morris, 6; Deborah, 5; and Wanda, 3. The others who died were Mrs Eula Mae Pitts, 24, from Miami, Florida; her daughter Jackie Lango, 8; and two children of Mrs Annie Mae Lowe, 27 – Sylvia, 8, and Vickie, 4. Mrs Lowe, who had been driving the car, and her daughter Faye, 6, were critically hurt and rushed to Crisp County Hospital.

Dooly County Sheriff H.C. Johnson said the train had hit the car dead centre and the bodies of most the victims had been scattered along the track. He did not know why no one had attempted to escape from the car when it stopped. He said, "I figure she {Mrs Lowe} probably got excited – panicked – and the car must have choked out on her. That train was just too close and there wasn't time. But we may never know what happened."

There was no protection at the crossing but there was a good view along the track in each direction. [1066] [1067] [1068]

A serious level crossing collision occurred in Czechoslovakia at 22:47 on the evening of Monday 28th July 1969. The accident happened at Bezděčín, in the Mladá Boslav district of Bohemia, about 50 kilometres north of Praha. The accident involved a ČSAD bus taking workers home from a factory late shift, 22 of whom were killed, as was the bus driver, and at least 20 more were injured, four seriously, when it was struck by a

1065 https://derforstwald.de/wp-content/uploads/2013/12/18_1989.pdf
1066 *The Herald-Tribune*, Sarasota, Florida, United States, 22nd June 1969
1067 *The Daily Independent*, Kannapolis, North Carolina, United States, 22nd June 1969
1068 *The Daily Times*, Salisbury, Maryland, United States, 24th June 1969

passenger train. The accident occurred on a gated level crossing and was apparently due to a mistake by the woman crossing keeper, who had failed to close the gates before the train arrived. [1069]

The Czech Communist Party newspaper *Rudé Právo* reported the bare details of the accident and the rescue work, then embarked on something of a diatribe about the irresponsibility of citizens in the workplace and the need for strict discipline among workers, especially those in occupations such as transport where individual carelessness could not only cost money but also lead to loss of life. [1070]

On, probably, Monday 4th August 1969, a goods train collided with a bus at an unstated location in the eastern Transvaal, South Africa. Six African bus passengers were killed and nine were seriously injured. [1071]

A Penn Central Railroad freight train from Maybrook, New York, to New Haven, Connecticut, consisting of four diesel units and 99 cars collided with a car on a rural level crossing south west of Highland, New York, United States, on the night of Sunday 28th September 1969. Four of the occupants of the car, two of them children, were killed more or less outright and a woman died later in hospital. The victims could not immediately be identified but the car bore Florida licence plates and they were believed to be migrant farm workers. The impact carried the car half a mile along the track. Protection at the crossing consisted of nothing more than a pair of crossbucks. [1072]

Eight people were killed more or less outright and another died in the following hours when an 80-car freight train collided with a car on a level crossing with Route 198 at Conneautville, Pennsylvania, United States, at 8.55 p.m. on the night of Sunday 12th October 1969. The car was carried 200 feet down the track.

Engineer William Thomas of the Bessemer and Lake Erie Railroad train stated that the bell of his engine had been ringing and he had whistled for the crossing. The car had stopped short of the crossing, hesitated, and had then moved forward again, at which point he had been unable to stop the northbound train, which had been moving at 45 miles an hour. "I believe I hit him right in the centre," the engineer said.

The victims were Denver Hamrick, 39, of Conneautville, the driver of the car; Dorothy Hanrick, a child, believed to be Denver's daughter; Franklin Charles Moats, of nearby Saegertown; Lucille Schrader, 34, of Conneautville; and five of Mrs Schrader's children: Sandra Sue, 9; James, 5; Allen, 3; Janet, 2; and Richard, five months.

Police said that Hamrick had borrowed the car, a 1963 two-door Rambler sedan, to pick up relatives from a service at the Pentecostal Church Revival Center a mile from the crossing.

'Eight of the victims, thrown from the car under the impact of the train, were

[1069] *Glenn Falls Times*, Glenn Falls, New York, United States, 30th July 1969
[1070] *Rudé Právo*, Praha, Czechoslovakia, 30th July 1969
[1071] *The Times*, London, England, 5th August 1969
[1072] *The Palladium-Times*, Oswego-Fulton, New York, United States, 29th September 1969

pronounced dead at the scene. The five-month-old baby died five hours after the 9 p.m. tragedy. Clothing on some of the victims was torn to shreds. Bodies were scattered throughout the area. The car was torn apart and thrown onto the east-side of the railroad right-of-way. The motor of the auto was completely separated from the chassis.

'Witnesses said the weather was clear and the crossing lights were operating. The lights were put out when the car hit a control box.

'The Conneautville Volunteer Fire Department set up portable lighting units to aid emergency work after the tragic accident.'

It emerged the following day that Hamrick did not have a driving licence. [1073]

A lorry carrying workmen was struck and crushed by a goods train on an open level crossing in the Bialystok region of north-east Poland on Tuesday 21st October 1969. Six people died and four more were seriously hurt. [1074]

On the morning of Tuesday 9th December 1969 a commuter train collided with road crane on Takane Crossing, Tatebayashi, Japan. The train was derailed and wrecked and its motorman, two train passengers, the crane driver, his mate, and the driver of a car stopped at the crossing were killed more or less outright. Over 100 people were injured, 24 seriously, and one of them, presumably a third train passenger, succumbed in Shoji hospital early the following morning. [1075] [1076]

A major level crossing accident occurred in Colombia, South America, at 12:25 on Sunday 11th January 1970 when a bus of the Sociedad de Buses Urbano de Atlantico was struck by the *Expreso del Sol* passenger train on an unprotected crossing at Mamatoco, just outside of the town of Santa Marta. The bus had left Barranquilla at about 06:00 on an excursion to El Rodadero, near Santa Marta.

A farmer, Joaquin Pinero, who witnessed the accident, told his story to a reporter. "The impact was brutal. I live very near the scene and was on my way to Mamatoco where I have some friends. I was looking forward to a pleasant Sunday. Suddenly I heard the whistle of a train but I didn't pay much attention to it, being used to such sounds. The next moment I saw a woman standing close to the line at the crossing frantically signalling to a bus which was close upon her, and then running towards it screaming 'Stop! Stop!' Then there was a huge impact like the sound of a large building collapsing. A train had hit the bus. People were shouting and others were imploring the saints for help and crying, 'Help me. I'm dying.' I didn't know what to do. Everything happened so fast. The lives of many people were ended in an instant."

The bus was completely demolished in the collision, with only its tyres surviving intact. Its engine was thrown several metres from its chassis. Around the remains of the bus lay the bodies of the dead and injured along with bloodstained clothes and

[1073] *The Derrick*, Oil City, Pennsylvania, United States, 15th October 1969
[1074] *The Times*, London, England, 22nd October 1969
[1075] *Pacific Stars and Stripes*, Tokyo, Japan, 11th December 1969
[1076] *The Dispatch*, Douglas, Arizona, United States, 10th December 1969

possessions. The first rescuers reportedly arrived at the scene only 15 minutes after the accident and the injured were taken to the San Juan de Dios hospital in Santa Marta.

The train had come from Bogotá and consisted of diesel locomotive No. 720 and 12 coaches, with 300 passengers. Driver Tomás Enrique Timoco had whistled several times for the crossing. The engine and leading five coaches were derailed by the impact and the coaches were left standing on the edge of a 30-metre-high embankment. No one on the train seems to have been hurt though.

Thirty-two occupants of the bus were known to have lost their lives by the day after the accident and it was feared that the bodies of another five small children might have been crushed beneath the wheels of the train, so as to be unrecognisable. At least another 14 people were injured. Those who died included the bus driver, Angel Maria Goethe, 33, and the organiser of the excursion, Segimundo Lleras Noriega, 52. Most of the others who died or were hurt were either young adults, teenagers, or children.

The accident was, it appeared, partly due to the recklessness of the bus driver, as several of the surviving passengers testified that the bus had been driven very fast, to the point where passengers had implored its driver to slow down, to no effect. However, visibility at the unguarded crossing was also poor due to lineside vegetation, including trees 'as high as telegraph poles.' A local official had written to the railway company towards the end of 1967 pointing out the restricted visibility at the crossing and predicting a tragedy if something were not done to improve it. Following his letter some efforts had been made to cut back the vegetation but clearly by 1970 it had grown back again.

The following day there was another level accident in Bogotá, resulting in serious injury to a car driver. This time it was apparently the fault of a crossing keeper who had failed to lower the barriers upon the approach of a train. [1077]

On the morning of Wednesday 28th January 1970 a disastrous school bus accident occurred in South Africa. At 14:15 an express train collided with a school bus carrying 72 children aged between six and 15 belonging to the Hoërskool DR Malan, an Afrikaner high school in nearby Meyerton. The accident occurred on a level crossing at Highbury, near Henley on Klip, a riverside resort town about 30 miles south of Johannesburg, named after Henley on Thames, England. The bus apparently stalled on the crossing.

Mrs Gracie Lamprechtes, who lived in a house near the crossing, told a reporter that she had heard "a tremendous bang." Running to the crossing she "saw children all over the place. They were hanging from the fences and were strewn over an embankment. I came running along and saw dead and dying children strewn for 25 yards up the track. I covered some of them with sheets." [1078]

Initially it was reported that there were 32 dead but later a railway police official stated that there were only 23 fatalities, with another 16 seriously hurt.

The train carried the bus a few yards. Eighteen ambulances took the dead and injured to local hospitals. [1079]

1077 *El Tiempo*, Bogotá, Colombia, 12th and 13th January 1970

1078 *The Times*, London, England, 29th January 1970

1079 *Glens Falls Times*, New York, United States, 29th January 1970

Johan le Roux, a matric (final year) student was credited with saving the lives of two of his schoolmates by pushing them out of a window at the last moment; he, himself, was killed. [1080]

After a basketball game in which Tracy Joint Union High School had beaten Atwater High School 76-48 a party of teenagers were driving around on the evening of Friday 13th February 1970, picking up others who wanted a ride home. There were 16 of them riding in what was variously described as an old truck and a minibus, when they reached a level crossing with the Western Pacific Railroad at Lathrop, California, United States.

The vehicle carrying the students apparently stopped, then restarted, moved forward, and stalled on the crossing before the warning lights at the crossing had begun to flash and its driver had then been unable to restart it despite repeated attempts. Eight teenagers had time to jump out, two were thrown out and killed when the freight train, travelling at an estimated 50 miles an hour, hit the truck, and the other six were trapped inside as it was carried sone distance down the line embedded on the front of the locomotive. All those who had remained in the truck were also killed.

The teenagers who lost their lives were Troy Bunner, 17, the driver; Joseph A. Church, 15; Phillis E. Handsome, Shirley Maddex, 16; Robin Moore, 16; Susan Pombo, 15; and Donnie Zimmerman, 17, all from Tracy; and Lisa Whiteman, 13, of Santa Clara. Of the survivors Shirly Dover, 16, was said to be "poorly" while Terry Knoll, 14, was slight hurt. Both were from Tracy.

Bruce Knoll, 16, said, "In a split second it was all over. I didn't see any flashing lights. I didn't see anything but the train. A couple of us managed to jump out. The others had to stay."

Tom Knoll, 14, brother of Bruce and Terry, said, "I just looked up and saw this light and saw that it was a train. We got one of the side doors open and I was kicking at the other. We were just driving to get out. I thought we had time to push it off the tracks." [1081]

Scott Houser, 18, related, "I jumped out and managed to haul two others out when the train hit. Eight others didn't make it. Why was my life spared? There must be a reason."

The National Transportation Safety Board (NTSB) sent investigators to the scene. Investigator Henry Wakeland declared, "We are looking especially at what caused the break-up of the van-type vehicle." He added that a study had begun the previous year to determine the feasibility of designing the front end of locomotives "so they would have less tendency to disintegrate vehicles struck. This case may be a prime example of the kind of disintegration that occurs in these types of accidents," he added. It has not proved possible to trace an NTSB report, if one was published. [1082]

The normal capacity of the vehicle was eight or ten people and other investigators theorised that it might have stalled because it was overloaded. Or possibly, as at Trent Road Crossing, England (see above), if it was an old vehicle perhaps its starter motor had been defective.

[1080] Wikipedia – Henley on Klip

[1081] *The Times Standard*, Eureka, California, United States, 16th February 1970

[1082] *The Argus*, Fremont – Newark, California, United States, 15th February 1970

On Monday 30th March 1970 a Japanese National Railways passenger train that had left Kawatana a few minutes earlier bound for Kyoto collided with a cement mixer lorry on a level crossing at Toyoura, Japan. The lorry driver had apparently missed or ignored the warning signals at the crossing and drove onto the line in the path of the train. The force of the impact derailed the leading two cars of the train and the first car overturned. The lorry driver and four train passengers were killed and 29 train passengers were injured. [1083] [1084]

Sixteen people died and at least 30 were injured when a diesel railcar collided with a double-decker bus on a level crossing at Roseworthy, South Australia, at about 7 p.m. on Sunday 12th April 1970. The bus had been carrying a party of picknickers to Elizabeth, about 15 miles away. Ambulances, police and rescue services from Adelaide, Elizabeth and Gawler went to the scene.

Mr Brian Leitchcock of Roseworthy, who arrived at the crossing half an hour after the accident, said, “It looked as if a bomb had been thrown into the bus. The train looked all right but the bus was a hell of a mess. Passengers from the train were standing around crying and some of them were helping to look after the injured. There were a lot of bodies around the bus. The police told me they had taken about 15 from the bus.” He said local people had been asked to rush blankets and sheets to the scene to use as bandages.

Mrs Otto Krieg, who lived on a nearby farm, spoke of there having been many children on the bus. “They were mostly English migrant families from Elizabeth who had been up here on a picnic. Some of the poor little kiddies were badly smashed up. I raced over with blankets and a few bandages but there was little we could do for many of them. I will never forget the sight. People screaming everywhere and poor little kiddies crying out for their parents. My son and husband tried to pull people from the bus but it was so twisted and smashed there was nothing they could do with their bare hands. People were still screaming inside the bus when the police and ambulance arrived. I did what I could and left. It was something I couldn’t take any longer.”

Farmer Ivan Kreig said he had heard the sound of the train followed by two loud thuds, which had made his dogs bark. “Even from my home I could hear people screaming out and crying,” he went on. “When I arrived dead and injured were lying everywhere. They were using crowbars, hacksaws, and anything they could to drag people from the wreckage. The bus was carried nearly 150 feet by the train, which was derailed. The bus’s engine was ripped right out and the front wheels were from the main wreckage. The bus was lying with its wheels in the air. It was flattened to the top floor window level. You couldn’t recognise it as a double-decker bus. It was more like a single decker. The accident happened right on dusk and it was dark by the time most people got to the wreckage. My wife was helping the injured lying on the ground as I helped drag others from what was left of the bus.”

The bus had been carrying about 100 members of the Gothic Hotel social club. There were only 15 passengers on the train, which was bound for Adelaide.

[1083] *Pacific Stars and Stripes*, Tokyo, Japan, 1st April 1970
[1084] *Athens News Courier*, Athens, Alabama, United States, 31st March 1970

Sixteen injured were taken to the Hutchinson hospital in Gawler with a dozen more being taken to the Royal Adelaide hospital. At Gawler hospital patients on stretchers were lined up in corridors to be dealt with as quickly as they could by overworked staff. One man on a stretcher asked repeatedly, "What happened to my boy? I just remember being on the bus singing. We were all half drunk, that's why we got a bus. The next thing I knew there was a crash. I was bleeding and my legs were pinned. I could smell burning. Listen, what happened to my boy?"

Acting railway secretary S.E. Hewitt stated that the Gladstone train had left Adelaide at 5.46 p.m. Driver K. Harrison had been taken to hospital, suffering from shock, he said, but had later been released and allowed to go home. Guard Phillips had not been hurt. The train passengers had been forwarded to Gladstone by bus. [1085]

A few days later the accident was the subject of a newspaper editorial complaining that it had taken a major tragedy to get the authorities to conduct a study into the safety of Australia's approximately 1,700 level crossings. The accident was the 42nd fatal level crossing accident in South Australia since 1966, while another had happened the same evening at Currency Creek Crossing.

Roseworthy (also known as Wasleys and 27 Mile) Crossing lay, the editorial pointed out, on a dead straight stretch of road. It had been dark at the time of the collision and the interior of the polished aluminium Bluebird railcar had been bright with white, fluorescent lighting. During the previous 15 years six other people had died in separate accidents at the same crossing.

State Roads and Transport Minister Hill had visited the scene and announced the opening of an inquiry but a previous inquiry in 1967 had little result. Under the 1961 Road Traffic Act the drivers of vehicles carrying eight or more passengers were required to stop at crossing not protected by warning lights, but few bus drivers were aware of the Act. South Australian Railways Signal and Telegraph Engineer R.I Carmichael believed that speed restrictions over the crossing would help; a previous speed limit of 20 miles an hour on the approach to the crossing had been removed, which he believed was a retrograde step.

The best solution would be the complete separation of road and rail traffic but this would be prohibitively expensive; even the cost of providing full signalling at all level crossings was too great for the State Government. Possibly wig-wag signals at the crossing might have helped.

The bus driver was amongst those killed and so could not explain why he had failed to stop but one theory, that he had suffered a heart attack, could be ruled out as there were short skid marks at the crossing indicating that he had applied his brakes at the last moment. [1086]

At the inquest on the victims of the accident Coroner Thomas Cleland apparently came to the conclusion that the bus driver had been intoxicated and that this was the explanation for his failure to see the train. [1087]

[1085] *The Canberra Times*, Canberra, Australian Capital Territory, Australia, 13th April 1970
[1086] Ibid, 23rd April 1970
[1087] https://www.abc.net.au/news/2019-07-01/wasleys-south-australias-worst-crash-remembered/11267518

A mother and seven of her eight children lost their lives in a level crossing accident as she was taking them to school at 7.40 a.m. on the morning of Wednesday 6th May 1970. The station waggon in which they were riding was struck by a Burlington Northern Railroad freight train being driven by Engineer J.S. Thompson on a crossing near Otis Orchards, Washington, United States, just east of Spokane.

Killed in the accident were Mrs Katherine Lorraine Krank, 36, the wife of an employee of the Inland Empire Paper Co.; and her children Rhonda G., 17; Audrey, 16; Kenneth J., 14; Dana J., 13; David T., 11; and twins Anita and Barbara, 6. The eighth child, Clara, 8, was seriously hurt and taken to the Valley General Hospital in Spokane.

The 53-car train was en route from Minneapolis, Minnesota to Auburn, Washington, and was travelling at 55 miles an hour. Engineer Thompson said that he had seen the station waggon when it was about 100 feet from the crossing and had been unable to pull up in time to avoid the collision.

By chance the accident was witnessed by Track Supervisor Raymond Kagele, who told police that the car had not hesitated at all before driving onto the tracks. The crossing was protected only by crossbucks, without even warning lights, let alone barriers.

State Trooper James Payne said that the station waggon had been carried 67 feet from the crossing and thrown down a 20 ft embankment. The bodies of three of the children were found at the top of the embankment, two children and the mother remained in the station waggon and the body of another child was found 100 feet west of the vehicle. The body of the other child was only located in the wreckage some time later. [1088] [1089]

There were many offers of support for the father, James Krank, and his surviving daughter after the accident, while local mothers and children planned to picket the crossing during the funeral of the victims in protest of its unsafe condition, and petitions were being circulated calling for the upgrading of protection at crossings in the area. Officials responded that there were warning signs in place at the crossing, visibility along the track was good and there had been no previous accidents recorded there at least since records had begun in 1959. [1090]

The crossing keeper and 28 of the 57 members of a wedding party were killed when a lorry in which they were riding was hit by a train on a level crossing between Hondon and Shree Mahavirjl, India, on the double-track line of the Western Railway between Bayana and Gangapur on the night of Tuesday 19th May 1970. Fourteen other members of the party were injured. [1091] [1092]

Another Burlington Northern Railroad freight train was involved in a major level crossing accident at about 10 p.m. on Friday 3rd July 1970. The 40-car train was on its

---

[1088] *The Billings Gazette*, Billings, Montana, United States, 7th May 1970
[1089] *Evening News*, Port Angeles, Washington, United States, 6th May 1970
[1090] *Walla Walla Union Bulletin*, Walla Walla, Washington, United States, 7th May 1970
[1091] *The Indian Express*, Madras, India, 20th May 1970
[1092] *The Lethbridge Herald*, Lethbridge, Alberta, Canada, 20th May 1970

way from Minneapolis-St Paul to Brainerd when it collided with a car on County Road 56 Crossing, five miles west of Anoka, Minnesota, United States. Again nine people, all but one of the family of Roger D. Smith, 37, of Osseo, Minnesota, were wiped out.

The victims were Roger Smith, the presumed driver of the car; his wife Mary, 32; and their children Susan, 12; Eugene, 10; Stephan, 8; Kevin, 7; Daniel, 6; Angela, 3; and Joanne, 11 months. A surviving child, Linda, 15, had been with her grandparents at Fergus Falls, Minnesota.

The train had been travelling at over 50 miles an hour when it hit the car, the front assembly of which was carried 100 yards down the track, with pieces of wreckage being carried another 300 yards, while parts of bodies were strewn over a wide area. [1093]

The crossing was marked by stop signs and crossbucks but lacked further protection. There was a good view along the line in each direction and the weather was cloudy but dry.

The train was delayed several hours by the accident.

A friend told the coroner that Smith was the superintendent of a Minneapolis construction company. He had been looking at property with a view to building in the area and the family had also stopped to visit a friend. The accident had occurred shortly after they left the friend's home. [1094]

A bus taking American airmen from the Don Muang air base in Thailand was struck near the airbase by the *Nong Khai Express* on the night of Tuesday 25th August 1970 on an unprotected level crossing. The bus was hurled 40 feet and five airmen were killed instantly, two died shortly afterwards and another died later in hospital. Two bodies were discovered later in an adjoining canal. Six other airmen, the bus driver and another Thai were injured.

A rescuer worker commented, "It looks like a war zone here." [1095]

Both the bus driver and the train driver were arrested, charged with 'reckless driving.' [1096]

A private bus carrying 70 people on a Sunday excursion was in collision with a freight train at Salta, Argentina, about 100 kilometres from the frontier with Bolivia, on 30th August 1970. The accident occurred on an unprotected level crossing when the bus driver tried and failed to get across the line in front of the train. The bus was dragged about 100 metres and the leading four waggons of the train were derailed. Thirteen bus passengers were killed and 42 were injured. [1097] [1098]

Early on Saturday 3rd October 1970 five young people on their way to the home of one

---

[1093] *La Crosse Sunday Tribune*, La Crosse, Wisconsin, United States, 5th July 1970
[1094] *The Sunday Tribune*, Albert Lea, Minnesota, United States, 5th July 1970
[1095] *The Herald Journal*, Logan, Utah, United States, 26th August 1970
[1096] *Pacific Stars and Stripes*, Tokyo, Japan, 30th August 1970
[1097] *Lebanon Daily News*, Lebanon, Pennsylvania, United States, 31st August 1970
[1098] *The Bee*, Danville, Virginia, United States, 31st August 1970

of them were killed when the car in which they were riding ran into the side of a southbound Chicago Milwaukee St Paul and Pacific Railroad freight train on an unprotected level crossing near Bellingham, Washington, United States. There was a heavy fog and the car was driven into the last waggon of the train, which chanced to be a tank waggon loaded with sulphur dioxide. The car was dragged 250 feet down the line and the bodies of the victims were found on its back seat.

Those who died were Paul Anderson, 20, the owner and presumed driver of the car, from Bellingham; Roy Carson, 20, also of Bellingham; Robert Dixon, 18; Anthony Eisenzimmer, 18; and Sheldon Querns, 19. The latter three were all Canadians from Vancouver, British Columbia. [1099]

In the second accident on Saturday 3rd October 1970 six people lost their lives when an Illinois Central Railroad passenger train collided with a car on a level crossing near Springfield, Illinois, United States. According to Engineer Jack Price the car had made a left turn from Route 54 onto the road leading to the crossing and had proceeded over the tracks without stopping. He thought that the car driver might have been blinded by the sun and hence failed to see the train.

Those who died were Mrs Verna Lee Dunlap, 34; her daughters Debra Sue, 14; Cynthia, 12; Laura, 9; and Lisa, 7; and the child of a neighbour, Charles Turnbull, 10, all from nearby Andrew. [1100]

A bus carrying excursionists who were on their way from Richmond to attend a church service at Porus, at which the Rev. Everard Allen, a Baptist minister, would be inducted, was struck by a passenger train on Job's Lane Crossing, near Spanish Town, Jamaica, a little after 3 p.m. on Monday 5th October 1970. The bus driver and eight passengers, including Mrs Maiselyn Gray, a sister of Rev. Allen, died at the scene, another passenger died on the way to hospital and a tenth soon after their arrival there. Another 45 of the approximately 60 people on the bus were injured, some badly, including Mrs Evelyn Allen, the mother of the Rev. Allen. Two more people died of their injuries some days later. [1101] Only one person on the train was slightly hurt.

The bus was a Leyland Master Key model, registration BV 209. It was halfway across the line when it was struck by the train, turned around and carried 25 yards along the track, while parts of it were attached to the front of the locomotive and borne 150 yards until the train stopped.

The train was on its way from Kingston to Port Antonio. It consisted of a diesel engine, two coaches and three waggons and was being driven by Driver Meikle, who had given the usual warning signal for the crossing, which was not heard by the bus driver.

A man named only as "Barvie," who lived near the crossing, witnessed the accident. He said, "I saw the bus enter Job Lane from Brunswick Avenue shortly after 3 o'clock, with people singing and clapping. I heard the train tooting, coming from Spanish Town. The bus did not stop at the crossing, which has no gates but is controlled by several

[1099] *The Daily Report*, Ontario-Upland, California, United States, 4th October 1970
[1100] *Dixon Evening Telegraph*, Dixon, Illinois, United States, 5th October 1970
[1101] *The Daily Gleaner*, Kingston, Jamaica, 21st October 1970

warning signs. I saw when the train hit the bus. The impact was so great that the bus turned to face the way from which it was coming. I shouted for help and ran to the spot. Other people bawled for mercy and also ran to the spot. When I got there I saw three dead bodies sprawled on the track. When the police and ambulances came some of the people helped to remove the trapped bodies and the injured persons from the wrecked bus." [1102] [1103] [1104]

An inquiry was held into the accident in mid-October. An important witness at the inquiry was a Mr Garricks, a Government Motor Vehicle Certifying Officer, who stated that he had been driving towards the crossing and 150 yards from it had observed the bus coming from the opposite direction. He had then stopped at the stop line of the crossing for the train, having heard its whistle, as had two vehicles coming towards him. The bus, travelling slowly and with its horn sounding, had then proceeded to overtake the two stationary vehicles and made no effort to stop, at which point, Mr Garricks said, he had remarked to the other person in his car, "It will happen now." Mr Garricks said he had heard no singing from the bus and could see no reason why the bus driver could not have seen the stop sign at the crossing, the stationary vehicles, and the train itself. He considered the bus driver responsible for the accident.

Mr Borris Gayle, the bus driver, maintained on the other hand that there had been loud singing and knocking of passengers' tambourines from the start of the trip at Richmond as far as Linstead, and again from Bog Walk to the crossing. He said the singing had been louder than the horn of his bus. He had never driven over that route before and was unaware of the crossing. He admitted overtaking two vehicles some yards from the crossing but denied hearing the train horn. He declined to answer any further questions on the advice of his lawyer.

Another witness at the inquiry, Mrs Miriam Johnson, a passenger on the ill-fated bus, stated that she had heard no singing and had seen no tambourines. She had seen the train approaching when the bus was 70 feet from the crossing. The bus had not stopped.

Farmer Titus Seville said that he lived part-time near the crossing. On the afternoon in question he had been approaching the crossing on foot with a friend, Ivan Richards. The farmer said that he had seen the car occupied by the motor vehicle examiner stopped on one side of the crossing and two vehicles stopped on the other side. He had then heard the train coming and watched the bus overtake the parked vehicles and enter the crossing. At this point he had exclaimed to his friend, "Watch yah, watch yah, watch yah, watch yah!" His friend had covered his face with his cap so as not to see what was about to happen. [1105]

The findings of the inquiry do not appear to have been made public.

The driver of a dump truck apparently attempted to beat a train over a level crossing between Hanasaki and Washinomya stations, Japan, on the night of Friday 9th October 1970 and in doing so caused an accident that killed five people, including himself, and injured another 148, 51 of whom were detained in hospital and 29 of whom were stated

---

[1102] Ibid, 6th October 1970
[1103] Ibid, 7th October 1970
[1104] Ibid, 10th October 1970
[1105] Ibid, 18th October 1970

to be very seriously injured.

The six-car Tobu Railway train had been carrying 1,300 passengers when the accident occurred at 20:19. The force of the impact derailed five of the train cars and threw passengers through windows into a paddy field and a shallow river bed. The leading car was described as having been smashed into an unrecognisable mass, while the second car fell into a river bed, the third jumped the track and two cars hung over a 20 ft embankment. A policeman remarked, "It's a miracle more people weren't killed."

Those who lost their lives were passengers Shoji Fujinuma, an employee of the government-owned NHK broadcasting network; his wife Hideko, Yoshiko Shinno, a 23-year-old Tokyo housewife; and a woman named as Kayoko Miwa from Yokohoma; plus the dump truck driver, Noboru Nishi, 25.

The dump truck, owned by the Toyo Busan Co. of Tokyo, had struck the leading car of the train. It was dragged 700 feet before becoming embedded in the river bed.

Local residents stated that they had heard the crossing signals operating before the crash. [1106]

On Wednesday 14th October 1970 a fleet of seven buses was carrying pupils from Kyungso Junior High School in Seoul who were returning from the Hyeonchungsa Shrine (built to commemorate a 16th-century Korean admiral who had won numerous naval battles against the Japanese) at the hot spring resort of Onyang, South Korea. The buses had only covered a short distance when they reached an unprotected level crossing at about 16:20. The first two vehicles passed over the crossing safely, the third followed them and was struck by a passenger train with disastrous results. The bus was carried about 50 metres and immediately burst into flames. The engine and three coaches of the train were derailed but no one on it was hurt.

The accident was witnessed by Korean Marine Sergeant Yu Sukmok, who was at a friend's house only 30 metres from the crossing. He said that when he rushed out of the house he had seen the train still pushing the bus along the track and several boys jumping off it. He and his friends had rushed to the bus and smashed windows so that the passengers could escape through them but none had. He told a reporter, "The bus was already ablaze. The boys seemed unable to help themselves. I picked up a rock, smashed some windows, and helped three of them out though." He said the heat had then become too intense for him to save any more.

Initially it was stated that 52 boys and the bus driver had lost their lives and another 24 boys had been injured, many suffering serious burns. The dead were all burnt beyond recognition and the wreckage was still smouldering five hours after the accident.

The police blamed the accident on the carelessness of the bus driver as all buses were supposed to stop before crossing the line. [1107]

The number of fatalities was later corrected to 45.

Following the disaster Mr Hong Jong-chul, the Education Minister, and Mr Paik Sun-

---

[1106] *Pacific Stars and Stripes*, Tokyo, Japan, 11th October 1970
[1107] Ibid, 16th October 1970

yup, the Transportation Minister, offered their resignations. [1108]

Nine people lost their lives and 53 were injured when a train collided with a bus on a level crossing near Moga, India, in the Punjab, on the night of Saturday 17th October 1970. [1109]

Five teenagers lost their lives in a level crossing accident at 10.15 p.m. on Saturday 7th November 1970 when the automobile in which they were riding was struck by a 107-car Baltimore and Ohio Railroad freight train on Holmes Street Crossing, Tiffin, Ohio, United States.

Engineer H.L. Pepple said that his train had been travelling at 50 miles an hour at the time but slowing for the town. He and Brakeman Michael Parvu stated that they had seen the car approaching the crossing and it had not appeared to reduce speed. The train carried the car 3,284 feet from the point of collision and its wreckage had to be pried from the front of the engine.

The victims, all from Tiffin, were Thomas Cummins, 15; Roger Faber, 16 (the driver of the car); Daniel Klepper, 16; Valerie Long, and Eugene Umlor Jr. All were pronounced dead upon arrival at Tiffin Mercy Hospital.

The crossing was marked only by a sign. [1110]

Belgrade Radio announced that a freight train had collided with a bus carrying workers to an iron ore mine at a location near Prijedor, Yugoslavia, on Tuesday 24th November 1970. Eight workmen were killed and 29 others were injured. [1111]

A Gulf Mobile and Ohio Railroad freight train making its final run before the beginning of a nationwide strike collided with a car on a level crossing with road Alabama 5 at Brent, Alabama, United States, on the night of Wednesday 9th December 1970. The accident led to the death of Mrs Willia Lockett, 39; her children Robert, 16; James, 14; Rose Mary, 12; Willie Dan, 9; and Michael, 8; and Mrs Lockett's sister, Louise Wilson, 50, from Chicago, Illinois.

Engineer Tossie Miller said that the automobile had hit his engine about 15 feet from its front end, derailing the engine and two of the 142 cars of the train. The train had been en route from Montgomery, Alabama, to Artesia, Mississippi.

Sheriff Harold Dailey pronounced the accident to be the worst that he had ever investigated. "We couldn't even tell for sure how many people had been involved for a good while," he said. He theorised that either Mrs Lockett had been attempting to beat the train or else she had failed to see the flashing red lights at the crossing because of a Christmas decoration atop the signal. He added that there had been many previous

[1108] *The Canberra Times*, Canberra, Australian Capital Territory, Australia, 16th October 1970
[1109] *The Daily Gleaner*, Kingston, Jamaica, 19th October 1970
[1110] *The Republican-Courier*, Findlay, Ohio, United States, 9th November 1970
[1111] *The Bakersfield Californian*, Bakersfield, California, United States, 25th November 1970

accidents at the crossing, but none that had resulted in fatalities.

Mrs Lockett's husband, Lee Marvin Lockett Sr., was working at a sawmill only a mile away while another son, Lee Marvin Jr., was at home.

The party had been returning from a visit to Mrs Lockett's mother. [1112]

Two young families had just collected their children from a babysitter at the end of an evening out and were on their way home when their car was struck by an eastbound 83-car Chicago Rock Island and Pacific Railroad freight train at 1.01 a.m. on Sunday 27th December 1970. The accident happened at the crossing with Kansas Avenue, Liberal, Kansas, United States.

The train was in charge of Conductor John Bearden with Engineer J.C. Henderson and Brakeman Thurman Herskell. One of the train crew estimated the speed of the car at 20 to 30 miles an hour while another put it at 40 miles an hour. There were no skid marks and the car was driven into the side of the leading locomotive and dragged 950 feet down the line, crushed with its occupants trapped inside. Orville Aldridge Jr., 23; his wife Mary Ann, 20; their daughter Wendy Sue, eight months; Leroy Peitz, 25; his wife Bonnie Ellen, 20; and their son Christopher, nine months, all lost their lives. Four of the victims were killed instantly, baby Peitz died on the way to hospital and Orville just over an hour after admission to the Southwest Kansas Medical Center.

The crossing was reportedly 'signalized,' meaning presumably that it was equipped with warning lights, which were said to have been working correctly. It had been the scene of several previous fatal accidents. [1113]

An empty (?) school bus was struck by a Baltimore and Ohio Railroad passenger train on about Saturday 13th February 1971 on Symmes Avenue Crossing, Fairfield, Ohio, United States. The glancing blow pushed the bus into a ditch and the bus driver and three maintenance workers were injured. [1114]

Bus driver Antonio Morja, 33, blamed faulty brakes for his failure to stop at a level crossing outside of Puebla, Mexico, on Sunday 21st February 1971. As a result the bus was in collision with a train and burst into flames. Twenty of its passengers were killed, including six children, and 40 were injured, many critically. Police said that many of the bodies were burnt beyond recognition. Morja himself escaped with slight injuries.

Another factor that might have contributed to the accident was the fact that the local bus was grossly overcrowded, as it was carrying 70 passengers, double its capacity. [1115]

Morja was subsequently arrested on a charge of homicide, the police alleging that he had tried to beat the train over the crossing and had never applied the brakes at all. [1116]

[1112] *Corpus Christi Times*, Corpus Christi, Texas, United States, 10th December 1970
[1113] *The Hutchinson News*, Hutchinson, Kansas, United States, 28th December 1970
[1114] *The Sunday Messenger*, Athens, Ohio, United States, 14th February 1971
[1115] *Lubbock Avalanche-Journal*, Lubbock, Texas, United States, 23rd February 1971
[1116] Ibid, 28th February 1971

A train of the Fuji Kyuko Railway was involved in a major disaster in Japan at about 08:25 on Monday 4th March 1971. Train No. 14 from Kawaguchiko to Otsuki, a two-car set, was travelling at 30 kilometres an hour when it collided with a small lorry on Midorigaoka No. 2 Crossing, between Kurechi (later renamed Kotobuki) and Mitsutoge stations. The lorry had smashed through the barriers of the crossing. Following the collision the train began to run away down an incline despite the motorman at once applying both the ordinary brakes and the hand brake. After running about four kilometres the train finally left the rails on a 240 metre radius curve and was wrecked. Seventeen people died and 69 people were injured, some seriously.

Investigation showed that the air reservoir of the train had broken when the lorry had caught the underside, and the supply of air to the brake cylinder had been interrupted.

Following the investigation measures were taken to install the brake equipment within the bogies and reinforced by a protecting wall so as to make a recurrence of the disaster unlikely. [1117]

Reckless driving on the part of the operator of a bus led to a level crossing accident at San Cristóbal Ecatepec, Mexico, on Saturday 10th April 1971. The driver of the second-class bus overtook another bus approaching the crossing about 20 kilometres north east of Mexico City. Having done so he then found himself too close to the crossing to be able to stop and the bus was struck by a Puebla-bound train. The vehicle was dragged 250 metres down the line and its fuel tank was punctured, resulting in an explosion and fire. Seven people died and 12 were injured. The bus driver was believed to be among the dead. [1118]

Tragedy struck a party of ten teenagers who were probably just riding around in a ten-year-old secondhand car that one of them had just bought on Sunday 2nd May 1971. Danny Henderson, 16, of Covington, Georgia, United States, had bought the car less than three weeks earlier. He was killed in the accident, as were Peggy Ann Fuller, 15; her sister Susie Ann, 12; Vickie Gilstrap, 14; Sheila Height, 14; Jimmy Norwood, 15; his sister Almethin, 12; Harold Tuggle, 16; and Dira Williams, 12. The only survivor, Sharon Belcher, 16, was critically hurt and taken to an Atlanta, Georgia hospital. The teenagers all lived close to one another in Covington.

The car was only about three miles from Covington when it was struck by a 65-car Georgia Railroad freight train travelling at 45 miles an hour on Duncan's Crossing, [1119] described as an 'elevated country road crossing.' The car had overtaken another vehicle that had already stopped at the crossing and came onto the line directly in the path of the train. The crossing was marked only by warning signs. [1120]

Sheriff Henry Odum said that the car had been bent nearly in half by the impact, with its right rear end lacking only five feet from touching the right front wheel. "They never knew what hit them," said Newton County Coroner San Cowan. "I was in the first

[1117] https://www.mlit.go.jp/jtsb/eng-rail_report/English/RI2021-1-1e.pdf
[1118] *The Brownsville Herald*, Brownsville, Mexico. United States, 11th April 1971
[1119] *The Daily Independent*, Kannapolis, North Carolina, United States, 3rd May 1971
[1120] *Albuquerque Journal*, Albuquerque, New Mexico, United States, 4th May 1971

ambulance to arrive. I didn't see how there could be any survivors." [1121]

Carelessness and driving too fast were the apparent causes of a level crossing accident late on Thursday 3rd June 1971. An automobile ran into the side of a St Louis-San Francisco Railway freight train about four cars back from the engine, knocking the fifth car from the rails, though it was dragged along by the rest of the train. The engineer told police that his train had been at about 40 miles an hour and he had not seen the car at all. The train stopped half a mile down the line. The car was carried about 75 yards along the track and the bodies of the victims were strewn along the right of way for the last 25 yards. The engine of the car was found 50 yards from the crossing, while its wheels remained beneath the train.

Killed in the accident on Greenwich Road Crossing, Wichita, Kansas, United States, [1122] were Mrs Sally Barnaby, 25, of Wichita; her children Christina Sue, 3, and Tammy Lynn, 23 months; Randall Wells, 17, of Piedmont, Kansas, a nephew of Mrs Barnaby; Franklin Bedigrew, 18, of Toronto, Kansas; and Deborah Waits, 16, of Rose Hill, Kansas.

Skid marks 160 feet long indicated that the car driver had seen the train and applied the brakes at the last moment, and the car had also swerved just before it hit the train.

The north side of the crossing was protected by an 'automatic crossbar,' but on the south side, the direction from which the car had come, there was only an unlit sign. [1123] [1124]

Local freight train Mg 17 655 collided with a car on a level crossing on the line between Münster and Gronau, West Germany, on Monday 26th July 1971, killing its five occupants. The car driver had failed to observe or had ignored the flashing warning lights protecting the crossing. [1125]

A bus carrying children to Sunday school at the First Baptist Church at East Gary, Indiana, United States, was struck by a Penn Central Railroad freight train on Grand Boulevard Crossing in the city, killing four children and injuring five others and the bus driver. The accident happened at 9.30 a.m. on Sunday 31st October 1971. It appeared that the bus had approached the crossing driving along Marquette Avenue, parallel to the railway, before the bus driver, Joseph Spanos, 34, had turned into Grand Boulevard and driven onto the line, having failed to notice the oncoming train. According to Jane Gaza, 18, who was a witness to the accident, he had failed to stop at the crossing, as he should have done, before entering it. The crossing was unprotected by either barriers or warning lights. It was less than two blocks from the crossing to the church.

The 74-car eastbound train was on its way from Chicago to Elkhart. Engineer E. Kmiecik said that the bus had crossed his path too late for him to be able to prevent the accident.

[1121] *The High Point Enterprise*, High Point, North Carolina, United States, 3rd May 1971
[1122] *The Hutchinson News*, Hutchinson, Kansas, United States, 4th June 1971
[1123] *Wellington Daily News*, Wellington, Kansas, United States, 4th June 1971
[1124] *The Waxahachie Daily Light*, Waxahachie, Texas, United States, 4th June 1971
[1125] *Katastrophen der deutschen Bahnen Teil II* by H.J. Ritzau, pg. 253

Donna Breckman, 13; Mary Lee Meler, 9; Elizabeth Miller, 17; and Becky Turner, 13, lost their lives. [1126] [1127]

A school bus carrying 60 students aged between seven and 18 was in collision with a St Louis Southwestern Railway (Cotton Belt) freight train on an unprotected level crossing at Lilbourn, Missouri, United States, at 3.25 p.m. on Tuesday 23rd November 1971. The westbound local train, which was being driven by Engineer James Partlow, was only moving slowly at the time and a tragedy was thus averted. The force of the impact knocked the bus into a ditch but it did not overturn. Thirty-six students and the bus driver were injured and taken to the Missouri Delta Community hospital; eight were admitted for further treatment. [1128] [1129]

Bus driver William Sherill, 61, had stopped the bus at the crossing and opened its door to look and listen for a train as he was legally required to do, but had then set off again, having failed, even then, to spot the train. He was cited by the Missouri Highway Patrol for failure to yield at a crossing. [1130]

Another minor school bus accident occurred at Punxsutawney, Pennsylvania, United States, at 1.45 p.m. on Monday 6th November 1971. An empty school bus being driven by Cordroy Wilson, 65, was unable to stop when it reached the lowered barriers at the crossing and smashed through them, striking the caboose of a Baltimore and Ohio Railroad train driven by Engineer John Ward. The bus was southbound on North Findlay Street. No one was hurt and total damage was only $80. [1131]

Five young people lost their lives in a level crossing at Toledo, Ohio, United States, on Saturday 11th December 1971. David Myles, 17; his brothers Dean, 16; Dennis, 13; and Dwight, 9, all from Toledo; and David's girlfriend Marilyn Donbrosky, 16, from Temperance, Michigan, had been returning home from a meeting at the Calvary Assembly of God Church. The car was driven into the side of a 40-car Penn Central Railroad train, bound for Toledo from Elkhart, Indiana, the engineer of which said he had not seen the car and had not been aware of the impending crash. [1132] [1133] The automobile was pushed nearly half a mile. [1134]

Mrs Lawrence Myles, the mother of four of the children, declared later, "I don't understand it, but it's God's will, and God doesn't make mistakes." [1135]

[1126] *Chicago Tribune*, Chicago, Illinois, United States, 1st November 1971
[1127] *The Dubois County Daily Herald*, Jasper, Indiana, United States, 1st November 1971
[1128] *The Daily Standard*, Sikeston, Missouri, United States, 24th November 1971
[1129] *The Courier News*, Blytheville, Arkansas, United States, 24th November 1971
[1130] *The Sedalia Democrat*, Sedalia, Missouri, United States, 24th November 1971
[1131] *Indiana Evening Gazette*, Indiana, Pennsylvania, United States, 7th December 1971
[1132] *The Fairborn Daily Herald*, Fairborn, Ohio, United States, 13th December 1971
[1133] *Journal-News*, Hamilton, Ohio, United States, 13th December 1971
[1134] *Norwalk Reflector*, Norwalk, Ohio, United States, 11th December 1971
[1135] *The Daily Jeffersonian*, Cambridge, Ohio, United States, 15th December 1971

At 07:30 on Monday 20th December 1971 a bus carrying 12 schoolchildren reached a level crossing with the Stockheim – Sonneberg at Burggrub, West Germany. From a point 120 metres from the crossing there was a speed limit on the road of 20 kilometres an hour but this, the bus driver ignored and he was still travelling at 35 kilometres an hour close to the crossing, which was too high a speed given the restricted visibility there. About eight metres from the crossing the bus driver spotted an approaching shunting engine, which was travelling at 14 kilometres an hour. He then accelerated in an attempt to get over the crossing in front of the train but failed to do so. The engine struck the bus and four children were injured.

The bus driver was subsequently convicted of a criminal offence, presumably something equivalent to dangerous driving. The court found that although he was familiar with the crossing and the limited view of the line he had failed to reduce speed as required, while the fact that he had the radio on, coupled with the noise of the engine of the bus and the children talking would have completely drowned out the noise of a train whistle. The train had not whistled at the first whistle as it should have done, but the whistle could not have been heard on the bus anyway, while the train had approached the crossing a lower speed than that permitted (20 kilometres an hour). [1136]

On Tuesday 25th January 1972 what was reportedly an empty school bus came into collision with an eastbound Minneapolis St Paul and Sault Ste Marie Railroad train on Airport Road Crossing, near Rhinelander, Wisconsin, United States. The bus driver, Charles Stroyny, 48, said that he had delivered his passengers, presumably schoolchildren, to the Rhinelander-Oneida County Airport near the crossing, having come from the Central Wisconsin Airport near Mosinee. He had been returning to Mosinee when he discovered a pair of skis that had been left on the bus, so he had been returning to the airport with them.

Approaching the crossing on a road covered in snow and ice Driver Stroyny was unable to stop his bus before it had skidded onto the line and been struck the train. He suffered head and face lacerations as well as possible internal injuries and was admitted to St Mary's hospital. The bus and the engine of the train were both somewhat damaged. Neither Engineer George Hurley nor anyone else on the train was hurt. [1137]

The conductor, driver, assistant driver, and a fourth railwayman were all killed when a Santander – Madrid electric train was in collision with a lorry on a level crossing between Cabezón and Corcos stations, Spain, on the Venta de Baños – Valladolid line. The accident happened at 18:55 on Tuesday 7th March 1972. The lorry driver also lost his life and eight train passengers suffered bruises. The train was not derailed and its passengers continued their journey to Madrid on another train. What protection was provided at the crossing was not stated in the fairly brief report on the accident. [1138]

The several minor level crossing collisions involving school buses in the United States

---

[1136] *Die Katastrophenszene der Gegenwart* by Hans Joachim Ritzau and Jürgen Hörstel, pg. 219
[1137] *Rhinelander Daily News*, Rhinelander, Wisconsin, United States, 26th January 1972
[1138] *ABC*, Madrid, Spain, 8th March 1972

documented above were followed on Friday 24th March 1972 by a considerably more serious one on Gilchrest Road Crossing, near Congers, New York. The accident was investigated by the National Transportation Safety Board (NTSB).

At about 7.55 a.m. on the day of the accident a school bus containing 49 passengers entered the crossing and was struck by Penn Central Railroad freight train No. WV-1. The bus was carried 1,116 feet down the track and disintegrated. Three schoolchildren were killed more or less outright, two more died later of their injuries and all of the remaining 44 pupils and the bus driver were injured.

Bus driver Joseph Larkin, 37, had started from the garage with the bus at 7.30 a.m. Encountering an unexpected detour he had been forced to make his journey in the reverse direction, an unfortunate circumstance that increased the severity of the accident, and he approached the crossing from the opposite side, 20 minutes later than usual, carrying 49 passengers instead of only six. When the bus was still 600 feet from the crossing one of the children saw the train, as did other pupils when the bus was closer to the crossing, and several shouted warnings to the driver. About 200 feet from the crossing the bus slowed momentarily but then resumed its normal speed, only braking at the last moment. [1139]

A newspaper reported, 'Patty Bundick, an 18-year-old senior at Nyack High School, was sitting near the back of the orange school bus packed with 49 teenage students when she glanced up and saw the approaching freight train. "I looked at the train and a few of the other kids must have too. I called out that it was coming…a few of us screamed out," she said. "He (the bus driver) hesitated to stop, but he kept on going. He was going too fast to stop," she said.' Patty was lucky enough to escape with fairly minor injuries. [1140]

When the train was about 150 feet from the crossing it became apparent to its crew that the bus was not stopping. Engineer Charles Carpenter related that he "frantically blew his whistle and set his emergency brakes when he saw the bus coming," adding that "he (the bus driver) never stopped at all." The train had been travelling at an estimated 25 miles an hour, well within the authorised line speed of 40 miles an hour. It consisted of three diesel units, 83 cars and a caboose.

The NTSB report stated that 58½ feet from the centre of the track was a standard stop sign, below which on the same post was a standard railroad crossing advance warning sign, while the word "STOP" was painted on the road. Nearer the crossing there should have been the usual crossbuck but this was lying on the ground. The speed limit on the road was 30 miles an hour.

A traffic census on 7th April showed an average of 451 eastbound and 452 westbound vehicles daily but the numbers might have been influenced by sightseers still visiting the scene. Eight freight trains used the line daily in each direction, one of them between 7 a.m. and 8 a.m.

Immediately after the accident the fireman of the train reported it by radio and the engineer and a brakeman began extricating children from the bus, assisted by several residents. A police car arrived at 7.58 a.m. and the police notified Nyack Hospital at 8 a.m., when a doctor and three nurses went to the scene. The last pupil left the site 55 minutes after the accident.

[1139] National Transportation Safety Board report

[1140] *The Daily Jeffersonian*, Cambridge, Ohio, United States, 25th January 1972

The NTSB stated that the bus driver was primarily employed as a fireman. He had been driving a school bus for Rockland Bus Lines since 1966, starting his morning run at 7.30 a.m. and finishing it at 8.25 a.m. to drive to the firehouse in New York City, 30 miles away, where he was due to start work at 9 a.m. Bus passengers described his driving in terms ranging from "average" to "unacceptable." He was charged with five counts of criminal negligence after the accident and declined to give evidence at the NTSB inquiry on the advice of his lawyer. [1141] At his subsequent trial he pleaded "Not Guilty" and stated that he had neither seen nor heard the train. He was convicted but escaped with a sentence of five years' probation. [1142]

Analysing the accident the NTSB noted that visibility at the crossing was 'unobstructed and more than adequate.' That the train's horn was audible within the bus was evident from the fact that several passengers had heard it. There was no evidence that the bus driver's hearing was impaired; possibly he had ignored the horn as he was not expecting to encounter a train at that time.

The fact that the bus had momentarily decelerated, then increased its speed again, implied that he had misjudged the speed and/or distance of the train and thought that he could safely cross in front of it.

A test run by a driver between the bus station and the firehouse, obeying the traffic regulations and by the best available route, took an hour and five minutes, implying that the bus driver would have been late for work every day that he drove the school bus. He would therefore have wished to finish his bus run without delay. There was no rule or regulation requiring employers to examine a bus driver's primary employment to see if there was any factor that might make them hurry, or otherwise affect their safe driving.

Had the warning signs been more appropriately placed and the crossbuck in position the bus driver and his passengers would have been better reminded of the proximity of the crossing. The driver was, however, familiar with the crossing. There was also a considerable controversy over the effectiveness of signs at crossings.

The engineer only braked when it had become obvious that the bus was not stopping. The dark green colour of the locomotive would have made it harder to see.

The protruding coupler at the front of the locomotive had penetrated the bus, which had been bent into a 'U' shape. Couplers were not required in mainline operations and could be recessed in future locomotive designs.

Just prior to the collision many of the passengers had got up and moved in order to try and avoid it, and were thrown about on impact. As a result of the collision the rear portion of the bus had separated from the remainder, and five or six passengers in it had been ejected from the vehicle. The front portion of the bus, locked to the locomotive, had meanwhile begun to disintegrate. Two passengers had been ejected from the forward section and two more had fallen through the buckled floor.

Most of the injuries were head injuries caused by passengers being thrown against one another and exposed metal parts of the bus. High-stress, well-padded seats would have reduced the injuries and seat belts would have prevented passengers being thrown about or ejected, though the use of seatbelts was controversial.

[1141] National Transportation Safety Board report

[1142] Wikipedia – Gilchrest Road, New York crossing accident

Although the nominal capacity of the bus was 66 passengers the seating of so many would be cramped for high school pupils. On the bus in question seven pupils had thus chosen to stand, two of whom had been killed.

The most important factor in the treatment of the injured was the fact that Nyack Hospital had an operational disaster plan.

A school bus route should be selected to avoid unprotected crossings, or else efforts made to get them protected.

The qualifications for school bus driver licensing in New York did not meet the minimal Federal standards for commercial drivers in interstate commerce.

Neither the contractor nor the district had complied with the regulations requiring references as to the bus driver's attitudes and personality, nor had the driver been examined by a physician during the previous year, and his primary employment had not been scrutinised.

A programme of mandatory and continuous training of bus drivers was required.

Supervision of bus drivers should include measures such as "check rides", periodic written reports on individual drivers and evaluation of his relationship with his passengers.

There were 4,300 level crossings in New York state, 1,170 on Federal highways (30% unprotected) and 3,130 on non-Federal roads (63% unprotected). A substantial improvement in safety could be achieved by the selective use of funds to eliminate crossings or provide better protection at them.

The NTSB came to 24 conclusions:

(1) The school bus driver and the train crew had an unobstructed view between vehicles when they were within 500 feet of the crossing.

(2) The train horn was audible for some time prior to the collision to several passengers in the bus as well as to witnesses who were near the crossing. It is not known whether the school bus driver heard the train horn.

(3) The dark colour of the locomotive increased the likelihood of the locomotive's blending into the surrounding environment. This factor was, however, not significant when the school bus was still within a safe stopping distance from the track.

(4) The school bus driver did not stop at the stop sign at the grade crossing. Because of the criminal charges placed against the driver as a result of the accident, his explanation of this failure to stop was not made available to the Board.

(5) The school bus driver would necessarily have been unable to report for work at 9 a.m. with the New York City Fire Department on the days he drove the morning bus route. At the time of the crash, he was already behind his normal schedule, due to a detour. If he had stopped for the train, he would have been further delayed.

(6) There is a need for a scrutiny of school bus drivers who hold other jobs which require them to be present at times which could adversely influence their driving behaviour.

(7) The mechanical condition of the school bus and train did not contribute to this accident.

(8) The geometry and condition of the road, track and crossing did not contribute to this accident.

(9) The train and its crew were in compliance with all applicable rules and regulations.

(10) The route for the school bus could have been established to eliminate the Gilchrest railroad/highway grade crossing without affecting the economy or the system.

(11) There is a need for the State of New York Department of Education to expand its pupil-transportation activities to assist in training, supervision, and management at the local levels.

(12) The administration and management of the pupil-transportation in the Nyack School were largely informal. With few exceptions, there were no written policies and procedures. Those provided by the State of New York were not always followed completely.

(13) The accident might have been prevented if the crossing had been protected by railroad gates and/or flashing lights. The installation of such active protection devices at crossings on non-Federal-aid highways is limited under existing economics in State and local government and in the railroad industry.

(14) The normal crossbuck sign was missing, and the advance-warning sign was placed too close to the crossing to serve as an advance warning.

(15) The driver was familiar with the existing crossing from the direction of travel usually followed in the completion of the school bus route. It is not known what, if any, his familiarity with the crossing was from the direction of approach on the day of the accident.

(16) The crash forces produced on the bus by the front surface of the train at about 25 m.p.h. were sufficient to produce direct injury only to passengers in the area directly impacted.

(17) Some of the fatal injuries to passengers were the result of abnormal dynamics and contacts which occurred when the bus structure disintegrated.

(18) At least two of the five fatalities were ejected as a result of a floor separation and a lack of availability and use of an occupant-restraint system in the bus.

(19) The controversy over the feasibility of installing active occupant restraint systems in school buses suggests the need for a demonstration project to obtain data which can be used to resolve the question.

(20) The structural disintegration of the bus displayed many examples of failures at joints assembled with relatively few fasteners; such construction is typical of current school bus construction practices.

(21) Several of the passengers experienced severe crash injuries from sharp metal edges which were exposed by the separation of structural parts of the bus body.

(22) The capacity ratings for school buses used to transport high school students do not always accurately reflect the actual seating capacity of the bus.

(23) The special status of innocent passengers transported by school bus drivers necessitates that driver qualifications be more than those presently required in many instances. Whenever applicable, the qualifications should be at least equal to those

for drivers of interstate motor carriers as found in the Motor Carrier Safety Regulations of the Federal Highway Administration (49 CFR 391)

(24) The intended effect of the stop sign was not clear, since a requirement to stop at all times had been abrogated by the absence of any history of local enforcement.

The NTSB concluded that the accident had been caused by the failure of the school bus driver to stop at the stop sign until the crossing was clear for an undetermined reason.

Contributing to the accident was the unnecessary routing of the school bus over an unprotected crossing.

Contributing to fatalities and severe injuries were:

(a) The lack of structural integrity of the body of the school bus.

(b) The absence of high-back padded seats and an occupant-restraint system in the school bus.

(c) The presence of standing students in the bus.

(d) The action of the coupler of the lead locomotive which had caused the crash forces to be concentrated on the bus.

The NTSB made eight recommendations:

(1) The National Highway Traffic Safety Administration should expeditiously adopt a Federal Motor Vehicle Safety Standard to control the strength of structural joints of school buses. In this connection careful consideration should be given to requirement 5.6 Body Structure of the Vehicle Equipment Safety Commission. This standard should also apply to the strengthening of the window columns of school buses.

(2) The Department of Transportation should seek legislation which would extend the use of Federal funds now available for grade crossing safety and improvement on the Federal Highway System to include those railroad/highway crossings on non-Federal-aid highways.

(3) The National Highway Traffic Safety Administration, in its revision of the Federal Highway Safety Program Standards, should continue the requirement which states that school bus drivers are to "be qualified as a driver under the Motor Carrier Safety Regulations of the Federal Highway Administration 49 CFR 391 if he or his employer is subject to the regulations" and to extend that provision (except regarding driver's age) to include all drivers of school buses regardless of whether they or their employer is subject to those regulations.

(4) The National Highway Traffic Safety Administration, in its revision of the Federal Highway Safety Program Standards, should include a provision under the title Pupil Transportation Safety, which would require those persons responsible for the hiring and supervision of school bus drivers to conduct pre-employment enquiries and continued surveillance to assure that other employment requirements of school bus drivers do not adversely influence their school bus driving.

(5) The State of New York and all other states should adopt and implement all the provisions of the present Highway Safety Program Standard No. 17, Pupil Transportation safety, with special emphasis on the provisions relating to the selection and training of Personnel (Personnel, Section IV, C1) the safe routing of

school buses, and the elimination of standees in school buses (Vehicle Operations, 1V, C5).

(6) The State of New York Department of Education should expand its pupil-transportation safety activities in order to provide liaison, management consultation, and supervision at the local level to assure compliance with its policies and procedures. The State's participation on the local level should also include more active assistance in training pupil-transportation personnel.

(7) The National Highway Traffic Safety Administration should assess the human factors involved in seatbelt usage in school buses through a demonstration project. The project should include a number of buses equipped with seatbelts and high back padded seats which are engaged in pupil transportation.

(8) The International Association of Chiefs of Police should use its influences and resources to redirect the attention of law enforcement agencies to the need for uniform enforcement of traffic laws pertaining to railroad/highway grade crossings (1963 IACP resolution F-18, Highway Safety Policies for Police Executives). Such enforcement should provide special emphasis on those crossings protected solely with stop signs. [1143]

Five people died when their car was struck by an Atchison Topeka and Santa Fe Railway passenger train on what was referred to as a 'marked crossing' at Strong City, Kansas, United States, at 4.30 a.m. on Friday 31st March 1972. Engineer Olen Orr stated that he had not seen the car until it came onto the track. The vehicle was thrown 30 feet and its occupants were all ejected from it.

Those who died were James Sparks, 31, from Emporia, Kansas; his wife Esther, age unknown; his brother Ned Sparks, 26, of Strong City; James Overstreet, 19, of Elmdale, Kansas; and Junior Warren, 17, of Strong City, who had been driving the car.

Twenty-five minutes before the accident Sheriff Frank Coburn had stopped the car in Emporia after it made an erratic turn. He said that all the passengers in the car had been drinking but its driver had apparently not and so he had not detained the vehicle. The party had said that they were bound for Strong City to spend what was left of the night there. Several beer cans were found in the wreckage of the car. [1144]

There were conflicting reports as to the severity of a level crossing collision between a train and a bus which occurred at Angeles, the Philippines, at 03:35 on Sunday 14th May 1972. The northbound diesel multiple-unit passenger train was carrying 300 people, mostly holidaymakers returning from the mountain resort town of Bagio and other places. The La Mallorca – Pambusco bus reportedly tried to beat the train over the crossing. As a result of the collision the fuel tank of the bus was punctured and a fire broke out that enveloped the bus, four cars of the train and a dozen lineside buildings, the latter being variously referred to as squatter shacks and as shops and houses. A Manila newspaper stated the next day that at least 10 people had died while another report spoke of 30

[1143] National Transportation Safety Board report

[1144] *The Hutchinson News*, Hutchinson, Kansas, United States, 31st March 1972 fatalities, but these reports were denied the same day by Nicanor Jimenez, the manager of Philippine National Railways, and by Angeles police chief José Mercado. Mercado stated that the train had carried the bus about 20 metres and the bus driver and four passengers had died. The train passengers had had to scamper to safety and some of them had been slightly hurt, while none of the squatters had been hurt. There had only been six passengers on the local bus and its conductor survived. [1145] [1146]

An accident reminiscent of the 1968 Hixon disaster in England (see above) occurred in Brazil on the evening of Wednesday 17th May 1972 in the São Paulo suburb of Lapa.

Ferrovia Paulista (Fepasa) passenger train U-75 had left Júlio Prestes station in São Paulo at 17:40 for Itapevi. It was in charge of Conductor João Pier Vieira with Driver Eduardo Fuglini and Assistant Driver Fiero Milanese and was carrying about 1,600 passengers in six coaches. Having called briefly at Lapa station, it had restarted and reached a speed of 40 kilometres an hour when, at 17:52, it collided with the low loader trailer of an articulated lorry, carrying, ironically, a railway locomotive. The engine and leading two coaches of the train were derailed and badly damaged. The accident could have been even worse as a second train from Itapevi to São Paulo was close at hand. Fortunately it was able to pull up just short of the wreckage.

Initially it was stated that 13 people had died in the accident and 60 had been injured; the following day the number of fatalities was corrected to eight, while the number of known injured had increased to 98, 19 of whom had been admitted to hospital, and it was believed that the true number of injured could be a good deal higher, as many victims had been treated by private doctors rather than go to the hospital emergency department.

Workman Gonçalo Expedito Simão, 34, said he had been taking the train to and from work for 10 years. "I'm saved by a miracle," he said, still panting and barely recovered from the fright. "The journey went as usual. I was riding in one of the front coaches, looking at the passing landscape and thinking of day-to-day problems. Suddenly I saw a red light on the level crossing and on the other side of the line an enormous lorry with an engine on its trailer, which seemed to be stopped on the tracks. More by instinct than anything else, I jumped up from my seat and ran towards the back of the train, which I think was going at 60 kilometres an hour. The crash was tremendous. The door of the coach into which I had fled was open. With the impact I was thrown from it. Incredible as it may seem, I ended up lying beside the track without a scratch."

Some of the injured victims were also interviewed by reporters. José Maria de Cruz, 18, who had suffered a broken right leg and injuries to both arms, was in a state of shock and repeatedly crying. In between sobs he kept repeating, "Please tell my wife that I am fine." He was worried because his wife, Catarina, 20, had given birth only 11 days previously and with his new-born son was awaiting his return in the shanty where they lived in Carapicuiba. "I'm afraid that she's going to lose milk and that's why I want her to know that I'm fine," he explained, adding that if his wife was unable to breastfeed their baby she would have a lot of difficulty obtaining powdered milk. "I didn't realise there was going to be an accident," he said. "I was in one of the first carriages and I didn't see anything. I just remember waking up in an ambulance. My head hurts from a blow but the

[1145] *The Manila Times*, Manila, The Philippines, 15th May 1972
[1146] *Pacific Stars and Stripes*, Tokyo, Japan, 16th May 1972

doctors don't think that it's anything serious. Ask the driver who comes from there to stop by my house and tell my wife that I'm fine," he appealed.

Maria Cicera Esteves, 34, was one of the first patients to be sent home from the casualty department. She had suffered bruises and had a lot of pain in her thigh. She said that she had been riding in the carriage immediately behind the engine and after feeling a violent shock, which had thrown everyone forward, she lost consciousness. "I didn't know anything else. When I came round I was lying on the floor and I don't even know how I got off the train," she said.

Manuel Arlindo Nunes, who owned the Dois Irmaos bar and café near the crossing, like a number of residents was alerted by the noise of the collision and going to investigate found that hundreds of men, women and children were leaving the derailed train, crying for help.

Alexo Gonçalo, who also lived near the crossing, likewise spoke of screaming passengers running from the wrecked train. Some, he related, in their desperation, had not even noticed the barbed wire fence that separated the railway from the Rua John Harrison, so he had seized a sledgehammer and with the help of others made a hole in the fence to form an exit. Then they had tried to stop cars to take the injured to hospital but some drivers refused to stop and they had been forced to stand in front of vehicles to force them to halt.

After a few minutes Captain Pellegrini of the Fire Department arrived with the first of 40 men, who would be working at the scene until dawn, removing the dead and injured, assisted by some of the uninjured passengers and nearby residents. Twenty minutes after the accident Colonel Teodoro Cabett of the Military Police was also at the scene and began to co-ordinate the work of the Fire Brigade, the Military Police, the Civil Police, and the medical staff.

At the Hospital das Clinicas 150 employees were mobilised to attend to 23 patients arriving at the emergency room and all employees who should have left duty at 19:00 were ordered to remain.

Transport Secretary Paulo Salim Maluf was early on the scene. Via the radio and television stations he requested that all buses passing through the area should approach the Municipal Market, near the crossing, to pick up the stranded passengers, saying that it was the most immediate measure he could take to lessen the crisis to the transport system that the blockage of the line had caused. Extra buses were provided to carry passengers – free of charge – between Júlio Prestes station and the station at Domingos de Moraes, in Lapa, but on the far side of the accident site. Señor Maluf added that he hoped the line could be cleared in 24 hours.

The IRGA company road transporter, a Scania Vabis lorry, had brought the Cosipa (São Paulo steel company) locomotive from Vila Leopoldina to the Estrada e Ferro Santos a Jundiai works at Lapa. The articulated transporter weighed 48 tons and the locomotive 70 tons, so the total weight of the obstruction blocking the railway was 118 tons. The lorry was being driven by Jair Gonçales Chaves, with assistant Válter Aparecido Medeiros and scout Heleno Luis dos Santos. The crossing leading into the works was a private road, provided with warning lights and was guarded by a crossing keeper, but it was without protecting signals the crossing keeper could have put to danger in an emergency. Before any unusual or special vehicles were to use the crossing the railway company should have been informed, so that special measures could be taken but this was apparently not done.

When the transporter entered the crossing the warning light was green but then the weight of the transporter caused it to ground on the crossing and it could not be moved. Driver Chaves, his assistant, and the scout began to try and raise the transporter with a jack while the crossing keeper ran to find a telephone to warn of the obstruction. Perhaps ten minutes later efforts to shift the transporter were still unavailing and the train had arrived at the crossing. Driver Fuglini, approaching the spot around a curve and only being alerted to the danger at the last moment by the sight of the crossing keeper waving his flag, was unable to do anything to avert the collision, but nevertheless stuck to his post. The Fepasa public relations department praised Driver Fuglini, who was seriously injured, for "his courageous and solid performance," adding that, "The driver, who could have left his cab before the collision, elected to remain at his post, doing what he could to brake the train, even knowing that he was putting his own life at risk, and he thus managed to reduce its speed and lessen the consequences of the collision." The men accompanying the transporter and the crossing keeper all got clear at the last moment and escaped. Afterwards Driver Chaves and his crew, in a state of shock, went to a nearby bar for a drink – sweetened water according to the newspaper but possibly something stronger.

Given that there were four and probably six people at the crossing, as a newspaper report indicates that there were also two General Motors mechanics accompanying the transporter, all of whom were obviously aware of the obstruction, and the apparent gap of ten minutes or more between the stalling of the transporter and the arrival of the train, it would appear that had at least one individual involved gone in either direction along the line to warn trains the disaster could have been averted. Probably they were too busy attempting to shift the grounded load to think of doing so. The line was worked by Centralised Traffic Control (CTC) with full track circuiting but the rubber tyres of the lorry would of course not activate the track circuits.

Fepasa stated that compensation of 10,000 cruzeiros would be paid to the families of those killed and to those permanently disabled, and lesser amounts to other victims depending on the nature and extent of their injuries. The company would also cover the funeral and hospital expenses of the victims.

Investigations were opened by the police and by Fepasa and the police had soon interviewed 25 people, but not the unnamed crossing keeper, the transporter crew or Driver Fugini.[1147] [1148]

Following the accident there were recriminations between IRGA, the transportation company, and Fepasa, the railway company. The day after the accident a spokesman for IRGA stated, "Fepasa was informed that the transporter carrying the locomotive was to use the crossing. Our driver only moved onto the crossing after receiving permission from the two crossing keepers present {only one crossing keeper was mentioned in the newspaper reports and two seems unlikely} and when the lorry became grounded there had still been time for a crossing keeper to have warned the train driver."

Fepasa in turn questioned why the transporter had been using a crossing that was not suitable for heavy vehicles, adding that the normal procedure was for notice to be given

---

[1147] *Jornal do Brasil*, Rio de Janeiro, Brazil, 18th and 19th May 1972

[1148] *O Estado do São Paulo*, São Paulo, Brazil, 18th May 1972 to the company that a heavy vehicle needed to use the crossing, so that a suitable time for the movement could be arranged. It added that the time when the transporter had entered the crossing, just before 18:00, was the evening rush hour, the time when the train service was most intensive.

According to the IRGA spokesman there should have been plenty of time for the lorry driver to have taken the jack from the roof of the transporter, placed it beneath an axle and begun lifting the trailer so that it was clear of the ground, while during that period of at least five minutes the crossing keepers could have taken measures to stop the rail traffic. Also the electronic signalling should have protected the transporter, the IRGA spokesman added, losing his temper somewhat, but it had not, and hence there was no safety at the crossing. It was clear to him that Fepasa was entirely to blame for the accident, as there had been plenty of time to prevent it. Calming down, the spokesman explained that the crossing was not used for ordinary traffic, but solely by large lorries and transporters bound for the Sorocabana yard, where locomotives and other equipment were taken for repair. “Everyone was aware that the locomotive was arriving,” the spokesman continued.

Two technicians from General Electric, the manufacturers of the locomotive, were at the site to assist in the unloading, as were Santos s Jundiai staff. The transporter driver had waited at the crossing for permission to cross and had only done so after receiving it from the crossing keeper. IRGA lorries, and those of other carriers, used the crossing many times a month. On the day of the accident there would have been time to move the stalled transporter, the spokesman repeated; work had been in progress to free the grounded transporter and there had been time to stop the trains. The transporter, he added, was a new 1971 model and had only been in use for six months. It had cost 200,000 cruzeiros and had third-party insurance; the locomotive loaded onto it had been fully insured by General Electric.

Fepasa stressed that the first matter to be investigated was whether there had been authorisation for the transporter to use the crossing. Visibility from the São Paulo direction was limited to 80 metres and a train required almost 200 metres to stop. The crossing was a private one for the Estrada de Ferro Santos a Jundiai and was manned by three crossing keepers of that company, working shifts. According to its information the crossing was used by an average of 185 vehicles a day, a total therefore of 1,100 over the six-day week. The crossing had been created to serve users of a canteen and was equipped with signs and a warning bell to warn the crossing keepers of an approaching train. Visibility from the opposite direction was greater, being 400 metres in clear weather.

Fepasa issued a statement regarding the accident: ‘The Board of Fepasa – Ferrovia Paulista S/A, has taken the following measures in relation to the accident which occurred at 17:52 yesterday, at the level crossing at kilometre 7 of the Sorocabana line with its U-75 train. which collided with the IP-7627 road transporter of the IRGA SA company, which was transporting a COSIPA locomotive to the yard of the 9th Division Santos-Jundiai – RFFSA.

(a) Immediate removal of the injured to the Hospital das Clinicas, Mater Dei Hospital – Lapa, Barra Funda Municipal Emergency Room, and Sorocaba Hospital.

(b) Removal of the bodies of those who had died to the Legal Medical Institute mortuary.

(c) Clearance of the site to allow rail traffic to resume, a task completed at 10:20 today.

(d) Appointment of a Committee of Inquiry composed of engineers Sergio Cruz Marelli, Norherto do Vale Ferraz and José Antonio Patricio to determine the causes of the accident.

(e) Fepasa has taken the necessary steps to cover the burial expenses of those who died and the hospital costs of the injured.

'In addition to the above measures Fepasa, considering the courageous actions of Driver Eduardo Fuglini, who had not been involved in a previous accident in 23 years of service, has decided to include a commendation in his work record. Whilst he could have abandoned the cab of his engine before the collision he chose to remain at his post, knowing that it might cost him his life, and had done what he could to stop his train, thus reducing its speed and diminishing the effects of the collision. He is at present in hospital, having suffered two broken legs as well as multiple bruises.'

A fatal mistake by a crossing keeper led to a collision between an express and a bus in Italy on Monday 11th September 1972. Both the railway and the road, which intersected at the crossing at Summaga, a kilometre and a half west of Portogruaro station, were busy. At 16:55 on the day of the accident the barriers of the crossing were lowered for a Trieste – Venezia train. By the time this train had reached the crossing, waiting on one side of it was an Autostradale company bus that had come from Milano, behind which were a road tanker and a police car. As the train cleared the barriers Crossing Keeper Bruno Mariotti, 26, raised them, forgetting or oblivious of the fact that train No. 185, a Venezia – Trieste direttissimo, was approaching from the opposite direction. This train was running either 40 minutes or two hours late according to different reports. Portogruaro carabinieri marshal Antioco Serra, a passenger in the police vehicle, declared, "My driver was just putting the car into gear ready to follow the tanker, which had just begun to move, when I heard the whistle of another train. By that time the bus was already on the crossing and the train struck it on its right side, dragging it like a twig, and it was thrown down the embankment. It rolled over and was torn open, with the bodies of its occupants tumbling out like puppets. It was a terrible sight."

Alvise Zanon, the driver of the tanker, said: "When the crossing carriers were raised after the passage of the rapido I put my lorry into first gear. I had moved forward a couple of metres when I heard the piercing whistle of another train and saw it approaching. I managed to stop just in time. It was too late for the bus though. From what I remember its driver tried in vain to accelerate to get clear of the crossing. What happened then I will never forget."

By the following day six people had lost their lives in the accident and another 14 were injured, six in a very serious condition. Five of those who lost their lives were named as Giuseppe Arpini, 58, from Milano, the bus driver; Remigio Piazzaq, 45, from Milano; Elizabatta Sartor, 63, from Torviscola; Stefano Timico, 11, from Dalmine; and Bruno Zanetti, 69, from Cervignano. The sixth fatality was a man aged apparently between 28 and 30.

The crossing keeper reportedly disappeared after the accident. The same evening the Ministry of Transport issued a statement that it was working to eliminate all level crossings and that 3,000 had been done away with. A Communist newspaper retorted sarcastically that another 14,000 remained. [1149] [1150]

Eleven Czechoslovak Army parachutists lost their lives on Thursday 21st September 1972 when the bus in which they were riding was struck by a train on a level crossing near Pernik, Bulgaria. They had been part of a group of competitors on their way to a parachute jumping competition for the soldiers of Eastern European armies being held in Bulgaria. [1151]

Twelve bus passengers were killed and 30 were injured, ten critically, when a crowded bus was struck by a train on an unprotected level crossing year Prizren, Yugoslavia, on Wednesday 4th October 1972. [1152]

Mrs Sylvia Marie Garcia, 27, was on her way to her mother's mobile home only 100 yards north of the Penn Central Railroad line with her four daughters when their car was stuck by an engine and caboose on Fisher Road Crossing, Ohio, United States, about five miles east of Lima, at 4.40 p.m. on Monday 23rd October 1972. Mrs Garcia and her children Christine Louise, 7; Veronica Sue, 6; Angela Marie, 3; and Rebecca Ann, four months, were all killed.

Engineer Arthur Lee Herron stated that he had sounded his horn when he had seen the car moving onto the crossing and made an emergency brake application when the car failed to stop. The train was travelling at 50 miles an hour and it carried the car about 400 yards down the line before it could be pulled up. It took rescuers more than an hour to remove the bodies from the wreckage of the vehicle.

The crossing was protected only by crossbucks.

Mrs Garcia only held a temporary driver's permit.

Having taken his engine and caboose into the yard at Lima and filed a deposition with the sheriff's deputies regarding the accident Engineer Herron picked up a train and continued towards Fort Wayne, Indiana. Just over the state line in Monroeville, Indiana, his train collided with a lorry that had stalled on a crossing. No one was hurt this time. [1153]

Five members of a family died when the pickup truck in which they were riding was struck by an Atchison Topeka and Santa Fe Railway train on a level crossing with FM 2301 Road, a mile and a half east of Aiken, Texas, United States. The accident happened at about 7.30 p.m. on Thursday 9th November 1972. Those who lost their lives were Pedro Diaz, 43; his wife Soledad, 35; their daughter-in-law Jobita, 22; and Jobita's children, Ignazio, 2; and Mario, 4 months. The train carried the pickup truck almost half a

[1149] *La Stampa*, Torino, Italy, 12th September 1972
[1150] *L'Unita*, Roma, Italy, 12th September 1972
[1151] *The Medicine Hat News*, Medicine Hat, Alberta, Canada, 22nd September 1972
[1152] *The Times*, London, England, 5th October 1972
[1153] *The Lima Times*, Lima, Ohio, United States, 24th October 1972

mile from the unprotected crossing. [1154]

A passenger train collided with an oil tanker on a level crossing at an unstated location in Mexico on Saturday 6th January 1973. An explosion occurred following the collision and the engine and two leading coaches of the train caught fire. Five people died and four more were injured. [1155]

A Lehigh Valley Railroad freight train consisting of a diesel locomotive and 25 cars collided with a car on Cayuga Road Crossing, Williamsville, New York, United States, just after 12 noon on Sunday 7th January 1973. Earl Brogan, 29, and his children Yvonne, 8; Keith, 5; and Yvette, 4; were killed, as was Debbie Nabb, 7, the daughter of a neighbour. Mrs Marion Brogan, Earl's wife, was taken to hospital suffering from shock and possible internal injuries. The crossing was protected by warning lights and bells that were working properly after the accident and it was theorised that the driver of car might have been blinded by the low sun. [1156] [1157]

On Tuesday 16th January 1973 a train collided with what was described as an excursion bus at a location somewhere in the vicinity of Hat Yai Junction, in Songkhla Province, southern Thailand. Fourteen women, nine men and two boys were killed outright or fatally hurt and another 35 people were seriously hurt, including the bus driver, whose chances of survival were said to be very slim. [1158]

A level crossing equipped with automatic half barriers, adjoining a station at which most trains stopped, was the scene of a major disaster at Kecskemét, Hungary, on Tuesday, 30th January 1973. Because the majority of trains stopped at the station the time between the lowering of the crossing barriers and the arrival of a train at the crossing was often as much as a minute and a half. This long interval led to the drivers of road vehicles, impatient at the delay, zig-zagging around the barriers. On the day of the accident cars and a service bus were driving around the barriers in the usual manner but the train for which they had descended was a non-stopping one, which reached the crossing in only 30 seconds. The train struck the bus, killing 38 people and injuring 18 more.

One solution to the problem of such crossings is the use of predictor circuits, by which the speed of the train is calculated and the barriers made to descend earlier or later, dependant on the speed of the train. Another solution is the use of stop signals protecting the crossing, which remain at danger with the barriers raised until the guard or conductor of a stopping train operates a plunger when the train is ready to start, which initiates the lowering of the barriers and the subsequent clearing of the signal. [1159]

[1154] *Lubbock Avalanche-Journal*, Lubbock, Texas, United States, 10th November 1972

[1155] *The Snyder Daily News*, Snyder, United States, 9th January 1973

[1156] *The Oneonta Star*, Oneonta, New York, United States, 8th January 1973

[1157] *The Bradford Era*, Bradford, Pennsylvania, United States, 8th January 1973

[1158] *Pacific Stars and Stripes*, Tokyo, Japan, 19th January 1973

[1159] Wikipedia – 1973 Kecskemét level crossing disaster

Tuesday 6th February 1973 saw another serious train/school bus accident in the United States. The accident happened on the outskirts of Littlefield, Texas, to a bus taking 29 pupils home from four schools in the town to the small farming community of Fieldton. Seven children died and another 17 were injured when the bus was struck by an Atchison Topeka and Santa Fe Railway freight train travelling at 58 miles an hour (the line speed limit was 60 miles an hour).

A local newspaper described the aftermath of the accident as it affected the residents of Fieldton: 'Fieldton community near here {Littlefield}, a town that can only hope for a brighter day, did not see the sun come up on Wednesday.

'Six school children, silent in the cloudy grey dawn, waited on the edge of the cotton fields to board bus No. 4 for the 10-mile trip south to Littlefield. On any other school day in the past another carrier, No. 7, would have been bringing 35 laughing and often rowdy youngsters to school. But Wednesday was different. No. 7 lay idle in the bus yard of the Littlefield School District, its wreckage a mute testament to one of the blackest moments in the history of Lamb County and its seat of 7,000.

'Seven of No. 7's Tuesday passengers will be buried later this week, and 16 others are in area hospitals. Three of these children are in critical condition.

'Only six children escaped injury in Tuesday's train bus condition in which the rear of the bus was severed by a fast-moving Santa Fe freight train at the rail crossing a few blocks from Main Street. As fate would have it, those six had remained in Littlefield to attend Boy Scout meetings, gatherings that caused one mother to say, "God was with us. He had a hand in it. We can't understand this thing, but we're so grateful today was Scout's day."

'But while some were counting their blessings, others were living out the horror of that black moment that rubbed out the lives of their friends.'

Surviving pupils described the moments leading up to the collision. "Everybody in the bus screamed that the train was coming," said 11-year-old Raul Ortiz from his hospital bed as he recalled the moments before the 58-car Santa Fe train sliced into the bus at the intersection of US 385 and the Santa Fe Depot.

"I saw the train coming in before the bus started after it stopped. Anita Martinez (one of the victims, a school drop-out who had decided to renew her education), screamed at Mr Johnson (bus driver)," said Jesse Diaz, 15, who was sitting in the middle of the bus as it stopped at the blinking light crossing. "When the bus stopped, everybody yelled that the train was coming. Everybody yelled and turned. He (the driver) tried to make it across," said Ortiz.

'Terry Foley, 13, thrown clear of the wreckage after the impact, told his mother that a little girl next to him was screaming. Terry said that, "The little boy next to me was dead."

'Mrs Jean Patterson, a substitute bus driver whose vehicle was stopped behind No. 7 driven by Artis Johnson, 68, said, "He pulled up there and stopped. Then he kept creeping across the tracks and all of a sudden the train came up from behind the depot."

'Another bus had just moved across the tracks earlier. Mrs Patterson, who crawled underneath the box cars when the train stopped, said a little girl was crying and saying, "That's my brother." And I could see that the kid was dead."

'Susie Moore, a Littlefield teacher, arrived at the scene minutes after the crash. "It was just horrible," said Mrs Moore, who lives in Lubbock. "We were on our way home. We pulled up to the scene about five minutes after it happened. The first person I saw was Rex Gilstrap running toward the wrecked bus, and the look upon his face was enough to scare anyone to death."

'Gilstrap, a business teacher at Littlefield High School, was driving a bus along the same route when the collision occurred at 4.05 p.m. Tuesday.

'Calm, but with an occasional crack in her voice, Mrs Moore said several school coaches arrived soon after the mishap and helped pick up books, children's purses, coats, and other items strewn about the area.

'Littlefield Hospital head nurse Barbara Symington, one of many nurses dispatched to the area in the town's four ambulances, said, "One of the nurses had to be given something to calm her down. She was terribly upset."'

It emerged after the accident that the bus driver had not held a chauffeur's licence required for a driver transporting people for hire, though he held a valid commercial driver's licence for a vehicle weighing over 2,000 pounds used to transport property and the test required for the two licences was the same, the only difference being that the chauffeur's licence cost $1.50 more. School officials were of the opinion that some leeway was permitted in the matter, stating that their regular driver had retired two weeks earlier and Johnson was only being employed temporarily as a bus driver. The licensing division of the Department of Public Safety agreed that this had been the case in the past but stated that it no longer was.

Johnson was only slightly hurt in the accident but was admitted to hospital suffering from shock, as was his wife. [1160]

A week after the Littlefield accident, on Tuesday 13th February 1973, yet another school bus accident happened in the United States, though this time it was only a minor one. A bus on loan from the Idaho Falls School District to the Eastern Idaho Child Development Center was struck by a train at the 19th Street Crossing in Idaho Falls. There were no children on the bus at the time, just the driver Charles Rankin, 37, and two other men, one of whom, Harold Black, 32, was injured, as was the bus driver, who was cited by police for failure to obey a railroad crossing signal. [1161]

Six Alberta, Canada, teenagers heading for a birthday party on a farm were killed when the car in which they were riding was hit by a Canadian National Railway rail-liner train on a level crossing near Stettler, Alberta. Gregory Ballinger, 19, of Ending, the driver; Virginia Baugh, 18; Cheryl Hauser, 17; and Evelyn Irene Thieme, 18; all of Stettler; Gail Evelyn Brooks, 18, of Botha; and Marilyn Louise Wasbel, 17, of Byemoor, were all killed instantly. Evelyn Thieme was the person for whom the party was being held; she would have been 19 the next day. The accident happened on Thursday 10th May 1973

[1160] *Amarillo Daily News*, Amarillo, Texas, United States, United States, 8th February 1973
[1161] *The Post-Register*, Idaho Falls, Idaho, United States, 16th February 1973

The 1984 automobile was pushed 200 feet from the unprotected crossing. It was not clear if the car driver had failed to see the train, perhaps through being blinded by the evening low sun, or if he had been trying to beat the train. [1162]

At least 23 bus passengers lost their lives when its driver attempted in foggy weather to negotiate an unprotected level crossing near Kyŏngju (also spelt Gyeongju), South Korea, at 07:25 on the morning of Wednesday 16th May 1973. Twenty-one passengers died on the scene and 62 were taken to hospital, where two more died soon afterwards, while 24 others were said to be in a serious condition. No one on the train was hurt. Although it was not strictly a school bus, most of the passengers on the vehicle were children on their way to school in the nearby city of Pohang-si. The train caught the rear quarter of the bus and sent it tumbling down an embankment into a stream. All that was provided to warn road users at the crossing was a stop sign, while the bus was also overloaded, carrying at least 83 passengers, considerably in excess of its seating capacity of either 60 or 70. A special police team was sent from Seoul to investigate the accident. [1163] [1164]

An apparent mistake by a crossing keeper combined with a fog led to a collision between a train and a small bus in the Valentin Alsina suburb of Buenos Aires, Argentina, on Saturday 9th June 1973. The train driver and the bus driver, both of whom were injured, agreed that the crossing barriers had been raised when the train reached the crossing, deceiving the bus driver into attempting to cross the line. Fourteen bus passengers died and 28 were injured. Travelling at speed, the train carried the bus 100 metres before it could pull up. [1165] [1166] [1167]

Unlike the situation in, for example, Great Britain, German schoolchildren start their days early in the morning and finish school correspondingly early, sometimes as early as 12:00. Thus it was that a bus from Kißlegg had already dropped off most of its pupils just before 13:00 on Monday 25th June 1973, though it had also picked up some children who wished to go into Kißlegg, where the bus was to return. The farthest point of the bus's route involved it negotiating a level crossing near Memmingen, West Germany, so as to drop off children who lived in Reipertshofen. As the bus was approaching the crossing, so was a diesel locomotive travelling at 90 kilometres an hour. The driver of the engine observed that the warning lights of the level crossing were operating and it appeared to him that the bus was going to continue straight ahead, along a road that paralleled the railway, when suddenly it instead turned to the right onto the crossing. It was too late for any effective braking and the engine stuck the bus, throwing it 15 metres into a meadow, while the engine travelled over 300 metres before it could be pulled up. Seven children were killed and 13 were injured, some seriously. Surviving pupils said that they had seen

---

[1162] *The News*, Medicine Hat, Alberta, Canada, 11th May 1973
[1163] *Pacific Stars and Stripes*, Tokyo, Japan, 18th May 1973
[1164] *The Times*, London, England, 17th May 1973
[1165] *Clarin*, Buenos Aires, Argentina, 10th June 1973
[1166] *The Cedar Rapids Gazette*, Cedar Rapids, Iowa, United States, 10th June 1973
[1167] *The Daily Review,* Hayward, California, United States, 10th June 1973

the flashing lights and shouted to warn the bus driver, without effect. [1168]

On Friday 14th September 1973 local freight train No. Nv 3018 collided with a car on a level crossing between Essen-Altenessen and Geisenkitrchen, West Germany, killing six people in the car. The barriers at the crossing had been raised prematurely. The crossing keeper pleaded both that he had not been advised of the approach of the train and that the crossing barriers had been defective. At the funeral of the victims and on the following night a demonstration was held by 300 members and supporters of the Communist Party who laid candles on the line and displayed banners reading: "Here angels are made." To the author it is not exactly clear what they were demonstrating against unless it was the unsafe nature of the crossing. The crossing keeper was subsequently given a six-month suspended prison sentence. [1169]

Eastbound Canadian Pacific Railway passenger train No. 2 collided with a car on a level crossing about three quarters of a mile west of Whitewood, Saskatchewan, Canada, at about 5 a.m. on Wednesday 14th November 1973. The car was demolished and its five teenage occupants, four men and a woman, were all killed. The train was delayed almost an hour. A pair of stop signs was the sole protection at the rural crossing. [1170] The victims were later identified as Eric Lindhal, 18; and David Tallentire, 19, both of Whitewood; Clifford Wilton, 18, of Langbank; Jack Korey, 18, of Esterhazy; and Joanne Nelson, 17, of MacLean. Slippery road conditions were a possible factor in the accident. [1171]

Six members of a family lost their lives in a level crossing accident not very far from their home in Kittrell, North Carolina, United States, on the afternoon of Sunday 9th December 1973. Sidney Roberson, 16, who had been driving the car; his brother Ricky, 14; a sister-in-law, Mrs Diane Roberson, 27; and Diane's children, Steven, 8; Scott, 7; and Suzanne, 3, were all killed. The accident happened two miles from Kittrell at 12.35 p.m. when the car in which they were riding was struck by a southbound Seaboard Coast Line Railroad freight train, consisting of an engine and 25 cars, and was dragged at least half a mile. According to one report it was pushed a mile and three-quarters. [1172]

Engineer Daniel E. Matthews Jr. said that the automobile had stopped at the crossing as his train had approached, then without warning had pulled forward onto the crossing. The group had been coming home from church. [1173]

A lorry carrying labourers to (from?) work on the evening of Tuesday 5th March 1974 was struck by an engine on a level crossing near Rajgir, Bihar, India. Nine people died

---

[1168] *Katastrophen der deutschen Bahnen Teil II* by H.J. Ritzau, pg. 255
[1169] Ibid, pg. 256
[1170] *Winnipeg Free Press*, Winnipeg, Manitoba, Canada, 14th November 1973
[1171] Ibid, 16th November 1973
[1172] *Statesville Record and Landmark*, Statesville, North Carolina, United States, 10th December 1973
[1173] *Playground Daily News*, Walton Beach, Florida, United States, 10th December 1973

and another 26 were injured. [1174]

Six people riding in what was referred to as a 'compact car' lost their lives in a collision at an unprotected level crossing near Fort Covington, New York, United States, on the evening of Thursday 28th March 1974. The automobile was hit by a 37-car Penn Central Railroad train, on its way from Massena, New York, to Montreal, Quebec, Canada. Mrs Sandra Jarvis, 26, the owner and driver of the car; her daughter Patricia, 6; Joseph Savage, 26; Constance LePage, 17; and Bernard LaLonde, 36, were all killed more or less outright. Steven LaLonde, 3, survived the initial impact but died early the following morning in a Plattsburgh hospital.

Engineer Eugene Gadway said that the car had initially slowed up, "but then increased speed in an apparent attempt to pass in front of the train." He had sounded his whistle for the crossing.

A trooper remarked, "I bet you could see a quarter mile down the tracks in either direction. It's an open area with just fields on either side." The weather had also been clear with good visibility. [1175]

Twenty-eight children, on their way home at about 12.35 p.m. on Friday 5th April 1974 for the start of the Easter holiday, were slightly hurt when the school bus in which they were riding was struck by a Southern Pacific Company train driven by Engineer James Melvin McCutcheon on a level crossing in Bakersfield, California, United States. There had been 56 pupils from the Rushmore Elementary School, aged between five and 13, on the bus, which was being driven by Daniel Sanchez Chairez, himself only 18.

The bus had failed to stop at the stop line of the crossing and a motorist who could see the approaching train through the bus windows sounded his horn when the bus began to move forward onto the track, but his warning was misunderstood or ignored. The train caused considerable damage to the front of the bus. One pupil suffered a broken arm, the remainder only cuts and bruises. They were taken to Kern General, Mercy and San Joaquin hospitals and all were released after treatment. [1176]

Five people from Idaho lost their lives just over the state line when a Burlington Northern Railroad freight train travelling at an estimated 60 miles an hour collided with their car on Idaho Road Crossing, near Otis Orchards, Washington, United States, on Thursday 4th July 1974. The victims were Laura Davey, 18, the driver, from Post Falls; her son Curtis, 2; Gloria Benham, 14; and her sister, Virginia, 4, also from Post Falls; and Ronald Copper, 16, of Coeur d'Alene. It was not stated if the accident occurred at the same crossing as that in the May 1970 accident noted above. [1177]

Six bus passengers were killed outright and 38 were injured when a freight train collided

---

[1174] *The Times*, London, England, 7th March 1974

[1175] *Adirondack Daily Enterprise*, Saranac Lake, New York, United States, 29th March 1974

[1176] *Bakersfield Californian*, Bakersfield, California, United States, 6th April 1974

[1177] *The Register*, Orange County, California, United States, 5th July 1974 with a bus at a level crossing in the suburbs of Mexico City, Mexico, on Thursday 25th July 1974. Later the same day 18 people were killed in a collision between a bus and a lorry north east of the city. [1178] Two of the people injured in the level crossing accident died later in hospital. The bus driver told the police that he had not seen the train.

After the accident irate families seized two buses and threatened to keep them until they received compensation. An official of the bus company met a spokesman for the families and promised to pay the equivalent of $12,000 per victim, to which the families responded that they would keep the buses until the compensation was in their hands. [1179]

Five occupants of a car lost their lives on the morning of Sunday 29th July 1974 when it was in collision with a train. The car was travelling on national road No. 430 from Badajoz to Valencia when the accident occurred in the vicinity of Almansa, Spain, at kilometre 306.500 of the railway. The car smashed through the lowered barrier of the crossing and entered the line directly in front of an Alicante – Madrid mail train, which struck the vehicle and dragged it a considerable distance along the line, killing its occupants instantly.

Those who died were Juan Díaz González, the driver, 23; his brother Alfonso, 21; their parents, Pablo Díaz Molina, 50; and Rosa González Agudo, 48; and the driver's wife, Enriqueta Rubio, 21.

The car could have taken an alternative route that would have avoided the level crossing altogether.

Some goods trains were delayed about an hour by the accident but no passenger services were affected. [1180]

A station wagon carrying eight people returning from a morning of blueberry picking was struck by a Minneapolis St Paul and Sault Ste Marie Railroad (Soo Line) freight train travelling from Gladstone, Michigan, to Sault Ste Marie, Ontario, Canada, on Tuesday 30th July 1974. The accident happened on an unprotected level crossing at Fibre, Michigan, United States. Sandra Jean Chamberlain, 34; Marie Cryderman, 34; her children Forest Jr., 11; and Sherri, 6; Charlotte Ann Darling, 26; her daughter Deane Renea, 6; and Carol Sauro, 30; were all killed outright or died soon afterwards at the military hospital at nearby Kincheloe Air Force Base. Mrs Darling had been eight months pregnant. Her other daughter, Tracy Marie, 3, was critically injured and succumbed at the hospital at following day. Mrs Chamberlain had been driving the station wagon northwards on Highway 40, back to Sault Ste Marie, where all the victims lived. The car was carried 1,254 feet by the train, which was not derailed and was later able to resume its journey. [1181] [1182]

[1178] *Independent*, Gallup, New Mexico, United States, 26th July 1974

[1179] *Vallejo Times-Herald*, Vallejo, California, United States, 29th July 1974

[1180] *ABC*, Madrid, Spain, 30th July 1974

[1181] *The Hillsdale Daily News*, Hillsdale, Michigan, United States, 31st July 1974

[1182] *Winona Daily News*, Winona, Minnesota, United States, 1st August 1974

An accident occurred to the *Freccia del Dolomiti* (*Dolomite Arrow)* express just after 16:00 on Wednesday 7th August 1974 at a level crossing between Carmigliano di Brenta and Fontaniva, Italy, on the Milano – Calalzo di Cadore line. The train had left Vicenza at 15:29 bound for Treviso. A minute or so after it had passed Carmigliano di Brenta station it collided with a lorry on the crossing, the barriers of which had apparently not been lowered by the crossing keeper. The force of the impact was sufficient to derail at least one of the coaches of the train, which then caught fire.

Firemen from Padua, Cittadella and Vicenza were quickly on the scene, as were numerous ambulances, railway officials, police and carabinieri. The dead and injured were taken to hospitals in Cittadella and Vicenza. Conductor Fausto Nicelli, 41, and four passengers lost their lives and another 18 passengers were injured. The passengers who died were named as Florio Candian, 47; Gerado Guglielmi, 23; Mario Pozzi, 51; and Simone di Sandri, 33.

Crossing Keeper Rino Fiorito, 53, was said to have been in a state of shock after the accident. He was arrested and taken to the police station at Cittadella. [1183] [1184]

A Philippines National Railway train bound for Manila collided with a bus on a level crossing in the town of Calumpit, in Bulacan Province, the Philippines, on Saturday 10th August 1974. Twelve bus passengers were killed more or less outright and another 51 were injured when the bus overturned and fell into a roadside ditch. The cause of the accident was unknown. [1185] Six of the most seriously injured victims died in hospital in the next day or so. [1186]

A 4-H Club (an American youth development organised by adults for 5- to 19-year-olds) outing ended in a tragic accident at 3.35 p.m. on Sunday 18th August 1974 when the children of four prominent local families and the mother of one of the children were killed in a level crossing accident half a mile from Wren, Ohio, United States. The party had been working on an exhibit for the Farm Focus-USA farm show, due to be held north west of Van Wert, Ohio.

The train that struck the car was a 78-car Erie-Lackawanna Railroad freight, in charge of Conductor Harvey Cone with Engineer Charles Milner and Head Brakeman Gordon Bates. Engineer Milner told police that his train had been travelling at 40 miles an hour. He had seen the car coming southwards; it had slowed to about 10 miles an hour as if it were about to stop at the crossing but then had proceeded across the line and its right rear had been struck by his engine. The impact had spun the car around so that the front of it struck the engine in turn. After the collision the vehicle came to rest in a ditch 52 feet from Clayton Road, the road along which it had been travelling.

Killed in the accident were Mrs Marcile Oechsle, 39; her son Scott Wayne, 10; Richard Alan Bowen II, 10; Kelley Leigh Ruskaup, 11; and Jeffrey Charles Sheets, 11. Critically injured was Mrs Oechsle's husband Lowell, 39; while their daughter Tamara,

---

[1183] *L'Unita*, Roma, Italy, 8th August 1974
[1184] *The Times*, London, England, 8th August 1974
[1185] *The Argus*, Fremont-Newark, California, United States, 11th August 1974
[1186] *Pacific Stars and Stripes*, Tokyo, Japan, 14th August 1974

16, was seriously hurt. Mr Oechsle was taken to Parkview Memorial Hospital in Fort Wayne while his daughter was taken to the Van Wert County Hospital.

It was unclear who had been driving the car. There had been a previous fatal accident at the same crossing in March 1966. [1187]

At 3.14 p.m. on the afternoon of Thursday 29th August 1974 a school bus was struck by a northbound six-car Norfolk and Western Railway freight train on a level crossing with what was described as an 'oil and dirt' road near Worden, Illinois, United States. In what was clearly a lucky escape neither the bus driver, Harold Loch, nor any of the 29 children on the bus was hurt, despite the train having sheared off the front of the bus.

Engineer Keith Feree stated that his train had been travelling at 55 to 60 miles an hour and he had sounded the horn a considerable distance from the crossing. The bus had apparently stopped short of the unprotected crossing, then inched forward to be struck by the train.

Schools superintendent John Stevenson described the crossing as "a bad one" because the road climbed on either side to the track, and he paid tribute to the bus driver for averting a more serious accident. "He has been driving for years and years for us," the superintendent said, "and the fact that nobody was injured is due to his driving ability."

As well as the bus the brake hoses of the train were damaged in the collision and another engine had to be sent from Decatur before it could be moved.

The crossing was used four times daily by the school bus because four children from one family lived on the far side of it. They had just been dropped off and the bus had been turned around to continue its journey when the accident had happened. [1188]

Less than a week after the Worden escape, on Wednesday 4th September 1974, there was another school bus level crossing accident in the United States, this time near Malad, Idaho. Again there were no serious consequences, though three of the 42 elementary school children going on a field trip were slightly hurt. The fact that the accident was not worse was doubtless due to the fact that the Union Pacific Railroad freight train involved was only moving at about 10 miles an hour at the time. In addition to the fifth grade pupils there were two adult passengers on the bus (teachers?).

The bus was being driven by Malad Elementary School principle Paul Gillies, who apparently belatedly became aware of the train, for there were skid marks 50 feet long on the approach to the crossing. He managed to stop the bus with its front end just foul of the line and then attempted to put it into reverse, but the bus was struck before he could do so. The train hit the front right side of the bus, which was pushed about 20 feet without overturning and was not badly damaged.

Sheriff Kenneth Wharton, who investigated the accident, commented that, "If the train had been going a little faster or the bus had been farther out on the tracks, it could have been bad."

[1187] *Times-Bulletin*, Van Wert, Ohio, United States, 19th August 1974

[1188] *Edwardsville Intelligencer*, Edwardsville, Illinois, United States, 30th August 1974

Two of the injured pupils, both girls, were taken to hospital and one was detained overnight but released the next day. The third injured pupil, a boy, only sustained a cut lip and was treated at the scene by a doctor.

Union Pacific general manager H.H. Brandt stated, "This is a prime example of disregard of simple grade crossing safety rules. Fortunately this train was travelling only 12 miles per hour, which lessened the impact. The responsibility of school bus operators cannot be overemphasised and only constant attention to grade crossing signals can prevent tragedy." There were in fact only crossbucks at the crossing anyway, while visibility along the track was said to be poor.

Meanwhile Union Pacific public relations officer Clarence Rockwell commented, "Railroad officials are concerned about this potential tragedy and the potentially grave consequences of this kind of accident." He added that the railroad had a "near miss policy" that required engineers to report incidents of the unsafe use of crossings by motorists where no accident resulted, and he too cautioned school bus operators to be especially careful. [1189]

Train No. 277, the Up Guntur – Gadang passenger, collided with a bus at Crossing No. 272/E between Santamagular and Savalyapuram stations, India, on the South Central Railway, at about 00:50 on Friday 6th September 1974. Eleven people on the bus were killed and four were seriously injured. No one on the train was hurt.

The train consisted of Class YB steam locomotive No. 30055 and seven coaches.

The motor bus belonged to the Andhra Pradesh State Road Transport Corporation and was on its way from Kurnool to Guntur.

The Class B crossing was manned by two gatemen working 12-hour shifts and provided with gates that swung away from the line and could be secured with chains and padlocks. The gates normally stood open for the road. Before the Savalyapuram stationmaster offered a Down train or accepted an Up one from Santamagular he was required to advise the gateman by telephone, then advise him again when the train was leaving one or other station. After the second telephone call the gateman must at once close the gates. If the gateman failed to answer the telephone trains were to be cautioned.

The accident was the subject of a full inquiry. Driver L.S. Devadanam told the inquiry that he had left Santamagular at 00:37 (00:42 according to the train register there). Approaching the crossing at 50 kilometres an hour he had seen the bus on it, had shut off steam and braked and then heard the sound of the train hitting the bus. Stopping, he had accompanied his guard back to the crossing, where the gateman told them that he had had no notice of the train from the Savalyapuram assistant stationmaster. The driver stated that his headlight had been lit and that he whistled at all crossings, including this one. He had never before found the gates of this crossing open.

Guard G. Veera Raju said that the collision had occurred at about 00:45. At about 01:10 he had found the gateman on the telephone and the latter alleged that he had had no telephone message regarding the train. Then he (the guard) had spoken to the bus driver, who said that the crossing gates had been open for him and the train had struck the rear of

[1189] *Idaho State Journal*, Pocatello, Idaho, United States, 5th September 1974 his bus.

Travelling Ticket Examiner R. Dhanapathy confirmed that the engine headlights had been lit and that the driver had whistled.

Railway Police Head Constable A. Suryanarayana had been a passenger on the train. He put the time of the collision at about 00:50 and said that he had gone to the gate lodge five or ten minutes later and had found two or three people trying to rouse the gateman, who had apparently been asleep.

Bus Driver D.C. Sreenivasulu testified that he had been travelling at 40 to 50 kilometres an hour approaching the crossing. Looking both ways he had seen no sign of a train and the gates had been open for him. Even as he had been entering the crossing he had not seen any train, but the rear of his bus had been struck by one, nevertheless.

At least two bus passengers, however, had apparently seen and heard the train, one of them saying that they had shouted to the bus driver to go ahead quickly.

Gateman A. Venkayya stated that since he had come on duty at 19:00 the evening before two trains had passed over the crossing, of which he had been duly advised by the assistant stationmaster. He had been aware that a train was due from the Narasaravupet side after midnight and had been on the lookout for it. No notice had been received of one and he had gone into the gate lodge five or six minutes before the accident. He had only heard the sound of a train whistle when the train had been very close, due to the noise of the heavy road traffic, while the engine headlight had been only dim, though he could still have seen it a kilometre away had he been outside. His attempt to belatedly close the crossing gates had then been too late, the bus had entered the crossing at considerable speed and been struck by the train. His mistake had been to rely on getting a telephone call. After the accident he had spoken to the assistant stationmaster on the telephone advising him of the collision and the assistant stationmaster told him that he had been busy issuing tickets and had forgotten to telephone. It was not true that he had fallen asleep. He had not himself asked the assistant stationmaster where the train was, as when he had done so previously he had been rebuked. On odd occasions previously trains had arrived of which he had not been notified.

Gateman Anandam, the other crossing keeper, denied that trains had ever arrived at the crossing unannounced while he had been on duty, and said that the telephone could be heard from outside the gate lodge.

Assistant Stationmaster M.C. Reddy at Santamagular stated that train No. 227 had been accepted by Savalyapuram at about 00:31 and left his station at 00:42, when he had sent the *Train Entering Section* signal. At about 00:55 the Savalyapuram assistant stationmaster had told him on the block phone that the train was not in sight and asked if it had a headlight. Yes, it had, he replied. At about 01:10 he had been informed of the collision.

Assistant Stationmaster J.B. Smith at Savalyapuram stated that he had telephoned the gateman both at about 00:26, when Santamagular asked for *Permission to Approach* for the train, and again when the train left Santamagular at 00:42. The gateman had informed him of the collision at 00:58 and the guard of the train had also informed him at 01:10. When asked about his query to Santamagular about the train having a headlight he explained that he had been anxious because he had a cash bag to put on the train.

Stationmaster M. Masthan Rao at Savalyapuram admitted that he had only recorded his first telephone calls to the gateman, when *Permission to Approach* was given, and not also the time of the *Train Entering Section* signal as per the rules. He was now recording both. The gateman had always responded to his telephone calls.

Additional Commissioner P.M.N. Murthy put the time of the accident at 00:50.

The train speed he decided had been 50 Km/h and that of the bus about the same.

The assistant stationmaster had testified that he advised the gateman of the train; the gateman denied it. It was difficult to believe that the assistant stationmaster could have completely forgotten to telephone the gateman, but the reason for his enquiry to Santamagular about the engine headlight could have been that he had hoped that the headlight might warn the gateman of the train, of which he had not been notified. Possibly the first but not the second telephone call had not been made and the gateman had positioned himself in the gate lodge awaiting this call.

Allegations that the gateman had been asleep in his lodge were unlikely. More likely he had retreated there after the collision to escape the wrath of the passengers.

It appeared that the engine headlight had been burning but had been extinguished by the force of the collision.

The train driver could not have averted the collision and had whistled for the crossing although there were no whistle boards. He was not responsible for the accident.

The gateman should have been outside his lodge looking out for the train as per General Rule 229(2) rather than inside it for five or six minutes before the collision. Had he been outside on the lookout he could have averted the collision despite not having received the second telephone call.

The bus driver, seeing the gates open for him, would have been concentrating on the road ahead and could not be blamed for the accident, though possibly he might have been more vigilant.

Additional Commissioner Murthy found that Gateman Venkayya was responsible for the collision because he had failed to keep a lookout for the train and thus to close the crossing gates in time. Assistant Stationmaster Smith had probably contributed to the accident.

The additional commissioner made seven recommendations:

(1) (a) A daily census of road traffic taken in 1969 had shown 669 motor vehicles, 153 bullock carts, 619 bicycles, 1133 head of cattle and 774 pedestrians. Another census in October 1974 had shown 251 motor vehicles, 11 bullock carts, 1 rickshaw, 34 bicycles, 61 cattle and 181 pedestrians, so the road traffic had markedly decreased. If, after rechecking, the figures were similar, possibly the normal position of the gates might be changed to across the road, or if this was not feasible signal protection might be provided.

(b) Meanwhile a system of exchanging private numbers should be introduced at this and other crossings with telephones. Gateman should be literate and able to read numbers.

(2) To prevent any delay in advising the gateman the *Train Entering Section* signal should be sent immediately from Santamagular to Savalyapuram for Up trains.

(3) Vegetation within 50 metres of the crossing should be removed and kept clear to improve visibility.

(4) (a) Comprehensive instructions for working crossings should be framed.

(b) Where the gates were normally open to the road the gateman should be instructed to display the danger signal to trains upon receiving the first telephone call, removing it only after having received the second telephone call and closed the gates.

(5) (a) Better upkeep of the gate lamps was required; the wick of one was not working.

(b) The gates needed repainting.

(6) The vacuum brake certificate of the train had been issued at Guntur with the vacuum having been insufficient.

(7) (a) The conductor had not been sat in his nominated seat next to the bus driver.

(b) There was some evidence that the bus had not been completely roadworthy and it had not even been carrying a First Aid box. [1190]

The level crossing collisions involving school buses in the United States continued on Wednesday 23rd October 1974 with another serious one, resulting in the death of seven pupils and the injury of the bus driver and of 71 of the remaining pupils (87 in all) who had been on the bus. The accident was investigated by the National Transportation Safety Board (NTSB).

The accident occurred on Bellview Road Crossing, Aragon, Georgia, when a Polk School District bus was in collision with a Southern Railway work train setting back southwards over the crossing.

Southern Railway Work Train No. 8284 consisted, in the words of the NTSB report, of 13 units, of which, when it was setting back, the caboose was at the leading end, followed by two panel cars, two gondola cars, a tank car, two coaches, a box car, the diesel locomotive, a boom car, a derrick car and a box car. The conductor was riding in the northernmost car, furthest away from the crossing, the engineer, a brakeman, and a road foreman of engines were on the locomotive, and a flagman was riding in the caboose at the leading end of the train. The flagman could communicate with the engineer via a radio telephone unit affixed to the wall in the middle of the caboose. The boxcar coupled to the locomotive interfered to some degree with the engineer's view of the crossing.

As the train approached the crossing the engineer told the NTSB inquiry that he had been fully occupied operating the train, including sounding the whistle, looking forward and back and communicating with the flagman.

Flagman R.W. Moree was looking out of the door at the south end of the caboose, then moving to the centre of the caboose to communicate with his engineer, then returning to the door. As the caboose reached the crossing he testified he had looked out

[1190] Government of India Ministry of Tourism and Civil Aviation (Commission of Railway Safety): Report on Collision of train No. 227 Up 'Guntur – Gadag Passenger' with motor bus No. APZ 6093 at Manned Level Crossing No. 272/E at Km. 607/13-14 between Santamagular and Savalyapuram Stations on the Guntur-Donakonda single line metre gauge section of the South Central Railway on 6th September 1974 but failed to see the eastbound bus and returning to the centre of the caboose he had told his engineer, "Cab on the crossing, crossing is clear." Almost at once the collision had occurred, which, he said, had knocked him down and stunned him, so that he had been unable to either tell the engineer what had happened or to apply the brakes from his caboose. As a result the train pushed the bus 315 feet instead of the 30 feet that it would have done had the brakes been applied.

Again chance played a part in the accident. The bus driver, Billy Kellet, 50, had been unable to cross the Southern Railway line via the usual level crossing because it had been blocked by a derailment, so had to switch to the Bellview Road Crossing instead. He had not driven a bus over this crossing before, though it was familiar to him. As he approached the crossing travelling at 30 miles an hour he had seen an engine south of the crossing but his view northwards was obstructed and he did not see the work train. About 750 feet from the crossing he said that he had changed down to fourth gear and slowed to 15 or 20 miles an hour. When within 100 feet of the crossing he had seen the top of the caboose of the work train but had thought it was stationary.

Driver Kellett stated that he had applied the brakes of his bus when it was about 60 feet from the crossing to no effect, then again 40 feet from the crossing when he had realised that the caboose was moving towards him. He said the brakes had still not responded, when he had attempted, instead, to accelerate. Witnesses put the speed of the bus at 5 to 10 miles an hour. The bus driver denied having heard the sound of the locomotive horn or having seen any of the trainmen before the collision.

The coupler of the train penetrated the bus, stopped its forward movement, and it was pushed along the track, unbeknown to anyone on the engine. As the bus was pushed along its right rear tires fell into a ditch, it rolled over to the right, its lower body came in contact with the rails of the track and it demolished a signal post at its concrete base. Further damage was caused to the bus as it struck lineside objects and the caboose was derailed before the train stopped. The train had finally stopped when the engineer had seen dust flying up and then the road foreman had seen the driver of a second bus, who had witnessed the accident, running along the track towards them waving his arms.

No one was ejected from the bus and a fire started from an overturned stove in the caboose was quickly extinguished. [1191]

A newspaper reporter managed to interview a couple of the children and a witness. '"The train stopped down there like it was going to let Billy go across," said Tommy Duke, 11, a sixth-grader at Euhallee Elementary School, who was sitting in the back of the bus. "When Billy started to go, it started back. Everybody started tumbling over the seats. They just started screaming and then it hit."

'Gerald Roper, an 11-year-old who suffered a broken arm, said he climbed out of a broken window when the bus came to rest. "They were all screaming and hollering," he said.

'Joan Miller, one of the first to reach the scene, said she saw, "all those little kids lying around. Some looked dead and they were covering them up. The bus looked like an accordion and some of the kids were lying under the bus. One kid was crying and looking

[1191] National Transportation Safety Board report

for two of his brothers." [1192]

'Ambulances rushed from nearby communities to take the injured to hospitals. Police and volunteers were called in to help as distraught parents rushed to the scene.

'Almost three hours after the accident, a large crowd of parents had gathered at the Rockmart-Aragon Hospital here {Aragon}, trying to determine the fate of their children.' [1193]

The NTSB report described the crossing. The road was an asphalted two-lane country road. There was a speed limit of 25 miles an hour at the crossing. Standard crossbucks were the only protection provided. An April traffic count showed 1,414 motor vehicles and 18 trains daily over the crossing, while an October count showed 2,328 vehicles and 26 trains. At least 20 of the motor vehicles were school buses during school term time. There had been four previous accidents at the crossing since 1966, none of them involving fatalities.

The bus had a Thomas-built body on a 1972 Ford frame. Its seating capacity was 66, though there had been 87 passengers on the bus at the time of the accident, with at least 22 of them standing. [1194] With regard to this a newspaper wrote, 'The state Department of Education said the state school board's regulations "don't recommend for any standees at all" on the buses. '"But being realistic," a spokesman for the department said, "we allow for 20 per cent standees because of the shifting populations" if there are "emergency conditions."

'Dr Guy Turner, superintendent of Polk County schools, said he was sure there had been some overcrowding on school buses but not as much as rumours had it.

'Under the state Board of Education's standards, the bus would have been considered overcrowded, but with the 20 per cent tolerance allowed for standees, the bus apparently was not overcrowded from a legal point of view.' [1195]

The NTSB stated that the bus driver had been doing the job for five years. None of the passengers had criticised his driving or his demeanour and he had no record of traffic violations, though he had not renewed his chauffeur's licence, which had expired 35 days before the accident. He had a history of cardiovascular disease but had passed a medical examination in August.

A transportation director for the school district stated that he had conducted a safety field check on school bus operation in response to complaints, including that buses had failed to stop at crossings, and had found such complaints unjustified. However the day after the accident NTSB investigators and Georgia State Police noted that another school bus had failed to stop at the same crossing.

Analysing the accident the NTSB noted that the bus had failed to stop short of the crossing. It rejected the suggestion that its brakes had failed and concluded that they had not been operated so as to stop the bus short of the track, for an unknown reason, perhaps because the bus driver thought that he could safely cross in front of the train.

[1192] *The Argus*, Fremont-Newark, California, United States, 24th October 1974

[1193] *The Berkshire Eagle*, Pittsfield, Massachusetts, United States, 23rd October 1974

[1194] National Transportation Safety Board report

[1195] *Times-Enterprise*, Thomasville, Georgia, United States, 24th October 1974

As regards the train crew the engineer would have had a better view of the crossing had he looked out of the window. The road foreman had not criticised the operation of the train, though he was aware that only he and the conductor had portable radio sets, while the flagman had not requested the use of either of these sets. Instead he had relied on the fixed radio equipment on the wall of the caboose, which was 19 feet from the point where he could observe the crossing. It appeared also that had the flagman gone out onto the leading platform of the caboose as he had stated, with the caboose on the edge of the crossing, before he had made his final report to the engineer, the caboose would have been over the crossing at the moment of impact and would not have been struck by the bus. Had he been on the leading platform of the caboose when it had become apparent that the bus was not pulling up, he could have stopped the train using the brake valve in the caboose.

The NTSB came to 11 conclusions:

(1) The driver of the school bus was not physically incapacitated prior to the collision.

(2) There is no evidence that the mechanical condition either of the school bus or the train contributed to the accident.

(3) The driver of the school bus, had he been looking, and the flagman, had he been on the rear platform of the caboose, had an unobstructed view between vehicles when they were within a safe stopping distance from their respective edges of the crossing.

(4) The driver on the school bus and the flagman on the caboose understood their respective responsibilities for the safe movement over the grade crossing. Neither the driver nor the flagman complied with those responsibilities.

(5) The requirement for school buses to stop short of railroad track(s) at a grade crossing is more than a legislated formality. It provides the time necessary to detect potential hazards and thereby ensures safe passage. School bus drivers need to be made aware of the necessity of stopping at all rail/highway grade crossings.

(6) It was essential for the safe operation of the train during its rearward movement that the actions of the train crew be monitored, and if necessary questioned, by the road foreman of engines. The road foreman took no exceptions to the manner in which any member of the crew was performing his duties.

(7) The number of students on the bus, which was in excess of the State of Georgia's and the manufacturer's rated passenger capacities of the vehicle, did not contribute to the accident.

(8) Had the caboose been equipped with a radio jack at the south end or the flagman been provided with a portable radio, the flagman could have remained on the leading platform during the entire movement. The use of hand signals by the flagman to direct the movement of the train would have permitted him to stay on the leading end of the caboose during the entire movement.

(9) If the flagman had elected to communicate with the engineer by hand signals, the engineer would have been able to see his signals.

(10) The failure of the flagman to notify the engineer of the collision and/or apply the air brake available to him in the caboose allowed the train to push the school bus over 300 feet from the point of impact before stopping. That movement contributed to the severity of injuries and losses of life.

(11) Formal guidelines had not been promulgated to assist train crews in the selection of the use of radio or hand signals to direct the movement of trains.

The NTSB determined that the probable cause of the accident was the failure of the school bus driver to stop his vehicle short of the track until it was safe to proceed, and the failure of the flagman of the train to guard the unprotected crossing.

Finally the NTSB made three recommendations:

(a) The Federal Railroad Administration (FRA) promulgate regulations governing railroad operating rules for the use of radio communications and flagging when trains are being pushed (shoved) across grade crossings.

(b) The FRA promulgate regulations to require that if radios are to be utilized for directing rearward movement of trains, the radios must be installed so that they can be used by an employee from the leading platform.

(c) The Polk School District, State of Georgia, establish a formal procedure for frequent field checks of school bus drivers to insure their compliance with the District's safety policies and the State of Georgia's uniform Rule of the Roads.

A month after the accident barriers, warning lights and bells were installed at the crossing. [1196]

A school bus that had collected four mentally handicapped children aged between eight and 12 to take them to school was struck by a train on a level crossing near the West Flanders town of Kortemark, Belgium, at about 07:00 on Tuesday 12th November 1974. The train, travelling at 80 kilometres an hour, caught the rear of the bus, which was carried 50 metres and thrown into a ditch. The bus driver, a teacher, and three of the children were killed outright and the fourth child, an eight-year-old boy, was seriously injured and apparently died later in hospital. Two of the children were reportedly orphans.

There were no barriers at the crossing but there were warning lights and bells, which were found to be in working order after the accident. [1197] [1198]

A northbound Seaboard Coast Line Railroad freight train collided with an automobile on an unprotected crossing in Columbia, South Carolina, United States, at 7.55 p.m. on Monday 18th November 1974. Paul Jones, 66, of Swansea, South Carolina, his son Walter, 33, of Columbia, and Walter's children Marvin, 11, Fridz, 10, and Kimberly, 7, were all killed, probably instantly.

Engineer Ray Gardner said his train had been travelling at about 30 miles an hour when it hit the car, which was welded to the front of the locomotive and carried a quarter mile down the track. It took an hour to remove the bodies of the victims.

There was a light, misty drizzle falling at the time of the accident, which might have

---

[1196] National Transportation Safety Board report

[1197] *The Times*, London, England, 13th November 1967

[1198] *Algemeen Dagblad*, Rotterdam, The Netherlands, 13th November 1974 reduced visibility. Standard crossbucks were provided at the crossing. [1199]

A disastrous level crossing collision occurred in the suburbs of Rio de Janeiro, Brazil, on the early morning of Friday 27th December 1974. At 06:07 that morning locomotive No. MK-3112 of the Rede Ferroviária Federal collided with bus FI-0150 of the Viação Esperança company on the level crossing at Caioaba, at kilometre 35 of the Linha Auxiliar from Belfort Roxo. The bus was carried 150 metres and immediately caught fire. Twenty bus passengers lost their lives and the same number and the bus driver were injured. Eighteen passengers died at the scene and the other two passed away at the Casa de Saúde Nossa Senhora de Fátima hospital, whence and to the Nova Iguaçu Hospital the injured had been taken in private cars and on another Viação Esperança bus. The bodies of the dead were so charred that by the evening only three had been identified.

The engine, in charge of Driver Altamiro Francisco Rosa, had been on its way from Engenheiro Pedreira to São Mateus.

The crossing was not provided with barriers. A surviving bus passenger, Paulo Lopes da Fonseca, told a reporter that he had been sat immediately behind the bus driver, Edgard de Oliverira Duque Filho, and had seen how the driver had proceeded without stopping over the crossing in the path of the rapidly approaching engine. Even with the engine only 10 metres away, Paulo said, the bus driver had tried to get across in front of it. It transpired that the bus driver had been in a hurry to reach the station at Nova Iguaçu in time to permit his passengers to catch the 06:20 train from there into the city. Crowded with passengers from Xavantes, its starting point, the bus was running seven minutes late, and was hurrying over the bumpy roads. The bus passengers were workers who if they arrived late at their jobs due to the bus failing to connect with the train risked losing a day's pay.

A newspaper wrote of the sad cases of two of the passengers who had lost their lives. Marlene Candido de Silva, 'a skinny girl of 18,' was supporting her whole family; her old and widowed mother, her 16-year-old nephew and four nephews, girls aged eight, five and three and a boy aged a year and seven months, the children of her sister who had died a few months earlier. She lived in Vila São Luis and had a job at São João de Meriti which meant catching a bus to Nova Iguaçu and then a second bus to her workplace.

José Fertado de Mendonça had been out of work but had just found a job. On the morning of the accident he had awoken feeling unwell and his wife had urged him to stay home but he insisted on going to work; that day he was due to collect his first pay packet.

Both Marlene and José had almost missed the bus but baker Antonio Pais dos Santos had seen them running to catch it and shouted to the bus driver, who had waited for them. They were the two victims who died in hospital. Two other intending passengers were luckier; they had flagged the bus down but its driver, anxious not to lose more time, had not stopped for them.

The railway company placed the blame for the accident entirely on the bus driver. There had been three previous accidents involving trains colliding with buses (only one of them belonging to the Viação Esperança company) at the same crossing. [1200]

[1199] *Statesville Record and Landmark*, Statesville, North Carolina, United States, 19th November 1973

[1200] *Jornal do Brasil*, Rio de Janeiro, Brazil, 28th December 1974

An article in the same newspaper the following day dealt with the efforts of the railway to eliminate various level crossings. The problem, as always, was the cost of doing so. In the previous year two crossings had been closed at Coelho da Rocha (in São João de Meriti) and at Nilópolis, but crossings remained in the Baixada Fluminense region at Olinda, Edson Passas, Mesquita, Juscelino Kubitschek, Belford Roxo, Comendador Soares, Austin, Queimados and Engenheiro Pedreira. There were plans to eliminate in the short term all of the level crossings on the main line between Rio de Janeiro and São Paulo so as to be able to run faster passenger trains to compete with road and air travel. [1201]

Reckless behaviour of the person driving a bus was clearly the cause of another disastrous level crossing collision at Cuautitlan, Mexico, in the suburbs of Mexico City, on the night of Saturday 8th February 1975. The bus was struck by the *Golden Eagle* passenger train, on its way from Mexico City to Ciudad Juarez on the Texas border, killing 26 or 29 persons and injuring 32. No one on the train was hurt and it resumed its journey after a short delay.

That negligence on the part of the bus driver was to blame was clear from the testimony of surviving bus passengers. Thus Pedro Solano, 22, related, "The train was very visible because it was very dark. I saw the light from afar even before we got to the railway crossing. I thought the driver was going to stop, but he didn't. Then I heard all this screaming. I started screaming: 'You stupid idiot, stop!' Then I heard what sounded like an explosion. Bright white light suddenly engulfed everything. Then I lost consciousness."

Another survivor, Lucila Chacon, 20, said, "I don't know how I survived. All I remember is grabbing the arm of a man next to me. Suddenly the train appeared. The passengers began shouting for the driver to stop and to let the train go by. But he paid no attention and instead accelerated."

The bus was cut in two and parts of it were scattered 500 metres along the track.

"I never saw anything so terrible," said Cuautitlan police chief Ignacio Silva Macedo. José Rodriguez Bacre, one of the first people on the scene, agreed, saying, "It was horrible to see the dead. The bodies were all mutilated. The screams of the injured were terrible."

Those who lost their lives were all from nearby Tultepec and had been returning from doing shopping in Cuautitlan.

The road along which the bus had been travelling was a gravel one running parallel to the main road, which was under repair. [1202]

The bus was overcrowded, to what degree varied according to different reports. It had a capacity of 40 or 44 and was carrying either 58 or 90 passengers. Who had been driving it was also in dispute. It should have been José Casimiro Patino, aged 38 or 47, who was killed, but according to some survivors 17-year-old Angel Peralto had actually been at the wheel at the time of the accident. [1203]

[1201] Ibid, 29th December 1974

[1202] *The Salt Lake Tribune*, Salt Lake City, Utah, United States, 10th February 1975

[1203] *Centralia Sentinel*, Centralia, Illinois, United States, 10th February 1975

A level crossing with Kraus-Maffei Strasse north of München-Allach station, West Germany, was over 50 metres wide. The main crossing, protected by lifting barriers, took the street over eight tracks, and a separate crossing with its own barriers took the street over a ninth track. Rail and road traffic over the crossings were both high; an average of 230 trains and 25 shunt moves passed over the crossing daily, plus an average of 500 pedestrians and 3,500 motor vehicles, including buses. The traffic moment of the crossing (number of train movements multiplied by the number of road vehicles) was therefore a very high 892,500. A crossing keeper operated the barriers from a building (Posten 5) placed between two of the tracks. An illuminated panel was provided in Posten 5 on which approaching trains were indicated by illuminated red lights, which changed to white once a train had cleared the crossing, while the arrival of a train was also announced by the sounding of a buzzer, which rang until cancelled by the crossing keeper. The announcement of each train was recorded in a train register and reported by the crossing keeper to the fahrdienstleiter. To deal with the eventuality of more than one train approaching the crossing at the same tine a cap was provided, which when this happened could be placed over the button operating the crossing barriers; it was marked 'Achtung, Zwei Züge' (Attention, two trains). The crossing barriers were, however, not interlocked with the nearest signals, so the signals could be cleared with the barriers raised and the barriers raised while the signals were clear for a train. No accidents had been recorded at the crossing between the end of the war and early March 1975.

On the morning of Friday 7th March 1975 the crossing keeper, who had come on duty at 06:00, was 35-year-old Josef Volk. At 07:07 train No. S 7221 was announced from Dachau (due at München-Allach at 07:13) and he lowered the crossing barriers. Shortly afterwards train No. N 4208 was announced from München Hauptbanhof (due through München-Allach at 07:17 in the opposite direction). Volk acknowledged the warning for this train and entered it in the train register (recording the time wrongly as 07:18) but he forgot to put the 'Achtung, Zwei Züge' cap over the crossing control button. Then, as soon as train No. S 7221 had passed over the crossing, with all thoughts of train No. N 4208 having passed out of his head, he raised the main crossing barriers.

Riding in the cab of train No. N 4208 that morning were its driver, an examiner who was stood to the left of the driver, and a trainee driver, who was stood behind the examiner. As the trainee was considerably taller than the examiner all three had a clear view of the crossing. The train was travelling at about 111 kilometres an hour (the maximum permitted speed at this point was 120 km/h). It was estimated that the train had been 141 metres from the crossing, approaching on track No. 7, when its driver realised the danger, applied his brakes, put down sand and sounded his horn.

The München-Allach fahrdienstleiter also saw the danger and shouted to the crossing keeper, 'Fünfer was machst du? Mach zu, mach zu!' (Five, what are you doing? Shut them, shut them!) With that, Volk began to close the barriers, but a No. 78 bus was already on the crossing and it was struck about its centre by the train and demolished. The train, which had still been travelling at about 108 kilometres an hour when it struck the bus, ran another 450 metres with three of its cars derailed, knocking down two posts supporting the overhead line equipment (OLE) and the Einfahrvorsignal (distant for the home signal) for Karlsfeld station. The damage to the OLE resulted in a loss of electric traction power in the area. Of the 14 people on the bus, including the driver, ten were killed outright and two more died of their injuries in hospital. Schoolboy Klaus Schwager, 13, and Clerk Margarete Gassner, 36, were thrown out of the bus and survived with serious injuries. Three people on the train were slightly hurt.

The blame for the accident clearly lay almost entirely with the crossing keeper, though it was suggested that if the bus driver had looked to his left and reacted very quickly he might have saved the situation.

It emerged that Crossing Keeper Volk had had a somewhat chequered railway service. He had initially served an apprenticeship as a baker, then worked in the construction industry before joining the railway in 1968. He had worked at first as a shunter, then between 1970 and 1974 had operated a small locomotive at Dinkelsherben. He had been found responsible for one or two minor mishaps, had domestic problems and was frequently off sick with nervous disorders, digestive problems, dizziness, eczema, and so on. In 1974 he transferred to Donauwörth, working there as a shunter, and, as his medical problems continued, he was examined by Dr. Heinz Weinand, who recommended his transfer to a post where he could eat regularly at home. He had then expressed a wish to work as a conductor or in the luggage department. The irregular hours of a conductor had ruled him out for that job, while there were no vacancies in the luggage department. Finally he was offered one of several positions as a crossing keeper; for personal reasons he chose that at München-Allach, and after training began working there on 18th January 1975.

Following an investigation that lasted a year and a half eight railwaymen found themselves in court charged with manslaughter and causing injury through negligence. Crossing Keeper Volk, however, was not one of them, the public prosecutor having decided that he had been overwhelmed by his job and was generally unfit, due to his physical and psychological condition, for operational service within Deutsche Bundesbahn. Instead the train driver, the trainee driver, Doctor Weinand, Bundesbahnoberatt Maximilian Kick and Amtmann Karl Saitner of Department 4A (human resources and careers) and Bundesbahndirektor Hermann Keilholz, Amtsrat Benedikt Steger, and Betriebsinspektor Anton Ostermajer of Department 5 (occupational welfare) all found themselves in court. The logic for prosecuting these officials was that they should have realised from Volk's record of disciplinary offences, illnesses, and mental state that he was unfit for such a responsible job as that of a crossing keeper and that having nevertheless agreed to his appointment in such a position they had been negligent. There does not seem to have been any sensible reason for prosecuting the two train drivers and it was found that the railway doctor who had examined Volk had no idea that he was to be employed as a crossing keeper. At the end of the initial trial the drivers, the doctor and the officials of Department 4A were all acquitted but the three officials from Department 5 were found guilty and each was sentenced to a suspended sentence of 8 months' imprisonment. The case then went to retrial and in February 1979, if the author's reading of the German text is correct, they too were acquitted.

Following the accident Volk left the railway and found a job as a bartender. The crossing was later abolished. [1204] [1205] [1206]

Before dawn on Tuesday 11th March 1975 an express train rammed a bus bound for Manila on a level crossing at Angeles, The Philippines. The bus driver had apparently

---

[1204] *Katastrophen der deutschen Bahnen Teil I* by H.J. Ritzau, pgs. 214-229

[1205] https://www.spiegel.de/politik/leere-im-kopf-a-d23d93b5-0002-0001-0000-000041001957

[1206] Wikipedia – Eisenbahnunfall von München Allach

tried to beat the train over the crossing. He failed and the train hit the bus, pushing it 300 yards before it overturned and crashed into a warehouse. Ten bus passengers died and 28 were injured. [1207]

Six kilometres from Bahia Blanca, Argentina, the driver of a bus on its way from Buenos Aires to Rio Negro province on Wednesday 26th March 1975 did as his counterpart in The Philippines had done a fortnight earlier and tried to beat a freight train over a crossing. Like his counterpart he failed and 14 bus passengers paid for his negligence with their lives, while 18 were injured, some very seriously. There was a heavy drizzle at the time, which might have made the train harder to see. The crossing lacked barriers but was equipped with warning lights and bells. [1208] [1209]

On Easter Monday, 31st March 1975, another of those fairly rare level crossing collisions that result in significant fatalities to train passengers occurred between Mjölby and Sya, Sweden, on the Södra main line. Express train No. Xt 15, the *Öresundspilen* from Stockholm to København, Denmark, driven by Driver Arne "Visslarn" Johansson, was already passing over the crossing when a small Volkswagen car was inexplicably driven into the side of it. Although the crossing lacked barriers, it was equipped with warning lights and bells and given the train driver's nickname, which translates as 'the whistler,' it seems certain that he had also sounded his horn for the crossing, while the lights and bells were found to be working normally after the accident. Possibly the car had skidded onto the crossing on the snow-covered and potentially slippery road. As a result of the collision the car was spun around through 180 degrees and the impact of its rear-mounted engine may have contributed to the derailment of the second coach of the train. The following carriages of the train were thrown off the line in turn and the second to seventh carriages were badly derailed, the second to fourth being thrown into a field, while the following three ending up diagonally across the track. The leading bogie of the eighth carriage was also derailed, the remaining six of the 14 coaches remained on the track. The engine and leading coach of the train came to a stand about 200 metres down the line. The second, third and fourth carriages were very badly damaged and the fifth coach somewhat less so. Thirteen train passengers lost their lives and 29 were injured. The car driver, the only occupant of the vehicle, also died. [1210] [1211]

No. 20 Down Mail train collided with a lorry on a level crossing between Deoband and Teleppari Buzurg stations, India, on Saturday 12th April 1975. Five people died and nine were injured in the accident on the Northern Railway, which was blamed on the negligence of the lorry driver. Damage to railway property was put at 550 rupees. [1212]

[1207] *Eau Claire Leader Telegram*, Eau Claire, Wisconsin, United States, 11th March 1975
[1208] *Clarin*, Buenos Aires, Argentina, 27th March 1975
[1209] *The Daily Review*, Hayward, California, United States, 27th March 1975
[1210] https://www.jvmv2.se/forum/index.php?mode=thread&id=220701
[1211] Wikipedia – The railway accident in Sya-Mjölby
[1212] https://indianculture.gov.in/flipbook/1471

On about Thursday 24th April 1975 a bus taking 31 industrial workers to a rest home in the Rodopi Planina mountains of central Bulgaria was struck by a freight train on an unguarded level crossing near Plovdiv, Bulgaria. Eleven of the bus passengers were killed and another 16 were seriously injured. One of the waggons of the train was reportedly derailed and hit the bus. Whether the derailment of the train was the initial accident, or whether the train struck the bus first, causing the derailment, is not clear from the brief newspaper reports on the accident. [1213]

No. 552 Down Secunderabad – Kurnool Town passenger train collided with a lorry carrying military personnel on an unmanned crossing between Balanagar and Gollapalli, India, on Wednesday 30th April 1975. Six soldiers died and three were hurt in the accident, which was blamed on the carelessness of the lorry driver. The South Central Railway estimated the cost of the damage to railway equipment at 300 rupees. [1214]

No. 13 Down Bombay – Madras *Janta Express* collided with a lorry on a level crossing between Patas and Daund, India, on Monday 19th May 1975, killing 61 people on the lorry and injuring another 17 lorry occupants and three engine crew.

In an answer given in parliament regarding the accident it was stated that it had been due to 'the driver of the motor truck having driven it into the unmanned level crossing in a rash and negligent manner in the face of the approaching train. The owner of the truck, who was actually driving it, has been held responsible for his failure to stop at the Warning Boards and to look out for the approaching train at the unmanned level crossing. He also carried excessive persons in his goods truck in contravention of the Motor Vehicles Act.'

In reply to a further question it was stated that the 40,000 unmanned level crossings in India were being upgraded on the basis of traffic density, the matter being reviewed annually. There were 14,000 manned crossings. [1215]

The investigation into the disaster recommended that 'Existing visibility should be improved by clearing soil dumps near the level crossing.' [1216]

An article on a level crossing catastrophe in India that appeared in a Canadian newspaper as late as the end of October clearly related to this accident: 'All work ceased in the tiny village of Moregaon, about 200 miles from Bombay. The doors of houses were closed and the muffled cries of women and children filled the air. Almost every family in Moregaon lost a member or a relation in India's most ghastly railroad accident in many years. Sixty-one residents were killed when the truck bringing them back from a wedding was hit by a speeding express train at a level crossing. The injured included Uttam Kumar, the bridegroom, and Asha Lata, the bride.

'Bagoo Jagtap, a 69-year-old farmer of Moregaon, said the tragedy could have been avoided if the level crossing had been manned by the railroad police or its safety ensued by electronic devices. "We shall not rest until all the crossings in the country are made

---

[1213] *Winnipeg Free Press*, Winnipeg, Manitoba, Canada, 1st May 1975

[1214] https://indianculture.gov.in/flipbook/1471

[1215] *Lok Sabha Debates* Volume 56 Issues 11-20

[1216] https://indianculture.gov.in/flipbook/1471

safe," said Jagtap. "If necessary, we will stage a march on Parliament in New Delhi. For the next year this issue will dominate Moregaon."

'But Junior Railroad Minister Mohammed Querishi said it was not possible to make every level crossing safe. "About 12,000 of the 42,000 crossings all over India now are manned for safety," he said. "We do not have the resources to man all of them. It is the duty of people using unmanned crossings to take care and watch out for trains."

'A senior official of India's government-owned railways said truck drivers are often careless when approaching level crossings. "Some are so drunk they can't even see or hear approaching trains."

'A report in the newspaper *Calcutta Patrika* said as many as 1,000 people die in level crossing accidents every year. Last year 200 careless commuters died under the wheels of speeding suburban trains in the Bombay area. Summer is the worst period for level tragedies in rural India because it is the traditional wedding season. It is a common sight to find dozens of oxcarts carrying wedding guests moving slowly across the tracks. Said the driver of the crack Bombay-Calcutta Mail train: "Often I have to keep sounding the locomotive whistle continuously for several minutes to scare away the oxcarts. I dread the coming of the wedding season."

'The federal railway ministry plans to spend $20 million this year on the construction of overpasses and underpasses at vulnerable crossings. Said a ministry official: "The problem is while more railroads are built to open up interior areas, peasants are also building more dirt roads across rail tracks to speed up marketing of their produce."' [1217]

An oil tanker loaded with domestic fuel reportedly broke down on an unprotected level crossing near Tarbes, in the south of France, on Thursday 22nd May 1975. Before it could be moved it was struck by an express from Bayonne to Toulouse. An explosion immediately followed the collision and three or four of the coaches of the train were set alight. At least five train passengers were burnt to death and another half dozen were injured. The lorry driver jumped to safety at the last moment and escaped. [1218] [1219]

On Friday 6th June 1975 No. 253 Up passenger train on the Northeast Frontier Railway collided with a lorry on an unmanned level crossing between Mahapur and Aunrihar Junction stations, India. Nine people died and eight were injured. As usual, the accident was put down to the negligence of the lorry driver. [1220]

A bad level crossing accident occurred on the South Central Railway between Satara Road and Koregaon stations, India, on Sunday 15th June 1975. No. 84 Up *Maharashtra Express* collided with a lorry, killing 20 people, and injuring 45, all of whom were on the lorry. Failure of the lorry driver to give way to the train was cited as the cause of the

[1217] *Winnipeg Free Press*, Winnipeg, Manitoba, Canada, 1st November 1975
[1218] *The Times*, London, England, 23rd May 1975
[1219] *The Lincoln Star*, Lincoln, Nebraska, United States, 23rd May 1975
[1220] https://indianculture.gov.in/flipbook/1471

accident. The damage to railway property was estimated at a mere 100 rupees. [1221]

Initial reports spoke of a train colliding with a bus carrying 'hundreds' of passengers on a level crossing 275 kilometres (170 miles) south of Seoul, South Korea, at about 07:30 on Saturday 14th June 1975. A more accurate second report started that there had been over 60 passengers on the crowded bus, many of them high school students on their way to school. Twenty bus passengers were killed and all or most of the remainder were injured in the accident, which occurred near Changsong. [1222]

The serious level crossing collisions in India continued with one on Tuesday 8th July 1975 between Anklar and Vased stations on the Western Railway. No. 688 Up goods collided with a lorry, killing five people, and injuring seven. Failure of the lorry driver was blamed. [1223]

On Friday 1st August 1975 train No. F 26 Down collided with a tractor trolley on a level crossing between Nawashar Doaba and Garh Shankar stations, India, on the Northern Railway due to the negligence of the tractor driver. Seven people lost their lives and the same number were hurt. [1224]

A southbound Amtrak passenger train, en route from Chicago Union to St Louis, Missouri, collided with a 1973 Ford LTD car on an level crossing with a gravel road three and a half miles south of Gardner, Illinois, United States, at 6.45 p.m. on Monday 22nd September 1975. The car was being driven by Mrs Antoniette Tweedt, 53, the principal Goodfarm Grade School, and was carrying five sixth-grade girl passengers who were returning from a softball game at Bonfield Grade School. They were returning later than expected, the game having gone into an extra innings. Mrs Tweedt and her pupils were all killed. The girls were identified as Laura Archer, 11; Becky Gearhart 11; Barbara Gschwendtner, 11; Carol Holm, 11; and Tonya Kincade, 14.

The train was travelling at 75 to 78 mile an hour. According to a spokesman, "The engineer said he saw the car nose down as if the brakes were being applied. Then it just came right onto the tracks slow, like she didn't see the train at all." The car was carried half a mile before the train stopped.

The accident was witnessed by Mrs Charles Pearson, following in a second car with the remainder of the softball team. "It all happened so fast. I don't know what happened," she said.

'About 70 area policemen, firemen, and emergency squad members were at the scene,' a newspaper wrote, 'using portapower and air chisels to cut the top out of the car, impaled on the train's engine. It took about 90 minutes to free the bodies which were then removed to the Gardner Fire Station for identification. The last victim was identified

[1221] Ibid.
[1222] *Pacific Stars and Stripes*, Tokyo, Japan, 15th and 16th June 1975
[1223] https://indianculture.gov.in/flipbook/1471
[1224] Ibid.

about 10.30 p.m.'

The engine of the train, 'a new French Turboliner,' was considerably damaged, but no one on the train was hurt. Its passengers were transferred to another train after a delay of four hours.

The only protection at the crossing with the Illinois Central Gulf Railroad line was a pair of crossbucks. It was bordered on two sides by cornfields and was described as a "blind crossing." [1225] [1226]

A bus crowded with holidaymakers was struck by a train on an unprotected level crossing at an unstated location in eastern Hungary, 240 kilometres from Budapest, on Tuesday 14th October 1975. Twelve bus passengers were killed, four were critically hurt and nine more were seriously hurt. [1227] It was stated that the bus driver had 'carelessly driven onto the line.' [1228]

Prompt action on the part of the school bus driver averted a potential tragedy when a bus stalled on Roddenberry Crossing, near Cairo, Georgia, United States, at 7.35 a.m. on Friday 17th October 1975. Mrs Linda Booth, 24, whose four- and eight-year-old daughters were two of the nine children, aged between four and ten, on board the bus, said afterwards, "It had been raining real hard. I'd been having trouble with my bus; it was wanting to drown out on me. It was real odd. The bus had been skipping some already and just before I got to the crossing the thought occurred to me, 'What if the bus stalls?' I got to the tracks and I stopped. I looked and the train wasn't coming. I started and gave it a lot of gas...the front tires were on the tracks...it choked down and went completely off." She tried to frantically restart the bus but failed. "I looked then and saw the lights from the train flashing among the trees; it was coming around the tracks." It was estimated that the train had then been a mile away as Mrs Booth had begun to shepherd the children off the bus.

"I knew right then it was either the bus or the children. I said, 'Kids, the train's going to hit the bus. Let's get out of here.' Luckily all the kids were near the front of the bus and nobody panicked. The kids were excited, they weren't scared at all," Mrs Booth went on. She managed to get the children to the relative safety of some bushes 200 feet away before the train hit the bus. "It slung the bus around. We were all in a huddle and screaming – the kids from excitement. I'd squeeze one, then go to another and squeeze him," the bus driver related.

The 156-car Seaboard Coast Line Railroad train, hauled by seven diesel units, was travelling at 40 miles an hour. It tore off the front wing, bonnet, and front wheels of the bus, which were thrown up to 60 feet. The passenger portion remained undamaged.

Mrs Booth said afterwards that she had thought of the Aragon school bus accident the year before (see above). "I just couldn't think about those children. But I wasn't scared a bit at the time. Now I'm all to pieces."

[1225] *Morris Daily Herald*, Morris, Illinois, United States, 23rd September 1975
[1226] *The Daily Leader*, Pontiac, Illinois, United States, 23rd September 1975
[1227] *The Dubois County Daily Herald*, Jasper, Indiana, United States, 14th October 1975
[1228] *Progress Bulletin*, Pomona, California, United States, 14th October 1975

Grady County School Superintendent Jack Willis commented that Mrs Booth had "acted with a great deal of maturity," while Georgia state trooper Jim Prine said that she had acted in a calm manner in a situation where many people would have lost their self-control.

Police took the children on to Southside Elementary School in Cairo.

One of the boy pupils, clearly not perturbed by what had happened, was more worried about what bus they were going to ride home in that evening. [1229] [1230] [1231]

Another serious school bus accident on a level crossing occurred on Wednesday 22nd October 1975 at an unprotected crossing at Untermallebarn, Austria, on the Nordwestbahn line of the Österreichischen Bundesbahnen, between Höbersdorf and Schönborn-Mallebarn. Six children and the bus driver lost their lives and 36 children were taken to hospital. [1232] [1233]

On the afternoon of Thursday 6th November 1975 farmer and school bus driver Keith Sizelove, 52, left his home near Alexandria, Indiana, United States, to pick up pupils from Alexandria Junior High School at the end of the day's lessons. His route to the school took him over an unprotected crossing with the Norfolk and Western Railway. When he reached the crossing he apparently slowed but failed to completely stop the bus and he failed to either hear the whistle of or observe a westbound freight train. His view was probably restricted by a line of box cars on an adjoining siding. The bus was pushed about 1,700 feet down the line and torn in two, seats and other parts being scattered along the line. The locomotive of the train was extensively damaged, a crossbuck at the crossing was torn down and six box cars were damaged. Driver Sizelove was thrown from his vehicle and killed.

Rescuers took nearly three hours to clear the wreckage from the line. Police were interviewing a maintenance crew who had been at the crossing and witnessed the accident, while the tachograph of the locomotive was removed for examination, though it was stated that there was no reason to believe that the train had been exceeding the 40 miles an hour restriction at the point. [1234] [1235]

Friday 12th December 1975 saw a serious level crossing accent in Canada, at the St Clair Avenue Crossing in Scarborough, Ontario, a suburb of Toronto. A Toronto Transit Commission (TTC) bus had stopped to pick up passengers just short of the crossing, then moved onto the crossing and stalled. The crossing, on a Canadian National Railway line

---

[1229] *Thomasville Times-Enterprise*, Thomasville, Georgia, United States, 17th October 1975
[1230] *Lincoln Evening Journal*, Lincoln, Nebraska, United States, Saturday 18th October 1975
[1231] *The Daily Times-News*, Burlington, North Carolina, United States, Saturday 18th October 1975
[1232] https://www.sierndorf.at/sites/sierndorf.at/files/files/pages/gemeindekurier_1987_12.pdf
[1233] https://de.wikipedia.org/wiki/Liste_von_Eisenbahnunf%C3%A4llen_in_%C3%96sterreich#1970%E2%80%931979
[1234] *The Anderson Herald*, Anderson, Indiana, United States, 7th November 1975
[1235] *Anderson Daily Bulletin*, Anderson, Indiana, United States, 7th November 1975

over which the GO Transit train involved had trackage rights (running powers), was fully protected with full barriers, warning lights and bells. No train was approaching when the bus entered the crossing but after it had stopped the crossing alarms began to operate and the barriers descended. Unable to restart the bus its driver, Donald Sine, told his passengers to get off and then ran along the line in a vain attempt to stop the oncoming train. Unfortunately westbound train passenger No. 967, non-stop from Toronto Union to Pickering, was running at 70 miles an hour and although its motorman was able to reduce speed to about 55 miles an hour the collision could not be avoided.

A number of the approximately 50 passengers on the bus had managed to get off before the train struck it and some of them stood behind it on the road, thinking that they were in a safe position. When the train collided with the front of the bus it spun the vehicle around and it struck some of the passengers standing next to it. One of the passengers who had time to alight, named Alice Mooney, told a reporter that a middle-aged woman had been thrown 20 feet into the air and landed face down on the frozen ground. "She might have smothered to death," Mrs Mooney said. "I turned her over and stayed with her until the ambulances came." In all eight bus passengers were killed outright in the accident and 20 were injured, six seriously, one of whom would later die.

Fire Brigade Captain Norman Shelton said that when his crew reached the scene ambulance crews were busy removing the dead and injured. "There was so much going on, we just moved from one to another, comforting them, giving them first aid and a little oxygen," he said.

No one on the train was hurt. The accident happened at 4.50 p.m. and evening rush hour traffic was disrupted. After the accident eastbound GO trains were running about an hour and a half late and two between 5 p.m. and 6 p.m. were cancelled. [1236] [1237]

The accident was discussed in parliament and there was discussion on the replacement of the crossing with a subway. Transport Secretary Otto Lang told the Commons that even if the proposal for a subway, estimated to cost $4.6 million, had been approved before the accident, it would not have been completed before 1977. Asked if more money was going to be available for the removal or upgrading of level crossings, he stated that he was always seeking more funding but the government had to decide on priorities. The previous year the government had approved $26 million in subsidies for 50 over or underbridges to replace level crossings and 194 signalling installations. [1238] The problem was a big one; there were 35 level crossings just in the borough of Scarborough. [1239]

An investigation determined that the cause of the accident was unusual. There had been no failure of the protection at the crossing or mistake by a crossing keeper, nor had the bus driver failed to observe the train or tried to beat it over the crossing. Instead the bus had stalled on the crossing and could not be restarted, due to a rather obscure technical defect.

In February 1976 Mr Gordon Hurlburt of the TTC was able to outline the sequence of events that led to the accident and detail its cause. When the bus had stopped at Laurel Avenue, just short of the crossing, to let a man and a woman off, its doors would not

---

[1236] *The News*, Medicine Hat, Alberta, Canada, 13th December 1975

[1237] *The Lethbridge Herald*, Lethbridge, Alberta, Canada, 13th December 1975

[1238] *Winnipeg Free Press*, Winnipeg, Manitoba, Canada, 16th December 1975

[1239] Ibid, 19th December 1975

open. Driver Sine had then operated an emergency override to open the doors but that had not worked either. The floor treadle normally used by passengers to open the doors was powered by a master electrical switch, which also powered the driver's emergency override. Because a tiny screw had fallen out two wires had been left hanging loose and these were only making intermittent contact with the master switch termini, hence the failure of both the door treadle and the emergency override to work.

At Laurel Avenue the doors had eventually been forced open manually by the male passenger and he and the woman passenger had alighted. Other passengers had then pulled the doors shut, whereupon Driver Sine had proceeded towards the crossing, but he had omitted to reset the emergency override switch. As the bus had bumped over the rough surface of the crossing the loose wires had again made contact with the termini and the emergency override had then opened the bus doors, which, ironically as a safety feature, had stopped the motor of the bus, so that it could not be driven with its doors open. [1240]

The crossing was eventually replaced by a bridge.

[1240] Ibid, 7th February 1976

CHAPTER 13

RAIL/RAIL CROSSING COLLISIONS

At locations where railways and roads intersected, the idea of having them cross on the level had obvious advantages in terms of economics over the cost of building an over or underbridge, especially if the level crossing were unmanned and the construction of a bridge would require the building of substantial approach embankments or possibly approach cuttings leading to an underbridge. With road users being limited in the early days to pedestrians, cyclists, and horse-drawn vehicles the dangers to a train from a collision on such a level crossing were also likely to be minimal. For railways crossing other railways on the level, though, the situation was very different. A collision with another train on a crossing was likely to be far more serious than a collision with, say, a horse and cart, and furthermore a train driver approaching a level crossing and seeing another train nearing or on it could not pull up in anything like the distance that a road user could. It is no surprise therefore that in most countries, where the railway system was either built directly by the state or under state supervision, that rail/rail crossings were not provided. It was only in countries where the railways were built by competing private companies without state interference that such crossings existed, most especially in the United States, where lines were often built as cheaply as possible and the economy of not needing to construct a bridge was doubtless an important consideration. Thus, the United States had many rail/rail level crossings, Great Britain at one time had about 20 or so and its colonies, such as Ireland and Canada, had a handful. There were also a number in Argentina, where the railways were constructed by British companies, such as Circunvalacion Crossing, Argentina, where the broad-gauge Buenos Aires Great Southern and the metre-gauge Province of Buenos Aires crossed, [1241] and one or two in other countries, such as at Istravoz (later Hilal), Turkey, where the Oriental Railway crossed the Smyrna – Cassaba Railway, and at Warthausen, Germany, where a major disaster had occurred in 1944. [1242]

As has been pointed out the American companies also seemed to have a perverse proclivity for building rail/rail crossings where they could easily have been avoided, perhaps because in the early days train speeds were low, trains infrequent and night operation unusual so that the dangers were not fully appreciated. [1243]

[1241] Facebook – Signal boxes and Signalling page.
[1242] Wikipedia – Hilal railway
[1243] *A History of Railroad Accidents, Safety Precautions and Operating Practices* by Robert B. Shaw, pg. 99

By 1961 every one of the 15 or so remaining rail/rail crossings in Great Britain was fully signalled and there were hardly any further collisions on them. In the United States, on the other hand, where some crossings remained with little or no signal protection, collisions on them continued to occur, though less frequently than before, doubtless due to improved protection at some locations and the elimination of other crossings either by replacing them with bridges or with the closure of one or both of the lines involved.

The first rail/rail crossing collision of any significance since the beginning of 1961 appears to have occurred at Barr, Illinois, United States, on Wednesday 7th August 1963. Here there was a crossing of the Chicago and North Western Railway and the Chicago and Illinois Midland Railway lines. Unlike many American crossings this one was protected by fixed signals, worked from a two-storey brick signal tower.

On the day of the accident a Chicago and Illinois Midland Railway was passing over the crossing and a heavy Chicago and North Western Railway train of 55 loaded and 83 empty cars was approaching it. The signals were of course against the North Western train but it failed to stop and cut through the Midland train, about 15 cars short of its caboose. Engineer John Ferguson, 69, of the North Western train attempted to jump at the last moment but was caught in the wreckage and killed.

Exactly why Engineer Ferguson had failed to stop was unclear. Head Brakeman C.L. Nicholson, who had also been on the engine, could only say that the engineer had applied the brakes in emergency at the stop signal protecting the crossing but by then it had been too late.

The derailed train also demolished the signal tower. Operator Bill Smith said that he "looked up, saw them down there, and ran." He escaped injury, as did the remainder of the railwaymen on the two trains.

About 40 cars formed a pile of wreckage, some stacked three high.

Because the accident had happened on the county line coroners from both Menard and Sangamon counties investigated it. [1244]

The problem with relying on hand signals when making shunting movements was illustrated by a very minor side collision that occurred at Granite City, Illinois, United States, on Friday 13th September 1963. A considerable amount of switching (shunting) was done over the crossing of the Illinois Terminal Railroad and the Alton and Southern Railway. Movements over the crossing from the Illinois Terminal were protected by a fixed signal but the practice had apparently arisen of an Alton and Southern employee signalling to an Illinois Terminal train that it was safe to cross, even though the signal was red. At 12.40 a.m. on the Friday in question the inevitable happened. The Alton and Southern employee gave a hand signal intended for his own train which was taken by the crew of an Illinois Terminal train as being intended for them. Both trains then moved slowly onto the crossing, the Illinois Central train setting back, and they collided. Four box cars and a caboose were derailed but no one was hurt. [1245]

[1244] *Daily Journal-Gazette*, Mattoon, Illinois, United States, 7th August 1963

[1245] *Edwardsville Intelligencer*, Edwardsville, Illinois, United States, 17th September 1963

A side collision at Sterling, Ohio, United States, early on Monday 11th January 1965 was considerably more serious than that at Granite City and was the subject of an Interstate Commerce Commission (ICC) inquiry.

At Sterling the Baltimore and Ohio (B&O) and Erie-Lackawanna (EL) Railroads crossed one another. Both lines were double track. The crossing was protected by fixed signals worked from Sterling tower. At 12.50 a.m. on Monday morning the B&O dispatcher telephoned Operator Sherman Heinz in Sterling tower to tell him that eastbound B&O freight Extra No. 6940 East, the *Virginian*, was to run over the westbound track, No. 1, from Sterling to Warwick, almost 11½ miles further east, and that he, the dispatcher, was about to issue a train order to that effect. Operator Heinz then displayed both a yellow flag and a yellow lantern as train order signals for the B&O train and notified the dispatcher that he had done so, and that dwarf semi-automatic signal No. 45 protecting the crossing of the two railroads was at danger. The dispatcher then issued train order No. 201, which in part read:

Extra 6940 East has right over opposing trains on No 1 one track Sterling to Warwick.

A little later the EL dispatcher advised Operator Heinz that Extra No. 6601 West had just passed Silver Creek, 10 miles east of Sterling. At about 1.02 a.m. an indicator in the tower advised the operator that the EL train was 6 miles away, while another indicator advised him that the B&O train was approaching. He decided to give the EL train priority over the crossing of the two lines and cleared semi-automatic signals L2 and 24-23 for it, then set the route for the B&O train to cross from track No. 2 to track No. 1 of its line, short of the rail/rail crossing. Some minutes later, while checking the train order, he heard Extra 6940 East approaching at 5 to 8 miles an hour and realised that it was not stopping at signal No. 45, nor had it sounded the prescribed whistle for the train order signal. [1246] As he told the ICC inquiry, "I was checking over the papers to make certain they were in order. Suddenly I heard the sound of the *Virginian's* lead unit, and I realised that it had gone past the red signal. I ran outside and waved my arms frantically, trying to stop the engineer with hand signals. I kept glancing over at the approaching Erie train, and back at the Virginian. I was so certain I could get him (the B&O engineer) to stop, but for some reason, he didn't stop." [1247] He stated that he had seen no one in the cab of the locomotive, had no response to his signals and the brakes of the train had not been applied. After the B&O engine had passed him he tried to flag the EL train, but doubted if its crew would have been able to see his hand signals. Almost at once the trains had collided on the crossing of the two lines, the diesel units and several cars had been derailed and a fierce fire had immediately broken out, enveloping the engines of both trains. Engineer Carl Duckworth, Fireman G.S. Williams and Head Brakeman F.C. Druschel of the B&O train and Engineer Don Narney and Fireman Carl Porter of the EL train were all killed. Head Brakeman Clyde Masters of the EL train survived the accident but was too badly hurt to give evidence at the inquiry. [1248]

The ICC noted that train Extra No. 6940 East had consisted of diesel units Nos. 6940, 5466B, 4521A and 6438 coupled in multiple control, 107 cars and a caboose. It had left

---

[1246] Interstate Commerce Commission report

[1247] *The Chronicle-Telegram*, Elyria, Ohio, United States, 26th January 1965

[1248] *The Daily Report*er, Dover, Ohio, United States, 12th January 1965

Willard at 11.55 p.m. after its brakes had been tested, passed Greenwich, 37¼ miles from Sterling, at 12.16 a.m., approached Sterling at 1 a.m. travelling at 54 miles an hour, and had still been moving at 18 miles an hour at the crossing, according to its tachograph. Conductor Arthur Llewellyn and Flagman John McGarey, in the caboose, the only surviving crew members, had been unaware that anything was wrong until the collision.

Train Extra No. 6601 West had consisted of diesel units Nos. 6601, 6552, 7143 and 7071, coupled in multiple control, 123 cars and a caboose. It had left Kent at 12.15 a.m. after its brakes had been tested and passed Silver Creek at 12.56 a.m. According to its conductor and flagman it had been travelling at 45 to 50 miles an hour approaching Sterling. About the time that it had passed signal No. 24-23 Engineer Narney had clearly seen that the B&O was fouling the crossing and had applied his brakes in emergency but the speed of his train had not been materially reduced before the collision. EL Conductor Fred Vestal and Flagman Nevil Tyson had likewise been unaware of any problem until the brakes had come on in emergency.

The signalling at Sterling was all thoroughly tested after the accident and found to be in order.

The ICC concluded that the accident had clearly been due to the B&O train running through signals and given the death of its engine crew it could not be determined why it had done so. [1249] There was newspaper speculation that possibly Engineer Duckworth might have suffered a heart attack [1250] but this possibility is not mentioned in the ICC report.

The ICC commented that Operator Heinz might have minimised the collision had he replaced signals L-2 and 24-23 to danger against the EL train immediately that he had realised that the B&O train was not stopping and that had he used the lighted red lantern and fusees in the tower to signal to the B&O train his signals might have been observed by its crew.

The ICC concluded that:

(1) The Sterling interlocking operator had properly established the route for the EL train to move over the crossing ahead of the B&O train.

(2) The EL train was operated in accordance with the rules.

(3) The B&O train failed to stop short of interlocking signal 45 as required and proceeded onto the crossing ahead of the EL train without authority.

(4) The accident was caused by failure of the engineer of the Baltimore and Ohio Railroad train to stop this train in accordance with the interlocking signal indication. [1251]

Two Chicago and North Western Railway (C&NW) freight trains collided on a crossing at Salem, South Dakota, United States, at 1.25 a.m. on Sunday 7th November 1965. The northbound train, driven by Engineer A.C. Olson, consisted of an engine and 56 cars bound from Sioux City, Iowa, to Huron and was running over the C&NW line. The eastbound train, driven by Engineer O.W. Curry, consisted of an engine and 29 cars

---

[1249] Interstate Commerce Commission report

[1250] *Hamilton Daily News Journal*, Hamilton, Ohio, United States, 12th January 1965

[1251] Interstate Commerce Commission report

en route from Mitchell to Sioux Falls. It was running over the Chicago St Paul Minneapolis and Omaha Railway (CMO) line, the latter company having been leased to the C&NW in 1957. It seems likely that the CMO train had reached the crossing first, as one of its cars was torn in two in the collision. Six cars of the C&NW train were derailed, as were two cars of the CMO train, while damage was also caused to the Salem station buildings, but no one was hurt. Damage to the two diesel engines was estimated at $30,000 and to the cars $20,000. The C&NW line was reopened later in the day and the CMO line the following day. [1252]

Another Chicago and North Western Railway (C&NW) freight train was involved in a side collision at about 6.30 p.m. on the evening of 22nd June 1968. This time the location was New London, Wisconsin, United States, and it had an encounter with a Green Bay and Western Railroad (GB&W) opponent. Again no one was hurt. GB&W freight train No. 2 was bound for Green Bay from Wisconsin Rapids and C&NW train No. 281 was running from Fond du Lac to Eland. The trains collided engine to engine and the C&NW locomotive was overturned, demolishing a small metal shed. Neither locomotive was much damaged. Two GB&W trains were diverted via Appleton but no diversions were planned for C&NW trains, although it was expected that the C&NW line would take longer to clear. Which train was at fault was not clear pending an investigation. [1253]

Details of a third rail/rail crossing accident involving a Chicago and North Western Railway train, again in the state of Wisconsin, this time at Watertown, on Tuesday 15th April 1969, are sketchy, but it apparently involved a C&NW track motor car that collided with a freight train of the Chicago Milwaukee St Paul and Pacific Railroad at the intersection of the two lines. The three men riding on the track motor car had time to jump clear before the collision. Elmer Gilbert suffered cuts to his head and face but his colleagues, Hilbert Hoot and Foreman Adolph Zietlow, escaped unhurt. [1254]

Early on Friday 2nd April 1971 a northbound Missouri Kansas Texas Railroad (Katy) freight train of 103 cars was halted over the crossing with the Missouri Pacific Railroad (MP) line at Paola, Kansas, United States, when it was run into by a slow-moving southbound MP of 93 cars. Two Katy box cars were overturned and four others were derailed, while four diesel units of the MP train were derailed, the leading two going down a steep 25 ft embankment. Engineer Rick Arner of the MP train was slightly hurt but declined medical assistance; he was taken home in an ambulance.

Damage was estimated at more than $40,000, while the rear of the MP train also blocked the adjoining rail/rail crossing between the MP and the St Louis-San Francisco Railway. A string of freight cars at the head of the MP train, loaded with new Chevrolet automobiles, came out unscathed, however. Wreck trains were sent from Kansas City and Coffeyville and it was estimated that it would take a day to clear the wreckage and repair the track. According to one railroad official the signals had been clear for the MP train,

---

1252 *The Daily Republic*, Mitchell, South Dakota, United States, 8th November 1965
1253 *Sunday Post Crescent*, Appleton, Wisconsin, United States, 23rd June 1968
1254 *Janesville Daily Gazette*, Janesville, Wisconsin, United States, 16th April 1968

which seems unlikely. [1255]

Paola, Kansas, United States, was the scene of a second rail/rail crossing accident on Saturday 20th May 1972, this time involving Missouri Pacific Railroad (MP) and St Louis-San Francisco Railway trains. Both freight trains were southbound on their respective lines. Several freight cars were derailed and hundreds of feet of track on both lines were ripped up. Engineer Guy Roberts of the MP train was the only person injured; he was taken to hospital in Jefferson City, Missouri, his home town. It was estimated that it would take 48 hours to clear the tracks. An investigation was opened into the cause. [1256]

Missouri Pacific Railroad (MP) and Kansas City Southern Railway (KCS) freight trains collided at the crossing of the two railroads at Texarkana, Texas, United States, early on Friday 6th April 1973. According to a witness 18 to 24 cars were derailed and the engines of the two trains were "torn all to pieces." It appears that the KCS train was on the crossing first and was run into by the MP train. MP Engineer Mark Rowe and Head Brakeman Ben Board jumped from their locomotive just before the collision and were injured. It was expected that it would take two or three days to clear the wreckage. [1257] [1258]

Chicago and Eastern Illinois Railroad (C&EI) and Illinois Central Gulf Railroad (ICG) freight trains collided on the crossing of their two lines at Kinmundy, Illinois, United States, at 12.15 a.m. on Thursday 24th July 1975. The ICG train, consisting of three diesel units and 27 cars, was passing over the crossing when its sixth car was struck by the leading locomotive of the C&EI train. Seventeen cars of the ICG train and two diesel units and two box cars of the C&EI train were derailed. Conductor L.E. Fry and Engineer P.H. Black of the C&EI train were taken to Salem hospital, treated for slight injuries, and released. [1259]

One southbound Amtrak train was terminated at Effingham and its passengers were bussed to Carbondale. The northbound ICG line was reopened at 5 p.m. in time for the northbound Amtrak *Panama Limited* passenger train, while the equivalent southbound train was diverted. The C&EI lines were reopened at 6 p.m.

Damage to the ICG line was estimated at $109,400 to 17 flatcars, $75,000 to 15 truck trailers, $14,000 to track and $10,000 to signals. Damage to the C&EI line had not been estimated at the time of the report.

The two railroad companies opened a joint formal investigation into the accident to which the Illinois Commerce Commission were invited. [1260]

[1255] *Ottawa Herald*, Ottawa, Kansas, United States, 2nd April 1971

[1256] *The Hutchinson News*, Hutchison, Kansas, United States, 21st May 1972

[1257] *Wichita Falls Times*, Wichita, Texas, United States, 6th April 1973

[1258] *San Antonio Express*/News, San Antonio, Texas, United States, 7th April 1973

[1259] *Southern Illinoisan*, Carbondale, Illinois, United States, 24th July 1975

[1260] *Centralia Sentinel*, Centralia, Illinois, United States, 25th July 1975

CHAPTER 14

FIRES

Fires often broke out on trains following derailments or collisions. While wooden coaches were still in use in passenger trains an overturned stove used for heating or spilt engine coals from a steam locomotive might set them ablaze. Later on, the substitution of gas for oil for train lighting proved to be another hazard, and a number of accidents were made much more severe by fires started by leaking or damaged gas cylinders until its dangers led to its replacement by electricity. This chapter will confine itself, however, to what might be termed primary fires, that is accidents where there was no precipitating derailment or collision but merely a fire. Serious primary fires were considerably rarer than what might be termed secondary fires, but they did occur from time to time from a variety of causes, while more minor fires were fairly common.

Smoke filled the dining car of the *Spirit of St Louis* passenger train of the Pennsylvania Railroad, en route from New York to St Louis, as it approached Brazil, Indiana, United States, at about 10.15 a.m. on Sunday 19th February 1961, travelling at about 60 miles an hour. The 20 passengers in the dining car evacuated it and the train was stopped at Knightsville, three miles short of Brazil, awaiting the arrival of the fire brigade. The 12-coach train was divided behind the burning kitchen car, to which the fire had by then spread, by the train crew, and about this time an explosion occurred, which slightly injured Cook Roland Wright and Waiter Otto Butler, the latter being blown from the train by its force onto the track. Butler said afterwards, "It felt as though two tough guys threw me off." Two passengers who had been standing at the rear of the coach next in front of the dining car were also injured when the blast tore off a steel door.

Due to a lack of water the firemen were unable to put out the blaze at Knightsville and the front of the train was instead towed to Brazil, where firemen extinguished the blaze after three hours. The kitchen car was considered a total loss and the interior of the dining car was badly damaged.

Fireman attributed the fire to a short circuit in a junction box and the explosion to the flames reaching a bottled gas cylinder in the kitchen car. [1261] [1262] [1263]

[1261] *The Terre Haute Star*, Terre Haute, Indiana, United States, 20th February 1961
[1262] *The Kokomo Tribune*, Kokomo, Indiana, Indiana, United States, 20th February 1961
[1263] *Logansport Pharos Tribune*, Logansport, Indiana, United States, 20th February 1961

Fire broke out in the box car of a Pennsylvania Railroad mail train from New York to Kansas City, Missouri via Philadelphia, and Pittsburgh, at about 7.30 p.m. on Friday 18th August 1961. The 17-vehicle double-headed train was stopped at Croydon, Pennsylvania, United States, for firemen to tackle the blaze. Crews from Croydon and Bristol attended and were in the process of fighting the fire when they had to temporarily vacate the burning car so that the current could be restored to the overhead line equipment. This having been done the two engines were then able to draw the leading two vehicles, including the burning car, the second vehicle in the train, clear of the remaining 15 vehicles, which were then worked forward by another engine. It proved impossible to detach the leading car from the second one. Parcel post and mail bags had to be thrown out of the box car by the firemen so they could access the fire; they were forwarded by road. Other trains were not delayed as they were allowed to pass on two adjoining tracks.

Firemen finally managed to bring the fire under control shortly after midnight but remained at the scene until 6 a.m. One fireman had to be taken to hospital suffering from smoke inhalation and about half of the contents of the burning car were lost.

The cause of the fire could not immediately be determined. [1264] [1265]

The Trans Europe Express was travelling at 140 kilometres an hour and had just passed through Dudelange, Luxembourg, on Christmas Day, Monday 25th December 1961, when passengers riding in a carriage in the middle of the train heard a dull explosion and saw flames licking the coach windows. Some degree of panic broke out but a passenger had the presence of mind to seize a fire extinguisher and begin to fight the fire, while another operated the emergency alarm. The train was brought to a stand a few hundred metres short of the French border, when the passengers were forced to break the coach windows and jump from the train.

Railwaymen detached the burning coach and firemen and other rescuers were soon on the scene. No one was hurt but the passengers lost all of their luggage which had been in the burning coach. It took the firemen three hours of effort to extinguish the blaze. The electric power has to be disconnected during this period and several international expresses were diverted, while many trains were delayed. The cause of the fire was not stated. [1266]

Fire broke out on a British Railways, London Midland Region, overnight express train from Glasgow Central to London Euston near Weaver Junction, England, early on Friday 31st August 1962. Sleeping Car Attendant Albert Burgess said, "I was first told about the fire by two passengers. My passengers behaved very well as I helped them down ladders onto the track. An invalid was helped off safely. The heat was intense and smoke was pouring from the carriages. After the danger was over I told the passengers they could board the train again." Driver William Warrington helped to evacuate the passengers and assisted by Driver Fisher and Guard Lee, both of whom had been travelling as passengers, divided the train on either side of the three coaches which were on fire. The leading coaches

[1264] *The Levittown Times*, Levittown, Pennsylvania, United States, 19th August 1961

[1265] *Leader Times*, Kittanning, Pennsylvania, United States, 19th August 1961

[1266] *La Stampa*, Torino, Italy, 27th December 1961

of the train were later worked forward to Crewe where they were reunited with the coaches from the rear of the train, which had been drawn back to Warrington and worked to Crewe via Chester. The train arrived in London many hours late.

The burning coaches, after firemen had extinguished the blaze, were stabled in sidings at Acton Bridge station, under police guard because they still contained mail. Two of them were burnt out and a third was badly damaged; some mail was lost. The West Coast Main Line to Scotland was blocked for several hours.

Initially arson was suspected as a cause for the fire [1267] but later it was put down to a discarded cigarette in a mail coach, which had at first smouldered then burst into flame as it was fanned by the rush of air through an open window, according to the official report as the train had passed through a tunnel, except there are no tunnels along that stretch of line. [1268]

A fire on a train which broke out in a tunnel near Sowerby Bridge, England, on Sunday 13th January 1963 has been detailed above in the Tunnel Accidents chapter.

A fire on a passenger train between Millbrook and Wilshamstead, England, on British Railways, London Midland Region, at about 1.43 p.m. on Saturday 10th October 1964 led to a fatality. The 12.35 p.m. London St Pancras to Bedford (Midland Road), consisted of an eight-car diesel multiple unit (DMU). It was travelling at 60 miles an hour on the Down Fast line between Millbrook and Wilshamstead signal cabins when fire broke out beneath the last coach. By the time the train had stopped just beyond Wilhamstead cabin several of the passengers, frightened by the smoke, heat and roar of the fire, had jumped from the train and an Italian woman, Mrs Camella Caanzano, was fatally injured in doing so. Mrs Canzano's husband, Lengthman A. Canzano, her five-year-old and her brother-in-law remained on the train and were unhurt but were taken to hospital suffering from shock. [1269] The Wilhamstead signalman called the fire brigade, who arrived within four minutes and the fire was quickly extinguished.

The accident was the subject of a full Ministry of Transport inquiry.

Running Foreman T. Shepherd said he had been a passenger on the train. As the train had passed Houghton Conquest cabin (between Millbrook and Wilshamstead) there had been a flash of flame outside and a blaze on both sides, with flames coming through open windows and heavy smoke. He had attempted to get into the driver's cab to apply the brakes and failing in this had pulled the communication cord twice, as the first application had not appeared to have any effect, shouting to the passengers to stay where they were as he was stopping the train. He doubted, though, if they heard him as a number of them had been screaming and shouting. After the train stopped he had been told by an Italian that his wife had jumped from the train. At that moment the Wilhamstead signalman had been coming towards him with a fire extinguisher, which he had taken from the latter and told the signalman to go back to his box and call for assistance.

[1267] *The Times*, London, England, 1st August 1962

[1268] Railway Accidents: Report to the Minister of Transport on the Accidents that occurred on the Railways of Great Britain during the year 1962.

[1269] *The Times*, London, England, 12th October 1964

Guard I.E. Lewis told the inquiry that he had been riding at the front of the last car. Seeing smoke he had rung the buzzer to stop the train and then gone to the brakes but found that they had already been applied. Once the train had stopped he had helped the passengers off and then used an extinguisher on the fire. He heard a man shout that his wife had jumped from the train and had been killed.

Driver A.T.R. Buxton said the train had been running normally when he had heard the fire bell sounding and stopped the train just beyond Wilhamstead cabin. He could not recall getting a bell from his guard.

Signalman G.A. Hughes at Millbrook said that the train had passed him at normal speed with nothing apparently wrong. About five minutes later the Wilhamstead signalman had telephoned him to call the emergency services, which he had done.

Relief Signalman A.G. Wallis at Wilhamstead stated that he had seen the train coming towards him; smoke had been coming from its rear and it had slowed and stopped. He had left his box with a fire extinguisher, which Mr Shepherd had taken from him, telling him to return to his box and call the emergency services, which he had done via Millbrook box, as the latter had a direct line to the railway exchange at Bedford, whereas his box was only on the omnibus circuit. He had also replaced all his signals to danger, though he had omitted to send the *Obstruction Danger* signal to the boxes on either side.

The inspecting officer, Colonel W.P. Reed, noted that the DMUs on the line were used intensively and had, on the whole, given good service, but they had been subject to various failures. The fastening of the engine to the underframe had required strengthening, the flange between the torque converter of the engine had opened out, the free wheel had jammed, the bearings in the final drive had worn, and the striking forks had broken. Detailed redesign had been carried out to strengthen all these weaknesses but they were still not as reliable as had been hoped. After the accident 40 cardan shafts on 60 units examined had been found to be slightly bent and were replaced.

There could be no doubt, Colonel Reed wrote, as to the cause of the accident. Out of balance forces caused by a bent cardan shaft revolving at speed had resulted in damage to other parts of the transmission and in the breakage of the shaft. The flailing ends of the broken shaft had then ruptured the fuel tanks and sparks had ignited the vaporised fuel. It might well be that the initiating cause of the failure on this occasion had been the one that had given rise to many of the previous troubles on this stock, and that serviceability would be very much better in the future as a result of the measures since taken.

He did not think that there had been any failure of the passenger communication cord. In the stress of the situation the brakes had probably not come on, in Mr Shepherd's perception, as quickly as they really had.

The robust and fireproof floor of the DMU had well withstood the briefly intense heat beneath it. [1270]

Eclipsing the previous train fires since the beginning of 1960 was a disaster that occurred to the Madrid – Barcelona passenger and mail train early on the morning of Wednesday 10th February 1965 between Grisén and Pinseque, Spain. As train No. 883

[1270] Ministry of Transport report

was passing through Grisén station the staff there saw that it was on fire and attempted to signal to its driver with a hand lamp, but the signal was not seen and the train continued for another two kilometres before it was stopped. By then, the fire, which had originated in a third-class coach, had spread to two adjoining coaches at the rear of the train, and fanned by a high wind and the draught from the motion of the train the wooden coaches were soon burning furiously. Some passengers had jumped from the train while it was still running at perhaps 80 kilometres an hour to escape the fire and rescuers found the bodies of four victims along the track, while an early report spoke of 26 more bodies being found on the train, with another 33 passengers injured, seven being critically hurt.

Civil Guard Marcelino Viegas helped to save a number of women and children, becoming badly burnt himself; he was taken to the military hospital in Zaragoza. He survived and many years later related his experiences. "I heard screams for help. It all happened in a moment. People did not want to get off the train but my colleague and I tried to help them to do so," he said. A soldier named Gregorio Torres, who was also active in saving passengers, himself died in the fire. [1271]

A number of the injured were also taken to Zaragoza in the front coaches of the train, after the burning coaches had been detached, while others, perhaps the less badly hurt, were treated at Grisén. Telegraph wires alongside the track were damaged by the fire and communications were interrupted. [1272] Most of the 250 passengers on the train had been asleep when the fire had started. Many of them were Andalusians on their way to northern European countries in search of better paid jobs.

A relief train was despatched to the scene from Calatayud. Fireman Jesús Piquero, who had accompanied it, spoke later of his horror at discovering the bodies of a woman and her three children in a toilet, where they had taken refuge in a vain attempt to escape the flames.

There were a number of theories as to the origin of the fire, the most credible of which was that a carelessly discarded cigarette end had lodged in the gap between a window pane and a seat, though some witnesses spoke of the fire having originated in the vestibule separating the coaches, while an electric fault and arson were suggested as other possible causes. [1273] The official version was that the fire had resulted from an overheated axle box ('hot-box') but a railwayman denied that there had been any problem with the axle boxes. One outlandish theory was that an individual named Francisco Gazo Sancerni from Husesca, who had been abusing his wife and three children, had started the fire to throw his family into it "to get them off his back." [1274]

The day after the fire Señor Jorge Vigón Diáz, the Minister of Public Works, presided over the burial of 30 unidentified victims in the local cemetery, while the bodies of four identified victims were sent to their home towns. Officially the number of victims was only 34 but officials documents appeared to indicate that the true number of dead was

---

[1271] https://historico-aragondigital-es.translate.goog/noticia.asp?notid=129171&_x_tr_sch=http&_x_tr_sl=es&_x_tr_tl=en&_x_tr_hl=en&_x_tr_pto=sc

[1272] *The Times*, London, England, 11th February 1965

[1273] Wikipedia – Accidente ferroviario de Grisén

[1274] https://www-heraldo-es.translate.goog/noticias/aragon/zaragoza/2015/02/10/medio-siglo-incognitas-sobre-incendio-del-tren-grisen-338506-2261126.html?_x_tr_sl=es&_x_tr_tl=en&_x_tr_h

between 60 and 70. The Franco government presumably wished to conceal the true magnitude of the disaster.

Red Nacional de los Ferrocarriles Españoles offered compensation to the families of the victims; 250,000 pesetas for those aged over 14, 100,000 pesetas for children between 3 and 14 and 50,000 for children under 3, though many relatives had difficulty claiming their money. [1275]

Panic spread through the whole of the *Northern Tablelands Express* on Tuesday 20th July 1965 as passengers from its rear carriage ran forward through it shouting, "Fire, fire, the train's on fire!" The blaze was in fact confined to the rear coach, from which thirty passengers made their escape, while train hostesses helped them retrieve their luggage.

Conductor H. Jones, the guard and an electrician managed to detach the burning coach from the remainder of the train and fight the fire at some risk to themselves so that after almost half an hour the fire was extinguished. By then the coach had been badly damaged.

Conductor Jones said afterwards that he could not understand why the burning carriage had not exploded, adding that, "There is a big oil tank underneath," presumably referring to a fuel tank.

The fire happened near Redfern, New South Wales, Australia. [1276]

Another diesel multiple unit (DMU) fire occurred on the British Railways, London Midland Region main line from London St Pancras on Wednesday 18th August 1965, this time at Sandridge, England, near St Albans. The 12.00 Down stopping passenger train from St Pancras to Luton consisted of a four-car set. It was travelling at about 35 miles an hour on the Down Slow line when the fire broke out about a mile north of St Albans. The communication cord was pulled, the train stopped and the passengers were evacuated. The crew could not extinguish the fire and by the time the fire brigade had done so the train was badly damaged. This time no one was hurt but the fire was again the subject of a Ministry of Transport inquiry.

Driver A. Longworth told the inquiry that he had driven the train from St Pancras to St Albans and had no particular problems.

Driver A.R. Swain said that he had taken over the train at St Albans. The departure from there had been normal until, as the speed reached 35 miles an hour, the brakes had suddenly come on, the fire bell had rung and the guard had informed him over the telephone that the train was on fire. Seizing a fire extinguisher he had got down from his cab and seeing that passengers were already alighting he had told them to go to the cess clear of the Up Slow line. He had stopped the engines on the leading power car but could not stop those on the rear power car by pressing the stop button. By crawling underneath the train he had managed to stop No. 1 engine by manually operating the shut-down solenoid, but the fire had been too intense for him to get near No. 2 engine. He had used the fire extinguisher but the fire had been too far gone for it to be much use and he asked

[1275] https://www-heraldo-es.translate.goog/noticias/aragon/zaragoza/2014/12/02/medio-siglo-una-tragedia-que-oculto-325300-2261126.html?_x_tr_sl=es&_x_tr_tl=en&_x_tr_hl=en&_x_tr_pto=s

[1276] *The Canberra Times*, Canberra, Australian Capital Territory, Australia, 21st July 1965 another railwayman, Driver F.I. Wright, to protect the Up lines.

Driver Wright, who had been learning the road, stated that he had gone back to assist the guard when he heard the fire bell ring, but neither he nor the guard had been able to operate the fire extinguisher. He had then gone forward to protect the train, putting down detonators, and had met a permanent way man, who had gone forward on a bicycle to report the fire to the signalman from a signal post telephone.

Passenger Guard A.G. Skinner testified that the journey had been quite normal until near the 20¾ milepost, at 12:44 by his watch, when he had felt the brakes coming on and looking out had seen clouds of thick smoke. Almost at once there had been an explosion, the fire bell had rung and flames had appeared. He said that his first concern had been to get the six or seven passengers off the train and to safety by the side of the line. He had then tried to use a fire extinguisher but had been unable to get its cap off, perhaps, he thought, because he had been unnerved by the suddenness and ferocity of the fire. He had then made his way to a lineside house where he asked a woman to telephone for the fire brigade and the police. By the time he had returned to the train the fire in the rear car had become, as he put it, 'an inferno.' He had then walked back over a mile to St Albans North box, forgetting to first collect some detonators from the driver's cab, where he had signed the train register at 13:15.

Mrs F.M. Radford, a passenger bound for Luton, said that she and a young man had been the only passengers in the non-smoking compartment of the train most affected by the fire. All had been normal until after leaving St Albans when she heard a sharp crack and noticed a burning smell resembling that of a burnt fuse. Two or three seconds later there had been a louder bang and the compartment had begun to fill with smoke, whereupon she had at once pulled the communication cord. By the time the train had stopped the compartment had completely filled with smoke. An attempt to alight from the Up Slow side had been frustrated by flames coming from beneath the coach but she and the young man had been able to get down from the other side.

Terminals Inspector W.C. Hasler stated that he had been travelling as a passenger. When the train stopped he had seen that it was on fire. He had assisted the passengers to alight and later helped Stationmaster G. Ward of St Albans load the passengers from the DMU onto the 12:35 train from St Pancras to Bedford, which had been specially stopped on the Down Fast line to pick them up.

Driver R. Moore of the 12:25 Down express from St Pancras to Manchester Central submitted a written statement to the effect that he had seen the DMU on fire as he had passed on the Down Fast line and reported it to the St Albans North signalman from the Down Intermediate Block Section (IBS) home signal telephone.

Relief Signalman A.B. Kirk at St Albans North testified that he had seen and heard nothing wrong with the 12:00 DMU as it passed him. When Driver Moore had telephoned him he had informed Control and the stationmaster. He had accepted a train on the Up Slow line but the Harpenden signalman had advised him that it was being held at the Up IBS home signal. He had also received the *Obstruction Danger* signal for the Down Fast line but had not sent it for the Up Fast line, though he realised subsequently that he should have done.

Signalman J.C. Wood at Harpenden stated that he had received a telephone call from a permanent way man at Sandridge ground frame (this was probably the former signal cabin) that the DMU was on fire. He had replaced his Up IBS signals to stop an Up goods train that had already passed his box, advised the stationmaster, and sent the *Obstruction Danger* signal to St Albans North on the Down Fast line.

Stationmasters Ward at St Albans and D.P. Lovatt at Harpenden spoke of being informed of the fire and proceeding by road to the scene.

Superintendent J.C. Marson at Cricklewood Diesel Depot described examining the DMU after the fire and discovering that a copper gasket in a bolted joint in the exhaust system adjacent to the No. 2 fuel tank had blown out over about one-sixth of its circumference. The missing portion of the gasket would have allowed hot exhaust gases to impinge directly on one corner of the fuel tank from a range of 6 inches. Mr. Marson was at first of the opinion that there would not have been a sufficiently thick layer of oily dirt on the outside of the tank to sustain combustion and allow the flames to climb to the top of the tank, but when he examined the other end of the vehicle he found that the felt strip used to prevent chafing between the fuel tank, and the angle iron frame by which it was supported, were saturated with fuel oil, and could have acted as a wick to sustain combustion. All traces of felt strips at the No. 1 end had disappeared in the course of the fire.

The inspecting officer, Lt. Col. I.K.A. McNaughton, asked Mr. Marson how recently the underside of this vehicle had been cleaned and whether its condition, judging by the amount of dirt he found at the end unaffected by the fire, could be regarded as normal for this type of vehicle. The superintendent said that the sets were cleaned every 6,000 miles, approximately monthly, and that this particular set had last been cleaned on 4th August, 14 days before the fire took place. He described its condition as normal in view of the time the car had been running since its last cleaning and added that, if a car was seen to be particularly oily, it was cleaned before waiting its normal turn. Mr. Marson pointed out that it was in fact extremely difficult to clean the underfloor equipment effectively, particularly the top sides of the main fuel tanks, the clearance above which was very restricted. Moreover, due to the design of the tank with its flat top and surrounding lip, it became quickly covered with dirt when in traffic and also retained any fuel spilt on the tank when refuelling.

Mr. Marson was also asked whether, in the course of his examination he had found any other defects which, in his opinion, could have been contributory to this accident. He stated that the car appeared to have been in good order mechanically and that the only other defect he had found was also in the exhaust system of the No. 2 engine, where the circular end plate at the engine end of the expansion chamber was fractured. This could possibly have been present before the fire, but he thought it more likely that it had occurred during the run back to Cricklewood as a result of excessive vibration in a long length of exhaust pipe, the flexible supporting brackets of which had been burnt away in the fire. In any case, he did not think that an exhaust leak in this position could have caused the fire.

In his conclusions, Lt. Col. I.K.A. McNaughton noted that for a fire to take place three things were necessary – a means of ignition, a flammable substance, and a suitable environment to sustain combustion. In this case the means of ignition had been a jet of hot exhaust gas, possibly carrying with it incandescent particles of carbon, directed towards the main fuel tank of the vehicle from a range of some 6 inches as a result of the deterioration and failure of a corrugated metal gasket in a bolted joint. The flammable substance that first caught alight, and without the presence of which the exhaust leak would in itself have not been a danger, had been spilt diesel fuel oil absorbed by the felt packing between the fuel tank and its supports, covering the sides of the tank in the form of a deposit of oily dirt. Once alight the oily dirt fire had been self-sustaining because the fuel tank had been hot, possibly in the region of 120°F, thus allowing the oil on its outer surface to vaporise more easily, and it had lain in a rapidly moving current of air caused by the motion of the train. The exhaust leak that caused this fire was not large enough to have made a noticeable noise and on account of its position was not easily visible. It was probable that it had only recently started and that no person could reasonably be held responsible for having failed to notice it.

The chain of occurrences that had caused the small oily dirt fire on the sides and leading end of the fuel tank to become a major fire involving the passenger accommodation had been, at every stage, the result of design inadequacies in the power-car and its equipment, each of which could well have been avoided in the first instance and could all be eliminated without major redesign. Once the fire had reached the tank top, which, because of its raised lip could retain several gallons of fuel oil, it could only have been a few seconds before the flames reached the rear end of the tank and ignited a jet of vaporised fuel oil issuing from the open breather vent pipe. The actual temperature of this vaporised oil could not be estimated exactly, but it would have been somewhere between the temperature of the bulk oil in the tank, say 120°F, and the temperature of 210°F at which the aerated fluid from the torque converter was returned to the top of the tank through a small pipe adjacent to the vent pipe. These temperatures should be considered in relation to the specified minimum flash point of the oil, 130°F, and the actual flash point of a sample taken at Cricklewood diesel depot on August 5th 1965, which was 170°F. A flashover into the interior of the tank, causing the explosion that split the upper seam of the tank, must have followed almost immediately, resulting in intense jets of flame being directed at the underside of the floor, eventually penetrating through the joints and round the trap doors.

The whole development of the fire could only have taken a matter of two minutes. There had been no sign of smoke or flames as the train passed St. Albans North signal box at 12:42 and the fuel tank explosion had already occurred by the time the communication cord was pulled by Mrs. Radford before the train had travelled one mile.

Turning to remarks and recommendations, the inspecting officer stated that the circumstances of the fire had been discussed in detail with the Chief Mechanical and Electrical Engineer of the London Midland Region, and many of the alterations necessary to prevent any occurrence had already been completed on the whole fleet of power cars used on the St. Pancras/Bedford services. Amongst the alterations put in hand as a matter of urgency were the following:-

(i) Replacement of all main fuel tanks with a modified design with a flush top.

(ii) Extension of the breather vent pipe over the side of the tank, terminating near the bottom of the tank.

(iii) Modification of the oil return pipe from the torque converter so that it discharged near the bottom of the tank below the surface of the oil.

(iv) Elimination of the felt packing strips between the tanks and its supports.

He believed that these four modifications on their own should effectively prevent another fire occurring in the manner described in his report, but there were a number of additional modifications at present under consideration by the British Railways Board that would go even further towards decreasing the fire hazard on these particular vehicles. Amongst these were the following:-

(i) Rerouting of the exhaust pipe from the No. 2 engine to avoid passing near the main fuel tank.

(ii) Extension of the area protected by the Graviner automatic fire fighting system to include the area of the main fuel tank.

(iii) Replacement of the inch light alloy sheet under the floor insulation by a mild steel sheet, which would be a more effective protection for the fibreglass insulation.

In view of the number of fires that had occurred on this design of power car in which an exhaust system defect had been established as the basic cause, the proposal to redesign and reroute the exhaust system was to be welcomed. In the meantime he had been informed that the need for special care in the examination of these exhaust systems had been brought to the notice of the staff concerned and that such examinations were now always carried out with the engines running, thus making it easier to detect leaks.

Apart from design considerations, the first line of defence against fire with diesel traction was cleanliness. It was, moreover, almost impossible to keep equipment in a proper state of maintenance unless it was kept clean. He was glad to be able to report, therefore, that he had been informed by the Chief Mechanical and Electrical Engineer, London Midland Region, that all DMU Depots had now been equipped with hand spraying equipment to enable the underframes to be sprayed with a detergent to emulsify the dirt and grease and subsequently rinsed off by hose, and that more sophisticated high-pressure rinsing equipment was in the process of being provided. He had no doubt, moreover, that the new design of fuel tanks would eliminate the spillage of oil over the top and sides of the tank and as a result reduce considerably the rate of build-up of the oily dirt deposit.

Steps were taken to ensure that all guards and drivers on the St. Pancras-Bedford services were familiar with, and able to operate, both types of fire extinguisher provided on the Rolls-Royce engined sets and that they were aware of the different kinds of fire for which the two types of extinguisher were suitable. In view of the importance of immediate correct action in dealing with an outbreak of fire on a train it was recommended that such practical instruction in the use of fire extinguishers should he included in the normal training of all guards and other staff who might have to use them. [1277]

A third serious DMU fire occurred on the British Railways, London Midland Region, on the main line from London St Pancras on Wednesday 9th February 1966 and for a third time it was the subject of a comprehensive investigation on behalf of the Ministry of Transport.

This time the accident happened between Napsbury and Radlett, England. The 08:07 Up passenger train from Harpenden Central to St Pancras, an eight-car train carrying about 700 passengers, was travelling at about 65 miles an hour on the Up Fast line when, as at Wilhamstead, a cardan shaft broke and punctured a fuel tank. No one was killed but

[1277] Ministry of Transport report

33 passengers required hospital treatment. The cardan shaft that broke had measuring instruments attached to it, and measurements of stress in the shaft together with associated measurements of torsional disturbances in the engine and torque converter were being carried out at the time.

The evidence presented in the Ministry of Transport report began with that of a passenger, Mr G.H. Powell, who said that as the train had begun to pass Radlett Airfield at about 50 miles an hour he had heard the sound of ballast being thrown up under the floor of the compartment. The sound had increased in intensity and showers of stones had passed the window at high speed. He had then stood up and pulled the communication cord and at the same time yellow flashes of flame appeared, and it had become impossible to see beyond them for smoke. The flashes had become one large flame covering the whole of the window area and licking in through the open top. The left side of his hair had been singed and his face burnt, the heat being intense and breathing difficult. The main flame had been immediately followed by dense black smoke that appeared to be pumped in through the heating ducts. He said that it had then been impossible to breathe or to see, but he had felt a general movement towards the east side of the car where the flames were less, and some confusion, though no panic. Some people had tried to open the doors while others had called for the windows to be shut to keep out the heat. After approximately a minute the door on the east side, which he thought had been jammed, appeared to have been opened and people had begun to jump out. He had also jumped through the smoke still thinking the train was moving and had been relieved to find that it was not. After collecting himself he had returned to assist other passengers and was eventually taken to hospital for treatment. He made the point that one of the engines under the car was still running when the airfield staff were applying foam to it and that it had speeded up at one time about 10 minutes or so after he alighted. Mr. Powell was strongly of the view that if people had not begun to leave the train before it stopped, some would have been suffocated.

A letter from another passenger, Mr R.A. Wing, read in part, "Undoubtedly some people jumped too soon but I doubt if most were able to accurately judge the situation. Some may not have done so if they could. Many did jump without serious injury or any injury. If no one had jumped I consider there would have been far more serious burn injuries, people suffocated with fumes, and possibly some accidents by the open doors. Many people have said to me 'You were sensible not to jump' – possibly, but I was also lucky because other people did jump and I was able to get out just in time as the train stopped. Any panic actions were taken not because of the fire outside but because of the thick choking and scorching fog inside, and the resulting confusion between people taking opposite actions to deal with the situation."

Relief signalman L. H. Copping at Napsbury said that the train passed his box in the normal fashion at about 50 miles an hour, under power he thought, and as there were no other trains in the vicinity he had watched it going away towards Radlett. When it was nearing the bridge he had seen what looked like a white wing on either side which he thought might be smoke, so he had sent the *Stop and Examine* signal to Radlett and had then telephoned the Radlett signalman. He had not sent the *Obstruction Danger* at this stage as he had not been sure whether the train was on fire. He had tried to telephone Control but the lines had been busy and he had not got through until 08:30 to report what he thought might have happened. Immediately afterwards he had been telephoned from Radlett Airfield and he had then informed all concerned and called out the emergency services. He was praised by the inspecting officer, Colonel W.P. Reed, for his vigilance and prompt actions.

Relief signalman T. J. Moriarty, on duty in Radlett box stated that he had just been able to stop a Down Slow train which was leaving the station when he got the *Stop and Examine* from Napsbury and he restored the signals on the Down Fast line which he had cleared for an express.

Driver G. A. Humphrys testified that he had driven the train concerned empty from the depot at Cricklewood to Harpenden, where it became the 08:07 hrs service to St. Pancras. He had been told of the measuring equipment but had not seen it himself. The train had responded normally on leaving Harpenden and again after the start from St. Albans, the speed at Napsbury had been just over 60 miles an hour and the train had been under power. He had shut off power at the Up Fast starting signal where the speed had been just over 65 miles an hour. He had then felt a fairly gradual brake application, had seen that he had lost the engine indicator light for the fourth car and heard the Graviner fire bell sounding. He had then made a full brake application. When he had got down after the train stopped he had met Superintendent J.C. Marson, from Cricklewood Depot, who being younger and more active, had volunteered to go ahead to protect the Down Fast line, while he, Humphrys, had done what he could to help in dealing with the fire. He said that the main fire seemed to be around the fifth car, with very dense smoke. He had left the engines running in his, the forward car, and had turned off the engines in the rear car by hand. He had not passed close to the intermediate power cars as the firemen had been dealing with them but he thought that the engines had been dead.

Guard S. Miller said that he had been in the guards' van in the rear car. He had heard noises under the train after passing Napsbury starting signal as if ballast were hitting the bottom of the car, had noticed the brake come on, and looked out to see smoke coming from underneath the two power cars in the middle of the train. He had applied the brake in his van and at the same time the fire bells had rung. When the train stopped he had taken a fire extinguisher and used it without much effect on the fire, which had been burning strongly, but the Handley Page firemen had arrived (from Radlett Airfield) by then and he had gone back to protect the train and the other lines.

Divisional Inspector D. Walker stated that he had accompanied the technical group in the instrumented brake van. He had taken quick action to get in touch with the Napsbury signalman and then busied himself in organising the rescue and relief arrangements. He had not noticed any bump or rough riding of the vehicle.

Superintendent Marson said that the failure had presented itself to him as a severe lurch and bang. As this happened the Rolls-Royce representatives in the instrumented van had reported a severe stress and loss of recording. The clattering had continued, the Graviner fire system had gone off with a crack, and smoke had entered the brake van. The fire bell had rung and he opened the guard's emergency brake valve and waited for the train to stop. He said that before it had done so the van became full of black/grey smoke and visibility had been virtually nil. His immediate concern had been to protect the Down Fast line as the Manchester express was due in a few minutes. He lost no time in carrying out the protection duties, which he had taken over from Driver Humphrys. On his return he had examined the damage and arranged for the train to be taken back to Cricklewood.

In his conclusions Colonel Reed wrote that the fire was the natural, though by no means inevitable, result of the cardan shaft breaking while the train was travelling at speed. The flailing ends, and subsequently the detached portions, had punctured the fuel tanks to release warm fuel oil, which then had ignited either from contact with hot exhaust pipes or more likely from the heavy sparking caused by the flailing end of the shaft in contact with other parts of the equipment and underframe.

The nuisance of the fire to the passengers, as distinct from the alarm that many of them must naturally have felt, appeared to have been caused in the main by the dense choking smoke forced into the compartments by the hot air heaters, and the inability of passengers to get away from it except through the carriage doors onto the track.

The cardan shaft had been rotating near to its whirling speed when it broke. In this condition it would have been unstable, and the final impulse to bow and break it need only have been some minor variation in stress incidental to the running of the car at speed. The whirling speed of the shaft had been reduced from the designed safe level to near the speed at which the shaft was turning, by the weight of instrumentation attached to it near its midpoint, but the effect of this weight on the whirling speed had apparently been overlooked. The responsibility for this oversight must rest with the management.

It was a matter for consideration whether the measurements should have been carried out on a service train. In the light of events it is clear that they should not, but it is only fair to point out that this was not a test in the normal meaning of the word in that no new engine or transmission equipment was being tried out, nor was existing equipment being intentionally stressed in some new way. The work was simply one of measurement of stresses and rotation speeds in standard equipment in normal running conditions. Had these measurements been successfully completed they might well have provided valuable data for British Railways and for Messrs. Rolls-Royce Ltd., for the improvement of transmission design. Instructions had been issued that if such an exercise was to be mounted again it must not be using a service train.

The fire had given rise, Colonel Reed continued, to some public feeling, particularly among passengers who travelled regularly on this line, because of previous breakdowns and fires on this stock, and he had been called upon to review the record of the St. Pancras/Bedford diesel multiple-unit stock. The points made by the public after this fire included the following:-

(a) Slow reaction of the train brakes to the operation of the communication cord.

(b) Excessive amount of scorching fumes and black smoke, apparently pumped through the heaters.

(c) Communicating doors required between compartments.

(d) Protection of fuel tanks against damage or repositioning them in a less vulnerable place.

(e) One engine of the burning car still continued to run after the train came to a stop.

In regard to item (a) he was satisfied that the brakes were in proper order. In circumstances such as this it would have been very easy for persons in the fourth and fifth cars to overestimate the length of the delay between pulling the communication cord and the brakes appearing to take effect, though in fact the rate of application was probably greater, because of the opening of the guards' brake valve, than it would have been if only the communication cord had been pulled. The communication cord application was designed to be more gradual than that of the guard's emergency brake, or of course of the driver's brake. The distance of about one mile from first signs of damage to where the train stopped was not unreasonable when it was realised that the train had been running at over 65 miles an hour. The time taken by the train to travel that mile would have been of the order of 30 seconds for the first half mile with little diminution of speed, and another 55 seconds to stop in the next half mile under the progressive brake application.

Items (b) (c) and (d) above were under examination with the Railways Board amongst a number of other modifications, and improvements could he made, particularly in respect of items (b) and (c). These were obviously important matters for the travelling public.

On item (e) the engine under the burning fifth car should have been shut down by the operation of the Graviner system, but the electrical circuit from the Graviner explosive fuse to the engine "shut down" solenoid was fused and the circuit might have been damaged or the fuse destroyed by the fire from the fourth car before the fire system operated. The Railway Officers were of the opinion that no additional danger was caused by the engine continuing to run because of the damage to the "shut down" circuit, but a modification to alter the method of operation of the solenoid from "energised to stop" to "energised to run" to ensure that the circuit will "fail safe" in the event of damage was under consideration.

Colonel McNaughton in his report on the accident at Sandridge on 15th August 1965 had tabulated the fires that occurred on this stock from its introduction in 1960 to that date, and the number that had been due to a faulty exhaust system. These were 33 and 18 respectively. The causes of the other 15 fires were various, five being attributed to fuel system defects, two to brake block sparks, two to a leaking fuel tank, one to the heater, and five to major mechanical defects. The fires had been for the most part minor ones. The trend showed an increasing incidence, culminating in 10 fires between January and August 1965. Since that date there had been three fires reported on this stock (up to July 1966) other than this one, one attributed to faulty exhaust, one to a heater defect, and one to the fuel system. There was another failure of a cardan shaft caused by the seizure of the final drive resulting in the main fuel tank being punctured, but there was no fire. Though the list of these fires and their causes appeared formidable, it must be remembered that this stock was intensively used, the normal monthly mileage of each car being about 6,000 miles. The cars had been designed to give the maximum amount of usable space, and all the machinery had been placed underneath. This had led to constrictions in design and a lack of accessibility for cleaning and maintenance. It might be on these grounds partly that reliability had not been as good as might be expected. The improvement since the Sandridge fire might be attributable to greater attention given to cleaning since that time, approved and under examination by the Railways Board.

The modifications suggested by Colonel McNaughton in his report on the Sandridge fire (see above) had been put in hand.

Further modifications had been discussed with the British Railways Board and were under examination by them.

(a) Strengthening of cardan shafts. The Board had decided to increase the wall thickness from one eighth of an inch to three fourteenths and new shafts were being ordered. Such shafts would have a slightly lower whirling speed, but considerably greater torsional strength which was considered to be necessary to stand up to torque fluctuations in service conditions.

(b) Strengthening or protection of fuel tanks from broken flailing cardan shafts. The Board considered that the tank could be made a little more robust by increasing the thickness of the metal from which they were constructed, but that it was not practical to armour them effectively against the potential attack of a broken, flailing, cardan shaft. A design for increasing the tank wall thickness from 18 SWG to 14 SWG had been approved and any new tanks would be constructed accordingly. Repositioning of the tanks away from the area of the cardan shaft was not a practical proposition.

(c) Re-routing of the exhaust system and improvement to the exhaust joints. The intention was to re-route No. 2 exhaust system to No. 1 end to keep it away from No. 2 fuel tank. Various ways of making the exhaust system itself more durable by the use of stainless steel, improvement in design of the flexible portion, improved gaskets, were under study, but instructions had been given, as an immediate measure, to provide shields between the exhaust system at joints and vulnerable parts of the underfloor equipment where oily fires might start. It could be hoped that the re-routing of this system would be approved and the work carried out at the next overhaul of each power car.

(d) Extension of the Graviner System. The Board's proposal was to extend the system in two respects, firstly by providing additional extinguisher bottles and circuits to protect each fuel tank, and secondly by installing duplicate extinguisher bottles which would discharge automatically through their respective circuits as soon as the train stopped. The delayed discharge was intended to extinguish any fire which had persisted due to the initial discharge being dissipated whilst the train had been travelling. These features would it appeared be added to this stock when it went through the shops.

(e) Cooling of torque converter return oil. From the fire-prevention point of view the object of this measure is to avoid heating the fuel oil in the tank so as to limit its volatility. The boiling point of fuel oil was in the region of 350° F and the flash point 160° F though the specification required not lower than 130°F. The return oil had a temperature in the region of 210° F and could heat up the fuel oil in certain conditions to near the flash point, which produced a dangerous situation if a fire source should be present. The Board were examining an alternative improvement with Rolls-Royce to isolate the torque converter system fluid from the main fuel supply. A non-inflammable fluid would then be used. The justification for this somewhat expensive modification would depend on the efficacy of the other measures. It was hoped that one of these alternatives would be introduced.

(f) Automatic shutdown of air heater intakes. The Board had prepared a design for incorporating such a control in the air heater system and a programme had been made to apply it to all the St. Pancras-Bedford stock. It was to be carried out as expeditiously as possible, within say three or four months. Additionally the instructions to guards would in future require them to shut down all heaters in case of fire, as a reinforcement of the automatic arrangements.

(g) Communicating doors between passenger compartments in power cars. This modification had been discussed with the Board's officers and it was hoped it would be carried out as each car went through works overhaul at Derby. The provision of gangways between cars was also discussed, but this affects the main structure of the car and its value was less significant than that of the communicating doors, which Colonel Reed considered provided sufficient freedom of movement.

(h) Fire resistance of floors. The Board held the view that the cost of this work, which involved much dismantling to give access to the underside of the floor, was not justified having regard to the other measures. This seemed reasonable.

The modifications being made to the St. Pancras-Bedford trains would also be incorporated in the 40 power cars on the Eastern Region's Lea Valley route, which were of substantially similar design.

An improvement of general value for all aspects of maintenance was the high-pressure detergent rinsing equipment now in the process of being provided by the Chief Mechanical and Electrical Engineer for depots where multiple-unit diesel sets were maintained. This would allow cleaning to he carried out quickly and thoroughly, so enabling the subsequent inspection and testing of engines and transmissions to be carried out in better conditions.

The measures outlined above, some of which were in hand or under examination before this accident took place, should go a long way towards reducing the risk of fire occurring in this stock. There was little doubt too that if fires should occur the measures would have the effect of minimising the consequences so that undue concern was not caused to the travelling public. [1278]

Fire broke out in a bogie of the *Banaras Express* bound for Bombay (now Mumbai) shortly after it had left Phulpur station, India, on Monday 25th April 1966. The alarm chain had been disconnected to prevent misuse and to escape the fire some passengers had jumped from the train. Two passengers died at the scene, a third died on the way to hospital, eight were seriously injured and 25 slightly hurt.

A question regarding the fire was raised in Parliament the following day, when the Railway Minister, Mr S.K. Patil, explained that alarm chains had been blanked off on the Allahabad – Banaras section four years previously, as well as in other sections where the incidence of episodes of misuse had exceeded a certain figure. In a reversal of this decision by the Government alarm chains would in future not be blanked off "no matter how much inconvenience is caused by their wrong use." Compensation would be paid to the dependants of those who had lost their lives. The cause of the fire was being investigated; sabotage had been ruled out, an electrical fault was a possibility. [1279]

Two valuable British thoroughbred racehorses, La Bage, a mare, valued at £5,000, and an unnamed foal, died, while a stallion, Wheel Wright II, valued at £4,000, was seriously injured in a train fire at Brühl, West Germany, on Friday 16th December 1966. The box car in which the horses had been riding was seen to be on fire as the train was passing through Brühl and the train was stopped, whereby the stallion jumped out into a field, covered in flames. The foal was burnt to death in the car and the mare was so badly hurt that she had to be put down. The horses had been brought privately by the Anglo-Irish

[1278] Ibid

[1279] *The Hindu*, Chennai, India, 27th April 1966

Bloodstock Agency for a Turkish purchaser and were on their way to Turkey. The cause of the fire was not stated; it was probably not immediately known. [1280]

Fire broke out in a Swiss coach forming part of international express D 207 from Basel (Switzerland) to Dortmund (West Germany) between Karlsruhe and Durlach, West Germany, on Saturday 4th February 1967. No one was hurt. The cause was believed to have been a faulty heater. [1281]

Two children died and five passengers were seriously and possibly fatally injured in a train fire near Multan, West Pakistan, on about Wednesday 12th April 1967. An overheated oil stove had caused the fire, probably one used by a passenger to cook meals. [1282]

Five minutes after it had left Parasnath station, India, on the Eastern Railway, on Saturday 13th April 1968, passengers in the fourth coach of train No. 17, the Up Sealdah – Pathankot Express noticed smoke coming from above a door. An attempt to stop the train by pulling the communication cord failed because it had been disconnected on the orders of the administration, except in the ladies' compartments, to prevent misuse. Later, flames appeared in the same area of the coach compartment. It was only after the train had passed through Chichaki that the assistant driver chanced to look back, saw flames, told his driver, and the train was pulled up at 18:47, before reaching Hazaribagh Road station, some 15 minutes after the smoke had first been seen. By then six panicking passengers had jumped from the train and been killed. In all, seven passengers lost their lives as a result of the fire and another 26 were admitted to hospital, nine of whom were seriously hurt.

Driver R.C. Singh told the inquiry into the fire that upon a shout from his colleague he had applied the brakes in emergency and stopped the train at Km. 341/7. With the help of the assistant driver he had then separated the fourth coach from the remainder of the train and played a hand-fired extinguisher on the blaze, but the fire had been too intense for it to have any useful effect. He had heard no complaints of defective lighting in the coaches.

Assistant Driver H.P. Yadaw described looking back as the train had been negotiating a right-hand curve, seeing a coach on fire, and alerting his driver. As he had gone back with the driver to separate the coach, flames had been coming from the right-hand lavatory window and the adjoining door. All the coach lights had been burning at that time.

Guard R.P. Mukkerjee stated that he had exchanged *all right* signals with the assistant stationmasters and switchmen at Chaudharibandh and Chichaki stations. He had observed the fire after the train stopped, tried to fight it with extinguishers and had given first aid to the passengers. He had tried to call the traction power controller at Asansol but had been unable to get the train's portable telephone to work. He had not seen any danger signals being made at Chichaki station nor any passengers jumping from the train there or elsewhere.

[1280] *The Times*, London, England, 17th December 1966

[1281] *Katastrophen der deutschen Bahnen Teil II* by H.J. Ritzau, pg. 139

[1282] *The Times*, London, England, 14th April 1967

Assistant Stationmaster R.N. Das at Chichaki testified that while exchanging *all right* signals from the platform he had seen smoke coming from the train, heard passengers shouting and had seen one fall from the train, at which point he had exhibited a danger signal and shouted to the guard, to no effect; he denied that the guard had exchanged the *all right* signal with him. When he had been able to contact the West cabin switchman the latter replied that he, too, had heard shouts and tried to stop the train by replacing the Advanced Starting signal, then he had advised his colleague at Hazaribagh Road East cabin. The assistant stationmaster added that he had told the switchman to hold the Down Andal Special back at Hazaribagh Road and had advised Control at 18:45.

Relieving Switchman R.L. Mukkerjee in Chichaki East cabin stated that the guard of the train had failed to acknowledge his *all right* signal, though he had not reported this to Control. He had not noticed any smoke coming from the train, or heard any shouting.

Relieving Switchman Gayana in Chichaki West cabin said that he had heard shouts coming from the train and at the same time seen smoke. He had shown a danger signal to the guard that had not been acknowledged, immediately replaced the Advanced Starting signal, and when this had not worked either, advised Hazaribagh Road. It seemed that the engine might have passed the Advanced Starting signal before it was replaced.

Assistant Stationmaster M.R. Ash at Chaudharibandh stated that the driver of train No. 17 had acknowledged the *all right* signal but not the guard. His two switchmen agreed.

Relieving Switchman Mohan Ram at Chegro Block Hut (between Parasnath and Chichaki) likewise stated that the guard had failed to acknowledge the *all right* signal. He had not bothered to report this to control as they usually took no notice of such reports.

Switchman Bandham Gope in Hazaribagh Road East cabin said that upon being advised by his colleague at Chichaki West he had informed his assistant stationmaster and Control.

The report on the fire continued with detailed evidence regarding the origin of the fire from electrical engineers and the accounts of a good number of the surviving passengers.

Discussing the accident, Additional Commissioner G.S. Pandor concluded that from the evidence the fire had begun at 18:28. The train had been overcrowded, with probably 75 passengers travelling in the affected compartment. The was no evidence of inflammable materials being carried by a passenger having started the fire, no evidence of an overheated axle box, while the train had been pulled by an electric locomotive and no steam-hauled train had passed in the opposite direction, ruling out the possibility of a live cinder from an engine firebox having started the fire. Likewise a discarded cigarette would not have caused a fire originating near the roof of the coach. A fault in the electrical wiring was therefore the likely cause of the blaze. Divisional Electrical Engineer B.K. Sarkar and Deputy Chief Electrical Engineer N.D. Gupte were, on the other hand, of the opinion that the fire had not been electrical in origin, but the additional commissioner disagreed with their views. The destruction of the coach made it impossible, he said, to establish the exact defect that had caused the fire, which could have been due to simultaneous leaks on the positive branch wire of the corridor light point of the coach and a wire of negative polarity in the same or another coach. The current so caused could have overheated the wire and ignited the wooden troughing in the vicinity.

The train lighting used a low-voltage system with aluminium conductors (which had been substituted for copper conductors to reduce thefts). Because of the low voltage heavier currents were required, leading to overheating if there was a bad connection or joint. Special tools and presses were required to make the joints and the task was difficult. High-resistance faults might not blow the protecting fuse and thefts of lighting equipment made loose and bare wires and thus faults more common.

Raising the system voltage to 110 volts, altering the arrangement of the wiring to reduce the load on individual wires and replacing the wooden battens with porcelain ones, were measures that might be undertaken to reduce the likelihood of such electrical fires.

The passengers had done their best to alert station staff by pulling the communication cord, but the latter had been disconnected. The railway administration were, he stated, within their rights to order the disconnection of the emergency cord (this seems surprising – they certainly would not be permitted to do so in, for example, Great Britain).

It was clear that the guard had failed to acknowledge the *all right* signals at different locations or the danger signals exhibited at Chichaki. Had he done so the casualties in the accident would have been minimised.

The additional commissioner concluded that no individual could be held responsible for the accident as the fault had developed while the train had been running and could not have been detected earlier. However, had Guard Mukkerjee exchanged signals with the station staff the train could have been stopped earlier.

The additional commissioner made five recommendations:

(1) As a short-term measure coaches should be intensively checked for electrical defects. In the longer term a higher voltage system, preferably 110 volts, should be used for train lighting.

(2) Wooden panelling should be coated with fire-retardant paints.

(3) The use of materials that were not fire resistant in the construction of the inner ceiling of coaches should be prohibited. The side panelling of coaches should likewise be fire resistant.

(4) The telephonic system of AC Electric Traction systems should be improved.

(5) The train communication should not be blocked off in long-distance express and mail trains except in exceptional circumstances, such as over notorious sections of line where misuse was very prevalent. [1283] A decision announced in Parliament two years earlier after a fire near Allahabad (see above) that alarm chains were not to be blanked off in any circumstances had apparently not been implemented.

The British Railways, London Midland Region main line from London St Pancras to Leicester again suffered a serious fire involving a diesel multiple-unit passenger train on Wednesday 12th June 1968, again at Sandridge, England. The train involved was the 07:40 Up train from Bedford to St Pancras, an eight-car set. Before the train had stopped a number of passengers jumped out of it to escape the fire, one of whom was killed and a second fatally injured. Ten other passengers were hurt, six of whom were detained in hospital.

[1283] Government of India Ministry of Tourism and Civil Aviation (Commission of Railway Safety): Report on Fire on No. 17 Up Sealdah – Pathankot Express Train between Parasnath and Hazaribagh Road Stations on Eastern Railway on 13th April 1968

Driver A.H. Dale told the Ministry of Transport inquiry into the accident that he had driven the 05:40 passenger train from St Pancras to Bedford, a four-car torque converter set, and there it had been coupled to two two-car gearbox sets for the return journey. He knew that the Up train should be driven as if it were composed of gearbox-type vehicles and had done so. The Up journey had been normal until after leaving Harpenden a Down express had passed him and he had seen its driver gesticulating to him. The fire alarm bell had rung almost at once and he had then applied the brakes in emergency. The train had been moving at 65 miles an hour and he had not noticed any brake application from the pulling of the communication cord, or by the guard. Looking back as soon as the train had stopped he had seen that its rear was shrouded in smoke. He had told Driver J.L. Daniel, an instructor driver riding with him, to go back and assist the guard while he went forward to protect the train, which he had done.

Driver R.H. Marshall said that he had been working the 08:02 Down express from St Pancras to Manchester and as he had rounded a gentle left-hand curve travelling at about 80 miles an hour he had seen the DMU and that the rear part was on fire. He had tried to alert its driver but did not think his gesture had been noticed in the brief moment available so he had stopped at Harpenden and told the signalman what he had seen.

Driver Daniel said that he had been on his way to take up duty at Cricklewood. He had also seen the express driver gesticulating and looking back had seen that the rear of his train was on fire. Going back with a fire extinguisher he had met the guard, whom he had told to go back to protect the rear of the train, while he used the extinguisher on what appeared to be the seat of the fire under the seventh vehicle, to little effect. He had been joined by Driver J.M. Geeson, who had been travelling as a passenger, and they used further extinguishers on the fire. The engine of the second vehicle had still been running and after the fire brigade arrived and sprayed it with water he had finally managed to shut it down by crawling under the vehicle and manually operating the solenoid, burning his hands in the process. With Driver Geeson he had then helped the passengers onto a train that had arrived cautiously on the Up Slow line.

Passenger Guard T.M. Wilson stated that between Harpenden and St Albans he had suddenly seen through the window that flames were leaping up from the seventh coach. He had jumped up to apply the brake as the fire alarm had rung but shortly smoke and fumes forced him to let go of the brake handle, the brakes being applied at about the same time by his driver.

Mr J.E.C. Palmer, a regular passenger on the line, said that after leaving Harpenden he had notice a vibration, which became more severe. As he had been considering pulling the communication cord, smoke had come into the compartment from the other end of the coach and soon had become so thick that he could not see the other end of the coach. Both the communication cords in the cross passage just ahead of where he had been sitting had been pulled, but he thought that the train started to slow down about a second before the cord was pulled. He himself had opened the sliding door to the small compartment at the front of the coach to warn the occupants and then had taken up a position by the near side door. He had opened the door and looked back and seen smoke and flames coming from the underside of both rear coaches. Not long after that, when the speed had dropped to perhaps 30 miles an hour, he had seen a passenger jump from the rear door of the seventh coach and others from the eighth coach. He had then seen them lying motionless on the track, having obviously jumped much too soon. By this time there were passengers crowding behind him and in order to speed up their escape through the limited number of doors available, he had decided to jump as soon as it was reasonably safe to do so. He had waited until the speed was down to about 8 to 10 miles an hour and then jumped out. He had landed on his feet and very soon afterwards the train had stopped.

Mr. J. Godfrey, travelling in the rear compartment of the seventh vehicle, regularly boarded this train at Bedford. He, too, had become aware of increasing vibration after leaving Harpenden, eventually becoming so bad that one could hardly sit on the seat at all. He had stood up and at once there seemed to be a minor explosion and within a few seconds the carriage had become full of choking smoke and fumes, apparently coming from a little forward of where he had been sitting. He had then become aware of the sound of breaking glass as passengers used their briefcases to break the windows. He had found and pulled the communication cord and it seemed to him about a minute before the train came to a stand. After the train stopped he had seen flames coming into the coach, apparently through a broken window. He had finally left the coach through the same door as Mr. Palmer and he stressed that everybody had jumped out in a more or less orderly fashion and that there had been no sign of panic at all.

Mr. J. V. A. Haggard had been travelling in the larger passenger compartment of the eighth vehicle. He related that the first sign of anything being wrong had been the sudden appearance of a sheet of flame outside the windows on the offside, accompanied by a roaring noise; there had been a concerted movement of passengers to the near side and a certain amount of confusion, with people asking where the communication cord was, and so on. He had two men struggling at the nearside door, adjacent to the guards' compartment, one trying to open the door, the other to keep it closed. He thought that the door in fact had not opened until the train came to a stand. He had been aware of a strong acrid smell but did not remember seeing any smoke in the compartment. He felt that the train had taken a long time to stop and before it had come to a stand he had seen a man jump through a broken window on the near side. He himself had been the last but one to leave the compartment, using the nearside door adjacent to the guards' compartment.

Mr. G. W. Webb said he had been sitting in the rearmost row of seats, facing the rear driving cab. Hearing a shout, he looked over his shoulder and had seen smoke and flames on the offside that had spread to cover all the offside windows. It had seemed about 10 to 15 seconds before the train started to slow down and then there was another outburst of flame that cracked the leading windows on the offside of the compartment. There had been no visible smoke inside the coach but passengers had been stood up in the gangway and cross passage making it impossible to see what was going on further forward. He had seen two men jump from the nearside door when the train was still running at 10-15 miles an hour but the remainder had waited until after the train had stopped; he had been the last to leave.

Mr. C. E. Parker of Harpenden said in a written statement that he had been travelling with his two sons, aged 12 and 14, in the large compartment of the eighth vehicle. Like the other witnesses travelling in this vehicle his first intimation that anything was wrong had been the sudden appearance of a sheet of flame outside the offside windows accompanied by a roaring noise. The windows had quickly become obscured and begun to crack. This had caused the passengers, particularly those on the offside, to jump up and make for the nearside doors. Because the seats were facing rearwards most had made for the cross passage at the rear of the compartment, but he, with his sons, had moved forward to the cross passage adjacent to the guards' compartment. He had heard someone shout, "Pull the communication cord!" and almost immediately the bell in the guards' compartment had started to ring continuously. In the belief that the fire was in the engine or fuel tank under the coach in which he was travelling, he had taken his sons through the communicating door into the guards' compartment; conditions there had been no better but there had been fewer windows. As the smoke and fumes had increased, he crouched down with his head near floor level and had instructed his sons to do the same. This had made breathing easier and he determined not to jump out until the train had stopped or conditions inside had become completely unbearable. When the train had stopped he jumped down from the nearside guards' door, and after ensuring that his sons were safely across the lines had returned to assist other passengers getting out of the train. He noted that, as soon as the train had stopped, the flames subsided and only a minor fire had remained under the seventh coach. He then re-entered the last coach to ensure that all the passengers had alighted, and none had been overcome by fumes or injured in the crush to escape but had found no one. He noted that amongst the passengers alongside the train there had been none with other than minor injuries, but he had seen, about 20 yards behind the train, a badly injured and unconscious young man lying in the "six-foot" between the Up Fast and Down Slow lines with his feet towards the direction of travel, his position being consistent with having jumped from the nearside of the train and landed on his head. Further back, some 50 yards behind the train, another man was lying in a similar position but with his head towards the direction of travel. He was unconscious and had obviously sustained a fractured skull.

In his conclusions the inspecting officer, Lt. Col. I.K.A. McNaughton, wrote, 'This fire was the direct result of the failure of the mechanical transmission on the seventh vehicle of the train. The failure arose from a lack of lubrication in an epicyclic gearbox, but there is insufficient evidence to show that any share of the responsibility for the failure and its consequences can reasonably be placed on either of the two fitters who most recently examined the vehicle prior to the accident. apart from the presumed failure on the part of Fitter Hughes, when carrying out a "B examination" on 4th June, to replace properly the gearbox oil filler plug, and of Fitter Whillis to notice that the plug was loose or missing during his quick visual examination on 10th June. The gearbox, which at the time of the fire had run 42,010 miles and was closely approaching the time at which it was due for an examination, may in fact have been using excessive oil but there is no definite evidence to show that this was so, despite Fitter Hughes' statement that he had found the oil level low when carrying out the "B examination" on 4th June, and put in a gallon from the dispensing trolley.

'As far as the action of the train crew is concerned. I am satisfied that the driver and the guard acted promptly and correctly as soon as they became aware of the fire and that there was no avoidable delay in bringing the train to a stand. It seems clear from the evidence found on the track in the neighbourhood of M.P. 22 that the actual time that elapsed between the gearbox seizure and the beginning of the fire can at the most have been some 3 seconds, and even if the brakes had been applied at the moment of failure the fire could not have been prevented.

'Although the "Graviner" automatic fire extinguishing system was found to have operated on both B the vehicles involved, it should be appreciated that it was not designed to deal with a major diesel fuel fire of the intensity that arose on this occasion, but only to smother minor fires in the immediate vicinity of the diesel engine itself. It was fortunate that by the time the train came to stand most of the fuel had run out of the ruptured tank and that only a residual fire remained to be dealt with, but I must nevertheless commend the exemplary action by Instructor Driver Daniel and the assistance given to him by Driver Geeson in fighting the fire prior to the arrival of the Fire Brigade.

'It is a matter of some concern that this fire led to the deaths of two passengers and injuries to a further ten. In the event, it was clear from the evidence given at my Inquiry that there would have been no injuries had all the passengers waited until the train stopped before jumping out. But in the circumstances. with the considerable confusion caused by the entry of smoke and fumes, the difficulties in locating the communication cord and the large numbers of people crowding at the small number of outside doors available. it is not surprising that, of those nearest the doors, some misjudged the earliest moment at which it was safe to jump from the moving train, not because they themselves were in any immediate danger from the fire, but because of the pressure of people behind them waiting their turn to escape from what might, at any moment, became an inferno. It is clear from the evidence of the passengers that, although there was confusion, understandable in the circumstances, there was no panic and that the evacuation of the two vehicles involved was carried out in an orderly manner.

'Whatever the circumstances, a failure of a part of the mechanical transmission of a diesel multiple unit power car should not lead, as in the present case, to a serious fire involving fatal casualties. That it did so on this occasion must be ascribed to the bad design of the vehicle involved. The location of the main fuel tank astride the cardan shaft meant that a failure of either of the universal joints, from whatever cause, or of the cardan shaft, could result in the shaft flailing and rupturing the fuel tank making a serious fire an almost inevitable consequence.

'Two previous instances are on record, in both of which the transmission failure was initiated by the loss or failure of a gearbox mounting suspension bolt, leading to misalignment of the gearbox and consequent failure of the universal joint at the leading end of the cardan shaft. The two cases. which were briefly reviewed in Colonel McMullen's Annual Report for 1967, occurred at Bay Horse (London Midland Region) on 19th January, and in Sough Tunnel near Darwen (London Midland Region) on 4th October. In each case, the fuel tank was ruptured and a serious fire ensued, leading in Sough Tunnel to the total gutting of 2 vehicles, fortunately without involving serious casualties.

'On this particular design of power car, fitted with only one engine and transmission, there seems no justification for this location of the main fuel tank, requiring as it does a complicated design of tank, when there was enough space at the other end of the underframe, between the engine and the non-driven bogie, for a simple rectangular tank of adequate capacity. The British Railways Board have accepted the need for these fuel tanks to be relocated in a less vulnerable position, and by August 1969. this modification had been completed on all except one of the power cars of this type to be retained in service.'

Turning to remarks the inspecting officer stated, 'In view of the public concern which attended this fire, it is opportune for me to review the fire record of the DMU services on the St. Pancras-Bedford line and to comment on the progress being made towards the completion of the modifications recommended in previous accident reports. As far as the Craven 2 car low density sets are concerned, the fire at Sandridge which is the subject of the present report is the only fire to be reported on this type of stock during its period of service in the London area, apart from a single incident at Radlett on 15th April 1968, when a carelessly discarded cigarette set light to waste paper under a seat without serious consequences, though as recorded there have been earlier fires involving this type of stock in service elsewhere.

'After this accident, however, I received a number of letters from members of the public, including some who had been travelling on the train involved, commenting on the use of low density DMU sets on the services between Bedford and St. Pancras. The general opinion expressed was that the limited number of outside doors, a time-consuming nuisance in any case on a busy commuter service, presented a real danger in the event of a fire or other emergency calling for the speedy detrainment of a full load of passengers. My attention was also drawn to the absence of lower step boards at any of the passenger doors and to the extremely limited number of places where the communication cord could be pulled, none of which were located in the actual passenger compartments.

'I discussed these points with the responsible officers of the British Railways Board and of the London Midland Region. As far as the use of low density designs of DMU on these services was concerned, the Board's Officers accepted that high density stock with a door to each bay of seating would be more appropriate, but in view of the limited clearances over the City Widened lines, none of the high-density vehicles then available on British Railways were permitted to run through to Moorgate at the time of the accident. Subsequently, however, with some operating restrictions, route availability was given for high density vehicles to run over the City Widened lines and I am able to report that all the Craven low density sets of the type involved in this accident were finally withdrawn from the Luton-Moorgate services and replaced by high density sets having a door to each bay of seating, with effect from 8th August 1968. Nine of the low density sets so displaced will continue in use on the Kentish Town-Barking service and on these eighteen vehicles additional lower footsteps are being provided at one passenger door on each side of each vehicle and also two additional points of access to the communication cord, one on each side, in each passenger compartment, i.e. 4 additional points are being provided in each motor-second brake and 6 in each motor lavatory composite.

'Three previous fires between St. Pancras and Bedford have been the subject of public Inquiries; these occurred between Millbrook and Wilshamstead in October 1964, at Sandridge in August 1965, and at Napsbury in February 1966. All these fires occurred on units of the standard St. Pancras-Bedford high density stock with Rolls-Royce engines and torque converter transmissions. Colonel Reed, in his Napsbury report, reviewed the fire record of these units and listed the modifications which the Railways Board had agreed to carry out on them and on the substantially similar units running at that time on the Lea Valley Services of the Eastern Region. Many of the modifications are extensive and can only be carried out as the units pass through main workshops for overhaul, with the result that the programme will not be finally completed until the end of next year and, in particular, it will be noted that the major modification to the torque converter, involving the provision of a self-contained fluid circuit to prevent the temperature of the fuel in the main fuel tank being raised by the oil circulated through the torque converter, has only been completed on one power car. This modification was originally planned to be completed by the end of this year, but the target date has now been put back because of delays in obtaining the redesigned components. There has already been a substantial reduction in the incidence of fires on this type of stock, however, and all the fires reported since that at Napsbury in February 1966 have been of a minor character confined to the underfloor equipment, the actual record being as follows: 1966 3 (excluding the fire at Napsbury) 1967 and 1968 2 Despite the improvements brought about by the various modifications made and the higher standards of cleanliness achieved as far as the underfloor equipment is concerned, the general fire record of the underfloor engined diesel multiple units on British Railways is not good, particularly when compared with that of the diesel-electric multiple-units in use on the Southern Region. I would like, therefore, to draw attention to a reference made in the British Railways Board's Annual Report for 1968 to a review that has been made of equipment for use in new designs of diesel multiple-units. Two basic schemes were proposed and are under consideration: high density trains for commuter services and high-speed Inter-City trains. Both schemes envisage the use of above-floor engines and electric transmission. I welcome this development.' [1284]

On Thursday 31st October 1968 fire broke out in a TLR (third-class, luggage and brake van composite) coach of No. 4 BRB Down passenger train between Hissar and Jakhod Khera stations, India, on the Northern Railway. Six passengers died, nine were seriously and 16 slightly injured.

The train consisted of Class YP steam engine No. 216 and seven coaches. The fire broke out a few minutes after the train left Hissar. The 50 passengers in the luggage compartment of the coach attempted to fight the fire by throwing the burning material out but in the panic there was no concerted effort. Some of the passengers also shouted to alert the driver, who looking back saw passengers jumping from the train and stopped it. He and the fireman managed to extinguish the fire with buckets of water from the engine, the train was set back a little and passengers who had jumped from it were picked up, then the train was set back again to Hissar station for assistance.

Driver Lachni Narain told the inquiry into the fire that he had left Hissar at 15:37. As he had approached Nyoli Kalan Halt he heard shouting and looking back had seen passengers jumping from the TLR coach next to his engine. Stopping, he had found the coach on fire but had been able to put the blaze out with water from his engine, then had reversed to Hissar station as already detailed. His engine was fitted with a spark arrester and no sparks had been emitted during the run. The alarm chain had not been pulled.

Grade B Fireman Shambhoo Dayal said that they had been travelling at about 30 kilometres an hour at the time. He had seen no sparks coming from the engine chimney.

Grade C Fireman Bharat Singh said that the engine fire damper had been closed after leaving Hissar and no sparks were being emitted.

Guard 'B' Sohan Lal agreed that his train had left Hissar at 15:37 and stopped at 15:48. Alighting, he had seen injured passengers and handed a first-aid box to Travelling Ticket Examiner Deep Singh. Going forward to the front of the train, the guard continued, he had found that a fire in the luggage compartment of the TLR had been extinguished by the driver and fireman. He pleaded that he had been unaware of a circular stating the empty luggage compartments should be locked or knowing that there had been passengers in said compartment.

[1284] Ministry of Transport report

Train Examiner in Charge J.P. Mittal at Hissar stated that he had examined the axle boxes of the train and they had been in a satisfactory condition with no sign of overheating.

Assistant Stationmaster Chanan Singh at Rewari testified that the stock of the train had been properly cleaned at Rewari.

Train Lighting Fitter Prem Nath at Rewari stated that the lighting and fans of the TLR coach had been checked and then switched off.

Deputy Chief Controller Audh Bahari Lal at Rewari said that he had been informed of the fire by the guard at 16:15.

Circle Fire Inspector B.M. Bharadwaj at Delhi stated that he had examined the luggage compartment of the affected coach. He ruled out the axle boxes and electrical connections as a cause of the fire, which he attributed to the ignition of krubi (a tropical plant) by a live bidi end (bidis were cheap cigarettes) or match.

Assistant Security Officer (Fire) S.P. Nag stated that he had also examined the coach and found burnt sirkanda (reeds), krubi and sweet potatoes on its floor plus a large number of half burnt bidi ends and matchsticks. Sirkanda, he added, caught fire very quickly and he thought that a lighted bidi end or matchstick had set it alight, the fire then spreading to the krubi. He ruled out the axle boxes, electrical system, or a spark from the engine as a cause of the fire.

Divisional Mechanical Engineer K. Reghunathan said that the engine was fitted with a fall type spark arrester and the Grade 1 Bengal coal being used was of the non-sparking type.

Divisional Electrical Engineer G.L. Vaishnavi testified that he had examined the coach thoroughly. He ruled out an electrical cause for the fire.

A number of passengers also gave evidence, several speaking of a number of people smoking bidis and one of a bundle of grass and fodder (the sirkanda and krubi) catching fire.

Additional Commissioner C.R. Sule noted that the train had stopped at 15:48 near Km. 146/3. According to the passenger's accounts the fire had begun five minutes earlier, thus at 15:43.

It was clearly established that it was the bundles of sirkanda and krubi that had ignited. Dried grasses such as these were inflammable but they were not substances passengers were prohibited from carrying. Other sources of ignition having been ruled out by the evidence, the inevitable conclusion must be that a lighted bidi or match carelessly thrown by a passenger was the cause of the fire. The severity of the accident was increased because the luggage compartment, not being meant for the carriage of passengers, had no communication chain.

No responsibility rested upon the railway for the occurrence of the fire but Guard Sohan Lal was responsible for not preventing passengers from riding in the luggage compartment.

The additional commissioner recommended that the rules preventing passengers riding in luggage compartments should be rigidly followed and it was desirable that the carriage of inflammable dried grasses such as sirkanda and krubi by passengers as luggage should be prohibited. [1285]

The disastrous fire in Vranduk Tunnel, Yugoslavia, on Sunday 14th February 1971 has been described in the Tunnel Accidents chapter, above.

Fire broke out in the sleeping car of a Ringling Brothers circus train travelling at nearly 60 miles an hour and bound for Venice, Florida, early on the morning of Tuesday 3rd December 1974, between Lake City and Lane, South Carolina, United States. Having finished its last performance at New Haven, Connecticut, on Sunday 1st the circus was on its way to its Florida winter quarters.

The fire had probably begun some miles further north, near Lake City, at about 5.30 a.m. and smoke was seen by residents who were early risers at Kingstree but the train was beyond Lane before it was discovered and the train was stopped. Notice of the fire was received by the fire department from the dispatcher at about 6.45 a.m. The place where the Seaboard Coast Line Railroad train had stopped was inaccessible and it had to be reversed three miles to Lane, where the fire could then finally be tackled and extinguished. Only the one sleeping car, in the middle of the train, was affected. Nine of the 13 occupants of the burning car managed to escape to adjoining cars but the charred and unrecognisable bodies of three men were discovered in the car after the fire had been put out. A fourth victim, a woman, had jumped from the burning train but she sustained fatal injuries in doing so as well as burns. None of the dead had been circus performers; the men had been part of its support staff and the woman was stated to have been a private contractor for one of the circus concessions. There had been 225 people on the train in all, plus circus animals and equipment, but no one else was hurt. The train later continued its journey after the badly damaged car had been detached. [1286] [1287] [1288]

The coach that had caught fire was divided into separate sleeping berths, each with a bunk, a lavatory, and a space for possessions. The fire had broken out in a berth at the front of the coach where the remains of an electric hot plate were found. The heat of the fire had caused windows to break, whereupon the draught of air caused by the fast-moving train had intensified the fire, as had the carriage door being left open by the woman who had jumped from the train. The exact cause of the fire could not be determined, it was believed to have been either due to the carelessness of the circus employee in the berth where it had begun or alternatively to an electrical fault. [1289]

Fire broke out on a southbound two-car diesel railcar near Eketahuna, New Zealand, at 7.30 p.m. om Sunday 24th August 1975. A passenger alerted the train crew, the train

---

[1285] Government of India Ministry of Tourism and Civil Aviation (Commission of Railway Safety): Report on Accident Caused by Fire in Coach No. NR 1704 TLR of 4 BTB Down Passenger Train Between Hissar and Jakhod Khera Stations, Northern Railway on 31st October 1968.

[1286] *Sarasota Herald Tribune*, Sarasota, Florida, United States, 4th December 1974

[1287] *Florence Morning News*, Florence, South Carolina, United States, 4th December 1974

[1288] *The Daily Freeman*, Kingston, New York, United States, 4th December 1974

[1289] https://scfirefighters.org/firewire/tragedies-of-the-circus-the-south-carolina-connection/

was brought to a halt, and Guard Len Cairncross and Driver Pat Hickton, finding that they were unable to control the blaze in the undercarriage of the front car, helped the 73 passengers, who included a disabled woman, descend the five feet to the ground. The two railwaymen then attempted to retrieve the passengers' luggage but were only able to retrieve five suitcases before smoke and flames forced them to abandon their efforts. It was a walk of about two miles to the nearest road and a local opossum trapper, Mr Barry Gaskin, took 12 passengers – children and elderly people – who were unable to walk that far, to the road in his Land Rover. No one was hurt.

A railway spokesman would not comment on the cause of the fire but stated that the company 'had their opinions and suspicions'. The 20-year-old Fiat railcar had been running on only one engine after the other had been shut down on the journey from Napier to Woodville. [1290]

On the evening Monday 29th September 1975 train A-41, a Central Railway Down Bombay VT to Ambernath local, left Kalyan station, India, at 20:29 and was between there and Vithalwadi station when there was a sudden flash on the roof of the leading coach. This burnt a hole in both the false roof and the main roof and burning material and molten metal fell on the passengers below in the crowded second-class compartment. The motorman heard an unusual noise, looking back saw smoke and fire, and heard cries for help from the passengers. He stopped the electric train and then lowered its pantograph. Many of the passengers jumped from the train. As a result of the accident 29 passengers were injured, eight seriously, five of whom later died in Kalyan hospital. There had been a severe thunderstorm on the evening of the accident but it was only raining at the time that it happened.

A number of passengers gave evidence at the inquiry into the fire, none of whom thought that the train had been struck by lightning. Nor had several witnesses who had seen the fire from the lineside seen any sign of lightning.

Guard M.A. Dabhoivala stated that the train had stopped suddenly and he had seen smoke and flames coming from the motor coach. A loud report had made him think at first that lightning must have struck the train but he had not seen any flashing light.

Motorman Abbas Haider said that his train had been travelling at 10 Km/h at the time. He thought that lightning had struck the train. On noticing the fire near the pantograph at the Kalyan end at about 20:32 he had applied the emergency brakes and lowered the pantograph.

Much technical evidence was also given at the inquiry as to the nature of the damage and its likely cause.

Additional Commissioner D.V. Divgi noted that a number of previous accidents in the roofs of EMU trains on the Central and Western Railways had been caused by short circuits due to stray pieces of wire.

The time of the accident, from the tripping of the feeder at the sub-station, he put at 20.33.

The train had started cautiously on an A marker at the automatic signal at Kalyan and passengers had jumped from it without sustaining serious injury, while it had taken four

---

[1290] *The Press*, Christchurch, New Zealand, 26th August 1975 minutes to cover a kilometre. Its speed at the outbreak of the fire must have been about 15 Km/h.

The formation of the hole in the roof bore a striking resemblance to past fires due to stray wires. That no stray wire had been found afterwards could be easily explained by it having been mixed up with the debris.

The lack of damage to the pantograph seemed to exclude a defect in the pantograph as a cause of the fire.

Though the motorman of the train thought he had seen a lightning strike five passengers, three lineside witnesses, the guard and a gateman all denied having seen any lightning.

There were many objects in the vicinity that lightning could have struck instead of the EMU.

Having reviewed all the evidence the additional commissioner concluded that the cause of the fire was a short circuit between the roof and live equipment above it due to an external object such as a stray piece of wire.

The location of the pantograph over the passenger compartment constituted a grave fire hazard, as shown by numerous previous accidents of the same type. This was an extremely unsatisfactory state of affairs from the point of view of the safety of the travelling public.

No fires had occurred in the roofs of pre-1951 rolling stock. This was because the roofs of earlier stock were of wood while those of more recent stock were of steel.

The absence of steps presented a grave hazard for passengers exiting the train in emergency, especially the aged, women and children.

In conclusion the additional commisioner was satisfied that an external object such as a stray piece of wire had caused an electrical short citcuit between the 1,500 volt overhead electric lines/pantograph and the roof of the coach.

No one was held to be responsible for the fire.

Additional Commissioner Divgi made three recommendations:

(1) The pantographs should be relocated as soon as possible onto the roofs of non-passenger compartments.

(2) A suitable means of roof insulation, especially in the pantograph zone, should be provided.

A means of alighting from EMUs in emergencies, such as steps, should be provided.[1291]

[1291] Government of India Ministry of Tourism and Civil Aviation (Commission of Railway Safety): Report on fire on Leading Coach No. 7606 (Motor Coach) of A-41 Down Bombay CT – Ambarnath Local Train at Km 54/5 between Kalyan and Vithalwadi Stations of Bombay Division on Central Railway on 29th September 1975.

CHAPTER 15

CHEMICAL EXPLOSIONS

Railway accidents involving explosions are principally of two types: those that might be defined as chemical explosions, and locomotive boiler explosions. Accidental chemical explosions, such as those due to mishaps with blasting powder or munitions, are rare, whereas in the early days in particular, locomotive boiler explosions were much more common, though by the mid-twentieth century they were becoming much rarer, with chemical explosions perhaps becoming a little more common. In wartime the latter often involved the accidental explosion of munitions. This chapter will deal with chemical explosions and also those involving electrical equipment, even though these are not strictly chemical in origin.

Disaster stuck *accelerato* train No. 1072 from La Spezia to Genova, Italy, on Saturday 16th January 1965. Despite what its name might seem to imply this was a local or stopping train. As it was a Saturday it was not quite as crowded as on a weekday, but it was still carrying a fair number of passengers, mostly a mixture of students and workers. It had been quite full to begin with but Conductor Pietro Quattordio said later that a fair number of passengers had got off at Levanto, while a few had got on there, including 18-year-old student Giuseppina Borgnotta, who had had to hurry to catch her train. Three other passengers were railwaymen, surveyors engaged in the doubling of the line; Renato Lagorio, 50; Giuseppe Loss, 51, both from Genova; and Giulio Rizono, 50, from Sanremo. Another passenger connected with the railway, and the only passenger riding in the first-class coach, was Ernesto Baldi, 79, retired chief secretary for the Ferrovie dello Stato, while other passengers not connected with the railway included Franco Graffigna, 36, a dental technician from Chiavari; Alfredo Salvadore, 26, a mason from Sermide da Mantova; and Franco Scisciò, 29, an optician from Chiavari. All the above named passengers were to lose their lives in the accident that was shortly to overtake them, together with Ticket Examiner Giovanni Bianchi, 45.

The train left Levanto at 13:08 and shortly entered the tunnel that separated that station from the one at Bonassola. In front of the passenger train was freight train No. 6178, which had arrived at Bonassola 12:57 and stopped both to unload freight and to give priority to the passenger train, before it continued to Sestri Levante, about 15 kilometres down the line.

When train No. 1072 arrived at Bonassola at 13:10 there was no one on the platform,

Stationmaster Claudio Ambrogi was in his office and a porter was working in the luggage room. Driver Angelo Donati applied his brakes as the train came into the station, passing the freight train on the adjoining line, and with that it happened. There was a terrific explosion that destroyed one coach of the passenger train and badly damaged two others, while three waggons of the freight train were reduced to a heap of scrap metal. The damage was not limited to the station either, houses within a radius of several hundred metres were more or less damaged with windows broken and tiles ripped from roofs, while those closer at hand suffered more extensive damage. The body of one victim was blown onto the roof of a house, another was found in the roadway and those of the three surveyors, who had been occupied playing cards, were found lying close to one another in a field. The initial number of dead was eight but student Giuseppino Borgnotto also died, shortly after her arrival at Levento hospital.

Driver Donati said, "I jumped down and ran into the stationmaster's office to telephone La Spezia and tell them that a disaster had happened."

The first ambulances from La Spezia were followed by others from Chiavari, Sestri and La Spezia, while private cars were also employed and two helicopters carrying doctors, nurses, and supplies of plasma flew from Genova, landing on the beach at Levanto. Fifty passengers were injured, ten of whom were admitted to Levanto hospital. Nicola Bercotti, age not stated but probably a student, and Vittorio Perfigli, aged 18, were said to be in a serious condition, as was a third unnamed person. The Minister of the Interior, Paolo Taviani, arrived at the accident scene by car at 18:00.

It transpired that one of the large waggons of the freight train, called a Carnera, had contained boxes of dynamite for use in the doubling of the railway, 16 boxes destined for the Colini company in Bonassola and another eight boxes for the Sogene company at Framura – in all 648 kilograms of explosives. The sticks of dynamite were wrapped in red greaseproof paper and were the standard length of about 20 centimetres. They did not have detonators and therefore should have been relatively safe from accidental explosion. They had been loaded onto the waggon, it was stated, taking all the usual precautions. [1292]

The following day a newspaper reported that the investigation into the causes of the disaster had been completed, though it would be some time before its results became known. It then went on to set out four theories as to how the explosion could have happened, listed in order of increasing probability:

(1) Sabotage had been rumoured but was considered the least likely theory. A potential saboteur was unlikely to have chosen an insignificant freight train as a target and could not have planted a time bomb such that it would explode at the exact time the passenger train arrived at the spot.

(2) Another waggon of the freight train had been carrying compressed natural gas. It was possible that a leakage of gas had occurred and a spark, perhaps from the brakes of the passenger train, had triggered an explosion of the gas.

(3) The cold weather could have caused some of the nitroglycerine in the sticks to partially separate from the absorbant; thus separated the nitroglycerine was much more unstable and much more liable to explode.

(4) The most likely theory seemed to be that during unloading a box of explosives had

[1292] *La Stampa*, Torino, Italy, 17th January 1965

been placed near to the track in an unstable position and the vibration of the approaching train had caused it to fall to the ground and under the wheels of the passenger train, the shock of the collision being sufficient to set off the dynamite. [1293]

Early on the morning of Wednesday 9th March 1966 an 80-car Missouri Pacific Railroad freight train was entering a siding at Corning, Arkansas, United States, when a boxcar loaded with 175mm artillery shells exploded. The explosion was powerful enough to be heard in the neighbouring states of Missouri and Tennessee. A hole 50 feet long and 15 feet deep was made by the blast. A house owned by Mr and Mrs Rappert was demolished but they fortunately had not been at home but were in St Louis, Missouri; the walls of two other houses were caved in and a natural gas line was breached, forcing the evacuation of nearby residents. The town of Corning itself was three miles away but the windows of its stores were broken.

The train was bound for the Red River Arsenal at Texarkana, Texas, when the explosion occurred at 2.48 a.m.

The siding adjoined a rice storage and drying elevator, the reason for the natural gas supply. John Hutchinson, secretary of the Corning Volunteer Fire Department, said that they had been hindered in fighting the fire that had followed by the explosion until the gas supply had been turned off, as well as for fear of further explosions from other munitions. Their efforts were supplemented by fire and civil defence crews from within a 50 mile radius.

Mrs Glendon McGuire, living half a mile from the siding, was awoken by the noise of the blast and her feet were cut by flying glass but she was not seriously hurt. No one else was injured.

The cause of the explosion was not immediately known. [1294]

Seven people were killed and 31 were injured when a munitions waggon exploded in a marshalling yard on the Central Railway at Bhusaval, India, on the evening of Monday 2nd April 1966. The explosion was presumably accidental in nature. [1295]

The Red River Army Depot at Texarkana, Texas, again featured in a train explosion on Wednesday 14th May 1969, this time as the starting point for an ammunition train bound for Port Chicago, California, whence the munitions were to be shipped to Vietnam. The 79-car Southern Pacific Company freight train, pulled by four diesel units, was standing in a siding at Palmdale, California, United States, while engine crews were changed when a boxcar was discovered to be on fire at about 4.50 p.m. It was loaded with 61 tons of ammunition and the heat soon caused live shells to explode. Eight homes near the scene were immediately evacuated and the Federal Aviation Administration banned all flights within a five-mile radius of the site. Later an evacuation of 400 residents within a two-mile radius was begun but before it had been completed the explosions and fire had diminished.

[1293] Ibid, 18th January 1965

[1294] *Winona Daily News*, Winona, Minnesota, United States, 9th March 1966

[1295] *The Cedar Rapids Gazette*, Cedar Rapids, Iowa, United States, 3rd May 1966

Road Foreman Charles Hastings was hailed as a hero for driving his pickup truck to within a short distance of the burning box car to uncouple the cars on either side so that engines could draw them clear. Cars close to the burning one were loaded with liquid sulphur and propane. All but 17 cars were initially detached and later all but four.

Six units of firemen were despatched to the scene as was an ordnance disposal team from the nearby Edwards Air Force Base. The firemen decided that there was little they could do to fight the blaze, as it was considered too dangerous to approach it too closely, so instead the fire was left to burn itself out, which it did after about four hours.

No one was hurt.

The fire was thought likely to have originated in an overheated journal ('hot-box'). [1296] [1297]

The guard and two passengers were killed more or less outright and 11 passengers were admitted to hospital, one of whom died later the same day, when a bomb apparently exploded at 02:30 on Wednesday 21st May 1969 in a rear third-class brake composite bogie of a Timsukia – Siliguri passenger train. The explosion occurred midway between Dharamtul and Auguri stations, India, on the North East Frontier Railway. That a bomb had been planted inside the coach was deduced from the fact that the coach had been all but wrecked but the track was undamaged. A relief train was sent from Gauhati with doctors and ten passengers were admitted to the hospital there, the other being taken to Chaparmukh hospital. Services on the line were suspended for some hours while an investigation was carried out. The train was later taken to Jagi Road station, near Gauhati, where it was checked by explosive experts, while a tracker dog was sent to the scene.

Inspector-General of Police G. Day stated that the explosion "is a definite case of sabotage," adding that, "our first thoughts go to the rebel Nagas." It would have been the first Naga attack after a lull of two years and observers believed that it had been designed to boost the morale of the Naga underground. The Chief Security Officer agreed, stating, "It is definitely a case of sabotage. The *modus operandi* is the same as that employed by Naga hostiles in April 1966."

Ex gratia payments were quickly made to the wounded passengers and it was announced that payments to the families of the deceased would follow. [1298]

Despite the apparent certainty of the initial reports that the explosion had resulted from a bomb, on the following day it was stated that an initial investigation had failed to find any evidence of an explosive device and that it was now believed that the explosion had been accidental, having resulted from the detonation of a substance carried by a passenger. The explosion had bored a large hole in the floor of the train and the substance had fallen onto the track, rebounded, and ripped open the side wings of the coach. Only three passengers and the guard were known to have died but six long seating benches had been blown off and it was thought that others of the 35 passengers in the compartment must have been blown to pieces. [1299]

[1296] *The Valley News and Valley Green Sheet*, Van Nuys, California, United States, 16th May 1969
[1297] *The New Mexican*, Santa Fe, New Mexico, United States, 16th May 1969
[1298] *The Indian Express*, Madras, India, 22nd May 1969
[1299] Ibid, 23rd May 1969

As through goods train Dg 57 767 passed Rethen station on Sunday 22nd June 1969 the signalman there saw sparks coming from a waggon and assuming that its brakes were binding informed the next station. After another signalman had observed smoke coming from the waggon the train was stopped at Hannover Linden, West Germany, and a railwayman attempted to extinguish the fire. At this point no one appreciated that there was any danger. A shunter uncoupled the Belgian waggon involved, in the middle of the train, from the following one, and the engine pulled the front of the train forward. The shunter and the conductor hurriedly examined the lading of the train. When they discovered what the burning waggon was carrying they immediately warned the signalman and via a loudspeaker the station staff that the burning waggon was carrying munitions. At that moment the waggon exploded, just as the Hannover fire brigade arrived with their fire engine. Eight of the firemen and four railwaymen were killed instantly and 30 people were injured, some seriously. Two waggons were destroyed and 71 were more or less damaged. The explosion left a crater 15 metres in diameter and there was considerable damage to surrounding buildings and equipment.

An investigation into the accident excluded an overheated axle box and decided that binding brakes were the probable, though not certain cause of the fire that had led to the explosion. The investigation thus closed after a year without any tangible result but the safety regulations regarding the transportation of munitions were tightened up. Waggons carrying munitions were to be fitted with spark protection plates, axles with roller bearings and sealed floors, must not have single brakes and the trains in which they were included must not have more than three unfitted waggons. The waggon markings for vehicles carrying explosives were also to be improved. [1300]

At 3.45 a.m. on Sunday 3rd August 1969 a 115-car Kansas City Southern Railway freight train en route from Kansas City to Shreveport, Louisiana, was stopped at Noel, Missouri, United States, when a brakeman noticed a glow and supposed that it was a vehicle on fire due to an overheated journal. The Noel Fire Department were alerted and on their way to the scene, while efforts were being made to detach the burning car from the remainder of the train, when there was a small explosion, quickly followed by a second massive explosion that devastated the town.

A newspaper wrote, '"It blew the hell out of things," said Kenneth Meader. One piece of wood ripped through a wall of Mrs Roxa Miller's house, killing the 47-year-old beauty parlor operator. An 800-pound railroad car wheel soared three blocks and smashed into the house of Virgil Bentley, seriously injuring Mrs Bentley.

'Roofs caved in, Walls buckled. Main Street was carpeted with shattered glass from store fronts. More than 40 persons were injured. Pieces of shattered railroad cars were thrown half a mile. At the site of the explosion there was a hole 15 feet deep and 50 feet across in the railway roadbed. "Practically every building in town is damaged," said Meader, who owns a gas station.'

The ammonium perchlorate being carried in the box that caught fire was apparently the cause of the first explosion, and the second larger blast had occurred when the fire

[1300] *Katastrophen der deutschen Bahnen Teil II* by H.J. Ritzau, pgs. 141-2 spread to an adjoining car loaded wth dehydrated alfalfa. The ammonium perchlorate, an oxidising agent, was being transported from Henderson, Nevada, to the Longhorn Army Ammunition plant at Marshall, Texas. It was used as an ammunition propellant. The blaze had begun, it was believed, some miles before Noel.

'The explosion severed some pipes on a propane gas tank standing beside the tracks, and a fire ignited which began burning off escaping fumes. Authorities feared the 12,500 gallon tank would explode and perhaps touch off two other tanks next to it. Police ordered the train evacuated at 2 p.m. while employees of the Empire Gas Co. rigged a 2½ inch pipeline to the tank's main valve. Just before 6 p.m. they rigged a valve allowing the propane to flow in liquid form through the pipe and ignited it. The fuel billowed into a 30 foot flame, which roared harmlessly over a ditch, and by 10.30 p.m. the tank was safely drained.' [1301]

At first it was thought that an explosion of a propane tank had caused the main blast but gas company officials denied that anything had happened to any of their tanks in the nature of an explosion.

The damage caused to property was very extensive. Thirty-eight homes and a church were demolished, 58 homes and a second church were heavily damaged and 55 homes were moderately damaged. The cost of the damage was estimated at almost a million dollars. [1302] [1303]

On the afternoon of Thursday 26th April 1973 a local freight train picked up a load of general purpose bombs loaded into 21 Department of Defence (DODX) box cars from the Hawthorne Naval Ammunition Depot in Nevada and took them to Sparks, Nevada. The bombs were destined for the Naval Weapons Station in Concord, California, and thence for Vietnam, where the United States was still engaged in its long, unwinnable, war against the forces of Communist North Vietnam. The cars were inspected at Sparks, where they remained in the yard until the night of Friday 27th April, when they became part of Southern Pacific Company Extra No. 9117 West, which had originated in Ogden, Utah. This train left Sparks at 11.45 p.m. bound for the Southern Pacific yard at Roseville, California, United States.

At Norden, California, just over 84¾ miles east of Roseville, the train began its long descent from the summit of the line in the Sierra Nevada Mountains. Slightly more than two hours was lost at Cisco, 73 miles from Roseville, when the train became divided due to a defective coupler lever. From Emigrant Gap, 64 miles from Roseville, the gradient became much steeper, and the engineer had to use both dynamic and automatic braking to control the train. At a point 50 miles from Roseville the train passed a 20-year-old college student who camped near the line and who was a keen railway enthusiast. He testified later that he had watched the train, had seen the box cars loaded with bombs, and had also observed:

(a) Sparks flying from the wheels of the train as a result of braking action.

(b) The rim of the front wheel on the south side of the leading truck of a DODX box car glowing red throughout its circumference; and

---

[1301] *The Hillsdale Daily News*, Hillsdale, Michigan, United States, 4th August 1969
[1302] *The Joplin Globe*, Joplin, Missouri, United States, 5th August 1969
[1303] *Monroe Morning World*, Monroe, Louisiana, United States, 4th August 1969

(c) Small flames flickering intermittently on the plate of the aforesaid wheel.

He said that he had noticed such sparks and small flames previously on descending trains and had not therefore considered them unusual. He had not noticed anything indicating an overheated axle box.

'Hot box' detectors west of Gold Run, 46 miles from Roseville, and at Rocklin, four miles from Roseville, detected no abnormality.

The occupants of a station waggon who saw the train as it arrived at Roseville at 6.05 a.m. saw nothing wrong with it, nor did any of the railroad employees as the train entered the receiving yard at about 7 a.m., or a car inspector at 7.30 a.m. At 7.40 a.m. and 7.45 a.m. a resident of the nearby community of Antelope and a switchman respectively noticed light smoke arising from a car. At 7.59 a.m. the resident and the car inspector both heard a low-order explosion. The car inspector saw smoke and fire coming from the area of the sound and reported his observation by radio with advice to call the fire brigade. He then saw thick black smoke and flames coming from one of the DODX box cars, reported what he had seen and began to flee the scene, when a high-order explosion knocked him to the ground. A supervisor and factory employees helped him to safety a few minutes later.

Two herders (pointsmen) in the yard and the Citrus Heights Fire Department at Antelope also heard the low order explosion. The firemen were discussing the explosion with the herders when most or all of the bombs in the DODX box car exploded at 8.03 a.m. in a high order explosion. Smoke and flames rose high into the air and fire spread to other cars at the west end of the yard. At various intervals until 10.30 a.m. 17 other DODX cars exploded. Bombs were strewn about the area and these continued to explode until 4.05 p.m. the next day.

A state of emergency was declared and about 18,000 people were evacuated from the surrounding area, while hospitals were placed on a disaster footing. [1304]

As a result of the explosion 15 railwaymen and 335 residents of Roseville, Antelope and Citrus Heights were injured, though mostly only slightly. No one was killed. About 5,500 buildings, mostly residential, were damaged. On the railway 169 freight cars were destroyed and another 98 cars and one engine were damaged. The total cost of the damages and compensation claims was estimated at $23 million.

The explosions were thoroughly investigated by the National Transportation Safety Board (NTSB) but they were unable to reach any firm conclusions as to their cause. Sabotage being considered most unlikely, there were, they decided, two possibilities:

(a) A fire had exposed one of the bombs to a sufficiently high temperature to cause it to explode. Given that two 'hot box' detectors had not alarmed and that no one had seen anything indicating one a 'hot box' was also ruled out. The observations of the college student and the sighting of smoke and flames immediately before the first explosion suggested that sparks or a burning brake shoe caused by the heavy braking descending from the Sierra Nevada summit could have set fire to the wooden floor of the DODX box car. The fire could have smouldered without producing any smoke until the floor had eventually burst into flame.

(b) Explosive experts consulted by the Southern Pacific favoured the alternative theory

[1304] *Daily Independent Journal*, San Rafael, California, United States, 28th August 1973 that an unstable bomb resulting from faulty assembly procedures at the Hawthorne Ammunition Depot had caused the explosion. Navy explosive experts disagreed with this conclusion, stating that there was no evidence of faulty manufacture or loading at the depot.

The NTSB could not determine which theory was the correct one. [1305] The Southern Pacific would obviously be keen to promote the faulty bomb theory as then the government rather than the railway would be liable for the substantial damages incurred.

Less than a month after the Roseville accident, on Thursday 24th May 1973, another Southern Pacific Company freight train, carrying as part of its load bombs bound for the Naval Weapons Station at Concord, California, suffered a series of explosions. The time the location was between Dragoon and Benson, Arizona, United States. The explosions fortunately occurred in a desert area; the nearest house was five miles distant.

Train 2nd BSM 22 had left El Paso, Texas, at 1.40 p.m. It consisted of five diesel units, 106 cars and a caboose, and cars 35 to 46 were loaded with bombs which had begun their journey at the Naval Munitions Depot at McAlester, Oklahoma. While it was crossing another train at San Simon, Arizona, the crew of that train had reported seeing smoke coming from it but the crew of train 2nd BSM could see no smoke as their train left there. A track crew inspected both sides of the train at 5.50 p.m., a train crew did the same at 6.05 p.m. and another track crew inspected it at 6.11 at Wilcox, Arizona. According to the National Transportation Safety Board (NTSB) report the train was in total inspected or observed by no less than 116 operating employees and 114 maintenance of way employees between San Antonio, Texas, and the site of the accident. It also passed four 'hot box' detectors and four dragging equipment detectors after Lordsburg, New Mexico, the last situated about 16½ miles before Dragoon.

At 6.43 p.m. the train was descending a 1% (1 in 100) gradient at Dragoon, Arizona, travelling at 45 miles an hour when there was an explosion in the 38th car. This was not severe enough to interfere with the train's progress and it went unnoticed by the train crew, but the occupant of a road vehicle saw "mustard yellow smoke" coming from the box car. As the caboose of the train passed the west switch of Dragoon siding its rear brakeman saw that a crosstie (sleeper) was smouldering and later he noticed grass fires and informed the conductor, who in turn informed the engineer. The latter looked back along the train, saw grey smoke, and began braking. Four and a half miles from Dragoon there was a second explosion in the 38th car that blew six bombs and part of a seventh from it. At this point the conductor saw flames and black smoke and applied the brakes in emergency from the caboose.

As the 98th car of the train passed the point where the second explosion had occurred one of the expelled bombs blew up and the train became divided in front of the 35th car, the leading one of those loaded with munitions. A series of further explosions then followed which continued until 1.15 a.m. the next morning.

The main series of explosions produced a crater with a maximum width of 115 feet and a depth of 25 feet and the desert was scorched for a quarter mile in all directions. As well as the destroyed munitions cars, following cars loaded with zinc ash, clays, fibreboard, water heaters, automobile tyres and asbestos were badly damaged and their

[1305] National Transportation Safety Board

contents scattered over the right of way. Of the 2,000 bombs being carried about 500 were recovered unexploded. Bombs were thrown as far as a mile. The nearest house, five miles away, had its windows shattered.

The damage caused was estimated at $274,600 for railway equipment, $70,000 for track, $47,000 for the load and $70,000 for the clearance operation. Total damages were put at $884,600.

The only casualties were the conductor and the rear brakeman, who were injured when they jumped from the caboose. [1306]

Newspaper reports suggested that the explosions had been due to sabotage and were being looked into by the Federal Bureau of Investigation [1307] but the NTSB did not consider that sabotage was the cause of the accident. It noted that the wooden floor of the 38th car had already been on fire 25 minutes before the first explosion. The evidence suggested that the fire had not come from a chemical reaction within a bomb below the level of an explosion but from an external source. There was no evidence of a 'hot box' as a cause of the fire but sparks from brake shoes due to heavy braking were, as a Roseville, very possible, especially as the wooden floor of the 38th car had been impregnated with sodium nitrate, from a previous load, and with oil, and had not been provided with spark shields. A fire within a box car would not be apparent from outside.

The NTSB came to a number of conclusions:

(1) The occurrence and control of car fires was a necessary consideration in the promulgation of safety requirements governing rail transportation of heat-sensitive hazardous materials shipments.

(2) Neither the DOT regulations for the transportation of "Class A" explosives by rail, nor DOD requirements for loading and shipment of MK 82, 500-lb. bombs, prohibited the use of cars with wood floors.

(3) Car 38 was susceptible to fire because it contained a contaminated wooden floor and because it was not equipped with spark shields to prevent sparks from striking the flooring or roller bearings which could preclude a hotbox condition.

(4) Sodium nitrate impregnation and possibly excess oil sprayed from the journal boxes made the wood more sensitive to ignition and accelerated combustion after the fire ignited.

(5) The accident events in this case began with a fire in Car 38. The physical evidence, suggesting the ignition of the car floor fire by brake-shoe sparks, though not conclusive, substantially outweighed the evidence of any other ignition sources.

(6) There was no evidence of arson or sabotage.

(7) None of the current visual or mechanical methods of surveillance of trains insured reliable detection of fires inside box cars.

(8) The wooden pallets, blocking, and bracing in Car 38 intensified the fire adjacent to heat-sensitive munitions.

[1306] Ibid.

[1307] *The Dispatch*, Douglas, Arizona, United States, 25th August 1973

(9) Existing Federal regulations did not address the problem of detection and suppression of fires in railcars transporting munitions.

(10) The explosion at MP 1052.6 was probably the bursting of the bomb casing caused by external heat, rather than the result of an internal chemical reaction.

(11) The consecutive placement of the cars carrying bombs in the train increased the probability of the en masse, multiple-car explosions and increased the size of the danger zone in the accident.

(12) The quality of the bombs deviated from the design specifications. Although the effects of these deviations on the safety of the bombs in transit could not be determined conclusively, the deviations probably did not initiate the fire in Car 38.

(13) The requirements in 14 CFR 170 - 174 for removal of potentially hazardous residues from railroad cars after unloading were not clear as to responsibility for, or method for, proper and timely cleaning.

(14) The circumstances of this accident indicated that improved coordination or integration of the compliance activities of all parties who prescribe safety requirements intended to assure safe transportation of hazardous materials could raise the overall level of safety. This warranted further study.

The NTSB decided that the probable cause of the accident was the exposure of heat-sensitive bombs in Car 38 to a fire inside the car. The fire had most likely originated from sparks off the brake-shoes, which had ignited the sodium nitrate impregnated floorboards.

The NTSB made several recommendations:

(1) The Secretary of Transportation should reassess the regulations applicable to the packaging, loading, storing and transportation of military munitions and develop a safety compliance program to unify and coordinate Federal activities relating to the shipment of military munitions as now authorized by the Transportation Safety Act of 1974.

(2) The Federal Railroad Administration in conjunction with the Secretary's program should develop and require the use of a mechanism for the timely detection of fire by crews of trains carrying Class A explosives.

(3) The Secretary of Defence should design, develop, and use special rail cars for the transportation of military munitions which were constructed of non-combustible materials, or are otherwise resistant to ignition from sources external to the car, and reduce or eliminate the use of wooden or other combustible sheathing, blocking, and bracing inside the cars.

As the result of this accident the Safety Board on 1st June 1973, made three recommendations to the Federal Railroad Administration which would require:

(1) the use of cars with roller bearings and either composition brake-shoes or spark shields in the transport of military munitions,

(2) the placement of "spacer" cars between cars transporting munitions, and

(3) increased inspection and surveillance of freight car loadings in the transportation of military explosives.

On 19th November 1974 the Federal Railroad Administration promulgated a regulation effective from 1st July 1975, establishing new requirements for selection, preparation, inspection, certification, and loading of rail cars transporting Class A explosives. [1308]

The next significant train explosion had a different cause and did not involve munitions. It occurred in the Burlington Northern Railroad yard at Wenatchee, Washington, United States, at 12.32 p.m. on Tuesday 6th August 1974, while shunting operations were being performed. This time there were two fatalities, Yard Foreman David Jones, [1309] and an unidentified transient, while another 113 people were injured. Again the accident was the subject of an investigation by the National Transportation Safety Board (NTSB).

The NTSB noted that the explosion had occurred in tank car DUPX 16009, which had contained about 10,000 gallons of monomethylamine nitrate, en route from E.I. DuPont de Nemours at Biwabik, Minnesota, to DuPont, Washington. The car had been taken from Biwabik to Superior, Washington, by the Duluth Missabe and Iron Range Railroad and thence transferred to the Burlington Northern. The car had arrived at Wenatchee at 6.55 a.m. on the morning of the accident and had been shunted several times in the Apple Yard there. Just before the explosion witnesses saw a column of grey-white smoke and orange fire, then a mushroom cloud of black smoke, then the explosion had followed.

Parts of the tank car were found a mile away from the site. Seventy-one cars and four containers were demolished by the explosion and 101 cars and 29 trailers were damaged. At least four buildings in the yard and a number of homes in the vicinity were more or less damaged. Damage to track and buildings was estimated at $1,444,000, to freight cars at $1,152,000 and to private property at $5,100,000.

The tank car had been built in April 1961. On 12th October 1973 it had been part of a train that was struck from behind by another but had not sustained any visible damage.

The load of the car was an 86% solution of monomethylamine nitrate in water, designated PRM. It was used to manufacture a water-gel explosive named Tovex. The NTSB noted that both PRM solutions and crystals could detonate, the crystals more readily, but there had been no history of previous accidental explosions. The PRM was classified as a flammable solid, the bulk transportation of which in tank cars was not normally permitted and the DuPont company had had to obtain a special permit for its shipment.

Analysing the accident the NTSB concluded that the events leading up to the explosion must have begun when a cut of five cars had collided with a stationary cut of cars on track 12 of the yard. According to the manufacturer a low-order or burning-type reaction could be initiated in three different ways:

(1) a coupler override which penetrated the tank of PRM,

(2) cavitation followed by recompression within the PRM solution,

(3) compression of an air bubble entrained in the liquid or entrapped in the air space at the top of the tank.

After the reaction had begun, it could escalate and the entire lading detonate.

[1308] National Transportation Safety Board

[1309] *The Daily Report*, Ontario-Upland, California, United States, 7th August 1974

The NTSB reached the following conclusions:

(1) PRM was known before its introduction into railroad transportation to be detonable under certain conditions when stimulated by heat, shock, or friction.

(2) The PRM solution was classified as a "flammable solid" under the Department of Transportation regulations, although it behaved like a detonating explosive in this accident.

(3) In view of classification of PRM as a flammable solid, the shipment did not receive adequate treatment as a material which, in that quantity and form, could explode in transportation. Quantity and form should be considered in the classification process.

(4) PRM crystals probably were present in the insulation and bottom outlet leg of DUPX 16009 before the accident.

(5) The PRM solution in DUPX 16009 was contaminated with iron in solution accumulated in the car during one or more earlier unloading cycles.

(6) The quality of the PRM in DUPX 16009 was not the same as the quality of PRM used by DOT to determine the hazardous material classification and transportation safety requirements or the material in the finished product. This was because of the absence of effective quality control procedures governing the shipper's manufacturing and loading processes.

(7) The Federal inspectors and carrier personnel were not likely to detect deviations from the quality of the PRM specified in Special Permit 5737 under existing surveillance practices.

(8) Any damages to the tank or components of DUPX 16009 during a train accident which occurred on an earlier trip remained undetected because the shipper did not know about the accident, because of the nature of the car and lading, and because of the loading and surveillance practices.

(9) The explosion of the PRM was triggered during the routine switching impact between the 5-car cut and the standing cut of cars, in a manner that could not be established.

(10) Because most of the evidence was destroyed in the explosion, it was impossible to determine the relationship between the escaped PRM, the contaminants, the quality deviations, and the explosion. Nevertheless, these problems required regulatory action.

(11) The methods of evaluation which led to the issuance of Special Permit 5737 did not expose safety problems which required safeguards.

The NTSB was unable to determine the probable cause of the accident. A partial list of the possibilities that could not be ruled out completely included a reaction of dried crystals, a reaction of spilled or leaking crystals in the insulation space, a reaction in one of the tank components, cavitation and recompression of the solution, compression of an air bubble entrained in the solution, a reaction of solution or crystals sensitised by contamination, or the ignition of escaped product by friction. The NTSB concluded that the classification of monomethylamine nitrate as a "flammable solid" permitted shipment of the chemical without proper safeguards against predictable hazardous conditions.

The NTSB made five recommendations, the first to the Federal Railroad Administration and the remainder to the Secretary of Transportation:

(1) Identify all liquids now transported in tank cars which are capable of detonation; determine whether detonation or other dangerous chemical reactions can be initiated by conditions and circumstances encountered by those liquids in railroad transportation and issue regulations to control the risks identified.

(2) Require applicants submitting proposals for transportation of detonable materials to make an examination of the transportation conditions for detonation risks and describe what they found.

(3) Publish guidelines describing methods available for conducting safety analyses that would facilitate the discovery of detonation risks and standards to be met in preparing the proposal.

(4) Amend 49 CFR 173 to establish appropriate explosives classification definitions and test procedures that address every known way in which detonable materials could explode accidentally in transportation.

(5) Establish regulations for quality specifications and quality control procedures in the manufacture, packaging, and loading of detonable hazardous materials. [1310]

The explosion of a package of fireworks being carried by a passenger was the apparent precipitating event that led to a disastrous fire in a second-class compartment of the *Upper India* express from Delhi to Sealdah (Calcutta) on Thursday 31st October 1974. A survivor related how the woman passenger carrying the package had objected to a man in front of her smoking, telling him that she had some dangerous material in her luggage. She had then removed something from her bag and a violent explosion had followed seconds later. According to some reports the man had flicked a lighted cigarette at the woman, perhaps in annoyance, and this had set off the fireworks. The blaze that followed the explosion swept through two carriages of the crowded train and about 20 passengers leapt from it to escape the fire, only to be killed in doing so. The train stopped near Mohanganj in Uttar Pradesh, India, about 25 miles from Allahabad.

Railway officials confirmed that preliminary reports indicated that fireworks carried by a passenger had caused the initial explosion and resulting fire. The deputy Railway Minister, Mr Buta Singh, went to the scene with senior officials.

A press report stated that there were 43 dead and about 60 injured, some very badly. [1311] [1312]

[1310] National Transportation Safety Board

[1311] *The Times*, London, England, 1st November 1974

[1312] *Charleston Daily Mail*, Charleston, West Virginia, United States, 31st October 1974

CHAPTER 16

BOILER EXPLOSIONS

By 1961 steam locomotives had given way to diesel or electric traction in many countries and the incidence of boiler explosions had drastically decreased. In the period between the beginning of 1961 and the end of 1975 the author has only been able to find three, so this chapter is a very short one.

Discovered since completion of the previous volume of this series was written is a boiler explosion that occurred on 13th March 1948, near Prunişor, Romania – a small station on the Bucharest – Timişoara line.

On the early evening of that day locomotive No. 150.001, from Turnu Severin shed, was pulling a passenger train when its boiler exploded. The five people on the engine were thrown about 20-30 metres from the destroyed locomotive. The force of the explosion derailed the engine and the first coach of the train, and a crater with a diameter of a metre and a half was made in the embankment. Traffic between Craiova and Turnu Severin was interrupted for 19 hours.

As usual the accident did not result in casualties among the train passengers but Driver Constantin Dogăroiu, Assistant Driver Constantin Colţatu and Fireman Dumitru Goga died on the spot. Firemen Ioan Chiriţoiu and Virgil Panduru, miraculously survived, though their bodies were scalded by steam and boiling water of temperatures exceeding 200 degrees Celsius.

One of the causes of this almost unique accident in the history of Romanian railways was two common pieces of wire, with diameters of 1 and 4 millimetres, bent in the shape of the letter U and inserted into the communication holes of the glass gauges indicating the water level in the locomotive's boiler. These small metal bodies had prevented the normal circulation of water, and sediment deposited around them had clogged the two holes, which meant that the water gauges no longer worked properly.

Even under these conditions the disaster that followed could have been avoided, but the safety valves did not work as they should have done either. The driver did not notice in time the fact that the firebox was no longer covered by water, and the special steel from which this part was made, heated to a temperature of over 500 degrees Celsius, lost its resistance, which led to the collapse of the entire assembly and the violent explosion of the locomotive boiler. [1313]

Early on Monday 9th October 1961 the boiler of the locomotive of the night express from Valencia to Madrid exploded soon after the train had left El Mugrón, Spain. The driver and fireman were killed and the conductor and five passengers were injured. [1314] [1315]

The British Railways, London Midland Region, 7.30 a.m. Up express passenger train from Holyhead to London Euston was approaching Bletchley, England, on Wednesday 24th January 1962 when there was a loud bang in the cab of the engine, Coronation Class 4-6-2 No. 46238 *City of Carlisle*. The driver and fireman were both scalded but were able to stop the train. Low water had caused steam at a pressure of 250 lb/sq. in. to bulge the crown sheet down at three points and had slightly warped the firebox tubeplate. The steam had forced the crown sheet over seven roof stay nuts, causing holes through which the scalding steam had escaped into the engine cab. Three lead safety plugs had melted.

Employment Inspector E.G. Brown investigated the accident. He discovered that the boiler had been generally in good condition but the right-hand gauge glass on the fireman's side of the footplate had been too dirty to allow the water level to be read. During the journey he had therefore used the left-hand gauge glass, which every time that he had looked at it appeared to be full, as the top of the water was not visible, but in fact it had been empty and the water level in the boiler had fallen until the crown sheet had become exposed.

This would be the last boiler explosion on British Railways, whose last steam locomotives were withdrawn in August 1968. [1316]

The boiler of Ty 51 Class locomotive No. 105 working a Rozwadów – Siedlice freight train exploded near Radzyń Podlaski station, Poland, on Friday 10th August 1973. [1317]

[1313] https://www-railnet-ro.translate.goog/viewtopic.php?f=119&t=4119&_x_tr_sl=ro&_x_tr_tl=en&_x_tr_hl=en&_x_tr_pto=sc
[1314] *The Times*, London, England, 10th October 1961
[1315] Federación Castellano Manchega de Amigos del Ferrocarril
[1316] *Locomotive Boiler Explosions* by C.H. Hewison, pgs. 140-1
[1317] Wikipedia – Railway accidents in Poland

CHAPTER 17

VANDALISM

'Accidents' due to vandalism and sabotage have always been a problem for the railways. They are obviously very closely related, inasmuch as both represent deliberate attempts to cause trains to crash, as opposed to accidents caused through non-deliberate carelessness, negligence or purely accidentally. Here they will, however, be treated separately, as they are slightly different. This separation is perhaps a little artificial, and if the perpetrator of a given act is not caught or his or her motives not sufficiently obvious, it may be difficult to decide which it was, but the distinction is nevertheless worth making.

Vandalism is usually, though not invariably, carried out by either children or young adults and its results are most commonly, though not always, less serious than those following acts of sabotage. The motives for train accidents resulting from vandalism are usually either simple curiosity, in the case of younger children – 'I wanted to see what would happen' – or in the case of older children and teenagers, a reflection of general anti-social behaviour; so that individuals who attempt to derail trains also carry out other acts of 'mindless vandalism,' and perhaps also commit burglaries or carry out arson attacks.

Sabotage on the other hand is usually undertaken by adults, who employ methods more likely to cause death and injury, so that while a vandal might for example just place an object upon the track or jam a stone into a set of points, a saboteur would be more likely to employ methods such as using explosives to blow up a section of track.

The engine of a Chesapeake and Ohio Railway passenger train from Chicago, Illinois, to Grand Rapids, Michigan, was derailed on a set of facing points at Michigan City, Indiana, United States, on the evening of Saturday 7th October 1961. The eight coaches of the train remained on the track and none of the train crew or 95 passengers was hurt. [1318] A couple of days later three boys, two aged 12 and one aged 13, after questioning by their parents, admitted to police that they had interfered with the points. No reason was given for their actions. [1319]

Two 15-year-old sophomores (second-year students) said that they had been playing when they opened a set of facing points at Birmingham, Alabama, United States, on

[1318] *The Brazil Daily Times*, Brazil, Indiana, United States, 9th October 1961
[1319] *The Holland Evening Sentinel*, Holland, Michigan, United States, 10th October 1961

Saturday 25th November 1961. About ten hours later, early on Sunday, an Atlantic Coast Line Railroad freight train arrived and was diverted onto the spur line. Five diesel units and 11 cars were derailed, about 900 feet of track was torn up and $200,000 of damage was caused.

The boys were the sons of a Fire Department lieutenant and a company foreman. On the Sunday one of them said that he and his friend had known nothing of the accident "until I heard the boys at Sunday School talking. I was mighty scared." By the time he had got home agents of the Federal Bureau of Investigation were there. Charged with train wrecking the boys were placed on $3,000 bonds.

One of the boys' fathers remarked, "If the spur had been in good condition the train could have stopped without derailing. A train official told me the tracks had split apart when the engines entered the spur. This caused the accident." [1320]

On Sunday 18th March 1962 a 115-car New York Central Railroad freight train was diverted onto a spur line 2,000 feet long at Nashville, Michigan, United States. Engineer C.R. Zane was unable to pull up before his train reached the end of the spur so that it collided with the buffer stops and ended up in State Street. [1321] A 14-year-old boy admitted opening the switch. Upon finding that he could not close it again he had just gone home and had not said anything about the matter to his parents. He explained that "he wanted to see what would happen," a typical motive for childhood vandalism on the railway. [1322]

At 11.15 a.m. on Wednesday 25th March 1964 a Down passenger train from Glasgow Central to Wemyss Bay consisting of a steam locomotive and six coaches was derailed in Cartsburn (New) Tunnel, Scotland, on British Railways, Scottish Region. Five coaches left the track but the tunnel walls kept them from overturning. None of the 59 passengers were more than slightly hurt but they were stuck in the tunnel for about an hour before they were rescued with the aid of ladders, conducted through the tunnel, and forwarded on buses.

Investigation showed that 75 bolts had been removed from a length of track in the 380-yard single-line tunnel and there had been similar interference in the adjoining 310-yard Cartsburn (Old) Tunnel. Tools had been stolen from a platelayer's hut that had been broken into.

Chief Constable David Williamson of Greenock told reporters, "This could not have been done by children. Whoever did it must have been an adult. It is very difficult to remove those bolts. They are big bolts, screwed into the sleepers. Any reasonable adult must have realized what the consequences were likely to be. This could have been a very serious accident. If the train had been travelling a little more quickly, or the driver had been a bit slower in stopping, people could have been killed. There was no real value in the things which were unscrewed – only a matter of coppers."

The chief constable proved to have been wrong in thinking that the derailment must have been the work of adults, for the day after the derailment two teenagers, John

[1320] *Times Democrat*, Cullman, Alabama, United States, 28th November 1961

[1321] *The Luddington Daily News*, Luddington, Michigan, 20th March 1962

[1322] *Traverse City Record-Eagle*, Traverse City, Michigan, 20th March 1962

McMenamie, aged 15, and Hugh Meredith Jones, aged 14, were arrested.

The following week the teenagers appeared in Greenock Juvenile Court, where they admitted stealing equipment and removing bolts, fishplates and (chair) keys from the tracks in the two tunnels. It appeared that the evening previous to the derailment they had loosened six bolts and a fishplate from the Up line Cartsburn (Old) Tunnel but desisted when a train had arrived. In the morning they had gone to Cartsburn (New) Tunnel and removed 75 bolts, fishplates and 20 keys, then moved a 60 ft length of rail using a pickaxe. A train had then appeared and they retreated to a tunnel recess. Despite the damage to the track the train had not been derailed and returning they had found that 20 remaining chairs were holding the rail secure, so they had prized it loose with the pickaxe, then left the scene. Later, realising the gravity of what they had done, they had gone back in the hope of repairing the damage, but the train reached the site first.

The Court was told that that the derailment had caused £538 worth of damage, excluding labour costs, while the line had been closed for 12 hours.

Mr R.M. Watson, for the prosecution, stated, "It is difficult to understand the mentality of boys who could do this. There could have been a serious loss of life, If the accident had happened outside the tunnel the coaches would have gone over the embankment."

The Court decided to send the offenders to an approved school (in Great Britain a residential institution for young offenders, in use between 1933 and the early 1970s). [1323] [1324]

The electric engine of a Veno – Amsterdam express was derailed on the afternoon of Monday 31st August 1964 near Amstel station, The Netherlands. Fortunately the train was only travelling at 20 kilometres an hour at the time and only two wheels of the engine left the rails, but it ploughed up the track for 200 metres. Driver J.A. Jansen suffered considerably from shock and was seen to be very pale after the derailment. Traffic was disrupted for two hours and the damage caused was estimated at more than 5,000 guilders.

The head conductor of the train saw three children running away from the scene, chased after them and was able to catch a ten-year-old girl, Grada K. Taken to the police station she confessed that she had placed three stones on the track, assisted by her seven-year-old sister Louise and their friend, eleven-year-old Carolina de J. The children were sent home after questioning. [1325]

A serious act of vandalism on British Railways, Eastern Region near Elm Park, England, on Monday 29th March 1965, resulted in the death of Driver Ernest Wibrew and one passenger and the injury of 18 passengers.

The 19:10 Down passenger train from London Fenchurch Street to Shoeburyness, an eight-car electric multiple unit, was travelling at 70 miles an hour when it ran into some

[1323] *The Times*, London, England, 26th and 28th March and 1st April 1964
[1324] Wikipedia – Approved School
[1325] *Die Volkskrant*, Amsterdam, The Netherlands, 1st September 1964

pieces of metal that had been placed on the track at 19:29. As a result the train was derailed and then struck the centre girder of an underbridge. The accident was the subject of a Ministry of Transport inquiry.

Guard T.W. Harper told the inquiry that he had been riding in the brake compartment in the sixth coach of the train, accompanied by Guard J.A. Banks. The train had been less full than normal. All had gone normally until near Elm Park the train had begun to sway. The swaying had become violent, had suddenly changed to an up and down movement and he had realised that the train was derailed. He did not recall any severe brake application.

Guard Banks explained that he had been helping Guard Harper with a large number of newspaper parcels. He did remember what he thought was an emergency brake application, accompanied by a severe oscillation and shuddering of the coach, which he had realised was off the line, and then "a series of leap frogs."

After the accident Guard Harper went forward and Guard Banks went back to protect the lines, using signal post telephones to contact Upminster and Barking signal boxes respectively.

Clerk D.E.H. Clark, who was travelling as a passenger, remembered what he recognised as an emergency brake application rather more than a quarter mile short of the point of the derailment. After the accident he said he had gone to a lineside house and telephoned Fenchurch Street control office, using an ex-directory telephone number he knew, and had explained the seriousness of the situation.

Driver Stanley Colville stated that he had been driving the 18:47 Special Up empty coaching stock from Southend Central to East Ham Depot. Passing the site of derailment at about 19:21 he had found objects on both lines and his train had run over those on the Up line. He had not stopped to report the matter before reaching the depot because such incidents were so common.

The drivers of Up trains at 18:00, 18:20, and 18:30 and of Down trains at 18:22 and 18:45 all likewise reported striking objects on the track, which they had reported, mostly at Barking and Upminster. The objects were described as pieces of steel and concrete and rows of stones.

Booking Clerk K.O. Newton at Elm Park described receiving a message from Fenchurch Street regarding children trespassing and calling the police.

Station Inspector E. Moody at Barking described receiving reports of objects on the line from the drivers of Up trains and reporting these to Control. He said that he had not advised the Barking signalman to caution Down trains because he thought that the signalman would have been listening in to his call to Control, while he admitted that it had not occurred to him that there might be objects on the Down as well as the Up line.

Station Inspector E.E. J. Williams at Upminster said that when the driver of the 18:22 Down train had reported striking pieces of iron near the River Beam bridge, he had telephoned Dagenham East station and Control. A little later Dagenham East had rung back and said that the location was not in their police district and could he ring Hornchurch or Romford. He had then rung Hornchurch and been told that a police car would be sent as soon as possible. The driver of the 18:45 Down train from Fenchurch Street had then reported running over ballast at the same place, he had advised Control and they told him to caution the 18:30 Up train from Shoeburyness but it had already left. He had not thought to advise the Upminster signalman.

The Barking and Upminster signalmen stated that they not infrequently received reports of vandalism at different locations either from the station staff or from drivers via signal post telephones, when they cautioned trains and advised their colleagues.

Deputy Chief Controller R. Cooper at Fenchurch Street testified that his first knowledge of the derailment had been at 19:29 from the Electrical Control office at Pitsea, who said that the circuit breakers on the feeders to the Up and Down lines had opened. Both breakers had been closed, whereupon that on the Down line had immediately reopened. At about 19:30 he had been advised by a section controller of Clerk Clark's telephone call and had at once instructed the Pitsea Controller to re-open the Up line breaker and keep it open.

Section Controller D.T. Hawkes described the various telephone calls with Barking and Upminster regarding obstructions on the line.

Ganger G.W. Smith said that his length did not usually get that much trespassing and was generally tidy, with no pieces of metal around.

Colonel D. McMullen, the inspecting officer, in his conclusions, stated that after examining the site and interviewing witnesses he had taken the unusual step of making an announcement:

"It is not usual at this stage of an Inquiry for a statement to be made on the cause of an accident, but I feel that in this case it is right that I should make one.

"From what I have seen at the site, from my examination of the track and equipment and from what I have heard today, I propose to advise my Minister that I have no doubt that this deplorable accident, in which two innocent people lost their lives and a number of others were injured, was caused by the wanton obstruction of the track by some metal objects. I have heard today, too, that this is not an isolated case either on this line or in this Region, nor is it an isolated case on British Railways, for I have received a number of reports of incidents of this nature from other Regions, some of them extremely serious.

"I propose, therefore, to discuss as urgently as I can with the Railway Managements what further steps can possibly be taken to prevent hooliganism, I think that's the only appropriate word, of this type, and the dangers which it presents to the travelling public and to train crews.

"Hooliganism itself is of course a social problem and will have to be dealt with as such."

The inspecting officer praised the actions of Guards Banks and Harper and of Clerk Clark, while criticising Station Inspectors Moody and Williams for not informing their respective signalmen so that trains could be cautioned.

Turning to remarks Colonel D. McMullen noted that the accident could easily have been more serious had a train on the adjoining District line also been involved. Forty-four accidents involving trains running into obstructions had been reported in 1962, 50 in 1963, 76 in 1964 and 45 already in the first five months of 1965, while there had been numerous other acts of vandalism such as stone throwing. In response the railway management had taken measures such as repairing fencing, working to keep linesides tidy, arranging for talks in schools in badly affected areas and experiments had been made with lineside closed circuit television and cine cameras on trains. Numerous suggestions had been made by the public, such as constant patrolling on foot, the use of helicopters, the provision of electric fences and so on, most of which were, however, impracticable for one reason or another.

The maximum penalties for trespass and vandalism on the railway were in the process of being raised from £5 to £25 under the railway bye-laws while there was already in the general law, under the Malicious Acts and the Offences Against the Person Acts of 1861, a maximum penalty of life imprisonment.

It would be useful if members of the public who observed acts of trespass telephoned the police or the nearest railway station. [1326]

Numerous reports of railway vandalism appeared in the newspapers in the month or so following the derailment, until eventually the Press lost interest in the subject. [1327]

At the inquest on the deaths Driver Wilbrew and passenger Eric Rayner the jury returned verdicts of 'Manslaughter by some person or persons unknown.' The perpetrators were never caught. [1328]

A Chicago and Eastern Illinois Railroad freight train consisting of three diesel units, three box cars and a coal car was travelling at about 40 miles an hour when it was derailed at 2.20 p.m. on Saturday 6th August 1966 near Marion, Illinois, United States. Engineer Frank Cruzen was trapped in the cab of his locomotive for almost two hours and was quite seriously injured. Conductor George Fields and Brakemen Arnold Bookout and Donald Huston were less badly hurt. [1329]

Inspection showed that the lock of a set of facing points a quarter mile north of Marion had been broken, the points had been opened and about 30 small rocks had been jammed in the toe of the points so as to hold them open, while the banner that would have indicated to the engineer that the points were open had been torn off. As a result the train had been diverted onto an out-of-use spur that had formerly served the Illinois Gas and Electric Co. The engines had run off the end of the spur and fallen into a ditch about 10 feet deep and 100 feet long. Damage was estimated at $500,000. [1330]

On Monday teenagers Carl Neafus, 17, and Jesse Watkins, 13, were arrested and the next day they appeared in court charged with causing criminal damage to property and felonious damage to railway locomotives. Because their parents were unable to pay the $8,000 bond they were being held in Williamson County Jail. [1331] [1332] [1333] The charge of causing damaged to locomotives was subsequently dropped on a legal technicality and in mid-August Neafus was sentenced to one to five years in Menard Penitentiary and Watkins was bound over to the Illinois Youth Commission for not more than five years,

[1326] Ministry of Transport report
[1327] *The Times*, London, England, 6th, 9th, 10th, 12th, 13th, 19th, 21st, and 22nd April and 11th, 14th and 21st May and 30th June 1965
[1328] Ibid, 5th June 1965
[1329] *Sunday Sentinel*, Centralia, Illinois, United States, 7th August 1966
[1330] *Southern Illinoisian*, Carbondale, Illinois, United States, 7th August 1966
[1331] *The Register News*, Mt. Vernon, Illinois, United States, 9th August 1966
[1332] *Centralia Sentinel*, Centralia, Illinois, United States, 10th August 1966
[1333] *Southern Illinoisian*, Carbondale, Illinois, United States, 10th August 1966

both having been found guilty of causing criminal damage to property. [1334]

SCOTS EXPRESS CRASHES

AT 100 M.P.H.

Six dead, 20 injured in midnight

derailment near Huntingdon

RESCUE TEAMS FREE TRAPPED

The above were the headlines in a national newspaper on the morning of Monday 6th March 1967. As usual the facts were not entirely accurate, only five people died and the train was only travelling at 75 miles an hour, but the accident was serious enough.

The train involved in the accident on British Railways, Eastern Region, was the 22:30 Down express from London King's Cross to Edinburgh Waverley, consisting of Deltic type 5 diesel electric locomotive D9004 *Queens Own Highlander*, a brake van, three Post Office sorting vans, a sleeping car, three second-class coaches, a first- and second-class composite coach, a second class and brake composite and another brake van, in charge of Guard J. Wright with Driver B Oakton and Secondman P.A. Wheaton. It had left King's Cross on time with 147 passengers and was due to run non-stop to Grantham, where it was due at 00:26 on Monday; it was derailed on a set of facing points at Connington South signal box, England, a few miles short of Peterborough, at 23:36 on Sunday night. The engine and leading five coaches came to a stand without leaving the track beyond the points, the sixth vehicle was partially derailed, the remaining vehicles were totally derailed and the eighth and ninth vehicles were overturned. Five passengers lost their lives and 18 were injured, two seriously. [1335]

'Ambulances, firemen, and police sped through this bleak fenland countryside to the scene of the crash where carriages lay drunkenly awry, doors swinging from their hinges.

'Rescuers helped out the injured and ambulances rushed them to Peterborough Memorial Hospital. Firemen worked under floodlights and smashed windows of the overturned coaches to free the injured. Passengers, many of them in their night clothes, stood by watching the rescue operation. The engine and the undamaged coaches were disconnected from the wrecked coaches and towed clear.

'Baggage, newspapers, and sandwich packets littered the empty compartments, the wrecked coaches blocking the up and down lines.

'A railway official said: "Some of the people in the bunks did not even know there had been a crash. They were still asleep."

'A crash crane was brought up and began to lift the derailed coaches while the rescuers with powerful hand torches made sweeps up and down the track. In the rear derailed coach, lying almost on its side, seats and cushions were piled against the smashed windows. A pack of playing cards was scattered down a corridor. Along the side of the track a ripped open blue suitcase and a trail of women's clothing was scattered. The crash took the 230 {sic} passengers by surprise, wakening people from sleep as they

---

[1334] Ibid, 17th August 1966

[1335] Ministry of Transport report

curled into the corner of their seats, settling down for the journey. Those lucky enough to escape were taken on to Peterborough, looking shaken and subdued, by coach.'

In another paragraph a reporter wrote, 'One man who saw the accident was Mr Alan Frost, who was on duty at the South signal box. He estimated that the train was travelling at more than 80 miles an hour on the straight stretch. Mr Frost said: "There was a rumbling and suddenly I saw the back of the last coach keel over. The train had gone by dead on time and everything seemed perfectly normal. Then I felt the ground shake and when I turned round I saw the red light on the last carriage tipping over. I got on the telephone and called headquarters for emergency." [1336]

The cause of the derailment was initially a mystery and the signalman was apparently merely an eyewitness to it, however the results of the investigation into it would show otherwise.

Driver Oakton told the Ministry of Transport inquiry that after a signal check at Huntingdon had brought his train down to 15 miles an hour, speed had increased to 80 miles an hour at Abbots Ripton. He had then closed the controller to allow the train to coast down the 1 in 200 falling gradient towards Conington, and speed had dropped to 75 miles an hour by there. The Connington South and North signals had been green but suddenly the brakes had come on and the vacuum had fallen to 5 inches and then almost at once to zero.

Secondman Wheaton confirmed his driver's statement and added that immediately after the accident he had gone forward to protect the Up line and advise the Conington North signalman of what had happened.

Guard Wright stated that he had been riding in the last vehicle of the train and had been somewhat bruised and dazed in the accident. With difficulty he had climbed out of the vehicle and made his way to a signal box 200 yards away (Connington South). Near the box he had shouted, "Are you there, Bobby?" and the signalman had come to the window. He had shouted to the signalman to call the emergency services and as he entered the box the signalman had been sending emergency bell signals, before speaking on the telephone to Control. The guard related that he had not noticed anything strange in the signalman's behaviour, though he added that he himself had been suffering from shock.

Relief Signalman H.G. Thaxter said that he had been on duty in Connington North cabin at the time of the accident. He had accepted the express from the South box at 23:28 and had received the *Train Entering Section* signal for it at 23:34. Shortly afterwards he had heard a noise and his signal repeaters had gone to "Wrong." At that point he had telephoned his colleague in the South box and asked him if anything was amiss, to be told that the signals were in order but that the train seemed to be stopping. Several minutes later he had received the *Obstruction Danger* signal from the South box, as the secondman of the express had come into his box to report that his train was off the road.

Signalman Frost in Conington South box testified that he was 20 years old. He had begun training as a signalman a year earlier and had worked Connington North box before moving to the South one. At the railway's internal inquiry he maintained that he had known nothing of the derailment until Guard Wright had come to his box, but questioned by the inspecting officer, Lt. Col. I.K.A. McNaughton, at the Ministry of

[1336] *The Times*, London, England, 6th March 1967

Transport inquiry, he admitted that he had both heard and seen it happen, but said he had been too shaken to do anything. He also admitted having told Signalman Thaxter that the train had just stopped, though he had known by then that it was derailed. He had taken no action to call the emergency services until the guard had arrived at his cabin four minutes after the accident, attributing this to his shock at what had happened. He insisted that he had not replaced his home signal prematurely, though he agreed that he had cleared the train back to Abbots Ripton box before it passed his clearing point.

Permanent Way Inspector J.G. Abbott stated that the derailment had apparently occurred at the Connington South facing points but he was unable to explain how it had happened, though there were slight marks on the switchblades.

The facing points, their locking and detection (the device that ensures that the points are correctly closed before the signal protecting them can be cleared) were examined immediately after the derailment by Chief Signal and Telecommunications Inspector A.D. Roulstone, who found the points lever, No. 27, and the facing point lock lever, No. 26, reversed, while the seals for the emergency releases for the electric locks were intact. He too was unable to account for the derailment.

Carriage and Waggon Assistant P.C. Barker stated that he had made a careful inspection of the derailed vehicles but was unable to find nothing wrong with them that could have caused the derailment.

At the end of the initial inquiry the cause of the derailment remained obscure and the fact that the facing points had been found locked and bolted in their normal position after the accident, with no damage to their locking mechanism, seemed to imply that they must have been so for the whole passage of the train and that the cause of the derailment must lie elsewhere. Exhaustive tests were therefore conducted to try and discover some fault with the track or with the vehicles of the train. The results were entirely negative and it was concluded that in some manner the mechanical and electrical locking of the facing points must have been circumvented and they must have been opened under the train. A series of experiments was then carried out, first in Arlesley signal box, which like Connington South had a Dutton's type lever frame, and then in Connington South box. It was also found that the electric locking at Connington South could be bypassed by the use of a table knife or a similar object.

The conclusion of the inspecting officer was that the derailment must have been caused by the unlocking and movement of the facing points by Signalman Frost. The signalling equipment was in order and this could only have been done by his deliberate and irregular action, making him solely responsible for the accident. The statement of the signalman that he had not restored his home signal until the train passed could not be accepted, nor could his later statement to a police officer that he had accidentally unlocked the points while swinging on the levers.

Lt. Col. McNaughton outlined what he considered to have been the likely sequence of events leading up to the derailment. After accepting the express from Abbots Ripton and having had it accepted in turn by Connington North box the signalman must have disabled the electrical lock on the points lever, No. 27, by inserting the blade of a table knife or similar in the locking apparatus. On the express reaching track circuit "G", 200 yards in the rear of the Down home signal, he had replaced the distant signal lever, No. 13, to normal and had then stood holding the home signal lever, No. 14, reversed, with its catch handle raised. As soon as the engine of the train had passed the home signal he must have restored its lever to normal, freeing the lever working the facing points lock, No. 26, which he had then reversed. At this moment the engine would just about have reached the facing points. Four seconds later, as the seventh vehicle of the train had reached the points, he had reversed them momentarily, derailing it and the remaining vehicles of the train. He had then reversed the facing points lever once more.

Turning to remarks the inspecting officer noted that there had only been one similar accident from the same cause in Great Britain, the derailment of a sleeping car express at Douglas Park, Scotland, in May 1949, when no serious consequences had resulted. Prior to joining the railway Signalman Frost had been discharged from the Royal Marines, reportedly because he had been below their required physical standard. It was later revealed, however, that Frost had been discharged from the marines because he had been judged to be suffering from 'hysteria and an immature personality' but this fact had not been made known to railway management earlier.

Signalman Frost subsequently stood trial on charges of manslaughter and of endangering the safety of passengers on the railway. After an 11-day trial the judge instructed the jury to acquit him of the manslaughter charge but he pleaded Guilty to the lesser charge of endangering passenger safety, for which he was sentenced to two years' imprisonment. [1337] The reason for his actions remains a mystery.

Stones placed on the track by children derailed a train at Adelsdorf, West Germany, on the Forchheim – Höchstadt line on Monday 10th May 1971. Two people were injured. [1338]

Seaboard Coast Line Railroad train No. 85, the *Champion*, operating between New York and St Petersburg, Florida, under contract to the recently formed National Railroad Passenger Corporation (Amtrak), was within seven miles of its destination at 6.10 p.m. on Thursday 23rd December 1971, when it was derailed near Cross Bayou, Florida, United States. It was running at about 55 miles an hour at the time and was about an hour late. Its three diesel engines were overturned, the baggage car was wrecked, with suitcases and clothes spread over the track, and 11 of its 14 passenger cars were derailed, the leading ones tilting over and the others just off the rails. Fuel leaking from one of the overturned engines caught fire and the flames spread to nearby woodland; it took two hours to put the blaze out. Nineteen of the 50 passengers on the train were injured but only two were admitted to hospital.

One of the two detained in hospital was Dr Lilian Salsman, 83, from Marblehead, Massachusetts, who said that she was scared of flying, "so when I heard that they had this wonderful train to St Petersburg, I took it." She had been planning a vacation in Florida with friends who were flying down and was riding in the rear car with Mrs Rita Manor of Boston. She had suffered a broken jaw in the accident. From her hospital bed she added, "All of a sudden there was this boom and we were both knocked down. I wasn't knocked

[1337] Ministry of Transport report

[1338] *Die Katastrophenszene der Gegenwart* by Hans Joachim Ritzau and Jürgen Hörstel, pg. 159

out, but I really saw stars. Then I felt sick." [1339]

Mrs Manor, sitting in the lobby of the St Petersburg General Hospital, pressing a tissue to her bleeding nose, told a reporter, "It felt like a big bang. I was in the last car. It was just myself and an 83-year-old woman. She got hit in the jaw."

Mr Al Nagy of New York City, another passenger who explained that he had taken the train because he was afraid of flying, related that he had been standing in the aisle of the second car. "All of a sudden the train braked hard and I was tossed back and forth. The door opened and hit me in the face. When I looked up, all the cars were helter-skelter." [1340]

Mr and Mrs E.R. Ross, returning home to Seminole, Florida, from New York City, were riding towards the rear of the train. Mr Ross said, "I walked out of the bedroom and all of a sudden it as all over. It was the fastest thing that ever happened to me." [1341]

Dining car employee A.L. Jackson said, "It was like running into a wall." The dining car was one of those that had tipped over. His colleague, Elra Ward, who sustained a back injury, commented, "The first thing I knew it had turned over. I fell over a chair." [1342]

An investigation determined that a switch had been tampered with, sending the train off the main line into a spur leading to a junkyard. Major Herman Vincent of the Pinellas County sheriff's office said, "The padlock that locks this switching station in place is missing. It would have taken quite an instrument to break it – or someone having a key."

Seaboard Spokesman Donald T. Martin said that the railroad had had problems with vandalism in the area previously, but not to such a serious degree. "It was things like a foreign object on the track." He said that a northbound freight train had passed the site without any problem three hours earlier. The passenger train had been running within the maximum permitted speed of 60 miles an hour, he added. [1343]

A Southern Pacific Company train, consisting of an engine and two empty cars was proceeding northwards along the main line near Ventura, California, United States, when it was struck by three runaway tank cars loaded with liquified petroleum gas (LPG) at about 3.20 p.m. on Tuesday 6th November 1973. Conductor Joe Manuel, Engineer Tony Emanuel, and Brakemen Gale Connor and Danny Seamans all jumped from their train before the collision. The leading tank waggon was ruptured in the ensuing collision and although there was no fire the conductor and engineer were splashed with the LPG and suffered severe cold burns. They were flown in a Navy helicopter to the Sherman Oaks Burn Center for treatment but the conductor died of his injuries early on Wednesday and the engineer likewise succumbed two days later. The two brakemen had landed in a ditch and escaped injury. [1344] [1345] [1346]

[1339] *The Daily Review*, Hayward, California, United States, 25th December 1971
[1340] *Sarasota Herald Tribune*, Sarasota, Florida, United States, 25th December 1971
[1341] Ibid, 24th December 1971
[1342] *News-Herald*, Panama City, Florida, United States, 24th December 1971
[1343] *The Progress Index*, Petersburg, Virginia, United States, 24th December 1971
[1344] *The Valley News*, Van Nuys, California, United States, 8th November 1973
[1345] *Press-Telegram*, Long Beach, California, United States, 7th November 1973

It transpired that five students from Ventura High School, aged between 13 and 16, were involved. As a prank they had released the brakes of the leading tank car and the vehicles began to run away. "As the cars started to pick up speed, the two older boys apparently panicked and leaped off," according to a detective. They had then run to their homes where they were taken into custody following an investigation.

The 15- and 16-year-old boys were charged with involuntary manslaughter and tampering with the brakes of a railroad car. The 13-year-old and two 14-year-old boys were charged with misdemeanour trespassing. [1347] It was suggested that the perpetrators were members of a Crips street gang.

Only four days after the Ventura accident, on the night of Saturday 10th November 1973, an Atchison Topeka and Santa Fe Railway freight train consisting of four diesel units and 52 cars was diverted into a siding at Lawndale, California, United States, and collided with a box car. All four engines and nine or ten cars were derailed. This time none of the railwaymen was hurt but the derailed engines and vehicles fell onto two automobiles stopped at a level crossing, crushing them. Margaira Eubanks, 36, occupying one of the automobiles, was fatally injured and a man and a woman in a second car were less badly hurt. [1348] [1349]

Engineer James Kerr said that he had not seen the open switch in time to stop. He told a reporter, "I threw on the emergency brake. Everything was happening so fast. It takes a while to stop these things, you know." Damage to rolling stock, track and signalling equipment was estimated at $80,000.

The day after the accident a 13-year-old local boy was arrested by detectives. It appeared that he had unlocked the padlock at the switch using a key stolen several weeks earlier. [1350]

In the aftermath of the accident it was proposed that the speed of trains passing through the Lawndale area should be limited to 15 miles an hour. The train involved in the accident had been travelling at 28 miles an hour. [1351]

A third derailment due to vandalism involving a fatality in the United States occurred on Thursday 22nd November 1973, at Windham, Ohio. A 27-car Erie-Lackawanna Railroad freight train from Chicago to New York was derailed at a set of points leading to a brick factory and partially embedded into the earth. Spilt diesel fuel precluded the use of oxy-acetylene equipment and the wreckage had to be pried apart to free Engineer Raymond Wilcox and a second railwayman, named as David Consider. Both were taken to the Robinson Memorial Hospital in Ravenna, where the engineer died two hours later.

[1346] *The Bakersfield Californian*, Bakersfield, California, United States, 7th November 1973
[1347] *The Press-Courier*, Oxnard, California, United States, 8th November 1973
[1348] *Star-News*, Pasadena, California, United States, 11th November 1973
[1349] *The Bakersfield Californian*, Bakersfield, California, United States, 11th and 12th November 1973
[1350] *The Press-Courier*, Oxnard, California, United States, 12th November 1973
[1351] *The Valley News*, Van Nuys, California, United States, 15th November 1973

[1352] Damage was estimated at over $1,000,000.

A couple of days later police stated that two boys aged seven and eight had told their parents that they had opened the switch, causing the derailment. The older brother of one of the boys had reportedly closed the same switch a week earlier to help a railwayman and had been "rewarded" with a dime.

"What can you do with babes?" an official asked. "You can't throw them in jail. There wasn't any malice intended. They'll probably get a good lecture, I just don't know."

No decision had been made, another official said, as to whether charges would be filed. [1353]

[1352] *Warren Times-Observer*, Warren, Pennsylvania, United States, 23rd November 1973

[1353] *The Titusville Herald*, Titusville, Pennsylvania, United States, 27th November 1973

CHAPTER 18

SABOTAGE

Four distinct motives for sabotaging trains may be distinguished (after Robert B. Shaw):

(a) As an aid to intended robbery, either of the train passengers or to steal, for example bullion being carried by the train.

(b) To satisfy a personal grudge against the railway company, such as by an ex-employee who feels that he has been wrongly dismissed.

(c) To hinder the operation of the railway, as for example during a strike.

(d) As an act of guerrilla warfare or terrorism, often against an occupying enemy power. Sabotage against the railways has been a feature of many conflicts from the American Civil War onwards.

By the period covered in this volume, acts of sabotage in countries such as Palestine, Algeria, Malaya, and the Dutch East Indies had diminished or ceased, with their becoming independent, but the lengthy conflict in Vietnam brought a new area of the world where numerous acts of sabotage were carried out for political motives, while various acts of sabotage were also carried out in other counties such as India, Burma, the United States, Uruguay, France, Mozambique. and so on.

In a departure from the method used in other chapters of this book acts of sabotage have been to some degree grouped together according to the country in which they occurred, rather than being listed in a strictly chronological order.

The first act of railway sabotage to occur in the 1961-75 period took place on Wednesday 4th January 1961, about 30 miles from the port of Albertville (renamed Kalemie 1971) in the soon to be independent Belgian Congo (now the Democratic Republic of Congo). A pay train that had left Albertville with wages for co-operative Baluba tribesmen working on the railway was derailed by a band of 30 or 40 rebels from the same tribe who had torn up the track. The rebels then attacked the train with bows and arrows and old muzzle-loading guns but were driven off by United Nations Nigerian troops who were guarding the train. Fourteen of the rebels were reportedly killed, one of the troops was wounded in the shoulder by an arrow. [1354]

[1354] *Corsicana Light*, Corsicana, Texas, United States, 6th January 1961

Karen rebels, fighting for independence, blew up a train on Saturday 25th March 1961, probably near The Daw, Burma, killing eight people. [1355]

In their second attack on a train in two days a band of about 200 Karen National Defence Organization rebels derailed eight coaches of a diesel-hauled Mandalay-Rangoon express with three land mines at Tawkywein, Burma, 190 kilometres north of Rangoon, early on Sunday 26th March 1961. They then attacked the overturned carriages with automatic weapons, shooting all those in uniform or suspected of being government employees. The other passengers were looted of about £8,000 of property and 60 or 80 passengers, including schoolgirls, were taken hostage, presumably for ransom. The daughter and son-in-law of Mr. W.D. Gunaratna, the Ceylon Minister to Burma, were passengers on the train, it was not stated if they were among the hostages. In total 23 people were killed and 100 wounded, 50 seriously and 30 critically, either in the initial derailment or in the post-derailment attack. [1356] [1357] [1358] [1359] [1360]

Karen rebels mined and then fired upon a passenger train on the night of Tuesday 27th June 1961 at an unstated location on the line between Rangoon (now Yangon) and Prome, Burma. The fireman of the train was killed and 41 passengers were wounded. [1361]

No. 6 Down train from Amingaon to Maniharighat on the North East Frontier Railway (NEFR), the *North Bengal Express*, was derailed at 10.10 p.m. on Wednesday 19th April 1961 between Sivok and Gulma stations, West Bengal, India. The engine and seven coaches left the track and capsized. The initial report on the accident stated that 19 passengers had been killed more or less outright, four had died soon after admission to hospital, and 81 people had been injured, 28 seriously. The injured were taken to railway, civil and military hospitals in nearby Siliguri.

Mr Jagjivan Ram, the Union Minister for Railways, accompanied by Mr Karnail Singh, the Chairman of the Railway Board, flew on a special plane from Delhi to Siliguri, while the General Manager of the NEFR and other senior officials likewise took an Indian Air Force plane from Gauhati to Siliguri.

The cause of the derailment was being investigated but Mr Shab Nawaz Khan, the Deputy Minister for Railways, stated in the Lok Sabha (the lower house of the Indian parliament) that it might be a case of tampering with the tracks. [1362]

A couple of days after the derailment the number of dead was stated to have increased to 34. A 56-foot length of rail had reportedly been found removed from the track at the site of the derailment but the Home Secretary told newsmen that it still could not be

[1355] *The Herald-Press*, St Joseph, Michigan, United States, 27th March 1961
[1356] *The Times*, London, England, 28th March 1961
[1357] *The Daily Republic*, Mitchell, South Dakota, United States, 27th March 1961
[1358] *Pacific Stars and Stripes*, Tokyo, Japan, 28th March 1961
[1359] *The Herald-Press*, St Joseph, Michigan, United States, 27th March 1961
[1360] *Red Bank Register*, Red Bank, New Jersey, United States, 27th March 1961
[1361] *The Cedar Rapids Gazette*, Cedar Rapids, Iowa, United States, 2nd July 1961
[1362] *The Indian Express*, Chittoor, India, 21st April 1961

confirmed that it was a case of sabotage, especially as it had not been a military train and there had been no looting after the derailment. If it was not a case of sabotage the displaced rail would presumably have been a result of the derailment rather than its cause. [1363]

Société nationale des chemins de fer français (SNCF) train No. 12, a Strasbourg to Paris express, was derailed at 15:10 on the afternoon of Sunday 18th June 1961 near Loisy-sur-Marne, France, while travelling at 130 kilometres an hour. The engine and four leading coaches remained on the rails and stopped several kilometres down the line, the next four coaches were derailed, one of them being badly damaged, and the rear ten coaches separated from the front of the train and fell down a 10-metre-high embankment, three of them landing in marshy ground at the foot of the embankment. Many passengers were stated to have been trapped in the wreckage.

Ambulances, firemen, police and private cars were rushed to the scene and the injured were taken to the hospital in nearby Vitry-le-François until it could take no more casualties, when the remaining victims were taken to a hospital at an American Air Force base. Three of the most seriously hurt were flown by helicopter to a hospital at Chalons-sur-Marne. The boggy ground increased the difficulty of the rescue operation and firemen were forced to cut through the roofs of the wrecked carriages to reach the trapped passengers.

Initially it was stated that 23 people had died in the accident and 130 were injured. By the following day it was stated that there were 24 dead, 99 seriously injured and at least 50 more slightly hurt. The final number of fatalities was 28.

M. Buron, the Minister of Transport, visited the scene of the derailment the following day and paid tribute to the work of servicemen from the American Air Force base, while two technical inquiries were opened and the *juge d'instruction* (local examining magistrate) began his investigation. [1364]

The day after the accident it was discovered that the track had been damaged by an explosive device that had detonated when the train passed over it. It also emerged that a few days prior to the derailment the Vitry-le-François stationmaster had received a threatening letter. Communist Party leader Jacques Duclos demanded an inquiry into the affair but instead the matter was hushed up by the French government and despite apparently overwhelming evidence the theory of sabotage as a cause of the disaster was ruled out.

The train had been derailed by members of the Organisation armée secrète (OAS), a far-right terrorist organisation opposed to the independence of Algeria. [1365]

An initial report stated that 30 people had been killed and 163 injured in the derailment at 2.30 a.m. on Friday 20th October 1961 of train No. 83, the Up *Ranchi Express*, between Ghatsila and Dalbumgarh stations, India, on the Howrah – Tatanagar main line of the South Eastern Railway. The engine and seven bogies (coaches) of the train left the rails. The four leading bogies capsized, the fifth and sixth, one of them a

---

[1363] Ibid, 23rd April 1961

[1364] *The Times*, London, England, 19th and 20th June 1961

[1365] Wikipedia – 1961 Vitry-le-François train bombing.

partly air-conditioned coach, telescoped one another, and the seventh, a first-class coach, also capsized. Driver A.V. Rao was one of those killed. The train was running two hours late and was crowded with holidaymakers.

Two relief trains were rushed to the scene from Kharagpur and rescue work was immediately begun. Forty-five of the injured were taken to the railway hospital at Kharagpur and 16 to Tatanagar hospital, while the General Manager of the railway and senior officials set off for the scene. Most of the passengers on the train had been bound for Tatanagar after visiting the Pooja celebrations (a Hindu festival) in Calcutta, which had ended on Thursday.

Train No. 34, the Down *Ranchi Express*, had passed the site of the accident about 15 minutes earlier without incident.

The line was not badly damaged and all traffic was resumed at 9 p.m. Meanwhile the Down Sambalpur passenger train and Bombay Mail bound for Howrah and the Up Howrah – Bombay express were diverted over the Adra – Kharagpur line.

Mr Jagjivan Ram, the Union Minister for Railways, and Mr Karnail Singh, the Chairman of the Railway Board, chartered a plane from Delhi for the scene, while Mr A.K. Gupta, the Government Inspector of Railways, planned to open an inquiry into the disaster at Ghatsila station, starting at 8 a.m. on Saturday 21st. [1366]

The following day it was announced that 44 bodies had been recovered from the wreckage. Fifty-three people had been admitted to hospital at Kharagpur, three of whom had died and one had been discharged, while 123 had been admitted to Tatanagar hospital, of whom 50 had been discharged. One of those who had lost their lives was Dr. B.S. Gulta, formerly a Director of the Anthropology and Tribal Research Institute at Ranchi.

Fishplates and bolts were found feet away from the track at the point of the derailment, indicating that sabotage was the cause.

Mr Jagjivan Ram made an *ex gratia* payment of 500 rupees to the widow of Driver Rao in addition to the 500 rupees also paid to the driver's family to cover the funeral expenses. The injured were each granted an immediate payment of 200 rupees and Mr Jagjivan Ram added that the entire cost of the hospital treatment of the injured would be met by the railway. [1367]

Two more victims died the following day [1368] and a couple of days later the toll had risen to 52. [1369]

At the end of October it was stated that a preliminary inquiry by the Government Inspector had confirmed that the derailment was due to sabotage. [1370] By whom, or to what purpose, was not stated, possibly was not known.

In what was apparently the second such attack in a month a 15 vehicle mixed train was mined and derailed in the vicinity of Mượng Mán, South Vietnam, at 04:00 on

[1366] *The Indian Express*, Chittoor, India, 21st October 1961
[1367] Ibid, 22nd October 1961
[1368] Ibid, 24th October 1961
[1369] Ibid, 30th October 1961
[1370] Ibid, 31st October 1961

Sunday 21st October 1961. The locomotive was reportedly wrecked by the explosion and five passenger coaches, three tank cars and four waggons loaded with rice were derailed. Only six people were wounded. The attackers made off with most of the rice. [1371]

Vietcong guerrillas mined and derailed a mixed train between Văn Phú and Phú Mỹ stations, South Vietnam, on Friday 3rd November 1961. Two people were reportedly killed and three seriously injured. [1372]

A passenger train was derailed by sabotage between Saigon and Nha Trang, South Vietnam, on Tuesday 5th December 1961. Passengers and a soldier guard were hurt and seven other soldiers were said to be missing. [1373]

Vietcong guerrillas opened fire with automatic weapons on a northbound train from Saigon at a point 20 kilometres from Quy Nhon, South Vietnam, on Wednesday 25th April 1962, wounding seven passengers. The train did not stop. [1374]

Three coaches of a train were 'blown up,' presumably by a mine, near Diêu Trì, South Vietnam, on Thursday 10th May 1962. Twenty-seven or 29 government civil guards were killed, 30 wounded and six captured. [1375]

On Tuesday 5th June 1962 a convoy of three trains was attacked by Vietcong guerrillas near Mượng Mán, South Vietnam. A mine derailed the leading train and the guerrillas then fired on the soldiers in it who had survived the derailment. Four vehicles of the train, two of them tank waggons loaded with petrol, were set ablaze. The rear train, an armoured escort, was driven back towards Sông Phan under intense fire from 57mm recoilless rifles carried by the guerrillas, while the middle train, carrying civilians, was not fired on but the guerrillas robbed its passengers. The casualties to the soldiers in the leading train were not known but it was believed that most of them had been killed or captured. [1376]

A passenger train was derailed by sabotage between Saigon and Mượng Mán, South Vietnam, on Monday 11th June 1962. About 45 metres of track had been torn up, and the engine and five coaches left the track. A child was killed and two passengers were wounded. [1377]

A patrolling armoured train exploded a mine between Nha Trang and Diêu Trì, South Vietnam, on Thursday 21st June 1962. The engine and about 30 metres of track were damaged but there were no casualties. [1378]

Captain Santiago Perez, the head of an anti-Castro raiding unit, claimed on Thursday 21st June 1962 that three of his men had escaped from the island after derailing trains at Central Hersey and Guanabacoa, Cuba, in Matanzas Province. Cuban radio confirmed the derailment of a Camaguey-bound freight train at Guanabacoa, which they said had killed

---

1371 *Pacific Stars and Stripes*, Tokyo, Japan, 24th October 1961
1372 *The Charleston Gazette*, Charleston, West Virginia, United States, 7th November 1961
1373 *The Portsmouth Herald*, Portsmouth, New Hampshire, United States, 6th December 1961
1374 *Pacific Stars and Stripes*, Tokyo, Japan, 29th April 1962
1375 *Beatrice Daily Sun*, Beatrice, Nebraska, United States, 13th May 1962
1376 *Pacific Stars and Stripes*, Tokyo, Japan, 7th June 1962
1377 *The Bee*, Danville, Virginia, United States, 14th June 1962
1378 *The Abilene Reporter-News*, Abilene, Texas, United States, 26th June 1962

the driver and injured the fireman. [1379]

Vietcong guerrillas mined and ambushed a train carrying what were described as 250 South Vietnamese paramilitary civil guardsmen who were moving to a new position further north on Friday 17th August 1962. The train was attacked at a location south of Diêu Trì, South Vietnam. Eight guardsmen were killed and 30 were wounded, either in the initial derailment or in the fire fight which followed. [1380]

A government train loaded with supplies and medicines for people in the northern provinces of Quang Tri and Thua Thien, which had been devastated by a tropical storm, was derailed by the Vietcong between Saigon and Mượng Mán, South Vietnam, on Friday 21st September 1962. The attackers had allowed an armoured escort train with 50 government troops to pass unmolested before attacking the relief train. A row of 12 electrically operated mines was set off beneath the train, killing two railwaymen and wounding one. [1381] Immediately after the derailment the attackers descended on the train and made off with 15,000 kilograms of supplies, including canned food and medicines. [1382] The engine and eight waggons were derailed and ten more waggons were badly damaged. A load of 180 tons of rice was scattered by the blast and damaged by fire. [1383]

On about Friday 5th October 1962 a Saigon and Hue express train was derailed at an unstated location south of Quang Ngãi, South Vietnam, guerrillas having ripped up the track. No one was hurt. [1384]

On about Sunday 2nd December 1962 over 100 Vietcong derailed a train with a mine and attacked it north of Da Nang, South Vietnam. They were beaten off by the so-called Tiger Force of American paratroopers, aided by South Vietnamese troops. [1385]

Seven passengers died and 15 were wounded when a train carrying holiday travellers was mined south of Huê, South Vietnam, on Sunday 23rd December 1962. Two coaches overturned and a third left the rails. [1386]

About 40 Communist soldiers and militiamen were wounded early on Thursday 17th January 1963 when a train exploded demolition mines near Po Kut, China, about 8 kilometres north of the border with Hong Kong. It was believed that Chinese Nationalist guerrillas had laid the mines. [1387]

Three passengers, one of them a child, reportedly died in the derailment of a Saigon – Hue train about ten kilometres south of Van Giã, South Vietnam, on about Thursday 17th January 1963. Whether the derailment was accidental or due to sabotage was not

1379 *Janesville Daily Gazette*, Janesville, Wisconsin, United States, 22nd June 1962
1380 *Pacific Stars and Stripes*, Tokyo, Japan, 19th August 1962
1381 *The Canberra Times*, Canberra, Australian Capital Territory, Australia, 24th September 1962
1382 *Sunday Journal and Star*, Lincoln, Nebraska, United States, 23rd September 1962
1383 *The Modesto Bee*, Modesto, California, United States, 23rd September 1962
1384 *Corsicana Daily Sun*, Corsicana, Texas, United States, 6th October 1962
1385 *The Cedar Rapids Gazette*, Cedar Rapids, Iowa, United States, 5th December 1962
1386 *The Daily Reporter*, Dover, Ohio, United States, 26th December 1962
1387 *The New Mexican*, Santa Fe, New Mexico, United States, 18th January 1963

started. [1388]

The derailment of a local train between Quy Nhon and Quang Ngãi, South Vietnam, on Monday 21st January 1963 was definitely due to sabotage. A mine laid by the Vietcong reportedly derailed 11 goods waggons and 8 passengers died and 15 were wounded – were they riding on the goods waggons or was it a mixed train with passenger coaches as well? [1389]

Travellers returning from Cuba reported an act of railway sabotage at Pina on about Sunday 10th February 1963 (very approximate date). Twenty-seven militiamen reportedly died in the derailment of a train and another 50 were taken prisoner. A variety of other attacks were being carried out by armed rebels opposed to Fidel Castro at the time including attacks on buses, electricity installations and so on. [1390]

Rebels derailed a train with a mine and then opened fire on it on Monday 18th March 1963, probably near Pyok Kwe, Burma. Four people died and four others were wounded. [1391]

A train struck a mine and was derailed near Long Khánh, South Vietnam, on Friday 5th April 1963, killing five government troops and wounding 20, who were taken to hospital in Long Khánh. [1392]

Seven mines exploded beneath a passenger train between Nha Trang and Quy Nhon, South Vietnam, on Wednesday 8th May 1963. Five coaches overturned and the armoured troop escort car was derailed. Five passengers were seriously injured and many were less badly hurt. Another 100 passengers were kidnapped by the guerrillas and taken into the jungle but were later released. [1393]

A Saigon – Hue passenger train was again mined between Nha Trang and Diêu Trì, South Vietnam, on Sunday 19th May 1963. This time five coaches overturned, five civilian passengers were killed according to one source, and six soldiers, five passengers and a railwayman were wounded in the derailment and subsequent fighting. [1394] Another source spoke of only two deaths. [1395]

A train was attacked, presumably derailed, probably near Mựợng Mán, South Vietnam, on Sunday 23rd June 1963. Nine government civil guards were killed and 14 wounded, while the guerrillas made off with three light machine guns, three submachine guns, two automatic weapons, six rifles, ten carbines, three cases of grenades and five radios. [1396]

[1388] *Pacific Stars and Stripes*, Tokyo, Japan, 19th January 1963
[1389] Ibid, 25th January 1963
[1390] Ibid, 19th February 1963
[1391] *The Bakersfield Californian*, Bakersfield, California, United States, 20th March 1963
[1392] Ibid, 7th April 1963
[1393] Ibid, 11th May 1963
[1394] Ibid, 23rd May 1963
[1395] *The Canberra Times*, Canberra, Australian Capital Territory, Australia, 23rd May 1963
[1396] *Independent Journal*, San Rafael, California, United States, 24th June 1963

By October 1963 measures taken to counter Vietcong sabotage included the running of early morning patrol trains to detect damaged track or mines laid during the night, the use of civilian track walkers, each covering about four kilometres, and the employment of Wickham trolleys – armoured cars manned by Civil Guard troops, as employed by the British in Malaya – which followed every patrol train and were there said to be very effective at deterring sabotage attempts. [1397]

Despite the above precautions the Vietcong staged what was described as a daring raid on an armoured train returning to Biên Hòa, just north of Saigon, after it had escorted a train of rice to an unnamed town on the coast, on Wednesday 30th October 1963. A mine exploded under the train near Mượng Mán, South Vietnam. Five of the 11 cars of the train were derailed. Eight soldiers were killed, 10 were wounded and 14 were described as missing. A haul of automatic weapons, submachine guns and other weapons was seized by the guerrillas. [1398] [1399] [1400] [1401]

A freight train from Saigon to Hue was derailed by a Vietcong mine on Friday 13th December 1963, near Gia Huỳnh, South Vietnam. The engine was wrecked and four tank waggons loaded with 120 tons of petrol were set ablaze. Nine other waggons loaded with 135 tons of beer and 20 tons of wood overturned. The train was reportedly raked with small arms fire after the derailment. No one was hurt. It was stated that an attempt to mine a train was made at least once a week but this was the first attack on a train carrying commercial goods. [1402]

Another freight train was mined less than 50 kilometres from Saigon, South Vietnam, on Wednesday 18th December 1963. The engine and ten waggons were overturned. No one was hurt and ground troops and aircraft prevented the Vietcong from looting the contents of the train. [1403]

At least 30 people died in an accident to a train attributed to sabotage about 12 kilometres from Montevideo, Uruguay, on the night of Friday 2nd August 1963. A three-coach train from Casupa carrying 200 passengers was travelling at 45 miles an hour when it was diverted into a little-used siding and collided with some freight waggons. Driver Nestor Martinez said he had seen that the points were set for the siding when his engine was 20 metres from them and had applied the brakes. His engine had collided with the waggons and the coaches had separated from the engine either before or after it struck the waggons and been smashed. Rescuers worked through the night to free victims from the twisted wreckage.

The president of the state railway, Juan Funes, stated that criminal hands had broken the lock of the facing points, opened them and then held them open with stones. [1404] [1405]

[1397] *Pacific Stars and Stripes*, Tokyo, Japan, 18th October 1963
[1398] *Corsicana Daily Sun*, Corsicana, Texas, United States, 31st October 1963
[1399] *The Victoria Advocate*, Victoria, Texas, United States, 1st November 1963
[1400] *The Citizen*, Prince George, British Columbia, Canada, 31st October 1963
[1401] *Pacific Stars and Stripes*, Tokyo, Japan, 2nd November 1963
[1402] Ibid, 15th December 1963
[1403] Ibid, 21st December 1963
[1404] *The Bakersfield Californian*, Bakersfield, California, United States, 3rd August 1963
[1405] *Daily Journal-Gazette*, Mattoon, Illinois, United States, 3rd August 1963

It was not clear who had been responsible for the sabotage, or even, given the method employed, that it was not in fact a case of vandalism by children or teenagers, but with more than usually serious consequences.

By Friday 10th January 1964 employees the Florida East Coast Railway had been on strike for almost a year in a bitter wage dispute, union members seeking a rise of 10.28 cents an hour in line with a national agreement and the railway company pleading that they could not afford it. On the evening of the Friday in question 18 cars of a 64-car southbound freight train were derailed near New Smyrna Beach, Florida, United States. Eleven of the cars were loaded and seven were empty. No one was hurt. A piece of rail had been wedged in a set of points. [1406]

Just before 2 a.m. on Sunday 9th February, in the most serious incident since the strike had begun, as a 91-car Florida East Coast Railway freight train was crossing a 15ft high trestle over a tidal creek there was an explosion of dynamite beneath it at North Miami, Florida, United States. The four diesel units and 33 cars were derailed and a fire broke out, fuelled by 20,000 gallons of methane in one of the cars, together with hay and paper products. Again no one was hurt. [1407]

New Smyrna Beach, Florida, United States, was again the location of an act of sabotage on Friday 14th February 1964 and again dynamite was used. The four diesel units and 14 of the 96 cars of a southbound Florida East Coast Railway freight train were derailed. A tank car of petroleum caught fire and exploded with a blast that could be seen and heard in Daytona Beach, 16 miles away. Two scab crew members were slightly hurt. [1408]

On Thursday 27th February five diesel units and 27 cars of a Florida East Coast Railway freight train were derailed by an explosion on the Jacksonville – Miami main line near Saybrook, Florida, United States. A few hours later two explosions went off under another freight train near Spuds, Florida. One explosion failed to derail the diesel units of the 190-car train but a second about 120 cars back derailed seven cars. Only about 15 miles away from the second attack President Lyndon B. Johnson was giving a speech. Learning of the attack he remarked, "This criminal action has got to stop." [1409]

On Friday 17th January 1964 the Vietcong blew up a bridge in Quang Ngãi Province, South Vietnam, about 625 kilometres north of Saigon. North of the bridge a train set off mines, wrecking an engine and eight cars and killing nine railwaymen. [1410]

Six coaches of a Saigon – Da Nang passenger train were derailed 15 kilometres north-east of Phú Mỹ, South Vietnam, in Binh Dinh Province, on Monday 24th February 1964. The explosion tore up 30 metres of track and ten passengers and one soldier of the armed escort were killed, and 18 passengers and four soldiers were wounded. It was estimated that the line would be blocked for 48 hours. [1411]

1406 *The Sunday News Tribune*, Fort Pierce, Florida, United States, 12th January 1964

1407 *Bluefield Daily Telegraph*, Bluefield, West Virginia, United States, 10th February 1964

1408 *Kenosha News*, Kenosha, Wisconsin, United States, 15th February 1964

1409 *The Cumberland News*, Cumberland, Maryland, United States, 28th February 1964

1410 *Pacific Stars and Stripes*, Tokyo, Japan, 20th January 1964

1411 *The Times Herald-Record*, Middletown, New York, United States, 26th February 1964

In another attack on Wednesday 26th February 1964 a section of track was removed, derailing the engine and two coaches of a passenger train near Tháp Chàm, South Vietnam. This time no one was hurt. [1412]

A train was derailed in Phú Yên Province, South Vietnam, on the main line to Huê, on Monday 20th April 1964. Only two vehicles were seriously damaged and no casualties were stated but the line was blocked for 12 hours. This was reportedly the 10th derailment in 20 days. The cause of this and the previous nine derailments was not stated, though sabotage was implied. [1413]

A passenger train on the Saigon – Huê main line was derailed by a Vietcong mine north of Biên Hòa, South Vietnam, on Friday 8th May 1964. The engine and eight of its 14 coaches overturned, two passengers were killed and six wounded. [1414]

Early on Sunday 24th May 1964 a Saigon – Hue passenger train was derailed by a mine laid by the Vietcong near Duy Xuyên, South Vietnam. Four coaches were derailed, eight passengers were killed and the same number were wounded. [1415]

The Vietcong managed to attack three trains within 12 hours on Friday 12th June 1964, as well as two lorries. The first train they derailed with a mine north of Nha Trang, South Vietnam. Two coaches were derailed and two militiamen and a child died. North of Quảng Trị, a mine badly damaged a locomotive. Later, south of Da Nang, another locomotive was damaged by a mine and 15 metres of track was destroyed. No one was hurt in the second and third attacks. [1416]

In what was stated to have been one of the worst acts of terrorism carried out by the Vietcong against the civilian population 20 Vietnamese died and 40 were injured when a crowded passenger train exploded a number of mines planted on the track at a point about 15 kilometres south of Nha Trang, South Vietnam, late on Thursday 18th June 1964. Three coaches overturned and a fourth was derailed. Most of those who died succumbed in the initial explosions, while the injured mostly suffered burns in the fire that had swept through two coaches after the derailment. Rescuers worked through the night to free the wounded, who were flown by helicopter to military hospitals. Many of the passengers were women and children travelling from Nha Trang to Saigon to visit relatives.

Commenting on the attack a newspaper wrote, 'Trains along Vietnam's north-south line are mined constantly, but never before have the Viet Cong gone deliberately after civilian passengers. American military men in Nha Trang express believe it was directed deliberately at civilians and may indicate an increase in terrorism throughout the countryside. The death toll has been rising in recent weeks as a result of civilian buses being mined on lonely country roads by communists.

'Officials in Nha Trang said the train was carrying no military supplies. American Special Forces men spaced in a nearby camp were the first to learn of the bombing. They radioed urgently to Nha Trang and hospitals were alerted. An American surgical team

[1412] *Wichita Falls Times*, Wichita Falls, Texas, United States, 27th February 1964
[1413] *Pacific Stars and Stripes*, Tokyo, Japan, 22nd April 1964
[1414] Ibid, 11th May 1964
[1415] Ibid, 27th May 1964
[1416] Ibid, 15th June 1964

based at Nha Trang raced to the scene to give first aid to the wounded.' [1417]

A fairly original act of sabotage was carried out between Mượng Mán and Tháp Chàm, South Vietnam, on Friday 10th July 1964. Two members of the Vietcong, disguised as track workers, flagged down a passenger train. They ordered the train crew and passengers off the train, uncoupled the engine and two coaches, ordered the driver back onto his engine and instructed him to accelerate and then to jump off, which he did. The runaway engine covered about five kilometres and then rammed the rear of an armoured patrol engine. The crew of this locomotive had time to abandon it and no one was hurt but the Vietcong also looted the passenger train. [1418]

On about Monday 28th September 1964 the Vietcong tried the same trick again and this time there was a fatality. A 'government train' was stopped, its engine was uncoupled and sent running away down the line, where after about 10 kilometres it collided with another train at a South Vietnamese junction station on the coast, killing one child and injuring another. The newspaper account unfortunately does not name the location. [1419]

A passenger train was derailed by a mine on Thursday 19th November 1964 between Biên Hòa and Binh Thuân, South Vietnam. Four Vietnamese were killed and 17 were wounded while an American 'adviser' was stunned and bruised. [1420]

Before dawn on Monday 30th November 1964 a large diesel ran away in Saigon, South Vietnam, and entered the station. The wooden buffer stops that had previously stood at the end of the track had been removed and so the train ran off the end of the line and struck a number of passengers waiting for an early-morning train before becoming derailed. There were conflicting reports as to the number of casualties, some accounts spoke of 17 people killed but other, perhaps more accurate accounts, stated that only four passengers and a policeman had lost their lives, while seven passengers and four soldiers had been injured. It was supposed that the engine had been run into the station as an act of sabotage. The casualties were probably lower than they might have been because the accident had happened before the lifting of a nightly curfew (the people already at the station had presumably either defied the curfew or perhaps more likely had arrived at the station the night before and had slept there as best they could). [1421] [1422]

There was a narrow escape from a potential major disaster in Italy early on Sunday 15th November 1964. The 14-vehicle *Alpen Express* from München, West Germany, to Roma was crowded with about 400 passengers, mostly Germans, when it arrived at Brennero, Italy, at 02:50. Here the Austrian staff were replaced by their Italian counterparts and a railway police officer boarded the luggage van, marshalled fifth in the train, for a routine check. There he became aware of a strange, continuous clicking noise coming from a parcel placed on the floor of the van. He began to investigate, flashing his pocket torch around, but at that moment a whistle from the conductor indicated that the

---

[1417] Ibid, 21st June 1964
[1418] *The Cedar Rapids Gazette*, Cedar Rapids, Iowa, United States, 11th July 1964
[1419] *The Lincoln Star*, Lincoln, Nebraska, United States, 29th September 1964
[1420] *The Salt Lake Tribune*, Salt Lake City, Utah, United States, 20th November 1964
[1421] The *News Journal*, Radford, Virginia, United States, 30th November 1964
[1422] *The Yuma Daily Sun*, Yuma, Arizona, United States, 30th November 1964

train, already over an hour late, was about to depart, and he had to abandon his search. Concerned at what might very likely be a bomb, however, he had reported the matter to his superiors, a message was sent down the line, and the train was stopped at Bressanone station at 04:10.

The likelihood of a bomb was considered sufficiently great for the luggage van to be detached at Bressanone and pushed into a siding well away from the station buildings. Meanwhile the train passengers were becoming annoyed at this further delay to their journeys, but seven minutes after the shunting operation had been completed a deafening explosion ripped out the side of the van. No one was hurt, though the postal parcels in the van were more or less destroyed.

It was found that a bomb containing 15 kilograms of dynamite and a clockwork timing device had caused the blast. The parcel containing the bomb, sent from a fictitious München resident to a likewise fictitious person in Italy, had apparently been placed in the luggage van at Innsbruck. Had the train been on time it would have arrived at Bolzano by 04:50, the time of the explosion, or if the explosion had occurred while the train had been in motion it would probably have been derailed. In either case there likely would have been a great loss of life.

The attack coincided with the opening of elections for the Trentino-Alto Adige Regional Council. [1423] The largely German-speaking population of South Tyrol, which had been annexed by Italy in 1919, previous to which had been part of the Austro-Hungarian empire, was seeking more autonomy, and there had been repeated terrorist bombings. [1424] [1425]

A Vietcong mine exploded beneath a mixed train at 06:15 on Sunday 17th January 1965, derailing four freight waggons at the head of the train (the passenger coaches were at its rear) and slightly injuring the driver and assistant driver. The derailment occurred just south of Thủ Đức, South Vietnam, and only about five kilometres outside of Saigon. [1426]

On Monday 5th April a passenger train was again mined close to Saigon, this time about ten kilometres out, near Dĩ An, South Vietnam. The diesel engine was completely thrown off the track and five of the following coach piled up in a zig-zag fashion across the track. The remaining sixth coach overturned onto its side. About 20 metres of track were ripped up but were soon repaired. Nine people were wounded in the dawn attack, four of them seriously, including two trainmen. [1427] The train had been bound for Phan Thiết and did not have a military escort. [1428] This was the second attack in three days, a train having been mined at an unstated location on the night of Saturday 3rd April, derailing eight cars and injuring three railwaymen. [1429]

An armoured military train exploded a mine 25 kilometres north of Da Nang, South Vietnam, in Quảng Nam Province, on Wednesday 7th April 1965. Three Vietnamese

---

[1423] *La Stampa*, Torino, Italy, 16th November 1964
[1424] *Appleton Post-Crescent*, Appleton, Wisconsin, United States, 16th November 1964
[1425] Wikipedia – South Tyrol
[1426] *Pacific Stars and Stripes*, Tokyo, Japan, 19th January 1965
[1427] *Kokomo Morning Times*, Kokomo, Indiana, United States, 5th April 1965
[1428] *The Winnipeg Tribune*, Winnipeg, Manitoba, Canada, 5th April 1965
[1429] *The Post Standard*, Syracuse, New York, United States, 6th April 1965

soldiers were killed and six soldiers and a civilian were wounded. [1430] [1431]

The Vietcong mined a train only 12 kilometres south of Da Nang, South Vietnam, on Thuersday 22nd April 1965. [1432]

On Friday 30th April 1965 a train was mined in Long Khánh Province, South Vietnam, 55 kilometres from Saigon. Two cars were derailed but the remainder of the train was able to continue. No casualties were mentioned. [1433]

At the end of July 1965 it was reported that rail traffic in South Vietnam had virtually ceased due to Communist mines. [1434] This was not entirely the case though, it would seem, for some trains continued to run and attacks continued.

On Tuesday 24th August 1965 a train was mined near Huê. Four civilians were killed and ten wounded. An aircraft was sent to the scene and the Vietcong withdrew. [1435]

Late on Saturday 2nd October or early on Sunday 3rd October 1965 an armoured train was derailed by sabotage near Mượng Mán, South Vietnam. About 500 guerrillas then stormed a station further north, possibly that at Mượng Mán. No casualties were reported. [1436]

On Monday 15th November 1965 a train was derailed by a mine just south of Huê, South Vietnam, overturning the locomotive. The attack occurred only about 5 kilometres from the American air base at Phu Bai. [1437]

A freight train was derailed by a mine on about Saturday 27th November 1965 between Huê and Quang Tri, South Vietnam. The engine and one waggon left the rails. [1438]

On Sunday 13th February 1966 a train travelling southwards from Huê to Da Nang was derailed by a mine north west of Phú Lôc, South Vietnam. In the afternoon an emergency train with troops was sent northwards from Da Nang but this in turn was mined 30 kilometres south east of Phú Lôc, sending two boxcars down into a stream adjoining the line. [1439]

On about Sunday 10th April 1966 a passenger train was mined near Vĩnh Xượng, South Vietnam, derailing an engine and a coach. No casualties were mentioned. [1440]

Twenty-seven people were initially reported as having been killed and 34 wounded when an explosion occurred on train No. 3, the Up Assam Mail, at about 21:20 on

[1430] *The Daily Telegram*, Eau Claire, Wisconsin, United States, 7th April 1965
[1431] *Lake Charles American Press*, Lake Charles, Louisiana, United States, 7th April 1965
[1432] *The Altoona Mirror*, Altoona, Pennsylvania, United States, 23rd April 1965
[1433] *The Pampa Daily News*, Pampa, Texas, United States, 30th April 1965
[1434] *The Modesto Be and News Herald*, Modesto, California, United States, 27th July 1965
[1435] *The Progress-Index*, Petersburg, Virginia, United States, 24th August 1965
[1436] *Kokomo Morning Times*, Kokomo, Indiana, United States, 4th October 1965
[1437] *Ogden Standard-Examiner*, Ogden, Utah, United States, 15th November 1965
[1438] *The Post*, Frederick, Maryland, United States, 27th November 1965
[1439] *Pacific Stars and Stripes*, Tokyo, Japan, 16th February 1966
[1440] *The Arizona Republic*, Phoenix, Arizona, United States, 10th April 1966

Wednesday 16th February 1966, between Farketing and Kamarbandhali, India, on the Dimapur – Mariani section of the Northeast Frontier Railway. The roof of a third-class bogie fifth from the engine was blown off. [1441]

By the following day it was stated that the number of known dead had risen to 37, though a final figure could not be given because of the complete dismemberment of some of the bodies. The number of wounded had risen to 52. Almost all the casualties were labourers from the Barauni area bound for Tinsukia in Upper Assam. Two of the three time bombs planted in the crowded coach by Naga terrorists had gone off, after the train had left Farketing.

The two bombs had not exploded simultaneously. Fourteen people had died in the first explosion, which was followed half an hour later by a second, more severe blast in the same coach, which killed more passengers. The train had been stopped after the first explosion, in the rearmost compartment of the coach, and the guard had been busy tending to the injured and setting up emergency lighting when the second explosion occurred in the leading compartment of the same coach. A third, live bomb, two boxes of explosives and some fuse wire were recovered from another compartment after a search had been carried out and seven suspects were then arrested on suspicion of having planted the bombs.

The train had been accompanied by two batches of the Railway Protection Force and one batch of the Assam Rifles, these forces riding in the leading and rear brake vans. This was the third attack on a train in three months in the section concerned, which ran through thick jungle very close to Nagaland. [1442]

The bombings were part of a long-standing fight for independence on the part of Nagaland from the rest of India. [1443]

Another explosion shattered several compartments of a Tinsukia – New Jalpaiguri passenger train as it stood in the station at Lumding, Assam, India, on the night of Wednesday 20th April 1966. Fifty-five passengers died in this attack and 127 were wounded. [1444]

Dr Ram Subhag Singh, the Minister of State for Railways, told the Lok Sabha (the lower house of the Indian parliament) that a passenger had reported seeing men placing crates on board the train in the place where a powerful explosion had ripped off the coach roof, as well as cutting down people standing on the platform. Members of Parliament accused rebel Naga tribesmen of carrying out the bombing. A separate state of Nagaland had already been formed at the end of 1963 but this was apparently not enough for one faction of the Naga independence movement and their terrorist attacks had continued. [1445]

Time bombs were believed to have been responsible for the second explosion on a passenger train in four days in Assam, India. This latest explosion occurred on the night of Saturday 23rd April 1966 when the train was stopped in Diphu station, just a few miles from the border of Nagaland. Again the toll was high – 39 people were reportedly killed and about 60 wounded, again including some on the platform awaiting the arrival of the

---

[1441] *The Indian Express*, Madras, India, 17th February 1966
[1442] Ibid, 18th February 1966
[1443] Wikipedia – Nagaland
[1444] *Southern Illinoisian*, Carbondale, Illinois, United States, 21st April 1966
[1445] *The Times*, London, England, 20th and 22nd April 1966

Tinsukia – New Jalpaiguri train. [1446] [1447]

Exactly a week after the Diphu bombing the Assam Mail was derailed at Panikheti, Assam, India, with an unknown number of casualties. Whether this was a case of Naga sabotage was not clear. [1448]

Train No. 203, the Up Bangalore City – Poona express, in charge of Guard S.G. Jog with Driver F.X. Barnabas, First Fireman S.R. Kale and Second Fireman H.N. Jadi, was derailed at 02:58 on Thursday 26th May 1966 between Khanapur and Desur stations, India, on the Southern Railway. The engine and leading four coaches completely left the rails and the second and third coaches telescoped one another. Twenty-one passengers and two railwaymen were killed and 40 people were injured, 11 seriously.

Driver Barnabas told the inquiry into the accident that his train had been travelling at 40 kilometres an hour when he had heard an unusual sound and the engine had suddenly swerved to the right, almost immediately being derailed. There had been no chance to apply the brakes.

The two firemen could not add much to their driver's evidence. They had had to dig him out from under a pile of coal.

Guard Jog spoke of feeling a very bad and heavy jerk and finding that the front of the train had been derailed. He had reported the accident using a mobile telephone and protected his train in the rear.

Stationmaster E.V. Venkatachalam of Castlerock stated that he had arrived at the accident scene at 06:05 on a railbus with doctors and medical equipment and helping rescue passengers. He had noticed that the track had been interfered with.

Keyman Laxman Honnappa alleged that he had slept on the night of the accident in Desur South cabin but this was denied by Leverman Sahadev Parasuram. It was not believed that he had anything to do with the derailment but that he had been lying to conceal the fact that he had actually gone home.

Gang Mate Babu Ishwar spoke of going to the accident site upon being informed of the derailment by Gangman Shivaram. He added that he was not aware of anyone having enmity with either himself or any other member of the permanent way gang.

Assistant Permanent Way Inspector T.F. Alvares, Permanent Way Inspector H.K. Kulkarni and Divisional Engineer Narasimhachar described the interference with the track they had found at the point of derailment.

Driver D. Velayudhan of train No. 3224 Down Goods and Driver P.V. Guttal of train No. 204, trains that had left Desur at 23:40 and passed through there at 00:48 respectively, stated that they had noticed no problems with the track, nor had they seen anyone in the vicinity of the accident.

Driver S.P. Francis of train No. 3293 Up goods, which had left Khanapur about 22:00, stated that about two miles from the scene of the derailment he had seen about five

---

[1446] Ibid, 25th April 1966

[1447] *Southern Illinoisian*, Carbondale, Illinois, United States, 24th April 1966

[1448] *The Sunday Tribune*, Albert Lea, Minnesota, United States, 1st May 1966 people alongside the track, who had shouted something to him.

Additional Commissioner Arya Bhushan described his examination of the track. A rail joint had been completely opened out, the track slewed to the left by 3⅞ inches and there was a longitudinal gap in the joint of about 2½ inches. Four fishplates had been removed. The indications were that the work of opening the joints had been done by untrained hands using non-standard tools. Two men could have completed the task in five or six minutes. Away from the point where it had been tampered with, the track was in a good state of maintenance.

The additional commissioner noted that the train had been running at a permissible speed and there had been no defects with the engine, rolling stock or permanent way that could have caused the derailment. It was clear that tampering with the track had caused the accident. There had been two hours since the passing of the previous train for the miscreants to have carried out an operation that would only have taken them a few minutes.

The sabotage had occurred on a dark night in a sparsely populated area that was hilly and had dense forest, hence an ideal spot for such an attack. It was sheer luck that trees and bushes had prevented the engine and carriages falling the whole of the way down the embankment.

The Press had surmised that the motive for the sabotage was a border dispute between Maharashtra and Mysore states.

Night patrols were provided by the railway during the monsoon season but it was not reasonable to expect them all year round to guard against sabotage. No culprits had been apprehended up to the time the report was made. [1449]

The report recommended the use of welded track in areas very prone to sabotage, the use of anti-telescoping or at least all-steel coaches, and the bringing of dog squads to locations where sabotage had occurred, to trace offenders. Further recommendations concerned the provision of emergency lighting on trains, the careful marking of materials before restoration work was begun, so as to enable reconstruction of the accident, and the removal of unused track material to prevent its use by anti-social elements.

A train exploded a land mine and was derailed about six kilometres from Poipet, Cambodia, on Sunday 12th June 1966, killing the driver and injuring a passenger. Cambodian radio blamed the sabotage on a recurrence of terrorist activities by neighbouring Thailand. [1450] Somewhat later, apparently, the Cambodians decided that it was pro-Red saboteurs directed from China, rather than Thailand, who had been responsible. [1451]

Twenty-seven people apparently died when a train was derailed due to sabotage at Pa-

[1449] Government of India Ministry of Tourism and Civil Aviation (Commission of Railway Safety): Report on Derailment of No. 203 Up Bangalore City – Poona Express between Khanapur and Desur stations on 26th May 1966

[1450] *Amarillo Daily News*, Amarillo, Texas, United States, 14th June 1966

[1451] *New Castle News*, New Castle, Pennsylvania, United States, 15th February 1967

Kong Bridge, China, on Thursday 11th August 1966. [1452]

Two electrically detonated mines exploded underneath a mixed train early on Saturday 10th September 1966 near Binh Lợi Bridge, on the outskirts of Saigon, South Vietnam, in what was described as a 'daring pre-election attack.' Two vehicles of the train were derailed, four were damaged and six people were wounded. [1453]

On Saturday 17th September 1966 a train was mined between Da Nang and Huê, South Vietnam. Five of the soldiers escorting the train and one passenger were killed and three passengers were wounded. [1454]

The Vietcong mined a train 15 kilometres from Saigon, South Vietnam, on Friday 14th October 1966, derailing the engine, an armoured car and several flatcars and wounding three Vietnamese guards. When US engineers sent six salvage vehicles to the scene they were attacked by the Vietcong with machine guns and small arms. According to the Americans they suffered no casualties in the 20-minute firefight. [1455]

A train exploded a mine between Saigon and Mượng Mán, South Vietnam, on Tuesday 18th October 1966, derailing the engine and three freight waggons and slightly injuring the driver. [1456]

A military train carrying ammunition and other goods was mined 10 kilometres from Saigon, South Vietnam, on Saturday 3rd December 1966. The engine and three box cars were derailed and three of the train crew were wounded. [1457]

Train No. 206, the Down Poona – Vasco Express, was derailed at 04:50 on Monday 10th October 1966 between Miraj and Mhaisal stations, Maharastra, India, the engine, tender and five bogies leaving the track. The train was in charge of Guard S. Hanumantha Rao with Driver L.G. Kangralkar, First Fireman C.G. Bhonsale and Brakeman V.T. Alavani.

State Home Minister D.S. Desal stated that, "It is undoubtedly a case of sabotage." He explained that two fishplates had been removed from the start of the bridge where the derailment had occurred and a chappal (type of leather sandal) had been found nearby. A special investigating officer and his team had left Bombay for the scene, he said.

Initial estimates were that nine people had lost their lives and 132 had been wounded, of whom 13 had been treated on the spot and allowed to proceed on their journeys and 110 had been admitted to the Civil Hospital at Sangli or the Mission Hospital at Miraj, though 85 of those admitted were not badly hurt.

The accident had occurred about three kilometres from Miraj on the Belgaum – Miraj main line, Hubli Division, of the new South Central Railway, formed only on 2nd October. According to an eyewitness two bogies fell 30 ft from a bridge to the river bed below, where the water was knee deep.

[1452] Author's notes, source mislaid.

[1453] *Albuquerque Journal*, Albuquerque, New Mexico, United States, 10th September 1966

[1454] *Wisconsin State Journal*, Madison, Wisconsin, United States, 19th December 1966

[1455] *The Bridgeport Post*, Bridgeport, Connecticut, United States, 14th October 1966

[1456] *The Daily Gazette*, Sterling-Rock Falls, Illinois, United States, 18th October 1966

[1457] *Tucson Daily Citizen*, Tucson, Arizona, United States, 3rd December 1966

Doctors from Miraj quickly reached the scene and breakdown trains were despatched from Miraj and from Hubli; the latter train carrying the Medical Officer and railway officials, while the General Manager and other officials chartered a plane to the site or nearby. [1458]

By the following day the death toll had risen to ten with the death of a one-year-old child.

Meanwhile, according to a newspaper report, a police dog named Sheru had traced the scent of the presumed saboteur to the home of a railway gangman, who, it seemed, had had a row with his superior over the allocation of duties. The police dog apparently picked the gangman out from a group of 20 people. [1459] The suggestion by the newspaper that this man, who would have been Gang Mate Laxman Bhima, had been responsible for the sabotage turned out, however, to be completely wrong. The police dog had picked up the scent of the gang mate for the simple reason that he had been working on the track at the site a few days earlier. The official report on the derailment confirmed that it had been due to sabotage but made no suggestion as to who had been responsible. It recommended the welding of rails and the provision of guard rails on bridges, while condemning the marshalling of the train with a passenger coach immediately behind the engine. Other recommendations concerned the provision of speedometers on engines and emergency lighting, the improvement of track recorder charts and the provision of quarters for a keyman nearer his place of work. [1460]

Early on the morning of Friday 11th November 1966 a train carrying Army personnel was derailed between Sivok and Pilanshat stations, West Bengal, India, North East Frontier Railway. An initial report spoke of 14 dead and 35 others were injured, including the driver and assistant driver. The derailment occurred at 03:20 about 25 kilometres from Siliguri. One bogie of the train, which had been bound for Assam, was reportedly completely smashed, while the diesel locomotive and five bogies were thrown from the track.

Some fishplates had reportedly been removed from the track and senior Army, railway and police officials went to the spot to further investigate the matter. The train service was, of course, disrupted.

The Additional Commissioner of Railway Safety stated that an inquiry into the accident would be opened at Siliguri Junction on Sunday 13th. [1461]

An Israeli goods train exploded a mine near Jerusalem and only a few hundred yards from the border with Jordan on the evening of Thursday 27th October 1966, and the driver was slightly hurt. Later handwritten posters in Arabic were found near the scene, bearing the name of the Palestine Liberation Front. Train services between Jerusalem and other

[1458] *The Indian Express*, Madras, India, 11th October 1966
[1459] Ibid, 12th October 1966
[1460] Government of India Ministry of Tourism and Civil Aviation (Commission of Railway Safety): Report on Derailment of 206 Down Poona – Vasco-da-Gama Express between Miraj and Mhaisal stations of South Central Railway on 10th October 1966
[1461] *The Indian Express*, Madras, India, 13th November 1966

parts of the country were suspended following the attack. [1462]

On Sunday 22nd January 1967 Communist guerrillas mined and then attacked a Rangoon (now Yangon) – Mandalay passenger train after staging a bloody ambush in which six policemen were killed and seven wounded. Ten people were injured in the mining and presumed derailment of the train 72 miles north of Rangoon, thus in the vicinity of Ka Toke, Burma. [1463]

In February 1967, in response to repeated acts of railway sabotage, the Indian government stated that it would introduce the death penalty for such acts. [1464]

Notwithstanding the government statement on Thursday 2nd March 1967 rebel Naga tribesmen derailed a military train that was acting as a pilot to the Assam Mail at an unstated location in Assam, India. They had blown up the track with explosives. After the train had stopped the rebels opened fire on it, killing one soldier and wounding five more. [1465]

A payroll train was derailed by guerrillas between Las Montoyas and El Opón stations, Colombia, at about 14:30 on Thursday 9th March 1967. Two dynamite charges exploded as the train was passing, demolishing the locomotive, and the guerrillas then opened fire on it with machine guns and rifles. Six railway policemen, Driver Jorge Coronel, and an assistant paymaster died either as a result of the initial explosion or from gunshot wounds; Conductor Luis Calvete, Brakemen Luis Felipe Ibarra and López Buitrago, Faustino Camargo, president of the Barrancabermeja zone of the railway; his wife Edith Sarmiento de Carmago and two workshop soldiers (?) were wounded. The train had consisted of an engine and a single coach and was en route from Puerto Olaya to Gamarra with the wages of Ferrocarril del Atlántico employees.

According to the police the 80-odd guerrillas involved in the attack had been seen in the area by locals the night before but no one had thought to alert the authorities.

The train had been attacked at a carefully chosen point in a deep cutting, from the tops of which the guerrillas fired upon it after it had left the rails. The surprise attack did not give the police escort time to properly resist and most of them died in the subsequent one-sided gunfight.

It was believed that the guerrillas had been under the command of Ricardo Lara Parada and were members of the Frente Camila Torres group, part of the Ejército de Liberación Nacional (ELN) (the National Liberation Army, a Marxist-Leninist group).

Paymaster Juan Crisóstomo León was the only person on the train who escaped death or injury, having been concealed beneath the bodies of other victims during the ten-minute attack. Interviewed later at Barrancabermeja he said that the attackers had declared that they only wanted to kill soldiers and they would continue their struggle until

[1462] *The Times*, London, England, 29th October 1966
[1463] *The Press Courier*, Oxnard, California, United States, 23rd January 1967
[1464] *The Humboldt Times*, Eureka, California, United States, 11th February 1967
[1465] *The Abilene Reporter-News*, Abilene, Texas, United States, 3rd March 1967

they seized power. He added that a few minutes earlier his train had crossed a railcar bound for Puerto Berria and that the guerrillas must have ignored the railcar, being clearly intent on attacking the pay train. Most of the wages had already been paid out before the attack and a briefcase taken by the guerrillas had only contained 50,000 pesos (equivalent to £9 Sterling at the 2023 exchange rate). The guerrillas had told the survivors of the attack that they were free to go and added that they greatly regretted the casualties to the civilians, as it was only those in uniform whom they had intended to kill. The attackers had then made their escape into the mountains, shouting, and firing their weapons into the air as they did so. [1466]

A few days later 15 leaders of the Communist Party, including Secretary General Gilberto Vieira, and 200 other known Communist and left-wing sympathisers, were arrested by the Colombian government's National Security Council. [1467]

On about Friday 19th May 1967 the Vietcong mined a train very close to Saigon, South Vietnam, and near the Tân Sơn Nhất air base. The engine and three waggons were thrown off the track. Casualties, if any, were not stated. [1468]

Sunday 21st May 1967 was meant to have been the beginning of a 48-hour truce between the South Vietnamese/American forces and the Vietcong, the 22nd May being celebrated as the anniversary of Buddha's birthday, but almost three hours after the truce was meant to begin at 07:00, at 09:45, a mine derailed a freight train near Mượng Mán, South Vietnam. The train was described as having been heavily guarded and included armoured cars manned by a platoon of South Vietnamese militia. Two cars were blown from the tracks and the militia suffered light casualties. It could not be determined if the mine had been detonated by Vietcong hiding nearby or if it had been a pressure device set off by the weight of the train passing over it. [1469]

A passenger train was pulling into the station at Shioya, Japan, on the Sanyo Electric Railway, on Sunday 18th June 1967, when a bomb placed in the luggage rack exploded, killing a 23-year-old woman and wounding 16 of the 50 passengers in the second coach of the two-car Hyogo – Himeji train. Investigators found the remains of a clock in the debris and it was believed that a hand-made time bomb had been involved. One of the wounded passengers apparently later succumbed. [1470]

In the aftermath of the Six-Day War between Israel and its Arab neighbours on about Saturday 1st July 1967 Egyptian saboteurs mined and derailed an Israeli freight train at an unstated location in the Israeli-occupied Sinai Peninsula. [1471]

[1466] *El Tiempo*, Bogotá, Colombia, 10th March 1967

[1467] *Glenn Falls Times*, Glenn Falls, New York, United States, 15th March 1967

[1468] *Austin Daily Herald*, Austin, Minnesota, United States, 20th May 1967

[1469] *The Daily Herald*, Biloxi, Mississippi, United States, 22nd May 1967

[1470] *Pacific Stars and Stripes*, Tokyo, Japan, 20th June 1967

[1471] *Walla Walla Union-Bulletin*, Walla Walla, Washington, United States, 7th July 1967

A train carrying 150 South Vietnamese soldiers and 100 civilians from Vũng Rô Bay, a port used by the American Army, to the South Korean base camp at Tuy Hòa, was derailed by a Vietcong mine near Tuy Hòa, South Vietnam, on about Friday 4th August 1967, killing the driver and his assistant, though all the passengers escaped. [1472]

On Friday 18th August 1967 Vietcong mines derailed a freight train at Biên Hòa, South Vietnam, just outside of Saigon. Six cars, including two armoured cars carrying security personnel, were derailed and the tracks to the large military camp there were damaged. No one was hurt. [1473]

Three large mines derailed the engine and two cars of a train near Tuy Hòa, South Vietnam, on Saturday 19th August 1967. Again no one was hurt. [1474]

In the first act of railway sabotage since the end of the Korean War 14 years earlier, a passenger train was derailed and overturned near Yeoncheon, South Korea, on the evening of Tuesday 5th September 1967. Ten passengers were wounded. Trinitrotoluene (TNT) had been planted under the rails. [1475]

In another act of sabotage on Wednesday 13th September 1967 explosives planted under the track derailed six waggons of a freight train carrying military supplies near Munsan, South Korea, 30 kilometres north of Seoul. At least six waggons were derailed, two were badly damaged and one railwayman was slightly hurt. Police stated that the explosives were of North Korean origin. [1476] [1477]

A mine exploded beneath an Israeli Lod – Haifa freight train near Tulkarm, Jordan, on Friday 15th September 1967. Three waggons were derailed but no one was hurt. [1478]

Train No. 4, the Down Assam Mail, was derailed at about 03:11 on Thursday 28th September 1967 between Samuktala Road and Salsalabari stations, India, on the North East Frontier Railway. Fourteen passengers and a conductor were killed and 19 were seriously and 11 slightly injured.

The train consisted of Class YDM4 diesel electric locomotive No. 6128 and 22 coaches. Following the derailment the engine separated from the train, the first eight coaches went off to the right, the first two capsized, the next three went down an embankment, the sixth coach was telescoped by those on either side, while the wooden dining car, ninth in the train, caught fire and was almost burnt out. The rear ten coaches of the train remained on the track.

Driver Bhola told the inquiry into the accident that he had left Gauhati at 19:35 the previous evening and passed Samuktala Road at 03:08. Exiting a curve at about 62

[1472] *The Evening Sun*, Hanover, Pennsylvania, United States, 5th August 1967
[1473] *Charleston Daily Mail*, Charleston, West Virginia, United States, 18th August 1967
[1474] The *Corpus Christi Times*, Corpus Christi, Texas, United States, 19th August 1967
[1475] *Moline Daily Dispatch*, Moline, Illinois, United States, 6th September 1967
[1476] *The Terre Haute Tribune*, Terre Haute, Indiana, United States, 13th September 1967
[1477] *Fairborn Daily Herald*, Fairborn, Ohio, United States, 13th September 1967
[1478] *Charleston Daily Mail*, Charleston, West Virginia, United States, 15th September 1967

kilometres an hour he had found the Salsalabari Distant signal at yellow and closed the throttle. Then he had heard a loud noise, had seen sparks, there had been a pull from the rear of his engine, and he had gone to apply the brakes but found that they had already been applied. He had not noticed any obstruction on the line. There was no speed recorder on his engine; none of the diesels had them.

Assistant Driver Sukha Prakash Ghosh stated that after passing out of the right-hand curve mentioned by his driver he had been filling out entries in his log book when he heard a loud sound, the engine had jumped ahead, then stopped. He had seen no sparks. His log book entry of the train speed had been 80, overwritten to 70, overwritten to 60.

Guard M.S. Bannerjee said his train had arrived at Jorai on time but had been held there 22 minutes to cross train No. 3 Up. As his train had come out of the curve in question he had felt a jerk, the train had oscillated, there had been a second jerk and the train had stopped. After the derailment he had examined the line for obstructions, without finding any.

Ticket Examiner N.R. Chakraborty described hearing an unusual sound, feeling a couple of heavy jerks as if something was being pulled along the track, then the train had stopped.

Ticket Examiner D.C. Chakraborty gave a similar account.

Railway Contractor R.K. Mitra, who said he had been a passenger on the train for a long time, considered that it had been travelling faster than usual.

Lt. Gen. M.M. Khanna related that on going to the bathroom just after 03:00 he had found it difficult to keep his balance, had felt that the track was uneven, and that the train was going faster. He had said to himself, "I hope there won't be a mishap." Back in his compartment he had said the same thing to himself again and with that, the train had left the track. At about 05:00 he had spoken to the driver and accused him of going too fast. The driver had answered that it was a case of sabotage. He had then examined the track, which to him as a layman appeared to be in poor condition, then he had seen a missing fishplate and a rail turned up and concluded that it was, indeed, a case of sabotage.

Assistant Stationmaster Binda Prasad Singh at Samuktala Road said that there had been nothing abnormal with the train as it passed his station.

Assistant Stationmaster J.T. Haldar at Salsalabari testified that his Down Distant signal had been green since about 03:00, before he had been given *Line Clear* for the mail train from the station ahead. He had received *Train Entered Section* signal from Samuktala Road at 03:13.

Train Examiner R.K. Dhol said that he had inspected the train before the accident.

District Mechanical Engineers S.M. Bammi and P.R. Narasimhan stated that they had examined the engine and rolling stock of the train respectively and found no cause for the derailment.

Driver Sumer of No. 667 Down goods, the last train before the Assam Mail, said that he had had a smooth run and had seen no one near the track.

Patrolman Santosh stated that he had taken over patrol duty at 22:00 with Patrolman Sarwan. Everything had been normal when they passed the accident spot about 23:30.

Patrolman Sarwan agreed that there had been nothing amiss when they passed the location of the derailment. He added that he had found fishbolts missing and loose dog spikes three or four times during the previous five years.

Gang Mate Baldeo said that the track had last been attended to on 7th September when it had been packed. He knew of no quarrel among the permanent way staff or with the public.

Assistant Permanent Way Inspector M. Prasad stated that the point of derailment was not a bad spot and had not required frequent attention.

Permanent Way Inspector G.C. Barman Rajkumar testified that the gang tools had all been correct after the derailment.

Assistant Engineer J.M. Gupta detailed the state of the track after the derailment. He said that security patrolling had been in operation in the section since the end of September 1965. In the recent past patrolmen had detected two cases of tampering between Alipur Duar (the station beyond Salsalabari) and Samuktala Road.

District Engineer D.H. Mukherjee agreed that the damage to the track suggested that tampering was the cause of the derailment. There had been five cases of tampering with the track in West Bengal since November 1966.

Masalchi Ramu explained that the fire in the dining car had originated in the spillage of burning coal from the chulha stoves when the dining car capsized in the derailment.

Dining Car Manager Taj Mohammed added that the fires in the chulha stoves were normally only dimmed at night and not extinguished.

Additional Commissioner P.M.N. Murthy calculated that the Assam Mail must have run through Samuktala Road at 76 to 80 kilometres an hour (the maximum authorised speed was 65 kilometres an hour). It had initially left the track at Km 245/5. Only the rear wheels of the engine had derailed at this point and it had continued thus to near Km 245/10. It appeared that the initial derailment had been due to the excessive speed of the train, combined with the poor condition of the track. The further derailment of the train had undoubtedly been the result of sabotage, a rail having been displaced upwards. A demonstration showed that two men armed with a crowbar and a spanner could carry out the sabotage in eight minutes.

The additional commissioner concluded that the initial derailment of the train at Km 245/5 was due largely to the excessive speed of the train, for which Driver Bhola was responsible. The track defects had contributed to the derailment. It was not possible to assign to any individual the responsibility for the track defects. Driver Bhola was not responsible for the final derailment, which he could not have averted. Detection of the sabotage in time was impossible.

That there had been an initial derailment due to excessive speed at Km 245/5 was not accepted by the Government.

The additional commissioner made five recommendations:

(1) It was essential that the speed recorders fitted to the locomotives of passenger trains be maintained to deter drivers from overspeeding.

(2) The train reports of guards should be checked to detect overspeeding.

(3) The standard of track maintenance between Samuktala Road and Salsalabari should be improved.

(4) The quality of track patrolling in the area should be improved.

(5) Fires in the ovens of dining cars should be completely extinguished at the end of the day's service, the manager being held personally responsible for this.

The additional commissioner made eight additional observations and recommendations:

(1) The coupling pivot pin at the rear of the locomotive had not been of the correct high-tensile steel.

(2) The vigilance control of the locomotive had not been working.

(3) The engine speedometer required repositioning so that the driver did not have to turn around to look at it when the engine was running with the long hood leading.

(4) The Salsalabari assistant stationmaster had cleared his Down Distant signal for the mail train before he had obtained *Line Clear* for it.

(5) The station clocks at Samuktala Road and Kamakhyaguri (the station before Samuktala Road) had both been out of order.

(6) Suggestions made by Lt. Gen. Khanna for improving the post-accident relief operations should be examined.

(7) No arrangements had been made to collect and record the evidence of witnesses.

(8) The location of the curve involved in the accident was incorrectly given in the working timetable. [1479]

A 51-car Southern Railway freight train from the Appalachia region to Birmingham, Alabama, [1480] was derailed at 1 a.m. on Sunday 29th October 1967, a dynamite charge exploding beneath it near Big Stone Gap, Virginia, United States. Six coal cars and a caboose were derailed; no one was hurt. [1481] Twenty employees of the Interstate Railroad (a subsidiary of the Southern railway company) who were members of the Brotherhood of Locomotive Engineers and Firemen, had been out on strike since 2nd August, and Southern Railway employees had refused to cross their picket lines. The Southern company offered a reward for the apprehension of those responsible for the sabotage, issuing notices worded "Ten thousand dollars reward…Do you know who wrecked our train?" [1482] By the beginning of December no one had informed on those responsible and the reward was doubled to $20,000. [1483]

Seven coaches of a Mexico City – Ciudad Juarez passenger train were derailed by sabotage on Wednesday 7th February 1968 near Celaya, Guadalajara state, Mexico. No

[1479] Government of India Ministry of Tourism and Civil Aviation (Commission of Railway Safety): Report on Derailment of Down Assam Mail Between Samuktala Road and Salsalabari Stations of N.F. Railway on 28th September 1967

[1480] *The News Journal*, Radford, Virginia, United States, 30th October 1967

[1481] *Kingsport Times*, Kingsport, Tennessee, United States, 30th October 1967

[1482] Ibid, 31st October 1967

[1483] *Middlesboro Daily News*, Middlesboro, Kentucky, United States, 1st December 1967 one was seriously hurt. [1484]

A 70-car empty cattle train, returning from Ciudad Juarez to Mexico City was diverted into a siding of the Aceros de Chihuahua steel company about 5 kilometres north of Chihuahua City, Mexico, and collided with a three-car train there on Friday 9th February 1968. The drivers of both engines jumped clear and escaped with minor injuries. The freight train was reportedly travelling at 110 kilometres an hour, though this seems unlikely. Six railwaymen in all were taken to hospital and damage was estimated at 1,000,000 pesos (about $80,000).

An investigation found that the lock of the set of facing points leading to the steel company yard had been broken. Only supervisors had a key to the lock. Mexican authorities blamed members of a 'peace march' protesting against American involvement in the Vietnam War for both of the acts of sabotage. [1485]

A bomb exploded on a crowded Yokosuko – Tokyo passenger train as it was entering Ofuna station, Japan, on Sunday 16th June 1968, killing one person and wounding 27 of the 50 passengers in the coach, four of them seriously. The victims had mostly been returning from weekend visits to the beach. Like the explosion the previous year (see above) the attack took place on Japan's Father's Day.

The blast broke most of the coach windows and blew several holes in the roof. Surviving passengers said that they had been momentarily dazed and stunned and the coach had quickly filled with thick white smoke. Then they could hear the wounded groaning. One of them, Hitoshi Nakajima, 19, said he had heard a bang that sounded like 100 firecrackers going off at once. Then he had noticed a sharp pain in his left leg and looking down had seen that his trousers were ripped and his leg was bleeding. He had been sat across the aisle from where the bomb was planted.

Other survivors spoke of having seen a small green package placed on the luggage rack at the rear of the carriage. Investigators recovered four dry cell batteries, pieces of newspaper and what was apparently the remains of a clock spring. [1486]

Isamu Hiroshima, 30, the passenger who died, had been returning from a hospital visit to his wife and first child. [1487]

By the beginning of August 1968 there were only two trains running in the whole of South Vietnam and on Monday 5th a Vietcong mine exploded beneath one of them, killing two people and wounding four near Phan Rang.

A reporter wrote, 'During the French administration it was possible to travel by train along the Vietnamese coast to Hanoi. But the railroads became a favourite target of guerrillas. Trains were blown up and tracks were hauled to build bunkers. Most of the lines are now overgrown with grass and the rails rusty. Saigon's railroad station in the

---

1484 *El Paso Herald Post*, El Paso, Texas, United States, 7th February 1968
1485 Ibid, 10th February 1968
1486 *Pacific Stars and Stripes*, Tokyo, Japan, 18th June 1968
1487 *The Lowell Sun*, Lowell, Massachusetts, United States, 17th June 1968

heart of the city has not heard an engine whistle in more than two years.' [1488]

A passenger train was derailed between Arbil and Kirkuk in northern Iraq on Wednesday 4th December 1968. Reports reaching Baghdad stated that the train had been attacked by a band of "rebellious agents" who had laid explosives on the track, derailing a number of carriages. One report added that "the attackers then descended on the passengers, killing old people, women and children." The number of casualties was not stated. The attack was reportedly linked to the dissatisfaction of the Kurdish minority in the northern part of the country and it was said that "the government is trying to solve the problem peacefully." [1489]

A passenger train was derailed by sabotage at an unstated location north of Rangoon (now Yangon), Burma, on about Friday 24th January 1969. Rebels then opened fire on the derailed train, killing four people and wounding five. Three rebels were killed by the train's police escort. [1490]

Train No. 6, the Down Allahabad Express, consisting of YP Class engine No. 2288 and ten coaches, was derailed at about 01:10 on Saturday 21st June 1969 just short of a bridge over the Mangai Nala between Jakhanian and Dulahpur stations, India, on the North Eastern Railway. The consequences were disastrous. The engine and three leading coaches fell to the bed of the nala (stream) and the next four coaches were derailed and more or less badly damaged. The rear three coaches remained on the track. Sixty-three people were killed outright and nine more died on the way to hospital or after admission. Forty-six others were seriously injured and 103 were slightly hurt. The driver and first fireman were among the fatalities.

Guard S.H. Singh told the inquiry into the accident that the train had left Varanasi at 23:19 the previous night, four minutes late. Further delays at Varanasi City to cross No. 17 Up passenger train and at Kadipur had made the train 24 minutes late. Soon after leaving Jakhanian there had been a jerk and the vacuum had dropped, leading him at first to think that the communication cord had been pulled, but then a roaring sound had told him that something else must be wrong. The train had stopped, he had alighted and had seen by the light of his torch that coaches were derailed. Efforts to send a message using a portable telephone had failed because the overhead lines were down, so he had sent the conductor guard back to Jakhanian on foot with the message, "The train met with an accident at 01:10 hrs and breakdown train and medical van urgently required." The guard said that he had then carried out a detailed examination of the wrecked train and in the presence of Railway Protection Force (RPF) sainiks (soldiers) who had been escorting a cash safe on the train he had also found bolts and keys that had been thrown into the bushes of the lineside embankment. He added that the train had been travelling at 65 to 66 kilometres at the time of the derailment and had been carrying about 600 passengers, with a sitting capacity for 442.

[1488] *The Daily Gleaner*, Kingston, Jamaica, 6th August 1968
[1489] *Charleston Daily Mail*, Charleston, West Virginia, United States, 5th December 1968
[1490] *The Times*, London, England, 25th January 1969

RPF Sainiks Ram Ugrah Singh and Ram Adhar Misra corroborated the statement of the guard.

Second Fireman C.B. Yadava stated that the engine had been working normally, its brakes functioning and headlight illuminated. When the train derailed he had been pulling coal. He had lost consciousness in the accident. When he came round he had shouted to his driver and fellow fireman but received no response. He had then walked to Dulahpur station with a severely scalded foot.

Assistant Stationmaster H.S. Lal at Jakhanian stated that the train had passed through his station at 01:05 under clear signals. When he had not received the *Train Out of Section* signal from Dulahpur after the usual time he had begun making enquiries but the Control phone had been out of order. He had then sent the gang mate down the line to investigate at 01:35 and at about 02:50 the conductor guard had arrived at his station with a memo regarding the accident, when he had advised Control through the block wires.

Assistant Stationmaster O.N. Vaish at Dulahpur testified that he had received the *Train Entering Section* signal from Jakhanian at 01:04. When the train had not appeared after the booked time he had tried to contact Jakhanian via the block wires, the Control phone being out of order. At 01:35 he had sent one of the gangmen to investigate and at 02:43 the second fireman had arrived and reported the derailment.

Train Examiner R.D. Rai of Allahabad City described some minor defects to the rolling stock of the train that he had attended to.

Train Examiners Jag Narain at Allahabad City and T. Prasad at Varanasi stated that they had found no defects in the train on the day before the derailment.

Loco. Foreman Ram Sarau at Varanasi detailed his examination of engine No. 2288 after the accident.

Driver Paras Ram said that he had experienced no problems with engine No. 2288 on its last run before that which had ended with the derailment.

Driver Purushottam of No. 71 Up passenger, Driver A.C. Niwas of No. 3 MB Up through goods and Driver Saquid Hussain of No. 76 Down passenger, who had passed the site of the derailment at about 21:12, 22:40 and 23:27 respectively on the 20th all denied having noticed anything wrong at Mangai Nala Bridge.

Permanent Way Inspector R.S. Lal described being called to the site of the derailment and finding a displaced rail and various missing fittings, which had been thrown down the embankment. The track had last been inspected by push trolley on the 18th and from the footplate on the 15th, no defects being found. In the previous year there had been frequent thefts of keys and jaws from the track and an attempt at sabotage at the bridge the previous July. Most permanent way equipment was kept in tool boxes at stations.

Assistant Permanent Way Inspector B.D. Tewari gave a similar account.

Gang Mate P. Singh spoke of having sent two gangmen to examine the line from the Dulahpur side on the orders of the assistant stationmaster and they having returned with the second fireman to report the accident. The bridge, which was in his section, he had last examined on the 17th, when all had been in order.

Keyman Sati Ram stated that the track and the bridge had been in perfect order on the morning of the 20th.

Assistant Engineer D.P. Gupta described the state of the track after the derailment in detail, as did Divisional Engineer R.C.S. Khichi and Senior Deputy Chief Engineer L.C. Agarawal.

Additional Commissioner G.S. Pandor noted that it was clear the derailment had occurred at 01:10. The speed of the train from the speed recorder was 62 kilometres an hour at the moment of derailment. There was no evidence of any defects in the locomotive or rolling stock and the track generally was in good condition. Evidence of the removal of a rail at the site of the derailment and of loose track components being found scattered about at the scene immediately after the accident was provided by the guard, two RPF personnel and three passengers, one of them a doctor. Negligence on the part of the permanent way gang could be excluded by the fact that seven trains had passed the site safely since 17:44. There had been an hour and 35 minutes since the passage of the last train to carry out the sabotage, a job that could be done in about nine minutes.

In conclusion the additional commissioner stated that the derailment had clearly been due to a deliberate act of sabotage, the 'dastardly act' being carried out by removing the joints and fastenings of a 39 ft length of rail. The offenders had planned to cause the maximum amount of damage by choosing a point on a 22 ft high embankment just short of a girder bridge. Damage to the track was impossible to detect at high speed in time to avoid an accident. The speed of the train had been well below the maximum permitted speed of 75 kilometres an hour at that point and Driver P.C. Bannerjee, who had lost his life, had been in no way responsible for the accident.

The additional commissioner made no recommendations. Under incidental observations and recommendations he remarked:

(1) The examination of locomotive No. 2288 had been overdue because of a large number of staff absences.

(2) The engine repair book had not been periodically scrutinised.

(3) Staff who had exceeded the speed limit through yards as shown by the speed recorder were not being disciplined. [1491]

Sixteen people died in the derailment of a train, possibly due to sabotage, at an unknown point between Rangoon and Mandalay, Burma, on Tuesday 23rd February 1971. [1492]

The Vietcong derailed a train between Da Nang and Huê, South Vietnam, on Tuesday 13th April 1971, killing 20 people and injuring another 23. [1493]

The derailment of a diesel-hauled passenger train near Potgietersraus (renamed Mokopane 2003), in the Transvaal, South Africa, on Friday 31st March 1972, was

[1491] Government of India Ministry of Tourism and Civil Aviation (Commission of Railway Safety): Report on Derailment of No. 6 Down Allahabad – Gorakhpur Express Train Between Jakhanian and Dulahpur Stations North Eastern Railway on 21st June 1969

[1492] Author's notes, source mislaid.

[1493] *Defiance Crescent-News*, Defiance, Ohio, United States, 14th April 1971 attributed by the police to sabotage. It was stated that signs of possible sabotage had been found at the scene. No explosives had been used, rather the rails had apparently been loosened on the approach to a bridge over a river. The derailed locomotive continued over the 60ft-long bridge but the crowded coaches fell from the bridge to the river, about 30 ft below. There was only a trickle of water in the river. The accident was a bad one, resulting in the death of 38 people and the injury of 174 more. All the passengers were described as 'non-whites,' most of whom had been on their way to a church meeting at Pietersburg, the next stop. The train had been bound for Messina, near the Rhodesian border. Ambulances going to the scene were delayed by congestion caused by buses and cars en route to the church meeting.

The following day the engine and seven waggons of a goods train were derailed near Springs, South Africa, supposedly because a set of points had been interfered with. [1494] [1495]

Frente de Libertacao Mocambique (Frelimo) guerillas carried out two attacks on trains on New Year's Eve, Monday 31st December 1973. A passenger train was mined near Marchipanda, Mozambique, on the Beita – Umtali line while a mail train was blown up near Vila Nova de Fronteira, Mozambique, on the Donso – Malawi line close to the Malawi border. News of the attacks was only announced a week later. No casualties were reported. [1496]

Frelimo guerillas mined a freight near Maforga, Mozambique, on the line connecting Ian Smith's Rhodesia with the Indian Ocean at Beira on the night of Saturday 19th January 1974. One African was killed. [1497] An Australian newspaper commented that the attacks on the railway, of which this was the second within a week, marked a major new phase of the fight for liberation in Southern Africa and that the railway was vital to landlocked racist (this term was used several times in the newspaper) Rhodesia. [1498]

Two Portuguese railwaymen were killed when a freight train was mined in an attack on the Beira railway at an unstated location in Mozambique on about Thursday 31st January 1974. [1499]

Two attacks on railways emanating from Beira were made by Frelimo guerillas on Sunday 17th February 1974. In the first and more serious of them a train on the line to Malawi was machine-gunned near Nangue, Mozambique, killing five passengers and wounding six others. In the second attack on the Umtali line a freight train was derailed at an unstated location. [1500] (Umtali was renamed Mutare 1982.)

On Wednesday 8th May 1974 a passenger and mail train was raked with machine-gun fire near Inhaminga, Mozambique, on the Malawi line. A mixed-race woman passenger was killed and several citizens were wounded. The following day, Thursday 9th May, 17 waggons of a southbound freight train were derailed by sabotage in the same area. No

[1494] *The Times*, London, England, 1st and 4th April 1972
[1495] *The New York Times*, New York, New York, United States, 1st April 1972
[1496] *The Canberra Times*, Canberra, Australian Capital Territory, Australia, 7th January 1974
[1497] *The Times*, London, England, 22nd January 1974
[1498] *Sydney Tribune*, Sydney, New South Wales, Australia, 22nd January 1974
[1499] *The Times*, London, England, 2nd February 1974
[1500] *Delaware County Daily Times*, Upper Darby, Pennsylvania, United States, 18th February 1974

one was hurt but the line was blocked. [1501] [1502]

On Friday 19th July 1974 Frelimo guerillas attacked the railway between Beira and Tete, damaging it in 74 places, in an attempt to disrupt supplies to the Portuguese Cabora Bassa Dam project. No trains were involved in these attacks. [1503]

Mozambique gained independence from Portugal on Wednesday 25th June 1975.

Only hours after an agreement between South Vietnam and the Vietcong regarding an exchange of prisoners a mine exploded under a train between Da Nang and Huê, South Vietnam, killing nine civilians and wounding another 26. This was probably the last act of railway sabotage of the Vietnam conflict, though the war lingered on until the end of April 1975. [1504]

[1501] *The Daily Gleaner*, Kingston, Jamaica, 10th May 1974
[1502] *Diario de Lisbõa*, Lisbõa, Portugal, 9th May 1974
[1503] *The Press-Courier*, Oxnard, California, United States, 19th July 1974
[1504] *The Daily Herald*, Provo, Utah, United States, 7th February 1974

CHAPTER 19

MISCELLANEOUS ACCIDENTS

The vast majority of railway accidents fall into one of a small number of categories such as derailments, collisions, bridge failures and so on. A few, though, do not easily fall into any obvious group, hence this chapter on miscellaneous accidents. Included in this chapter are a good number of accidents involving the release of hazardous or toxic chemicals, a type of accident that has only become widespread in modern times.

As noted above a feature of railway traffic that has greatly increased since the 1960s, especially in the United States, is the transport, usually in tank waggons, of a variety of explosive, flammable, or poisonous chemicals. Although the majority of these have been carried safely, inevitably from time to time trains carrying them have been involved in accidents and leakages of chemicals have occurred. Due to the large volume of the traffic there have, in total, been quite a few accidents. A few of those involving explosions have already been described in the Chemical Explosions chapter, the remainder that occurred in the period covered by this book will be individually described here.

On Tuesday 31st January 1961 a Louisiana and Arkansas Railroad freight train, running on the Texas and Pacific Railway, was derailed at La Barre, Louisiana, United States. Northbound train No. 42 from New Orleans consisted of three diesel units working in multiple, 113 cars and two cabooses and was moving at 42 miles an hour when the derailment occurred. The 81st to 96th cars left the track, the 89th car was ruptured and chlorine gas escaped from it and covered an extensive area. As a result an 11-month-old baby was fatally poisoned and 168 residents and the front brakeman of the train were injured, some very seriously. [1505] The Harris family were the worst affected, baby James Harris III being the fatality. The 24-bed local hospital was overwhelmed with patients and many had to be treated in corridors. A large number of farm animals also died. [1506]

The accident was investigated by the Interstate Commerce Commission, who ascribed the derailment to a broken journal resulting from overheating (a 'hot box') on the 80th car, which had not been detected prior to the derailment. Another car in the train with an overheated journal had previously been removed at North Baton Rouge, Louisiana. The

[1505] Interstate Commerce Commission report.
[1506] *The Humboldt Times*, Eureka, California, United States, 1st February 1961

chlorine was en route from Baton Rouge to Evadale, Texas. [1507]

Early on Saturday 25th March 1961 26 cars of a Chicago Rock Island and Pacific Railroad freight train were derailed at Hazen, Arkansas, United States. Five of the derailed cars were carrying military explosives and two others were carrying ammonia, some of which escaped. Two daughters of Mr J.W. Baxter, whose house was only about 100 feet from the site of the derailment, were slightly affected by the fumes, and he took them and his wife away from the scene. Police stated that no one else was in danger either from the ammonia or from the explosives. It was believed that the derailment had been due to a broken journal. [1508]

Thirty-one people were reportedly killed and 35 were injured, 12 seriously, when a one-car electric train was buried beneath a muddy landslide between Oita and Beppu, in southern Japan, at about 15:00 on Thursday 26th October 1961. There had been 66 passengers on the train. Heavy rain for over 24 hours had preceded the landslide. [1509] The reinforced concrete wall bordering the track had given way under the weight and pressure of the water-sodden soil behind it. Most of the passengers on the train were high school students. [1510] The landslide, which was reportedly 15 metres high, 30 metres wide and 20 metres long, was embedded with pine trees and covered both the railway and the adjoining highway skirting Beppu Bay. [1511]

The full details of something of a freak accident near Trebbin, East Germany, which was also apparently a major catastrophe, only emerged in late 1989, 27 years after it happened. On Thursday 1st March 1962 a Soviet Army train was carrying a tank battalion from a training area at Jüterborg to East Berlin. The train consisted of one or more locomotives, about 15 flat waggons, each of which was loaded with two T-55 tanks, seven or eight boxcars with wooden seats, each carrying 30 to 50 soldiers, more boxcars doubtless loaded with equipment, and flat waggons loaded with wheel vehicles at the rear. Half of the cannon barrels of the tanks were presumably orientated in the direction of travel and the other half in the opposite direction.

Travelling in the opposite direction at 120 kilometres, bound for Leipzig, was an express from Berlin.

The gun barrel of one of the tanks cannot have been properly locked, or the lock was faulty, and the vibration of the train caused the barrel to swing to the side, fouling the opposite running line. This must have happened at some point before the troop train and the express passed one another just before 19:00, near a level crossing with the F 101 road (Crossing Post No. 39) at Kliestow, close to the town of Trebbin. The gun barrel first struck the locomotive of the express, tearing off its water pump, swung away, then swung back and tore out the side of the third or fourth coach of the express, killing a

[1507] Interstate Commerce Commission report.
[1508] *Northwest Arkansas Times*, Fayetteville, Arkansas, United States, 25th March 1961
[1509] *Pacific Stars and Stripes*, Tokyo, Japan, 28th October 1961
[1510] *Pharos Tribune*, Logansport, Indiana, United States, 27th October 1961
[1511] *The Times*, London, England, 27th March 1961

passenger who was thrown from the train. As chance would have it this was the corridor side of the compartment coach, or the casualties on the express would have been higher. The force of the collisions caused the tank to overturn and fall from the flat waggon onto the track, derailing the following vehicles of the troop train, which were wedged together and piled up.

The accident was briefly reported at the time but the only fatality made public was that of the unfortunate passenger on the express, information as to casualties among the soldiers on the troop train being hushed up. According to eyewitnesses who had partaken in the subsequent rescue operation about 70 or 90 soldiers were also killed outright and many others were injured, some from splintered wood from the broken waggons. [1512] The injured soldiers were initially taken to the civil hospital in the town of Luckenwalde before later being transferred, on the orders of higher military officials, to military hospitals. An eyewitness stated many years later that the injured soldiers had been treated quite brutally by some of the Soviet military involved in their rescue, leading to angry protests from the senior doctor at Luckenwalde hospital, who refused to allow seriously hurt soldiers to be moved. [1513]

Seven waggons of a freight train were derailed by an earthquake between Kogota and Tajiri stations, Japan, on Monday 30th April 1962. The Tohoku trunk line was blocked. [1514]

Six men, two women and a child were killed and 114 people were injured when No. 3232 Down Nagpur – Tatanagar passenger train was completely derailed near Kumhari station, India, on the South Eastern Railway, at about 7.20 p.m. on Friday 8th June 1962. The train had been struck by a cyclone running from north to south, and which dissipated after about half an hour. Twenty-eight of the injured were admitted to Raipur hospital while the others were treated and sent home. Apart from the damage done to the train the roofs of many homes were torn off and waggons were derailed in a nearby marshalling yard. [1515]

Four diesel units and 35 cars of a Delaware and Hudson Railroad freight train were derailed on Friday 4th January 1963 a mile and a half south of Starucca, Pennsylvania, United States. Many of the waggons overturned and fire broke out after the derailment, consuming two box cars loaded with hay. The fire brigade attended but was unable to extinguish the blaze due to a lack of water. More to the point, as far as this book is concerned, a tank car loaded with an unspecified acid broke in two, spilling over the roadbed. No one was hurt and the cause of the derailment was not immediately ascertained. [1516]

[1512] Wikipedia – Eisenbahnunfall von Trebbin

[1513] https://www.maz-online.de/brandenburg/ruecksichtslose-offiziere-und-mutige-mediziner-DENRNMDFJW3UBCPHFE7MWJHRDY.html

[1514] *Pacific Stars and Stripes*, Tokyo, Japan, 1st May 1962

[1515] *The Sunday Standard*, Chittoor, India, 10th June 1962

[1516] *The Daily Record*, Stroudsburg, Pennsylvania, United States, 5th January 1963

A Chesapeake and Ohio Railway freight train was derailed at Fayette station, West Virginia, United States, on Monday 7th January 1963 and a tank car loaded with epichlorohydrin fell into the New River. About 30% of the 8,000 gallons of the chemical reportedly leaked into the river, despite which a spokesman for the Union Carbide Chemicals Co. stated that there had been no pollution. Nevertheless, pumping of water at the towns of Alloy and Montgomery, downstream of the accident, was suspended for some time. [1517] [1518] [1519]

As a 60-car Grand Trunk Western Railroad freight train was passing the station at Stillwell, Indiana, United States, on Monday 14th January 1963 Operator E.E. Cash spotted a hot-box on its 31st car and attempted to signal it to stop. Before it could do so 18 cars had become derailed, demolishing telegraph wires and blocking the line. One of the derailed vehicles was a tank car loaded with irritant, carcinogenic and flammable formaldehyde, some of which escaped and firemen stood by while work was in progress to plug the leak. [1520]

At about 10.30 a.m. on Friday 1st March 1963 20 cars of a Pennsylvania Railroad freight train were derailed near Bowerston, Ohio, United States. During the derailment the valve of a tank car loaded with liquefied petroleum gas (LPG) broke off and some of its contents escaped. Harrison County School had to be closed and the area around the derailment sealed off until the danger of a fire or explosion had passed. [1521] [1522] [1523]

A westbound Canadian National Railway freight train collided with an articulated lorry on a level crossing near Portage du Fort, Ontario, Canada, at about 7.30 a.m. on Thursday 11th April 1963. Three diesel locomotives and 27 cars of the train were derailed, including three cars carrying highly volatile naptha gas, 30,000 gallons of which were spread over a wide area. Engineer William Callan said afterwards, "It's a peculiar feeling when you know you are going somewhere, but don't know where you end up. I thought my fireman was dead when I saw him. He and the brakeman were lying right on top of me after the train turned over on its side." In fact he, Fireman Edward Rock and Brakeman Raymond Markle were only slightly hurt, though all were taken to hospital. Conductor Ernest Beddington and the lorry driver were unhurt. [1524]

The derailment of a Western Maryland Railroad train running over the Reading Company line near Brandtsville, Pennsylvania, United States, at 4.45 a.m. on Sunday 28th April 1963 reportedly resulted from a large hole having been punched in the wheel of a tank car loaded with liquefied petroleum gas (LPG) by an axle. Twenty-three of the 45 cars left the track and five hours later three of the LPG tank cars exploded over ten minutes. Fire broke out after the initial derailment; the fire department was called and

---

[1517] *The Charleston Gazette*, Charleston, West Virginia, United States, 8th January 1963
[1518] *Bluefield Daily Telegraph*, Bluefield, West Virginia, United States, 8th January 1963
[1519] *The Athens Messenger*, Athens, Ohio, United States, 8th January 1963
[1520] *The Anderson Herald*, Anderson, Indiana, United States, 15th January 1963
[1521] *Steubenville Herald-Star*, Steubenville, Ohio, United States, 1st March 1963
[1522] *The Daily Jeffersonian*, Cambridge, Ohio, United States, 2nd March 1963
[1523] *Altoona Mirror*, Altoona, Pennsylvania, United States, 2nd March 1963
[1524] https://churcher.crcml.org/circle/Wreck%20Details/1963PortageduFort.html

they had thought they had the fire under control, when the first explosion occurred at 9.25 a.m. In addition there was a leakage of chlorine from another tank car. Twenty-two people, included a number of firemen, were injured, either from burns or from chlorine poisoning and 80 families had to be evacuated. [1525] [1526]

Fifteen cars of a Chicago Milwaukee St Paul and Pacific Railroad freight train were derailed at Seymour, Iowa, United States, on the afternoon of Sunday 2nd June 1963. Ten of the cars were loaded with road oil, which was spilt over a road at a level crossing but fortunately did not catch fire. The derailment was blamed on a broken axle or wheel. [1527]

There appears to have been a sudden increase in accidents involving the leakage of chemicals in 1963 for some reason and another happened at Eureka, Missouri, United States, on Sunday 18th August 1963. Twenty cars of a 111-car Missouri Pacific Railroad were derailed due to a 'hot-box' and one of them began to leak liquefied petroleum gas (LPG). Over 400 residents of the town had to be evacuated for fear of an explosion. No one was hurt. [1528] [1529]

Over 100 residents of Rochester, New York, United States, had to be evacuated on Wednesday 18th December 1963 after 26 of the 67 cars of a New York Central Railroad freight train were derailed. Twelve of the cars fell down a 40 ft embankment and there was a reported leakage of 21,000 gallons of styrene monomer from one tank car and 10,000 gallons of vinyl chloride monomer from a second tank car (another account states only 8,000 gallons were lost from each car). Both chemicals are toxic and carcinogenic. [1530] [1531] [1532]

Nineteen of the 66 cars of a Chicago Rock Island and Pacific Railroad freight train from Houston to Fort Worth and Denver, Texas, left the track on Wednesday 8th January 1964 near Deco, Texas, United States, about 17 miles from Houston. The accident occurred on the Burlington – Rock Island Railroad, jointly owned by the Chicago Burlington and Quincy and Chicago Rock Island and Pacific companies. The derailment occurred on a 300ft wooden trestle over Cypress Creek and petrol and diesel fuel that spilt from five tank cars set fire to both the trestle and to two box cars, while there were reportedly several small explosions. A 'hot box' was thought to have caused the derailment. No one was hurt. [1533]

Both main lines of the Baltimore and Ohio Railroad were blocked on the afternoon of Friday 17th April 1964 when 42 cars of an eastbound freight train, the *Advanced Baltimorean*, left the rails half a mile west of Orleans Road, West Virginia, United States, at 4.47 p.m. The 109-car train was bound for Cumberland, Maryland, and was in charge of Conductor C.J. Phillips with Engineer C.J. Walker. As a result of the derailment 20,000 gallons of caustic soda (sodium hydroxide) and an undetermined quantity of fuel oil were lost into the Potomac River. Downstream communities were warned not to take

---

[1525] *The Gettysburg Times*, Gettysburg, Pennsylvania, United States, 29th April 1963
[1526] *The Evening Standard*, Uniontown, Pennsylvania, United States, 29th April 1963
[1527] *Atlantic News-Telegraph*, Atlantic, Iowa, United States, 3rd June 1963
[1528] *Alton Evening Telegraph*, Alton, Illinois, United States, 19th August 1963
[1529] *Mansfield News-Journal*, Mansfield, Ohio, United States, 19th August 1963
[1530] *The Daily Messenger*, Canandaigua, New York, United States, 19th December 1963
[1531] *The Post Standard*, Syracuse, New York, United States, 19th December 1963
[1532] *The Citizen-Advertiser*, Auburn, New York, United States, 19th December 1963
[1533] *Galveston Tribune*, Galveston, Texas, United States, 9th January 1964

water from the river. [1534]

A misunderstanding between a signalman and a driver resulted in a light engine being set back into collision with a 48-waggon train of petrol tank waggons early on Friday 14th August 1964 at Didcot North Junction, England, on British Railways, Western Region. Petrol leaked from the damaged waggons, 12 of which were overturned, and a major fire eventually affected 14 waggons, with dense smoke and large sheets of flames. Fortunately it was possible to pull the remaining waggons clear with another engine. It took firemen about three and a half hours to get the blaze under control. No one was hurt but apart from the damage to the rolling stock, all the tracks in the area were destroyed, a footbridge was damaged, a cornfield was set alight, and there were fears of petrol polluting the River Thames. [1535] [1536]

Fourteen cars of a Delaware and Hudson Railroad freight train derailed at 1.45 a.m. on Wednesday 23rd September 1964 and piled up "like a deck of cards." The former depot at Central Bridge, New York, United States, was demolished and there was a spillage of merchandise and perishables and also of ethylene glycol. Firemen stood by to avert a possible fire. A hot-box was again blamed for the derailment. [1537] [1538]

Firemen were called after a New York Central Railroad freight train derailed and caught fire at Columbus, Ohio, United States, on Sunday 13th December 1964. A tank car loaded with ethylene oxide was involved when 12 of the 33 cars of the train left the track. An hour after the derailment there were reportedly 500ft-high flames (doubtless a considerable exaggeration) but there was then a small explosion and the firemen were forced to beat a hasty retreat. After that they let the ethylene oxide drain into the ground and left the fire to burn itself out as there were fears of further explosions from a second tank car loaded with a petroleum product. [1539]

In contrast to air crashes and in earlier years shipwrecks it is very unusual for an accident to a passenger train carrying a number of people to have no survivors, and the author is only aware of two accidents in which this has been the case. One was the famous Tay Bridge disaster in Scotland on 28th December 1879, dealt with in the first volume of this series. The second was an accident that happened almost exactly 85 years later off the coast of southern India.

Pamban or Rameswaram Island is situated between mainland India and Sri Lanka (named Ceylon at the time of the accident). In 1964 the island was linked to the mainland by a long railway bridge to Mandapam. At Pamban village at the western end of the island the railway divided into two with an 11-kilometre branch to the town of

---

[1534] *The Cumberland News*, Cumberland, Maryland, United States, 18th April 1964

[1535] Ministry of Transport: Railway Accidents, Report to the Ministry of Transport on the Accidents that occurred on the Railways of Great Britain during the year 1964.

[1536] *The Times*, London, England, 15th August 1964

[1537] *The Oneonta Star*, Oneonta, New York, United States, 24th September 1964

[1538] *The Kingston Daily Freeman*, Kingston, New York, United States, 23rd September 1964

[1539] *The Greely Daily Tribune*, Greeley, Colorado, United States, 15th December 1967

Rameswaram and a 25½-kilometre-long branch to Dhanushkodi, a pilgrimage centre and a ferry port for Ceylon. Towards its far end the latter branch ran on a long sandy strip of land with the sea on either side.

On Tuesday 15th December the Meteorological Department of India had warned of a developing cyclone south of the Andaman and Nicobar Islands in the Bay of Bengal. The cyclone moved slowly towards India and by Tuesday 22nd it was approaching that country from Ceylon.

The guard and driver of No. 660 Up Mixed train of the Southern Railway, which arrived at Mandapam at 20:30 from Dhanushkodi on Tuesday evening, warned that a large amount of sand had accumulated on the track between Dhanushkodi and Rameswaram Road stations and Madurai Control were informed at 20:35. At 21:00 the wind speed had increased to 64 kilometres an hour and the Pamban Bridge operator advised Control that it was not safe for trains to cross the bridge. It was therefore not possible for the permanent way inspector to come from Mandapam over the bridge with his staff to clear the line. Instead the Pamban bridge inspector and some khalasis accompanied Down Passenger train No. 653, which left Pamban at 23:10 for Dhanushkodi. The train consisted of engine No. B.31378, a third-class, brake and luggage composite coach, a first-, second- and third-class composite, three third-class and a first-, second- and third-class composite coach. It reached Rameswaram Road station at 23:45 and left there at 23:55. By then the intensity of the storm had increased and communication between Control and Dhanushkodi failed at 00:05 and with Rameswaram Road at 01:35.

The train did not arrive at Dhanushkodi. The station there was flooded by the sea and the severity of the storm prevented the staff there or at Rameswaram Road from learning what had happened to the train. The staff at Dhanushkodi hoped that the train had been held back at Rameswaram Road and those at the latter place similarly hoped that the train had reached its destination. It was only on the morning of Christmas Eve, Thursday 24th, that the chimney of the engine was seen poking out of the water near the Dhanushkodi Down Outer signal. For some reason the assistant stationmaster at Pamban had counted the number of passengers on the train, which was probably 110; in addition 18 railwaymen lost their lives.

Stationmaster M.C. George at Dhanushkodi told the inquiry that at 02:30 on the morning of the accident he had been woken by water flooding his bedroom and wading through to the station he had found that train No. 653 had not arrived. All communication had failed and he thought that the train might have returned to Rameswaram Road. No weather warning had been received before he had left duty.

Assistant Stationmaster K. Perriannan at Dhanushkodi stated that a message had been received from Rameswaram Road that the train crew of train No. 660 had reported sand on the track and he had advised the assistant permanent way inspector. There had also been a weather warning from Control at 21:00. He had given *Line Clear* to Rameswaram Road at 23:35 for train No. 653 and had been informed that the bridge inspector and his men were piloting the train to clear the sand. The train had been reported 'out' at 23:55 but at 00:40 could still not be seen and communications had failed. He had intended to proceed up the line to investigate but the intense gale made it impossible to get beyond the station platform. He too had hoped that the train had retreated to Rameswaram Road.

Driver N. Karrupiah of train No. 660 testified that he had left Dhanushkodi at 18:40. Three kilometres out he had found sand on the track and proceeded cautiously to Rameswaram Road, arriving there at 20:06, when he had reported the matter. Cyclones and sand on the line, he added, were nothing unusual. There had only been a breeze even on arrival at Pamban at 20:50 but the anemometer there had shown an impending storm and the train had been cancelled over the bridge. Control had then wanted him to return to Rameswaram Road for water but he had refused to do so on account of the weather.

First Fireman V. Marieppan, Second Fireman T. Mani, and Guard A.M. Masilomoney corroborated their driver's account.

Assistant Stationmaster Mathalainathan at Rameswaram Road spoke of receiving a message from the staff of train No. 660 that there was sand being blown onto the line between Kms. 676 and 679, due to the wind. He had duly cautioned train No. 653 and two men from his station had accompanied it. He had received a weather warning from Madurai at 00:30 but it had only started raining at 01:30 and the wind intensity had not increased until 02:30. At about 01:25 he had received a message on the block telephone from the assistant permanent way inspector at Dhanushkodi that 100 kilometre an hour winds and heavy rain were preventing him from going beyond the Outer signal there. Passing the message on to Control they had asked about train No. 653. At 01:30 Dhanushkodi had replied that it was not in sight. Five minutes later telephone communication with Control had failed.

Pointsman A. Valli at Dhanushkodi stated that he had lowered his signals for train No. 653 to go to No. 1 road on the instructions of the assistant stationmaster. The backlights of the signals had then been visible but later disappeared. Cyclonic conditions had gradually developed and water had engulfed his cabin, rising to half its height. He had neither seen the headlight of train No. 653 nor heard its whistle. He had remained in his cabin until noon on the 23rd when the water had finally subsided and he returned to the station.

Section Controller T.N. Ramaswamy at Madurai testified to receiving a message from Rameswaram Road that there was sand on the line. Such reports were received now and then. The marine superintendent at Mandapam not being available he had asked the bridge inspector to proceed with men and material on train No. 653 to clear the line and the assistant permanent way inspector at Dhanushkodi to introduce a night patrol. At 21:00 the operator of the Scherzer lift bridge at Pamban had reported that the wind speed was over 40 miles an hour (65 kilometres an hour). Train No. 660 had thus been held at Pamban and train No. 183 at Mandapam. Train No. 653 had left Pamban at 23:10 and Rameswaram Road at 23:55. From 00:05 he had only been able to contact Dhanushkodi via Rameswaram Road. At 01:00 Rameswaram Road had advised him that train No. 653 had not cleared Dhanushkodi. At about 01:30 the assistant permanent way inspector and stationmaster at Dhanushkodi had reported that because of the weather they could not go out beyond the Outer signal there, while at 01:40 all communication beyond Pamban had been lost.

Assistant Permanent Way Inspector D. Chellappan at Dhanushkodi spoke of being informed that sand was blocking the track between Kms 676 and 679. The wind had been too strong to use a trolley and walking to the site would have taken too long; instead it had been arranged for the line to be cleared by the bridge inspector and his men using train No. 653. At about 22:15 Control had asked him to introduce a night patrol in view of the weather and he had sent two men out. One had been 'lost' (drowned), the other, unable to keep up with his companion, had eventually returned to the station. At midnight he had, himself, attempted to walk to Rameswaram Road with five more men but after a furlong (220 yards/200 metres) the wind and rain had forced them back. He had not heard the whistle of train No. 653, nor seen it approaching.

Stationmaster R. Kulandaisamy at Rameswaram Road stated that train No. 653 had already left when he started duty at midnight. When it had not cleared Dhanushkodi by 01:30 he deputed Pointsman Alagan to investigate, but the wind and rain had prevented the latter from getting further than the Up Home signal and he returned. By then all communication had failed. The wind had progressed into a cyclone that lasted until 14:00. There had been a break in the storm between 06:00 and 08:00 when he had sent a gangman to examine the line. The gangman had returned at 09:00 and reported that there was a breach at Km 670 1-2. A pointsman sent out at 14:00 had said the same thing. On the night of 23rd a person who had swum across from Dhanushkodi said that the train had been washed away by a tidal wave.

Relieving Stationmaster R.K. Nair at Pamban said that train No. 653 had been due to depart at 20:40 but the bridge inspector had not been ready until 23:00, so the train had not left until 23:10.

Shunting Jamelar M. Kasruppan and Shuntingman G. Krishnan at Dhanushkodi said they had not heard the engine of train No. 653 whistling, nor seen its headlight.

Pointsman V. Alagan at Dhanushkodi (this must be a different person from the pointsman at Rameswaram Road – a brother perhaps), on duty from 13:00 to 22:00, spoke of seeing the signals lowered for train No. 653 and having been washed towards the loco shed as he had proceeded towards the station. He had been marooned by the water all night.

Railway Police Constable No. 560 R. Rengarathinam described hearing the engine of train No. 653 whistling continuously for about 45 minutes from about midnight and seeing its headlight.

Gangman M. Lakshmanan testified that he had been sent out on the night patrol from Dhanushkodi with Gangman N. Lakshmanan. With the increasing wind and rain he had lagged behind the other patrolman. At the Outer Home he had sat down, exhausted, and later had crawled back towards the station and taken shelter in a stores room.

Driver L. Thirumalaswamy said he had worked No. 101 Down train (the Boat Mail) to Dhanushkodi. Lying in bed in the running bungalow there he had heard an engine whistling.

Assistant Stationmaster A.M. Bellie at Pamban explained that someone at Control, he thought the section controller, had asked him to count the number of passengers on train No. 653. Being busy shunting Train No. 660 he had asked the guard of train No. 653 to count them but later had done so himself, anyway, counting 100 to 110 passengers. The section controller denied having given him any such instruction.

Permanent Way Inspector K.V. Radhakrishnan at Mandapam said that at 21:30 he had been told to proceed to Pamban and take train No. 653 from there with some men to clear the line but the wind had made the bridge impassable. The assistant permanent way inspector at Dhanushkodi having said that the wind was too strong to use his trolley, arrangements had finally been made for the bridge inspector at Pamban to take some men on train No. 653.

Assistant Engineer G.P. Ramachandran stated that he had arrived at Ramanathapuram at 21:40 on Rail Car No. 213 Down. A weather warning at 16:00 had mentioned only heavy rain. He had been advised of the report of sand on the line, and of the impossibility of sending staff to clear it from either Mandapam or Dhanushkodi, when he had arranged for the bridge inspector to proceed from Pamban with whatever men he could find. Night patrols had been discontinued from 13th December due to continued fine weather.

Box Boy T. Thavamani at Dhanushkodi said that he had gone to meet train No. 653 and it not appearing and the wind getting stronger, he had taken shelter in the waiting room. During the night he had heard an engine whistling. At about 07:00 on the 24th he had accompanied Driver Thirumalaswamy to a point near the Outer Signal where they found the bodies of the driver and fireman of train No. 653 and that of a woman. The driver denied having seen any bodies and the boy's account was not believed at the inquiry.

Bridge Operator V. Abdullah Koya at Pamban Bridge stated that when the Pamban assistant stationmaster had asked for his private number at 21:10 to start train No. 660 Up, he had told the stationmaster that the wind was blowing at 40 miles an hour and the anemometer was indicating danger. The train was then cancelled. The wind power had increased further and he informed Control of this at midnight. At about 01:30 the wind had been shaking his cabin and he could contact neither Pamban station nor Control. At 04:00 the roof of his cabin had been blown off, water had entered the cabin and he could see that four girders of the bridge on the Pamban side had been washed away. When no rescue boat had appeared by 15:00 on the 23rd he had begun swimming towards Pamban and been picked up by a boat after swimming part of the way.

Steam-man V. Ramayya Thevar at Dhanushkodi denied hearing any engine whistling. Neither of the engines in his shed had been in steam, he said.

Railway Police Constable No. 418 N. Narayanan Nair alleged that he had travelled to Rameswaram Road on train No. 660 Up, returning to Dhanushkodi on train No. 653. There had, he said, been 30 to 35 passengers in his bogie, including children. The train had proceeded at 10 miles an hour part of the way, then gone slower. It had stopped at the Dhanushkodi Outer signal, which had been at danger and showing a red light, and its driver had whistled. After 15 minutes he had alighted, had seen that the sand was blown away beneath the sleepers, and had told the driver. Someone on the engine had remarked that this often happened. He had then walked to the station, the wind being so strong that the signal posts were shaking. He had not realised that the train was in any danger. If he had been a passenger on the train from its last stop the police constable would, effectively, have been a survivor of the catastrophe (but see below). According to Press reports a Dhanushkodi vegetable merchant had likewise alighted from the train and survived but despite publishing appeals in different languages the inquiry was unable to trace him and the additional commissioner discounted the story.

Additional Commissioner Arya Bhushan noted that the normal running time between Rameswaram Road and Dhanushkodi was 34 minutes but train No. 660 Up Mixed had taken an hour and 26 minutes and he did not believe that train No. 653 could have taken any less time. Nor did he accept P.C. Nair's statement that the train, running, as it had been, under a caution order and piloted by the bridge inspector, had not stopped at any point en route. The accident must have occurred at some time after 01:00.

The engine regulator had been closed, the reversing lever notched up and the vacuum ejector handle had been in the running position, implying that the engine had been coasting at the time of the accident. Even if it had been stopped at the Outer signal it would not have remained there very long without someone on the engine going forward to the station. It seemed clear that the train must have been struck by a sudden tidal wave as had the water level risen gradually someone on the engine would have informed the station.

The additional commissioner reconstructed the likely course of events. Train No. 653 had been due to leave Pamban at 20:40 but had been delayed awaiting Bridge Inspector Kumaraswamy and his men, who were to examine the line, finally leaving at 23:10. The run to Rameswaram Road had been normal and *Line Clear* had been obtained there from Dhanushkodi in the usual fashion. After the train had left Rameswaram Road at 23:55 with a caution order conditions must have considerably deteriorated. The train had been struck by a tidal wave, an unprecedented event in living memory, which could not therefore have been anticipated. The assistant stationmaster at Dhanushkodi could not be faulted for giving *Line Clear* for train No. 653 and was not in any way responsible for the accident. The steps taken to examine the line had been in order. There was no evidence of defects in the track, rolling stock or engine or of mismanagement by the driver.

The statement of P.C. Nair that he had been on train No. 653 was unbelievable, as it contained a number of discrepancies. His story that the train had not stopped anywhere between Rameswaram Road and Dhanushkodi was improbable. That he had spoken to the driver after 15 minutes and had then walked to the station, without the driver asking him to tell the station staff where the train was seemed highly unlikely. That after some time at the station he had heard the train whistling outside, implying it had just been stood at the signal all that time, was unbelievable. If, as he had said, the sand had been blown away from under the sleepers the train would not have remained standing on the track to begin with. His timings lacked authenticity, his statement as to the signals could not be reconciled with any other, and so on. It seemed that if he had ever gone to Rameswaram Road he had not returned to Dhanushkodi that night, as he should have done to take up duty, and that he had been attempting to cover up his failure by alleging that he was on the train.

The additional commissioner dismissed Press criticisms that warnings of bad weather had been ignored and noted that night patrols could not have predicted a sudden tidal wave. Two patrolmen had in any case been sent out from Dhanushkodi, one of whom had been forced back and the other had lost his life.

Concluding his report, Additional Commissioner Bhushan wrote that the catastrophe had been an Act of God and there had been no negligence on the part of any railway staff.

Pamban Bridge was repaired after the accident but the branch to Dhanushkodi was abandoned. [1540]

On Tuesday 4th May 1965 three American Naval aircraft raided unstated locations in North Vietnam, during the course of which a train believed to be carrying ammunition

[1540] Government of India Ministry of Tourism and Civil Aviation (Commission of Railway Safety): Report on the Wash-away of No. 653 Down Pamban – Dhanushkodi Passenger Train between Kms. 681 and 682 near Dhanushkodi Station on 23rd December 1964 bound for the Vietcong in the south was blown up. [1541]

On about Saturday 27th November 1965 American Naval aircraft from the aircraft carrier USS *Midway* attacked a supply train north of Vinh, North Vietnam, setting it ablaze. [1542]

A signalman was the principal victim of a train accident at Coton Hill, England, at 05:50 on Monday 11th January 1965. Coton Hill was a marshalling on just north of Shrewsbury on the former Great Western Railway, but now British Railways, London Midland Region, main line between there and Chester. Approaching Coton Hill from the north the Up main line fell for about a mile at 1 in 100 (Hencote incline), while there was an Up Goods loop slightly less than half a mile long between Coton Hill North and South signal boxes. At the beginning of the incline there was a stop board, at which all goods trains were required to halt and pin down brakes, a requirement reinforced by a sign just in the rear of the stop board, which read, "All Up Goods and Mineral Trains must Stop Dead Here."

On the morning of the accident the 03:56 Class 5 Up goods train from Saltney to Pontypool Road consisted of Type 4 Brush Sulzer diesel locomotive No. D1734, 46 waggons and a brake van, in charge of Goods Guard J.J. Butler, with Driver George Pike and Fireman Kenneth Clorley. The train failed to stop at the board and then ran out of control down the incline. It was diverted into the goods loop, ran the length of the loop, failed to stop at the signal the end, was derailed on the trap points protecting the main line and the engine demolished Coton Hill South signal box, killing Signalman T. Ferrington. [1543] Driver Pike and Fireman Clorley were injured, the driver seriously. Eleven waggons were derailed, one of them a tank waggon from which a quantity of hydrochloric acid was spilt. The accident was the subject of a Ministry of Transport inquiry.

Goods Shunter A. Rose at Saltney marshalling yard told the inquiry that 22 waggons fitted with the vacuum brake had been the last to be added to the train, which was therefore a partly fitted one. He was sure that the vacuum pipes had been properly connected and free of blockages.

Carriage and Waggon Examiner E.P. Powell confirmed that the vacuum pipes had been properly connected but said he had been unable to carry out a brake test before the train engine was attached and he had had no time to do so afterwards.

Driver A.S. Bevan stated that he had had no problems with the brakes of engine No. D1734 on a previous trip.

Signalman S.M. Upson at Weston Rhyn said that the driver of the goods train appeared to be normal and had raised his hand as the train passed his box.

Signalman G. Davies at Leaton stated that he had offered the goods train to Coton Hill North box at 05:39, had it accepted and sent it on (given the *Train Entering Section* signal) at 05:45. He had received the *Train out of Section* signal already at 05:50. He knew that goods trains were, according to the local instructions, not meant to cover the distance between the two boxes in less than ten minutes but said that he would have been surprised if the train had taken that long, as they did not usually do so. The train had

[1541] *Kenosha News*, Kenosha, Wisconsin, United States, 4th May 1965
[1542] *Idaho State Journal*, Pocatello, Idaho, United States, 30th November 1965
[1543] *The Times*, London, England, 12th January 1965

passed at its usual speed of 40 miles an hour; he had not noticed the driver.

Signalman F.H. Foxall at Coton Hill North testified that he had offered the goods train to the South box on the Up Goods loop as soon as Leaton had sent it on and cleared his signals for it at once. Then he had been tidying up the box ready to go home at 06:00 and had thus not noticed the approach of the goods train until it had been close to his box. Although this was only four minutes after the *Train Entering Section* signal, and on average trains took ten minutes to arrive, it had not struck him at the time as being particularly quick. Sending the train on to the South box he had gone to the box window to acknowledge the guard's hand signal that the train was in the loop. It was only at this point that he had realised the train was going much too fast, when he put down his lamp and gave the *Emergency Call Attention* signal to the South box (a special bell code peculiar to the Great Western Railway – any number of beats given in rapid succession – the signal was not authorised on the London Midland Region but was one familiar to ex-Great Western signalmen). The next thing he knew was that the lights of the yard had gone out and the needle of his block instrument had dropped (because the derailment had brought the telegraph wires down). He put the speed of the train as it had passed him at 50-60 miles an hour with no sign of braking. The engine horn had been sounded as if the driver was surprised that he was entering the loop. He said it was left to the signalman's discretion if the train was put into the loop, though this train usually was. The signal box instructions did not require him to check trains going into the loop.

Signalman G.H. Jones at Shrewsbury Crewe Junction spoke of having heard a thud, which he had later found was the noise made by the accident.

Driver Pike was too badly hurt to give evidence until over a month after the accident and in any case he suffered retrograde amnesia and could remember nothing of it.

Fireman Clorley had only a patchy recollection of the events leading up to the accident. He could not recall if Driver Pike had carried out a brake test at Saltney. Nor could he recall much of the journey to Coton Hill, which included a steep descent at Ruabon. He remembered, however, the last part of the journey and Driver Pike working the handle of the vacuum brake repeatedly after nearing the stop board and the brake not appearing to work. Repeated questions to his driver as to what was amiss had not been answered. As the engine was becoming derailed at the trap points his driver still had his hand on the brake handle. His driver usually stopped at the stop board with Class 5 trains.

Goods Guard Butler stated that no brake test had been carried out before leaving Saltney. One was normally only done when there was time to spare, which was unusual. The journey to Leaton, including the descent of Ruabon bank, had been normal. He had not been concerned when the train did not stop at the stop board, as not all drivers of Class 5 trains did so. The Coton Hill North Outer Home had been at Danger when he had first seen it and he applied the brake in his van. He did not remember hearing the engine horn at all.

Head Goods Shunter J. Clarke did, however, remember the engine horn as the train entered the goods loop, at a speed he put at 50 miles an hour and said that he had seen sparks indicating braking and realised that something was amiss.

Yard Inspector A.T. Davies, said he had likewise heard the engine horn, seen signs of braking, and had known that something was wrong.

Driver E.J. Worrall, in charge of a steam engine on the Down Main line at Coton Hill

South, said that the horn of the goods train engine had been sounding continuously and described sparks coming from the engine brake blocks as having been "like Catherine wheels."

Fireman G.A. Jones on diesel light engine No. D 1684, running on the Third Line at Coton Hill South, described seeing the signal box being struck and having shouted to his driver to stop.

Rolling Stock Inspector R.L. Norman stated that after the accident he had found that the Automatic Warning System (AWS) in the leading cab of the goods engine was isolated and the driver's vacuum brake handle was jammed in the applied position.

Assistant Carriage and Waggon Engineer J. Colbert stated that there had been flats only on the wheels of the brake van. Only one defective waggon brake had been found but the damage to the waggons made it impossible to tell if the brakes had been applied on them.

Signal box train registers showed the average speed of the goods train to have been 28 miles an hour between Gobowen and Whittington (Low Level), rising to 40 miles an hour between Whittington and Baschurch and to 45 miles an hour between Baschurch and Leaton.

An analysis of the running of Class 5 trains suggested, stated the inspecting officer, Colonel J.R.H. Robertson, that some drivers were ignoring and others were paying scant attention to the instructions to stop at the stop board. During the previous year when in charge of Up trains Driver Pike had been routed at Coton Hill North box via the Main line 21 times and the Goods Loop ten times. His average speed between Leaton and Coton Hill with Class 5 trains was seven minutes, his fastest five minutes. With Class 6 trains it had been always ten minutes or more.

The inspecting officer went on, in his report, to make a detailed examination of the possibility of a partial or complete brake failure as a cause for the accident and to consider other possibilities such as the driver having thought that he was being routed along the main line or having fallen asleep or at least become drowsy.

In conclusion, Colonel Robertson noted that had the train been stopped at the stop board as per the instructions it could have stopped short of the Goods loop by the action of the waggon brakes, even if there had been a failure of the brakes on the engine. He (the inspecting officer) was satisfied, however, that there had been no brake failure and that Driver Pike was responsible for the accident. It did not seem likely that the driver had dozed off. What was more probable was that he had forgotten that he was to be relieved at Coton Hill South (the reason for his train being routed via the Goods loop), had been in the habit of only checking Class 5 trains at the top of Hencote Incline, and with his train being over half an hour late he had not checked it at all, assuming that he would be running via the Main line. This assumption might have been confirmed by his seeing at a long distance the Coton Hill North Outer Home signal at clear and he might not have realised that he was being routed into the Goods loop, until his train was actually entering it.

Turning to remarks and recommendations Colonel Robertson expressed his surprise that the routine checking of signal box train registers had not brought to light the fact that Class 5 goods trains were repeatedly taking less than ten minutes between Leaton and Coton Hill.

The inspecting officer rejected the suggestion made by the jury at the inquest on the dead signalman that the stop board should be resited much further out, as this was impracticable and, moreover, contradicted basic signalling principles.

Although it appeared not to have had any bearing on the accident the isolation of the AWS equipment had been unnecessary and deprived the driver of an important aid. Such isolations were apparently frequently made and drivers should be reminded that they must not make them without good reason.

A control on the Coton Hill North Outer Home signal to prevent it being cleared prematurely when the road was set for the Goods loop would probably have prevented the accident. One had subsequently been installed.

A derailment serious enough to wreck Coton Hill South signal box could not reasonably have been forseen but the replacement box would be positioned elsewhere.

A duplication of the pipe connection between the vacuum train pipe and the triple valve on engines was recommended.

The instructions for the testing of the brakes of partially fitted goods trains should be amended to make it clear that the guard was always responsible for testing the brakes of a train before starting unless he could obtain assurance that the Carriage and Waggon staff had already done so.

The inspecting officer was not impressed with the Coton Hill North Up fixed Distant signal, which he said conveyed nothing to the driver other than where he was. [1544]

Almost half of a 150-car Southern Railway freight train left the rails at Taylor, South Carolina, United States, on Monday 31st May 1965, after its engines had collided with a bulldozer working too close the line. The derailment happened close to a 150ft-high trestle over the Saluda River, above which the five derailed engines were left perilously balanced. The train crew scrambled off the trestle and the bulldozer driver also had time to jump to safety. Butane gas leaked from two tank cars and one of the engines, situated 200 yards from the escaping gas, caught fire. For fear of an explosion ten families living nearby and a large number of spectators were evacuated by the police, who cordoned off the area, and no one was allowed to approach the leaking cars until the gas dissipated. [1545] [1546]

A Baltimore and Ohio Railroad freight train became derailed west of Garrett, Indiana, United States, on Saturday 19th June 1965, probably due to a broken journal, and a tank car carrying either acetic or acetic anhydride was breached and leaked. Residents in the area had to be evacuated. [1547]

Eight of the nine cars of a Southern Pacific Company freight train were derailed at 3.15 a.m. on Wednesday 25th August 1965 at Berenda, California, United States, six miles north of Madera. A tank car loaded with 8,000 gallons of liquefied sulphur dioxide was punctured and fumes covered the adjoining Highway 99. Lorry driver Reuben Stewart, who had stopped to offer assistance, was overcome by the fumes, and taken

---

[1544] Ministry of Transport inquiry

[1545] *Kingsport Times*, Kingsport, Tennessee, United States, 1st June 1965

[1546] *Florence Morning News*, Florence, South Carolina, United States, 1st June 1965

[1547] *The Sunday Pharos Tribune and Press*, Logansport, Indiana, United States, 20th June 1965 unconscious to Madera Hospital, where the next day he was stated to be in a fair condition. The road was closed until 6 a.m. The train crew were not hurt. The cause of the derailment was not immediately known. [1548] [1549]

Thirty-one cars of a Norfolk and Western Railway freight train from Frankfort, Indiana, to Buffalo, New York, were derailed on the edge of the yard at Tipton, Indiana, United States, at 11.20 p.m. on Wednesday 13th October 1965, after a wheel broke on one of the leading cars. The train was carrying a wide variety of goods such as lumber, plywood, whiskey, canned goods, powdered plastic, paper, and fire brick clay but the problem load was a tank car of acrylonitrile, a toxic, flammable, potentially explosive and carcinogenic chemical. A valve on the tank car, which was the fifth vehicle in the train, was damaged and the acrylonitrile began to escape. At least two families living nearby had to be evacuated while workers wearing gas masks transferred the chemical to five road tankers. [1550] [1551]

A baggage car at the rear of the *Southwesterner* passenger train of the New York Central Railroad from St Louis, Missouri, to New York, became derailed near De Graff, Ohio, United States, on the night of Friday 28th January 1966. The derailed vehicle fouled the opposite running line and was struck by a 68-car freight train from Buffalo, New York, to St Louis. As a result of the collision three diesel units and 42 loaded cars of the freight train were derailed. Some of the cars fell down an embankment and one of them began leaking an initially unidentified chemical, forcing a local evacuation. The chemical as later identified as alcohol. Three of the crew of the freight train were slightly hurt. [1552]

A long Kansas City Southern Railway freight train of 183 cars became derailed on Thursday 31st March 1966 at Shreveport, Louisiana, United States. Seventeen cars in all left the rails and gasoline (petrol) escaped from a damaged tank car and caught fire. Eleven fire engines attended and firemen sprayed water onto three adjoining tank cars containing lubricating oil to prevent them also becoming involved in the blaze. Five hundred feet of track were also ripped up. Engineer Bill Van Hook said that one of the wheel boxes had frozen, presumably meaning that it had had a ‘hot box’. [1553]

Twenty-three cars of a Norfolk and Western Railway freight train in charge of Conductor P.P. Dietrick with Engineer Lowell Dumechey left the rails on the edge of Lima, Ohio, United States, on the night of Friday 15th April 1966. Two tank cars loaded with fuel oil ruptured and their contents were spilt but fortunately there was no fire or explosion. [1554]

The derailment of 18 cars of a 110-car a Texas and Pacific Railway freight train at 6.30 a.m. on Thursday 5th May 1966 at Gordon, Texas, United States, resulted in the death of two hobos, transients, or itinerants, as they were variously known. The men were riding on a flat car next to two tank cars carrying sulphuric acid. When the cars left the

---

[1548] *The Bakersfield Californian*, Bakersfield, California, United States, 25th August 1965
[1549] *Humboldt Standard*, Eureka, California, United States, 25th August 1965
[1550] *Kokomo Morning Times*, Kokomo, Indiana, United States, 15th October 1965
[1551] *The Logansport Press*, Logansport, Indiana, United States, 15th October 1965
[1552] *Ogden Standard-Examiner*, Ogden, Utah, United States, 29th January 1966
[1553] *The Daily Star*, Hammond, Louisiana, United States, 1st April 1966
[1554] *The Lima News*, Lima, Ohio, United States, 16th April 1966

rails an estimated 20,000 gallons of acid escaped and covered the men. Local residents pulled the victims, named as Charley Jones, 56, from Clarksville, Mississippi, and Johnny Kale, 48, from Helena, Arkansas, from the wreckage and an ambulance took them to hospital, where they died several hours later. The derailment was blamed on a common cause – a broken journal due to a 'hot box'. Bulldozers covered the spilt acid with soil and the railway arranged for alkaline chemicals to be brought to the site to neutralise the acid. Other derailed cars were loaded with fertiliser, furniture, and lorry trailers. [1555] [1556]

Thirteen of the 120 cars of a New York Central Railroad freight train became derailed on the night of Tuesday 28th June 1966 near Clarendon, Michigan, United States, on Michigan Central Railroad tracks. The wheels of a box car were the first to leave the rails and the following vehicles at the end of the train followed it off the track, several piling up on one another. A car of 'non-flammable alcohol' ruptured, spilling liquid over the tracks. [1557]

Planes from the aircraft carrier *Oriskany* attacked a train about midway between Thanh Hóa and Nam Dinh, North Vietnam, on Tuesday 16th August 1966. They damaged four boxcars before they were attacked in turn by MIG jets and forced to withdraw. [1558]

On about Saturday 20th August 1966 American Naval aircraft bombed and strafed a camouflaged ten-car freight train near Thanh Hóa, North Vietnam, and reportedly destroyed the locomotive and five cars and damaged two more cars. [1559]

American aircraft attacked a train between Hanoi and Yên Bái, North Vietnam, on about Monday 22nd August 1966, destroying five rail cars, damaging five more and cutting the line. [1560]

A 15-car train on the line between Hanoi, North Vietnam, and the Chinese border was bombed and strafed by American aircraft on Tuesday 6th September 1966. Five box cars were reportedly destroyed, one was damaged and the line was cut in three places. [1561]

Carrier-based Navy bombers attacked a 58-car train and a railway bridge at Ninh Binh, North Vietnam, on about Sunday 18th September 1966. Pilots reported that they had destroyed at least three box cars and damaged the bridge and its approaches. [1562]

On about Tuesday 20th September 1966 US Air Force F105 Thunderchief aircraft attacked a train in the vicinity of Đống Đăng, North Vietnam, north east of Hanoi on a main line to China. One of the pilots, Major Neil Eddins, spoke of the destruction of the train as, "total, with all cars exploding, one fire and derailed." [1563]

[1555] *Lubbock Avalanche-Journal*, Lubbock, Texas, United States, 6th May 1966
[1556] *The Abilene Reporter-News*, Abilene, Texas, United States, 6th May 1966
[1557] *Marshall Evening Chronicle*, Marshall, Michigan, United States, 29th June 1966
[1558] *The Gettysburg Times*, Gettysburg, Pennsylvania, United States, 16th August 1966
[1559] *Port Angeles Evening News*, Washington, United States, 21st August 1966
[1560] *Waukesha Freeman*, Waukesha, Wisconsin, United States, 22nd August 1966
[1561] *Fond du Lac Commonwealth Reporter*, Fond du Lac, Wisconsin, United States, 7th September 1966
[1562] *Alamogordo Daily News*, Alamogordo, New Mexico, United States, 18th September 1966
[1563] *Winona Daily News*, Winona, Minnesota, United States, 20th September 1966

US Navy A4 Skyhawk aircraft from the Coral Sea aircraft carrier attacked a train about 15 kilometres north east of Thanh Hóa, North Vietnam, on Saturday 24th September 1966. Five box cars were reportedly destroyed and ten were damaged. [1564]

The derailment at about 7.45 a.m. on Sunday November 1966 of part of a 45-car Missouri Pacific Railroad freight train led to the evacuation of over 1,000 residents of Basile, Louisiana, United States. Four tank cars laden with ethylene dibromide and a caboose left the track. Ethylene dibromide, used at the time as a pesticide, fumigant, and an additive to petrol, is both highly toxic and carcinogenic, and it leaked from one of the cars. Conductor Charles Ray Henagan, sat at his desk in the caboose, was thrown about but not much hurt. He said, "God was riding with me in the caboose." Engineer M.B. Wilkinson, Fireman J.W. Adams, and Brakemen I. Buck and J.L. Cooks, all on the diesel engine, were also unhurt.

The residents were allowed to return home at about 10 p.m. and the line was reopened at 12.15 a.m. on Monday. No official cause was given for the derailment but it was theorised that a load of metal weighing about 1,000 pounds that was being carried in a box car might have bounced about, broken through the floor of the car, and ended up on the track. [1565] [1566]

Two separate derailments involving the spillage of different dangerous chemicals happened in Ohio, United States, on Friday 11th November 1966. In Toledo on the Detroit and Toledo Shore Line Railroad, a tank car with a capacity of 20,000 gallons, en route from Texas City, Texas, to the Monsanto Chemical Co. at Trenton, Michigan, overturned and began leaking vinyl acetate, highly flammable, explosive, and potentially toxic. Fireman sprayed foam over the chemical to dilute and dissipate it and stood by in case it should ignite.

In Homer, on the Michigan Central Railroad, the derailed vehicles of the train jack-knifed into a lineside oil storage tank, leading to a fire fed by 60,000 gallons of petroleum. In turn, the heat from this fire caused the rupture of a tank car loaded with vinyl chloride and the release of its contents. Vinyl chloride is flammable, potentially explosive and toxic and burns to produce hydrogen chloride, carbon monoxide and a small amount of phosgene, though Robert Wright, a chemical expert flown to the scene, stated that the small amount of gas released did not represent a problem. Nevertheless most of the 1,600 inhabitants of Homer were evacuated for some hours. [1567] [1568]

There was a fear of contamination to water supplies after the derailment of 18 cars of a Western Maryland Railway freight train near the station at Loy's, Maryland, United States, on Tuesday 20th December 1966. Five of six tank cars loaded with liquid asphalt that left the rails were ruptured and the liquid ran into a creek, a tributary of the Moncacy River. The cause of the derailment was not stated. [1569]

[1564] *Pharos Tribune and Press*, Logansport, Indiana, United States, 25th September 1966
[1565] *Lake Charles American Press*, Lake Charles, Louisiana, United States, 7th November 1966
[1566] *Wichita Falls Times*, Wichita Falls, Texas, United States, 7th November 1966
[1567] *The Times Recorder*, Zanesville, Ohio, United States, 12th November 1966
[1568] *Van Wert Times-Bulletin*, Van Wert, Ohio, United States, 12th November 1966
[1569] *The News*, Frederick, Maryland, United States, 22nd December 1966

Raw turpentine and a small quantity of chlorine were spilt when a Southern Railway train was derailed two miles east of Rock Hill, South Carolina, United States, on Tuesday 17th January 1967. Fifteen cars in all left the rails from an undetermined cause. Firemen covered the chlorine with foam and stood by all day lest the turpentine should catch fire. [1570]

Two diesel units and ten cars of a 106-car Delaware and Hudson Railroad freight train from Binghamton to Mechanicville left the rails at Albany, New York, United States, on Sunday 19th March 1967, and a tank car loaded with toluene was ruptured. Toluene is flammable, potentially explosive, and toxic. Firemen spent the night flushing the leaking toluene into the sewers and played their hoses on another derailed car loaded with liquefied petroleum gas. Other derailed cars were carrying caustic soda (sodium hydroxide). There was no fire, but a fireman remarked, “Any kind of heat, sparks, or open flame could set this thing off.” A policeman stated, “If that thing ever blows up, it’ll take three blocks with it.” There seems, however, only to have been a very limited evacuation. The derailment was blamed on a broken rail. [1571]

Twenty cars of an Illinois Central Railroad freight train derailed at Gates, Tennessee, United States, on Wednesday 14th June 1967, jack-knifed and blocked both running lines. Firemen again stood by to deal with a leakage of liquid ammonia. Trains were diverted over the Louisville and Nashville Railroad while the wreckage was being cleared. Witnesses reported having seen a ‘hot box’ on one of the cars just before the derailment. [1572]

At 12.10 p.m. on Saturday 17th June 1967 12 cars of a 111-car Norfolk and Western Railway freight train were derailed three miles from Lima, Ohio, United States. The derailed vehicles included a tank car loaded with 190,000 pounds of highly toxic and flammable formaldehyde. An adjoining box car knocked a two-foot hole in the tank car and the gas spewed out into a nearby field. Fortunately there were no houses nearby but a nearby road was closed off until the gas dispersed. A ‘hot box’ was blamed for the derailment. [1573]

The next derailment to be considered occurred on the Atchison Topeka and Santa Fe Railway near Cave, Kansas, United States, about eight miles west of Satanta. About 3 p.m. on Tuesday 27th June 1967 eight cars in the middle of a 27-car freight train left the track, two tank cars containing liquefied petroleum gas (LPG), three tank cars of helium and three cars of wheat. One of the LPG cars was punctured and there were fears of a massive explosion of the LPG and the inert but pressurised helium. Luckily the derailment occurred in a cutting in an unpopulated area. Water had to be played on the LPG car and its remaining contents were later transferred to road tankers brought to the scene. [1574]

In another derailment in Kansas, United States, on Sunday 6th August 1967, 27 cars of a Chicago Rock Island and Pacific Railroad train were involved. A tank car containing naphtha gas was breached and a second car containing the gas and a third loaded with

---

[1570] *Florence Morning News*, Florence, South Carolina, United States, 18th January 1967

[1571] *The Citizen-Advertiser*, Auburn, New York, United States, 20th March 1967

[1572] *Kingsport Times*, Kingsport, Tennessee, United States, 15th June 1967

[1573] *The Lima News*, Lima, Ohio, United States, 18th June 1967

[1574] *Garden City Telegram*, Garden City, Kansas, United States, 28th June 1967 liquefied petroleum gas overturned but were not ruptured. The area was cordoned off for fear of an explosion and firemen from the nearby Forbes Air Base spread foam over the wreckage. The derailment occurred near Willard. [1575]

The derailment of about nine cars of a Texas and Pacific Railway freight train due to a 'hot box'/broken journal led to a fire, an explosion, and the evacuation of almost 6,000 residents of Texarkana, Texas, United States. The accident happened late on Tuesday 22nd August 1967. The fire involved at least four cars but a car of granulated plastic burnt most fiercely. The explosion happened at about dawn the following morning and involved a tank car of muriatic (hydrochloric) acid. Hydrochloric acid is not explosive but if it reacts with metals hydrogen is produced and this must have been the cause of the explosion, which shook windows several miles away. Other derailed cars contained nitric acid and naphtha gas. Seven firemen and three Red Cross workers suffered cuts and burns evacuating residents and fighting the blaze. [1576]

Twenty-six cars of a Missouri Kansas Texas Railroad freight train were derailed a mile from Selma, Kansas, United States, early on Monday 9th October 1967. Ammonia escaped from one of the derailed cars and spread over half a mile of countryside but there was no habitation nearby and the gas, which stayed close to the ground, was stated to present no hazard to residents, though it was a "handicap" to workmen clearing the site. [1577]

Early on Saturday 28th October 1967 a Missouri Pacific Railroad freight train derailed from an undetermined cause near Danbury, Texas, United States. Twenty-five cars left the rails and a tank car loaded with wood alcohol (methanol) ruptured. Fire then broke out and spread along the train and into a rice processing plant, which was totally destroyed. Other cars contained naphtha gas and chlorine and the risk of an explosion led to the evacuation of 1,500 people while firemen fought the fire. Danbury Fire Chief W.R. Wollam had to be taken to hospital suffering from nausea and smoke inhalation. [1578]

American Phantom jets attacked a 60-car freight train on Thursday 20th April 1967, probably between Kép and Đống Đăng, North Vietnam. The first aircraft reportedly attacked a box car in the middle of the slowly moving double-headed train, while following aircraft "walked" their bombs forward from one box car to another towards the leading engine, throwing six to ten box cars off the track and jack-knifing others. [1579]

On Saturday 17th June 1967 American Navy Skyhawk aircraft from the carrier Bon Homme Richard attacked a 20-car freight train, apparently carrying high explosives, between Hanoi and Thanh Hóa, North Vietnam. [1580]

Collisions with animals trespassing do not usually have serious consequences, except to the animals themselves, but there have been exceptions, such as was shown by a disaster in

---

1575 *The Columbus Daily Telegram*, Columbus, Nebraska, United States, 7th August 1967
1576 *The Pampa Daily News*, Pampa, Texas, United States, 23rd August 1967
1577 *Great Bend Daily Tribune*, Great Bend, Kansas, United States, 9th October 1967
1578 *The Danville Register*, Danville, Virginia, United States, 29th October 1967
1579 *Carthage Press*, Carthage, Missouri, United States, 21st April 1967
1580 *Pacific Stars and Stripes*, Tokyo, Japan, 20th October 1967

Italy on Thursday 9th November 1967. The accident happened near the small station of San Nicola Varco, in an area near Salerno in the southern part of the country, where buffalo were commonly reared for both their meat and their highly sought after milk.

At about 04:55 on the day of the accident Guardablocca (Signalman) Pietro Narducci at San Nicola Varco was alerted by the bellowing of a herd of buffaloes that had apparently made their way onto the railway from the adjoining fields. *Direttissimo* (express) No. 904, which had left Palermo at 17:20 for Roma, consisting of an engine and 17 coaches, had just been reported to him from Albanella, 6 kilometres to the south, and in the opposite direction the *Conca d'Oro* express from Milano to Palermo was near Battipaglia. He replaced his signals in both directions and Driver Luigi De Gaetano, in charge of the engine of train No. 904, obeying the signals, duly pulled up at the station. Signalman Narducci warned the driver that there might be animals on the line and the train then proceeded at about 30 kilometres an hour. He had only proceeded 300 metres when he was confronted by a herd of 50 buffaloes on the line. He then braked violently but still struck three of the animals, while the violent traction shock was sufficient to throw two of the coaches of the train off the track. The 11th was diverted harmlessly towards the cess on the side of the line but the 12th was thrown in the opposite direction foul of the opposite running line.

Driver Michel Del l'Aquila on the engine of the *Conca d'Ora* was also slowing down, alerted by the signals, but the obstruction was well on the Battipaglia side of the station signal and before he could stop his engine had run into 29 beasts which were on his line. The train was half an hour late and the stretch of line involved was dead straight, so the train was travelling at 120 kilometres an hour. Its engine demolished about 100 metres of a lineside wall, was thrown into the air, and landed facing the way it had come. Behind it the leading four carriages of the 13 in the train were partially crushed into one another, the fifth went down the embankment and the sixth collided with the derailed coach of the train from Palermo.

Twelve passengers died in the accident and 72 were injured, 48 seriously. All but one of those who died had been on the *Conca d'Ora*.

A following train was stopped at Battipaglia. Upon news of the disaster being received railwaymen and carabinieri went to the scene from Battipaglia, firemen attended from Salerno and police and ambulances were sent from different towns. A newspaper report spoke of firemen working for hours with oxy-acetylene torches to free the dead and injured from the wreckage. The bodies of those who had died were temporarily lined up in a field while the injured were transported to hospitals in Battipaglia and Eboli by ambulances and private cars. The corpses of the dead buffaloes were quickly buried in a large grave excavated by a bulldozer. Traffic over the line was, of course, suspended until it was cleared but no passenger trains were cancelled, travellers being transhipped by railway busses between Battipaglia and Albanella.

Speaking after the disaster Assistant Conductor Ernesto Puntillio, of train No. 904, said, "It took more than two hours to be able to regain some calm. I have seen bloodied people screaming endlessly from pain, crying desperately, begging for help, calling out the names of relatives, of loved ones, while all the time the buffaloes, maddened by the noise, by the smell of blood and by the confusion, bellowed threateningly."

Assistant Driver Mario Ruono of the *Conca d'Oro* told a reporter, "We had just left

Battipaglia and were travelling at a good speed when we saw the red signal. We tried to pull up but it was useless. There was a deafening noise and our locomotive was thrown into a field. I managed to break a window with a pickaxe and climb out to safety. It was pitch dark all around and you couldn't see anything, just hear cries for help."

The sad scenes in the mortuary at Eboli, where the bodies of those who had lost their lives were taken, were described by a newspaper. Two distraught women arrived, supported by weeping relatives, and dressed all in black. They were Domenica and Marianna, the daughters of 76-year-old Vito Russotto, who had stayed with them in Milano for two months and had been on his way home to Cammarata. Another woman, Renza Reginella, from Tuscany, sat in a chair and murmured to herself, "The beasts took him from me." Her husband, Sebastiano, had died in the accident, along with his brother Ignazio. They had come from Prato, where they had been working, the previous evening, to visit their dying mother at Calatafimi. Each brother had left behind two young children. For others there was somewhat better news. A man in his sixties arrived, asking for his son. "He is not here," he was told. "Perhaps he is among the injured."

A judicial investigations were launched into the accident, accompanied by a separate inquiry by the Ferrovie dello Stato. Signalman Narducci, 45, was taken into custody after the disaster, for failing to take sufficient measures to warn the drivers of the two trains, as was the man who should have been supervising the buffaloes, herdsman or farmer, Raffaele Gaeta, aged 35. Narducci had apparently accepted the *Conca d'Oro* after having become aware that there were buffalo on the line, while train No. 904 had already been in the section. Described as 'a thin man, of medium height, with an alert look,' he was said to have been in severe shock for hours after the accident. The two men were charged with three different offences: causing a disaster (article 430 of the penal code – penalty five to 15 years' imprisonment); creating the conditions for such a disaster to occur (article 449 – penalty two to ten years' imprisonment); and manslaughter (article 589 – up to 12 years' imprisonment).

Gaeta said that he could not explain how the animals had ended up on the railway. He said that they were usually taken for milking at 04:00 and told a story of having had to leave them at 03:30 because one of them had kicked him in the leg. However an examination of him by a doctor, ordered by a magistrate, failed to reveal any sign of a leg injury.

Various theories were advanced to explain how the buffaloes had got onto the line. One was that they had been taken across the line to avoid a flooded ford. Gaeta himself suggested that an attempt had been made to steal animals but this seemed very unlikely as the thieves would have had to have taken the animals past San Nicola Varco station, brightly lit by neon lights, where the signalman could have seen them. Such livestock thefts were also unknown in the area. Another rumour was that there had been a nocturnal transport of buffaloes between fields via the railway to avoid the expense of using a lorry to take them via a longer road route. Yet another possibility was that the animals had been frightened by something. It was unusual for the buffaloes to stray from the pastures where they were feeding, though their enclosure was only separated from the railway by a low stone wall, in most places one and a half metres high but near the station only 20 centimetres high and at one point entirely missing. [1581] [1582]

Questions regarding the disaster were raised in Parliament, Socialist deputy Martuscelli asking Minister of Transport Scalfaro to reassure the public that the dangers to train users had not increased, especially in view of the recent introduction of trains travelling at over 200 kilometres an hour on the Roma – Napoli line. Deputy Canestrari called for the introduction of radio-telephones on trains, which he said would have prevented the disaster. The Minister of Transport responded that the railway was already undertaking various measures to make travelling safer, such as the extension of automatic signalling. Radio-telephones, which were useful in a country such as the United States, with its greater distances, were not considered to be so important in Italy. Experiments were in progress, however, on the repetition of lineside signals on locomotives. Deputies Canestrari and Montandi also drew attention to the abolition or upgrading of unattended level crossings and the need for adequate lineside fencing. The Minister of Transport replied that the provision of improved fencing along 16,000 kilometres of railway (the length of the national network) would entail enormous expense with scarcely appreciable advantages, while measures were already in place to improve the safety of unattended level crossings. [1583]

Early on Wednesday 15th November 1967 26 cars of a Louisville and Nashville Railroad freight train left the track at East Chattanooga, Tennessee, United States, due to a broken rail. There was a spillage of 10,000 gallons of sulphuric acid and 500 feet of track was ripped up. No one was hurt. [1584]

Thirty cars of an Illinois Terminal Railroad freight train were derailed at Springfield, Illinois, United States, on Christmas Day, Monday 25th December 1967. A quantity of vinyl chloride escaped and most of it ran into a small stream. No one was hurt and there was no evacuation but nearby roads were closed. [1585]

The year 1968 was less than a day old when an accident of major proportions in terms of the number of people evacuated and the damage caused occurred at Dunreith, Indiana, United States. At 9.30 p.m. on Monday 1st January westbound Pennsylvania Railroad freight train No. PR-11A (Extra 2210 West), consisting of five diesel units and 98 cars, was passing Dunreith, travelling at 42 miles an hour, when its 88th car was derailed by a broken rail and fouled the opposite running line. At the same time eastbound freight train No. SW-6 (Extra 2217 East), consisting of five diesel units and 106 cars, was passing in the opposite direction moving at 32 miles an hour. A collision occurred between the derailed car or cars of the westbound train and those of the eastbound train and then a general derailment. The rupturing of several cars of flammable chemicals resulted in a large fire, followed after 45 minutes by the violent explosion of a tank car loaded with ethylene oxide. All 236 residents of Dunreith were then evacuated. The accident was the

[1581] *Railway Accidents of Great Britain and Europe* by Ascanio Schneider and Armin Masé, pgs. 300-302

[1582] Wikipedia – Battipaglia railway accident

[1583] *La Stampa*, Torino, Italy, 10th and 11th November 1967

[1584] *Kingsport News*, Kingsport, Tennessee, United States, 16th November 1967

[1585] *The Post-Standard*, Syracuse, New York, United States, 26th December 1967 subject of a full report by the National Transportation Safety Board (NTSB). [1586]

According to the NTSB report the only (minor) injuries resulting from the accident were to three fireman and two policemen but it appears that at least one civilian, David Fairchild, of New Castle, Indiana, was also hurt. He told his story to a reporter: 'My wife and I were visiting my mother when the first explosion occurred. We ran to the front door to see what happened. We were about 75 or 100 feet from where the fire broke out. We knew something had happened to the train. The railroad cars were one on top of another. But I didn't know what was on fire. There were several explosions – you couldn't tell what was on fire. But the canning factory was burning right away.

'There was a semi-truck sitting crossways on the road like he was trying to get out. The highway was blocked right away. He backed the truck out and drove away.

'I got my mother and wife out right away. We ran to the car – it was parked in front. After I got them out, I went back. I was trying to get back to the house to get some medicine for mother. I was in the back of the house, trying to go in the back door when the big explosion occurred. A big ball of flame shot up in the air and spread. The fire just looked like it was 10 or 15 feet above me and all around me. It looked like an atomic explosion. The heat felt like a blowtorch on me. The heat was so great it actually scorched the leather jacket on my back. I was blistered on my head and hands. I tried to dive under the porch, but it was so hot I got up and ran for my car. It was about half a block away.

'I went to the hospital and they put some stuff on the blisters.' [1587]

The westbound train had left Columbus, Ohio, at 5.22 p.m. and passed Newman, 29 miles from Dunreith, at 8.54 p.m. The eastbound train had left Pine (Indianapolis) at 8.31 p.m. and passed Thorne, 32 miles from Dunreith, at 8.45 p.m. As the trains approached one another the engineer of train Extra 2217 East saw sparks and fire coming from a car of train Extra 2210 West and alerted its engineer to stop his train. Before the latter could do anything the brakes of his train went on in emergency, as did the brakes of the eastbound train following the collision and derailment of both. Fire erupted immediately after the derailment and engulfed a lineside cannery.

The crew of train Extra 2217 East called the fire service by radio at 9.35 p.m. while a railwayman from the caboose of train Extra 2210 West, after first attempting to use a lineside telephone, which had been damaged in the accident, then proceeded to the adjoining US Route 40 to stop road traffic and finding a State Patrolman warned him that there were tank cars involved in the fire, which presented a grave danger. Another railwayman went into the town to warn people to stay away because of the danger from hazardous materials. At 9.45 p.m. railroad officials requested the assistance of the chemical warfare unit at Fort Benjamin Harrison, Indianapolis, and breakdown trains were sent from Indianapolis, Richmond and Logansport, Indiana, and from Columbus, Ohio. Between midnight and 8 a.m. the local Sheriff's Department requested gas masks and airborne equipment to fight the fire but no gas masks could be found and the only airborne firefighting equipment was on the West Coast tackling forest fires.

The fire had begun in a tank car loaded with vinyl chloride, which had spilt from the damaged car and immediately ignited. Other cars containing acetone cyanohydrin

---

[1586] National Transportation Safety Board report

[1587] *Greensburg Daily News*, Greensburg, Indiana, United States, 2nd January 1968 (flammable and toxic), methyl methacrylate monomer (flammable) and ethylene oxide then became involved and the ethylene oxide car exploded at about 10.15 p.m., just as firemen and bystanders were beating a hasty retreat after warnings shouted by railwaymen and police. Firemen and police then evacuated the town. The explosion spread the fire and by 3.15 a.m. the canning factory, four migrant workers' homes, and three residences had been destroyed. The intensity of the fire prevented local firemen from getting close enough to fight it and matters were complicated by ignorance as to exactly what hazardous chemicals were involved, the cardboard "Dangerous" placards on five cars having been destroyed.

Clearance of the wreckage began at 5.15 a.m. on 2nd January but had to be discontinued at intervals due to the danger of further explosions or poisoning. The fire was finally extinguished by the afternoon of 2nd January, the line was cleared at 6 p.m. on 3rd January and Dunreith residents began returning home at 6.30 p.m. the same evening. A further problem was the leakage of some of the unburnt acetone cyanohydrin into Buck Creek, which had poisoned and killed farm animals drinking from it and contaminated householders' private water wells.

Damage to Dunreith property, including the tomato cannery, its major industry, was estimated at $500,000, and to railroad property, rolling stock and so on about the same amount.

After its usual very thorough analysis of the accident the NTSB came to 22 conclusions:

(1) Trains PR-11A and SW-6 were being operated in compliance with all applicable rules and regulations.

(2) A broken rail in a compromise joint in the north rail, 258 feet west of Dunreith Interlocking Station, caused the initial derailment of the trailing truck of AESX car 850, an empty tank car in westbound train PR-11A.

(3) The break in the rail was a progressive fracture originating at a bolt-hole.

(4) After the settlement of the crossties, the four-hole compromise joint did not resist vertical deflection sufficiently to prevent the development of the fracture in the rail. The "General Requirements of a Rail Joint," as found in the Manual of Recommended Practice of the American Railway Engineering Association and the Pennsylvania Railroad's "Specifications for Construction and Maintenance of Track," are intended to serve as guides for the economic maintenance of track. However, they are vague and non-specific as to the required performance of the joint in relation to the other factors of track condition. This precludes any judgment by the Board as to whether the four-hole compromise joint was adequate for the required performance in relation to the deteriorated condition of the track support. The Federal Government does not have regulatory authority over track conditions.

(5) An unknown factor in the compromise joint, combined with defects in track maintenance, led to the development of the bole-hole crack and broken rail.

(6) Electronic equipment for the detection of hidden flaws and defects in rail ends was used by the Pennsylvania Railroad but its use was not required on the joint where the broken rail occurred. The bolt-hole crack progressed undetected to the extent that a 5-inch piece of rail broke out.

(7) Visual track inspections that were made did not identify the track defects that required attention.

(8) Although rules of the Pennsylvania Railroad and recommendations of the American Railway Engineering Association apply to track maintenance and rail joint features of this case, these documents were ineffective as part of the railroad's system of risk management because of indefinite language and relief clauses which made such rules unenforceable by the railroad. Thus, no written regulation governing the features of track condition critical to this accident was constructively in effect.

(9) The separation of the truck from AESX car 850 led to the break in train PR-11A and the subsequent general derailment. If there had been some positive means of keeping the truck and car properly attached, possibly the train would have remained coupled and probably would have been brought to a halt without disastrous results after the derailment was reported by radio from Train SW-6. The radios were ineffective in preventing the wreck because of the rapid development of the derailment situation after the initial derailment.

(10) The derailment of train SW-6 was caused by a side collision with derailed cars in train PR-11A.

(11) The secondary derailment in train SW-6 was the result of slack action.

(12) The intense fire, explosion, and contamination of water resulted from the large volume of flammable and poisonous liquids which flowed rapidly out of the tanks after they were mechanically punctured in the collision and resulting derailment.

(13) The safety valve on the tank car of ethylene oxide operated for about 45 minutes but did not prevent the explosion.

(14) The scope of the fire and explosion and the technical complexity of the hazardous materials involved were beyond the normal capabilities of local fire departments. Specialized information, equipment, and experience were required to combat safely the fire and were not readily available.

(15) Firefighting personnel from military installations which were called in to assist could not effectively fight the fire because of lack of knowledge of the nature of the materials burning. The train consists, waybills, and notices to the crews did not furnish sufficient information to permit adequate handling of the hazardous materials in this wreck. The "Dangerous" placards and "Red Labels" were ineffective after the derailment because they were destroyed by fire.

(16) The use of large amounts of water contributed to the decomposition of the acetone cyanohydrin which drained cyanides into the ground water system. The ground water system drained into streams, resulting in the poisoning of farm animals, before the Indiana Board of Health could initiate preventive and corrective measures.

(17) No toxicity tests were performed before allowing personnel to work in the area contaminated by spilled and burned chemicals.

(18) The railroad clearing operation and the transfer of acetone cyanohydrin to tank trucks were not well coordinated by the railroad or the Bureau of Explosives, resulting in unnecessary hazard to the workmen and community.

(19) Federal regulations for the safe transportation of dangerous articles by rail were complied with, however, this did not serve to prevent the fire, explosion, and contamination of water.

(20) Title 49 CFR 174.506, which attempts to assign the handling of emergency matters related to the transportation of hazardous materials by rail for the protection of the public to the Bureau of Explosives of the Association of American Railroads, does not require that any report be made by the Bureau of Explosives concerning its handling of the emergencies. Although the Safety Board requested such a report of its action from the Bureau of Explosives in this case, no report was provided.

(21) The reporting of this accident to the Bureau of Explosives for handling of the emergency, as required for the protection of the public by 49 CFR 174.506, did not produce immediate attendance at the scene by any representative of the Bureau of Explosives until about 10 hours after the Bureau of Explosives representative was notified. The Federal regulation was intended to produce protection of the public by assigning the handling of the emergency to the Bureau of Explosives but the regulation proved ineffective in this respect.

(22) The Pennsylvania Railroad complied with the requirements of the Accident Reports Act. Because there were neither serious nor fatal injuries involved in the accident, regulations did not require a report to the Federal Railroad Administration until 30 days after the end of the month in which the accident occurred (January). In this case, only a monthly report was required and it was not due until 60 days after the occurrence.

Turning to the probable cause of the accident the NTSB stated:

The probable cause of the initial derailment in train PR-11A was the broken rail within the compromise joint where two different sizes of rail were joined.

A contributing causal factor was the inadequate track maintenance that left the joint unsupported and allowed the development of the break in the rail. This initial derailment and the design of the lift-off type of center-pin connection between the truck and body of AESX car 850, which allowed the truck to separate from the car under impacts of a simple derailment, led to the secondary collision and general derailment.

The collision and general derailment led to the intense fire, explosion, and contamination of water by the poisonous chemical. Contributing causal factors were, (a) the large volume of flammable and poison liquids in tank cars, which were located next to one another in the train, and flowed rapidly out of the tanks after they were mechanically punctured, (b) the shortage of information and lack of firefighting equipment necessary to extinguish promptly the fire around the ethylene oxide tank, (c) the destruction of the heat insulation of the tank, and (d) the failure of pressure relief valves to vent sufficiently to prevent the very rapid buildup of pressure consequent to uncontrolled heating of the ethylene oxide.

Finally the NTSB made ten recommendations:

(1) The Safety Board recommends that the Penn Central Transportation Company and the American Railway Engineering Association revise their track inspection and maintenance standards or recommended practices for track construction and maintenance so that they provide objective measures of conditions and definite

criteria for correction. Decisions as to adequacy of track conditions should not be solely dependent upon variable judgment or interpretation of individuals but should be made according to objective measurements required by written enforceable rules.

(2) The Board recommends that the American Railway Engineering Association and the Association of American Railroads initiate a research program to improve the present rail-joint design. The improved joint should be as strong and dependable as the rails to which applied and also prevent the development of rail defects and failures which now commonly occur in the joint area.

(3) The Board recommends that the Penn Central Transportation Company and the railroad industry in general employ to a greater degree the available rail-flaw detector equipment. It is further recommended that complete use be made of the available technical knowledge to insure the development of more dependable means of detecting rail defects within the joint areas with a greater degree of accuracy.

(4) The Board recommends that the Federal Railroad Administration include in its current study of an improved coupler design, the problem of keeping cars coupled and in line with the track and with each other after a derailment occurs. In order to attain an integrated organization of track and rolling stock features that would limit the after effects which can now follow a simple derailment, the Federal Railroad Administration should also study other related technical approaches to control interference with traffic on adjacent tracks and wayside structures during derailments, such as means of limiting the lateral excursion of wheels, and the separation of trucks from the cars.

(5) The Board recommends that the Department of Transportation, through its Assistant Secretary for Research and Technology, and the Federal Railroad Administration, in cooperation with the Association of American Railroads, study the problem of proper management of hazardous materials in train accidents and take appropriate action. This study should include the matter of whether it is feasible to provide vehicles that will resist mechanical puncturing, the problem of controlling the flow of flammable and poisonous liquids out of punctures in tank cars, and whether it is feasible and preferable to separate cars of hazardous materials with cars of inert materials. The problem of adequate capacity of safety valves should also be reviewed.

(6) The Board recommends that the Federal Railroad Administration review and correct the applicable regulations to provide markings at multiple locations which can survive a fire and which can be identified from a distance during fire and smoke (such as a large silhouette or see-through). Such markings should be legible to inform constructive actions by persons looking for them in cases of leakage, threat of fire, or during a fire. The problem of insufficient data relative to hazardous materials on the manifests and notices to the crew should be reviewed and appropriate action taken to establish a system that will provide those requiring it with the necessary information to initiate corrective action in emergency situations.

(7) The Board recommends that the Federal Railroad Administration review 49 CFR 174.506, which intended to protect the public against fire or explosion resulting from railroad accidents by the assignment of the handling of the emergency to the Bureau of Explosives of the Association of American Railroads. This regulation appears to place responsibility for public safety in the hands of a private organization

representing only one of the interests involved, and which may not be able to handle expeditiously emergencies which may develop. The Board is aware that the practice of delegating responsibility for hazardous materials regulations to private agencies is under study by the Department of Transportation.

(8) The Board recommends that the Department of Transportation study means of improving the training methods available to local fire departments so that they can upgrade their skills in their handling of emergencies created by the increasing transportation of hazardous materials. The problems of controlling such accidents are especially troublesome because of the daily introduction into commerce of numerous new kinds of hazardous materials. The Board believes that local emergency organizations cannot be expected to be conversant with necessary procedures to handle situations involving the many possible emergencies involving the transportation of hazardous materials unless some form of assistance in training is provided such as a model type training course.

(9) The Board recommends that the Federal Railroad Administration amend its requirements for the reporting by railroads of accidents to include the immediate reporting of any accident involving cars containing hazardous materials which constitutes a current or potential hazard to the carrier, the passengers, its personnel, or the local environment.

(10) The Board recommends that the Secretary of Transportation undertake a study of the feasibility of establishing a National Hazardous Materials Advisory Data Center. It is envisioned that such a national data center, through the use of computer-stored data would be able to furnish emergency information on flammable, explosive, highly reactive, and poisonous substances, and locally available sources of expertise and specialized emergency equipment on a 24-hour-a-day basis. The Board is of the opinion that such data center is now necessary to aid Federal, State, and local authorities faced with the varied and complex problems involved in combating actual or potential disasters involving hazardous materials.

There are working programs throughout the country which could be used as pilot models for such a center. Some of these are the Coast Guard's programs in their Hazardous Materials Division in Washington, D. C., and in the various districts and the "poison control centers" in the various States.

At about noon on Sunday 21st January 1968 seven of the 123 cars of a Western Maryland Railway freight train became derailed at Chambersburg, Pennsylvania, United States. A car loaded with zinc fell down an embankment into the Conococheague Creek but of more concern was leakage from a tank car carrying 10,064 gallons of ethylbenzene, a highly flammable, potentially explosive, somewhat toxic, and carcinogenic chemical used in the manufacture of polystyrene and for several other purposes. A large number of firemen went to the scene and nearby streets were cordoned off while three volunteer firemen plugged a hole in the tank car at great personal risk and a fireman warned newspeople not to use flash bulbs to take photographs, saying that a single spark could cause an explosion. The remaining chemical was then transferred to two road tankers, an operation that could not begin until large amounts of foam concentrate had been obtained, which took until 1.45 a.m. the following morning. About

1,000 gallons of ethylbenzene was lost and formed a pool by the lineside. Several families evacuated the area until the danger was over. No cause was given for the derailment. [1588]

A 66-car Atchison Topeka and Santa Fe Railway freight train was derailed at about 4.30 a.m. on Friday 8th March 1968 near Miles, Texas, United States, possibly because something hanging from one of the cars had caught in a crosstie. Brakeman C.F. England said afterwards, "The train went into emergency, and we thought we had broken in two. Sure enough we had." At first it had seemed a minor mishap but he and Brakeman J.A. McClaskey went towards the rear of the train and "when we got back there they were scattered all over the country." Only three of the 66 cars were salvageable and one of them had spilt liquid sulphur into a ditch while other cars caught fire. Neither the brakemen, Conductor Alex Bradshaw nor Engineer T.F. Alexander was injured. [1589]

Thirty-six of the 65 cars of an Erie-Lackawanna Railroad freight train from Marion, Ohio, to Meadville, Pennsylvania, left the track at Akron, Ohio, United States, on the afternoon of Sunday 10th March 1968 from an undetermined cause. Leakage of acetic acid from three tank cars hampered the clean-up operation until firemen hosed away the acid. No one was hurt and there was no evacuation. [1590]

The derailment of a Chicago and North Western Railway freight train at Kelley, Iowa, United States, at about 3 a.m. on Sunday 21st April 1968 led to the escape of a large quantity of ammonia from a damaged tank car and the evacuation of 240 residents. Fortunately the wind blew the gas fumes away from most of the residences but two people suffered from the effect of inhaling fumes and had to be taken to hospital. A bull, a cat, a dog, and a number of birds were fatally poisoned. Damage to another car carrying diesel oil led to a fire that spread to cars of grain, flour and magazines and a fireman suffered first- and second-degree burns. Residents began to return home at 6 a.m. after the gas had dispersed; elementary school was cancelled on Monday. The line was reopened about 6.30 p.m. on Monday. [1591]

On Thursday 5th September 1968 an eastbound 126-car Union Pacific Railroad freight train became derailed, 43 cars towards its rear leaving the rails at Thayer Junction, Wyoming, United States. A tank car loaded with petrol burst open and a fire erupted and spread to nine other cars. Firemen from Green River and Rawlins prevented it spreading further. The *City of Los Angeles* and *City of St Louis* streamlined passenger trains were held at Cheyenne while cranes, also from Green River and Rawlins, built a shoo-fly diversion around the wreckage. No one was hurt. [1592] The derailment was possibly due to a 'hot box.' [1593]

When 33 cars of a Norfolk and Western Railway freight train became derailed at South Whitley, Indiana, United States, on Tuesday 10th September 1968 and a tank car lost its contents, there were fears at first that either ammonia gas or some potentially explosive chemical were involved. A total of 1,300 residents were evacuated before it

---

[1588] *The Record Herald*, Waynesboro, Pennsylvania, United States, 22nd January 1968
[1589] *The Abilene Reporter-News*, Abilene, Texas, United States, 9th March 1963
[1590] *The Republican-Courier*, Findlay, Ohio, United States, 11th March 1968
[1591] *Ames Daily Tribune*, Ames, Iowa, United States, 22nd and 23rd April 1968
[1592] *The Billings Gazette*, Billings, Montana, United States, 6th September 1968
[1593] *The Herald Journal*, Logan, Utah, United States, 6th September 1968

was found that rather less hazardous anhydrous acetic acid had escaped. Firemen washed away the spillage and the residents returned home after a few hours. [1594] [1595] [1596]

A broken rail caused the derailment of four diesel units and 41 of the 135 cars of northbound Kansas City Southern Railway freight train No. 42, driven by Engineer Woodrow Jones, near Dalby, Missouri, United States, at 2.30 p.m. on Wednesday 13th November 1968. A tank car of ammonia was ruptured and fire broke out involving another tank car containing liquid oxygen. Firemen from Neosho, Saginaw and Diamond fought the fire and played water from Shoal Creek on the leaking ammonia. There were no residences nearby. [1597] [1598] It was expected to take several days to clear the wreck, meanwhile trains were diverted via the St Louis-San Francisco Railway. [1599]

Southern Railway train No. 154 was derailed at Laurel, Mississippi, United States, at about 4.15 a.m. on Saturday 25th January 1969. It consisted of four diesel units, 139 cars and a caboose and was proceeding at about 30 miles an hour when a wheel of the 62nd car broke as it was passing over the crossing with the Gulf Mobile and Ohio Railroad (GMO). This and the 14 following tank cars loaded with liquefied petroleum gas (LPG) were derailed, a large quantity of LPG escaped from the damage cars, fire immediately broke out and a series of explosions followed over the next 40 minutes, setting fire to nearby dwellings. In addition pieces of tank cars were blown up 1,600 feet, damaging and setting fire to more buildings. [1600]

The engine crew pulled away the underailed front cars of the train, which included a tank car of hydrocyanic acid, the underailed rear cars were uncoupled at considerable personal by Switchman Frank Wells Jr., who was off duty at the time, and a road foreman of engines together with the rear end crewmembers then pulled them away with another locomotive. [1601]

A large number of residents had to be evacuated. No less than 54 residences and several businesses were destroyed or badly damaged, 1,350 residences, six schools and five churches were damaged to a lesser degree and 35 people were badly hurt, two of whom later died. An unspecified number of people were slightly hurt.

The fire had been brought under control by Sunday morning, the residents were able to return at about noon on Sunday and the line was reopened about 5.30 p.m. that day. Total damage was put at $3,000,000.

The National Transportation Safety Board (NTSB) investigated the accident.

Train No. 154 had started from Oliver Yard, New Orleans, Louisiana, at 12.10 a.m., picking up 26 LPG tank cars at Dragon, Mississippi, then had made a normal run to

---

[1594] *The Anderson Herald*, Anderson, Indiana, United States, 11th September 1968
[1595] *Tipton Daily Tribune*, Tipton, Indiana, United States, 11th September 1968
[1596] *The Pharos Tribune* and Press, Logansport, Indiana, United States, 11th September 1968
[1597] *The Kansas City Times*, Kansas City, Missouri, United States, 14th November 1968
[1598] *The Joplin News Herald*, Joplin, Missouri, United States, 14th November 1968
[1599] *The Chillicothe Constitution-Tribune*, Chillicothe, Missouri, United States, 14th November 1968
[1600] National Transportation Safety Board report
[1601] *Delta Democrat Times*, Greenville, Mississippi, United States, 27th January 1969

Laurel. The operator at Laurel had watched the train as it passed over the crossing with the GMO and immediately after the 61st car had passed he heard a loud noise and saw sparks and a fire. Unable to signal to the caboose he then went into his tower to notify the Hattiesburg dispatcher to stop the train but before he could do so the derailment had occurred and all his telephones were put out of action.

The NTSB reached 28 detailed conclusions regarding the accident as well as general conclusions. The detailed conclusions were:

(1) The speed of the train at the time of the derailment was between 28 and 35 miles per hour. There was no device on the locomotive to record the speed of the train.

(2) The variations in the track condition in the accident area were not abnormal for track restricted to speeds of 30 miles per hour. There was no evidence to indicate that the track in general was unsafe for this speed. The Southern Railway follows the recommended practices for track maintenance of the American Railway Engineering Association which do not provide a specific objective measure of track cross level. Decisions as to adequacy of track condition on the Southern Railway are dependent upon the judgment and interpretation of track standards by individuals. The Board has noted in the report of the accident at Dunreith, Indiana, January 1, 1968, that the standards did not provide objective measures of condition.

(3) Telephones in the tower and hand signals are not adequate means of communications to convey information quickly to the crew that an operator has noted an unsafe condition in the train.

(4) The L-3 wheel of the 62nd car, ACSX 932003, broke as it was traversing the G. M. & O. crossing and resulted in the initial derailment at a point about 256 feet north of the center line of the G. M. &O. crossing.

(5) The gross weight on rail for loaded tank car ACSX 932003 did not exceed the maximum allowable load under A.A.R. Interchange Rules.

(6) Although the Laurel Fire Department and fire departments from neighboring communities responded rapidly and performed in an excellent manner, they could not control the massive LP-gas fire and explosions.

(7) In this disaster, telephones alone proved inadequate for communications. The use of radio and television was necessary to mobilize forces required to protect the public.

(8) Local volunteer groups and other public and commercial groups responded to the demands of the situation and did an effective job in meeting the immediate needs of the displaced and injured persons.

(9) Although venting the intact tank in the vapor space by explosives was successful, experts disagree as to the safety of using this procedure.

(10) The tread-worn-hollow condition of the L-3 wheel did not require it to be condemned under the Rules of Interchange of the Association of American Railroads, which are based upon measurement of flange height. There is no A. A. R. gauge to indicate a tread-worn-hollow contour for wrought steel freight car wheels, although the need for such a gauge has been recognized by A.A R. for many years.

(11) Though not condemnable by high-flange condemning limits, wheels with substantial hollow-worn treads may produce, under certain normal circumstances, significant

stresses which were not considered in the development of the wear-limit rules.

(12) The A. A. R. specification for finish is not specific as to surface finish because the words "workmanlike finish" are undefined and their interpretation merely a matter of opinion. This specification does not function as a safety control.

(13) The chemical composition of the steel of the L-3 wheel met the requirements of the A. A. R. specification.

(14) The corrective machining of the plate of the L-3 wheel resulted in a torn surface with resulting discontinuity in the area of the break.

(15) The discontinuities in the machining of the plate of the L-3 wheel made the wheel more subject to impact fracture in the machined area at low metal temperatures similar to air temperatures existing on the day of the accident.

(16) The first fracture of the L-3 wheel originated near the hub in the area of rough machining on the back of the plate.

(17) The L-3 wheel failed in the back plate near the hub because of suddenly applied lateral forces. The suddenly applied lateral forces came from the change in stress when the tread-worn-hollow wheel traversed the 35° 20' crossing frog, and from the pressure on the back of the flange by the guardrail because of the depressed east rail. The rough machining of the plate provided a notch effect and a stress concentration which reduced the ability of the wheel to withstand impact. The FRA reported the failure of a wheel for a similar reason in the accident at Cold Springs, Ohio, January 8, 1968, but has no authority to require any changes.

(18) Although the A. A. R. became aware, through inspection, that Armco's Butler plant over a period of years was producing wheels whose finish, in A. A. R.'s opinion, did not comply with the specifications, similar wheels were sold and put into service on A. A. R. member roads, some of which had no knowledge that such wheels had been criticized. A large number of these wheels may be in use.

(19) Lacking an objective standard for surface finish, there is no current basis by which to inspect wheels and determine which are too rough. Essentially all wrought steel freight car wheels are subject to rough finish and potential breakage because of the nonspecific specifications which have been in effect for many years.

(20) Title 49 CFR 174. 506 delegates to the Bureau of Explosives of A. A. R. the responsibility for protecting the public from the dangers of hazardous materials accidents but fails to specify the Bureau of Explosives' authority or relationship with the carriers or Federal agencies regarding this function.

(21) All tanks involved in the wreck were constructed in compliance with DOT and A. A. R. specifications.

(22) At present the A. A. R. Committee on Tank Cars develops all new proposed tank car tank specifications for handling hazardous materials and submits them to the Department of Transportation which has the statutory responsibility for issuing and enforcing regulations regarding these specifications. Relative to this accident, recommendations by the A. A. R. Committee on Tank Cars to construct a tank car without external heat insulation to transport LP-gas, and a car without continuous center sills, were approved by the Interstate Commerce Commission without change as proposed by the A. A. R.

(23) Much of the initial mechanical damage to the tanks was caused by couplers of cars which became uncoupled in the derailment. The use of interlocking couplers on these cars would have increased the probability of the couplers remaining in line after the derailment and decreased the probability of puncture damage.

(24) According to the Certificates of Construction, the tanks involved in the wreck were constructed of the material meeting the specifications. Test results indicate the need for additional and continued research and tests to determine the best combination of design and materials to withstand the effects of derailments.

(25) Tests and service experience indicate that the design of the 30, 000-gallon tank car without a continuous center sill is adequate for normal conditions of transportation. However, research or analysis to determine the effect of well-known classes of accidents on this kind of tank loaded with flammable compressed gas have not been accomplished.

(26) The flight of two of the tanks over long distances resulted from the exhausting of the rapidly expanding and burning gas from the open end of the ruptured tank in which the liquid gas served as an energy reservoir, and the open ruptured end approximated the functions of a rocket nozzle. The flight of these tanks was thus rocket-propelled.

(27) The investigation did not develop sufficient evidence to determine conclusively the full scope of causes of the numerous tank ruptures. However, it is evident that mechanical damage to the tanks reduced substantially the bursting strength of the tank shells in a number of cases. Intense heat on the metal around the vapor space seriously weakened and thinned the metal in some cases. Under these conditions, the abnormal pressure produced by overheating the vapor space, or the sudden increase in pressure when large volumes of LP-gas flashed to vapor, could have caused some of the tank ruptures.

(28) The basis upon which safety relief valves are sized for undamaged tanks has been proven to be sound by research and experience; however, in actual accidents, safety valves sized in this manner do not account for changes in the strength of tanks due to structural damage or localized heating of tanks. Such safety valves did not have sufficient venting capacity to accommodate the physical damage or fire-caused weakness which the tanks received in this accident.

The general conclusions were:

'This accident illustrates the consequences which may ensue when, in the carriage of large quantities of hazardous materials through populated areas, supposedly effective safety controls do not work. Many of the failures of safety controls are attributable to ineffective planning design, and management of safety controls involving Government and private industry.

'Safety controls of private industry were involved in the factors surrounding the initial derailment. These controls relied upon voluntary specifications, voluntary communications between agencies, and implementation by economic self-interest of an informed purchaser. Some specifications involved in this phase of the accident were specific, were followed and were not causally involved. However, the controls over the most determinative factors, namely, the specifications for wheel surface finish and wheel tread wear limits failed to be specific. A method of numerical specification of surface finish was available but had not been employed. The gauging method for tread wear had been determined to be unsatisfactory but the appropriate method, already in use on similar wheels, had never been specified.

'When an attempt was made to enforce the nonspecific surface finish specification by an opinion judgment of an inspector, the attempt was ineffective. The approach made to the manufacturer as a voluntary recommendation did not produce the desired changes and no communication was effected with users of the wheels, or with members of the trade association whose economic interest might have prevented the wheels from going into use.

'The slight out-of-level condition of the crossing, a routinely measured and easily specified matter which might have been included in the industry self-regulatory recommendations of the American Railway Engineering Association, is not included in that industry document. As pointed out by the Board in the report of the accident at Dunreith, Indiana, that AREA document cannot be a basis for any forceful industry mechanism of enforcement.

'Thus, in this case, three different forms of safety control involved in the wheel breakage were inadequate as to engineering specificity, although the state of the art allowed specificity, and two of the controls (cross level of track and surface finish of wheel) were unenforceable within the industry pattern of self-regulation, even if they had been adequately specified.

'The writing of these safety-controlling documents and the arrangements for their application involved a number of professional engineering task committees and review groups of the industry whose work output was required to be technically competent and properly organized into a safety control structure in order to be effective. The documents produced under committee organization, however, must be judged incompetent as pieces of professional engineering work because they did not employ the available necessary engineering specifications. Also, the fact that these documents could not be enforced further raises the question whether reliance upon these voluntary methods to control safety of the public should be continued.

'The Safety Board has noted other instances of vagueness, non-specificity, and loose control in the self-regulated areas of railroad safety in its reports of the New York Central Railroad collision which occurred May 22, 1967, and the Pennsylvania Railroad accident at Dunreith, Indiana, January 1, 1968.

'The second phase of this accident involved the large-scale spread of fire, rupture of tanks, and the extension of fire over long distances through rocket propulsion of the ruptured tanks. The field of tank car safety has long been a statutory Federal responsibility – first, under the Interstate Commerce Commission, now, under the Department of Transportation – as enacted under the Transportation of Explosives and Other Dangerous Articles Law (sections 831-835 of Title 18, United States Code, as amended). The law allows, but does not require, the assistance of carriers and shipper associations in the formulation of regulations. In practice, virtually all the regulations relating to tank cars originate in the Committee on Tank Cars of the Association of American Railroads, and the specifications of that organization are cited in regulations to be the controlling specifications for certain purposes not detailed by the regulations.

'Two of the changes in tank car regulations related to this accident (a preceding abandonment of requirements for external heat insulation, and approval of new design high-capacity tank cars without a continuous center sill) were originated by the

Committee on Tank Cars of the A. A. R and adopted as Federal regulations without change. The abandonment of heat insulation is related to the development of hot spots and metal thinning in a fire which can produce explosive rupture. The high-capacity tank cars without a continuous center sill have shown the phenomenon, new to railroads, of tanks which become rockets when a fire occurs.

'The responsibility of the Federal Government, when such regulation changes are proposed, should extend to the determination of possible dangerous side effects produced, regardless of the supposed competence of the source of the proposal. As the Board has recommended, these determinations should include the effects produced in full-scale tests which simulate accidents. However, neither the Interstate Commerce Commission nor the Department of Transportation has had sufficient technical capability to make such independent technical tests or analyses to guide regulations, nor have they ever had the necessary expertise to evaluate independently hazardous materials regulations in general. In effect the tests were made at Laurel, Mississippi. The observations of new phenomena and recommendations, which could have come far earlier, have now been developed as investigative findings of a major accident. This procedure, of course, involves far higher cost to the public not only in dollars, but in death, disability, and suffering.

'The costly results of this accident and the weakness of preceding efforts at safety control by both Government and private industry indicate that regulatory authority is necessary, but authority alone is not enough. It is also crucial that there be adequate fund support for research and development in support of regulations and sufficient budgeted technical capability in Government to insure that all regulatory actions will improve safety and none will, through ignorance, approve new hazards.'

Regarding the cause of the accident the NTSB stated:

'The cause of the initial derailment in Southern train 154 was the broken L-3 wheel on the 62nd car in the train.

'Contributing causal factors were the rough machining on the plate of the wheel, the tread-worn-hollow condition of the wheel and the sudden lateral loading on the wheel as it crossed the G. M. &O. crossing. The sudden lateral loading was caused by the tread-worn-hollow condition with an unknown degree of contribution of the slight out-of-level condition of the crossing under load.

'The cause of part of the spread of the fire beyond the immediate site was the projection of large parts of the tank cars long distances by rocket propulsion as the liquefied gas changed to vapor and was expelled through the ruptured open ends of the tanks. A contributing causal factor to this jet propulsion effect was the unhindered propagation of circumferential cracks in the rupture of two of the tanks which configured the open end of the tanks in a manner similar to that of a wide rocket nozzle.

'Contributing causal factors to the fire and explosions at the site were the large volume of propane released and set fire by the punctured tanks, tank shells weakened by mechanical damage and abnormal heat, and the absence of a pressure-relief system that would provide adequate safety release under such conditions.'

Finally the NTSB made 12 recommendations:

(1) The Safety Board recommends that the Federal Railroad Administration take the necessary steps to impose regulations requiring all mainline trains to be equipped with devices to record the speed of trains.

(2) The Safety Board reiterates the recommendation made in its Dunreith report issued December 18, 1968, that " ... the American Railway Engineering Association revise their track inspection and maintenance standards or recommended practices for track construction and maintenance so that they provide objective measures of conditions and definite criteria for correction."

(3) The Safety Board recommends that employees required by carriers to observe passing trains for defects be provided with means of rapid direct communications with personnel on the train.

(4) The Safety Board reiterates the recommendation made in its report covering the derailment and collision of Pennsylvania Railroad train PR-11A and SW-6 in Dunreith, Indiana, on January 1, 1968, "... that the Department of Transportation study means of improving the training methods available to local fire departments so that they can upgrade their skills in their handling of emergencies created by the increasing transportation of hazardous materials. The problems of controlling such accidents are especially troublesome because of the daily introduction into commerce of numerous new kinds of hazardous materials. The Board believes that local emergency organizations cannot be expected to be conversant with necessary procedures to handle situations involving the many possible emergencies involving the transportation of hazardous materials unless some form of assistance in training is provided, such as a model type training course."

(5) The Safety Board recommends that the Association of American Railroads and the American Short Line Railroad Association develop plans that will result in the fire chief of each community through which the track of a member road passes knowing where immediate information can be obtained, describing the location and characteristics of all hazardous materials in any train involved in a train accident that affects a community. This recommendation can be accomplished in a relatively short time regardless of the level of training which may be achieved later by fire departments.

(6) The Safety Board recommends that the A. A. R. make a study of the stresses developed in freight car wheels with hollow-worn treads while moving over frogs, switches, and crossings. If increased impact stresses are being developed as a result of the wear and the stresses under the worst possible combination of dimensional and material variations approach the design stress, consideration should be given to changing the wear limits or the wheel design. Further, if new wear limits are determined, beyond-limit wheels should be removed from service as quickly as practicable.

(7) The Safety Board recommends that the A. A. R. conduct physical tests on specimens of wheel steel having a range of surface finishes to determine the best surface finish for a wrought steel wheel and, further, that the resultant surface finish be incorporated in the specification for wheel finish as a specified surface texture measured in microinches.

(8) The Safety Board recommends that the American Railway Engineering Association study the design of railroad crossings to produce a crossing design that will lessen impact to wheels and require less maintenance.

(9) The Safety Board recommends that the A. A. R. take the necessary steps to give proper notice to the purchases of wheels when inadequacies in manufacturing practices are found, thereby improving enforcement of the specifications to the degree possible under industry self-regulation.

(10) The Safety Board recommends that the A. A. R. review the function of the Bureau of Explosives regarding its performance in protecting the public from danger resulting from railroad accidents involving hazardous materials and take the necessary action to develop an effective, cooperative program with the carriers to accomplish the intended purpose of the responsibility delegated to the Bureau of Explosives by Title 49, Code of Federal Regulations, Section 174.506. The Board endorses the FRA's proposed amendment of the regulations which will provide that reports of incidents and accidents involving hazardous materials presently made to the Bureau of Explosives by rail carriers will also be filed with the FRA.

(11) The Safety Board reiterates and emphasizes the recommendation made in its report of the railroad accident which occurred on the Pennsylvania Railroad at Dunreith, Indiana, on January 1, 1968, which reads as follows " ... that the Federal Railroad Administration include in its current study of an improved coupler design, the problem of keeping cars coupled and in line with the track and with each other after a derailment occurs. In order to attain an integrated organization of track and rolling stock features that could limit the aftereffects which can now follow a simple derailment, the Federal Railroad Administration should also study related technical approaches to control interference with traffic on adjacent tracks and wayside structures during derailments, such as means of limiting the lateral excursion of wheels, and separation of trucks from cars."

(12) The Safety Board recommends that the Department of Transportation develop a cooperative program with the A. A. R., manufacturers of tank cars, and producers and shippers of hazardous materials aimed at determining a full range of technical improvements for railroad transportation of liquefied petroleum gas and other hazardous liquids. This program should include a comprehensive study of the causes of the tank rocketing phenomenon, causes and weakening effects of accident damages to tanks, and means of guarding against explosive ruptures of tanks under the types of conditions actually encountered in accidents. The study should be supplemented by engineering development work and full-scale testing under conditions known to exist in service. The Board also recommends that the FRA develop and impose suitable regulations to correct any identified deficiencies. The Board further recommends that regulations for hazardous materials tank cars require, in all cases, a demonstration of satisfactory performance under test conditions which reflect the full scope of accident conditions known to be encountered in service. [1602]

At about 6.30 a.m. on Tuesday 18th February 1969 Chicago Burlington and Quincy Railroad freight train No. 64 was entering Crete, Nebraska, United States, moving at

[1602] National Transportation Safety Board report

about 52 miles an hour, when it became derailed. The derailed cars stuck cars standing on a siding south of the main line and the cars of train No. 824 standing on a line north of the main line. A tank car in train No. 824 was shattered and 29,200 gallons of ammonia were released from it and formed a gas cloud that blanketed the area for a considerable time. Three transients on train No. 64 were killed in the derailment, six people in Crete were killed and 53 were injured due to exposure to the ammonia. The accident was investigated by the National Transportation Safety Board. [1603]

A newspaper told the stories of some of the Crete residents who had survived the accident: "I opened the front door and I couldn't see anything. The gas just knocked me backwards." Those were the words of Mrs Mamie Littler, 60, of Crete, one of hundreds of Crete residents forced to evacuate their homes early Tuesday when ammonia fumes, loosed by a train derailment and explosion, spread over part of the city. Rescuers ran from door to door after the explosion, warning residents to flee. Many, like Mrs Littler, had already noticed the smell. Mrs Littler was helped from the area by national guardsmen. But she didn't leave empty handed. She brought her two chihuahua dogs, Red and Poddy, along. "I would rather have stayed and died than left them," she said.

'Les Parks, a volunteer fireman, said he was on a rescue unit with a pickup truck and four other men. He said the victims were in a state of shock. They were not saying too much, just 'Help us, Help us.' "Most of them were bleeding from the mouth and nose. For some of them we didn't have enough masks to go around by far. Guys were walking ahead of the truck. We just were kind of feeling our way as we drove down the street. We tried to go to each house to check if the people got out."

'"It was just like fog except thicker, said John Hoesche, one of those whose home was attacked by the heavy, poisonous fumes. Although many persons tried to walk out of the area with wet towels wrapped around their faces, most gave it up, and waited for evacuation by rescue workers. Many residents outside the area hit by the fumes volunteered their services and automobiles to shuttle persons out of the area. Cars became almost impossible to use near the derailment because visibility dropped to zero.

"We left the house with towels wrapped around our faces," Mrs Raymond Foreman said. "But we had to give it up after we got in the car. We couldn't see." Her family returned to their home and blockaded themselves in a "newer section of the house, with better insulation. But the ammonia began to seep in here too. Finally some rescue workers came by and got us," Mrs Foreman said, "and we got out." [1604]

Train No. 824, consisting of a diesel engine and 49 cars, had arrived at Crete at 2 a.m. from the Wymore branch and had subsequently left 11 box cars on the south siding.

Train No. 64, operating as No. 624 East, consisted of seven diesel units, 92 cars and three cabooses. Approaching Crete it had been travelling at 60 miles an hour. Speed was then reduced to about 47 miles an hour but crept up to about 52 miles an hour on the falling gradient nearer the station. There was no appreciable slack action. The first warning that its engineer had of any problem was the brakes coming on in emergency. He remained unaware of the derailment until informed by Conductor Frank Coffman by radio. The conductor and Flagman Chester Weber had fled the caboose when they had seen the gas cloud and smelt the ammonia. Both were among those hospitalised and were

---

[1603] Ibid.

[1604] *The Lincoln Star*, Lincoln, Nebraska, United States, 19th February 1969 seriously injured.

The NTSB reached 15 detailed conclusions regarding the accident and some general conclusions. The detailed conclusions were:

(1) The track was not maintained to standards normally acceptable for freight train operation at 50 m.p.h. The irregularities in the track contributed to the development of undue lateral forces on the rails.

(2) The guardrail protecting the frog of the turnout met specifications, but it was not designed to withstand impacts with the derailed wheels of the cars.

(3) The uncentered geometry of the trucks on the U-28C locomotive unit had no significant detectable effect on the lateral forces produced by the trucks in the circumstances of the accident.

(4) While No. 64 was moving on the descending grade approaching Crete, the control of the speed by braking was not performed in the manner prescribed by the carrier.

(5) The use of the independent brake on the locomotive, in the manner described by the engineer, was shown by calculations to have produced only about 10 percent as much lateral force on the rails as was produced by track irregularities.

(6) Train No. 64 slightly exceeded the maximum authorized speed as it approached the accident area, but the excess speed was of little significance.

(7) The breaking of the coupler yoke on the west end of B&M 77924, the 14th car, was a result of the derailment, not a cause.

(8) Damage to the 14th car, (B&M 77924), the 16th car (GATX 1827), and the 18th car (SP 209451) was the result of the falling of the follower block of the 14th car to the track structure when the coupler yoke broke because of forces resulting from the derailment.

(9) Most fractures of the tank were of a brittle nature. At the ambient temperature of +4° F., the steel plates of the tank car were far below their nil-ductility transition temperatures of 75° to 80° F.

(10) The original fracture occurred in a defective weld in the tank head where there was a gap between two plates. It could not be determined by the Board whether a complete weld would have prevented the fracture.

(11) The steel plates of the tank car met the specification required by DOT 112A340W. These standards do not, however, include microstructural or general toughness (e.g., NDTT) specifications, and thus did not serve to control the low temperature brittleness.

(12) The welds between the tank car shell plates were found to be of satisfactory quality and were not causally involved with the propagation of the fractures.

(13) Tank cars specified by Federal regulations were used for the loading of the anhydrous ammonia and the loading procedures had no influence upon the failure of the tank.

(14) All persons directly exposed to the cloud of ammonia gas received either fatal or serious chemical injury to the respiratory tract; whereas people within the area of the

gas concentration who remained in their homes or who had adequate protection, escaped with either no injuries or minor injuries.

(15) The citizens of Crete were able to cope with the situation because they had developed adequate plans for emergency procedures before the accident occurred.

The general conclusions were:

'Federal responsibilities are involved in two major areas of this accident. The shattering of about one-half of the tank, which produced rapid vaporization of the ammonia, is a Federal responsibility in that brittleness is now potentially controllable but was not controlled by the specification for DOT 112A340W tank cars. Control of track irregularities and of the integrity of track construction was not subject to Federal regulation at the time of the accident but is now subject to such regulation under the Federal Railroad Safety Act of 1970.

'The brittleness of the tank car material and the extent of the shattering of the tank in cold weather were surprising findings of great significance for potential hazards. This finding of brittle failure by the AAR was reported to the Federal Railroad Administration by the Association of American Railroads in July 1969. The fact that the deformation which resulted in shattering was not sufficient to have produced a puncture at higher temperatures means that the ability of this tank car to resist accident damage was compromised at the +4° F. temperature to a serious degree. Furthermore, any other tank car employing this same type of steel and metallurgical treatment would be seriously compromised at that low temperature.

'No tank cars of the DOT 112A340W class assigned to hauling anhydrous ammonia or liquefied petroleum gas (LPG) have been subjected to control of low-temperature brittleness. This is not the same as saying that all tank cars of this type are subject to low-temperature brittleness. Some of the cars of this class may not exhibit low-temperature brittleness, but if they do not, it would be a result of some significant difference in the metallurgy of the steel employed, as compared to this car. Thus all cars of specification DOT 112A340W assigned in this service must be considered under suspicion of having low-temperature brittleness. These tank cars could potentially fracture in other accidents in the same manner as occurred at Crete, releasing ammonia or other hazardous materials in the same manner.

'As of June 1969, there were more than 11,000 cars of this type in service, most of them carrying hazardous materials. Most of these cars are used to ship liquefied petroleum gas. The result of an accident producing a similar large scale failure with LPG in the tank would be sudden release of more than 30,000 gallons (liquid) of flammable gas, probably with fire and rocketing of large tanks similar to the events which have occurred at Laurel, Mississippi, at Crescent City, Illinois, and at other localities. Thus, the implication of the brittleness found in this accident is that, in low-temperature conditions, the consequences of accidents involving some large LPG tank cars are more likely to be catastrophic than in warm weather. Although the very cold-weather accident occurred to a car carrying anhydrous ammonia, a conclusion may properly be drawn that the risk of transporting similar cars carrying LPG is also increased at very low temperature.

'In the report of the LPG tank car accident at Laurel, Mississippi, the Safety Board noted that full-scale testing of this type of tank car by fire had not been accomplished prior to approval of the car and that the full-scale testing which disclosed the rocketing phenomenon had been, in effect, performed at Laurel, Mississippi, at very high cost to the public. The accident at Crete was, in effect, a low-temperature crash test which disclosed the need for control of low-temperature brittleness of tank materials. This test was also performed at very high human and economic cost. Both of these tests were performed long after the tank cars had gone into service, and after the pattern of rates and other economics of the use of these cars had been fixed. Under these circumstances, the correction of the existing cars may now be too costly to accomplish. There is also a danger that changes to cars not yet built, which would have seemed reasonable before the economic pattern was established and it may be regarded as costly and profit-reducing, and therefore questionable improvements rather than corrections of an overlooked matter.

'The conclusion is inescapable that obtaining knowledge of tank car hazards by analysis and testing, is far less costly to the public than "testing" by service and would produce more practical corrections. The omission of thorough testing is not only false economy, but an injustice to persons living along the tracks. The "testing" by service use of the Specification 112A340W or 112A400W tanks revealed hazards which in three accidents resulted in 11 fatalities, 152 injuries, including many disfiguring burns, and approximately $5,300,000 in property damage. There is a history of budgetary requests by the Interstate Commerce Commission for research and development funds which might have shown the need for such testing, and testing by manufacturers could have been required by regulation.

'This accident again points up the difficulties in attempting to control safety of track and train operation by regulations based upon current track standards concepts. The Safety Board has now found and pointed out the results of vague or absent track maintenance rules and the substitution of "experienced judgment" in three accidents: those at Dunreith, Indiana; Glenn Dale, Maryland; and Franconia, Virginia. In two of these cases, experienced track or engineering personnel of the same railroad gave conflicting testimony of the requirements of track maintenance. In two of the accident cases, Crete and Franconia, experienced railroad personnel could not explain the accident in terms of any existing definitions of track conditions. In the Glenn Dale accident, railroad personnel felt they knew the cause of the track's buckling, but none could say that any corrective measures known to them would prevent a recurrence. Such problems indicate that many of the static and dynamic relationships within the track and train system are genuinely unknown. In the instant accident, even the available track recommendations of AAR would not have controlled this cyclic oscillation that was found. If the relationships are unknown, of course, they cannot be logically controlled.

'In this accident analysis, an effort was made to determine analytically which of the various forces could have shifted the rail. The analysis revealed unsuspected sources of cyclic forces in the configuration of the track, and it questioned an impression that a braking action was an important factor. However, the analysis could not positively provide the answers to a comparison between force and ability to resist force. The amount of lateral force capable of being resisted by a railroad spike has never been defined, even though such spikes have been used to hold railroad track together for more than a century. As indicated, the holding power of spikes is highly variable, and this is the probable reason why it has not been defined. The same variability is found in many other aspects of track structure. The freezing of the roadbed having alignment and surface irregularities changed the dynamic behavior of the track, and in ways that are now essentially unpredictable.

'The unpredictable and non-analytical approach to track maintenance is thus partially rooted in the present methods of track construction which allow such wide variations. Thus the Crete accident implies again the need for changes in track construction to reduce the number of variables as a step toward objective safety control.

'The analytical method of the Board's analysis employed well-known track measurements to produce force indications, thus interrelating the action of train and track on an instantaneous basis. Several assumptions were necessary to compensate for ignorance of some facts. Nevertheless, the method came close to describing the track condition on the basis of what forces the track exerts on the train (and vice versa) rather than on the basis of track dimensional measurements judged on a pass-or-fail basis, which does not reflect forces directly.

'It should be pointed out that these calculation methods could be employed on a computerized basis, employing track measurements taken not from laborious hand methods, but from electronic test devices such as the capacitor gage car of the Department of Transportation and those being used by some railroads. Determination of forces and car motions by a fast-moving test car might offer a very definite way of confirming safe alignment and surface conditions.'

As to the probable cause of the accident the NTSB stated:

'The National Transportation Safety Board determines that the derailment was caused by the movement of a rail of the turnout due to lateral forces produced by the locomotive as it moved over track which had alignment and surface deficiencies. The cause of the complete failure of the tank car which released the poisonous gas, following derailment, was a heavy blow delivered to the head of the tank car by the coupler of another car and the brittleness of the tank car steel at the ambient temperature of 4° F.'

The NTSB made six recommendations:

(1) Because statutory authority has now been conferred by the Railroad Safety Act, the Federal Railroad Administration should establish as soon as possible, objective, and enforceable standards for the inspection of condition of track and maintenance of track, such as gauge, surface, alignment, and securement. Such standards should be based upon the effects produced upon trains and should rely upon subjective opinion as little as possible. Such standards should consider the effects of roadbed freezing.

(2) The Federal Railroad Administration revise the specifications for material for the construction of future tank cars to include transition temperatures for steels so that tank cars hauling hazardous materials will be adequate to operate through the full range of ambient temperatures encountered in service.

(3) The Federal Railroad Administration make a study of the steels used to construct existing tank cars to determine the ranges of transition temperatures and the ability of these cars to resist brittle fracture in various temperatures so that the scope of the existing problem can be determined and adequate corrective action taken.

(4) The Federal Railroad Administration in cooperation with the Association of American Railroads make a study of the analyses and tests of the forces produced by the various components of a train while traversing track over the full range of operating conditions which would serve as a basis for the design and maintenance of track according to objectively determined requirements.

(5) The Office of Hazardous material in cooperation with Civil Defense and State agencies conduct a study of methods that could be employed rapidly to disperse or absorb clouds of toxic vapors. The study should also determine the suitable and practical means of protecting the public which may be exposed to the hazard on an emergency basis. Arrangements should be formulated for the dissemination of this information.

(6) The Safety Board reiterates the recommendation made in its Glenn Dale, Maryland, report of the derailment of Penn Central train 2nd No. 115, issued July 13, 1970, that the: "Federal Railroad Administration, in cooperation with the Association of American Railroads and the American Railway Engineering Association, conduct studies, including tests, to determine desirable combinations of track and equipment components required to act as a system to keep derailed cars upon and in line with the track structure." [1605]

Ten of a 105-car Chicago Indianapolis and Louisville Railway (Monon) freight train travelling at about 40 miles an hour were derailed on the edge of Rensselaer, Indiana, United States, on Saturday 1st March 1969, probably due to a 'hot box' leading to a broken journal. A considerable amount of sulphuric acid was lost from a ruptured tank car and a small quantity of chloroform from a second. Nearby streets were cordoned off but the danger was not considered sufficient to require an evacuation. No one was hurt. Damage, excluding the lost cargo, was put at $125,000. [1606] [1607]

Eleven of 32 cars of Chicago Rock Island and Pacific Railroad freight train No. 997 from Liberal, Kansas, to Amarillo, Texas, were derailed at Pringle, Texas, United States, north of Stinnett, at about 9.30 a.m. on Wednesday 5th March 1969. Two of the derailed cars were jumbo cars loaded with liquefied petroleum gas (LPG) and in all there were nine tank cars of LPG in the train. Some of them were damaged in the derailment, fire broke out and an explosion followed. As the accident occurred in an isolated area there was no danger to persons or property.

Engineer Potter and Fireman Fristoe (the roles of these two men have been assumed as lines are missing in the newspaper report) took the two diesel units and 11 undamaged cars at the front of the train south to Stinnett. Fristoe told a reporter that having checked that Conductor R.E. Neidhardt and Brakeman Aaron Lancaster were unhurt, "I went back and cut the angle cock, or air hose (separating the last car from the derailed cars) and got out. We got pretty excited and didn't bother to look at our watches until we got on out. I think it happened between 9.15 and 9.30." A following freight train pulled the undamaged rear cars and the caboose back clear of the wreckage. Firemen left the scene at noon on Wednesday, leaving the fire to burn out and not apparently expecting any further explosions, but three further explosions nevertheless followed on Thursday night. Stinnett residents were described as 'jittery' on Friday but the decision was apparently taken just to let the fire burn itself out, traffic being diverted in the meantime.[1608] [1609]

There was a lucky escape from more serious consequences when a Seaboard Coast

---

[1605] National Transportation Safety Board report
[1606] *The Times*, Hammond, Indiana, United States, 2nd March 1969
[1607] *The Evening News*, Jeffersonville, Indiana, United States, 1st March 1969
[1608] *Amarillo Daily News*, Amarillo, Texas, United States, 6th and 8th March 1969
[1609] *Amarillo Globe Times*, Amarillo, Texas, United States, 7th March 1969

Line Railroad freight train was derailed at Powder Springs, Georgia, United States, on Monday 17th March 1969. Two diesel units and 17 cars of train No. 298 from Birmingham, Alabama, to Atlanta, Georgia, jumped the track due to a broken rail and a tank car carrying 30,000 gallons of liquefied petroleum gas (LPG) was ruptured. Escaping gas led to the evacuation of 500 townspeople together with 625 children and 25 teachers from Clarksdale elementary school. Police cordoned off the area, and electrical power to the area was cut for fear that a spark would set off an explosion. The derailment occurred just after the train had crossed a trestle over a deep ravine. Had it happened on the trestle and the cars fallen from the structure and exploded, the accident would have been much more serious. [1610]

The metal band securing gas pipes on a 96-car Penn Central Railroad freight train from Logansport, Indiana, to Effner, Illinois, failed as the train was passing through Wolcott, Indiana, United States, on Thursday 20th March 1969 and fouled a set of facing points, derailing three tank cars loaded with anhydrous ammonia. A small quantity of gas escaped but there was reported to be no danger. [1611]

Six cars of a Penn Central Railroad freight train were derailed at Simsbury, Connecticut, United States, on Thursday 15th May 1969. Four of them were tank cars carrying 10,000 gallons of crude oil, some of which was spilt. Two railwaymen riding in the caboose of the train were shaken up. No cause was stated for the derailment. [1612]

The collision of a 19-car southbound Western Pacific Railroad freight train with an automobile on Hammer Lane Crossing, Stockton, California, United States, at about 11.15 p.m. on Thursday 28th May 1969, led to a considerably more serious accident when the train became derailed as a result. The derailed vehicles struck the cars of a second train parked on a siding, a tank car containing 33,600 gallons of butane was damaged, the gas escaped and erupted into flame. The flames reportedly rose hundreds of feet into the air and could be seen from all parts of the city for several hours. With the possibility of an explosion present local residents had to be evacuated. The fire burnt for over 20 hours.

The car driver, William Silva, 46, of Stockton, and Engineer E.A. Tibbedeaux, 58, were injured, the car driver seriously. They were taken to St Joseph's Hospital. A third casualty was Helen Azevedo, 44, who had been watching television in a nearby house and suffered from shock when a freight car fell into her backyard. She related that it had sounded as if there had been an earthquake. According to the police Silva had failed to notice the flashing warning lights at the crossing and drove into the side of the train. [1613] [1614] [1615]

Ten tank cars of a 26-car Korean National Railway freight train en route to Seoul were derailed in the early hours of Monday 30th June 1969 near Yongchon, South Korea. Gasoline (petrol), kerosene and oil escaped from five of the cars, a fire broke out at once and enveloped a number of lineside farmhouses. Four sleeping children were burnt to

[1610] *The Holland Evening Sentinel*, Holland, Michigan, United States, 18th March 1969
[1611] *The Pharos Tribune and Press*, Logansport, Indiana, United States, 21st March 1969
[1612] *Naugatuck Daily News*, Naugatuck, Connecticut, United States, 16th May 1969
[1613] *The Modesto Bee*, Modesto, California, United States, 29th May 1969
[1614] *The Times*, San Mateo, California, United States, 29th May 1969
[1615] *The Daily Review*, Hayward, California, United States, 29th May 1969

death, one person was missing and between seven and ten people were injured. At least eight farmhouses were destroyed and 1,000 chickens, 12 pigs and other livestock were killed. Firemen from Yongchon and from the US Army base at Taegu fought the fire for eight hours. The cause of the derailment was unknown. [1616]

Six cars of an Atchison Topeka and Santa Fe Railway freight train left the track and piled up like dominoes at Marion, Kansas, United States, on Saturday 26th July 1969. A 240-ton tank car was ruptured an liquefied petroleum gas (LPG) escaped. Sheriff's deputies and highway patrolmen kept the area clear until all the gas had dispersed. [1617]

Part of an Erie-Lackawanna Railroad freight train bound for Wayne, New Jersey, from Marion, Ohio, was derailed at Cambridge Springs, Pennsylvania, United States, on Sunday 17th August 1969, apparently due to a 'hot box' leading to a broken journal. A tank car containing toxic, carcinogenic, and flammable acrylonitrile fell down an embankment and was breached, losing about 5,000 gallons of the 20,000 that it contained, the chemical running into the marshy ground adjoining the track. Some of the acrylonitrile found its way into French Creek, the main source of water for the towns of Cambridge Springs and Franklin, so the two places stopped taking water from the stream temporarily and water was rationed. The remaining chemical was transferred into a road tanker. [1618] [1619]

A 74-year-old man walking along the track was the cause of a major accident at Glendora, Mississippi, United States, at 2.35 p.m. on Thursday 11th September 1969. In an effort to avoid striking the trespasser the engineer applied his brakes in emergency, whereupon the train buckled towards its rear and 15 cars were derailed, including eight loaded with vinyl chloride. One car was punctured by the coupler of another, vinyl chloride escaped and at 8.30 p.m. it was ignited from an unknown source and several explosions followed. About 20,000 people were evacuated due to fears of poisoning by phosgene gas produced by the burning vinyl chloride, though the NTSB pointed out that the quantity of phosgene produced by the burning of vinyl chloride was very small and the hydrogen chloride so produced was a greater danger. The pedestrian and a second person were injured and four houses, other buildings, cars, and equipment were damaged by the fire and explosions. No one was killed. The National Transportation Safety Board (NTSB) inquired into the accident.

Illinois Central Railroad train Second No. 76 had left Gwin, Mississippi, at 12.10 p.m. for Memphis, Tennessee and consisted of four diesel units and 149 cars. It was travelling at about 42 miles an hour approaching Glendora. When the train was braked hard by its engineer a witness saw a box car, 108th in the train, buckle, and the 106th to 119th cars were derailed in turn. Immediately after the derailment the conductor advised his engineer via a portable radio what had happened, then saw a white vapour escaping from a tank car and consulting his waybills realised that it was vinyl chloride. The dispatcher then advised the conductor to warn local inhabitants of the possible danger the gas posed, which was done. The evacuation of inhabitants due to fear of phosgene poisoning was

---

[1616] *Pacific Stars and Stripes*, Tokyo, Japan, 2nd July 1969
[1617] *The Salina Journal*, Salina, Kansas, United States, 27th July 1968
[1618] *Somerset Daily American*, Somerset, Pennsylvania, United States, 20th August 1969
[1619] *The Bradford Era*, Bradford, Pennsylvania, United States, 18th August 1969

begun at 7.30 p.m. and completed by 11 p.m.

Concluding its report into the accident the NTSB came to the usual specific and general conclusions. Its specific conclusions were:

(1) The train was being operated in compliance with prevailing regulations and instructions at the time of the accident.

(2) The curvature of the track and vegetation growing adjacent to the track prevented the engineer and brakeman on the locomotive from seeing the pedestrian at a distance greater than about 500 feet.

(3) After the engineer applied the brakes in full emergency, the head end of the train stopped more quickly than would have been expected. The exact distance that the train traveled and the reason for the shorter than usual stopping distance were not developed in the investigation.

(4) The empty MILW car buckled because of undue pressure resulting from uneven braking of the train under full emergency conditions. The front part of the train decelerated at a rate greater than the rear part.

(5) The arcing electric line did not ignite the vinyl chloride vapor because the vapor never accumulated to that depth.

(6) Phosgene gas was not a hazard to the population because it was not developed in sufficient quantity to be measurable.

(7) After the derailment, the site was not protected in a manner to prevent injury to railroad employees and others. At one time, employees climbed upon the wrecked tank cars to make observations.

(8) There was an inadequate assessment by persons on scene of the dangers to the public in this accident.

As to its general conclusions the NTSB stated:

'This accident is a good example of what can happen when a normal freight train on a level track is decelerated by a properly initiated emergency brake application. At best, a properly initiated emergency brake application is transmitted rearward successively from the locomotive through each car, resulting momentarily in greater deceleration on the head end than on the rear. The unequal decelerating forces established within a train by the variance of braking power of the individual cars contributes to the problem. The braking system is designed to provide maximum braking of an empty car without sliding the wheels; consequently, when the car is loaded, the braking efficiency is reduced.

'It is not uncommon for empty cars to buckle under the pressures created in normal emergency braking. Train speeds have increased, and car capacity has increased. In turn, the loaded car braking ratio is lower; therefore, dependable, consistent deceleration is harder to attain than it was 25 years ago. An emergency application of the brakes can result in damage to the train due to slack action or separations in the train because of broken couplers or knuckles. The action of a given freight train when brakes are applied is not always predictable. The proper service braking of a freight train is not a science but an art. The result depends upon the skill and experience of the engineer. When train brakes are applied in emergency, however, the only control that an engineer can exert is to release the brakes on the locomotive.

'These aspects of train braking constitute a problem of long standing. An engineer's knowledge of these conditions sometimes deters his use of the train brakes in preference to other means of controlling the train. For instance, an engineer of a train, approaching a grade crossing occupied by a motor vehicle knows that full emergency braking may damage his train and he hopes that the vehicle will vacate the crossing, therefore, the engineer sometimes makes a full emergency brake application only as a last resort. At Waterloo, Nebraska, on October 2, 1967, a freight train struck a school bus, resulting in the death of four children and injuries to the other nine. The bus stopped, and then moved slowly onto the crossing in full view of the engineer; however, the engineer did not apply the train brakes until he was sure that the school bus was actually going to occupy the crossing in the path of the train. The engineer at a grade crossing faces the dilemma of whether to make a full emergency application and risk damaging the train, or to make a full-service application and lessen the chances of avoiding the vehicle or pedestrian. Engineers should not be placed in situations such as this where they have to decide which course is better – the one which places the risk on the train or the one which places the risk on the other party. This dilemma would not exist if the freight train braking system were consistently capable of maximum emergency braking without damaging the train.

'Stopping distances achieved in the 1887 Burlington Brake Tests are comparable to those achieved in actual practice today. It was concluded after the tests that the best braking system was one operated by air and controlled with electricity. Several other tests conducted since that time also have concluded the electropneumatic brake was most desirable. With the exception of the radio-controlled remote systems, the electropneumatic brake has not been developed for use on freight trains.

'It is often not possible to determine the stopping distance of brake application history of a train involved in an accident. Proper data of this type, however, could indicate the frequency with which less than ideal braking is responsible for or contributes to accidents. Statistical justification from such accident data could indicate the need for train braking improvements. Thus, the lack of braking history in accidents or in railroad operations in general is a key problem in determining how much improvement is necessary.

'The first approach to the overall braking question should be to develop equipment capable of recording the braking effects in accidents. The degree of need for improvements to benefit the general public could then be determined from the accident data. It is now evident, however, that improved braking equipment is already desirable to avoid accidents and damage resulting from the combination of rapid deceleration and slack action. This situation in railroad emergency brake application is one of the few remaining in transportation technology in which an unpredictable accident can be caused by normal use of an operating control.'

Regarding the probable cause of the accident the NTSB wrote:

'The Safety Board determines that the derailment was caused by the buckling of the underframe of the 108th car when the engineer made a full emergency brake application in an attempt to avoid striking a pedestrian who was walking on the track. The car buckled because of excessive and uncontrollable compression in the train which developed when the full emergency brake application created greater braking force on the head of the train than on the rear.

'The fire and explosions resulted from the rupture of a tank car of vinyl chloride by the coupler of one of the derailed cars. The absence of interlocking couplers and other means of preventing separation and jackknifing allowed the cars to jam up together. The pileup resulted in additional mechanical damage to the tanks and allowed the fire from the leaking tank to impinge on the others.'

As to its recommendations the NTSB wrote:

(1) The Federal Railroad Administration, in cooperation with the railroad industry, initiate research and development which will result in practical locomotive instrumentation capable of recording data relating to the braking behavior of trains. The data obtained should be of sufficient scope and accuracy to allow evaluation of the braking performance of the train and to record the manner in which the brakes were applied. The project should contemplate initially the instrumentation on sufficient numbers of locomotive units to insure a sampling of freight train accidents and should include provision for cooperative evaluation of records by Government and industry.

(2) The Federal Railroad Administration initiate research and development to provide prototype models of freight train braking systems.

(a) Capable of providing shorter stopping distances which nearly approach the theoretical limit under all conditions of loading and length of trains;

(b) Capable of stopping a train in the emergency applications now required by regulations without internal collisions, train separations, or damage to the train or its lading;

(c) Capable of propagating brake application, both service and emergency, throughout the length of a train more expeditiously and surely;

(d) Capable of more rapid application of the full intended stopping force to the rails at each car after the application signal is received at each car.

(3) The Illinois Central Railroad Company initiate a program to train employees in the proper procedures to insure adequate protection to employees and bystanders at accidents involving hazardous materials.

The Safety Board reiterates and emphasizes the following recommendations made in the report of the accident which occurred on the Southern Railway at Laurel, Mississippi, on January 25, 1969:

"5. . . . the Association of American Railroads and American Short Line Railroad Association develop plans that will result in the fire chief of each community through which the track of a member road passes knowing where immediate information can be obtained, describing the location and characteristics of all hazardous materials in any train involved in a train accident that affects a community. This recommendation can be accomplished in a relatively short time regardless of the level of training which may be achieved later by fire departments.

"10. . . . the A.A.R. review the function of the Bureau of Explosives regarding its performance in protecting the public from danger resulting from railroad accidents involving hazardous materials and take the necessary action to develop an effective, cooperative program with the carriers to accomplish the intended purpose of the responsibility delegated to the Bureau of Explosives by Title 49, Code of Federal

Regulations, Section 174.506. The Board endorses the FRA's proposed amendment of the regulations which will provide that reports of incidents and accidents involving hazardous materials presently made to the Bureau of Explosives by rail carriers will also be filed with the FRA.

"11. . . . the Federal Railroad Administration include in its current study of an improved coupler design, the problem of keeping cars coupled and in line with the track and with each other after a derailment occurs. In order to attain an integrated organization of track and rolling stock features that could limit the after effects which can now follow a simple derailment, the Federal Railroad Administration should also study related technical approaches to control interference with traffic on adjacent tracks and wayside structures during derailments, such as means of limiting the lateral excursion of wheels, and separation of trucks from car." [1620]

Five cars of a Missouri Kansas Texas Railroad freight train were derailed about three miles south of Rockwall, Texas, United States, at about 8 p.m. on Wednesday 26th November 1969 and anhydrous ammonia escaped from one of two tank cars that was ruptured when they slid down an embankment, leading to the evacuation of 200 people from nearby homes and a trailer court. The tank car contained 30,000 gallons of chemical. The cloud of gas spread through a wooded area alongside of Lake Ray Hubbard and south of US 66, and fog and light rain held it close to the ground. It was believed that it might take 24 hours for it all to escape. The evacuees were able to return home on Thursday in time for their Thanksgiving Day dinners. [1621] [1622]

An Illinois Central Railroad freight train became derailed early on Wednesday 4th February 1970, at Sarah, Mississippi, United States, 14 cars leaving the rails. A tank car loaded with flammable and mildly toxic methyl acrylate leaked and both ends of a ditch running into the Coldwater River were dammed up to prevent contamination by the chemical. No one was hurt. [1623]

Five cars of a Penn Central Railroad freight train were derailed due to a 'hot box' leading to a broken journal at Muncie, Illinois, United States, on Thursday 12th February 1970. Highly flammable naphtha leaked from a tank car and 400 pupils of St Lawrence Elementary School and the residents of 15 houses were evacuated as a precaution. Firemen pumped the remaining chemical out of the car and diluted that on the ground with water. The residents were allowed home after about an hour; the pupils were sent home for the day. [1624]

Thirty of 101 cars of an Atchison Topeka and Santa Fe Railway freight train were derailed near La Rose, Illinois, United States, at 3.26 a.m. on Wednesday 18th February 1970. Two tank cars of liquified petroleum gas (LPG) ruptured, spilling their contents onto a siding and nearby fields, the LPG ignited and there was at least one reported explosion. Fearing further explosions state police evacuated the 400 or 500 residents of La Rose but they were allowed to return home after about four hours after another train

[1620] National Transportation Safety Board report

[1621] *The Port Arthur News*, Port Arthur, Texas, United States, 27th November 1969

[1622] *The Victoria Advocate*, Victoria, Texas, United States, 27th November 1969

[1623] *Laurel Leader-Call*, Laurel, Mississippi, United States, 5th February 1970

[1624] *The Anderson Herald*, Anderson, Indiana, United States, 13rd February 1970 had pulled away the underailed LPG cars. Unable to approach the burning cars because of the intense heat, firemen left the blaze to burn itself out. [1625] [1626]

Track defects were stated to have been the principal cause of the derailment of a Markham Colliery to Cottam Power Station merry-go-round coal train at Worksop, England, on British Railways, Eastern Region, on Tuesday 17th March 1970, though the riding characteristics of a worn 32-ton waggon contributed. Twenty-four waggons were derailed in all and were run into by a Gainsborough to Ellesmere Port oil train, the engine, two runner waggons and 10 tank waggons of which were derailed in turn. Spilt oil caught fire and 30 families had to be evacuated from their homes. The driver, fireman and guard of the oil train were taken to hospital where the driver was detained. [1627] [1628]

Three of the 28 cars of a Penn Central Railroad freight train bound for Pottstown were derailed at Phoenixville, Pennsylvania, United States, on Sunday 17th May 1970 as the train was travelling at about 15 miles an hour. A reported 135,000 gallons of vinyl chloride were lost from a ruptured tank car and 20 families were evacuated. Firemen washed away the chemical and a water treatment plant was briefly closed for fear of contamination of the water supply. No one was hurt. [1629]

Eighteen cars of a northbound Missouri Pacific Railroad freight train derailed at about 4 p.m. on Sunday 7th June 1970 two miles east of De Soto, Illinois, United States. The cars were carrying ammonium nitrate, anhydrous ammonia, nitric acid, oil, glue and asphalt and the following morning it was stated that the first three chemicals and naphtha were all leaking. A cloud hanging over the area was believed at first to be either ammonium nitrate or anhydrous ammonia but later it was stated to be the contents of a 24,500 gallon tank car of corrosive, toxic and combustible diethylenetriamine. An adjoining road was closed off and one family living nearby was evacuated. No one was hurt. [1630] [1631]

The dangers of transporting liquefied petroleum gas (LPG) were again demonstrated by an accident at Crescent City, Illinois, United States, on Sunday 21st June 1970. The National Transportation Safety Board (NTSB) investigated it.

Eastbound train No. 20 of the Toledo Peoria and Western Railroad consisted of four diesel units, 108 cars and a caboose. It had left East Peoria Yard, Illinois, at 3.30 a.m. for Effner, Illinois, passed Forest, 61¾ miles from East Peoria, at 5.55 a.m., and Gilman, about 83½ miles from East Peoria, at about 6.30 a.m. Here the line crossed the Illinois Central Railroad and an operator was on duty there, who observed the train as it passed

---

[1625] *Stevens Point Daily Journal*, Stevens Point, Wisconsin, United States, 18th February 1970
[1626] *Dixon Evening Telegraph*, Dixon, Illinois, United States, 18th February 1970
[1627] *The Times*, London, England, 18th March 1970
[1628] Ministry of Transport: Railway Accidents, Report to the Ministry of Transport on the Accidents that occurred on the Railways of Great Britain during the year 1970.
[1629] *The Evening Sun*, Hanover, Pennsylvania, United States, 18th May 1970
[1630] *Southern Illinoisian*, Carbondale, Illinois, United States, 8th June 1970
[1631] Ibid, 9th June 1970

and indicated to the train crew that he had observed no problems with it. A car driver stopped at a level crossing at Gilman did, however, see smoke coming from the journal box of a hopper car and attempted to signal to the train crew but like the operator received no reply. He was public spirited enough to then stop and report the matter to the police from a nearby restaurant, to be told that the accident had already happened and the police were on their way to Crescent City, 7¼ miles further on.

The speed of the train increased from 30 to 43 miles an hour between Gilman and Crescent City and an initial derailment occurred at a level crossing just short of Crescent City. Several residents of the town observed smoke and fire coming from beneath a car and a car bouncing up and down. General derailment followed at a set of facing points and the coupler of the 27th car struck the 26th, a tank car loaded with LPG, which was punctured. The LPG escaped and was immediately ignited, probably either by a spark from the derailment or from the hot-box. Flames erupted to several hundred feet, people suffered burns and buildings were set on fire.

The residents of Crescent City were alerted by the noise of the derailment and the fire, the fire alarm was set off and volunteer firemen began to fight the blaze and at first appeared to get it under control. Meanwhile the police, finding that cars containing propane gas were involved in the fire, began to evacuate residents. The intensity of the fire then increased and the local fire chief called for assistance from neighbouring communities. At 7.33 a.m. the 27th car exploded, the 28th exploded at 9.40 a.m., and the 32nd and 33rd at 10.55 a.m., damaging and releasing propane from the 34th and 35th cars. The fires were then left to burn out and the residents who had been evacuated finally returned at 2.30 p.m. on Tuesday 23rd June after 56 hours. No one was killed but 66 people were hurt, 11 seriously. Twenty-five homes and 16 businesses were destroyed and a good number of other homes and businesses were damaged. The total cost of the damage was estimated at $1,700,000. [1632]

A newspaper reporter wrote, 'The cars moved slowly out of Crescent City today loaded with children, pets, and a few hastily gathered possessions. Billowing black smoke rose from the town behind them. For a time even the firemen abandoned the center of this little Iroquois farm town to the fire that raged out of control after a railroad car derailed and tank cars carrying liquid propane exploded. The intense heat and the constant threat that another tank car would explode at any moment forced the firemen to stand back and watch and fight only a defense action around the perimeter of the fire in an attempt to keep it from spreading.

'The fire, with smoke rising high enough for it to be seen for miles across the flat cornfields, had devastated the two-block main street and devastated 15 homes. The explosions had blown sections of tank cars into front yards a number of blocks away. Ennen's blacksmith shop was gone – so was Jenkin's barber shop, Hofmeister's hardware and Leo's tavern – 15 businesses in all. It had come quickly – without warning.

'Orvel Carlson, Crescent City fire chief, was sitting on the edge of his bed two blocks away when he heard the train going thru town. "I heard a rumbling noise, then there was this big flash like the world was coming to an end," he said. He immediately blew the fire siren from his house and requested all available equipment from Iroquois and Kankakee counties. "It was like an atom bomb had just gone off and then just a wall of fire," he

[1632] National Transportation Safety Board report

said. Carlson's head and hands were swathed in thick, white, bandages. He had been among those fighting the blaze when a tank car blew up a block and a half away. "The explosion knocked off my hat and the flames swept at me and up under my raincoat," he said.

'George Gocken, a Crescent City volunteer fireman, was behind the blacksmith shop when that car blew up. His head and hands also were damaged. "There was fire in a ring all around me," Gocken said, "I ran like hell."

'H.J. Sterrenberg, the town mayor, was one of the many standing helplessly in the Crecent-Iroquois High School waiting as the fire swept through more and more of the town. He had been nearby when the second tank blew. "It was just a miracle we weren't all killed," he said. "Pieces of the car came flying thru the air. I put my hands over my head and ran for the fire station. This is the worst thing that has ever happened in the area. We'd talked at firemen's meetings about how this kind of thing could happen. But we never really thought it would." Sterrenberg said he was proud of his town and the way its people had responded to the emergency. "Everybody tried to help," he said. "I guess we'll have to start all over now."

'Emergency relief operations began as soon as word of the tragedy spread. The radio station in nearby Watseka broadcast pleas for cots, blankets, and food. Relief centers were set up in two schools and at the fairgrounds outside Crescent City. Dozens of families in the area offered to take in Crescent City residents and by shortly after noon so much food had been donated that persons were asked to wait several hours before bringing more.' [1633]

After its usual very thorough analysis of the accident the NTSB reached 19 conclusions:

(1) CB&Q 182544, the 20th car of train No. 20, on Jun 21, 1970, was loaded with sand within its carrying capacity and without prior inspection or servicing of its journal boxes.

(2) No exceptions were taken to the condition of the car during inspections made at various points between the loading and the accident points.

(3) The journal boxes of CB&Q 182544 were serviced en route, but the adequacy of the servicing could not be determined.

(4) There was sufficient time between the beginning of the overheated journal and the derailment in which to detect the defective journal if some method of surveillance had been in use.

(5) The lack of curves on the track made observations of the train by crew members more difficult.

(6) Increased use of 'hot box' detectors could have detected the defective journal before it broke.

(7) The journal at L-4 location of CB&Q 182544 broke, due to excessive overheating.

(8) The cause of the overheating of the journal could not be determined.

[1633] *Chicago Tribune*, Chicago, Illinois, United States, 22nd June 1970

(9) The placement of the 12 rank cars loaded with propane, coupled together, greatly increased the severity of the accident by allowing the flames from one car to play upon another.

(10) The separations of the tank cars on derailment resulted in the puncturing of one tank, escape of propane, and fire.

(11) The jumbling of the tank cars after they separated exposed the tanks to excessive heating which, in turn, resulted in their exploding.

(12) The overloading of several of the tank cars did not increase the severity of the accident nor materially reduce the time between the accident and the explosion of the tanks.

(13) Information as to the lading in the derailed cars was not readily available to the firemen and the other emergency crews.

(14) The absence of a properly protected power source, or the failure to provide an emergency source of power for the town's water supply, resulted in a reduction of water to fight the fire.

(15) The lack of a central source of command to oversee, advise, and direct the activities of the assisting fire departments resulted in confusion and might have caused injuries to some of the firemen.

(16) The lack of adequate training for the firemen who were called to fight this fire resulted in their being exposed unnecessarily to danger.

(17) Some of the firemen's protective clothing was not adequate to resist the impingement of the fire.

(18) The fire-fighting forces in this case were not provided with adequate equipment to handle fires of this type and size.

(19) The evacuation of the town and the warnings to firemen of the hazard in fighting the fire were given by the Illinois State Police. This warning resulted in substantially lower casualties than otherwise might have occurred.

As to the probable cause of the accident the NTSB wrote:

'The National Transportation Safety Board determines that the probable cause of this accident was the breaking of the L-4 journal of CB&Q 182544, the 20th car, due to excessive overheating, which permitted the truck side to drop to the track and derail the leading wheels of the car. The cause of the overheating could not be determined.

'The cause of the initial fire was the puncturing of one tank during the derailment, the jumbling of the derailed cars, and the large volume of propane released which immediately ignited and subjected the other tanks to impingement of fire.

'The cause of the explosive rupture of several tanks was localized heating which weakened the steel of the rank so that it could no longer resist the pressure of the propane. Contributing to the explosive rupture were (1) the placement of a number of tank cars together in the train which permitted interaction between cars, (2) the speed of the train which allowed jumbling of cars to occur and, (3) the absence of heat insulation of the cars which was formerly required.

'The injury to the fireman and spectators was due to lack of appreciation by fireman of the large scope of the fire and explosion which could occur in a fire of this type.'

The NTSB made eight recommendations:

(1) The Federal Railroad Administration concluded the proposed rulemaking regarding the transportation of liquefied flammable gases in tank cars of DOT Specifications 112A and 114A having capacities exceeding 25,000 gallons. This rulemaking was initiated with the Hazardous Materials Regulations Board's "Request for Public Advice on Speed Restrictions on Tank Cars," Docket No. HM-60, published in the Federal Register on October 15, 1970.

(2) The Association of American Railroads and the Federal Railroad Administration encourage the expeditious completion of the testing of insulating material for the coating of the outside of tanks, and, if a satisfactory material is found, require its use on all tank cars transporting flammable liquids (liquefied flammable gases).

(3) The Federal Railroad Administration encourage the expeditious completion the RPI-AAR Railroad Tank Car Safety Research and Test Project and take steps to ensure that the results of the research are applied to enhance the safety of transportation of hazardous materials.

(4) The Federal Railroad Administration promptly publish their current study of the failure of freight car journals, and if necessary continue such studies so that regulations can be promulgated to establish standards for freight car journals and their maintenance.

(5) All railroads which proposed action in response to the July 7, 1970, letter to the Chairman of the National Transportation Safety Board advise the Federal Railroad Administration of the action taken by them and that these railroads and all others who transport liquefied flammable gas in DOT Specifications 112A and 114A having capacities exceeding 25,000 gallons, as a minimum precaution, initiate such additional action for full compliance with the recommendation of the General Committee, Operating Transportation Division, Association of American Railroads, made in response to the above letter.

(6) The Safety Board reiterates and emphasizes the following recommendations made in previous accident reports:

Railroad Accident Report, Pennsylvania Railroad Company, Dunreith, Indiana, January 1 1968:

The Board recommends that the Department of Transportation study means of improving the training methods available to local fire departments so that they can upgrade their skills in their handling of emergencies created by the increasing transportation of hazardous materials. The problems of controlling such accidents are especially troublesome because of the daily introduction into commerce of numerous new kinds of hazardous materials. The Board believes that local emergency organizations cannot be expected to be conversant with necessary procedures to handle situations involving the many possible emergencies involving the transportation of hazardous materials unless some form of assistance in training is provided such as a model type training course.

Railroad Accident Report, Southern Railway Company, Laurel, Mississippi, January 25 1969:

5. The Safety Board recommends that the Association of American Railroads and the

American Short Line Railroad Association develop plans that will result in the fire chief of each community through which the track of a member road passes knowing where immediate information can be obtained, describing the location and characteristics of all hazardous materials in any train involved in a train accident that affects a community. This recommendation can be accomplished in a relatively short time regardless of the level of training which may be achieved later by fire departments. [1634]

Thirty-five of the 128 cars of a Chesapeake and Ohio Railway freight train at a bridge at Carrs Run, near Omega, Ohio, United States, on Tuesday 14th July 1970. Eight cars went into a stream and one of them was ruptured, leaking an unspecified toxic acid into the water. [1635]

A Grand Trunk Western Railroad freight train suffered a derailment on a level crossing at Lapeer, Michigan, United States, on the night of Tuesday 21st July 1970, eight of 23 cars leaving the rails. A small quantity of ammonia escaped from a tank car and local residents were warned against lighting matches. The situation was brought under control within an hour. [1636]

Naphtha gas reportedly escaped from one of five cars of a 96-car Canton to Cleveland Penn Central Railroad freight train which was derailed near Canton, Ohio, United States, on Saturday 25th July 1970. A wheel had apparently come off a car. Firemen stood by as a precautionary measure. [1637] [1638]

Twenty-six of a 170-car Southern Railway freight train left the rails near Boligee, Alabama, United States, on Thursday 6th August 1970. Three of the derailed cars contained ammonia, some of which escaped, but no danger was anticipated. [1639]

Much more serious than the last three accidents was the derailment of an Atchison Topeka and Santa Fe Railway freight train at Melvern, Kansas, United States, at about 4.55 p.m. on Friday 14th August 1970. Twenty-seven of 113 cars derailed, two cars containing liquefied petroleum gas (LPG) left the track in a cutting on the edge of town as the train was travelling at about 50 miles an hour.

Wendell Allison, 53, told a reporter, "I've lived in that house near the tracks all my life, and they've had a lot of accidents there, including some that took lives. But I never saw anything like that. I was watching TV. I heard the train coming. I looked out the door and saw a boxcar coming up over the embankment. Then there was a big explosion. It was a ball of fire. I ran out the other door and warned my sister and another woman get away. I looked back and saw stuff flying in the air."

Gary Vogler, who said he had also been watching TV at the time, related, "All of a

---

[1634] National Transportation Safety Board report
[1635] *The Chronicle-Telegram*, Elyria, Ohio, United States, 16th July 1970
[1636] *The Holland Evening Sentinel*, Holland, Michigan, United States, 22nd July 1970
[1637] *The Times Recorder*, Zanesville, Ohio, United States, 26th July 1970
[1638] *The Sunday Chronicle-Telegram*, Elyria, Ohio, United States, 26th July 1970
[1639] *The Anniston Star*, Anniston, Alabama, United States, 7th August 1970

sudden all you could see was fire coming over the houses. I grabbed the two kids and my wife ran out of the house. The heat could almost knock you down. It shook the whole house. My wife screamed and hollered. When we got outside all the neighbours were running out of the houses, getting into their cars, and leaving. I never had anything that scared me so much in my life."

The LPG tank waggons had ruptured in the derailment, the gas had escaped, caught fire and the explosion had followed. The Melvern Fire Department were assisted by firemen from Lebbo, Lyndon, Quenemo, Pomono, Topeka and Waverly.

Ten people were hurt, one of whom was in a serious condition with second- and third-degree burns. The injured included Wendell Allison and Evelyn Schultze, both of whose homes were destroyed, as were the wooden Methodist church (which had been rebuilt once already after burning down in 1910) and several other buildings. Property damage was estimated at $200,000. The cause of the derailment was not immediately clear. [1640] [1641]

Three tank cars containing vinyl chloride were among 28 of a 111-car Penn Central Railroad train that derailed in a rural area near Batavia, New York, United States, on Thursday 27th August 1970. The chemical leaked from at least one car, caught fire and there was an explosion. It was not known if a second vinyl chloride car was involved in the fire. The surrounding area was cordoned off but there was no evacuation and no one was hurt. The cause of the derailment was unknown. [1642] [1643]

At 6.30 a.m. on Friday 8th January 1971 14 of a westbound 63-car Burlington Northern Inc freight train were derailed near a bridge over a 350ft-long Charlton River, three miles west of Callao, Missouri, United States. The derailed cars collided with the bridge abutments, a car containing anhydrous ammonia exploded, the bridge caught fire and collapsed and eight cars, including five containing ammonia, fell into the river.

Resident Bernard Fiedler had just left home for work in New Cambria. He said he "saw a ball of fire going up in the air and shake the house. It kept kind of roaring and re-exploding, rolling up. I saw sparks of fire going up the railroad tracks. I guess that was the rest of the train that had broke loose. The sparks of fire were where the brakes were dragging. I tried to call the patrol, but all the neighbours were on the line."

Residents were warned by police to be on the alert for ammonia fumes and to leave the area if they smelt any until they cleared. No one was hurt but there were expectations that ammonia leaking into the river would kill all the fish downstream. The line was expected to be blocked for at least a fortnight while the bridge was rebuilt. [1644] [1645]

[1640] *Ottawa Herald*, Ottawa, Kansas, United States, 15th and 17th August 1970
[1641] *Lawrence Daily Journal-World*, Lawrence, Kansas, United States, 15th August 1970
[1642] *Wellsville Daily Reporter*, Wellsville, New York, United States, 27th August 1970
[1643] *The Daily Messenger*, Canandaigua, New York, United States, 27th August 1970
[1644] *Moberly Monitor-Index*, Moberly, Missouri, United States, 8th January 1971
[1645] *The Columbian Missourian*, Columbia, Missouri, United States, 10th January 1971

Twenty-two cars of an Illinois Central Railroad freight train derailed om the edge of Ashkum, Illinois, United States, on the afternoon of Wednesday 10th February 1971. A tank car of methanol was ruptured, its contents emptying into a ditch and it, a car of vinyl chloride and a third car caught fire. Fumes from the fire spread northwards towards Clifton, then the wind changed and the fumes shifted towards Ashkum. In the end the residents of both places were evacuated for fear of an explosion and the fire was left to burn itself out. [1646] [1647]

"I should have been standing there out on the platform and should have been killed," declared Agent Louis Hulpman, at Slinger, Wisconsin, United States, adding, "but fortunately I was somewhere else, inspecting some other cars, when it happened." [1648] A 25-car Minneapolis St Paul and Sault Ste Marie Railroad train from Chicago, Illinois, to Fond du Lac, Wisconsin, had 19 of its cars leave the rails and a tank car containing 7,980 gallons of phosphoric acid overturned and spilt several thousand gallons of its contents. Parts of Slinger were cordoned off while a dike was constructed to contain the acid, which was then neutralised with powdered lime and limestone. Oil and gas leaked from other cars and started a small fire which was quickly extinguished. One fireman suffered an acid burn. [1649]

What Mayor Louie Welch declared the worst disaster in the history of the city of Houston [1650] occurred at 1.44 p.m. on Tuesday 19th October 1971 at Mykawa, Texas, United States, two miles within the city limits. The accident occurred on a line of the Atchison Topeka and Santa Fe Railway (ATSF) but a Missouri Pacific Railroad (MP) train was involved, the MP having trackage rights (running powers) over the ATSF.

The accident began when 20 cars of northbound MP train No. 94, which consisted of four diesel units and 84 cars, left the track. Six of the derailed vehicles were tank cars containing vinyl chloride monomer, two of which were ruptured, the chemical escaped and ignited. Houston Fire Department attended and fought the fire. About 45 minutes after the derailment, a tank car exploded and another "rocketed" about 300 feet. A fireman, Thurston Hathaway Jr., was killed and about 50 people, mostly firemen but also newsmen and three spectators, were injured. [1651]

A newspaper wrote, '"The whole dammed world turned orange." That's the way Houston Post reporter saw the tank car explosion Tuesday that killed a fireman and injured 37. "It hurt like hell, Oh God, it hurt like hell," Bishop, 27, said in Memorial Baptist Hospital shortly after the explosion. He and Post photographer Dan Hardy were reported in fair condition at Memorial Baptist Hospital Southeast today. "I was talking with a deputy fire chief," Bishop said. "He was telling his men all they could do was cool the cars around the burning tanks before the other cars ruptured. It didn't look like much water to me. I think there were some derailed cars piled up, but before I could take many

---

[1646] *The Dubois County Daily Herald*, Jasper, Illinois, United States, 11th February 1971

[1647] *Journal Gazette*, Mattoon, Illinois, United States, 11th February 1971

[1648] *Stevens Point Daily Journal*, Stevens Point, Wisconsin, United States, 14th May 1971

[1649] *Oshkosh Daily Northwestern*, Neenah, Wisconsin, United States, 15th May 1971

[1650] *The Cougar*, Houston, Texas, United States, 20th October 1971

[1651] National Transportation Safety Board report

notes, everything went haywire. It was an instant holocaust. I ran and ran and ran. I never lost consciousness, but kept thinking, God, why was I here?"

'Bishop and Hardy, 44, were two of the first newsmen on the scene. They were standing and talking about 100 yards from the burning cars when an explosion threw a mushroom cloud of fire hundreds of feet in the air. "I was taking pictures, and I heard an explosion," Hardy said. "Suddenly there was fire everywhere. Before I knew what happened, my shirt caught fire and I dove for the nearest mud puddle." Hardy was burned on his back and both arms, but while lying in the puddle he grabbed a camera and tried to take more pictures.

'Ray Covey, 44, another Post photographer, received minor burns on both arms and a severe neck burn. He was in satisfactory condition today at Gulfway General Hospital. "I was standing 50 or 60 yards from the tank on fire," Covey said. I was just behind three firemen who were talking about not having enough water. They were watching almost as if hypnotized, but still no water came. I climbed up on their fire truck to shoot an overall. I turned toward the truck, and suddenly there was an explosion. Everything just blew up. Everything kept getting louder and louder and I started to run. I was scared, damn scared. I thought I'd had it," Covey said. "I wanted to save the cameras and the film and the film, but I didn't know I had them until later. As I ran, debris was flying everywhere. But I guess I'm fortunate. If my back hadn't been turned, I might not be able to tell you about it."

'A less fortunate photographer, Othell Owensby, of the Houston Chronicle, received burns over 80 per cent of his body. He was in satisfactory condition today in the intensive care unit of the Gulfway General.' [1652]

The National Transportation Safety Board investigated the accident.

The train had left Angleton, Texas, at 11.25 a.m. without a proper air brake test having been performed. After several miles it was stopped by an emergency brake application, due to a ruptured air hose. It then restarted but after 13 miles was stopped by a second emergency brake application initiated by the rear brakeman when he thought that he had seen something hanging down. Restarting again, the trip passed without incident till the train reach Mykawa, where a third automatic brake application occurred and the train was derailed.

The NTSB report is very lengthy, much too long to be given in full here, so we will content ourselves, as usual, with considering its conclusions, assessment of probable cause, and recommendations. Twenty-four conclusions were listed:

(1) Inspections of the train and track before and after the accident did not reveal any apparent cause of the derailment.

(2) The rocking tank car observed by one witness, the broken rails, and the loose wheel discovered in the wreckage all resulted from the derailment.

(3) The speed of Train 94 was within the Santa Fe's speed limits, even though the train crew did not have the correct information to determine the authorized speed.

(4) The speed of Train 94 contributed to the seriousness of the accident because it directly affected the kinetic energy that was dissipated by braking and derailment.

[1652] *Amarillo Globe Times*, Amarillo, Texas, United States, 20th October 1971

(5) When the train approached the accident site, all switch ties were in place and tamped. Alternate switch ties were spiked.

(6) The stability of the track had been affected adversely by work in progress on the installation of continuous, new switch ties, which involved the removal of rail anchors, and the spiking of alternate switch ties.

(7) The derailment of the 38th car and the 52nd car occurred at almost the same instant.

(8) The cause of an emergency brake application could not be established, but it was determined that the brake application was propagated before the 48th car passed over the newly installed switch ties.

(9) The unusual noise that various witnesses heard just before the derailment may have been the "chattering" associated with a brake application.

(10) Federal regulations for testing airbrakes were not observed when Train 94 was assembled at Angleton, although a Missouri Pacific supervisor was overseeing the work.

(11) A block of empty cars preceded a block of heavily loaded cars, and this increased the probability of excessive impact occurring within the train.

(12) The location of the 53rd car between a series of empty cars and a series of heavily loaded cars, coupled with the short truck spacing of the 53rd car, contributed to the development of high lateral forces during the emergency brake application. The track fastenings were not able to withstand these forces, which caused the rails to spread and overturn.

(13) Railroad personnel were on the scene when the accident occurred, but they did not help control the emergency.

(14) Everyone who was injured entered the danger zone after the derailment and was exposed to the injury-producing fireball and rocketing tank which followed the abrupt, violent rupture of tank car ESMX 4804.

(15) Information was inadequate for on-scene identification and assessment of the hazards, potential injury-producing events, and consequent response options.

(16) If firemen had known the names of the contents in the wrecked cars, it would not have lessened the injuries sustained.

(17) There were no violations of current, regulatory hazard-information requirements.

(18) A commodity-oriented placarding requirement, similar to that proposed in Hazardous Materials Regulation Board Docket HM-103, would not have affected the outcome of this accident.

(19) The full range of emergency response alternatives and the information needs of parties at risk in transportation accidents involving hazardous materials have not yet been set forth in a comprehensive document.

(20) Firemen reacted to the emergency promptly and followed procedures based on their prior experience and training, as well as on current, recommended firefighting practices.

(21) The absence of a logical classification theory in the current hazardous materials regulatory system of the Department of Transportation is an underlying problem

which contributes to information deficiencies experienced in transportation accidents.

(22) The number of firemen injured indicates that current, recommended firefighting practices are unsatisfactory in some transportation accidents.

(23) Recommended firefighting practices, which were not effective in this accident, are consensus recommendations.

(24) There are no documented quality standards for such recommendations.

As to the probable cause of the accident the NTSB wrote:

'The National Transportation Safety Board determines that the probable cause of this accident was an unexplained emergency brake application which induced lateral forces in the area of the 52nd car as the train's slack was taken up. These lateral forces exceeded the holding capability of the track fasteners and a rail overturned, which caused the following cars to jackknife. Tank cars ESMX 4803 and 4804 were punctured. The retardation of the empty cars which preceded the 52nd car was greater than the retardation of the more heavily loaded cars which followed. This variation in retardation contributed to the development of high lateral forces.

'The severity of the accident was increased by the abrupt, violent rupture of the tank car which contained a pressurized, flammable gas; the lack of adequate training, information, and documented procedures for on-scene identification and assessment of the threats to public safety; and firemen's reliance on firefighting recommendations which did not take into account the full range of hazards in such situations.'

Turning to its recommendations the NTSB stated:

(1) The Missouri Pacific Railroad Company take the necessary action to ensure that affected employees understand and comply with Federal regulations pertaining to the operation of trains.

(2) The Atchison, Topeka and Santa Fe Railway Company and the Missouri Pacific Railroad Company initiate programs to train affected employees in proper procedures to follow at accidents involving hazardous materials. These procedures should include methods for identification of and communication with local emergency service personnel.

(3) The Hazardous Materials Regulations Board hold in abeyance the establishment of a hazard-information system (Docket HM-103) until it develops the criteria which a regulatory threat-information system should address. These criteria should be solicited from and documented by the various parties at risk.

(4) The Secretary of Transportation initiate the development of a logical hazardous materials classification theory which will establish a rational basis for classifying hazardous materials, handled under normal transportation conditions and in transportation emergencies as well.

(5) The National Fire Protection Association, and similar standard-setting organizations, establish documented quality standards and quality control procedures for developing recommended practices to combat transportation emergencies which involve hazardous materials. These standards and procedures should include a

requirement for a technical safety analysis of their applicability and the risks associated with their use.

The benefits that will be derived through the implementation of Recommendations 3,4 and 5 will be continuous, but also will require some time to become fully effective. Recognizing this, the Safety Board reiterates and emphasizes the importance of the following recommendations made in previous accident reports:

Railroad Accident Report, Southern Railway Company, Laurel, Mississippi, January 25, 1969

5. "The Safety Board recommends that the Association of American Railroads and the American Short Line Railroad Association develop plans that will result in the fire chief of each community through which the track of a member road passes knowing where immediate information can be obtained, describing the location and characteristics of all hazardous materials in any train involved in a train accident that affects a community. This recommendation can be accomplished in a relatively short time regardless of the level of training which may be achieved later by fire departments." (Recommendation No. R-69-22)

Railroad Accident Report, NTSB-RAR-70-2, Illinois Central Railroad Company, Glendora, Mississippi, September 11, 1969

2. "The Federal Railroad Administration initiate research and development to provide prototype models of freight train braking systems.

(a) capable of providing shorter stopping distances which nearly approach the theoretical limit under all conditions of loading and length of trains,

(b) capable of stopping a train in the emergency applications now required by regulations without internal collisions, train separations, or damage to the train or its lading,

(c) capable of propagating brake application, both service and emergency, throughout the length of a train more expeditiously and surely,

(d) capable of more rapid application of the full intended stopping force to the rails at each car after the application signal is received at each car." [1653]

For a fortnight in December 1971 India and Pakistan were at war for the third time since the division of colonial India and the formation of what was then West Pakistan in 1948. During this brief conflict Indian aircraft attacked a passenger train near Narowal, Pakistan, on Monday 6th December 1971, killing five people and wounding 30. [1654]

An Illinois Central Railroad freight train was derailed at Corbin, Louisiana, United States, on Saturday 11th December 1971, it was believed due to vandals tampering with a switch. The engine and 17 cars left the rails and a tank car carrying chlorine was ruptured, the poisonous gas escaping. The population of the small settlement was evacuated until Monday, Sunday being spent transferring the remaining chlorine to a road

---

[1653] National Transportation Safety Board report

[1654] *Miami Daily News-Record*, Miami, Oklahoma, United States, 6th December 1971 tanker. There were also some small explosions and a fire. [1655] [1656]

The excessive speed of cars during hump shunting operations at the Gateway Yard, East St Louis, Illinois, United States, of the Alton and Southern Railway, at about 6.20 a.m. on Saturday 22nd January 1972, led to the collision of a tank car containing liquefied petroleum gas (LPG) with a hopper car, the puncturing of the tank car by the coupler of the hopper car and the release of a large cloud of LPG. Fire broke out, followed by a small explosion and then a second, more powerful explosion. No less than 223 people were injured, 19 of whom were admitted to hospital. Between 870 and 1,000 homes and other buildings, including a church and a school, were damaged and the total damage was estimated at $7,500,000. The National Transportation Safety Board investigated and made four recommendations:

(1) The Alton and Southern Railroad Company review the design of the hump at Gateway Yard and make those changes necessary to insure that all cars brought to the hump for classification will be handled without overspeed crashes. This review should include a systems analysis which covers grades, retarders, control for switching and maintaining programmed speed, provisions for backup, decelerating, and stopping devices, constant monitoring of overspeeds, and hump procedures in general.

(2) The Federal Railroad Administration establish a requirement that railroad carriers handle switching operations of cars containing large shipments of hazardous materials, with a danger range beyond railroad property boundaries, in the same manner as they handle switching operations of cars containing explosives.

(3) The Federal Railroad Administration develop requirements for the collection, documentation, and technical analysis of the principal ways in which the hazardous materials present produce injuries and damages in railroad accidents. Such documentation should identify the range of distances at which injuries or damages occurred.

(4) The Federal Railroad Administration review and, if necessary, revise its reporting requirements for accidents involving hazardous materials to obtain more accurate reporting of casualties adjacent to railroad premises. [1657]

The day after the East St Louis accident, Sunday 23rd January 1972, 24 of the 75 cars of a Louisville and Nashville Railroad freight train were derailed at Waitman, Kentucky, United States, and liquefied petroleum gas began to leak from one of the three cars carrying it. This time there was no fire and no one was hurt. A defective set of points were thought to have caused the derailment. [1658]

Also on Sunday 23rd January 1972 a Norfolk and Western Railway freight train was derailed south west of Clarksfield, Ohio, United States. Two diesel engines and 36 of the 160 cars of the train left the track at 4.30 p.m. and 18 of the cars fell from the trestle into the Vermilion River. A car containing 19,000 gallons of 32% hydrochloric acid came to rest on the river bank and acid leaked from a safety valve. It was hoped that the acid

---

[1655] *The Daily Star*, Hammond, Louisiana, United States, 13th December 1971

[1656] *The Monroe News-Star*, Monroe, United States, 13th December 1971

[1657] National Transportation Safety Board report

[1658] *The Vidette Messenger*, Valparaiso, Indiana, United States, 24th January 1972 would simply be absorbed into the ground but residents were warned not to use water from the river for human or animal consumption until it had been shown to be safe, for fear of possible contamination. The cause of the derailment was unknown. [1659]

Liquefied petroleum gas (LPG) was again involved in the derailment of a Toronto to Montreal freight train of the Canadian National Railway near Morrisburg, Ontario, Canada, at 8.15 a.m. on Friday 11th February 1972. Twenty cars left the road and in the usual sequence of events LPG gas escaped from a ruptured tank car, a fire broke out, with reportedly 100ft-high flames, and then three LPG tank cars exploded, one after the other. Alex English, living six miles away, thought that it was an earthquake. No one was hurt, though a dozen men cutting dead elm trees as part of a federal winter work project had to flee on their hands and knees to escape flying pieces of metal. [1660] [1661]

An estimated 2,100 people were forced to flee on Monday 29th May 1972 after 22 of the 80 cars of an eastbound Penn Central Railroad train from St Louis, Missouri, to New York were derailed near Ashmore, Illinois, United States, on the Cleveland Cincinnati Chicago and St Louis Railway. A 40-square-mile evacuation of the rural area was ordered after the overturning and rupturing of a tank car containing over 15,000 gallons of nitric acid, 90% of which was lost. Bright orange nitric acid fumes spread over the area. Twelve people complained of burning sensations in their throats and lungs from inhaling the corrosive fumes and were treated at Charleston Community Hospital and released, while it was feared that others might suffer a delayed reaction to the fumes. Dams were bulldozed around the derailment site in an effort to prevent the acid entering the water supply and farmers were advised not to let their livestock drink from possibly contaminated creeks. [1662]

Shortly before midnight on Thursday 1st June 1972 a 66-car Missouri Pacific Railroad freight was leaving the Texas and Pacific Railway yard at Dallas, Texas, United States, for Minneola, Texas, when five of its cars left the track, two of them loaded with liquefied petroleum gas (LPG). The LPG leaked from a damaged car, there was an explosion and a fierce fire. Fifteen people suffered minor injuries from the blast or fire, over 200 people fled to a Red Cross shelter, five homes were destroyed and 40 or 50 were damaged. An estimate of the financial cost of the damage put it at over $100,000. Fireman declared the blaze under control at 7.30 a.m. on Friday and decided to let it burn itself out rather than attempting to extinguish it and possibly leaving pockets of residual gas, which might explode and involve other LPG tank cars. [1663] [1664]

Twelve cars of an 83-car eastbound Western Maryland Railway freight train left the track near Mount Savage, Maryland, United States, on Monday 26th June 1972. Three of the derailed vehicles were tank cars, each containing 40,000 gallons of corrosive and toxic carbolic acid (phenol). At least one of the tank cars was ruptured and its contents

---

[1659] *Norwalk Reflector*, Norwalk, Ohio, United States, 24th January 1972
[1660] *Winnipeg Free Press*, Winnipeg, Manitoba, Canada, 11th February 1972
[1661] *The News*, Medicine Hat, Alberta, Canada, 12th February 1972
[1662] *Daily Republican Register*, Mount Carmel, Illinois, United States, 30th May 1972
[1663] *Lubbock Avalanche-Journal*, Lubbock, Texas, United States, 3rd June 1972
[1664] *McKinney Daily Courier-Gazette*, McKinney, Texas, United States, 2nd June 1972

flowed down a hillside towards the small community of Slabtown, leading to the evacuation of five families. Trains were rerouted while 40 volunteer firemen from three towns washed down the spillage under specialist direction. No one was hurt and no cause was given for the derailment. [1665]

Twenty cars of a Penn Central Railroad freight train derailed at Wolcott, Indiana, United States, on Monday 3rd July 1972 and a tank car loaded with 190 degree proof (95%) grain alcohol shed its contents over the right of way. Sheriff's deputies kept spectators away, warning that the slightest spark could set off a tremendous blaze. The alcohol evaporated without incident. [1666] [1667]

A Chesapeake and Ohio Railway freight train was in trouble near Bath, Indiana, United States, on Friday 7th July 1972, and there was a small spillage from a tank car of hydrogen peroxide and another car carrying activated carbon. Firemen were called and one stated that "there was a small bit of fire but not much danger of an explosion." The fire was quickly extinguished and work began to re-rail the cars. [1668]

Either late on Saturday 29th July or early on Sunday 30th 1972, 13 cars of a Penn Central Railroad train were derailed at Jamestown, Indiana, United States, and there was a leakage of carbon disulphide. About 45 people within a half-mile radius were evacuated for several hours as a precaution and traffic was rerouted, carbon disulphide being flammable and fairly toxic. No one was hurt. The cause of the derailment was not immediately known. [1669]

Several cars of an eastbound Southern Pacific Company train were derailed near Thousand Palms, California, United States on Wednesday 26th July 1972 and there was an escape of yet another flammable and toxic chemical, this time mesityl oxide (4-Methylpent-3-en-2-one). An unstated number of the repair crew and a sheriff's deputy had to be treated for inhalation of the gas and heat exhaustion and some needed hospitalisation. There was no evacuation. [1670]

East River Crossing, Mahoningtown, Pennsylvania, United States, was the location of the derailment of the Baltimore and Ohio Railroad *New Yorker* freight train at 2.30 p.m. on Friday 18th August 1972. Thirty-three of its 90 cars left the track. The train was carrying automobile parts, furniture, lumber, lube oil and caustic soda (sodium hydroxide). The lube oil and the sodium hydroxide were reportedly spilt. A broken axle was proposed as the cause of the derailment. It was said that if any road vehicles had been on the crossing at the time, "they would have been squashed like pancakes." [1671]

A loose wheel was said to have been the cause of the derailment of a Southern Pacific Company freight train travelling at 40 miles an hour near Jaynes, Arizona, United States, north of Tucson, on Monday 20th November 1972. Twelve cars left the rails, including a tank car loaded with 30,000 gallons of toxic and flammable methanol, which ruptured

---

[1665] *The Daily Times*, Salisbury, Maryland, United States, 27th June 1972
[1666] *Amarillo Globe Times*, Amarillo, Texas, United States, 4th July 1972
[1667] *The Anderson Herald*, Anderson, Indiana, United States, 6th July 1972
[1668] *Anderson Daily Bulletin*, Anderson, Indiana, United States, 8th July 1972
[1669] *Seymour Daily Tribune*, Seymour, Indiana, United States, 31st July 1972
[1670] *Independent*, Long Beach, California, United States, 27th July 1972
[1671] *New Castle News*, New Castle, Pennsylvania, United States, 19th August 1972

and lost its contents. About 200 residents and 344 children and teachers of the Laguna Elementary School were evacuated. The only person hurt was the driver of a pickup truck stopped at a level crossing whose vehicle was crushed by the derailed cars. The remaining methanol was transferred to a road tanker. [1672] [1673]

In another derailment of a westbound 18-car Southern Pacific Railroad freight train at Tucson, Arizona, United States, at 9.25 a.m. on Thursday 7th December 1972 flammable and harmful methyl methacrylate was spilt. Two box cars and the tank car containing the chemical just behind the engine left the rails. Only a small amount of chemical was lost and firemen sprayed foam on it to reduce the likelihood of a fire. There were no buildings within a mile of the derailment but spectators were kept at bay. A broken drawbar was given as the cause of the derailment. [1674] [1675]

The train derailments involving the spillage of dangerous chemicals continued with one on the Louisville and Nashville Railroad near La Follette, Tennessee, United States, on Tuesday 12th December 1972. Twelve cars left the track near Duff, Tennessee, United States, 12 miles north of La Follette, and the chemical to escape was a small amount of toxic and carcinogenic orthotoluidine. The residents of nearby Cotula were warned not to drink the water from Davis Creek or let their livestock do so, and a ditch was dug to contain the spill. No one was hurt and the cause of the accident was unknown. [1676]

The next accident, on Thursday 4th January 1973, involved a fatality. Twenty-one units of a 60-car Erie-Lackawanna Railroad freight train were derailed in the middle of Harrod, Ohio, United States. A tank car of pentane was ruptured, the pentane escaped, caught fire and the intense heat caused three portable lineside tanks of anhydrous ammonia to explode. Betty Sherard, 14, standing on the front porch of her house, was struck and killed by a piece of flying debris, in effect shrapnel. Her father was also struck and his leg was broken. Firemen spread foam over the remaining ammonia tanks and then let the fire burn itself out. Damage, mostly to railroad equipment, was estimated at $1,000,000. [1677]

A 'hot box' was put down as the cause of the derailment of a Penn Central Railroad train on Monday 5th February 1973 near Downingtown, Pennsylvania, United States. At least 7,000 gallons of a mixture of nitric and sulphuric acid escaped into Brandywine Creek when four cars fell from a bridge and thick yellowish clouds of fumes filled the area. Squads of police and firemen went from door to door to alert residents and 1,500 people had to evacuate, though officials stated that there was no danger of contamination of the water supply. It was planned to neutralise the acid with two tons of soda ash (sodium carbonate). At least 20 people were treated for minor injuries. [1678]

According to news reports the Town Marshal at Andrews, Indiana, United States, Ed Bigelow, attempted to telephone the offices of the Norfolk and Western Railway to report

---

[1672] *Tucson Daily Citizen*, Tucson Arizona, United States, 21st and 22nd November 1972
[1673] *The Arizona Republic*, Phoenix, Arizona, United States, 21st November 1972
[1674] Ibid, 8th December 1972
[1675] *Tucson Daily Citizen*, Tucson Arizona, United States, 8th December 1972
[1676] *Northwest Arkansas Times*, Fayetteville, Arkansas, United States, 13th December 1972
[1677] *The Chronicle-Telegram*, Elyria, Ohio, United States, 5th January 1973
[1678] *Lebanon Daily News*, Lebanon, Pennsylvania, United States, 6th February 1973

a broken rail there on Wednesday 7th February 1973, but they had refused to accept the collect call so he had said, "To hell with you," and hung up on them. About half an hour later 20 vehicles of a 75-car train were derailed near the spot and there was a leakage of crude oil and asphalt. Railroad officials blamed the derailment on a faulty wheel bearing rather than the track. [1679]

Three engines and 24 of 27 cars of a Canadian National Railway freight train were derailed at Loos, British Colombia, Canada, either late on Saturday 3rd or early on Sunday 4th March 1973. One of three tank cars containing chlorine was ruptured, releasing the gas, and there were conflicting reports as to whether sodium hydroxide had escaped from other cars. Dow Chemicals experts were being flown to the scene to help with the clear up. The nearest residence was a mile and a half away and there was no evacuation, though residents were alerted as to the danger. Two railwaymen were treated for slight chlorine poisoning and released. [1680] [1681] [1682]

A Southern Railway freight train was derailed at Lovick, Alabama, United States, on Sunday 1st April 1973. Cars loaded with ammonia and caustic soda (sodium hydroxide) were involved and the ammonia escaped, leading to the evacuation of 300 people. Most were allowed to return home later in the day but those who lived closest to the scene were advised to stay away longer. Nobody was hurt but nearby trees were stripped of their leaves. [1683] [1684]

About 300 residents of Lufkin, Texas, United States, were evacuated after the derailment of 12 cars of a Southern Pacific Company freight train late on Thursday 3rd May 1973, and the leakage of vinyl chloride from a ruptured tank car, while roads were closed. By morning the toxic and flammable gas had all evaporated and the residents were allowed to return. No one was hurt. [1685] [1686]

It was ammonia that was again involved in the derailment of a train at Arnold, Missouri, United States, late on Wednesday 9th May 1973. Thirteen cars of a St Louis-San Francisco Railway train left the rails and a tank car of anhydrous ammonia was ruptured, leaking to a small leak. Thirty residents were evacuated until the leak was capped after two hours. Two people were slightly hurt. [1687] [1688]

Early on Wednesday 23rd May 1973 600 residents of Mentor, Ohio, United States had to be evacuated after the derailment of 14 cars of a 69-car Penn Central Railroad freight train from Cleveland, Ohio, to Selkirk, New York, and the leakage of corrosive

---

[1679] *New Castle News*, New Castle, Pennsylvania, United States, 9th February 1973
[1680] *Escanaba Daily Press*, Escanaba, Michigan, United States, 5th March 1973
[1681] *Fairbanks Daily News-Miner*, Fairbanks, Alaska, United States, 5th March 1973
[1682] *The Evening News*, Sault Ste Marie, Michigan, United States, 5th March 1973
[1683] *Athens News Courier*, Athens, Alabama, United States, 3rd April 1973
[1684] *The Anniston Star*, Anniston, Alabama, United States, 2nd April 1973
[1685] *The Odessa American*, Odessa, Texas, United States, 4th March 1973
[1686] *The Ennis Daily News*, Ennis, Texas, United States, 4th March 1973
[1687] *The Joplin Globe*, Joplin, Missouri, United States, 11th May 1973
[1688] *Jefferson City Post-Tribune*, Jefferson City, Missouri, United States, 10th May 1973

chlorosulphonic acid from one of four tank cars carrying it. No one was hurt. [1689]

The derailment of an Atchison Topeka and Santa Fe Railway freight train on a trestle near Chanute, Kansas, United States, on Friday 6th July 1973 resulted in one tank car containing acetic anhydride falling into the lake and a second tank car overturning on the trestle. The corrosive chemical escaped into the water from both and by Sunday had killed a large number of fish. [1690]

On Friday 20th July 1973 13 cars of an eastbound Penn Central Railroad freight train were derailed on the edge of Fort Wayne, Indiana, United States. Vinyl chloride leaked from a ruptured tank car, and there was a fire that involved a second tank car carrying the same chemical. About 4,000 people were evacuated from a two-mile-square area and three hours after the initial derailment there was a small explosion that demolished a frame house. No one was hurt and the decision was taken to let the fire burn itself out. [1691] Most residents were able to return home the next day, except those who lived closed to the accident site. [1692]

Illinois Central Gulf Railroad train No. 93 from Centralia to Cairo became derailed north of Du Quoin, Illinois, United States, on Friday 17th August 1973. Nine box cars, a piggyback car, a coal car, a rack car, and a tank car left the rails and the rank car leaked fuel oil. [1693]

A semi-trailer (articulated lorry) was driven into the side of a Union Pacific Railroad freight train on a level crossing with Highway 61 south of Cortland, Nebraska, United States, on the afternoon of Tuesday 21st August 1973. The lorry struck the train about six cars back from its engine, derailing about 30 cars. A tank car of anhydrous ammonia burst open and another was leaking. The lorry driver was killed but no one was hurt by the escaped ammonia, though its fumes burned several acres of crops. The lorry was presumably either being driven too fast or its brakes had failed. [1694]

Early on Monday 17th September 1973 a Toledo, Ohio, to Detroit, Michigan Chesapeake and Ohio Railway train was stopped at a signal at Dearborn, Michigan, United States, when it was run into from behind by a yard engine pulling 25 cars. A westbound train leaving the yard then collided with the wreckage. A tank car began leaking harmful formaldehyde following the double collision and several families living nearby were evacuated. Seven railwaymen were injured in the collision, two of whom were admitted to hospital. Firemen washed down spilt chemical. [1695]

A 181-car northbound Illinois Central Gulf Railroad freight train as derailed at about 3.50 a.m. on Tuesday 25th September 1973 near Ruby, Mississippi, United States. Sixteen cars left the track, five of which were tank cars, four containing vinyl chloride, one of which was breached; and a fifth car carrying a different unidentified gas. About 50 residents were evacuated and nearby roads were closed but the danger posed by the leakage

---

[1689] *The Derrick*, Oil City, Pennsylvania, United States, 24th May 1973
[1690] *The Chanute Tribune*, Chanute, Kansas, United States, 9th July 1973
[1691] *The Times-Bulletin*, Van Wert, Ohio, United States, 21st July 1973
[1692] *Greensburg Daily Times*, Greensburg, Indiana, United States, 21st July 1973
[1693] *Southern Illinoisian*, Carbondale, Illinois, United States, 19th August 1973
[1694] *The Lincoln Star*, Lincoln, Nebraska, United States, 22nd August 1973
[1695] *Marshall Evening Chronicle*, Marshall, Michigan, United States, 18th September 1973

was played down as it had occurred in a rural location. The derailment was blamed on a blamed on a square lead ingot either falling from or through the floor of a car. [1696]

A 119-car Seaboard Coast Line Railroad freight train from New Orleans, Louisiana, to Jacksonville, Florida, had about 10 cars derail at about 3.30 a.m. on Wednesday 24th October 1973 near Ochlocknee, Florida, United States. The accident occurred in a rural area and no one was hurt but police sealed off a small area near the accident site and the few people living nearby were alerted but not evacuated. Workmen wearing gas masks transferred the remaining ammonia to an empty tank car. The unaffected part of the train continued its journey, only to suffer a second derailment near Madison about 8 a.m. [1697]

The accident at Ventura, California, United States on Tuesday 6th November 1973, which involved a leakage of liquefied petroleum gas, has already been dealt with in the Vandalism chapter, above.

About noon on Sunday 9th December 1973 21 vehicles of a 117-car Burlington Northern Inc freight train from Staples to Minneapolis, Minnesota, became derailed at Elk River, Minnesota, United States, and an estimated 90,000 gallons of No. 4 fuel oil poured onto Highway US 10-52, congealing as it came into contact with the cold air. No one was hurt. Sand barricades were used to contain the oil and more sand was poured onto the oil to form a mixture that could be removed. A little oil entered the Mississippi River. [1698] [1699]

A Southern Railway freight train was derailed near Gilbert, South Carolina, United States, on Friday 28th December 1973. One of three tank cars of fuel oil ruptured and at least 20,000 gallons escaped. There was no fire and no one was hurt. [1700]

Forty-four cars of another Southern Railway freight train were derailed near Columbia, South Carolina, United States, on Monday 31st December 1973, and sulphuric acid leaked from two tank cars into a ditch. No one was hurt. [1701] [1702]

On Thursday 10th January 1974 14 cars of an 81-car Penn Central Railroad freight train were derailed in a wooded area near Quarryville, Pennsylvania, United States, and petroleum naphtha was lost from a ruptured car. Firemen were called but were not needed. No one was hurt and the leak was soon patched up. [1703]

Ten cars of a 67-car Lehigh Valley Railroad freight train left the rails on the night of Monday 14th January 1974 in a sparsely populated area near White Haven, Pennsylvania, United States. Four of the derailed cars were loaded with compressed chlorine gas, some of which escaped, and police evacuated residents within half a mile of the accident site.

[1696] *Delta Democrat-Times*, Greenville, Mississippi, United States, 25th September 1973
[1697] *News-Herald*, Panama City, Florida, United States, 25th October 1973
[1698] *The Evening Tribune*, Albert Lea, Minnesota, United States, 10th December 1973
[1699] *The Daily Journal*, Fergus Falls, Minnesota, United States, 10th December 1973
[1700] *Florence Morning News*, Florence, South Carolina, United States, 29th December 1973
[1701] Ibid, 1st January 1974
[1702] *The Gastonia Gazette*, Gastonia, North Carolina, United States, 1st January 1974
[1703] *Lebanon Daily News*, Lebanon, Pennsylvania, United States, 11th January 1974

[1704] [1705]

Twenty-one cars of a 52-car Colorado and Southern Railway freight train from Cheyenne to Casper, Wyoming, were derailed near Horse Creek, Wyoming, United States, on Monday 14th January 1974. Carbolic acid (phenol) was spilt from as tank car. Eleven workers were slightly injured in clearing the wreckage and had to be treated at the Laramie County Memorial Hospital in Cheyenne. [1706]

At about 4.20 p.m. on Tuesday 12th February 1974 Delaware and Hudson Railway freight train No. NWB-4 was derailed just north of Oneonta, New York, United States. The derailed cars included seven carrying liquefied petroleum gas (LPG), one of which was ruptured and the escaping gas ignited. By about half an hour after the derailment four other LPG cars had ruptured. Fifty-four firemen and press reporters were injured in the fire and explosions.

The train consisted of three locomotive units and 122 cars in charge of Conductor Harvey Boone, and had left Binghampton, New York, at 1.55 p.m., bound for Mechanicsville, New York. [1707]

The accident was investigated by the National Transportation Safety Board (NTSB), whose report listed 16 conclusions:

(1) D&H train NWB-4 was being operated according to company procedures as it entered the 3° 30′ curve.

(2) No. 1 track in the area of the derailment probably met Federal track safety standards.

(3) Inspections made in Binghampton and in Oneonta and observations en route did not detect the defect in Car 4, because the cracked centre sill was not visible to inspectors.

(4) The fracture in the centre sill started at a rivet hole and progressed across the vertical web of the sill.

(5) The center sill was deflected downwards by weight and vibration in the vicinity of the crack which raise the coupler end and allowed it to pull over the coupler of car No. 3. Consequently the train separated.

(6) The emergency brake application following the train separation as the normally expected action under the provisions of 49 CFR 232. The force which overturned the east rail was the lateral component of a longitudinal compressive force in the train, generated by more effective emergency brake application at the locomotives.

(7) Whether two or three spikes were used to secure the rail on the 3° 30′ curve could not be determined.

(8) The number of defective ties, if any, on track No. 1 in the 3° 30′ curve could not be determined.

[1704] *The Evening Standard*, Uniontown, Pennsylvania, United States, 15th January 1974

[1705] *The Evening Independent*, Massillon, Ohio, United States, 15th January 1974

[1706] *The Billings Gazette*, Billings, Montana, United States, 16th January 1974

[1707] *The Oneonta Star*, Oneonta, New York, United States, 13th and 14th February 1974

(9) The east rail of No. 1 track was canted eastward as the locomotive of NWB-4 moved northward around the curve, which resulted in wide gage.

(10) The fire in the derailment resulted in the ignition, from an unknown source, of escaping LPG from the crushed tank of the 21st car.

(11) The violent rupture of four tanks was the result of impingement of fire on uninsulated tank sheets which heated and weakened the steel.

(12) Regulations have not been promulgated by the FRA {Federal Railroad Administration} which would eliminate or reduce the severity of the problem which arises when tank cars loaded with hazardous materials are exposed to an accident or a fire.

(13) Recommendations have not been made by the RPI-AAR {Railway Progress Institute-Association of American Railroads} Railroad Tank Car Safety Research Test Project that would help to eliminate or reduce the severity of accidents and fires which involve tank cars loaded with hazardous materials.

(14) Firemen had not received information from the D&H on matters that would have aided them in combatting the fire.

(15) Firemen had not been trained to combat fires of such magnitude. Further, they had not been trained for or exposed to information on the hazards that can result when tank cars loaded with hazardous materials are exposed to fire.

(16) The propulsion of large portions of tank cars 1,200 feet from the site was a result of rocketing action already displayed by type 112A and 114A tank cars in previous accidents.

The NTSB decided that the probable cause of the accident was 'the inability of the track to resist the lateral forces which canted the outside rail outward and widened the gage of the track. These forces which were induced at the third locomotive unit resulted from the emergency application of the brakes when the train was separated between the third and fourth cars as it entered the 3° 30′ curve. The train separated as a result of the broken center sill on the fourth car.'

The NTSB made four recommendations:

(a) That the FRA ensure that the inspections for defects in other cars with a similar center sill design be continued until all such cars had been inspected.

(b) The FRA should re-evaluate the Federal Track Standards to determine whether the spiking and crosstie requirements for tracks on curves were adequate to prevent the rails from spreading when subjected to lateral forces during an emergency brake applications of trains of various consists travelling at various speeds.

(c) Using the information from their own tank cars studies and those of the RPI-AAR the FRA should expedite the promulgation of regulations to eliminate or reduce the severity of the losses to the community when type 112A and 114A tank cars were exposed to accident or fire.

(d) The FRA should promulgate regulations to require railroads to provide pertinent lading information to the train crews of trains carrying hazardous materials as they were for explosives. These notices should include the name of the commodity and its hazard rating.

Recommendations made after the Laurel accident (see above) regarding ensuring local fire chiefs were able to quickly obtain information regarding hazardous materials being carried on trains passing through their areas, and after the Glendora accident (see above) regarding freight train braking systems, were reiterated. [1708]

Seven tank cars of a 113-car Southern Pacific Company freight train bound for Pine Bluff, Arkansas, were derailed at Colmesneil, Texas, United States, late on Saturday 16th February 1974. The contents of the tank cars, reportedly liquefied petroleum gas (LPG) and vinyl chloride, were spilt across US Highway 69 and caught fire. Three of the tank cars exploded and flames were said to have risen almost 1,000 feet into the air. An unoccupied house was engulfed by fire and destroyed but fortunately there were no vehicles on the highway at the time. An area of about half a mile radius around the accident site was evacuated and firemen from several towns fought the fire. No one was hurt and people began to return home on Sunday as soon as the danger of further explosions was past. Conductor H.A. Gardner of the train declared, "It was bound to have been a broken rail. An air hose broke and it disconnected the seven cars from the front of the train." [1709]

Another Southern Pacific Company freight train of 47 cars derailed just before 11 p.m. on Thursday 14th March 1974 near Dayton, Texas, United States, while negotiating a shoo-fly temporary track that had been laid after the derailment of an Amtrak passenger train at the same spot on the previous Monday. At least one of two tank cars of vinyl acetate leaked its contents, there was an explosion and a fire that also involved two box cars loaded with paper. No one was hurt and the fire was left to burn itself out. [1710] [1711]

Vinyl chloride was again involved in the derailment of a Penn Central Railroad freight train at Terre Haute, Indiana, United States, on the evening of Sunday 17th March 1974. Seven cars left the track, two of which carried 110 tons of the chemical, some of which at any rate escaped, and a thick white cloud hovered over the scene. Several hundred residents had to be evacuated as power supplies were cut to reduce the risk of an explosion and fireman wearing gas masks laid down foam. No one was hurt. [1712]

On the night of Friday 12th April 1974 a named Chesapeake and Ohio Railway freight train of 83 cars, *The Manhattan*, was passing through Clarksburg, West Virginia, United States, moving at about 15 miles an hour, when it became derailed from an unknown cause. Ten cars were derailed, two of which contained ethyl alcohol (ethanol), one of which was ruptured. The escaped alcohol caught fire and there were at least eight explosions. Two or three families were evacuated and at least 14 firemen were slightly hurt fighting the fire. Property damage was estimated at $40,000. [1713]

A Chesapeake and Ohio Railway freight train en route from Wheeling, West Virginia, to Pittsburgh, Pennsylvania, was derailed at Eighty-Four, Pennsylvania, United States, on

[1708] National Transportation Safety Board report
[1709] *The Victoria Advocate*, Victoria, Texas, United States, 18th February 1974
[1710] *The News*, Port Arthur, Texas, United States, 15th March 1974
[1711] *The Baytown Sun*, Baytown, Port Arthur, Texas, United States, 15th March 1974
[1712] *The Brazil Daily Times*, Brazil, Indiana, United States, 18th March 1974
[1713] *The Dominion Post*, Morgantown, West Virginia, United States, 14th April 1974

Saturday 13th April 1974 and a tank car of hydrochloric acid ended up in Little Chartiers Creek. Twenty-seven of the 65 cars of the train derailed in all, mostly carrying grain but one liquefied petroleum gas (LPG) so for a while there were fears of a possible explosion. The acid got into the stream and five bags of soda (sodium carbonate) were tipped in to try and neutralise it. Evacuation of 300 residents was considered but was not done. [1714]

In the third derailment involving dangerous chemicals in three days just after 11 p.m. on Sunday 15th April 1974 a 52-car Penn Central Railroad freight train bound for Flemington, New Jersey, was derailed a mile from 30th Street station, Philadelphia, Pennsylvania, United States. Only three cars left the rails but one contained toxic vinyl chloride, which spilt onto West River Drive, a major artery. The thoroughfare was blocked off by police while firemen hosed down the spillage. No one was hurt but police predicted that there would be a massive traffic jam during the Monday morning rush hour. The tank car held 38,000 gallons of the chemical but it was not known if it had been full at the time of the accident. [1715]

Another Penn Central Railroad freight train derailed at Milford, Ohio, United States, on Thursday 30th May 1974, 18 cars of the 104-car train leaving the rails, and a large part of the 20,000 gallons of oil in a tank car was lost into the Little Miami River. The remainder of the oil was siphoned from the tank car and efforts were made to skim some of it from the river. [1716] [1717]

Six cars of a Chicago Milwaukee St Paul and Pacific Railroad freight train left the rails late on Tuesday 4th June 1974 near Taft Tunnel, Idaho, United States, close to the state line with Montana. Two tank cars ruptured and 25,000 gallons of caustic soda (sodium hydroxide) and 10,000 gallons of aluminium sulphate were spilt into Cliff Creek, a tributary of the St Joe River. No ill effects to humans were anticipated but it was not known if any fish had suffered. [1718]

The power of the wind was demonstrated on Monday 10th June 1974 when a westbound Southern Pacific Company freight train was derailed at about 6.55 p.m. near Vail, Arizona, United States. Hollow-core aluminium plates loaded onto a flatcar were lifted up by the wind, the restraining bands were broken and the car tipped over. Seventeen other cars were derailed in turn and oil from a number of 55 gallon drums was spilt into a ditch. No one was hurt. [1719] [1720]

On Wednesday 12th June 1974 35 cars of an Atchison Topeka and Santa Fe Railway freight train left the rails near Norman, Oklahoma, United States. Among them was a box car loaded with 55 gallon drums of toxic phosphorous trichloride. At least some of the drums were damaged and by the next day a dense white cloud of fumes had spread over a 12-mile-square area. The police and the National Guard evacuated 7,000 people from Norman, Moore, and Oklahoma City and 150 people were treated for poisoning, though only five had to be admitted to hospital. The all-clear was announced at 5.30 p.m. by

---

[1714] *The Evening Standard*, Uniontown, Pennsylvania, United States, 15th April 1974
[1715] *The Evening Sun*, Hanover, Pennsylvania, United States, 15th April 1974
[1716] *Lancaster Evening Gazette*, Lancaster, Ohio, United States, 1st June 1974
[1717] The *Journal-News*, Hamilton, Ohio, United States, 2nd June 1974
[1718] *Idaho State Journal*, Pocatello, Idaho, United States, 6th June 1974
[1719] *Tucson Daily Citizen*, Tucson, Arizona, United States, 11th June 1974
[1720] *The Yuma Daily Sun*, Yuma, Arizona, United States, 18th June 1974

which time all the drums were either intact, empty, or buried. Returning residents were cautioned to thoroughly cleanse everything in their homes, especially food and cooking utensils. The damage was estimated at $3,500,000. [1721]

A St Louis Southwestern Railway freight train running on Missouri Pacific Railroad tracks was derailed near Wolf Lake, Illinois, United States, at about 4.30 p.m. on Friday 21st June 1974. A good quantity of acetic anhydride was spilt, most of it going into the six-mile-long lake, which it was expected would dilute the chemical and render it harmless. There was no evacuation even though at least two people were injured by exposure to the chemical, lineside vegetation at the site and water lilies in the lake were harmed and it was thought that some fish would be killed. Damage was estimated at over $500,000. [1722]

At noon on Thursday 27th June 1974 a northbound Penn Central Railroad train was derailed at Adams, Massachusetts, United States, and sulphuric acid leaked from one of three 10,000 gallon capacity overturned tank cars. There was a brief evacuation of 250 workers at the Petricca and the General Electric Ordnance Co. Plant 3 businesses but they were able to return at 1.30 p.m. Two secretaries and a policeman needed treatment after inhaling the acid fumes. The poor condition of the track was blamed for the derailment. [1723]

The derailment of a Norfolk and Western Railway freight train at Portsmouth, Ohio, United States, on Sunday 30th June 1974 involved three tank cars containing toxic (ortho)phosphoric acid, two of which began to leak. The situation was soon being described as "well under control," a ditch being dug to contain the spillage. No one was hurt. [1724]

Distortion of the track on a hot day due to insufficient expansion gaps between lengths of rail caused the derailment of a 54-car Chicago Milwaukee St Paul and Pacific Railroad freight train just after 3 p.m. on Tuesday 16th July 1974 at Lake Beulah, Wisconsin, United States. Eighteen cars in all left the rails and one of them, containing carbolic acid (phenol) was punctured. No one was hurt. [1725]

A collision during shunting operations at Decatur Yard, Illinois, United States, on the Norfolk and Western Railway, on Friday 19th July 1974, resulted in a disastrous explosion which fatally burned seven railwaymen nearby, injured 33 railwaymen and 316 residents of Decatur, and resulted in property damage estimated at $18,400,000.

During the shunting operations five jumbo boxcars loaded with isobutane were shunted onto track No. 11 and rolled into an empty boxcar. As a result of the collision a 26-inch puncture hole was made in the leading loaded boxcar, from which liquid isobutane escaped at an estimated rate of about 5,000 gallons a minute. The isobutane vapourised and was spread south-eastwards by a north-west wind for about 8 or 10 minutes. Unaware of what had happened the switching crew continued to shunt cars on tracks Nos. 10 and 12, when they noticed a dense grey cloud near a railroad dormitory.

[1721] *The Times-Standard*, Eureka, California, United States, 14th June 1974
[1722] *Southern Illinoisian*, Carbondale Illinois, United States, 23rd June 1974
[1723] *The Berkshire Eagle*, Pittsfield, Massachusetts, United States, 28th June 1974
[1724] *Logan Daily News*, Logan, Ohio, United States, 1st July 1974
[1725] *The Janesville Gazette*, Janesville, Wisconsin, United States, 17th July 1974

At that moment an unknown source caused the escaped gas to explode at 5.03 a.m.

The fire and concussion demolished 283 freight cars in the yard and damaged 312 others. The dormitory was demolished and those who lost their lives had been stood outside of it. About 700 residences, 31 commercial residences and 11 schools, especially Lakeview High School, which adjoined the yard, were damaged. Sixty-seven residences were declared uninhabitable and part of Lakeview School had to be demolished.

The National Transportation Safety Board (NTSB) investigated the accident. Its report came to eleven conclusions:

(1) The head of tank car GATX 41623 was punctured by the coupler of N&W boxcar 49203.

(2) The coupler of N&W 49203 overrode the coupler of GATX 41623 because of an attempted coupling at an undetermined speed which was equal to or exceeded 12.7 m.p.h.

(3) The overspeed impact resulted because switchmen released the five-car cut of tank cars at an excessive speed.

(4) The blast resulted from the explosion in open air of vaporized isobutane which had escaped from the punctured tank car, GATX 41623

(5) The crew that switched the tank cars was not trained and instructed adequately about risks involved in the switching of tank cars of flammable compressed gases.

(6) The N&W did not have a plan or a system that made firefighting materials convenient and readily available to combat fires in Decatur Yard.

(7) Head shields, or top and bottom shelf couplers, on GATX 41623 might have prevented the puncture of the tank head; however, head shields and top and bottom shelf couplers cannot be relied upon to eliminate completely all tank car punctures.

(8) Requirements for limiting losses when the heads of tank cars containing hazardous materials are punctured need to be considered in the regulations.

(9) GATX 41623, its lading, and its handling complied with requirements of Federal Regulations and railroad interchange rules.

(10) Reporting requirements for DOT {Department of Transportation} form F-5800.1 need to be amended to provide information to support implementation of the regulatory approach that the degree of protection should reflect the degree of severity in hazardous materials accidents.

(11) Federal requirements for reporting railroad accidents which involve hazardous materials distort intermodal safety comparisons and do not require sufficient information to assess accurately the hazards to the public from various materials.

The NTSB determined that the probable cause of the accident was 'the overspeed impact between the heavy cut of tank cars and the uncoupled light boxcar, which resulted from the release of the tank cars at a higher-than-acceptable switching speed. The lack of written guidelines to assist the switchman in determining the proper switching speed contributed to the accident. The crewmembers' lack of understanding of the risks involved in switching hazardous materials was also a contributing factor.'

The NTSB report made five recommendations:

(a) The FRA, in co-operation with the RPI and the AAR, should promulgate regulations to limit losses in accidents involving the transportation of hazardous materials by rail.

(b) The FRA, RPI and AAR should issue regulations requiring that DOT 112A and 114A tank cars be equipped with the best possible combination of top and bottom shelf couplers and head shields.

(c) The Norfolk and Western company should ensure that yard employees involved in the handling of hazardous materials were cognizant of the risks associated with the switching of them and require the switching to be done accordingly.

(d) The Norfolk and Western company should establish a plan to deal with emergencies such as that which had occurred at Decatur Yard including an adequate fire control system.

(e) The Secretary of Transportation should revise form F-5800.1 to obtain information required to support the rulemaking approach so that the degree of protection reflected the degree of severity of specific commodities in accidents. Such changes should at least address the delineation of the danger zone, and the types and degrees of injury or damages suffered by the various kinds of parties at risk.

The recommendation to the FRA issued after the Gateway Yard accident (see above) that it review its reporting requirements for accidents involving hazardous materials (which, the NTSB noted, had not been fully implemented), was reiterated and re-emphasised. [1726]

Eight cars and the caboose of a 117-car Illinois Central Gulf Railroad (ICG) freight train from New Orleans, Louisiana, to Chicago, Illinois, were derailed on the night of Sunday 4th August 1974 at Peotone, Illinois, United States, and a tank car spilt poisonous ethylene glycol, leaking to the evacuation of 500 people. Some of the chemical evaporated but most of it soaked into the ground and was covered with foam by firemen. The train resumed its journey after the derailed cars had been detached. No one was hurt. [1727] The derailment was later attributed to harmonic rocking due to a low spot in the track and the speed of the train. Locals clamoured for the speed of trains through Peotone to be reduced while an ICG official blamed the introduction of new, very large tank cars and the fact that the development of improved track had not kept pace with their introduction. [1728]

Salvadore Navarro, a Mexican trespasser, lost his life in the derailment of a Southern Pacific Company train at Buckeye, Arizona, United States, on Sunday 8th August 1974. His companion, another Mexican named Jesus Torrez, was trapped for 11 hours by the steel bars of the waggon in which they had been riding and was seriously hurt. Torres was found up to his knees in hot liquid asphalt, which it was thought had probably saved his life by preventing him bleeding to death, while liquid oxygen was also spilt. The 64-car train from West Colton, California, to Phoenix, Arizona, had been derailed by a 'hot box' leading to a broken journal. Amtrak passengers were taken by bus between Phoenix

---

[1726] National Transportation Safety Board report

[1727] *Dixon Evening Telegraph*, Dixon, Illinois, United States, 5th August 1974

[1728] *Chicago Heights Star*, Chicago, Illinois, United States, 8th August 1974 and Casa Grande until the line could be cleared. [1729]

An accident similar to that at Decatur Yard (see above) occurred at about noon on Saturday 21st September 1974 at Englewood Yard, Houston, Texas, United States, of the Southern Pacific Transportation Company. This was a gravity hump yard. Two jumbo cars loaded with butadiene, Nos. 17 and 18 of a 145 car complement, were uncoupled as a unit at the crest of the hump. They passed through the hump master retarder and group retarder without being slowed and then accelerated as they ran down the slope into bowl track No. 1. Moving at 18 or 20 miles an hour they collided with an empty tank car, whereupon the coupler of the empty car overrode the coupler of car No. 17 and punctured its tank head. Butadiene escaped and formed a vapour cloud that spread over the area. Two or three minutes later there was a violent explosion. Engineer J.A. McKnight, who had been operating a locomotive 600 feet away, was seen running from it with his clothes on fire. He died in hospital of burns two days later. [1730] No less than 235 other people were injured, 231 cars were destroyed, 282 more were damaged and the total cost of the damage was estimated at $13,364,085. About 1,700 people were evacuated.

The accident was investigated by the National Transportation Safety Board (NTSB), who in their report on it listed 17 conclusions:

(1) The failure of the retarders to control the cars was not caused by human error or the malfunction of the equipment.

(2) The foreign substances discovered on the brake shoes of the master retarder and the group retarder which served bowl tracks 50 and 51 were deposited there by the wheels of cars 15 and 16. The foreign substances discovered on the braker shoes of the group retarder which served bowl track 1 were carried there by the wheels of cars 17 and 18, which had picked up the material from the brake shoes of the master retarder.

(3) The foreign substances on the wheels of cars 15 and 16 were picked up from the flangeways of the track adjacent to the warehouse in the Shell plant where the epoxy resins were loaded.

(4) Although one of the seven sections of the master retarder had been removed from service and the brake shoes on the various retarders were worn, sufficient capacity remained to retard cars 15 through to 19 if no foreign substances had been on the retarder brake shoes and car wheels.

(5) The two rank cars containing butadiene (cars 17 and 18) collided with the empty tank car (car 10) while it was freestanding or moving slowly. The collision speed was 18 to 20 mph, as estimated by witnesses.

(6) The puncture of the head of the loaded tank car by the coupler of an empty car was generally consistent with the circumstances of accidents that occurred at Decatur, Illinois, and East St Louis, Illinois.

(7) The damage done by the explosion was consistent with a detonative open-air explosion rather than deflagration.

[1729] *The Yuma Daily Sun*, Yuma, Arizona, United States, 9th September 1974

[1730] *Wichita Falls Times*, Wichita Falls, Texas, United States, 24th September 1975

(8) Shell Oil Company management failed to heed the clear warnings of the Southern Pacific representatives issued in 1973 that the spilled epoxy resin at Shell's plant offered the potential for an accident involving hazardous materials at Englewood Yard.

(9) The Southern Pacific management recognized that foreign substances on car wheels could nullify the braking capabilities of the retarders; however, actions taken to prevent a recurrence were ineffective. The various requests and instructions initiated were either not received, not understood, or were ignored by the recipients. No follow-up action was taken by the Southern Pacific management although there was reason to suspect that these various communications were not being obeyed.

(10) Southern Pacific car inspectors had not been taking exception to foreign substances on car wheels even though such substances had contributed to previous accidents in Englewood Yard.

(11) The retarders were not able to brake cars with the frictional characteristics that were said to be within the retarder's capability. The coefficient of friction had not been the basis for judging retarder performance, as an occasional runaway car was tolerated.

(12) Car journal oil, grease, epoxy resins, and various combinations of these mixed with other contaminants exhibited similar coefficients of friction when subjected to simulated retarder conditions. The exact effect on retarder operations of these materials and many other substances that are common to car wheel surfaces is not known.

(13) There were no violations of Federal regulations by the carrier or shipper.

(14) Top and bottom shelf couplers probably would not have prevented the tank head puncture since the coupler on car 10 broke.

(15) A head-shield on the car which released the butadiene probably would have prevented the puncture of the tank head by the overriding coupler of the empty tank car.

(16) Although there had been instances of overspeed cars in switching operation at Englewood Yard, there were no emergency procedures or facilities for handling these overspeed incidents.

(17) The efficient handling of the fire emergency resulted from the joint pre-accident planning and training by the Southern Pacific and the Houston Fire Department.

At to the probable cause of the accident the NTSB wrote that 'the probable cause of the overspeed impact was the failure of the retarding system to slow the two coupled tank cars and the absence of a backup system to control cars which pass through the retarders at excessive speeds. The failure of the retarding system was caused by foreign substances on the wheels of the two cars that preceded the two tank cars through the retarders.

'Contributing to the accident was the failure of the Southern Pacific Transportation Company to enforce procedures to exclude cars with a foreign substance on their wheels from the humping system, and the Shell Oil Company's failure, after notification of the hazard, to eliminate spilled epoxy resin from the flangeways of their track.'

The NTSB made three recommendations:

(a) That the Southern Pacific Transportation Company review the design of the hump and make changes to ensure that all humped cars were properly controlled. The review should include a systematic analysis of the interrelated effects of grades, retarders, controls for switching and maintain programmed speed, with separate back-op provisions to stop overspeed cars, constant monitoring of overspeeds, car inspection procedures, and a prediction of the reliability of the system with cars actually using it.

(b) The Federal Railroad Administration should research the minimum performance standards for retarding systems in gravity switching yards in co-operation with the Association of American Railroads.

(c) The Federal Railroad Administration should issue regulations superseding Emergency Order No. 5 requiring railroads to handle the switching of cars carrying large amounts of hazardous materials which posed a danger beyond railroad property boundaries in the same manner as they handled explosives. [1731]

A Maine Central Railroad train from South Portland to Bangor, Maine, was derailed at 4.20 a.m. on Friday 8th September 1974 at North Belgrade, Maine, United States. Twenty-four of the 109 cars of the train pulled by three diesel units left the track and bunker oil being carried in two tank cars spilt into Messalonkee Lake. There were no immediate fears of contamination of the water supply. The train was travelling at 41 miles an hour at the time, within the speed limit, and the cause of the derailment was not known. Damage to the rolling stock and track was extensive. [1732]

A 32-car Southern Pacific Company freight train from Lafayette to New Orleans, Louisiana, was derailed between Boutte and Paradis, Louisiana, United States, on the morning of Sunday 12th January 1975. The train was carrying both hydrochloric acid and liquefied petroleum gas (LPG) and the acid spilt from one of the 15 cars that left the track. Families living nearby were evacuated and a nine-mile stretch of us Highway 90 between Des Allemands and Boutte was closed for several hours while firemen diluted the acid with water and spread lime (calcium hydroxide) on it to neutralise it. No one was hurt. [1733]

Sulphuric acid was spilt in a derailment west of Dora, Arkansas, United States, at about 6.45 a.m. on Saturday 17th May 1975, when 11 cars of a southbound Missouri Pacific Railroad train left the track and a tank car was overturned. Six families whose homes were in low-lying areas near the site of the accident were evacuated. No one was hurt. [1734]

Twenty cars of a 51-car Chesapeake and Ohio Railway freight train were derailed on Friday 30th May 1975 at Lawton, Kentucky, United States, and a small quantity of sulphuric acid escaped from a tank car. A ditch was dug to contain the spillage and no great problems were anticipated, nor was anyone injured. [1735]

A Missouri Pacific Railroad freight train was derailed at Tillar, Arkansas, United States, probably early on Friday 18th July 1975 and chlorine and an unspecified acid

[1731] National Transportation Safety Board report
[1732] *Greenfield Recorder*, Greenfield, Massachusetts, United States, 8th November 1974
[1733] *Monroe News-Star*, Monroe, Louisiana, United States, 13th January 1975
[1734] *Northwest Arkansas Times*, Fayetteville, Arkansas, United States, 18th May 1975
[1735] *Beckley Post-Herald*, Beckley, West Virginia, United States, 31st May 1975

leaked from derailed cars. Thirty residents were evacuated, most of whom were allowed to return later in the day after the chlorine leak had been stopped, only those whose homes were closest to the accident site being kept away longer. The acid continued to leak for a longer period. No one was reported hurt. [1736]

A 62-car Chesapeake and Ohio Railway freight train from Hinton to Charleston, West Virginia was derailed at 2.25 p.m. on Saturday 2nd August 1975 at Marmet, West Virginia, United States, while travelling at 48 miles an hour. Twenty-nine cars left the rails, only six of which were loaded, two with acetic acid, which was spilt. It was stated not to pose a serious threat but several people reported irritation to their eyes, noses, and throats, while a prize walking horse was crushed by a derailed car and killed. [1737]

The derailment of a 78-car Erie-Lackawanna Railroad freight train from Chicago, Illinois, to New York, claimed a fatality at Youngstown, Ohio, United States, on Tuesday 5th August 1975. Alberta Ceryak, 64, was sat in her camper van at a level crossing waiting for the train to pass when it was crushed by a derailed car. Another driver in a tractor trailer (articulated lorry) had a lucky escape from the same fate. A tank car of ethyl alcohol (ethanol) was ruptured, the alcohol was spilt and caught fire and there were several small explosions, preventing firemen from reaching the camper van for several hours. Five other people, three of them fireman, were slightly hurt. [1738]

Mrs Leslie Stout and her husband were on their way to a friend's home along Interstate 35-80 on Monday 1st September 1975 when suddenly "there were flames above us and all around us. We knew our only chance was to leave that car and try to outrun it," she related from her hospital bed in Des Moines, Iowa, United States. A 69-car southbound Chicago Rock Island and Pacific Railroad freight train had derailed just as it passed under the bridge, 11 cars leaving the track including three loaded with liquefied petroleum gas (LPG) which escaped from the damaged cars and in the usual fashion caught fire, forming a huge fireball, while a series of explosions followed.

Riding just in front of the Stouts' car was a motorcyclist, Von Ray Abbott, who caught the full force of the first explosion and whose motorcycle was swept from the bridge. He was in critical condition in hospital the next day with burns over the upper half of his body. Four other people were injured to a lesser degree. Homes, factories, and warehouses within a 2-square-mile radius were evacuated and the highway was closed. The noise of the blast was heard 40 miles away. A hundred firemen from Des Moines and the surrounding area kept control of the fire but did not attempt to extinguish it for fear of forming explosive pockets of propane which could have devastated part of Des Moines. [1739] [1740]

At 3.10 a.m. on Tuesday 2nd September 1975 a Chesapeake and Ohio Railway freight train passing through Defiance, Ohio, United States, had 31 of its cars derailed, one of them a tank car carrying corrosive chlorosulphuric acid, which was ruptured. About 400 residents in the Holgate and Harrison Avenue were evacuated by the police at 4 a.m. as

[1736] *Northwest Arkansas Times*, Fayetteville, Arkansas, United States, 19th July 1975
[1737] *Sunday Gazette-Mail*, Charleston, West Virginia, United States, 3rd August 1975
[1738] *Washington C.H. Record-Herald*, Washington Court House, Ohio, United States, 6th August 1975
[1739] *Ottumwa Courier*, Ottumwa, Iowa, United States, 2nd September 1975
[1740] *Daily Capital News*, Jefferson City, Missouri, United States, 3rd September 1975

acid fumes spread over the area and Slocum Elementary and St John's Catholic schools were closed for the day. Later firemen spread soda ash (sodium carbonate) to neutralise the acid. The train had been travelling at 54 miles an hour – within the railroad speed limit but well above the 35 miles an hour laid down in a city ordinance. The cause of the derailment was unclear. [1741]

Three engines and 11 tank cars containing No. 4 industrial fuel oil of a Penn Central Railroad freight train left the rails at Kingwood, New Jersey, United States, and about 1,000 gallons of fuel oil escaped into the Delaware River on the morning of Wednesday 3rd December 1975. It was hoped that the oil would dissipate before it reached the water supply of Trenton, New Jersey. [1742]

[1741] *The Crescent-News*, Defiance, Ohio, United States, 2nd and 3rd September 1975
[1742] *The Daily Register*, Shrewsbury, New Jersey, United States, 4th December 1975

INDEX

147th Street, New York City, New York, United States, 257
19th Street, Idaho Falls, Idaho, United States, 515
27th Street, Chicago, Illinois, United States, 219
Abejera and Sarracín de Aliste, Spain, 247
Abomey, Dahomey, 236
Accous, France, 315
Acropolis Express (named train), 276
Acting Foreman P. Dunphy, 325
Acting Leading Car Examiner Frederick Wager, 370
Acting Running Superintendent Hand, 325
Acting Stationmaster S. Dubey, 148
Adams, Massachusetts, United States, 708
Adana and Elâziğ, Turkey, 328
Additional Commissioner A.C. Khastgir, 44
Additional Commissioner Arya Bhushan, 132, 205, 620, 644
Additional Commissioner C.R. Sule, 573
Additional Commissioner D.V. Divgi, 79, 575
Additional Commissioner G.S. Pandor, 126, 127, 149, 211, 215, 565, 632
Additional Commissioner H.S. Hart, 139, 203
Additional Commissioner P.B. Aibara, 118
Additional Commissioner P.M.N. Murthy, 524, 627
Additional Commissioner S.K. Mojumder, 156, 162, 288, 320
Additional Commissioner V Venkataramayya, 310
Additional Fireman Raghunath Prasad, 160
Additional First Fireman E.V. Selvaraj, 139
Adelsdorf, West Germany, 601
Advanced Baltimorean (named train), 639
Afyon, Turkey, 480
Agent Esperidião Jonas de Prado, 152
Agent Jair de Lima Ferreira, 153
Agent José Carlos da Costa, 152
Agent Louis Hulpman, 692
Agnew, Nebraska, United States, 424
Água Branca, Brazil, 217, 470
Aiken, Texas, United States, 512
Airport Road Crossing, near Rhinelander, Wisconsin, United States, 500
Airport Way Crossing, California, United States, 398
Aitrang, West Germany, 65
Akibamachi, Japan, 377
Akron, Ohio, United States, 664
Albany, New York, United States, 653
Albany, Wisconsin, United States, 381
Albertville, Belgian Congo, 605
Alby, Sweden, 52
Alexandria, Indiana, United States, 539
Aligarth, India, 482
Allahabad, Uttar Pradesh, India, 124
Almansa, Spain, 519
Almedabad Express (named train), 117
Almeida Brandão, Brazil, 234
Alpu, Turkey, 230
Altamirano, Argentina, 182
Alton and Southern Railway, 543, 697
Amatilán, Guatemala, 306
Amstel, The Netherlands, 594
Amtmann Stephan Dorfner, 298
Amtrack, 31
Anaparthy, India, 124
Anatolian Express (named train), 78
Andimeshk, Iran, 374
Andrews, Indiana, United States, 700
Angeles, The Philippines, 506, 533
Angleton, Texas, United States, 380
Anklar and Vased, India, 537
Ann Arbor Railroad, 391
Annemasse, France, and Chêne-Bourg, Switzerland, 301
Anse-à-Gascon, Quebec, Canada, 399
Apennine Base Tunnel, Italy, 364
Apóstoles and Pindapoy, Argentina, 6
Appayya Gate, between Venkatachallum and Manoboholu, India, 471
Aquilon (named train), 24
Arbil and Kirkuk, Iraq, 630
Archbold and Pettisville, Ohio, United States, 471
Arcoverde, Brazil, 400
Area Assistant District Engineer J.R. Stockdale, 28
Area Assistant to the Civil Engineer P.G. Rawling, 61
Area Maintenance Engineer B.M. Jordan, 38
Area Manager J. Leonard, 332
Arnold, Missouri, United States, 701
Arroyo Seco, Santa Fe, Argentina, 440
Ashkum, Illinois, United States, 692
Ashmore, Illinois, United States, 698
Assam, India, 623
Assistant Block Signal Inspector P. Chinnoyya, 287
Assistant Booking Clerk B.C. Mukherjee, 214

Assistant Booking Clerk P.K. Paul, 214
Assistant Carriage and Waggon Engineer J. Colbert, 648
Assistant Chief Engineer of Maintenance of Way Ernest Schultz, 5
Assistant Chief Operating Manager C.H. Cope, 371
Assistant Civil Engineer K.C. Haysom, 71
Assistant Conductor Ernesto Puntillio, 655
Assistant District Signalman's and Traffic Inspector M.A. Bush, 93
Assistant Divisional Civil Engineer T.F. Grisdale, 61
Assistant Divisional Manager T.B. Savin, 372
Assistant Driver Antonino Barbera, 353
Assistant Driver Antonio Avila Lloret, 167
Assistant Driver António Barriga Branco, 240
Assistant Driver Antonio Romeo, 31
Assistant Driver Charles Calvez, 357
Assistant Driver Constantin Colțatu, 590
Assistant Driver Eric Denis Grenhoefer, 257
Assistant Driver Fabio Benjumea, 244
Assistant Driver Faivre, 116
Assistant Driver Fiero Milanese, 507
Assistant Driver G. Punniah, 203
Assistant Driver H.P. Yadaw, 564
Assistant Driver João Modesto Azevedo, 151
Assistant Driver Juerg Kuhn, 482
Assistant Driver Leonel Montoya, 244
Assistant Driver Luciano Ruiz Méndez, 247
Assistant Driver Manuel Méndez Alvarez, 247
Assistant Driver Mario Ruono, 655
Assistant Driver Nicolás Fernández Fernández, 167
Assistant Driver Otello Manzi, 45
Assistant Driver Pasquale Pugliatti, 352
Assistant Driver R.K. Purl, 210
Assistant Driver Ramón Lamaspa Rubio, 256
Assistant Driver Riccardo Scarpellini, 347
Assistant Driver Roberto Giardini, 232
Assistant Driver S.K. Chakravarti, 205
Assistant Driver Soriano Fabbri, 100
Assistant Driver Stjepan Varga, 82
Assistant Driver Sukha Prakash Ghosh, 626
Assistant Driver Vignier, 25
Assistant Engineer A.K. Ganguly, 156
Assistant Engineer D.P. Gupta, 632
Assistant Engineer G.P. Ramachandran, 644
Assistant Engineer J.M. Gupta, 627
Assistant Engineer Thakur Singh, 44
Assistant Mechanical Engineer Mohammad Murtaza, 319
Assistant Medical Officer Dr. G. Subramanyam, 203
Assistant Motorman Gabriel Hernández Castillo, 284
Assistant Motorman Masahiro Miyata, 359
Assistant P.W. Section Supervisor Simpson, 87
Assistant Permanent Way Inspector B.D. Tewari, 631
Assistant Permanent Way Inspector D. Chellappan, 642
Assistant Permanent Way Inspector M. Prasad, 627
Assistant Permanent Way Inspector S.R. Mahato, 287
Assistant Permanent Way Inspector T.F. Alvares, 619
Assistant Permanent Way Inspector W.S. Evans, 86
Assistant Permanent Way Supervisor P.R. Batkin, 86
Assistant Shift Supervisor G. Abbas, 38
Assistant Signal Inspector R.K. Biswas, 156
Assistant Stationmaster A.K. Guha, 213
Assistant Stationmaster A.K. Monda, 214
Assistant Stationmaster A.M. Bellie, 643
Assistant Stationmaster B. Adhikari, 131
Assistant Stationmaster B. Stephen, 118
Assistant Stationmaster B.B. Das, 130
Assistant Stationmaster B.C. Chakravarty, 214
Assistant Stationmaster B.D. Sahai, 178
Assistant Stationmaster Bheema Naraslugh, 202
Assistant Stationmaster Binda Prasad Singh, 626
Assistant Stationmaster Chanan Singh, 573
Assistant Stationmaster D Kar, 204
Assistant Stationmaster D. Padmanaban, 309
Assistant Stationmaster D.K. Majumdar, 213
Assistant Stationmaster D.P. Sharma, 210
Assistant Stationmaster G.N. Sharma, 210
Assistant Stationmaster Ganpat Singh, 118
Assistant Stationmaster H.S. Lal, 631
Assistant Stationmaster J.B. Smith, 523
Assistant Stationmaster J.J. Brady, 325
Assistant Stationmaster J.T. Haldar, 626
Assistant Stationmaster K. Perriannan, 641
Assistant Stationmaster L.M. Das, 127
Assistant Stationmaster M.B. Ramachandra Rao, 202
Assistant Stationmaster M.C. Reddy, 523
Assistant Stationmaster M.L. Yadav, 118
Assistant Stationmaster M.R. Ash, 565
Assistant Stationmaster M.V. Patil, 201
Assistant Stationmaster Mathalainathan, 642
Assistant Stationmaster O.N. Vaish, 631
Assistant Stationmaster R.N. Das, 565
Assistant Stationmaster R.S. Gnanavadivelu, 309
Assistant Stationmaster S. Subramanian, 309
Assistant Stationmaster S.K. Saha, 287

Assistant Stationmaster T.G. Balaraman, 139
Assistant Stationmaster Vincent Cyril McRae, 385
Assistentenwärter Harald Kruse, 99
Astorga, Spain, 449
Atakpame, Togo, 212
Atchison Topeka and Santa Fe Railway, 380, 506, 512, 514, 603, 653, 664, 680, 684, 690, 692, 702, 707
Atlantic Coast Line Railroad, 403, 406, 421, 593
Bahia Blanca, Argentina, 534
Bakersfield, California, United States, 518
Balanagar and Gollapalli, India, 535
Baltimore and Ohio Railroad, 336, 428, 495, 496, 499, 544, 639, 649, 699
Bamio, Spain, 90
Banaras Express (named train), 563
Baoule River Bridge, Mali, 328
Barr, Illinois, United States, 543
Baseline Road Crossing, near Mesa, Arizona, United States, 411
Basile, Louisiana, United States, 652
Batavia, New York, United States, 691
Bath, Indiana, United States, 699
Baudpur, India, 183
Beckingen, West Germany, 65
Beit Dagon, Israel, 392
Bell's Mills Crossing, Pennsylvania, United States, 428
Bellflower, Missouri, United, 343
Bellingham, Washington, United States, 492
Bellview Road Crossing, Aragon, Georgia, 525
Benavidez and General Pacheco, Argentina, 134
Berenda, California, United States, 649
Beska, Yugoslavia, 169
Bessemer and Lake Erie Railroad, 484
Bestwig and Frankenberg, West Germany, 397
Bezděčín, Czechoslovakia, 483
Bhusaval, India, 579
Bialystok, Poland, 485
Biên Hòa and Binh Thuân, South Vietnam, 615
Biên Hòa, South Vietnam, 614
Biên Hòa, South Vietnam, 625
Big Creek Bridge, near Lavender, Washington, United States, 330
Big Stone Gap, Virginia, United States, 628
Birdtail Creek Trestle, Manitoba, Canada, 343
Birmingham, Alabama, United States, 592
Biskra, Algeria, 68
Blanton, Florida, United States., 403
Bletchley, England, 591
Block Maintainer Jotilal Mahato, 287
Blue Dragon (named train), 129
Boligee, Alabama, United States, 690
Bonassola, Italy, 577
Bongsandong Tunnel, South Korea, 354
Booking Clerk H. Hunter, 195
Booking Clerk K.O. Newton, 595
Booking Clerk W.R.A. Thorley, 75
Booking Clerk-in-Charge Jagannath Prasad, 44
Borgallo Tunnel, Italy, 351
Boston and Maine Corporation, 436
Boutte and Paradis, Louisiana, United States, 713
Boutte, Louisiana, United States, 402
Bowden, Alberta, Canada, 273
Bowerston, Ohio, United States, 638
Box Boy T. Thavamani, 644
Bradford Exchange, England, 302
Brakeman A. Nelston, 381
Brakeman A.E. Walker, 291
Brakeman Aaron Lancaster, 678
Brakeman Arnold Bookout, 597
Brakeman C.F. Chandler, 444
Brakeman C.F. England, 664
Brakeman Chester Lee Wilburn, 342
Brakeman Danny Seamans, 602
Brakeman Donald Huston, 597
Brakeman Edwin Mitchell, 440
Brakeman Gale Connor, 602
Brakeman Ghlee Wilson, 281
Brakeman Herbert Fero, 259
Brakeman I. Buck, 652
Brakeman J.A. McClaskey, 664
Brakeman J.L. Cooks, 652
Brakeman Jams Schulthies, 259
Brakeman João Lima, 248
Brakeman José Luis Maroto Echevarria, 167
Brakeman Leonard Sullivan, 328
Brakeman López Buitrago, 623
Brakeman Luis Felipe Ibarra, 623
Brakeman Melvin Black, 291
Brakeman Michael Parvu, 495
Brakeman Nathan Carter, 312
Brakeman Peter Carter, 312
Brakeman R.E. Arthur, 342
Brakeman Raymond Markle, 638
Brakeman Robert McGraw, 280
Brakeman Sergio Goria, 251
Brakeman Thurman Herskell, 496
Brakeman V.T. Alavani, 621
Brakeman Walter Wendell, 291
Brakeman Weldon Nybbelin, 450
Brakeman William Habeck, 328
Brakemen Gary Parks, 291
Brakesman H. Lal, 147
Brandtsville, Pennsylvania, United States, 638
Brecha, India, 474
Brent, Alabama, United States, 495
Bressanone, Italy, 616
Bridge Inspector Kumaraswamy, 645

Bridge Operator V. Abdullah Koya, 644
Brigadier C.A. Langley, 95
Brisbane Express (named train), 384
British Railways, Eastern Region, 26, 59, 594, 598, 685
British Railways, London Midland Region, 85, 91, 105, 193, 458, 549, 550, 553, 557, 566, 591, 646
British Railways, North Eastern Region, 349
British Railways, Scottish Region, 334, 593
British Railways, Southern Region, 9, 69
British Railways, Western Region, 35, 640
Brühl, West Germany, 563
Bucharest and Mogosoia, Romania, 48
Buckeye, Arizona, United States, 710
Buena Vista, Guatemala, 104
Burggrub, West Germany, 500
Burlington – Rock Island Railroad, 639
Burlington Northern Inc, 691, 703
Burlington Northern Railroad, 35, 280, 363, 490, 518, 587
Businessmen's Special (named train), 413
Byblos, Lebanon, 415
Bydgoszcz, Poland, 20, 400
Cabezón and Corcos, Spain, 500
Cabin Assistant Stationmaster A.N. Roy, 125
Cabinman Amar Chand, 117
Cabinman Angnoo, 148
Cabinman C.P. Singh, 179
Cabinman Ram Autar, 178
Căile Ferate Române, 56
Caioaba, Brazil, 530
Calamba, Laguna, The Philippines, 478
Callao, Missouri, United States, 691
Calumpit, The Philippines, 520
Cambridge Springs, Pennsylvania, United States, 680
Caminhos de Ferro de Moçambiqe, 235
Caminhos de Ferro Portugueses, 51, 84, 143, 189, 398
Canadian National Railway, 308, 343, 383, 391, 392, 399, 410, 431, 455, 478, 515, 539, 638, 698, 701
Canadian Pacific Railroad, 273
Canadian Pacific Railway, 379, 399, 413, 440, 517
Canton, Ohio, United States, 690
Car Examiner J. James, 370
Carmigliano di Brenta and Fontaniva, Italy, 520
Carriage and Waggon Assistant P.C. Barker, 600
Carriage and Waggon Examiner E.P. Powell, 646
Carriage and Waggon Examiner H. Portus, 28
Carriage and Waggon Examiner J.W. Hathaway, 28
Carriage and Waggon Examiner S.F. Bunn, 61
Carriage and Waggon Examiner T.H. Walker, 28
Carriage and Waggon Fitter T. Rathinam, 309
Carriage Foreman S.K. Patra, 319
Cartsburn (New) Tunnel, Scotland, 593
Cassano d'Adda, Italy, 22
Castel Bolognese, Italy, 45
Castelfranco Veneto, Italy, 3
Cave, Kansas, United States, 653
Cayuga Road Crossing, Williamsville, New York, United States, 513
Celaya, Mexico, 628
Celle, West Germany, 17
Central Bridge, New York, United States, 640
Central do Brasil, 427
Central Hersey, Cuba, 609
Central of Georgia Railway, 327
Central Railway, 78, 191, 575, 579
Cesário Alvim, Brazil, 20
Ceylon Government Railway, 26
Chairman Bishan Narain, 179
Chambersburg, Pennsylvania, United States, 663
Champion (named train), 601
Chanute, Kansas, United States, 702
Chełmek, Poland, 114
Chemin de fer Dakar-Niger, 328
Chemins de fer de fédéraux suisses, 24, 262, 272, 358, 750
Chesapeake and Ohio Railway, 342, 382, 402, 417, 592, 638, 690, 699, 702, 706, 713, 714
Chicago and Eastern Illinois Railroad, 547, 597
Chicago and Illinois Midland Railway, 543
Chicago and North Western Railway, 384, 401, 543, 545, 546, 547, 664
Chicago Burlington and Quincy Railroad, 343, 456, 477, 672
Chicago Indianapolis and Louisville Railway, 678
Chicago Milwaukee St Paul and Pacific Railroad, 330, 381, 386, 481, 492, 546, 639, 707, 708
Chicago Rock Island and Pacific Railroad, 404, 416, 422, 496, 636, 653, 678, 714
Chicago St Paul Minneapolis and Omaha Railway, 546
Chief Civil Engineer A Paterson, 13
Chief Civil Engineer A.H. Cantrell, 13
Chief Guard Abdul Hamoed, 138
Chief Inspector J.L. James, 38
Chief Inspector Luigi Marazzo, 338
Chief Mechanical and Electrical Engineer S.R.D. Power, 38
Chief Medical Officer P.A.B. Raffle, 371
Chief Permanent Way Inspector T. Davenport, 12
Chief Signal and Telecommunications Inspector A.D. Roulstone, 600

Chief Signal and Telecommunications Technician D.L. Turner, 37
Chihuahua City, Mexico, 629
Chillicothe, Missouri, United States, 481
Chivasso and San Sebastiano Po, Italy, 251
Chonan, South Korea, 129
Chopin (named train), 6
Chorzów, Poland, 164
Chualar, California, United States, 404
Cincinnati Special (named train), 385
Circar Express (named train), 471
Circle Fire Inspector B.M. Bharadwaj, 573
Církvice, Czechoslovakia, 178
Cisalpin (named train), 24
City of Carlisle (named engine), 591
City of Denver (named train), 388
City of Los Angeles (named train), 664
City of New Orleans (named train), 31, 378
City of St Louis (named train), 664
Clarendon, Michigan, United States, 651
Clarksburg, West Virginia, United States, 706
Clarksfield, Ohio, United States, 697
Clayton Road Crossing, near Wren, Ohio, United States, 520
Clerk D.E.H. Clark, 595
Clerk Teodor Petrişor, 263
Cleveland Cincinnati Chicago and St Louis Railway, 698
Colmesneil, Texas, United States, 706
Colombo, Ceylon, 402
Colonel D. McMullen, 14, 29, 108, 196, 596
Colonel J.R.H. Robertson, 63, 71, 303, 472, 648
Colonel W.P. Reed, 551, 559
Colorado and Southern Railway, 704
Columbia, South Carolina, United States, 529, 703
Columbus, Ohio, United States, 640
Combe aux Loup, France, 24
Comendador Soares, Brazil, 54
Conca d'Ora (named train), 655
Conductor Alex Bradshaw, 664
Conductor Andrés Fernández Trapote, 113
Conductor Arthur Llewellyn, 545
Conductor B.R. Porterfield, 341
Conductor Biagio Bonifacio, 352
Conductor Blondi, 347
Conductor Bruno Serra, 251
Conductor Bryce Morehouse, 422
Conductor C.J. Phillips, 639
Conductor Carroll Fletcher, 31
Conductor César Hugo Barrios y Barrios, 104
Conductor Cesáreo González Maroto, 247
Conductor Charles Ray Henagan, 652
Conductor D'Andrea, 347
Conductor Ed Burnley, 291
Conductor Edward Cook, 291
Conductor Edward McMullin, 414
Conductor Edward Shaw, 425
Conductor Edwin Hunt, 436
Conductor Edwin Losinski, 328
Conductor Emilio Villaban Miranôn, 249
Conductor Ernest Beddington, 638
Conductor Ernest Burnal, 391
Conductor Ernest R. Cummings, 221
Conductor Fausto Nicelli, 520
Conductor Fernando Faccio, 134
Conductor Frank Coffman, 673
Conductor Fred Vestal, 545
Conductor G.F. Ryan, 224
Conductor G.W. Weidenhammer, 281
Conductor Gabriel Godínez Campos, 104
Conductor George Fields, 597
Conductor Gino Bizzarri, 45
Conductor Guard J. Wells, 36
Conductor H. Jones, 553
Conductor H.A. Gardner, 706
Conductor Harvey Boone, 704
Conductor Harvey Cone, 520
Conductor Howard Henry, 386
Conductor Ignacio Olivera, 408
Conductor James Christiansen, 396
Conductor João Pier Vieira, 507
Conductor João Romano Proença, 240
Conductor Joe Manuel, 602
Conductor John Bearden, 496
Conductor José Canoso, 142
Conductor José Guilherme Dias, 248
Conductor José Maria Pallejá Sangenis, 167
Conductor José Vicente Sousa, 21
Conductor Joseph Larabell, 382
Conductor Kjell Åge Kristiansen, 296
Conductor L.E. Fry, 547
Conductor Luigi Aristodemo, 337
Conductor Luis Calvete, 623
Conductor Mariano Espiniel, 244
Conductor Mario Antonangeli, 100
Conductor Mario Illenghi, 251
Conductor P.P. Dietrick, 650
Conductor Pedro Pérez, 133
Conductor Pietro Quattordio, 577
Conductor R.E. Neidhardt, 678
Conductor Ricardo Bolanos, 244
Conductor Sadahiro Mitojl, 176
Conductor Scott Newby, 342
Conductor Shoju Kiribara, 176
Conductor Teodoro Latorre, 256
Conductor Ulisses da Silva, 151
Conductor V. Diesche, 409
Conductor V.A. Marsh, 382
Conductor W.K. Besteder, 4

Conductor Władysław Czapla, 114
Conductor Z. Azana, 248
Conneautville, Pennsylvania, United States, 484
Connington South signal box, England, 598
Continental (named train), 383
Controller Dagfinn Dyresen, 295
Controller Frank McDonnell, 264
Controller P. McKay, 324
Córas Iompair Éireann, 326, 330
Corbin, Louisiana, United States, 696
Cordoba, Colombia, 57
Corning, Arkansas, United States, 579
Corrigan Road Crossing, near Noble Park, Victoria, Australia, 403
Cortland, Nebraska, United States, 702
Cosney-sur-Loire, France, 56
Coton Hill, England, 646
Cotonou, Dahomey, 207
County Road 9 Crossing, Harris, Minnesota, United States, 479
County Road No. 29 Crossing, near Waterloo, Nebraska, United States, 450
Crăciunel and Bucurdea Grânoasă, Romania, 263
Crescent City, Illinois, United States, 685
Crete, Nebraska, United States, 672
Crew Foreman R.D. Fritz, 5
Croesus, South Africa, 120
Cross Bayou, Florida, United States, 601
Crossing Keeper Alfonso Jesús de Aragón, 393
Crossing Keeper Alvaro Silva, 217
Crossing Keeper Andrés Dorado Villa, 446
Crossing Keeper Armandina Ramos de Carvalho, 398
Crossing Keeper Bruno Mariotti, 511
Crossing Keeper Candido de Sousa Martins Jacob, 398
Crossing Keeper Desmond Stewart, 479
Crossing Keeper Domenico Posca, 337
Crossing Keeper Josef Volk, 532
Crossing Keeper Jovelin Pereira da Costa, 427
Crossing Keeper Miguel Andrade, 420
Crossing Keeper Monty, 430
Crossing Keeper Piotr Kowalski, 399
Crossing Keeper Reginald Bagshaw, 385
Crossing Keeper Rinalda Perotta, 251
Crossing Keeper Rino Fiorito, 520
Crossing No. 137, Hammerum, Denmark, 428
Crossing No. 272/E between Santamagular and Savalyapuram, India, 522
Crossing PN 349, Andancette, France, 430
Croydon, Pennsylvania, United States, 549
Crozet Tunnel, France, 356
Cuautitlan, Mexico, 531
Cuba, 423
Cucamonga, California, United States, 380
Cuernavaca, Mexico, 408
Custóias, Portugal, 50
Da Nang and Huê, South Vietnam, 621, 632, 634
Da Nang, South Vietnam, 610, 614, 616, 617
Dahlerau, West Germany, 273
Dalby, Missouri, United States, 665
Dallas, Texas, United States, 698
Danbury, Texas, United States, 654
Danske Statsbaner, 428
Day Stationmaster Wenceslao Robledo Martin, 250
Dayton, Texas, United States, 706
De Graff, Ohio, United States, 650
De Soto, Illinois, United States, 685
Dearborn Michigan, United States, 702
Decatur Yard, Illinois, United States, 708
Deccan Express (named train), 198
Deco, Texas, United States, 639
Defiance, Ohio, United States, 714
Dekéleia, Greece, 151
Del block station, near Bowie, Maryland, United States, 40
Delaware and Hudson Railroad, 637, 640, 653
Delaware and Hudson Railway, 704
Delta Lake, Texas, United States, 422
Denizil, Turkey, 430
Deoband and Teleppari Buzurg, India, 534
Department of the Environment, 36, 60, 71, 369
Department of Transport, 86
Department of Transport and Power, 322, 331
Depok and Tjitajam,Java, Indonesia, 263
Depot Engineer S.J. McManus, 372
Depot Manager Raffaele Morrone, 338
Deputy Chief Controller Audh Bahari Lal, 573
Deputy Chief Controller R. Cooper, 596
Deputy Chief Electrical Engineer N.D. Gupte, 565
Deputy Chief Inspector Luigi Calvelli, 338
Deputy General Manager Mario Cirillo, 23
Deputy Stationmaster Giovanni Nicoletti, 340
Deputy Yard Master R.M. Gandhi, 120
Derveni, Greece, 128
Des Moines, Iowa, United States, 714
Deštnice, Czechoslovakia, 429
Detroit and Toledo Shore Line Railroad, 652
Deutsche Bundesbahn, 66, 67, 99, 160, 186, 239, 243, 273, 275, 297, 533
Deutsche Reichsbahn, 188, 286, 447
Dhanbad, Bihar, India, 392
Dhanushkodi, India, 641
Dharamtul and Auguri, India, 580
Dĩ An, South Vietnam, 616
Diablo Heights, Panama, 477
Dichwa Daboli and Sidhwalia, Uttar Pradesh, India, 415
Didcot North Junction, England, 640

Diesel Assistant B. Peter, 132
Diesel Assistant R. Raju, 138
Diêu Trì, South Vietnam, 610
Diêu Trì, South Vietnam, 609
Dining Car Manager Taj Mohammed, 627
Diphu, Assam, India, 618
Diredawa, Ethiopia, 8
Dispatcher F.J. Carrigan, 258
District Engineer D.H. Mukherjee, 627
District Engineer D.L. Carden, 11
District Manager V. Ryan, 325
District Mechanical Engineer P.R. Narasimhan, 626
District Mechanical Engineer S.M. Bammi, 626
District Signalman's and Traffic Inspector J. Redmond, 93
Divisional Civil Engineer J.L. Newing, 71
Divisional Electrical Engineer B.K. Sarkar, 565
Divisional Electrical Engineer G.L. Vaishnavi, 573
Divisional Engineer Madan Gopal, 44
Divisional Engineer Narasimhachar, 619
Divisional Engineer R.C.S. Khichi, 632
Divisional Inspector D. Walker, 559
Divisional Mechanical Engineer J.L. Soni, 44
Divisional Mechanical Engineer K. Reghunathan, 573
Divisional Operating Superintendent Shahid Ali Khan, 179
Divisional Signal and Telecommunications Engineer Harbans Lal, 156
Divisional Superintendent P.B. Rasmusson, 35
Dol-de-Bretagne, France, 81
Don Muang, Thailand, 491
Dora, Arkansas, United States, 713
Dorion, Quebec, Canada, 431
Dovreekspressen (named train), 295
Downingtown, Pennsylvania, United States, 700
Doxaras and Orfana, Greece, 276
Dragoon and Benson, Arizona, United States, 584
Driver A. Grundy, 473
Driver A. Jansen, 145
Driver A. Longworth, 553
Driver A. O'Reilly, 323
Driver A.C. Niwas, 631
Driver A.H. Dale, 567
Driver A.R. Swain, 553
Driver A.S. Bevan, 646
Driver A.T.R. Buxton, 551
Driver A.V. Rao, 608
Driver Ahmed Ali, 79
Driver Alfonso Fernindez, 249
Driver Alfredo Amoroso, 134
Driver Alfredo Luis de Freitas, 49
Driver Alres Teixeira, 198
Driver Altamiro Francisco Rosa, 530
Driver Andrés Pavón Carpena, 247
Driver Angelo Donati, 578
Driver Annibale Cannalire, 347
Driver Antonio Martiño Echevarris, 249
Driver Antonio Saglimbeni, 353
Driver Arne "Visslarn" Johansson, 534
Driver B Oakton, 598
Driver B.V.S. Rao, 203
Driver Bård Gundersen, 295
Driver Benigno Angado Antón, 247
Driver Bhagita, 45
Driver Bhola, 625
Driver Bindeshwari, 147
Driver Blijkerk, 146
Driver Burchi, 347
Driver C.W. Ireson, 87
Driver Caetano Lemes Filho, 217
Driver Caron, 25
Driver Catalon, 430
Driver Charles Neyrat, 116
Driver Chevalier, 430
Driver Chinnappan, 139
Driver Chitta Bhanumurty, 131
Driver Clodomiro Vilches, 393
Driver Cobb, 194
Driver Constantin Dogăroiu, 590
Driver Cyril Cordell Percy, 257
Driver D. Simkin, 257
Driver D. Velayudhan, 619
Driver Deonarayan, 125
Driver Diego Cabrera, 104
Driver Diego Di Benedetto, 231
Driver Domanico Fiorani, 232
Driver E. O'Grady, 323
Driver E.J. Worrall, 647
Driver E.S. Chatterjee, 43
Driver Eduardo Fuglini, 507, 511
Driver Elio Mossino, 251
Driver Emilio Andrade da Salva, 198
Driver Ennio Covacci, 45
Driver Erich Röhr, 185
Driver Ermenegilo Castillo, 407
Driver Ernest Wibrew, 594
Driver Eufrosino Manuel dos Santos, 234
Driver F.I. Wright, 554
Driver F.X. Barnabas, 619
Driver Fanie van Niekerk, 78
Driver Fernando Benavente Molero, 167
Driver Fernando Ruiz Noriego, 167
Driver Filadelfia Di Leo, 352
Driver Fisher, 549
Driver Francisco José dos Santos, 248
Driver G. A. Humphrys, 559

Driver G. McMahon, 473
Driver G.C. Das, 131
Driver G.H. Gott, 302
Driver Gaston Mathieu, 357
Driver George Francis William Hedgcock, 106
Driver George Fyfe, 334
Driver George Pike, 646
Driver Geraldo Couto, 151
Driver Gerritsen, 184
Driver Giovanni Bilardi, 31
Driver Grantley Small, 479
Driver H.D. Godayya, 199
Driver H.J. Lillystone, 194
Driver Hans Zehnder, 482
Driver Harold Garnett, 193
Driver Harpassad, 210
Driver Helio Bodo, 217
Driver Hernando Orozco, 412
Driver Iqbal Hussain, 43
Driver Ivadir Santana, 428
Driver J Mahato, 127
Driver J. Atkinson, 460
Driver J. Lawler, 331
Driver J. Walker, 26
Driver J.A. Jansen, 594
Driver J.E. Wilkinson, 72
Driver J.L. Daniel, 567
Driver J.M. Geeson, 567
Driver Jack Bowden, 264
Driver Jan Parzyka, 268
Driver Jean-Pierre Bieri, 482
Driver Jesús Madrigal, 412
Driver Jesús Rocha Serna, 317
Driver Joaquin Passan, 249
Driver John Arthur Smith, 303
Driver John Evans, 26
Driver John McKay, 85
Driver John Russell, 107
Driver Jorge Coronel, 623
Driver José Alves de Sousa, 418
Driver José Augusto Monteiro Lourenço, 240
Driver José de Azevedo, 470
Driver José Grané, 112
Driver Josef Feldrin, 212
Driver Joseph O'Neill, 330
Driver Juan B. Gómez, 244
Driver Juan Diozkez, 134
Driver Julio Pérez Rodriguez, 446
Driver K. Harrison, 489
Driver K.L. Purohit, 154
Driver Kamal Ram, 319
Driver Kart Krause, 187
Driver L. Braganza, 178
Driver L. Thirumalaswamy, 643
Driver L.G. Kangralkar, 621
Driver L.S. Devadanam, 522
Driver Lachni Narain, 572
Driver Lalta Prasad, 126
Driver Lanfranco Pigiani, 100
Driver Laszlo Papp, 206
Driver Lawrence Rosevear, 264
Driver Leslie Byers, 60
Driver Lourival Nunes Duarte, 234
Driver Luigi De Gaetano, 655
Driver M.L. Sharma, 119
Driver Manohar Lal, 147
Driver Manuel de Costa Leitão, 240
Driver Manuel Ferreira, 50
Driver Manuel Moreira, 190
Driver Mário Augusto Esteves, 189
Driver Meikle, 492
Driver Melchor Sánchez Echeverría, 316
Driver Michel Del l'Aquila, 655
Driver Miguel Sanchez Cruz, 115
Driver Miodrag Nestorović, 275
Driver Mohamad Abbas, 160
Driver Mohammad Siddique, 178
Driver Murath, 214
Driver N Karrupiah, 642
Driver N.C. Lakshmanan, 138
Driver N.C. Roy, 319
Driver Nestor Martinez, 612
Driver Nicholas Erasmus, 52
Driver Nikola Kneževiċ, 82
Driver Noribumi Minakami, 176
Driver P. Rogers, 323
Driver P.C. Bannerjee, 632
Driver P.S. Stevens, 72
Driver P.V. Guttal, 619
Driver Paras Ram, 631
Driver Pat Hickton, 575
Driver Paul Zowchuk, 431
Driver Purushottam, 631
Driver R. Moore, 554
Driver R.C. Singh, 564
Driver R.H. Marshall, 567
Driver Rahn, 66
Driver Ram Lall, 117
Driver Ramanand Kurmi, 155
Driver Robert Turner, 87
Driver Robert Wilsdon, 71
Driver Roberto Arbeláez, 244
Driver S. Booth,, 302
Driver S. Chinnappan, 138
Driver S.K. Ojha, 205
Driver S.P. Francis, 619
Driver S.S. Gupta,, 117
Driver Saquid Hussain, 631
Driver Saulo Salguera, 243
Driver Sebastião Nogueira da Silva, 20

Driver Shankar, 204
Driver Silvino Orellana, 104
Driver Stanley Colville, 595
Driver Stanley Turner, 460
Driver Stephen Kipp, 306
Driver Sumer, 626
Driver Svetomir Arsenijević, 275
Driver T.D. Owen, 36
Driver Thanu Mahato, 319
Driver Tomás Enrique Timoco, 486
Driver Van Leer, 170
Driver Venkatamuni Naidu, 201
Driver Vicente Rodriguez, 427
Driver Victor Ceroli, 133
Driver Villarreol, 182
Driver Vishwanath, 199
Driver W. Scanlon, 322
Driver Wieslaw Kolarczyk, 6
Driver Willem H, 145
Driver William Sharpes, 87
Driver William Warrington, 549
Driver William West, 192
Driver Williams, 124
Drummondville, Quebec, Canada, 392
Du Quoin, Illinois, United States, 702
Dube, South Africa, 117, 141
Dudelange, Luxembourg, 549
Duff, Tennessee, United States, 700
Dugald, Manitoba, Canada, 383
Duivendrecht and Amsterdam Amstel, The Netherlands, 144
Dumraon, India, 178
Duncan's Crossing, near Covington, Georgia, United States, 497
Dunreith, Indiana, United States, 657
Duy Xuyên, South Vietnam, 614
Dzialdowo, Poland, 8
Đồng Đăng, North Vietnam, 651
Eagle (named train), 390
Eagle Rock, North Carolina, United States, 328
East Chattanooga, Tennessee, United States, 657
East River Crossing, Mahoningtown, Pennsylvania, United States, 699
Eastern Railway, 124, 126, 153, 160, 178, 204, 212, 318, 564
Eastern Transvaal, South Africa, 484
Echeconnee Creek, Georgia, United States, 327
Ederland (named train), 159
Effingham Junction, Natal, South Africa, 54
Eighty-Four, Pennsylvania, United States, 706
Eketahuna, New Zealand, 574
El Arahal, Spain, 419
El Arenal, Mexico, 397
El Cuervo and Lebrija, Spain, 284
El Luciérnaga (named train), 182
El Mixto, *(named train)*, 134
El Mugrón, Spain, 591
El Rosarino (named train), 165
El Shangai (named train), 112
El Zarateño, (named train), 133
Elazig, Turkey, 78
Eldena, Illinois, United States, 430
Electrical Technician Lodovico Fiumana, 48
Electrical Technician Stafando Montando, 347
Elk River, Minnesota, United States, 703
Elm Park, England, 594
Eltham (Well Hall), England, 68
Embry Crossing, near Lincoln, Alabama, United States, 418
Employment Inspector E.G. Brown, 591
Encarnación de Diaz, Mexico, 20
Encinita Avenue Crossing, Temple City, California, United States, 396
Engineer A. A. Adams, 421
Engineer A.C. Olson, 545
Engineer A.E. Lewis, 273
Engineer A.G. Harnish, 401
Engineer Adlbert White, 391
Engineer Arl Booher, 387
Engineer Arthur Lee Herron, 512
Engineer B. Patterson, 411
Engineer B.E. Lee, 425
Engineer Bill Van Hook, 650
Engineer Braxton Gilbert, 456
Engineer C.C. Blakeslee, 401
Engineer C.J. Logan, 426
Engineer C.J. Walker, 639
Engineer C.R. Zane, 593
Engineer C.T. Kelley, 422
Engineer Calvin Cearley, 440
Engineer Carl Duckworth, 544
Engineer Carl Martin, 291
Engineer Charles Carpenter, 501
Engineer Charles Guidice, 386
Engineer Charles Milner, 520
Engineer Daniel E. Matthews Jr., 517
Engineer Dave Baskett, 291
Engineer David Morris, 476
Engineer Don Narney, 544
Engineer E. Johnson, 384
Engineer E. Kmiecik, 498
Engineer E.P. Estes, 341
Engineer Earl Knorr, 396
Engineer Ed Thygeson, 35
Engineer Eugene Gadway, 518
Engineer Francis English, 383
Engineer Frank Cruzen, 597
Engineer Fred Aden, 424
Engineer Frederick Mitchell, 280
Engineer George Hurley, 500

Engineer George Jeshner, 476
Engineer George O'Leary, 417
Engineer Glenn Peterson, 450
Engineer Glenn Way, 281
Engineer Gordon Chase, 312
Engineer Guy Roberts, 547
Engineer H.C. Billing, 278
Engineer H.L. Pepple, 495
Engineer Harry Bierce, 412
Engineer Harry Brown, 409
Engineer Henry Adams, 445
Engineer Herbert Sommers, 388
Engineer Howard Bruso, 258
Engineer Howard Horner, 40
Engineer Hugh Gross Merriman Jr., 421
Engineer I.J. Jacobson, 428
Engineer J. F. Shue, 5
Engineer J.A. McKnight, 711
Engineer J.C. Henderson, 496
Engineer J.D. Logan, 363
Engineer J.S. Thompson, 490
Engineer James Bell, 407
Engineer James Benham,, 342
Engineer James Case, 380
Engineer James Kerr, 603
Engineer James Melvin McCutcheon, 518
Engineer James Partlow, 499
Engineer Jimmy Livingston, 378
Engineer Jimmy Mayfield, 415
Engineer John Arden Simpson, 414
Engineer John Bernell, 416
Engineer John Ferguson, 543
Engineer John Houchard, 481
Engineer John Meusser, 382
Engineer John Prentice, 400
Engineer John Ruscio, 330
Engineer John Ward, 499
Engineer Keith Feree, 521
Engineer L.J. Rollins, 422
Engineer Lacy Haney, 31
Engineer Lowell Dumechey, 650
Engineer M. Stranahan, 444
Engineer M.B. Wilkinson, 652
Engineer Mario Rinaldo, 398
Engineer Mark Rowe, 547
Engineer Matthew Calvin, 443
Engineer O.A. Whittington, 381
Engineer O.W. Curry, 545
Engineer Olen Orr, 506
Engineer Oren Inman, 382
Engineer Orville McRay, 391
Engineer P.H. Black, 547
Engineer Potter, 678
Engineer R.G. Bowman, 471
Engineer R.H. Burns, 477
Engineer Ray Gardner, 529
Engineer Raymond Wilcox, 603
Engineer Rick Arner, 546
Engineer Robert Cripe, 404
Engineer Robert Emerson, 343
Engineer Robert Orr, 258
Engineer Robert Otto, 278
Engineer Rudy Pflanze, 328
Engineer T.F. Alexander, 664
Engineer T.T. Kelly, 427
Engineer Thomas Norman Bickers, 395
Engineer Tony Emanuel, 602
Engineer Tossie Miller, 495
Engineer Verna Zander, 386
Engineer Wayne Roberts, 445
Engineer William Callan, 638
Engineer William Mills, 403
Engineer William Norman, 410
Engineer William Thomas, 484
Engineer William Thomson, 483
Engineer William Witt, 411
Engineer Woodrow Jones, 665
Englewood Yard, Houston, Texas, United States, 711
Erie-Lackawanna Railroad, 427, 476, 520, 544, 603, 664, 680, 700, 714
Essen-Altenessen and Geisenkitrchen, West Germany, 517
Eßlingen Hauptbahnhof, West Germany, 241
Estaire, Ontario, Canada, 379
Estrada de Ferro Santos-Jundiaí, 470
Eureka Drive Crossing near Sterling, Florida, United States, 424
Eureka, Missouri, United States, 639
Everett, Massachusetts, United States, 436
Expreso del Sol (named train), 485
Expresso Campista (named train), 20
Fahrdienstleiter Alfred Messer, 98
Fahrdienstleiter Franz Habart, 297
Fahrdienstleiter Franz Reichmann, 212
Fahrdienstleiter Gottfried Sengbart, 274
Fahrdienstleiter Ludwig Hiergeist, 297
Fairview Avenue Crossing, Barberton, Ohio, United States, 426
Fanfa, Brazil, 198
Farketing and Kamarbandhali, India, 618
Fayette, West Virginia, United States, 638
Federal Railroad Administration, 271, 449
Ferrocarril de Panamá, 477
Ferrocarril del Atlántico, 623
Ferrocarriles Argentinos, 6, 25, 133, 165, 393, 408
Ferrocarriles de Cataluña, 436
Ferrocarriles Nacionales de Colombia, 57, 243, 245

Ferrocarriles Nacionales de México, 115, 316, 427, 435
Ferrocarrils de Catalunya, 351
Ferrovia Paulista, 507
Ferrovia Paulista S/A, 217
Ferrovie dello Stato, 3, 23, 47, 103, 232, 346, 351, 354, 577, 656
Fettler Thomas Dennis Finnigan, 385
Fexhe-Le Haut-Clocher, Belgium, 197
Fibre, Michigan, United States, 519
Fireman Alfred Varga, 343
Fireman Arnfried Brendecke, 264
Fireman Arumugam, 199
Fireman Bhimayya, 124
Fireman Brown DuVall, 429
Fireman C.E. Goodwin, 342
Fireman Carl Porter, 544
Fireman Charles Meuser Jr, 258
Fireman Colin Gradley, 273
Fireman D.H. Khan, 319
Fireman D.L. Samuel, 401
Fireman David Wolfenden, 306
Fireman Dumitru Goga, 590
Fireman Edward Rock, 638
Fireman Felipe Ulibarre, 113
Fireman Franklin Jenkins, 258
Fireman Fristoe, 678
Fireman G.A. Jones, 648
Fireman G.S. Williams, 544
Fireman Garcia Carrizales, 317
Fireman Georg Wensch, 187
Fireman Guilliaem (Rusty) Aertsen, 312
Fireman H. Homballayya, 199
Fireman Ignacio Gonzáles, 317
Fireman Ioan Chirițoiu, 590
Fireman J.W. Adams, 652
Fireman Jesús Piquero, 552
Fireman John Collins, 378
Fireman Johnson, 124
Fireman José Antonio Simón Campos, 249
Fireman Juan Carlos Miranda, 393
Fireman Kenneth Clorley, 646
Fireman Kyles Andes, 402
Fireman Le Roy King, 409
Fireman Leslie Hockenhull, 460
Fireman Lorenzo Martin Guillén, 249
Fireman Mario Duffini, 251
Fireman Melvin Swanson, 388
Fireman Mervyn Coulthard, 264
Fireman Molina, 133
Fireman N Sannappa, 199
Fireman Octavio Cárdenas, 104
Fireman Ramón Casasola, 104
Fireman Robert Warford, 395
Fireman Roy Rickhoff, 328
Fireman Thomas McMahon, 410
Fireman Truman Ellard, 387
Fireman Victor McCallum, 107
Fireman Virgil Panduru, 590
Fireman William Wyatt, 40
First Fireman Arjun Prasad, 154
First Fireman C.G. Bhonsale, 621
First Fireman Ganga Ram, 117
First Fireman Gurbachan Singh, 44
First Fireman Hamid Hussain, 147
First Fireman Narayan Mandal, 204
First Fireman P. Velappan, 139
First Fireman R. Bhaskaran, 308
First Fireman S.R. Kale, 619
First Fireman S.S. Panda, 131
First Fireman Satya Narayan, 125
First Fireman V. Marieppan, 642
Fisher Road Crossing, Ohio, United States, 512
Fitter Chargeman C. Sundaravadivelu, 309
Fitter Chargeman-in-Charge M.S. Gopal, 309
Fitter Hughes, 569
Fitter Jagannath, 79
Fitter Whillis, 569
Fitter-in-Charge M. Satyanarayana, 320
Fiumarella Viaduct, near Catanzaro, Italy, 337
Flagman Chester Weber, 673
Flagman John McGarey, 545
Flagman Nevil Tyson, 545
Flagman R.W. Moree, 525
Flagman R.W. Woodkirk, 281
Florida East Coast Railway, 613
Fornos de Algodres, Portugal, 240
Forsthaus Haltepunkt, West Germany, 482
Fort Covington, New York, United States, 518
Fort Wayne, Indiana, United States, 702
Foyil, Oklahoma, United States, 457
Franco-Ethiopian Railway, 8
Frankfurt Oder, East Germany, 56
Freccia del Dolomiti (named train), 520
Freeport, Illinois, United States, 412
Fruitridge Road Crossing, Sacramento, California, 440
Frunzeasca, Romania, 56
Fuji Kyuko Railway, 497
Fukuoka, Japan, 405
Fürnitz, Austria, 212
Futog, Yugoslavia, 445
Gádor and Santa Fe, Spain, 249
Gaira Crossing, Colombia, 407
Galmiz, Switzerland, 482
Gang Mate Babu Ishwar, 619
Gang Mate Baldeo, 627
Gang Mate Bhairon Ghasi, 119
Gang Mate Laxman Bhima, 622
Gang Mate P. Singh, 631

Ganger (Acting Sub-Inspector) R. Watson, 28
Ganger G.W. Smith, 596
Ganger M. Morris, 331
Gangman Kamaruddin, 43
Gangman M. Lakshmanan, 643
Gangman N. Lakshmanan, 643
Gangman Shivaram, 619
Gangpur, India, 126
Gardner, Illinois, United States, 537
Garmsār, Iran, 394
Garrett, Indiana, United States, 649
Gateman A. Venkayya, 523
Gateman Ali Raza, 148
Gateman Anandam, 523
Gateman Kanauji, 43
Gateman Mangal, 148
Gateman Moharman, 44
Gateman Ram Singh, 118
Gates, Tennessee, United States, 653
Gateway Yard, East St Louis, Illinois, United States, 697
General Manager Leo Cusick, 352
General Rodriguez, Argentina, 417
Geneva, Ohio, United States, 402
Georgia Railroad, 497
Ghamra, United Arab Republic, 408
Ghatsila and Dalbumgarh, India, 607
Gia Huỳnh, South Vietnam, 612
Gilbert, South Carolina, United States, 703
Gilchrest Road Crossing, near Congers, New York, United States, 501
Gioia Taura, Italy, 29
Glasgow Road Crossing, near Jackson, Michigan, United States, 400
Glendora, Mississippi, United States, 680
Golden Eagle (named train), 531
Goldküstenexpress (named train), 272
Gonda Junction, India, 146
Goods Guard C.R. Gooden, 195
Goods Guard E. Fenton, 91
Goods Guard H.P. Wake, 26
Goods Guard J. Gray, 9
Goods Guard J.J. Butler, 646
Goods Guard Muhammed Azad Khan, 473
Goods Shunter A. Rose, 646
Gordon, Texas, United States, 650
Gorey, Eire, 330
Gormanston, Eire, 322
Gostivar and Tetovo, Yugoslavia, 314
Grade B Fireman Shambhoo Dayal, 572
Grade C Fireman Bharat Singh, 572
Grand Boulevard Crossing, East Gary, Indiana, United States, 498
Grand Forks, North Dakota, United States, 410
Grand Trunk Western Railroad, 477, 638, 690
Granite City, Illinois, United States, 543
Great Northern Railway, 435
Greeley, Colorado, United States, 387
Green Bay and Western Railroad, 328, 546
Greenville, North Carolina, United States, 406
Greenwich Road Crossing, Wichita, Kansas, United States, 498
Grisén and Pinseque, Spain, 551, 747
Großen Linden, West Germany, 403
Guanabacoa, Cuba, 457, 609
Guard 'B' Sohan Lal, 572
Guard A.M. Masilomoney, 642
Guard A.S. Abdul Hameed, 138
Guard Abdul Satar, 132
Guard B.A. Friar, 371
Guard B.B. Chakravarty, 319
Guard B.K. Asthvana, 43
Guard B.N. Bhaduri, 155
Guard Biswanath, 127
Guard C.R. Ekambaram, 308
Guard D.C. Dey, 214
Guard D.L. Simcoe, 88
Guard E.A. Shore, 85
Guard E.V. Shah, 178
Guard G. Veera Raju, 522
Guard G.C. Day, 125
Guard G.D. E. Theron, 120
Guard G.N. Saha, 287
Guard Gordon Pereia, 306
Guard H.K. Mathur, 147
Guard Henry Atterbury, 71
Guard I.E. Lewis, 551
Guard J. Monahan, 322
Guard J. Wright, 598
Guard J.A. Banks, 595
Guard J.K. Paul, 160
Guard J.N. Mishra, 204
Guard J.T. Catney, 370
Guard J.W. Smith, 194
Guard Juan Picón Alvarado, 317
Guard K McRae, 107
Guard K.M. Lal, 209
Guard L.D. Mishra, 79
Guard Lee, 549
Guard Len Cairncross, 575
Guard M. Lee, 323
Guard M.A. Dabhoivala, 575
Guard M.S. Bannerjee, 626
Guard Munilal, 157
Guard O.P. Varma, 147
Guard P. Salmon, 323
Guard Phillips, 489
Guard R. Hewitt, 60
Guard R. Wilkinson, 106
Guard R.P. Mukkerjee, 564

Guard R.P. Ram, 178
Guard Robert Harris, 370
Guard S. Hanumantha Rao, 621
Guard S. Miller, 559
Guard S.C. Moulik, 127
Guard S.G. Jog, 619
Guard S.H. Singh, 630
Guard S.L. Sharma, 117
Guard S.R. Chowdhury, 125
Guard S.S. Sarkar, 205
Guard Suraj Narain, 117
Guard T. Burlong, 330
Guard T.N. Bhattacharjee, 160
Guard T.S. Rangasami Achari, 201
Guard T.W. Harper, 595
Guard V.R. Sankaran, 139
Guard V.S. Nim, 210
Guard Vincente Martinez Torres, 317
Guard W.C. Final, 460
Guard W.T. Veale, 257
Guard William Wyer, 264
Gulf Mobile and Ohio Railroad, 495
Guntershausen, West Germany, 159
Hachenburg and Unnau-Korb, West Germany, 276
Haiya, Sudan, 306
Hamburg Berliner Tor and Rothenburgsort, West Germany, 98
Hamburg Hausbruch, West Germany, 239
Hammer Lane Crossing, Stockton, California, United States, 679
Hanasaki and Washinomya, Japan, 493
Hannover Linden, West Germany, 581
Hanoi and Thanh Hóa, North Vietnam, 654
Hanoi and Yên Bái, North Vietnam, 651
Hanoi, North Vietnam, 651
Harmelen, The Netherlands, 170
Harrah, Oklahoma, United States, 404
Harrod, Ohio, United States, 700
Hastedt, West Germany, 186
Hat Yai Junction, Thailand, 513
Havelock, Ontario, Canada, 440
Havre St Pierre, Quebec, Canada, 182
Hazen, Arkansas, United States, 636
Head Brakeman Amil Tekale, 382
Head Brakeman Ben Board, 547
Head Brakeman Bob Starkovick, 330
Head Brakeman C.L. Nicholson, 543
Head Brakeman Clyde Masters, 544
Head Brakeman Emil Serhan, 278
Head Brakeman F.C. Druschel, 544
Head Brakeman Gordon Bates, 520
Head Brakeman H.L. Whitney, 384
Head Brakeman W.P. Kennedy, 278
Head Conductor Franz Gerling, 184
Head Goods Shunter J. Clarke, 647
Head Shunter R.H.J. Tolster, 194
Head Shunter Romano Corelli, 46
Heating Manager José Gutiérrez Giménez, 167
Heimemkirch, West Germany, 426
Hellas Express (named train), 329
Herndon, Pennsylvania, United States, 277
Herrliberg-Feldmeilen and Meilen, Switzerland, 272
Hicksville, New York, United States, 445
Highbury, near Henley on Klip, South Africa, 486
Highland, New York, United States, 484
Hindupur, India, 68
Hissar and Jakhod Khera, India, 572
Hither Green, England, 9
Hixon, England, 458
Hogebrug, The Netherlands, 481
Hokuriku Tunnel, Japan, 362
Holfontein, South Africa, 49
Holmes Street Crossing, Tiffin, Ohio, United States, 495
Holmsveden and Kilafors, Sweden, 252
Homer, Ohio, United States, 652
Hondon and Shree Mahavirjl, India, 490
Horse Creek, Wyoming, United States, 704
Horto Florestal, Brazil, 49
Hotar and Magra Hat, India, 212
Hsinchu, Taiwan, 137
Huayihuaica, Chile, 315
Huê and Quang Tri, South Vietnam, 617
Huê, South Vietnam, 610, 617
Hungary, 538
Hurricane, Alabama, United States, 341
Idaho Road Crossing, near Otis Orchards, Washington, United States, 518
Illinois Central Gulf Railroad, 219, 538, 547, 702, 710
Illinois Central Railroad, 31, 269, 378, 412, 430, 653, 680, 684, 692, 696
Illinois Terminal Railroad, 543, 657
Iloilo, The Philippines, 478
Imperial (named train), 387
Indianapolis, Indiana, United States, 385
Inhaminga, Mozambique, 633
Inspector Alexandru Cainaru, 191
Inspector Sousa, 191
Interstate Commerce Commission, 40, 544, 635
Iswepe, South Africa, 48
Italia (named train), 212
Italicus Express (named train), 364
Jaipur-Keonjhar, India, 130
Jajinci, Yugoslavia, 113
Jakhanian and Dulahpur, India, 630
Jamestown, Indiana, United States, 699

Janta Express (named train), 535
Japanese National Railways, 7, 176, 362, 488
Jaratjt, Iran, 387
Järna, Sweden, 230
Jaynes, Arizona, United States, 699
Jedliczach Lódzkich, Poland, 208
Jerico Park Road Crossing, Bowie, Maryland, United States, 429
Jerusalem, Israel, 622
Jessore, Bangladesh, 218
Job's Lane Crossing, near Spanish Town, Jamaica, 492
Joramow block hut and Madhupur, India, 153
Ka Toke, Burma, 623
Kabanovce, Yugoslavia, 49
Kaffir, Texas, United States, 417
Kaithalkurichi, Assam, India, 186
Kalamazoo, Michigan, United States, 440
Kalyan and Vithalwadi, India, 575
Kamrup Express (named train), 186
Kamrup Express(named train), 394
Kansas Avenue Crossing, Liberal, Kansas, United States, 496
Kansas City Southern Railway, 344, 420, 445, 547, 581, 650, 665
Kansas City, Kansas, United States, 435
Karachi, Pakistan, 344
Karampada and Kiriburu, India, 286
Karlsruhe and Durlach, West Germany, 564
Karola (named train), 285
Kartal, Turkey, 91
Kasapovac and Lipe, Yugoslavia, 275
Kaštel Stari and Sadina, Yugoslavia, 307
Katoomba and Wentworth Falls, New South Wales, Australia, 306
Kayseri, Turkey, 160
Kazakhstan, USSR, 399
Kecskemét, Hungary, 513
Kelkheim-Hornau and Unterliedbach, West Germany, 307
Kelley, Iowa, United States, 664
Kellmünz, West Germany, 19
Kép and Đống Đăng, North Vietnam, 654
Keyman Laxman Honnappa, 619
Keyman Sati Ram, 631
Khanapur and Desur, India, 619
Khandvara, India, 56
Khurja Junction, India, 209
Kingwood, New Jersey, United States, 715
Kinki Nippon Railway, 359
Kinmundy, Illinois, United States, 547
Kitaguni (named train), 362
Kittrell, North Carolina, United States, 517
Klington, South Africa, 181
Knightsville and Brazil, Indiana, United States, 548
Kogota and Tajiri, Japan, 637
Korat, Thailand, 446
Korean National Railway, 679
Korean National Railways, 129
Kortemark, Belgium, 529
Kościerzyna and Skorzewo, Poland, 268
Koshigaya, Japan, 431
Kota, Rajasthan, India, 188
Kumhari, India, 637
Kuppam, India, 308
Kyee Taw, Burma, 188
Kyogle, New South Wales, Australia, 384
Kyŏngju, South Korea, 516
L'Hospitalet de Llobregat, Spain, 167
La Barre, Louisiana, United States, 635
La Delfina and Cisneros, 244
La Grande, Oregon, United States, 329
La Louviere, Belgium, 207
La Palma, Arizona, United States, 387
La Picada, Argentina, 408
La Rose, Illinois, United States, 684
Lac Mégantic, Quebec, Canada, 399
Ladpura, Rajasthan, India, 117
Lake Beulah, Wisconsin, United States, 708
Lake City and Lane, South Carolina, United States, 574
Lam Lam, Senegal, 209
Langalanga, Nigeria, 63
Langenweddingen, East Germany, 446
Langhagen, East Germany, 187
Langlaagte, South Africa, 58
Langsa, Sumatra, Indonesia, 344
Lapa, Brazil, 507
Lapeer, Michigan, United States, 690
Larimore, North Dakota, United States, 35
Las Montoyas and El Opón, Colombia, 623
Lathrop, California, United States, 487
Laurel, Mississippi, United States, 665
Lawndale, California, United States, 603
Lawton, Kentucky, United States, 713
Leading Fireman Mohamad Salimuddin, 160
Leading Fireman Mukdeo Ram, 154
Leading Fireman Nur Mohomad, 79
Leading Railman F.T. Fleming, 72
Leading Railman Ralph Akehurst, 74
Leading Trackman A. Corri, 71
Leading Trackman A. Sandy, 71
Leading Trackman D. Costello, 87
Leave Reserve Assistant Stationmaster K.D. Pandey, 148
Lebanon, Missouri, United States, 401
Lehigh Valley Railroad, 513, 703
Leiferde and Meinersen, West Germany, 16

Leinhausen, West Germany, 182
Leith Docks, Scotland, 334
Lengthman A.J.P. Terras, 97
Leoben, Austria, 434
Leonard, Ontario, Canada, 413
Leptokarya, Greece, 329
Leuce Road Crossing, Alma, Michigan, United States, 382
Leverman Hari Gopa, 205
Leverman Sahadev Parasuram, 619
Leverman Sheik Nabi Bux, 131
Leverman Subal Naik, 131
Lewisville, Arkansas, United States, 421
Liaquatpur, Pakistan, 219
Lieut. Col. I.K.A. McNaughton, 38, 373, 555, 569, 599
Lilbourn, Missouri, United States, 499
Lilulah and Howrah, India, 124
Lily Lake Road Crossing, near Peterborough, Ontario, Canada, 478
Lima, Ohio, United States, 650, 653
Linfengyin and Longtien, Taiwan, 379
Litorina (named train), 151
Littlefield, Texas, United States, 514
Loboguerro and Puerto Dagua, 244
Loco Foreman M.K. Bose, 320
Loco Foreman M.S.W. Hasan, 319
Loco. Foreman Ram Sarau, 631
Locsin, The Philippines, 57
Lodhran, Pakistan, 430
Loisy-sur-Marne, France, 607
Lombardie Express (named train), 115
London Transport, 365
Long Island Railroad, 445
Long Khánh, South Vietnam, 611, 617
Loos, British Colombia, Canada, 701
Lop Buri, Thailand, 480
Los Santos de Maimona and Villafranca de los Barros, Spain, 300
Louisville and Nashville Railroad, 341, 657, 697, 700
Lovick, Alabama, United States, 701
Loy's, Maryland, United States, 652
Loyasse Tunnel, France, 350
Lucknow Express (named express), 186
Lufkin, Texas, United States, 701
Lumding, Assam, India, 618
Luttre, Belgium, 345
Lutzville, Pennsylvania, United States, 443
Ma'an, Jordan, 392
Mae Tha, Thailand, 299
Maforga, Mozambique, 633
Magno, Brazil, 89
Magnolia, Mississippi, United States, 378
Magude, Mozambique, 236
Mahapur and Aunrihar Junction, India, 536
Maharashtra Express (named train), 536
Maiduguri and Jos, Nigeria, 78
Maine Central Railroad, 713
Mainpuri and Bhongaon, Uttar Pradesh, India, 42
Major C.F. Rose, 88
Malad, Idaho, United States, 521
Malaga Road Crossing, Newtonville, New Jersey, United States, 475
Mamatoco Crossing, Colombia, 485
Mandini, South Africa, 158
Maquon, Illinois, United States, 280
Marchipanda, Mozambique, 633
Marion, Illinois, United States, 597
Marion, Kansas, United States, 680
Marl-Sinsen, West Germany, 233
Marmet, West Virginia, United States, 714
Marmi e Salici Tunnel, Italy, 346
Marmolejo and Arjonilla, Spain, 150
Marslev, Denmark, 121
Masalchi Ramu, 627
Massachusetts Bay Transportation Authority, 351
Matunga, India, 191
Maverick, Massachusetts, United States, 352
Maxton, North Carolina, United States, 421
McManus, Louisiana, United States, 269
Mechanical Engineer T.J. Lowe, 372
Medina del Campo and Gomeznarro, Spain, 263
Medjez-Sfa, Algeria, 78
Megara, Greece, 78
Melvern, Kansas, United States, 690
Memmingen, West Germany, 516
Mende and Sülysáp, Hungary, 206
Mendota, California, United States, 391
Menomonie, Wisconsin, Wisconsin, 384
Mentor, Ohio, United States, 701
Merritt, Oklahoma, United States, 416
Meteor (named engine), 182
Mexico, 513
Mexico City, Mexico, 406, 435, 519
Michigan Central Railroad, 651, 652
Michigan City, Indiana, United States, 592
Midday Congressional (named train), 429
Mid-day Scot (named train), 105
Middleton Street Crossing, South Kempsey, New South Wales, Australia, 479
Midland City, Alabama, United States, 406
Midland, Michigan, United States, 417
Midorigaoka No.2 Crossing, Japan, 497
Midway Road Crossing, near Dixon, California, United States, 444
Mikawashima, Japan, 174
Miles City, Montana, United States, 386
Miles, Texas, United States, 664
Milford, Ohio, United States, 707

Millbrook and Wilshamstead, England, 550
Minghsiung, near Chiayi, Formosa, 382
Ministry of Transport, 10, 27, 93, 107, 194, 302, 459, 550, 553, 557, 567, 595, 599, 646
Minneapolis St Paul and Sault Ste Marie Railroad, 409, 500, 519, 692
Minnesota 400 (named train), 401
Mińsku Mazowieckim, Poland, 399
Miraj and Mhaisal, Maharastra, India, 621
Missouri Kansas Texas Railroad, 546, 654, 684
Missouri Pacific Railroad, 380, 390, 422, 445, 546, 547, 579, 639, 652, 654, 685, 708, 713
Mjölby and Sya, Sweden, 534
Moatize, Mozambique, 188
Mobile Ganger P.J. Cartland, 13
Moga, India, 495
Mogosoia, Romania, 191
Mohanganj, Uttar Pradesh, India, 589
Monapo, Mozambique, 219
Montevideo, Uruguay, 612
Moorgate, England, 364
Moose Jaw and Archydal, Saskatchewan, Canada, 391
Moradabad, Uttar Pradesh, India, 235
Morley, Louisiana, United States, 334
Morpeth, England, 59
Morrisbureg, Ontario, Canada, 698
Moscow, USSR, 177
Motive Power Supervisor W.C.T. Reeves, 73
Motorman A.S. Sarkar, 125
Motorman Abbas Haider, 575
Motorman Alfonso Sánchez Martínez, 165
Motorman António do Espírito Santo Tavares, 142
Motorman Aristides do Espirito Santo, 247
Motorman Aurino Miguel da Silva, 254
Motorman B. Rosario, 370
Motorman B.C. van Tonder, 120
Motorman Bruno Montaldo, 251
Motorman C.A. Gladding, 370
Motorman Carlos Fernández Sánchez, 166
Motorman Ciro Miceli, 337
Motorman D.S. Purves, 9
Motorman Dreyer, 358
Motorman Durval Ribeiro da Costa, 254
Motorman Edouard Mottaz, 262
Motorman Enrique Cuartero, 375
Motorman Ernst Schröter, 239
Motorman Fictor, 170
Motorman Gopal Sharma, 160
Motorman Guillermo Camejo, 408
Motorman H. Lang, 370
Motorman Hiroyuki Okumara, 359
Motorman J.P. Bray, 370
Motorman J.P. Singh, 128
Motorman James A. Watts, 221
Motorman Katsumi Nishiyama, 359
Motorman Leslie Newson, 365
Motorman Mauroce Gaillex, 208
Motorman Paul Goelhen, 207
Motorman Pritam Singh, 127
Motorman Raymond Deadman, 371
Motorman Robert W. Cavanagh, 220
Motorman S.P. Banerjee, 125
Motorman Salvador Salcedo Fernández, 284
Motorman Thomas Bagley, 436
Motorman Thomas Shaw, 91
Motorman Vilmar Valentin Limasse, 90
Motorman W.D. Burton, 192
Motorman Yoshinori Sugimoto,, 444
Motorman Yukio Haga, 176
Mount Savage, Maryland, United States, 698
Mount Washington Cog Railway, 311
Mozambique, 633
Mr Brian Gibbens, Q.C, 458
Mr J.V. Feehan, 322, 332
Multan, West Pakistan, 564
München-Allach, West Germany, 532
Muncie, Illinois, United States, 684
Munsan, South Korea, 625
Münster and Gronau, West Germany, 498
Mượng Mán and Tháp Chàm, South Vietnam, 615
Mượng Mán, South Vietnam, 612
Mượng Mán, South Vietnam, 608, 609
Mượng Mán, South Vietnam, 617
Mượng Mán, South Vietnam, 624
Mustang, Oklahoma, United States, 290
Mykawa, Texas, United States, 692
Namwon, South Korea, 315
Nangue, Mozambique, 633
Nankai Electric Railway, 444
Napsbury and Radlett, England, 557
Narowal, Pakistan, 696
Nashville, Michigan, United States, 593
National Transportation Safety Board, 33, 220, 259, 277, 281, 291, 440, 450, 487, 501, 525, 583, 584, 587, 658, 665, 673, 680, 685, 693, 697, 704, 709, 711
Nawashar Doaba and Garh Shankar, India, 537
Nederland, Texas, United States, 344
Nederlandse Spoorwegen, 145, 172, 185
New Hampshire Public Utilities Commission, 313
New London, Wisconsin, United States, 546
New Smyrna Beach, Florida, United States, 613
New York Central Railroad, 257, 385, 400, 440, 593, 639, 640, 650
New York Chicago and St Louis Railroad, 402
New Yorker (named train), 699

Newark, Arkansas, United States, 445
Nha Trang and Diêu Trì , South Vietnam, 609
Nha Trang and Diêu Trì, South Vietnam, 611
Nha Trang and Quy Nhon, South Vietnam, 611
Nha Trang, South Vietnam, 614
Night Stationmaster Antonio León Cornejo, 250
Nilópolis and Olinda, Brazil, 253
Ninh Binh, North Vietnam, 651
Nocturno Campista (named train), 21
Noel, Missouri, United States, 581
Nong Khai Express (named train), 491
Nordjyden (named train), 121
Nordpilen (named train), 52
Norfolk and Western Railway, 521, 539, 650, 653, 664, 697, 700, 708
Norfolk Southern Railway, 328, 407
Norges Statsbaner, 295
Norman, Oklahoma, United States, 707
Norrköping, Sweden, 238
North Belgrade, Maine, United States, 713
North Bengal Express (named train), 606
North Bihar Express (named train), 153
North Coast Daylight Express (named train), 479
North East Frontier Railway, 169, 186, 580, 606, 622, 625
North Eastern Railway, 146, 415, 630
North Miami, Florida, United States, 613
North Vietnam, 645
North Western Railway, 181
Northeast Frontier Railway, 536, 618
Northern Express (named train), 299
Northern Pacific Railroad, 410, 479
Northern Railway, 42, 209, 482, 534, 537, 572
Northern Tablelands Express (named train), 553
Norvalspont, South Africa, 77
Novska, Yugoslavia, 191
Nuneaton, England, 85
Ochlocknee, Florida, United States, 703
Odernheim, West Germany, 65
Ofuna, Japan, 629
Oita and Beppu, Japan, 636
Okucani, Yugoslavia, 133
Omega, Ohio, United States, 690
Oneonta, New York, United States, 704
Onyang, South Korea, 494
Operator Bill Smith, 543
Operator E.E. Cash, 638
Operator P.F. Naumann, 281
Operator Sherman Heinz, 544
Oreitia, Spain, 419
Oreshetz, Bulgaria, 133
Öresundspilen (named train), 534
Orleans Road, West Virginia, United States, 639
Österreichischen Bundesbahnen, 539
Otis Orchards, Washington, United States, 490
Ozaki and Tarui, Japan, 444
Ozumi, Japan, 377
Padang Besar, Thailand, 174
Pa-Kong Bridge, China, 621
Palestina Crossing, Colombia, 412
Palmdale, California, United States, 579
Panama Limited (named train), 547
Panel Supervisor O.G. Hawkins, 37
Panhandle and Santa Fe Railway, 417
Panhandle and Sante Fe Railway, 395
Panikheti, Assam, India, 619
Pannonia Express (named train), 211
Panorama (named train), 344
Paola, Kansas, United States, 546, 547
Parada Modelo, Brazil, 417
Paraíba do Sul and Engenheiro Vieira Cortes, Brazil, 247
Parasnath and Hazaribagh Road, India, 564
Pas Viejo and Pas Centro, Nicaragua, 48
Passed Fireman F.W. Kelly, 106
Passed Fireman G.A. Cliffe, 302
Passed Fireman J. Dean, 91
Passed Fireman J.G. Walker, 474
Passed Fireman K.S. Craven, 97
Passenger Guard A.G. Skinner, 554
Passenger Guard G.R. Bradshaw, 91
Passenger Guard Joseph Hansell, 303
Passenger Guard T.M. Wilson, 567
Passenger Guard William Anderson, 26
Patas and Daund, India, 535
Patrolman H.I. Suttle, 12
Patrolman Santosh, 626
Patrolman Sarwan, 626
Penn Central Railroad, 277, 471, 484, 498, 499, 501, 512, 518, 679, 684, 685, 690, 691, 699, 700, 701, 702, 703, 706, 707, 708, 715
Pennsylvania Railroad, 3, 40, 429, 443, 548, 549, 638, 657
Pennsylvania-Reading Seashore Lines, 449, 475
Peotone, Illinois, United States, 710
Perambur, India, 137
Permanent Way Inspector D. Donovan, 332
Permanent Way Inspector G.C. Barman Rajkumar, 627
Permanent Way Inspector H.K. Kulkarni, 619
Permanent Way Inspector J.G. Abbott, 600
Permanent Way Inspector J.N. Wilkinson., 93
Permanent Way Inspector K.N. Verna, 44
Permanent Way Inspector K.V. Radhakrishnan, 643
Permanent Way Inspector R.S. Lal, 631
Permanent Way Inspector S. C. Maheshwari, 79
Permanent Way Inspector T.E. Watson, 28
Permanent Way Sub-Inspector P.G. Dawes, 12
Permanent Way Supervisor D. Mellen, 61

Permanent Way Supervisor P.G. Dawes, 71
Pernik, Bulgaria, 512
Phan Rang, South Vietnam, 629
Philadelphia, Pennsylvania, United States, 707
Philippine National Railways, 55, 57, 248, 478, 507
Philippines National Railway, 520
Phoenixville, Pennsylvania, United States, 685
Phú Lôc, South Vietnam, 617
Phú Mỹ, South Vietnam, 613
Phú Yên, South Vietnam, 614
Phulera and Sambhar Lake, India, 393
Phulpur, India, 563
Pina, Cuba, 611
Piotrków Trybunalski, Poland, 6
Placedo, Texas, United States, 390
Platelayer J. Doyle, 331
Platform Assistant Stationmaster J.N. Purushottam, 201
Platform Assistant Yard Manager S.R. Bhownick, 127
Plaza de España, Barcelona, Spain, 351
Plovdiv, Bulgaria, 535
Po Kut, China, 610
Pointsman A. Valli, 642
Pointsman Alagan, 643
Pointsman Alexandru Cristea, 263
Pointsman Devi Singh, 118
Pointsman E. Gazi, 214
Pointsman M. Ramaraju, 202
Pointsman M. Viswanathan, 202
Pointsman Siddappa, 201
Pointsman Thyagarajan, 202
Pointsman V. Alagan, 643
Poipet, Cambodia, 620
Pont D'Héry, France, 115
Port Barre, Louisiana, United States, 380
Portage du Fort, Ontario, Canada, 638
Porter Antonio León Martin Gallego, 247
Porter Bruno Martínez, 112
Porter Giovanni Malli, 252
Porter Molenaar, 171
Porter Signalman Máximo Blanco, 137
Porter/Booking Clerk D.A. Leigh, 109
Portsmouth, Ohio, United States, 708
Potchefstroom, South Africa, 236
Potgietersraus, South Africa, 632
Potgietersrus, South Africa, 429
Powder Springs, Georgia, United States, 679
Power Controller A. Aiyasamy, 309
Power Controller A.S. Krishnamurty, 309
Praha-Hloubětin, Czechoslovakia,, 117
Presidente Illia (Villa Soldati), Argentina, 393
Prijedor, Yugoslavia, 495
Pringle, Texas, United States, 678
Prizren, Yugoslavia, 512
Prunişor, Romania, 590
Puebla, Mexico, 480, 496
Puente Moreno, Mexico, 316
Puntalara, Argentina, 416
Punxsutawney, Pennsylvania, United States, 499
Purwokerto, Java, Indonesia, 186
Pyok Kwe, Burma, 611
Qalyūb, Egypt, 84
Quang Ngãi, South Vietnam, 610, 613
Quarryville, Pennsylvania, United States, 703
Quebec Iron and Titanium Company, 182
Quincieux and Saint-Germains-au-Mont d'Or, France, 58
Quipaw, Oklahoma, United States, 395
Quitaque Tunnel, Texas, United States, 363
Quy Nhon and Quang Ngãi, South Vietnam, 611
Quy Nhon, South Vietnam, 609
Radkowice, Poland, 158
Radzyń Podlaski, Poland, 591
Railman T. Andrews, 371
Railway Police Constable No. 418 N. Narayanan, 644
Railway Police Constable No. 560 R. Rengarathinam, 643
Railway Police Head Constable A. Suryanarayana, 523
Railway Protection Force Sainik J.N. Sarkar, 214
Railway Protection Force Sainik Ram Adhar Misra, 631
Railway Protection Force Sainik Ram Ugrah Singh, 631
Raiwind Junction, Pakistan, 480
Rajgir, Bihar, India, 517
Ranchi Express (named train), 607
Randfontein, South Africa, 52
Randspruit and Bume, South Africa, 122
Rangoon and Mandalay, Burma, 632
Rangoon and Prome, Burma, 606
Rangoon, Burma, 630
Reading Company, 638
Red Nacional de los Ferrocarriles Españoles, 300, 420, 553
Red River (named train), 435
Redditt, Ontario, Canada, 410
Rede Ferroviária Federal, 20, 54, 89, 151, 198, 247, 253, 400, 417, 530
Redfern, New South Wales, Australia, 553
Regulator F.H.A. Perry, 196
Regulator J. Marlow, 88
Relief Clark C. Brown, 370
Relief Signalman A.B. Kirk, 554
Relief Signalman A.G. Wallis, 551
Relief Signalman C.F. Pickering, 473
Relief Signalman Cecil Jackson, 107

Relief Signalman H.G. Thaxter, 599
Relief Signalman L. Cunningham, 324
Relief signalman L. H. Copping, 558
Relief signalman T. J. Moriarty, 559
Relief Signalman W. Martin, 324
Relief Signalman Walter Wade, 371
Relief Stationmaster Donaldo Deffino, 248
Relief Stationmaster Pires de Almeida, 190
Relieving Assistant Stationmaster Mohammed Hafiz Khan, 147
Relieving Assistant Stationmaster P.K. Biswas, 214
Relieving Stationmaster R.K. Nair, 643
Relieving Switchman Gayana, 565
Relieving Switchman Mohan Ram, 565
Relieving Switchman R.L. Mukkerjee, 565
Renigunta, India, 201
Rensselaer, Indiana, United States, 678
Rheinweiler, West Germany, 66
Rias Baixas (named train), 375
Řikonín, Czechoslovakia, 211
Río Luján, Argentina, 165
Robledo de Chavela and Santa Maria de la Alameda, Spain, 311
Rochester, New York, United States, 639
Rock Hill, South Carolina, United States, 653
Rockwall, Texas, United States, 684
Roddenberry Crossing, near Cairo, Georgia, United States, 538
Rolling Stock Engineer G.H. Hafter, 372
Rolling Stock Inspector R.L. Norman, 648
Roseville, California, United States, 582
Roseworthy, South Australia, Australia, 488
Ruby, Mississippi, United States, 702
Ruess Road Crossing, near Owosso, Michigan, United States, 391
Running Foreman T. Shepherd, 550
Ruse and Kaspichen, Bulgaria, 170
Rust Stasie, near Malmesbury, 77
Rzepin, Poland, 56
Saginaw, Michigan, United States, 402
Saigon and Mượng Mán , South Vietnam, 610
Saigon and Mượng Mán, South Vietnam, 609, 621
Saigon and Nha Trang, South Vietnam, 609
Saigon, South Vietnam, 612, 615, 621, 624
Salem, South Dakota, United States, 545
Sallisaw, Oklahoma, United States, 420
Salta, Argentina, 491
Sambo Road Crossing, near Blackfoot, Idaho, United States, 411
Samuktala Road and Salsalabari, India, 625
San Cristóbal Ecatepec, Mexico, 497
San Fabian and Damortis, The Philippines, 248
San Joaquin Daylight (named train), 396
San Jose, California, United States, 415
San Nicola Varco, Italy, 655
Sandridge, England, 553, 566
Sant Antonio Tunnel, Italy, 353
Santa Apolónia, Portugal, 83
Sanyo Electric Railway, 624
São João de Talha and Póvoa de Santa Iria, Portugal, 142
São Miguel, Mozambique, 235, 236
Sarah, Mississippi, United States, 684
Sarajevo, Yugoslavia, 169
Sasson, Iran, 272
Satara Road and Koregaon, India, 536
Savannah Avenue Crossing, Port Arthur, Texas, United States, 445
Saybrook, Florida, United States, 613
Schofield, South Carolina, United States, 456
Schoolcraft, Michigan, United States, 477
Schweinsburg-Culten, East Germany, 285
Schweiz Express (named train), 66
Scientific Officer J.B.C. McCann, 28
Scott's Crossing, near Corry, Pennsylvania, United States, 476
Seaboard Air Line Railroad, 425
Seaboard Coast Line Railroad, 456, 517, 529, 538, 574, 601, 679, 703
Second Fireman Abdul Salam, 79
Second Fireman B. Parthasarathy, 308
Second Fireman B.B. Singh, 147
Second Fireman C.B. Yadava, 631
Second Fireman Gaffur Ansari, 319
Second Fireman H.N. Jadi, 619
Second Fireman Kashmiri Lal, 209
Second Fireman N.K. Dhal, 131
Second Fireman Nandji Choudhury, 204
Second Fireman P. Nachimuthu, 139
Second Fireman P.C. Nandy, 125
Second Fireman R.K. Sinha, 319
Second Fireman R.S. Ramdan, 117
Second Fireman T. Mani, 642
Secondman B.K. Sugden, 195
Secondman C. Graham, 60
Secondman Dennis Smith, 26
Secondman James Toghill, 460
Secondman P.A. Wheaton, 598
Secondman Paul Stokes, 71
Secondman R.P. Woolnough, 36
Section Controller A.N. Sen, 156
Section Controller A.P. Gupta, 178
Section Controller D.T. Hawkes, 596
Section Controller S.A. Hameed, 309
Section Controller S.B. Bhattacharjee, 213
Section Controller S.R. Roy, 162
Section Controller T.N. Ramaswamy, 642
Section Controller V.N. Marghabandhu, 309

Seguin, Texas, United States, 426
Seibu Railway, 377
Selma, Kansas, United States, 654
Semizovac, Yugoslavia, 216
Senior Cabin Assistant Stationmaster S.P. Chowdhury, 126
Senior Deputy Chief Engineer L.C. Agarawal, 632
Senior Secondman F.S. Roddy, 87
Senior Secondman Leslie Norman, 85
Senior Train Examiner P.S. Sripathy, 310
Seoul. South Korea, 480
Seymour, Iowa, United States, 639
Shallowater, Texas, United States, 395
Shioya, Japan, 624
Shopman C.O. Pitter, 38
Shopman F.A. Barret, 38
Shortcreek Trestle, between Richmond and Boston, Indiana, United States, 342
Shreveport, Louisiana, United States, 650
Shunter (Passed Driver) P.V. Rao, 286
Shunter (Passed Driver) Premdhan, 287
Shunter Guglielmo Balbi, 100
Shunter L. Millar, 325
Shunter Lislo Barbieri, 100
Shunter N. Doran, 325
Shunter Stafano Rossi, 232
Shunting Jamadar Khadar Sahib, 202
Shunting Jamelar M. Kasruppan, 643
Shuntingman G. Krishnan, 643
Shwenyaung, Burma, 55
Sidi Yahia du Rharb, Morocco, 84
Signal and Telegraph Engineer R.I Carmichael, 489
Signal Khalasi D. Pullah, 202
Signalman Alan Frost, 599
Signalman Alberto Basualdo, 183
Signalman B.S. Register, 460
Signalman C. Searle, 302
Signalman C.F. Duffin, 37
Signalman C.H. Curtis, 194
Signalman Derek Harrison, 94
Signalman F.D. Obee, 74
Signalman F.H. Foxall, 647
Signalman F.M. Whiteside, 94
Signalman Fenrez Szantai, 206
Signalman Franco Sarchi, 100
Signalman G. Davies, 646
Signalman G.A. Hughes, 551
Signalman G.E. Young, 88
Signalman G.H. Jones, 647
Signalman H. McKenna, 324
Signalman H. Monks, 93
Signalman H. Oteman, 184
Signalman H. Scott, 302
Signalman Herbert Sutton, 107
Signalman J. Collier, 324
Signalman J. Russell, 28
Signalman J.C. Wood, 554
Signalman José Dias, 191
Signalman Kwakkenbus, 170
Signalman M. Fraiser, 27
Signalman M.J. Woodcock, 460
Signalman Makoto Ohashi, 176
Signalman Nicolau, 254
Signalman P. Garrahan, 324
Signalman P. Matthews, 325
Signalman Pietro Narducci, 655
Signalman R.W. Bradbury, 194
Signalman Rudolfo Luis Flores Guriérrez, 165
Signalman S.M. Upson, 646
Signalman Stasys Urbonavičius, 238
Signalman T Ferrington, 646
Signalman T. Davidson, 61
Signalman T.M. Halbert, 61
Signalman T.P. Carroll, 302
Signalman Wally Hartslief, 55
Sillipur and Birpur, India, 78
Simplon I Tunnel, Italy, 354
Simplon Tunnels, Switzerland, 357
Simsbury, Connecticut, United States, 679
Sinai Peninsula, Egypt, 624
Singleton Bank, England, 91
Sintra and Algueirão Mem Martins, Portugal, 189
Sion/Sitten and St-Léonard, Switzerland, 262
Sirajganj, Bangladesh, 233
Sistema de Transporte Collective, 165
Sivok and Gulma, West Bengal, India, 606
Sivok and Pilanshat, West Bengal, India, 622
Skavstaby, between Upplands Väsby and Roserberg, 314
Skultorp, Sweden, 115
Skyline Siding (Mount Washington), New Hampshire, United States, 312
Sleeping Car Attendant Albert Burgess, 549
Slinger, Wisconsin, United States, 692
Sociedad de Ferrocarrils y Tranvias de Bilbao, 271
Societa Italiana per le Strade Ferrate del Mediterraneo, 337
Société nationale des chemins de fer français, 81
Société nationale des chemins de fer français, 24
Société nationale des chemins de fer français, 314
Société nationale des chemins de fer français, 341
Société nationale des chemins de fer français, 350
Société nationale des chemins de fer français, 361
Société nationale des chemins de fer français, 607
Sofia, Bulgaria, 58
Somerdale, New Jersey, United States, 449
Son Nagar, India, 204
Sotani Tunnel, Japan, 359

South African Railways, 236
South Australian Railways, 489
South Bihar Express (named train), 153
South Central Railway, 124, 471, 522, 535, 536
South Eastern Railway, 130, 183, 286, 607, 637
South Whitley, Indiana, United States, 664
Southern Pacific Company, 387, 391, 398, 402, 404, 411, 415, 426, 443, 444, 518, 579, 582, 584, 602, 649, 699, 701, 706, 707, 710, 713
Southern Pacific Railroad, 327, 700
Southern Pacific Railroad of Mexico, 397
Southern Pacific Transportation Company, 711
Southern Railway, 137, 198, 201, 308, 419, 483, 525, 619, 628, 641, 649, 653, 665, 690, 701, 703
Southwesterner (named train), 650
Sowerby Bridge Tunnel, England, 349
Spirit of St Louis (named train), 548
Spoorwegongevallenraad, 173
Springfield, Illinois, United States, 657
Springs, South Africa, 633
Spuds, Florida, United States, 613
St Clair Avenue Crossing, Scarborough, Ontario, Canada, 539
St Joseph, Minnesota, United States, 435
St Louis Southwestern Railway, 421, 499
St Louis-San Francisco Railway, 290, 395, 401, 457, 498, 547, 701
Stanley, Wisconsin, United States, 409
Starucca, Pennsylvania, United States, 637
Statens Järnvägar, 52, 231, 252
Station Foreman G. Fox, 371
Station Inspector E. Moody, 595
Station Inspector E.E. J. Williams, 595
Station Manager B L.H. Arundell, 73
Station Supervisor K.R. Henderson, 73
Stationmaster Angelo or Giuseppe Venuto, 353
Stationmaster Angelo Timpano, 340
Stationmaster Artur Braga, 49
Stationmaster Carmelo López, 300
Stationmaster Chhuttan Lal Jain, 118
Stationmaster Claudio Ambrogi, 578
Stationmaster D.P. Lovatt, 555
Stationmaster E.V. Venkatachalam, 619
Stationmaster Francesco Sinopoli, 103
Stationmaster G. Ward, 554
Stationmaster Giovanni Tarlazzi, 46
Stationmaster Goncalor Correia, 84
Stationmaster J. Reed, 28
Stationmaster João Biu, 153
Stationmaster José Manuel Vallejo Gárate, 112
Stationmaster Juan Abando Amechuzarra, 271
Stationmaster M. Masthan Rao, 523
Stationmaster M.C. George, 641
Stationmaster Mario Podestà, 348
Stationmaster N.G. Choudhury, 286
Stationmaster Nikolaos Gekas, 276
Stationmaster R. Kulandaisamy, 643
Stationmaster R.P. Gautam, 43
Stationmaster Stroe Tiberiu, 57
Stationmaster T.O. Staley, 110
Stationmaster U.S. Srivastava, 210
Stationmaster Van Zevenberg, 171
Steam-man V. Ramayya Thevar, 644
Stechford, England, 192
Steelton, Pennsylvania, United States, 3
Stellwerksleiter Guenther Zimmermann, 212
Sterling, Ohio, United States, 544
Stettler, Alberta, Canada, 515
Stillwell, Indiana, United States, 638
Stratford, Ontario, Canada, 455
Strong City, Kansas, United States, 506
Stuart, Iowa, United States, 422
Sub-Ganger D. Semple, 94
Sub-Inspector R. Coldwell, 28
Sub-Inspector T. Dagger, 93
Substitute Porter Abdul Latif, 155
Substitute Token Porter Jugal Kishore Mahato, 287
Sud Express (named train), 240
Sudexpreso (named train), 249
Sukabumi, West Java, Indonesia, 54
Summaga Crossing, Italy, 511
Sumulong, The Philippines, 56
Sunchon, South Korea, 412
Sunset Limited (named train), 396
Super Continental (named train), 383
Super-Continental (named train), 344
Superintendent J.C. Marson, 555, 559
Supervisor Motiejus Šiško, 238
Surabaya, East Java, Indonesia, 49
Suzano and Jundiapeba, Brazil, 151
Switchman B. Lal, 210
Switchman B.B. Tanti, 156
Switchman Bandham Gope, 565
Switchman Frank Wells Jr., 665
Switchman Kishan Lal, 210
Switchman M.M. Mukherji, 155
Switchman Muhammad Khalil, 155
Switchman Ram Chandra Ram, 204
Switchman S.N. Mandal, 205
Sydwestjyden (named train), 121
Symmes Avenue Crossing, Fairfield, Ohio, United States., 496
Szolnok, Hungary, 113
Taben-Rodt, West Germany, 238
Tábor, Czechoslovakia, 173
Tacotalpa, Mexico, 114
Taft Tunnel, Idaho, United States, 707
Tagekapu, West Java, Indonesia, 301
Tahona, Argentina, 25

Tainan, Formosa, 423
Taipei, Formosa, 385
Taipei, Taiwan, 231
Takane Crossing, Tatebayashi, Japan, 485
Talaimannar Express (named train), 25
Tamouth, Egypt, 65
Tarbes, France, 536
Tardienta and Almudévar, Spain, 112
Tarrasa, Spain, 436
Tatuapé, Brazil, 427
Tawkywein, Burma, 606
Taylor, South Carolina, United States, 649
Tebing Tinggi, Sumatra, Indonesia, 470
Technical Assistant (Waggons) G.A. Appleby, 28
Technician Cllief, 171
Technician E. Bickley, 460
Telegraphist Benjamin Batista Santana, 234
Teltar, India, 169
Teotihuacan, Mexico, 423
Terminals Inspector W.C. Hasler, 554
Terre Haute, Indiana, United States, 706
Texarkana, Texas, United States, 547, 654
Texas and Pacific Railway, 334, 635, 650, 654, 698
Texoco, Mexico, 427
Tezrao Express, 219
Thanh Hóa and Nam Dinh, North Vietnam, 651, 652
Thanh Hóa, North Vietnam, 651, 652
Tháp Chàm, South Vietnam, 614
Thayer Junction, Wyoming, United States, 664
The Aberdonian (named train), 60
The Daw, Burma, 606
The Manhattan (named train), 706
The Southern Aurora (named train), 264
Theil, France, 169
Thirsk, England, 26
Thousand Palms, California, United States, 699
Thủ Đức, South Vietnam, 616
Ticket Collector Tarapada Guha, 162
Ticket Examiner D.C. Chakraborty, 626
Ticket Examiner Giovanni Bianchi, 577
Ticket Examiner Johan Viljoen, 58
Ticket Examiner N.R. Chakraborty, 626
Ticket Examiner Nandagopal, 199
Ticket Examiner S.H. Roets, 58
Ticket Inspector Orlando Pereira Mendonça, 50
Tillar, Arkansas, United States, 713
Tilrah, Bihar, India, 394
Tipton, Indiana, United States, 650
Tirreno (named train), 346
Tobu Railway, 494
Token Porter Narayan Patro, 286
Token Porter Padmanah Parida, 131
Toledo Peoria and Western Railroad, 685
Toledo, Ohio, United States, 499, 652
Tonti, near Salem, Illinois, United States, 31
Tornillo, Texas, United States, 415
Torrecon, Mexico, 427
Torricola, Italy, 231
Toyoura, Japan, 488
Track Chargeman Harold Merrick, 87
Track Supervisor Raymond Kagele, 490
Traction Driving Inspector A. Maitra, 319
Traction Inspector A.E. Holder, 37
Traffic Brakesman A.K. Raa, 131
Traffic Brakesman B. Latchayya, 131
Traffic Inspector P.B. Sarkar, 127
Traffic Inspector S.L. Kackar, 43
Traffic Regulator H.L. Hickmott, 74
Train Examiner C.K. Jha of Gujhandi, 320
Train Examiner in Charge J.P. Mittal, 573
Train Examiner Jag Narain, 631
Train Examiner R.D. Rai, 631
Train Examiner R.K. Dhol, 626
Train Examiner T. Prasad, 631
Train Examiner V.C. Kurien, 309
Train Lighting Fitter Prem Nath, 573
Trainee Engineer Charles Kenison, 312
Trainman Herbert Degerstedt, 343
Trans-Europa-Express (named train), 354
Transportation Inspector S.D. Singh, 318
Travelling Ticket Collector D. Mulhearn, 109
Travelling Ticket Examiner Deep Singh, 572
Travelling Ticket Examiner R. Dhanapathy, 523
Travelling Ticket Inspector Samuel Elliot, 59
Trebbin, East Germany, 636
Treno del Sol (named train), 29
Trent Road Crossing, Beckingham, England, 472
Tretten, Norway, 294
Tsurumi and Shin-Koyashu, Japan, 7
Tucson, Arizona, United States, 700
Tulkarm, Jordan, 625
Tunis, Tunisia, 115
Tunnel City, Wisconsin, United States, 401
Tunnel No. 66, between Laza and Villar de Barrio, Spain, 375
Tuy Hòa, South Vietnam, 625
Twilight Limited Flyer (named train), 440
Twin Cities 400 (named train), 384
Ultadanga Road, India, 160
Ultrasound Ganger D.V. Gateland, 13
Umeshnagar, India, 181
Union Pacific Railroad, 329, 388, 411, 424, 435, 450, 521, 664, 702
Untermallebarn, Austria, 539
Upper India (named train), 589
Urdúliz and Plencia, Spain, 271
Vail, Arizona, United States, 707
Valašské Meziříčí, Czechoslovakia, 209

Valdivia, Chile, 40
Valentin Alsina, Argentina, 516
Van Giã, South Vietnam, 610
Văn Phú and Phú Mỹ, South Vietnam, 609
Varigotti and Finale Ligure Marina, Italy, 358
Velars-sur Ouche, France, 341
Venchan, Bulgaria, 231
Ventura, California, United States, 602
Viaducto, Mexico, 165
Vicentica, Cuba, 344
Vienna, Georgia, United States, 483
Vierzy Tunnel, France, 359
Vila Nova de Fronteira, Mozambique, 633
Villabilla, Spain, 446
Villafranco del Campo and Santa Eulalia del Campo, Spain, 256
Villar-de-los-Álamos, Spain, 249
Vĩnh Xương, South Vietnam, 617
Vinh, North Vietnam, 646
Violet Town, Victoria, Australia, 264
Virgen Peregrina (named engine), 150
Virginian (named train), 544
Virrey Amat, Spain, 375
Voghera, Italy, 99
Volta de Ferradura Crossing, Portugal, 398
Vranduk Tunnel, Yugoslavia, 355
Vranje, Yugoslavia, 329
Vrbno pod Pradědem, Czechoslovakia, 299
Waitman, Kentucky, United States, 697
Warngau, West Germany, 296
Washington, North Carolina, United States, 407
Watertown, Wisconsin, United States, 546
Waβmannsforf, East Germany, 8
Weaver Junction, England, 549
Wells Creek, Pennsylvania, United States, 336
Wenatchee, Washington, United States, 587
Wendlingen and Kirchheim-Ötlingen, West Germany, 273
West Ealing, England, 35
West Java, Indonesia, 49
West Point, Texas, United States, 327
Western Maryland Railway, 652, 663, 698
Western Pacific Railroad, 487, 679
Western Railway, 120, 188, 474, 490, 537
Western Talisman (named engine), 36
Westervoort, The Netherlands, 183
Wever, Iowa, United States, 456
Whakaki, New Zealand, 257
White Haven, Pennsylvania, United States, 703
Whitewood, Saskatchewan, Canada, 517
Willard, Kansas, United States, 653
William Robinson, 241
Wilwatte, Ceylon, 25
Windham, Ohio, United States, 603
Windsor and Sherbrooke, Quebec, Canada, 308
Winsford and Coppenhall Junction, England, 105
Wolcott, Indiana, United States, 679, 699
Wolf Lake, Illinois, United States, 708
Worden, Illinois, United States, 521
Worksop, England, 685
Yadugram block hut and Gurpa, India, 318
Yalvigi, India, 198
Yard Foreman David Jones, 587
Yard Inspector A.T. Davies, 647
Yardmaster Wayne Barber, 291
Yeoncheon, South Korea, 625
Yilmaz, Turkey, 407
Yongchon, South Korea, 679
Yongkang, South Korea, 395
Youngstown, Ohio, United States, 714
Yugoslavia Express (named train), 133
Yutan, Nebraska, United States, 477
Zagreb, Yugoslavia, 82
Zahana, Algeria, 65
Zákolany, Czechoslovakia, 306
Žaslial, Lithuanian Soviet Socialist Republic, USSR, 237
Zemun Polje, Yugoslavia, 164
Zeytinlik, Turkey, 235
Zidani Most, Yugoslavia, 188

TABLE OF MAJOR RAILWAY ACCIDENTS 1961-1975

Date	Location	Railway Company	Description	Passenger Fatalities
9 Jan 1961	L'Hospitalet de Llobregat, Spain	Red Nacional de los Ferrocarriles Españoles	Head-on collision	20
11 Jan 1961	Valdivia, Chile		Derailment	30
13 Jan 1961	Akibamachi, Japan		Collision with lorry on open level crossing, derailment, collision, fire	6
1 Feb 1961	Near Cassano d'Adda, Italy		Derailment, defective vehicle	13
2 Feb 1961	Del Block Station, near Bowie, MS, USA	Pennsylvania	Derailment, excessive speed	6
8 Mar 1961	Teltar, India	North East Frontier	Head-on collision	11
25 Mar 1961	Near The Daw, Burma		Derailment, sabotage	8
26 Mar 1961	Tawkywein, Burma		Derailment, sabotage	23
30 Mar 1961	Marmi e Salici, Italy	Fellovie dello Stato	Fire in tunnel	5
Mar 1961	Castelfranco Veneto, Italy	Fellovie dello Stato	Derailment, defective points	29
19 Apr 1961	Between Sevak and Gulma, India	North East Frontier	Derailment, sabotage?	34
30 Apr 1961	Kartal, Turkey		Rear collision	13
13 Jun 1961	Near Eßlingen Hbf, West Germany	Deutsche Bundesbahn	Head-on collision during temporary single-line working	33
18 Jun 1961	Loisy-sur-Marne, France	Société nationale des chemins de fer français	Derailment, sabotage by the OAS	28
8 Jul 1961	Near Mae Tha, Thailand		Runaway train, derailment	19
16 Jul 1961	Singleton Bank, England	British Railways, London Midland Region	Rear collision	6
23 Aug 1961	Beska, Yugoslavia	Zajednica Jugoslovenskih Zeleznica	Train diverted onto wrong line, head-on collision	11
27 Aug 1961	Near Vrbno pod Pradědem, Czechoslovakia		Runaway waggons, collision, fire	18
29 Aug 1961	Theil, France	Société nationale des chemins de fer français	Collision with light engine	8

16 Sep 1961	Between Los Santos de Maimona and Villa France de los Barros, Spain	Red Nacional de los Ferrocarriles Españoles	Runaway waggons, collision	15
3 Oct 1961	Near Sarajevo, Yugoslavia		Collision	6
5 Oct 1961	Between Hamburg Berliner Tor and Rothenburgsort, West Germany	Deutsche Bundesbahn	Rear collision	26
20 Oct 1961	Between Ghatsila and Dalbhungarh, India	South Eastern	Derailment, sabotage	51
26 Oct 1961	Between Oita and Beppu, Japan	Oita Kotsu Corporation	Train struck by landslide	31
29 Oct 1961	Between Mainpuri and Bhongaon, Uttar Pradesh, India	Northern	Derailment, excessive speed	18
23 Dec 1961	Fiumarella Viaduct, near Catanzaro, Italy	Societa Italiana per le Strade Ferrate del Mediterraneo	Derailment, excessive speed, part of train fell from viaduct	71
6 Jan 1962	Between Ruse and Kaspichen, Bulgaria		Collision?	12
8 Jan 1962	Harmelen, Near Woerden, The Netherlands	Nederlandse Spoorwegen	Head-on collision	91
29 Jan 1962	Tábor, Czechoslovakia		Buffer stop collision at end of siding	11
22 Feb 1962	Between La Delfina and Cisernos, Colombia	Ferrocarriles Nacionals de Colombia	Head-on collision, single line	35
1 Mar 1962	Near Trebbin, East Germany	Deutsche Reichsbahn	Collision with load projecting from train on adjoining line, derailment	c70
8 Mar 1962	Castel Bolognese, Italy	Fellovie dello Stato	Derailment, excessive speed	14
21 Mar 1962	Iswepe, Transvaal, South Africa	South African	Derailment	10
1 Apr 1962	Between Pas Viejo and Pas Centro, Nicaragua		Derailment	9
17 Apr 1962	Padang Besar, Thailand		Head-on collision	7
3 May 1962	Mikawashima, Japan	Japanese National	Buffer stop collision at end of siding, double collision	160
10 May 1962	Near Dieu Tri, South Vietnam		Derailment, sabotage	29
10 May 1962	Moscow, USSR		Sidelong collision?	?
31 May 1962	Voghera, Italy	Fellovie dello Stato	Rear collision	63
8 Jun 1962	Near Kumhari, Italy	South Eastern	Train struck by tornado	9

6 Jly 1962	Cirkvice, Bohemia, Czechoslovakia		Double collision	16
21 Jly 1962	Dumraon, Bihar, India	Eastern	Head-on collision	74
22 Jly 1962	Between Bucharest and Mogosoia, Romania		Derailment, excessive speed	32
23 Jly 1962	Velars-sur-Ouche, France	Société nationale des chemins de fer français	Derailment, track distorted by the heat, part of train fell from viaduct	38
28 Jly 1962	Steelton, PA, USA	Pennsylvania	Derailment, track distorted by the heat	19
17 Aug 1962	Near Dieu Tri, South Vietnam		Derailment, sabotage	8
18 Sep 1962	Near Horto Florestal, Brazil		Derailment, excessive speed, driver intoxicated?	9
2 Oct 1962	Holfontein, Orange Free State, South Africa	South African	Derailment	17
5 Oct 1962	Combe Aux Loup, France	Société nationale des chemins de fer français	Derailment, broken axle, collision	7
5 Oct 1962	Klington, South Africa	South African	Head-on collision	8
9 Oct 1962	Piotrków Trybunalski, Poland	Polskie Kokeje Państwowe	Derailment, broken rail, collision	34
5 Nov 1962	Kabanovce, Yugoslavia	Zajednica Jugoslovenskih Zeleznica	Derailment	23
21 Nov 1962	Near Tahona, Tucuman, Argentina	Ferrocarriles Argentinos	Derailment, excessive speed over track under repair? Broken axle?	9
20 Dec 1962	Buena Vista, Guatemala		Rear collision	22
23 Dec 1962	'South of Huê,' South Vietnam		Derailment, sabotage	7
26 Dec 1962	Between Winsford and Coppenhall Junction, England	British Railways, London Midland Region	Rear collision	18
4 Jan 1963	Umeshnagar, Bihar, India	North Western	Head-on collision, fire	42
21 Jan 1963	Between Quy Nhon and Quang Ngãi, South Vietnam		Derailment, sabotage	8
c10 Feb 1963	Pina, Cuba		Derailment, sabotage	27
24 Feb 1963	Between Tardienta and Almudévar, Spain	Red Nacional de los Ferrocarriles Españoles	Rear collision	5
26 Feb 1963	Between Apostoles and Pindary, Argentina	Ferrocarriles Argentinos	Derailment, track subsidence	7
5 Apr 1963	Near Long Khanh, South Vietnam		Derailment, sabotage	5

11 Apr 1963	Near Tagekapu, West Java, Indonesia	Perusahaan Jawatan Kerata Api	Runaway train, derailment	36
19 May 1963	Between Nha Trang and Diêu Tri, South Vietnam		Derailment, sabotage	5
23 Jun 1963	Mượng Mán, South Vietnam,		Derailment? Sabotage?	9
13 Jly 1963	Near Surabaya, East Java, Indonesia	Perusahaan Jawatan Kerata Api	Derailment	21
2 Aug 1963	Montevideo, Uruguay		Derailment, sabotage	30
26 Aug 1963	Loyasse Tunnel, France	Société nationale des chemins de fer français	Rear collision in tunnel	7
19 Sep 1963	Near Fukuoka, Kyushu, Japan		Collision with lorry on open level crossing, derailment, collision	8
9 Nov 1963	Tsurumi, Japan	Japanese National	Derailment, track and vehicle defects, collision	161
25 Nov 1963	Leinhausen, Hannover, West Germany	Deutsche Bundesbahn	Sidelong collision	5
24 Dec 1963	Near Szolnok, Hungary		Rear collision in fog	45
4 Jan 1964	Jajinci, Yugoslavia		Rear collision in fog	61
25 Jan 1964	Zákolany, Czechoslovakia		Runaway train? Collision	14
1 Feb 1964	Altamirano, Argentina	Ferrocarriles Argentinos	Head-on collision, fire	32
24 Feb 1964	'15 Km North East of Phu My', South Vietnam		Derailment, sabotage	11
9 Mar 1964	Baudpur, India	South Eastern	Train diverted onto wrong line, head-on collision	22
30 Apr 1964	West Java, Indonesia		Derailment	21
9 May 1964	Chełmek, Poland	Polskie Kokeje Państwowe	Rear collision?	12
24 May 1964	Near Duy Xuyen, South Vietnam		Derailment, sabotage	8
18 Jun 1964	'15 Km South of Nha Trang, South Vietnam		Derailment, sabotage, fire	20
26 Jly 1964	Custoias, Portugal	Caminhos de Ferro Portugueses	Derailment, 'faulty loading', excessive speed?	88
29 Jun 1964	Near Randfontein, South Africa	South African	Derailment	21
21 Aug 1964	Leonard, Ontario, Canada	Canadian Pacific	Collision with lorry on open level crossing, derailment, fire	7

31 Aug 1964	Westervoort, The Netherlands	Nederlandse Spoorwegen	Sidelong collision, fire	5
5 Sep 1964	Alby, Sweden	Statens Järnvägar	Derailment, excessive speed through crossover	8
19 Sep 1964	Kaithalkurichi, Assam, India	North East Frontier	Head-on collision	8
25 Sep 1964	Zákolany, Czechoslovakia		Runaway train, collision	14
11 Oct 1964	Near Purwokerto, Java, Indonesia		Collision with waggons	5
17 Oct 1964	Vilwatte, Ceylon		Derailment, defective vehicle	25
24 Oct 1964	Hastedt, West Germany	Deutsche Bundesbahn	Sidelong collision	7
1 Nov 1964	Langhagen, East Germany	Deutsche Reichsbahn	Buffer stop collision at end of trap siding, collision	43
20 Dec 1964	Tacotalpa, Mexico	Ferrocarriles Nacionales de México	Rear collision in fog, fire	39
23 Dec 1964	Near Dhanushkodi, India	Southern	Train struck by tidal wave during cyclone	110
16 Jan 1965	Bonassola, Italy		Explosion	8
24 Jan 1965	Near Amatilán, Guatemala		Runaway train, collision	33
1 Feb 1965	Near Sukabumi, West Java, Indonesia		Derailment, faulty loading	12
10 Feb 1965	Between Grisén and Pinseque, Spain.	Red Nacional de los Ferrocarriles Españoles	Fire	65
20 Feb 1965	Skultorp, Sweden	Statens Järnvägar	Rear collision	10
26 Feb 1965	Near Haiya, Sudan		Runaway train, collision, fire	46
28 Mar 1965	Comendador Soares, Brazil		Derailment, track defects, collision	19
30 Mar 1965	Between Moatize and Motarara, Mozambique		Collision	9
5 Apr 1965	Between Paraíba do Sul and Engenheiro Vieira Cortes, Brazil	Rede Ferroviaria Federal	Head-on collision, single line?	21
23 Apr 1965	Near Tunis, Tunisia		Rear collision	9
16 Jun 1965	Between San Fabian and Damortis, The Philippines	Philippine National	Head-on collision, single line	14
24 Jun 1965	Waßmannsdorf, East Germany	Deutsche Reichsbahn	Derailment, track defects	9
16 Aug 1965	Between Gádor and Santa Fe, Spain	Red Nacional de los Ferrocarriles Españoles	Head-on collision, single line	13
28 Aug 1965	Pont d'Hery, France	Société nationale des chemins de fer français	Rear collision	13

21 Sep 1965	Praha Hloubětin, Czechoslovakia		Rear collision	14
4 Oct 1965	Effingham, South Africa	South African	Derailment	89
30 Nov 1965	Zidani Most, Yugoslavia		Collision	7
8 Dec 1965	Kyee Taw, Burma		Head-on collision	80
18 Dec 1965	Villar-de-los Alamos, Spain	Red Nacional de los Ferrocarriles Españoles	Collision in fog	34
20 Dec 1965	Sintra and Algueirão Mem Martins, Portugal,	Caminhos de Ferro Portugueses	Head-on collision	17
9 Jan 1966	Novska, Yugoslavia		Collision in blizzard	5
14 Feb 1966	Near Shwenyaung, Burma		Derailment, excessive speed?	23
16 Feb 1966	Between Farketing and Kamarbandhi, Assan, India	Northeast Frontier	Bombs exploded on train	37
16 Feb 1966	Between Kaštel Stari and Sadina, Yugoslavia		Runaway waggons, collision, fire	31
1 Apr 1966	'40 Kilometres east of Diredawa, Ehiopia	Franco-Ethiopian	Derailment, track washed away	9
4 Apr 1966	Sumulong, The Philippines	Philippine National	Derailment	9
20 Apr 1966	Lumding, Assam, India		Bomb exploded on train	57
23 Apr 1966	Diphu, Assam, India		Bomb exploded on train	39
24 Apr 1966	Between Holmsveden and Kilafors, Sweden		Head-on collision, single line	6
26 May 1966	Between Khanapur and Desur, India	Southern	Derailment, sabotage	21
30 May 1966	Dube, South Africa	South African	Rear collision	10
31 May 1966	Mogosoaia, Romania		Head-on collision	38
13 Jun 1966	Near Matunga, India	Central	Sidelong collision	60
19 Jun 1966	Ladpura, Rajasthan, India	Western	Rear collision	15
1 Aug 1966	Croesus, South Africa	South African	Rear collision, fire (arson)	6
11 Aug 1966	Pa-Kong Bridge, China		Derailment, sabotage	27
17 Sep 1966	Between Da Nang and Huê, South Vietnam		Derailment, sabotage	6
10 Oct 1966	Between Miraj and Mhaisal, India	South Central	Derailment, sabotage, part of train dell from bridge	10

21 Oct 1966	Cosney-sur-Loire, France	Société nationale des chemins de fer français	Derailment	10
11 Nov 1966	Between Sivok and Pilanshat, India	North East Frontier	Derailment, sabotage	14
16 Nov 1966	Between Nilopolis and Olinda, Brazil	Rede Ferroviaria Federal	Head-on collision during temporary single-line working	33
17 Nov 1966	Between Kelkheim-Hornau and Unterliederbach, West Germany	Deutsche Bundesbahn	Runaway train, collision	7
17 Dec 1966	Near Khandvara, India		Derailment	13
18 Dec 1966	Between Villafranco del Campo and Santa Eudalia del Campo, Spain	Red Nacional de los Ferrocarriles Españoles	Head-on collision single line in fog, fire	24
19 Dec 1966	Near Frankfurt Oder, East Germany	Deutsche Reichsbahn	Derailment	6
28 Dec 1966	Everett, MA, USA	Boston and Maine	Collision with oil tanker on open level crossing, derailment, explosions, fire	11
29 Dec 1966	Near Rzepin, Poland	Polskie Kokeje Państwowe	Derailment	6
28 Feb 1967	Stechford, England	British Railways, London Midland Region	Sidelong collision	8
5 Mar 1967	Connington South, England	British Railways, Eastern Region	Derailment, points moved under train	5
9 Mar 1967	Between Las Montoyas and El Opón, Colombia		Derailment, sabotage	?
12 Mar 1967	Frunzeasca, Romania	Căile Ferate Române	Derailment, excessive speed	10
1 Apr 1967	Between Ozarki and Tarui, Japan	Nankai	Collision with lorry on open level crossing, derailment, part of train fell from bridge	5
21 May 1967	Kuppam, India	Southern	Runaway train, derailment in sand drag at end of trap siding	58
30 May 1967	Locsin, The Philippines	Philippine National	Derailment	6
3 Jly 1967	Dzialdowo, Poland	Polskie Kokeje Państwowe	Derailment, track distorted by the heat? Broken rail?	7
6 Jly 1967	Near Langenweddingen, East Germany	Deutsche Reichsbahn	Collision with petrol tanker on FLB level crossing, fire	93
31 Jly 1967	Thirsk, England	British Railways, Eastern Region	Collision with derailed waggons	7

10 Aug 1967	Marsley, Denmark	Danske Statsbaner	Rear collision	11
17 Sep 1967	Skyline Siding (Mount Washington), NH, USA	Mount Washington	Derailment on points, runaway train (cog railway)	8
28 Sep 1967	Between Samuktala Road and Salsala, North Bengal, India	North East Frontier	Derailment, excessive speed and track defects/sabotage, fire	13
4 Oct 1967	Fexhe-le-Haut-Clocher, Belgium		Double collision	11
5 Nov 1967	Near Hither Green, England	British Railways, Southern Region	Derailment, broken rail	49
9 Nov 1967	Near San Nicola Varco, Italy		Collision with buffaloes, derailment, traction shock, collision	12
30 Dec 1967	Near Cordoba, Colombia	Ferrocarriles Nacionales de Columbia	Derailment, excessive speed?	7
1 Jan 1968	Between Randspruit and Bume, South Africa	South African	Rear collision	12
6 Jan 1968	Hixon, England	British Railways, London Midland Region	Collision with exceptional load on AHB level crossing, derailment	8
27 Jan 1968	Fanfa, Brazil	Rede Ferroviaria Federal	Sidelong collision	42
15 Mar 1968	Between Robledo de Chavela and Santa Maria de la Alameda, Spain	Red Nacional de los Ferrocarriles Españoles	Runaway engine, collision, fire	24
19 Mar 1968	Yalvigi, Mysore, India		Head-on collision	40
30 Mar 1968	Near Allahabad, India		Rear collision	20
13 Apr 1968	Between Parasnath and Hazaribagh Road, India	Eastern	Fire	7
9 May 1968	Langsa, Sumatra, Indonesia		Derailment, train fell from bridge	15
28 May 1968	Renigunta, India	Southern	Train diverted onto wrong line, collision	14
20 Jun 1968	Between Liluah and Howrah, India	Eastern	Rear collision	5
24 Jun 1968	Between Sion/ Sitten and St Leonard, Switzerland	Chemins de fer de fédéraux suisses	Head-on collision single line	10
1 Jly 1968	Between Quincieux and Saint-Germains-au-Mont d'Or, France		Derailment	6

13 Jly 1968	Between Medina del Campo and Gomeznaro, Spain	Red Nacional de los Ferrocarriles Españoles	Head-on collision during temporary single-line working	9
18 Aug 1968	Gangpur, India	Eastern	Rear collision	6
20 Sep 1968	Between Depok and Tjitajam, Java, Indonesia		Head-on collision single line	18
30 Sep 1968	Near Derveni, Greece		Rear collision	34
7 Oct 1968	Betweeen Crăciunel and Bucurdea Grânoasă, Romania,	Căile Ferate Române	Head-on collision single line	18
31 Oct 1968	Between Hissar and Jakhod Khera, India	Northern	Fire	6
15 Nov 1968	Son Nagar, Bihar, India	Eastern	Sidelong collision	9
4 Dec 1968	Between Arbil and Kirkuk, Iraq		Derailment, sabotage	?
22 Dec 1968	Mende and Sülysáp, Hungary		Train diverted onto wrong line, head-on collision in fog	41
31 Jan 1969	Chonan, South Korea	Korean National	Rear collision in blizzard	41
7 Feb 1969	Violet Town, VA, Australia		Head-on collision single line, fire	5
10 Feb 1969	About 20 kilometres from Cotonou, Dahomey		Collision	10
17 Feb 1969	Near Langlaate, South Africa	South African	Derailment, collision, leakage of petrol, fire	13
25 Mar 1969	La Louvière, Belgium		Head-on collision	13
1 May 1969	Near Sofia, Bulgaria		Derailment	25
7 May 1969	Morpeth, England	British Railways, Eastern Region	Derailment, excessive speed	5
15 Jun 1969	Between Kościerzyna and Skorzewo, Poland		Head-on collision single line	5
15 Jun 1969	Sant Antonio Tunnel, Italy		Head-on collision single line, in tunnel	5
21 Jun 1969	Between Jakhanian and Dulahpur, India	North Eastern	Derailment, sabotage, train fell from bridge	70
14 Jly 1969	Jaipur-Keonjhar, Orissa, India	South Eastern	Rear collision	86
28 Jly 1969	Between Leiferde and Meinersen, West Germany	Deutsche Bundesbahn	Derailment, track distorted by the heat	5
30 Jly 1969	Between Gostivar and Tetovo, Yugoslavia	Zajednica Jugoslovenskih Zeleznica	Runaway waggons, collision	29
26 Oct 1969	Okucani, Yugoslavia		Rear collision	15
31 Oct 1969	Jedliczach Lódzkich, Poland		Sidelong collision in fog	5

31 Dec 1969	Near Lam Lam, Senegal		Collision	16
2 Jan 1970	Oreshetz, Bulgaria		Rear collision?	12
1 Feb 1970	Berween Benavidez and General Pacheco, Argentina	Ferrocarriles Argentinos	Rear collision	139
12 Feb 1970	Near Hsinchu, Taiwan		Rear collision	9
16 Feb 1970	Langalanga, Nigeria		Derailment, excessive speed? Driver intoxicated	105
23 Feb 1970	Valašské Meziříčí, Czechoslovakia		Collision	11
29 Mar 1970	'25 Miles East of Karachi,' Pakistan		Train fell from trestle	20
11 May 1970	Beckingen, West Germany	Deutsche Bundesbahn	Derailment, train ran into landslide, collision	5
5 Jun 1970	Celle, West Germany	Deutsche Bundesbahn	Derailment, track distorted by the heat	5
22 Jly 1970	Gioia Tauro, Italy	Fellovie dello Stato	Derailment, defective vehicle	6
9 Aug 1970	Between Urduliz and Plencia, Spain	Ferrocarriles y Tranvias de Albao	Head-on collision single line	31
16 Aug 1970	Zahana, Algeria		Derailment	10
22 Aug 1970	Tamouh, Egypt		Derailment	13
17 Sep 1970	Baoule River Bridge, Mali	Chemin de fer Dakar-Niger	Bridge collapsed under train	27
17 Oct 1970	Bongsandong Tunnel, South Korea		Head-on collision in tunnel	14
26 Oct 1970	Khurja Junction, India	Northern	Head-on collision	7
31 Oct 1970	Perambur, India	Southern	Rear collision	14
31 Oct 1970	Dube, South Africa	South African	Rear collision	14
11 Dec 1970	Řikonín, Czechoslovakia		Double collision	32
11 Dec 1970	Between São João de Talha and Póvoa de Santa Iria, Portugal	Caminhos de Ferro Portugueses	Rear collision	11
31 Dec 1970	Sasson, Iran		Head-on collision, single line?	70
1 Jan 1971	Atapamé, Togo		Collision	13
18 Jan 1971	Between Herrliberg-Feldmeilen and Meilen, Switzerland	Chemins de fer de fédéraux suisses	Head-on collision, single line	6
9 Feb 1971	Aitrang, West Germany	Deutsche Bundesbahn	Derailment, excessive speed	26
14 Feb 1971	Vranduk Tunnel, Yugoslavia	Zajednica Jugoslovenskih Zeleznica	Fire in tunnel	34
23 Feb 1971	Between Rangoon and Mandalay,		Derailment? Sabotage	16

	Burma			
4 Mar 1971	Midorigaoka No. 2 Crossing, Japan	Fuji Kyuko	Collision with lorry on FLB level crossing, derailment	17
11 Mar 1971	Huayihuaica, Chile		Runaway train, derailment	12
13 Apr 1971	Between Da Nang and Huê, South Vietnam		Derailment, sabotage	20
1 May 1971	Fürnitz, Austria		Train diverted onto wrong line, collision	8
18 May 1971	Near Kellmünz, West Germany	Deutsche Bundesbahn	Derailment, track distorted by the heat	5
27 May 1971	Near Dahlerau, West Germany	Deutsche Bundesbahn	Head-on collision, single line	44
30 May 1971	Between Dulvendrecht Asi and Amsterdam Amstel, The Netherlands	Nederlandse Spoorwegen	Rear collision	5
5 Jun 1971	Gonda Junction, India	North Eastern	Rear collision	8
10 Jun 1971	Tonti, IL, USA	Illinois Central	Derailment, engine defect, fire	10
6 Jly 1971	Between Hotar and Magra Hat, India	Eastern	Double collision	13
17 Jly 1971	Semizovac, Yugoslavia	Zajednica Jugoslovenskih Zeleznica	Head-on collision	15
21 Jly 1971	Rheinweiler, West Germany	Deutsche Bundesbahn	Derailment, excessive speed	22
22 Jly 1971	Simplon Tunnel, Switzerland		Derailment on points in tunnel	5
4 Aug 1971	Between Kasapovasc and Lipe, Yugoslavia	Zajednica Jugoslovenskih Zeleznica	Head-on collision, single line	40
20 Sep 1971	Near Biskra, Algeria		Derailment	22
13 Oct 1971	Namwon, South Korea		Runaway train, collision	18
25 Oct 1971	Sotani Tunnel, Japan	Kinki Nippon	Head-on collision, single line in tunnel	21
30 Nov 1971	Near Água Branca, Brazil	Ferrovia Paulista	Head-on collision	8
6 Dec 1971	Near Narowal, Pakistan		Train attacked by Indian aircraft	5
23 Dec 1971	Dekéleia, Greece		Rear collision?	6
31 Dec 1971	Between Hachenberg and Unnau-Korb, West Germany	Deutsche Bundesbahn	Head-on collision, single line	6
16 Jan 1972	Between Doxaras and Orfana, Greece		Head-on collision, single line, fire	19
31 Mar 1972	Potgietersraus, Transvaal, South	South African	Derailment, part of train fell from bridge,	38

	Africa		sabotage?	
26 Apr 1972	Hindupur, India		Derailment	21
17 May 1972	Near Lapa, Brazil	Ferrovia Paulista	Collision with articulated lorry on FLB level crossing, derailment	8
3 Jun 1972	Near Bydgoszcz, Poland		Derailment, track distorted by the heat	12
4 Jun 1972	Jessore, Bangladesh	Bangladesh	Train diverted onto wrong line, collision	76
8 Jun 1972	Between Suzano and Jundiapeba, Brazil	Rede Ferroviaria Federal	Rear collision in fog	21
11 Jun 1972	Eltham (Well Hall), England	British Railways, Southern Region	Derailment, excessive speed, driver intoxicated	5
16 Jun 1972	Vierzy Tunnel, France	Société nationale des chemins de fer français	Collisions with rubble that had fallen from tunnel roof	103
3 Jun 1972	Near Norvalspont, South Africa	South African	Derailment	11
21 Jly 1972	Between El Cuervo and Lebrija, Spain	Red Nacional de los Ferrocarriles Españoles	Head-on collision, single line	77
6 Aug 1972	Liaqatpur, Pakistan	Pakistan	Train diverted onto wrong line, rear collision	90
28 Aug 1972	Near Larimore, ND, USA	Burlington Northern	Derailment, broken wheel	5
29 Sep 1972	Rust Stasie (Malmesbury), South Africa	South African	Derailment	48
5 Oct 1972	Puente Moreno (Saltillo), Mexico	Ferrocarriles Nacionales de México	Derailment, excessive speed (fire), engine crew intoxicated?	?
29 Oct 1972	Near Monapo, Mozambique		Collision	10
30 Oct 1972	Near Schewinsburg-Culten, East Germany	Deutsche Reichsbahn	Head-on collision, single line in fog	25
30 Oct 1972	27th Street, Chicago, IL, USA	Illinois Central Gulf	Double collision	45
31 Oct 1972	Alpu, Turkey	Türkiye Cumhuriyeti Deviet Demiryollari	Head-on collision, fire	30
6 Nov 1972	Hokuriku Tunnel, Japan	Japanese National	Fire in tunnel	29
10 Dec 1972	Venchan, Bulgaria		Rear collision	26
26 Dec 1972	Between Maiduguri and Jos, Nigeria		Derailment	6
8 Jan 1973	Near Encarnación de Diaz, Mexico		Derailment, inadequate protection of PW work?	5
1 Feb 1973	Medjez-Sfa, Algeria	Société nationale des transports ferroviares	Derailment	35

2 Feb 1973	Near Cesário Alvim, Brazil,		Derailment, track distorted by the heat	8
16 Feb 1973	Megara, Greece		Derailment	10
16 Mar 1973	Vicentica, Cuba		Derailment, part of train fell from bridge	23
11 May 1973	145 Kilometres South of Taipei, Taiwan		Head-on collision	12
7 Jun 1973	Torricola, Italy	Fellovie dello Stato	Double collision	5
27 Jly 1973	Between Joramow Block Hut and Madhupur Junction, India	Eastern	Rear collision	19
27 Aug 1973	Near Radkowice, Poland		Rear collision	16
16 Sep 1973	Sirajganj, Bangladesh	Bangladesh	Head-on collision, fire	?
26 Sep 1973	Mandini, South Africa	South African	Rear collision	17
5 Oct 1973	Marl Sinsen, West Germany	Deutsche Bundesbahn	Double collision	5
5 Nov 1973	Guntershausen, West Germany	Deutsche Bundesbahn	Rear collision	13
17 Dec 1973	Near Almeida Brandão, Brazil,	Rede Ferroviaria Federal	Head-on collision	17
19 Dec 1973	Near West Ealing, England	British Railways, Western Region	Derailment, engine defect	10
7 Jan 1974	Zeytinlik, Turkey		Train diverted onto wrong line, head-on collision	25
7 Feb 1974	Between Da Nang and Huê, South Vietnam		Derailment, sabotage	9
17 Feb 1974	Near Nangue, Mozambique		Train machine-gunned by guerillas	5
21 Feb 1974	Near Moradabad, Uttar Pradesh, India		Collision	40
27 Feb 1974	Near Elazig, Turkey		Derailment, train ran into landslide	10
9 Mar 1974	Between Silipur and Birpur, India		Derailment, excessive speed	5
27 Mar 1974	São Miguel, Mozambique		Head-on collision, explosion, fire	60
4 Aug 1974	Apennine Base Tunnel, Italy	Fellovie dello Stato	Bomb exploded on train in tunnel	12
4 Aug 1974	Dol-de-Bretagne, France	Société nationale des chemins de fer français	Derailment, excessive speed, driver intoxicated	8
15 Aug 1974	Luttre, Belgium		Derailment, track distorted by the heat? Part of train fell from bridge, fire	15
30 Aug 1974	Zagreb, Yugoslavia	Zajednica Jugoslovenskih	Derailment, excessive speed	153

		Zeleznica		
31 Oct 1974	Near Monanganj, Uttar Pradesh, India		Explosion, fire	43
6 Nov 1974	Near Abomey, Dahomey		Head-on collision, fire	80
13 Dec 1974	Near Kayseri, Turkey		Rear collision	7
27 Dec 1974	Near Santa Apolonia, Portugal	Caminhos de Ferro Portugueses	Derailment, collision	9
18 Jan 1975	Near Qalyub, Egypt		Derailment, excessive speed	27
29 Jan 1975	Ultadanga Road, India	Eastern	Rear collision	35
17 Feb 1975	Near Potchefstroom, South Africa	South African	Collision	14
22 Feb 1975	Near Tretten, Norway	Norges Statsbaner	Head-on collision, single line	25
28 Feb 1975	Moorgate, England	London Transport	Collision with end wall of tunnel	42
28 Mar 1975	Magude, Mozambique		Head-on collision, single line?	33
31 Mar 1975	Between Mjölby and Sya, Sweden	Statens Järnvägar	Collision with car on open level crossing, derailment	13
4 Apr 1975	Near Žaslial, Lithuanian Soviet Socialist Republic		Fouling collision, fire	20
22 Apr 1975	Taben-Rodt, West Germany	Deutsche Bundesbahn	Collision	5
15 May 1975	Near Vranje, Yugoslasvia	Zajednica Jugoslovenskih Zeleznica	Bridge collapsed under train	13
15 May 1975	Norrköping, Sweden	Statens Järnvägar	Sidelong collision	8
22 May 1975	Near Tarbes, France	Société nationale des chemins de fer français	Collision with petrol tanker on open level crossing, explosion, fire	5
22 May 1975	Near Sidi Yahia du Rharb, Morocco	Office national des chemins de fer du Maroc	Derailment	47
8 Jun 1975	Near Warngau, West Germany	Deutsche Bundesbahn	Head-on collision, single line	38
17 Jly 1975	Magno, Brazil	Rede Ferroviaria Federal	Derailment, excessive speed?	13
19 Jly 1975	Zemun Polje, Yugoslacia	Zajednica Jugoslovenskih Zeleznica	Rear collision	16
22 Jly 1975	Hamburg Hausbruch, West Germany	Deutsche Bundesbahn	Head-on collision	11
12 Aug 1975	Between Bamio and Vilagarcia de Arousa, Spain	Red Nacional de los Ferrocarriles Españoles	Derailment	6

1 Sep 1975	Near Chorzów, Poland		Rear collision?	6
29 Sep 1975	Between Kalyan and Vithalwadi, Indis	Central	Fire	5
29 Sep 1975	Rio Lujan, Argentina	Ferrocarriles Argentinos	Rear collision	32
20 Oct 1975	Viaducto, Mexico		Rear collision	34
8 Nov 1975	Tunnel No. 66, between Laza and Villar del Barrio, Spain	Red Nacional de los Ferrocarriles Españoles	Derailment in tunnel, broken rail	5
14 Dec 1975	Fornos de Algodres, Portuga	Caminhos de Ferro Portugueses	Train diverted onto wrong line, head-on collision	7
31 Dec 1975	Near Gorey, Eire	Córas Iompair Éireann	Derailment, underbridge stuck and damaged by high road vehicle	5

ABOUT THE AUTHOR

The author has been interested in railways for many years, ever since he was invited up a signal box at the age of 14 back in 1966. Not long afterwards he 'cabbed' the engine of a goods train that was shunted out of the way of a faster train and its driver showed him a copy of *Red for Danger*, the late L.T.C. Rolt's book about British railway accidents. This was the first railway book that he ever bought, several months' saved-up pocket money going towards a paperback copy of it. Later he came across a copy of *A History of Railroad Accidents, Safety Precautions and Operating Practices* by Robert B. Shaw, dealing with American railroad accidents, and his interest in the subject grew.

When the local signal boxes were all done away with in a modernization scheme in the mid-1980s he began a more detailed study of world railway accidents. There was no internet back then of course and many hours were spent in the local reference library slogging through bound volumes of *The Times*, writing out longhand in large notebooks details of every accident found. Later it was possible to view the paper on microfilm and print pages, which made things a little easier, and with the coming of the internet the amount of information available vastly increased. Some years ago, the author contributed to *Railway Disasters of the World* by P.W.B. Semmens (which deals only with 20th-century accidents) and more recently the idea came of writing his own book or books on world accidents, starting at the beginning.

Made in United States
Cleveland, OH
06 November 2026

47673669R20412